21st Century Nanoscience – A Handbook

T0133484

21st Century Nanoscience – A Handbook: Nanophysics Sourcebook (Volume One)

21st Century Nanoscience – A Handbook: Design Strategies for Synthesis and Fabrication (Volume Two)

21st Century Nanoscience – A Handbook: Advanced Analytic Methods and Instrumentation (Volume Three)

21st Century Nanoscience – A Handbook: Low-Dimensional Materials and Morphologies (Volume Four)

21st Century Nanoscience – A Handbook: Exotic Nanostructures and Quantum Systems (Volume Five)

21st Century Nanoscience – A Handbook: Nanophotonics, Nanoelectronics, and Nanoplasmonics (Volume Six)

21st Century Nanoscience – A Handbook: Bioinspired Systems and Methods (Volume Seven)

21st Century Nanoscience – A Handbook: Nanopharmaceuticals, Nanomedicine, and Food Nanoscience (Volume Eight)

21st Century Nanoscience – A Handbook: Industrial Applications (Volume Nine)

21st Century Nanoscience – A Handbook: Public Policy, Education, and Global Trends (Volume Ten)

21st Century Nanoscience – A Handbook

Nanopharmaceuticals, Nanomedicine, and Food Nanoscience (Volume Eight)

Edited by

Klaus D. Sattler

CRC Press is an imprint of the
Taylor & Francis Group, an **informa** business

MATLAB® is a trademark of The MathWorks, Inc. and is used with permission. The MathWorks does not warrant the accuracy of the text or exercises in this book. This book's use or discussion of MATLAB® software or related products does not constitute endorsement or sponsorship by The MathWorks of a particular pedagogical approach or particular use of the MATLAB® software.

CRC Press
Taylor & Francis Group
6000 Broken Sound Parkway NW, Suite 300
Boca Raton, FL 33487-2742

First issued in paperback 2022

© 2020 by Taylor & Francis Group, LLC
CRC Press is an imprint of Taylor & Francis Group, an Informa business

No claim to original U.S. Government works

ISBN-13: 978-0-815-35707-0 (hbk)
ISBN-13: 978-1-03-233651-0 (pbk)
DOI: 10.1201/9780429351587

This book contains information obtained from authentic and highly regarded sources. Reasonable efforts have been made to publish reliable data and information, but the author and publisher cannot assume responsibility for the validity of all materials or the consequences of their use. The authors and publishers have attempted to trace the copyright holders of all material reproduced in this publication and apologize to copyright holders if permission to publish in this form has not been obtained. If any copyright material has not been acknowledged, please write and let us know so we may rectify in any future reprint.

Except as permitted under U.S. Copyright Law, no part of this book may be reprinted, reproduced, transmitted, or utilized in any form by any electronic, mechanical, or other means, now known or hereafter invented, including photocopying, microfilming, and recording, or in any information storage or retrieval system, without written permission from the publishers.

For permission to photocopy or use material electronically from this work, please access www.copyright.com (http://www.copyright.com/) or contact the Copyright Clearance Center, Inc. (CCC), 222 Rosewood Drive, Danvers, MA 01923, 978-750-8400. CCC is a not-for-profit organization that provides licenses and registration for a variety of users. For organizations that have been granted a photocopy license by the CCC, a separate system of payment has been arranged.

Trademark Notice: Product or corporate names may be trademarks or registered trademarks, and are used only for identification and explanation without intent to infringe.

Publisher's Note

The publisher has gone to great lengths to ensure the quality of this reprint but points out that some imperfections in the original copies may be apparent.

Library of Congress Cataloging-in-Publication Data

Names: Sattler, Klaus D., editor.
Title: 21st century nanoscience : a handbook / edited by Klaus D. Sattler.
Description: Boca Raton, Florida : CRC Press, [2020] | Includes bibliographical references and index. | Contents: volume 1. Nanophysics sourcebook—volume 2. Design strategies for synthesis and fabrication—volume 3. Advanced analytic methods and instrumentation—volume 5. Exotic nanostructures and quantum systems—volume 6. Nanophotonics, nanoelectronics, and nanoplasmonics—volume 7. Bioinspired systems and methods. | Summary: "This 21st Century Nanoscience Handbook will be the most comprehensive, up-to-date large reference work for the field of nanoscience. Handbook of Nanophysics, by the same editor, published in the fall of 2010, was embraced as the first comprehensive reference to consider both fundamental and applied aspects of nanophysics. This follow-up project has been conceived as a necessary expansion and full update that considers the significant advances made in the field since 2010. It goes well beyond the physics as warranted by recent developments in the field"—Provided by publisher.
Identifiers: LCCN 2019024160 (print) | LCCN 2019024161 (ebook) | ISBN 9780815384434 (v. 1 ; hardback) | ISBN 9780815392330 (v. 2 ; hardback) | ISBN 9780815384731 (v. 3 ; hardback) | ISBN 9780815355281 (v. 4 ; hardback) | ISBN 9780815356264 (v. 5 ; hardback) | ISBN 9780815356417 (v. 6 ; hardback) | ISBN 9780815357032 (v. 7 ; hardback) | ISBN 9780815357070 (v. 8 ; hardback) | ISBN 9780815357087 (v. 9 ; hardback) | ISBN 9780815357094 (v. 10 ; hardback) | ISBN 9780367333003 (v. 1 ; ebook) | ISBN 9780367341558 (v. 2 ; ebook) | ISBN 9780429340420 (v. 3 ; ebook) | ISBN 9780429347290 (v. 4 ; ebook) | ISBN 9780429347313 (v. 5 ; ebook) | ISBN 9780429351617 (v. 6 ; ebook) | ISBN 9780429351525 (v. 7 ; ebook) | ISBN 9780429351587 (v. 8 ; ebook) | ISBN 9780429351594 (v. 9 ; ebook) | ISBN 9780429351631 (v. 10 ; ebook)
Subjects: LCSH: Nanoscience—Handbooks, manuals, etc.
Classification: LCC QC176.8.N35 A22 2020 (print) | LCC QC176.8.N35 (ebook) | DDC 500—dc23
LC record available at https://lccn.loc.gov/2019024160
LC ebook record available at https://lccn.loc.gov/2019024161

Visit the Taylor & Francis Web site at
http://www.taylorandfrancis.com

and the CRC Press Web site at
http://www.crcpress.com

Contents

Editor

Klaus D. Sattler pursued his undergraduate and master's courses at the University of Karlsruhe in Germany. He earned his PhD under the guidance of Professors G. Busch and H.C. Siegmann at the Swiss Federal Institute of Technology (ETH) in Zurich. For three years he was a Heisenberg fellow at the University of California, Berkeley, where he initiated the first studies with a scanning tunneling microscope of atomic clusters on surfaces. Dr. Sattler accepted a position as professor of physics at the University of Hawaii, Honolulu, in 1988. In 1994, his group produced the first carbon nanocones. His current work focuses on novel nanomaterials and solar photocatalysis with nanoparticles for the purification of water. He is the editor of the sister references, *Carbon Nanomaterials Sourcebook* (2016) and *Silicon Nanomaterials Sourcebook* (2017), as well as *Fundamentals of Picoscience* (2014). Among his many other accomplishments, Dr. Sattler was awarded the prestigious Walter Schottky Prize from the German Physical Society in 1983. At the University of Hawaii, he teaches courses in general physics, solid state physics, and quantum mechanics.

Contributors

Luis Almeida
Centre for Textile Science and Technology
University of Minho
Guimarães, Portugal

M. Angelakeris
Department of Physics
Aristotle University
Thessaloniki, Greece

Alejandro Baeza
Dpto. Materiales y Producción Aeroespacial, ETSI Aeronáutica y del Espacio
Universidad Politécnica de Madrid
Madrid, Spain

Stefano Bellucci
Instituto Nationale de Physica Nucleare (INFN)
Laboratory Nationali di Frascati
Frascati, Italy

Linda Böhmert
Department of Food Safety
German Federal Institute for Risk Assessment (BfR)
Berlin, Germany

Albert Braeuning
Unit Effect-Based Analytics and Toxicogenomics
German Federal Institute for Risk Assessement (BfR)
Berlin, Germany

Alexandra Nava Brezolin
Department of Food Engineering
URI - Campus of Erechim
Erechim, Brazil

Jeffrey M. Farner
Department of Chemical Engineering
McGill University
Montreal, Canada

Matthew Y. Chan
Institute for Critical Technologies and Applied Science, Virginia Tech
Blacksburg, Virginia
and
Charles E. Via Jr. Department of Civil and Environmental Engineering
Virginia Tech
Blacksburg, Virginia

Zhen Cheng
Molecular Imaging Program at Stanford, Department of Radiology
Stanford University
Stanford, California

Subhasree Roy Choudhury
Habitat Centre
Institute of Nano Science and Technology
Mohali, Punjab

Jyotirekha Das
School of Biotechnology
National Institute of Technology Calicut
Kerala, India

Arnaud Desrosiers
Laboratory of Biosensors & Nanomachines, Département de Chimie
Université de Montréal
Montréal, Canada

Sourav Ghosh
Laboratory of Toxinology and Experimental Pharmacodynamics, Department of Physiology
University of Calcutta
Kolkata, India

Antony Gomes
Laboratory of Toxinology and Experimental Pharmacodynamics, Department of Physiology
University of Calcutta
Kolkata, India

Michael R. Hamblin
Wellman Center for Photomedicine
Massachusetts General Hospital, Harvard Medical School
Boston, Massachusetts

Jouni Hirvonen
Drug Research Program, Division of Pharmaceutical Chemistry and Technology, Faculty of Pharmacy
University of Helsinki
Helsinki, Finland

Michael F. Hochella, Jr.
Department of Geoscience
Virginia Tech
Blacksburg, Virginia
and
Energy and Environment Directorate
Pacific Northwest National Laboratory
Richland, Washington

Yu Hou
Guizhou Provincial Key Laboratory for Information Systems of Mountainous Areas and Protection of Ecological Environment
Guizhou Normal University
Guiyang, China

Jiwei Hu
Guizhou Provincial Key Laboratory for Information Systems of Mountainous Areas and Protection of Ecological Environment
Guizhou Normal University
Guiyang, China
and
Cultivation Base of Guizhou National Key Laboratory of Mountainous Karst Eco-environment
Guizhou Normal University
Guiyang, China

Mahdi Karimi
Department of Medical Nanotechnology
School of Advanced Technologies in Medicine, Iran University of Medical Sciences
Tehran, Iran

Surajit Karmakar
Habitat Centre
Institute of Nano Science and Technology
Mohali, Punjab

Silke Krol
Nanomedicine Laboratory
Fondazione I.R.C.C.S. IStituto Neurologico Carlo Besta
Milan, Italy
and
I.R.C.C.S. "S. De Bellis"—Ente Ospedaliero Specializzato in Gastroenterologia
Bari, Italy

Alfonso Lampen
Department of Food Safety
German Federal Institute for Risk Assessement (BfR)
Berlin, Germany

Janine Martinazzo
Department of Food Engineering
URI - Campus of Erechim
Erechim, Brazil

Marek Martinec
Department of Environmental Chemistry
University of Chemistry and Technology Prague
Prague, Czech Republic

Lenka McGachy
Department of Environmental Chemistry
University of Chemistry and Technology Prague
Prague, Czech Republic

Arielle C. Mensch
Environmental Molecular Science Laboratory
Pacific Northwest National Laboratory
Richland, Washington

Lucas B. Naves
CAPES Foundation, Ministry of Education of Brazil
Brasília, Brazil
and
Centre for Textile Science and Technology
University of Minho
Guimarães, Portugal
and
Center for Nanofibers & Nanotechnology, Department of Mechanical Engineering
National University of Singapore
Singapore

Chris U. Onuegbu
Research and Technology Development Centre
Sharda University
Greater Noida, India
and
Department of Agric
Bioresources Engineering Federal Polytechnic
Oko, Nigeria

Leena Peltonen
Drug Research Program, Division of Pharmaceutical Chemistry and Technology, Faculty of Pharmacy
University of Helsinki
Helsinki, Finland

Tonya R. Pruitt
Nanoearth
Institute for Critical Technologies and Applied Science, Virginia Tech
Blacksburg, Virginia

Aaron J. Prussin II
Charles E. Via Jr. Department of Civil and Environmental Engineering
Virginia Tech
Blacksburg, Virginia

Jimei Qi
Guizhou Provincial Key Laboratory for Information Systems of Mountainous Areas and Protection of Ecological Environment
Guizhou Normal University
Guiyang, China

G. K. Rajanikant
School of Biotechnology
National Institute of Technology Calicut
Kerala, India

Seeram Ramakrishna
Center for Nanofibers & Nanotechnology, Department of Mechanical Engineering
National University of Singapore
Singapore
and
Hongkong-Macau Institute of CNS Regeneration (GHMICR), Jinan University
Guangzhou, China

Michael Riley II
Graduate Program in Plant Cellular and Molecular Biology
University of Florida
Gainesville, Florida
and
UF Genetics Institute
University of Florida
Gainesville, Florida

Barbara Sanavio
Nanomedicine Laboratory
Fondazione I.R.C.C.S. IStituto Neurologico Carlo Besta
Milan, Italy

Mohammed Nadim Sardoiwala
Habitat Centre
Institute of Nano Science and Technology
Mohali, Punjab

Jayeeta Sengupta
Laboratory of Toxinology and Experimental Pharmacodynamics, Department of Physiology
University of Calcutta
Kolkata, India

Holger Sieg
Junior Research Group Nanotoxicology
German Federal Institute for Risk Assessement (BfR)
Berlin, Germany

N. B. Singh
Research and Technology Development Centre
Sharda University
Greater Noida, India

Mayank Singhal
Drug Research Program, Division of Pharmaceutical Chemistry and Technology, Faculty of Pharmacy
University of Helsinki
Helsinki, Finland

Radek Škarohlíd
Department of Environmental Chemistry
University of Chemistry and Technology Prague
Prague, Czech Republic

Anup K. Srivastava
Habitat Centre
Institute of Nano Science and Technology
Mohali, Punjab

Clarice Steffens
Department of Food Engineering
URI - Campus of Erechim
Erechim, Brazil

Juliana Steffens
Department of Food Engineering
URI - Campus of Erechim
Erechim, Brazil

Alexis Vallée-Bélisle
Laboratory of Biosensors & Nanomachines, Département de Chimie
Université de Montréal
Montréal, Canada

Wilfred Vermerris
Graduate Program in Plant Cellular and Molecular Biology
University of Florida
Gainesville, Florida
and
UF Genetics Institute
University of Florida
Gainesville, Florida
and
Department of Microbiology & Cell Science
University of Florida
Gainesville, Florida

Gonzalo Villaverde
Dpto. Química en Ciencias Farmacéuticas, Facultad de Farmacia
Universidad Complutense de Madrid
Madrid, Spain

Xionghui Wei
Department of Applied Chemistry, College of Chemistry and Molecular Engineering
Peking University
Beijing, China

Volkmar Weissig
Department of Pharmaceutical Sciences & Nanomedicine Center of Excellence in Translational Cancer Research
College of Pharmacy Glendale, Midwestern University
Glendale, Arizona

Yiqiu Xiang
Guizhou Provincial Key Laboratory for Information Systems of Mountainous Areas and Protection of Ecological Environment
Guizhou Normal University
Guiyang, China

Pengfei Xu
Institute of Clinical Pharmacy & Pharmacology, Jining First People's Hospital
Jining Medical University
Jining, China

1

Review of Nanopharmaceuticals

Volkmar Weissig
Midwestern University College of Pharmacy Glendale & Nanomedicine Center of Excellence in Translational Cancer Research

1.1 Introduction

Based on substantial progress made in material science at the nanoscale level during the last three decades of the 20th century, the National Institute of Health (NIH) launched in 2000 a federal government program aimed at supporting, coordinating and advancing research and development of nanoscale projects. The impact of this new program named the National Nanotechnology Initiative (NNI) on research and development in various scientific disciplines related to human health became quickly noticeable. Extensive governmental financial support triggered the start of new and deepened already existing interdisciplinary research. Nanoscience quickly merged with medicine which resulted in the conceptualization of nanomedicine as a new scientific subdiscipline. Subsequently, nanoscience terminology was swiftly and, as I will explain in this chapter, largely uncritically adapted by pharmaceutical scientists leading to the advent of so-called nanopharmaceuticals. In general, the term "nano" became quickly associated with "cutting-edge", and the pharmaceutical science community devotedly embraced this new word. Colloidal systems were renamed nanosystems. Colloidal gold, a traditional alchemical preparation, became known as gold nanoparticles (GNPs), and colloidal drug delivery systems (Kreuter 1994) were sanctified as nanodrug delivery systems (Babu et al. 2014). The investigation of colloidal systems, i.e., systems containing nanometer-sized components, for biomedical applications, however, begun already more than 50 years ago, i.e., long before the launch of the NNI (Bangham et al. 1965a, b, c), and attempts to utilize colloidal (nowadays nano) particles for drug delivery date back 40 years (Marty et al. 1978). The perhaps best known example for such an early development of a nanopharmaceutical (but without using the term "nano" to describe it) is Doxil®, which is doxorubicin, a widely used anthracycline with anticancer activity, encapsulated in phospholipid vesicles commonly known as "liposomes". Efforts to utilize liposomes for reducing the cardiotoxicity of anthracyclines began at the end of the 1970s (Forssen 1981, Forssen and Tokes 1983). The 1980s saw a boom of investments into US start-up companies dedicated to the commercialization of liposome-based anthracycline formulations. Three companies, Vestar in Pasadena, CA, The Liposome Company in Princeton, NJ, and Liposome Technology Inc. in Menlo Park, CA, were competing with each other to bring liposomal anthracyclines into the clinic. In 1995, the FDA-approved Doxil® as the very first liposomal formulation of an already clinically used anticancer agent, i.e., doxorubicin. There was no mentioning or use of the term "nano". Actually, in the entire liposome literature comprising over 50,000 publications and books, the term "nano" was essentially absent until the year 2000. However, simply based on their size, which lies between about 50 and 200 nm for commercially used liposomes, they are nowadays known as nanolipid vesicles or nano-liposomes, and Doxil® was re-christened by one of his inventors as "the first FDA-approved nanodrug" (Barenholz 2012).

In two papers (Weissig et al. 2014, Weissig and Guzman-Villanueva 2015), we have recently evaluated all FDA-approved drug products (at that time, a total of 43, see Table 1.1, which had been publicized and referred to in at least one peer-reviewed publication or company press release as "nanopharmaceutical" or "nanodrug" or "nanomedicine".

We have argued (Weissig et al. 2014, Weissig and Guzman-Villanueva 2015) that for the vast majority of all

TABLE 1.1 Approved Drugs Commonly Referred to as "Nanopharmaceuticals"

Name	Description	Approval	Mechanism of Action	Indication
LIPOSOMES				
AmBisome®	Amphotericin B (AMB) encapsulated in small unilamellar liposomes (60–70 nm) composed of hydrogenated soy phosphatidylcholine, cholesterol and distearoylphosphatidylglycerol (2/0.8/1 molar) (Hiemenz and Walsh 1996)	FDA 1997	**MPS targeting**. Liposomes preferentially accumulate in organs of the mononuclear phagocyte system (MPS), as opposed to the kidney. Selective transfer of the drug from lipid complex to target fungal cell with minimal uptake into human cells has been postulated (Juliano et al. 1987a, b). Negative charge contributes to MPS targeting	Systemic fungal infections (i.v.)
DaunoXome®	Daunorubicin citrate encapsulated in liposomes (45 nm) composed of distearoylphosphatidylcholine and cholesterol (2/1 molar) (Eckardt et al. 1994, Guaglianone et al. 1994)	FDA 1996	**Passive targeting via enhanced permeability and retention (EPR) effect**. Concentration of the available liposomal drug in tumors exceeds that of the free drug, liposomal daunorubicin persists at high levels for several days. Distinct mechanisms for the localization of free and liposomal daunorubicin have been suggested (Forssen et al. 1996)	HIV-related Kaposi's sarcoma (i.v.)
DepoCyt®	Cytarabine encapsulated in multivesicular liposomes (20 μm)[a] made from dioleoyl lecithin, dipalmitoyl phosphatidylglycerol, cholesterol, and triolein (Kim et al. 1987)	FDA 1999/2007	**Sustained release**. Sustained-release formulation of cytarabine that maintains cytotoxic concentrations of the drug in the cerebrospinal fluid (CSF) for more than 14 days after a single 50-mg injection. (Glantz et al. 1999)	Lymphomatous malignant meningitis (i.t.)
DepoDur®	Morphine sulfate encapsulated in multivesicular liposomes (17–23 μm)[a] made from dioleoyl lecithinum cholesterol, dipalmitoyl phosphatidylglycerol, tricaprylin, and triolein	FDA 2004	**Sustained release**. After the administration into the epidural space, morphine sulfate is released from the multivesicular liposomes over an extended period of time (Alam and Hartrick 2005, Pasero and McCaffery 2005)	Pain reliever (administered into the epidural space)
Doxil®	Doxorubicin hydrochloride encapsulated in Stealth® liposomes (100 nm) composed of *N*-(carbonyl-methoxypolyethylene glycol 2000)-1,2-distearoyl-sn-glycero3-phosphoethanolamine sodium (MPEG-DSPE), fully hydrogenated soy phosphatidylcholine (HSPC), and cholesterol (Barenholz 2012)	FDA 1995	**Passive targeting via EPR effect**. Extravasation of liposomes via by passage of the vesicles through endothelial cell gaps, which are present in solid tumors. Enhanced accumulation of doxorubicin in lesions of AIDS-associated KS after administration of PEG-liposomal doxorubicin (Northfelt et al. 1996)	AIDS-related Kaposi's sarcoma, multiple myeloma, ovarian cancer (i.v.)
Inflexal®V	Influenza virus antigens (haemagglutinin, neuraminidase) on surface of 150 nm liposomes	Switzerland 1997	**Mimicking native antigen presentation**. Liposomes mimic the native virus structure, thus allowing for cellular entry and membrane-fusion (Herzog et al. 2009). Retention of the natural presentation of antigens on liposomal surface provides for high immunogenicity (Mischler and Metcalfe 2002, Bachmann and Jennings 2010)	Influenza vaccine
Marqibo®	Vincristine sulfate encapsulated in sphingomyelin/cholesterol (60/40, molar) 100 nm liposomes	FDA 2012	**Passive targeting via EPR effect**. Extravasation of liposomes through fenestra in bone marrow endothelium. No enhanced interaction with scavenger receptors on macrophages due to the absence of negative charge	Acute lymphoid leukemia, Philadelphia chromosome-negative, relapsed or progressed (i.v.)
Mepact™	Synthetic muramyl tripeptide-phoshatidylethanolamine (MTP-PE, mifamurtide) incorporated into large multilamellar liposomes (via reconstitution of dry powder) composed of 1-palmitoyl-2-oleoyl-sn-glycerol-3-phosphocholine and 1,2-dioleoyl-sn-glycero-3-phospho-L-serine (Frampton 2010)	Europe 2009	**MPS targeting**. Drug (an immune stimulant) anchored in negatively charged liposomal bilayer membrane via PE	Non-metastasizing resectable osteosarcoma (i.v.)
Myocet®	Doxorubicin encapsulated 180 nm oligolamellar liposomes composed of egg phosphatidylcholine/cholesterol (1/1, molar)	Europe 2000	**MPS targeting** (though not negatively charged). Forms "MPSdepot", slow release into blood circulation resembles prolonged infusion(Batist et al. 2002)	Metastatic breast cancer (i.v.)
Visudyne®	Verteporfin in liposomes made of dimyristoyl phosphatidylcholine and egg phosphatidylglycerol (negatively charged); lyophilized cake for reconstitution	FDA 2000	**Drug solubilization**. Rendering drug biocompatible and enhancing ease of administration after intravenous injection. No other apparent function of liposomes. Liposomal formulation instable in the presence of serum. Data of spin column chromatography indicated fast transfer of Verteporfin from Visudyne to lipoproteins(Ichikawa et al. 2004)	Photodynamic therapy of wet age-related macular degeneration, pathological myopia, ocular histoplasmosis syndrome (i.v.)

(*Continued*)

TABLE 1.1 (*Continued*) Approved Drugs Commonly Referred to as "Nanopharmaceuticals"

Name	Description	Approval	Mechanism of Action	Indication
Lipid-Based (Non-liposome) Nanoformulations				
Abelcet®	Amphotericin B complex 1:1 with DMPC and DMPG (7:3), >250 nm, ribbon like structures of a bilayered membrane (Hiemenz and Walsh 1996)	FDA 1995	**MPS targeting**. Selective transfer of the drug from lipid complex to target fungal cell with minimal uptake into human cells has been postulated (Janknegt et al. 1992, de Marie et al. 1994)	Systemic fungal infections (i.v.)
Amphotec®	Amphotericin B complex with cholesterylsulfate (1:1). Colloidal dispersion of disc-like particles, 122 nm × 4 nm (Hiemenz and Walsh 1996)	FDA 1996	**MPS targeting**	
PEGylated Proteins, Polypeptides, Aptamers				
Adagen®	PEGylated adenosine deaminase (Hershfield 1993)	FDA 1990	**Increased circulation time and reduced immunogenicity** PEGylation generally increases hydrodynamic radius, prolongs circulation and retention time, decreases proteolysis, decreases renal excretion and shields antigenic determinants from immune detection without obstructing the substrate-interaction site (Davis 2002, Wang et al. 2013)	Adenosine deaminase deficiency—severe combined immunodeficiency disease
Cimzia®	PEGylated antibody (Fab' fragment of a humanized anti-TNF-alpha antibody)	FDA 2008		Crohn's disease, rheumatoid arthritis
Neulasta®	PEGylated filgrastim (granulocyte colony-stimulating factor, G-CSF)	FDA 2002		Febrile neutropenia, in patients with non-myeloid malignancies; prophylaxis (s.c.)
Oncaspar®	PEGylated L-asparaginase	FDA 1994		Acute lymphoblastic leukemia
Pegasys®	PEGylated interferon alfa-2b	FDA 2002		Hepatitis B and C
PegIntron®	PEGylated interferon alfa-2b	FDA 2001		Hepatitis C
Somavert®	PEGylated human growth hormone receptor antagonist	FDA 2003		Acromegaly, second-line therapy
Macugen®	PEGylated anti-VEGF aptamer	FDA 2004		Intravitreal Neovascular age-related macular degeneration
Mircera®	PEGylated epoetin beta (erythropoietin receptor activator)	FDA 2007		Anemia associated with chronic renal failure in adults
Nanocrystals				
Emend®	Aprepitant as nanocrystal	FDA 2003	**Increased bioavailability due to increased dissolution rate.** Below about 1000 nm, the saturation solubility becomes a function of the particle size leading to an increased saturation solubility of nanocrystals which in turn increases the concentration gradient between gut lumen and blood and consequently the absorption by passive diffusion (Junghanns and Muller 2008)	Emesis, antiemetic (oral)
Megace ES®	Megestrol acetate as nanocrystal	FDA 2005		Anorexia, cachexia (oral)
Rapamune®	Rapamycin (sirolimus) as nanocrystals formulated in tablets	FDA 2002		Immunosuppressant (oral)
Tricor®	Fenofibrate as nanocrystals	FDA 2004		Hypercholesterolemia, hypertriglyceridemia (oral)
Triglide®	Fenofibrate as IDD-P (insoluble drug delivery microparticles)	FDA 2005		
Polymer-Based Nanoformulations				
Copaxone®	Polypeptide (average MW 6.4 kDa) composed of 4 amino acids (glatiramer)	FDA 1996/2014	No mechanism attributable to nanosize. Based on its resemblance to myelin basic protein, glatiramer is thought to divert as a "decoy" an autoimmune response against myelin	Multiple sclerosis (s.c.)

(*Continued*)

TABLE 1.1 (*Continued*) Approved Drugs Commonly Referred to as "Nanopharmaceuticals"

Name	Description	Approval	Mechanism of Action	Indication
Eligard®	Leuprolide acetate (synthetic GnRH or LH-RH analog) incorporated in nanoparticles composed of block copolymer D,L-lactide-glycolide	FDA 2002	**Sustained release** (Sartor 2003)	Advanced prostate cancer (s.c.)
Genexol®	Paclitaxel in 20–50 nm micelles (Oerlemans et al. 2010) composed of block copolymer poly(ethylene glycol)-poly(D,L-lactide)	South Korea 2001	**Passive targeting via EPR effect**.	Metastatic breast cancer, pancreatic cancer (i.v.)
Opaxio®	Paclitaxel covalently linked to solid nanoparticles composed of polyglutamate	FDA 2012	**Passive targeting via EPR effect.** Drug release inside solid tumor via enzymatic hydrolysis of polyglutamate	Glioblastoma
Renagel®	Cross-linked poly allylamine hydrochloride (Chertow et al. 1997) (sevelamer), MW variable	FDA 2000	No mechanism attributable to nano size. Phosphate binder	End stage renal disease—hemodialysis—hyperphosphatemia (oral)
Zinostatin stimalamer®	Conjugate protein or copolymer of styrene-maleic acid (SMA) and an antitumor protein neocarzinostatin (NCS) (Ishii et al. 2003). Synthesized by conjugation of one molecule of NCS and two molecules of poly(styrene-*co*-maleic acid)(Masuda and Maeda 1995)	Japan 1994	**Passive targeting via EPR effect (Greish et al. 2003)**	Primary unresectable hepatocellular carcinoma
Protein–Drug Conjugates				
Abraxane®	Nanoparticles (130 nm) formed by albumin with conjugated paclitaxel (2013, Saif 2013)	FDA 2005	**Passive targeting via EPR effect.** Dissociation into individual drug-bound albumin molecules, which may mediate endothelial transcytosis of paclitaxel via albumin-receptor (gp60)-mediated pathway (Desai et al. 2006, Foote 2007)	Metastatic breast cancer, non-small-cell lung cancer (i.v.)
Kadcyla®	Immunoconjugate. Monoclonal antibody (against human epidermal growth factor receptor-2)—drug (DM1, a cytotoxin acting on microtubule) conjugate, linked via thioether	FDA 2013	No mechanism attributable to nanosize	Metastatic breasts cancer
Ontak®	Recombinant fusion protein of fragment A of diphtheria toxin and subunit binding to interleukin-2 receptor	FDA 1994/2006	Fusion protein binds to interleukin-2 receptor, followed by receptor-mediated endocytosis, fragment A of diphtheria toxin then released into cytosol where it inhibits protein synthesis (Foss 2001)	Primary cutaneous T-cell lymphoma, CD25-positive, persistent or recurrent disease
Surfactant-Based Nanoformulations				
Fungizone®	Lyophilized powder of amphotericin B with added sodium deoxycholate. Forms upon reconstitution colloidal (micellar) dispersion	FDA 1997	**Drug solubilization**. Rendering drug biocompatible and enhancing ease of administration after intravenous injection. No other apparent function of micelles, which dissociate into monomers following dilution in circulation	Systemic fungal infections (i.v.)
Diprivan®	Oil-in-water emulsion of propofol in soybean oil/glycerol/egg lecithin	FDA 1989	**Drug solubilization**. Rendering drug biocompatible and enhancing ease of administration after intravenous injection	Sedative-hypnotic agent for induction and maintenance of anesthesia (i.v.)
Estrasorb™	Emulsion of estradiol in soybean oil, polysorbate 80, ethanol, and water	FDA 2003	**Drug solubilization.**	Hormone replacement therapy during menopause (transdermal)

(*Continued*)

TABLE 1.1 (*Continued*) Approved Drugs Commonly Referred to as "Nanopharmaceuticals"

Name	Description	Approval	Mechanism of Action	Indication
Metal-Based Nanoparticles				
Feridex®	Superparamagnetic iron oxide (SPIO) nanoparticles coated with dextran. Iron oxide core 4.8–5.6 nm, hydrodynamic diameter 80–150 nm	FDA 1996	**MPS targeting.** 80% taken up by liver and up to 10% by spleen within minutes of administration. Tumor tissues do not take up these particles and thus retain their native signal intensity (Wang et al. 2013)	Liver/spleen lesion MRI imaging (i.v.)
NanoTherm®	Aminosilane-coated superparamagnetic iron oxide 15 nm nanoparticles	Europe 2013	**Thermal ablation.** Injecting iron oxide nanoparticles exposed to alternating magnetic field causing the nanoparticles to oscillate, generating heat directly within the tumor tissue	Local ablation in glioblastoma, prostate, and pancreatic cancer (intratumoral)
Virosomes				
Gendicine®	Recombinant adenovirus expressing wildtype-p53 (rAd-p53)	China 2003	"... the adenoviral particle infects tumor target cells and delivers the adenovirus genome carrying the therapeutic p53 gene to the... nucleus... The expressed p53 gene appears to exert its antitumor activities". (Peng 2005)	Head and neck squamous cell carcinoma
Rexin-G®	Gene for dominant-negative mutant form of human cyclin G1, which blocks endogenous cyclin-G1 protein and thus stops cell cycle, inserted into retroviral core (replication-incompetent retrovirus) devoid of viral genes. About 100 nm particle	Philippines 2007	**Targeted Gene Therapy**. This retrovirus-derived particle targets specifically exposed collagen, which is a common histopathological property of metastatic tumor formation (Gordon and Hall 2010, 2011)	For all solid tumors

Source: Reprinted with permission from Weissig et al. (2014).

[a] Classifies as nanopharmaceutical based on its individual drug containing "chambers".

of these "nanodrugs", only the apparent size served as the basis for their classification as nanopharmaceuticals. We found that out of these 43 approved nanopharmaceuticals, 15 received FDA (or related foreign agency) approval before the year 2000. Further, assuming a development time of at least 10 years before a new drug gets approved for clinical use, it becomes evident that another 22 drugs in Table 1.1 were already under development long before the year 2000. Out of the total of the 43 "nanodrug" products listed in Table 1.1, only four have been approved for marketing after the year 2010. We concluded that "attributing the successful development of the fast majority of the products listed in Table 1.1 to the widely advertised and NNI-supported promotion of nanoscience and nanotechnology appears questionable" (Weissig et al. 2014), and we expressed our belief that "the undoubted promise of nanoscience and nanoengineering for the development of unique and highly efficient therapeutics has still to materialize" (Weissig et al. 2014). Before I will elaborate on any of the "nanopharmaceuticals" listed in Table 1.1, I shall explore in more detail what is actually so exciting or so special about any material at the nanoscale level.

1.2 What Is "Nano"? A Few Noteworthy Facts

The term "nano" is derived from the Greek νᾶνος (nânos), which translates into English as "dwarf". It is used as a prefix for metric units of time, mass and length. "Nano" denotes a factor of 10^{-9} or 0.000000001; i.e., one nanometer is one billionth of one meter, or one nanosecond is a billionth of one second. Such small dimensions are elusive to unaided human senses. Under optimal light conditions and at about 15 cm distant from the face, humans can distinguish between two objects only when they are separated by at least 26,000 nm. When closer to each other, these objects blur into one. The smallest objects a healthy human eye unaided by any magnifying tool can see are as small, or in this particular example as thin as a single human hair with a diameter of around 17,000 nm. A single sheet of paper is roughly 100,000 nm thick. Three single atoms of gold lined up would amount to 1 nm since a single gold atom has a diameter of roughly 0.3 nm. A single so-called low-molecular-weight compound is made up of molecules with a size of around 1 nm, while a macromolecule such as a protein can be 10 nm in size. Single cell bacteria have dimensions around several 1,000 nm. Human fingernails grow approximately 1 nm every second, which means (24 h × 60 min × 60 s) they grow roughly 86,400 nm in a day, certainly still too small for anyone to notice.

Within the dimensions of the "nanoworld" (namely between 1 and 100 nm) falls the size of adenosine triphosphate (ATP) synthase, a protein complex essential for the conversion of "food energy" into chemical energy in form of ATP. This protein complex can actually be thought of as a nature-built nanomachine.

Nanotechnology is all about the design and construction of man-made nanomachines, nanostructures and nanodevices. For this, two fundamentally different approaches are being explored. The so-called "bottom-up approach" involves the assembly of smaller components into larger structures but still within nanoscale dimensions. That's exactly what nature does. ATP synthase is a protein complex which nature assembles starting from single amino acids. On the other hand, the "top-down approach" starts with bulky material which is being scaled down to the nanometer range. Independent on which approach we are pursuing, we need to meet two major challenges. First, for manipulating matter on such a small scale, i.e., the nanoscale, we need appropriate tools. Any progress of nanotechnology is irrevocably associated with the design and building of appropriate tools enabling us to operate on the nanoscale. Second, we need to understand the laws of physics governing the nanoworld, which are very different from the laws of physics ruling our familiar and observable macroworld. These different laws of physics between the macro- and the nanoworld will be subject of the following section.

1.3 Unique Nanoscale Properties—Physics at the Nanoscale

This section is neither meant to be comprehensive nor meant to be written in depth for physicists or material scientist. It is intended mainly for pharmaceutical scientists and pharmacy graduate students, and only those unique nanoscale properties which might eventually be relevant for drug formulation development shall be considered.

1.3.1 Band Theory

Perhaps the most important difference between the macro- and the nanoworld involves the replacement of the classical Newton mechanics by quantum mechanics. The laws of motion discovered and described by Isaac Newton at the end of the 17th century fail to describe what happens to matter on the nanoscale. Newton's three laws of motion state the following: first, a body remains at rest or in uniform motion, i.e., with a constant speed and direction unless acted on by an external force; second, in response to that external force, a body accelerates proportional to the external force and inverse proportional to its mass; and finally, third, for each force, there exist an equal but opposite force. Based on Newton's laws, classical mechanics is being considered as "deterministic", which means when the position and velocity of a particle are known, one can calculate its past and future positions.

On the nanoscale and below, small matter obeys the laws of quantum mechanics, rather than the familiar laws of Newtonian mechanics that rules the macroscopic world. Simply put, Newton's laws of motion are useless for properly describing the movement of an electron around the nucleus. A single electron (or any other elementary particle) can be pictured as a "classical" particle or as a "classical" wave, a phenomenon known as "wave–particle duality". The discovery of the ambiguous behavior of elementary particles like electrons began towards the end of the 19th century when several experimental findings suggested that electromagnetic radiation can also behave like particles. The subsequent trailblazing work of Max Planck, Albert Einstein and Louis de Broglie at the beginning of the 20th century leads to the development of Erwin Schrodinger's fundamental wave equation for matter, the detailed discussion of which is beyond the scope of this chapter. In brief, this wave equation allows to interpret electromagnetic waves as being made up of particles, or the other way around, one can say that under certain circumstances particles, i.e., electrons can be wavy. Furthermore, we cannot know the exact position and velocity of an electron at the same time. Heisenberg's uncertainty principle states that the more precisely we determine the position of some particle, the less precisely we can measure its momentum, and vice versa. Therefore, in contrast to Newton's "deterministic" mechanics, quantum mechanics is considered as being "probabilistic". With respect to an electron Heisenberg's uncertainty principle, we can only calculate a certain probability, likelihood of detecting that electron somewhere around the atom's nucleus. All possible solutions of the corresponding mathematical functions form the basis for "atomic orbitals" (1s, 2p, 3d and 4f). When atoms combine to form molecules, their atomic orbitals combine to form molecular orbitals. The related mathematical function then can be used to calculate chemical and physical properties of that molecule as determined by its electrons. When we extend this concept to bulk material like a big piece of metal, the difference between successive molecule orbitals decreases, and as a result, their energy levels merge into a continuous band, a phenomena which gave rise to the development of the "Band Theory", which in essence postulates the existence of a "bonding band" at a lower energy level and a "antibonding band" at a higher energy level. Both are also called valence band and conducting band, respectively. The electrical conductivity of a metal depends on how easy electrons can transition from the valence band to the conducting band. In metals, both bands overlap. Upon the application of an electric field, electrons increase their energy, therefore easily transition into the conducting band and can thereby act as charge carriers. In semiconductors and insulators, there is a gap ("band gap") between those two bands. Insulators cannot conduct electrify since the energy for overcoming the difference between valence band and conducting band; that is the band gap is too big for any electrons to make the transition from the lower to the higher band. In semiconductors, however, energy provided by light absorption is sufficient to "excite" an electron, i.e., to elevate an electron form the valence band to the conduction band. As a result of this transition, the elevated electron leaves behind a "hole". The electron and the hole combined form a so-called "exciton". When the electron "falls back" from the conduction band into the hole on the valence band, the energy of that exciton is emitted as light, which is called fluorescence. When we now drastically decrease the size of the semiconductor (bulk) material by forming nanocrystals, its electronic and optical properties will significantly change and subsequently deviate from those of the bulk material as I discussed next.

1.3.2 Quantum Confinement

Reducing the dimension of a bulk crystal down to the nanoscale, energy levels become discrete; that is, to say, the quasi-continuous bands of the large macroscopic crystal disappear, and the differences between the energy levels increase. For semiconductors, this means that the band gap increases as the size of the nanocrystal decreases. This unique phenomenon appearing on the nanoscale is called quantum confinement, which generally can be observed once the diameter of a material is of the same magnitude as the de Broglie wavelength of the electron wave function. For most materials, quantum confinement effects appear at dimensions below 10 nm. Quantum confinement subsequently allows the manipulation of the optoelectronic properties of material such as the band gap by controlling the size of that material, though shape and material composition also play an important role. Essentially, the photoluminescence of nanocrystals, i.e., the absorption and emission wavelengths can be tuned to whichever values by choosing a certain particle diameter. Quantum dots with a radius above 4–5 nm, for example, emit light with a longer wavelength making them appear orange or red, while decreasing the diameter down to around 2–3 nm results in the formation of blue or green appearing quantum dots. Simply put, the larger the nanocrystal, the redder its absorption onset and fluorescence spectrum, and contrariwise, the smaller the bluer. These unique and above all controllable photophysical properties of quantum dots (or semiconductor nanocrystals) are the basis for currently ongoing revolutionary developments in the broad area of biomedical science, and specific diagnostic and therapeutic applications thereof are under preclinical development.

1.3.3 Localized Surface Plasmon Resonance

While quantum confinement is a phenomenon of matter at the nanoscale, the exploration of which has just begun. Localized surface plasmon resonance is a nanoscale property which has been utilized by mankind for many centuries, though, of course, unknowingly so. Namely, all the beautiful colors in stained glass windows from medial times are being caused by localized surface plasmon resonance of gold and silver nanoparticles. Noble metals like gold change their appearance, i.e., color, when we dramatically reduce their or size. Putting a golden-shiny piece of gold metal in water and cutting it down to smaller and smaller pieces until we reach the nanoscale, we end up with a red appearing solution of GNPs, during the last century better known as "colloidal gold". The first who actually prepared colloidal gold was the German physician and alchemist Paracelsus (1493–1541). He called this purple-reddish solution of gold Aurum Potabile, which is Latin and means "drinkable gold", and he advocated its assumed healing power for a variety of physical and mental ailments. Not surprisingly, Paracelsus is being credited as the found of toxicology. The fact that gold possesses biological activity earned Robert Koch (German physician, 1843–1910) the Nobel Prize for having demonstrated that compounds made with gold are able to inhibit the growth of the bacillus that causes tuberculosis. In present days, the antibacterial and antiviral activity of colloidal silver or silver nanoparticles has found many applications. So, how do we explain the red color of colloidal gold? Let's go back to the earlier discussed Band Theory. Metallic gold as an electrical conductor has "free, movable" electrons in the conduction band. These electrons can be considered as delocalized since they extend outside of just one atom; that is, they are populating the valence/conduction band. When irradiated with light, i.e., exposed to electromagnetic waves with certain frequencies of oscillation and wave lengths, these free conduction band electrons collectively start to oscillate relative to the stationary positions of the positively charged atoms in the metal lattice (the term "lattice" refers to the orderly arrangement/position of all positively charged atoms/ions in the bulk metal). We can say that we have a wave of fluctuating electron density in the metal lattice. Such collective oscillation of "free" electrons or of an "electron cloud" in conducting material is called a "plasmon", which is more precisely defined as a quasiparticle associated with a collective oscillation of charge carriers (in our case electrons). Or in other words, the collective excitation of electrons in the conduction band of bulk metal is called "plasmon oscillation". Upon exposure to light, the wave of oscillating electrons propagates along the surface of that conducting bulk material. This interaction between light and electron density, or between light and surface plasmons, forms the basis for a large variety of potential revolutionary new technologies. Plasmonics actually has emerged as a whole new scientific subdiscipline of physics. There is broad area in which the utilization of surface plasmons has already begun to significantly improve existing technology. Surface plasmons have been found to be extremely sensitive to any substance in close proximity to the conducting band, i.e., to the surface of the metal or the metal film. Very low amounts of molecules, proteins, viruses, or bacteria which are getting in contact with the thin metal surface, in many cases a thin film of gold, change the properties of its plasmons, thereby allowing to sense and quantify extremely low concentrations of any agent "touching" the gold surface which is conducting or propagating surface plasmons. Going back to the question where colloidal gold gets its red color from: irradiation of a gold surface with light causes the oscillation of electron density in the conducting band, called plasmon oscillation. When we now reduce the size of the bulk material, i.e., the large piece of gold metal until we get particles which are in diameter smaller than the wavelength of the light shining upon them (as a reminder, the wavelength of visible light ranges from 390 to 700 nm), the free conduction band electrons literally become confined. As a result, the oscillating electrons or surface plasmons cannot propagate anymore along the surface of the bulk metal. Instead, the electron cloud becomes polarized on one side of the particle and oscillates in resonance with the frequency of the incident light. Light consists of a continuous range of frequencies or wavelengths (as a reminder, frequency and wavelength of an electromagnetic wave are linked by the equation $c = \lambda\nu$, where c is the speed of light which is a constant, λ the wave length and v the frequency). In order for the electron cloud to oscillate, it has to be excited with light of the proper wavelength. This means the "natural intrinsic frequency" of the electron cloud's oscillation has to resonate with the frequency of the incoming light (as a reminder, resonance is the increase in amplitude of a system's oscillation when exposed to an external oscillation with equal frequency). When such resonance happens, the light waves with that resonance frequency or wavelength are being selectively absorbed from the incident light by the particle and never again to be released as light, while all other remaining light waves are being reflected. The selective absorption of light by a specific GNP (or other noble metal particles) occurs because the selected frequency of the light wave matches the "natural intrinsic frequency" at which the electron cloud of that nanoparticle is able to oscillate. Now, the "natural intrinsic frequency" of the electron cloud at the surface of a GNP, i.e., of plasmons, depends strongly on the diameter of the particle, its shape and its immediate environment, and in particular on the dielectric constant of that environment. Consequently, varying the size and/or shape allows the design of nanoparticles with tailored optical properties suited for different applications. For example, spherical GNPs with a diameter of around 30 nm absorb light in the blue-green portion of the electromagnetic spectrum, which is around 450–500 nm. In other words, the plasmons localized (confined) at the surface of that nanoparticle are able to oscillate with a frequency $v = c/\lambda$, with λ between 450 and 500 nm. In other words, when these localized surface plasmons find a "match" with an incoming oscillation of equal frequency, we get what we call localized surface plasmon resonance (LSPR). The light reflected from the above 30 nm GNP is now missing its blue-green portion; i.e., it is a mixture of all of the wavelengths of light that are not absorbed by the nanoparticle, and therefore, we see an orange-reddish solution of GNPs. When we increase the particle size, the light being absorbed by the surface plasmon oscillation shifts to longer wavelengths; i.e., reddish light is then being absorbed and consequently, the color of the colloid particle solution appears pale blue or purple.

1.3.4 Surface Area-to-Volume Ratio at the Nanoscale

Clinically approved drugs like Emend®, Rapamune® and Tricor® (see Table 1.1) are active drug agents (Aprepitant, rapamycin and Fenofibrate, respectively) which have been formulated as nanocrystals. Such "nanocrystal drugs" possess a significantly increased bioavailability due to their increased dissolution rate, which in turn increases the concentration gradient between gut lumen and blood and consequently the absorption by passive diffusion (Junghanns and Muller 2008). Please note that these drugs have been developed for oral administration. If we were to inject, for example, rapamycin and Fenofibrate intravenously, a monomolecular solution of these drugs would has been the preferred dosage form. The unique nanoscale property utilized for the development of nanocrystal drugs is the exponential increase of the surface-to-volume ratio at the nanoscale. Let's imagine a cube with a volume of 1 m^3 and a surface area of 6 m^2. Cutting this cube down into eight smaller cubes increases the surface area to $8 \times (6 \times 0.5 \times 0.5) = 12$ m^2, while the total volume of all eight cubes combined obviously remains constant at 1 m^3. Cutting these 8 cubes down to 64 cubes further increases the total surface to 24 m^2, but which again does not change the total volume. Generally, decreasing the particle size of any solid material not only "gradually", i.e., in a linear way, but exponentially increases the surface-to-volume ratio. The immediate consequence of the increase of the surface-to-volume ratio is that more and more atoms or molecules of that particle are being exposed to the outer surface versus being "hidden" inside the particle. In other words, with the decrease of the particle size, the fraction of atoms/molecules on the particle surface increases exponentially, while the fraction of so-called bulk atoms/molecules exponentially decreases. For example, about 15% of all atoms in a 10 nm spherical particle occupy "surface" positions, while in a particle a 100 times that size, i.e., in a 1,000 nm particle, only 0.0015% of all atoms are exposed to the surface. Such differences between bulk material and nanoparticles have significant impact on all material properties which are mostly "surface driven". Surface-dependent properties of nanoparticles are fundamentally different from the corresponding bulk material. Reactivity, solubility, affinity, melting point, vapor pressure, and more are all material properties which significantly change with a decrease in the particle size. The key for understanding the fast differences in surface-dependent properties between bulk and nanosized material lies in understanding the phenomenon of surface energy, to be discussed next.

1.3.5 Surface Energy at the Nanoscale

When discussing any kind of specific features of nanoparticles, surface energy is one of the key properties to be considered. Surface energy is always positive, it is like additional energy, i.e. surplus energy which all atoms positioned at the surface possess. Where is this extra energy coming from? By definition, surface energy is the energy required to create a unit area of "new" surface. Simply imagine cutting cheese. You have to push the knife through the cheese, which means you have to apply energy. In doing so, you create two new surfaces, i.e., the surfaces of the two new slices of cheese. Why do you need your muscle power, i.e., energy to do that? Because you break bonds between the atoms you are separating with your knife. It does not matter whether the bonding is covalent as in metals, ionic as in solid salt or noncovalent as in any liquid. As a result, these atoms are now under-coordinated, and they have tangling bonds which originally connected them with the atoms you were separating them from with your knife. Because of those free, unengaged bonds, surface atoms or molecules are "trying to find companionship" inside the material, as if an inwardly directed force acts on them. As result, the distance between the surface atoms and atoms next to them at the subsurface is smaller than that between all atoms in the interior of that material. Obviously, such shortened distance between surface and subsurface atoms in a piece of metal, i.e., in bulk material matter, hardly matters. But when we significantly decrease the size of the material and thereby exponentially increase the ratio between surface and interior atoms or molecules, such differences in distance turn significant and become the basis for new and unique properties of such nanosized particle. In the case of solid material, we use the term "lattice parameters", which generally refers to the physical dimensions of single unit cell within a crystal lattice. Quite obviously, three-dimensional lattices have three lattice constants, usually named a, b and c, and all of them are being affected when downsizing the bulk material to a nanoparticle. In other words, due to the increased surface energy, due to the dangling bonds of the surface atoms and due to the shortened distance between surface and subsurface atoms, the whole nanolattice in a nanoparticle becomes kind of compressed, since an unproportionally large portion of all atoms are actually surface atoms.

1.3.6 Melting Point Depression of Solid Materials at the Nanoscale

In the macroworld, melting points are considered a constant. Most of the metals in bulk form melt at temperatures above 1,000°C. For example, a one ounce gold bar melts at 1,064°C (1,948°F). A sand-sized piece of gold of approximately 1 mm in diameter still would melt at exactly 1,064°C. When further reducing the size of this sand-sized grain of gold down into the micrometer range, i.e., into the size range of human cells, the melting point starts to change very slightly, a cell-sized "piece" of gold would melt near to 1,064°C. However, from here on, when going further down the size scale, the effect of size on the melting point becomes dramatic with an about 1 nm gold particle having a melting point of approximately only 20°C; i.e., it would melt already at room temperature.

Such major melting point decrease of solid materials at the nanoscale has significant practical consequences. Likewise, turning water-insoluble drug particles into nanoparticles can improve their pharmacokinetics based on the increased dissolution velocity nanoparticles possess in comparison with microparticles. Obviously, any melting point depression going along with size reduction may require practical considerations with respect to the manufacturing process and storage conditions for the final drug product.

So, why do nanoparticles melt at lower temperatures as larger particles of the same material? When a solid melts, the bond between atoms or molecules of that solid material has to be dissolved. The energy needed to dissolve that bond is equal to the binding or cohesive energy. Per definition, the cohesive energy of any substance (this applies to solids and liquids) is the energy required to break all bonds associated with an atom or molecule of that substance. It is the energy needed to liberate a single component (atom or molecule) from interacting with its neighboring components. Each bond, be it covalent or ionic, an atom shares with its neighbors contributes cohesive energy. As I discussed earlier, nanoparticles have a significantly larger surface-to-volume ratio than bulk materials. This means that the ratio of surface atoms or molecules to bulk atoms or molecules increases. The more surface atoms we have (in comparison with bulk atoms), the more the physical and chemical properties of nanoparticles are determined by surface atoms and not by the bulk atoms. Since surface atoms interact with fewer neighboring atoms, they bind to the solid phase with a lower cohesive energy. Subsequently we need less energy to overcome the cohesive energy, and as a result, this nanomaterial melts at a lower temperature.

1.3.7 Brownian Motion

When looking at the surface of a lake at a very calm day, the water appears to be at a total standstill. Yet, if you were able to follow the movement of single water molecules, your head would spin. At room temperature, the average speed of water molecules has been measured to be about 300 m/s equal to 671 mph. In a very irregular way, all water molecules collide with each other at a very high speed and repel each other, thereby causing each other to change their path. Any large macroscopic object, like a boat floating on water, is also subject to this "high-speed bombardment" with water molecules, yet due to the enormous size difference between a single water molecule and a boat, the effect on that boat is not detectable, which means the boat is not being moved at all, and only some wind or a current could accomplish that. Now, let's make that boat smaller and smaller, let's create a "nanoboat". Would such a tiny boat also be resistant against the high-speed bombardment with water molecules? That's obviously not the case. What would happen to our nanoboat was observed for the first time by Robert Brown, a Scottish botanist in 1827. When Brown examined plant pollen grains suspended in water under a microscope, he saw tiny particles moving continuously and in a quite erratic fashion. We know today that what he saw were amyloplasts, membrane-bound vesicles containing starch grains, which have been ejected from the pollen grains he put into the water. We also know today that with an average size of 1–2 μm, amyloplasts apparently are small enough to be effected by the high-speed bombardment with water molecules. This phenomenon Brown observed two centuries ago is now known as "Brownian motion" or "pedesis" (from ancient Greek meaning "leaping"). It is defined as the random motion of particles in a liquid or a gas which is driven by the particle's collision with fast moving atoms or molecules of the liquid or gas the particle is suspended in. Nitrogen molecules have an approximate diameter of 0.3 nm, and at room temperature, they move around in the air at a speed of about 300 m/s (671 mph). Their sporadic and extremely fast movement affects particles up to 1 or 2 μm in size, but not anymore any particle larger than that. Dust particles in the air caught by a sunbeam or even pollen grains which are between 6 and 100 μm in diameter are already too big to exhibit Brownian motion. So, let's imagine a nitrogen molecule of 0.3 nm in size hits a solid object small enough to show Brownian motion, which requires that object to be around or smaller than 1,000 nm in size. Proportionally upscaling this situation, it would be the same as if a Boeing 777 with a length of 73.9 m and a wingspan of 60.9 m sitting on the ground would be constantly bombarded from all directions with a tremendous amount of spherical objects (nitrogen molecules) approximately 1 in. in diameter with a speed of 671 mph. Would you even think about getting onto this plane?

This situation illustrates the problems molecular nanotechnology, where mechanical engineering moves from the macroworld into the nanoworld, is facing. Any design and construction of mechanical systems like motors, gears and so forth has to consider Brownian motion, since none of these nanomachines is intended to function in a vacuum, and they all are exposed to the "high-speed bombardment" by atoms or molecules making up their environment. Reading about nanomachines, for sure Isaac Asimov's "Fantastic Voyage" comes to mind. A nanosized submarine able to navigate through the entire circulatory system inside a human body detecting and treating any kind of damages or intruders might be the holy grail of medical nanotechnology. But even if one could construct such a "Nanosub", what would be its chances to stay on course? As Georg Whitesides eloquently writes: "Any effort to steer (such Nanosub) a purposeful course would be frustrated by the relentless collisions with rapidly moving water molecules". Navigators on the nanoscale would have to accommodate their Nanosub to the "Brownian storm that would crash against their hulls" (Whitesides 2001).

1.3.8 Superparamagnetism

Superparamagnetic nanoparticles (NanoTherm®, Table 1.1) are able to produce a lot of heat when exposed to a relatively weak external electromagnetic field. Injecting such iron oxide nanoparticles into a tumor tissue followed by exposing the tumor to an alternating magnetic field causes the nanoparticles to generate heat directly within the tumor tissue leading to its demise, a medical procedure called thermal tumor ablation. Since superparamagnetism is one the few unique nanoscale properties which makes, for example, NanoTherm® a "true" nanopharmaceutical (more about the definition of "true" nanopharmaceutical below), I am now going to explain the property of superparamagnetism in more detail.

Magnetism is a basic property of all matter associated with the movement of electric charges. Any wire conducting an electric current, independent on whether it is a direct current (DC) or alternating current, (AC) is surrounded by an invisible field called a magnetic field which exerts an unseen force on any object exposed to it. It is said that in 1820, during an evening lecture about electricity, the Danish scientist Hans Oersted discovered by chance that a compass needle was deflected from its position once a wire carrying electric current came close to it. Turning off the current but keeping the wire in place allowed the compass needle to return back to its original position. You can also send electric currents in the same direction through two parallel wires, and you will see both wires are attracted to each other; they will move towards each other as if driven by an invisible force. Oersted's discovery triggered the development of the theory of electromagnetism, which in general describes the production of a magnetic field by an electric current flowing through a conductor. If a DC is sent through a wire, the strength of the induced magnetic field around that wire remains constant; in the case of AC, the produced electromagnetic field is periodically growing and shrinking due to the constantly changing current in the wire.

The motion of charges creating such a magnetic field can take many forms. As in the above example, it is the movement of an electric current through a conductor, but it can also be the movement of charged particles through space. The magnetic field of our own planet is thought to be caused by the flow of the liquid iron in the outer layer of earth's core, thereby producing an electric current which in turn generates a magnetic field encompassing the entire planet. On the atomic level, it is the movement of electrons which is responsible for magnetic properties of all matter. Individual electrons, in very simple terms, rotate around themselves; i.e., they possess a so-called "spin", and in addition, they orbit around the nucleus, which means they possess a so-called "orbital angular momentum". A "momentum" is a vector which means it has a direction and a magnitude. The velocity of the car you are driving is a vector, and it describes how fast your car is driving and in what direction you are going. "Angular momentum" means the movement follows a circle around a center, which is the nucleus in case of the electron's orbital angular momentum. The spin of the electron around itself and the orbiting of the electron around the nucleus combined are described as the "total angular momentum". You can picture such momentum as a real physical force. The angular momentum of a spinning top, for example, keeps it standing on its tip as long as it is spinning. Without the existence of an angular momentum, figure skaters wouldn't be able to make their high-speed rotations around themselves called pirouettes, and bicycling would plainly be impossible.

Back to the electron, the total angular momentum of all electrons combined accounts for the total magnetic moment of an individual atom, which possesses a specific direction. The momenta of paired electrons with opposite spins cancel each other out. The oxygen molecule has two unpaired electrons in its outer shell which "spin in the same direction"; subsequently, molecular oxygen has magnetic properties. The direction of any magnetic moment is said to point from the "south" to the "north pole" of the magnet. Subsequently, any magnetic moment is also being referred to as a magnetic dipole moment. You can imagine such a magnetic dipole as a tiny bar magnet.

In bulk material, like in a piece of metal, all atoms are arranged in a lattice structure and each of the atoms occupying a lattice position possesses a magnetic moment, i.e., represents a magnetic dipole. However, of uttermost importance is the direction of each individual dipole. Putting only a limited number of atoms together, the likelihood that all magnetic dipole moments are aligned with each other, i.e., point into the same direction is very high; we call this group of atoms a magnetic domain. By definition, a magnetic domain is a region in solid material and in particular in magnetic material in which all individual magnetic moments are aligned with each other, therefore pointing into the same direction. Depending on the material, particles with a size below about 50 nm are usually single-domain particles. Larger particles, i.e. bulk material, possess multiple domains, the magnetic dipoles of which point into random directions, consequently canceling each other out.

When exposing magnetic dipoles to an external magnetic field, they are subjected to a torque which can lead to their alignment with the magnetic moment of the external magnetic field, like the needle of a compass which aligns with the magnetic field of the earth. The property of matter to respond to an external magnetic field is called magnetic susceptibility. The magnetic susceptibility depends on the size of the individual magnetic dipole moments, i.e., on the magnetic dipole moments of all atoms in that material, and second, it depends on the extent all dipole moments are aligned relative to each other. Remember, each individual domain possesses a magnetic dipole moment which produces a magnetic field around that domain. Under normal circumstances, i.e. in the absence of an external magnet, these magnetic fields cancel each other out. But once the dipole moments of all domains are aligned with an external magnetic field pointing into the same direction, a large magnetic field is being produced. In other words, the external magnetic field has magnetized, the material which was placed into it.

When the direction of the induced magnetic field, i.e., the field of the material placed into the external field, points into the same direction as the external field, the material is being attracted in the direction of that external field. According to what happens to the induced magnetic field when the external or "driving" field is being switched to zero, we distinguish between paramagnetic and ferromagnetic materials.

In paramagnetic materials, removal of the external field leads to the entire loss of the induced magnetism, while in ferromagnetic materials, the induced magnetic field persists, though at a lower strength. Such remaining magnetism is called remanence; it is defined as the remaining magnetization after the external or "driving" field goes back to zero. The phenomenon of remanence in ferromagnetic materials is based on the persistent alignment of individual domains with each other. One can picture this as a cross-talk between individual domains. The magnetic fields of neighboring individual domains force each other to stay pointed into the same direction. Consequently, ferromagnetism is linked to the presence of multiple domains.

In order to eradicate the remaining magnetic field, i.e., to eliminate the remanence, the ferromagnetic material has to be exposed to an external magnetic field with opposite direction. The strength of such reverse field needed to drive back the magnetization to zero is called coercivity. Increasing further the strength of the reverse external field leads again to magnetization until all domains are aligned, which is called saturation point. No matter how much further we increase the strength of the driving magnetic field, we cannot increase the strength of the induced magnetic field, i.e., the field of the ferromagnetic material exposed to the external driving field. What we can do is to switch again the direction of the driving field. Now the domains start re-aligning again into the other direction. When the external field reaches zero, we have again remanence but of opposite direction as we had in the beginning. Further increasing the external field eventually will lead again to a saturation point.

Increasing the temperature literally agitates all domains leading to the loss of their identical alignment which in turn results in the disappearance of those materials' magnetic properties. The temperature at which ferromagnetic materials, i.e., permanent magnets lose their magnetism, is the so-called Curie temperature named after Pierre Curie who showed that magnetism can be destroyed by increasing the material's temperature. It is not wrong to say that above the Curie temperature ferromagnetic materials acquire paramagnetic properties. When the external magnetic field induces a magnetic field in materials with a direction opposite to the external field, those materials are being repelled by the applied field. Such materials are called diamagnetic materials.

So, what is superparamagnetism? How is superparamagnetism different from ferromagnetism and paramagnetism? Let's take a piece of ferromagnetic material like a large particle of iron which possesses countless magnetic domains largely aligned with each other and reduce its size dramatically. At a certain point, one can easily imagine ending up with only one single domain per particle. This happens when we reduce the size of the iron particle down to below 50 nm. Remember, ferromagnetism is based on the mutual impact all domains have on each other. The magnetic field of each domain forces the magnetic field of its neighboring domain to stay aligned. Having now only one domain left, such interaction between domains obviously is not possible anymore. Subsequently, ferromagnetism disappears and transitions into what we call superparamagnetism. Exposing such a single-domain particle to an external field causes its magnetic moment (the sum of all total moments of all atoms composing that particle) to align with the direction of the external field, which means it acts like a paramagnetic particle, and it becomes magnetized. Since each particle possesses only one single magnetic domain, and in other words, since these individual domains are literally on their own, removal of the external field results in the loss of the induced magnetism. While a large piece of iron, i.e., ferromagnetic material, stays magnetic after the removal of the external field, i.e., it "remembers" the external field due to individual domains forcing each other to stay aligned, the small single-domain particle (depending on the temperature) quickly loses its original magnetic dipole orientation. Therefore, one can say the nanosized single-domain particle derived from a larger ferromagnetic particle has lost its magnetic memory.

We discussed earlier that above a certain temperature, the Curie temperature, the synchronous alignment of individual domains in ferromagnetic material gets lost, which means its magnetic properties are destroyed. Since superparamagnetic particles have only one domain, the concept of the Curie temperature or Curie point obviously does not apply. Nevertheless, in the absence of an external magnetic field and still depending on the temperature, the direction of the dipole of the single magnetic domain can randomly flip between different orientations. The time between such flips is called the Neel relaxation time. Now imagine we suspend a large number of single-domain magnetic particles, i.e., superparamagnetic nanoparticles in a liquid medium, and apply an external electromagnetic field produced by a DC. Two things will happen: first, the magnetic moments of all particles will align with the moment of the external field (something you cannot see with your eyes), and second, the particles themselves will line up in the direction of the external field (something you can easily observe provided you have either a microscope with sufficient magnification or a large enough concentration of nanoparticles in that suspension). Now, turn the current off, and all nanoparticles literally will "relax". First, all individual magnetic moments will flip into random directions, since they are not forced anymore by the external field to align into the same direction. This is the so-called (already introduced above) Neel relaxation. Second, the nanoparticles themselves are not forced anymore to align with the external field, and they can now freely move in the medium they are suspended in, only exposed to the "high-speed bombardment" by atoms or molecules making up their environment. Rightly so, this is called Brown relaxation.

Let's now repeat the same scenario using an electromagnetic field induced by an AC. Each time the direction of the current switches, the external magnetic field loses its strength into one direction and builds up its strength again in the opposite direction. At the same time, all nanoparticles exposed to the alternate magnetic field "briefly relax" and are then immediately forced to align their magnetic moments as well as their physical localization in the medium along with the new direction of the external magnetic field. All nanoparticles are magnetized, loose their magnetism and are being magnetized again in the opposite direction. Likewise, they physically move into one direction, loose this directed movement and immediately are being moved into the opposite direction. At this point, it should be mentioned that the smaller the particle is, the less important is the physical movement of that particle. Or we say, Neel relaxation dominates over Brown relaxation in small enough nanoparticles, while Brown relaxation may prevail in larger nanoparticles. The exact size at which one or the other relaxation regime dominates depends on the nature of the material the nanoparticle is made of, the shape of the nanoparticle and its surface characteristics, as well as the nature of the medium and the concentration of the nanoparticles in that medium.

Now, let's look at our scenario, i.e., at our magnetic nanoparticles suspended in a liquid medium and exposed to an external field from the perspective of energy conversion. Whether our nanoparticles are single-domain or multi-domain particles doesn't matter yet. In our setup, electric energy of the AC is transformed into electromagnetic energy of the magnet producing the external field which in turn is transformed into magnetic energy of the nanoparticles. Energy transformations never occur with 100% efficiency. When converting one particular energy form into another one, a certain amount of that energy is always lost, i.e., dissipated as heat energy. Upon combustion, chemical energy of gasoline is converted to kinetic energy of expanding gas which in turn is converted into the kinetic energy of the moving piston. If that energy conversion would be 100% efficient, we wouldn't need a cooling system in our cars, which is necessary to handle the "lost" energy that is dissipated as heat energy. The general inefficiency of all energy conversions is also the reason for magnetic particles to heat up when exposed to an AC electromagnetic field. Energy is dissipated as heat energy during the repeated magnetization reversal the nanoparticles go through. In particular, heat is generated in any magnetic particle exposed to an alternating magnetic field through hysteresis loss, Neel relaxation loss and Brown relaxation loss. Hysteresis losses can only occur in ferromagnetic multi-domain magnetic particles. Neel relaxation dominates increasingly over Brown relaxation with decreasing particle size. The energy amount (of the AC electromagnetic driving field) which is converted into heat energy (i.e. heats up the magnetic nanoparticles) is described by the term "specific absorption rate (SAR)". Many particle parameters determine the value of SAR, like size, shape, chemical nature of the particle, surface characteristics as well as the possible highest magnetization degree achievable, i.e., the saturation magnetization. However, most important of course is the frequency and amplitude of the external driving AC electromagnetic field and here lays the major difference between ferromagnetic (multi-domain) and superparamagnetic (single-domain) particles. To maximally utilize the heat generated by hysteresis losses of multi-domain particles, i.e., to push the particle repeatedly through the entire hysteresis loop, a very high external magnetic field is required which in turn poses significant technical and physiological restrictions for any potential medical applications. Subsequently, multi-domain magnetic particles possess only low SAR values. In contrast, single-domain superparamagnetic nanoparticles generate heat mainly via Neel relaxation. Consequently, superparamagnetic nanoparticles potentially can generate more heat (higher SAR) at much lower field strength in comparison with ferromagnetic multi-domain particles. Exactly, this feature of superparamagnetic nanoparticles, their ability to produce a lot of heat when exposed to a relatively weak external electromagnetic field, makes them so interesting for medical applications (see NanoTherm®, Table 1.1).

1.4 What are "True" Nanopharmaceuticals?

In its NNI, the NIH defines nanotechnology as "the understanding and control of matter at dimensions between approximately 1 and 100 nanometers (nm), where unique phenomena enable novel applications not feasible when working with bulk materials or even with single atoms or molecules". In my interpretation, the emphasis in this definition lies on the unique phenomena directly associated with this particular size range. As I elaborated on in detail above, nanosized materials can display unique and distinctive physical, chemical, and biological properties. Appropriately, Schug et al. (2013) stated that *a* "true nanomaterial … possess properties, which neither the bulk material nor the atoms or molecules of that same material display. In extension, engineered nanomaterials (ENMs) of near infinitive variety of sizes and shapes produced from almost any chemical substance have the potential of exhibiting unique optical, electrical and magnetic properties". Logically, therefore, Rivera et al. proposed that "true" nanopharmaceuticals should be defined as "pharmaceuticals engineered on the nanoscale, i.e. pharmaceuticals where the nanomaterial plays the pivotal therapeutic role or adds additional functionality to the previous compound" (Rivera Gil et al. 2010). In accordance, we have suggested that any drug product has to meet two major criteria in order to be classified as a nanopharmaceutical: "First, nanoengineering has to play a major role in the manufacturing process. Second, the nanomaterial used has to be either essential for the therapeutic activity or has to confer additional and unique properties to the active drug entity" (Weissig et al. 2014).

Applying this strict definition to all of the drug formulations listed in Table 1.1, it should become obvious that the majority actually might not "deserve" the label "nano" even if indeed all of these drug formulations are truly "nanosized". To paraphrase a statement by the NIH in its NNI, simply working at ever smaller dimensions does not constitute nanotechnology, but utilizing the unique physical chemical, mechanical, and optical properties of materials present at that scale does. Nevertheless, and this shall be the subject of the last section in this chapter, already the nanosize alone confers to the drug formulations listed in Table 1.1 properties which are of significant therapeutic benefit.

1.5 Impact of the "Nanosize" of a Drug Formulation on the Therapeutic Effect of the Drug

Going back to the 43 listed drug formulations in Table 1.1, with the exceptions of nanocrystals and of metal-based nanoformulations (both of which I have discussed above under Sections 3.4 and 3.8., respectively, as "true" nanopharmaceuticals), most of all other drug preparations are either therapeutically active low-molecular-weight compounds formulated into a nanosized dosage forms (e.g., liposomes, surfactant-based nanoformulations, polymer-based nanoformulations, protein–drug conjugates) or biologically active macromolecules which have undergone a simple chemical modification reactions (e.g. PEGylated proteins, polypeptides and aptamers). Both strategies, i.e., the incorporation of small drug molecules into a larger, but still nanosized carriers and the chemical modification of, for example, proteins with polyethylene glycol (PEG) chains significantly, alter the pharmacokinetics (PK) of the low-molecular-weight drug or of the therapeutic protein, thereby, which is the ultimate goal, improving the drug's therapeutic efficiency. All basic pharmacologic properties of a drug molecule may be affected: the drug's absorption (A) may be enhanced (mainly for orally administered drugs); the distribution (D) of the drug within the human body, for example, after intravenous administration can be significantly altered; the metabolism (M) of the drug may be slowed down, altered or even prevented; and consequently the drug's elimination (E) from the human body via the intestinal or renal route may be changed. From Table 1.1, it can be seen that specific nanoformulations affect a drug's ADME properties in different ways, see "Mechanism of Action", which does not refer to the pharmacologic activity of the active drug molecule but to the impact the drug's nanoformulation has on of drug's ADME properties. In the following, I shall briefly discuss a generally recognized mechanism by which a nanoformulation of a drug can change its ADME properties, even independent on the specific nature of the nanoformulation. Any drug molecule, once it has entered the blood circulatory system, is potentially able to extravasate, i.e., to cross the endothelial cell linen in any tissue or organ, depending on the drug's physico-chemical properties (mainly its lipo-or hydrophilicity). Subsequently, the drug can act inside the human body anywhere where there is an appropriate target, i.e., a fitting drug receptor. This in turn is the main reason for side effects. Best examples are anticancer drugs. A drug designed to inhibit the replication of DNA does not distinguish between the DNA in a healthy cell and the DNA in a cancer cell. Unless there is a specific drug target only cancer cells possess (and this aspect is far beyond the scope of this chapter), the only way to make anticancer drugs cancer specific is to target them specifically to cancer cells. And here, the unique pathophysiology of solid tumors opened a way, which was described for the first time over 30 years ago (Matsumura and Maeda 1986) and which has quickly come to be known as the EPR effect (Maeda et al. 2016). In normal blood vessels, the endothelial cell lining provides a first barrier for any circulating particle to escape from the circulation. Blood vessels in solid tumors formed during tumor-specific angiogenesis, however, are "leaky" due to gaps between the endothelial cells. Such gaps have been found to be to some extent tumor-specific and to be in the size range roughly between 50 and 800 nm. Subsequently, any nanoformulation of an anticancer drug, be it a liposome (phospholipid vesicles), solid nanoparticle or a polymer-aggregate carrying a drug, will eventually "fall" through such gaps and thereby enter the tumor interstitium. The fact that solid tumors also lack lymphatic drainage eventually leads to a tumor-specific drug accumulation, hence the name enhanced permeability and "retention" effect. Prerequisite for such EPR effect is a sufficient long half-life of the drug formulation in circulation, which can be achieved by, for example, attaching long polyethylene glycol chains to the liposome or nanoparticle surface (as done for all PEGylated proteins, polypeptides and aptamers, see Table 1.1). The very first drug formulation based on this principle, liposomal-encapsulated doxorubicin called "Doxil®", was FDA-approved in 1995 (Barenholz 2012). Going back to our definition of true nanopharmaceuticals, one could of course now argue that the nanomaterial (i.e. liposomes) "adds new functionality" to the drug (i.e., changes its ADME properties) and therefore, all liposome-based drug formulations are indeed true nanopharmaceuticals. But then one would have to consider naturally occurring phospholipids as nanomaterials and their entropy-driven self-assembly into bilayer structures, i.e., liposomes as "nanoengineering", which I believe would be quite a stretch.

1.6 Summary and Conclusion

Without any doubt, nanoscience and nanotechnology has already impacted and will even much more change and improve many aspects of our life. The merger of nanoscience with medicine has spawned nanomedicine and material science at the nanolevel, and nanotechnology/nanoengineering will undoubtedly create unforeseen opportunities for both therapy and diagnostics of human diseases. But here, we are just at the beginning. For an outlook towards what the near future might bring us in the area of drug development, I would like to refer to our recent review about "Nanopharmaceuticals (part 2): products in the pipeline" (Weissig and Guzman-Villanueva 2015).

References

Alam, M. and C. T. Hartrick (2005). Extended-release epidural morphine (DepoDur): An old drug with a new profile. *Pain Pract* **5**(4): 349–353.

Babu, A., A. K. Templeton, A. Munshi and R. Ramesh (2014). Nanodrug delivery systems: A promising technology for detection, diagnosis, and treatment of cancer. *AAPS PharmSciTech* **15**: 709–721

Bachmann, M. F. and G. T. Jennings (2010). Vaccine delivery: A matter of size, geometry, kinetics and molecular patterns. *Nat Rev Immunol* **10**(11): 787–796.

Bangham, A. D., M. M. Standish and N. Miller (1965a). Cation permeability of phospholipid model membranes: Effect of narcotics. *Nature* **208**(5017): 1295–1297.

Bangham, A. D., M. M. Standish and J. C. Watkins (1965b). Diffusion of univalent ions across the lamellae of swollen phospholipids. *J Mol Biol* **13**(1): 238–252.

Bangham, A. D., M. M. Standish and G. Weissmann (1965c). The action of steroids and streptolysin S on the permeability of phospholipid structures to cations. *J Mol Biol* **13**(1): 253–259.

Barenholz, Y. (2012). Doxil(R)–the first FDA-approved nano-drug: Lessons learned. *J Control Release* **160**(2): 117–134.

Batist, G., J. Barton, P. Chaikin, C. Swenson and L. Welles (2002). Myocet (liposome-encapsulated doxorubicin citrate): A new approach in breast cancer therapy. *Expert Opin Pharmacother* **3**(12): 1739–1751.

Chertow, G. M., S. K. Burke, J. M. Lazarus, K. H. Stenzel, D. Wombolt, D. Goldberg, J. V. Bonventre and E. Slatopolsky (1997). Poly[allylamine hydrochloride] (RenaGel): a noncalcemic phosphate binder for the treatment of hyperphosphatemia in chronic renal failure. *Am J Kidney Dis* **29**(1): 66–71.

Davis, F. F. (2002). The origin of pegnology. *Adv Drug Deliv Rev* **54**(4): 457–458.

de Marie, S., R. Janknegt and I. A. Bakker-Woudenberg (1994). Clinical use of liposomal and lipid-complexed amphotericin B. *J Antimicrob Chemother* **33**(5): 907–916.

Desai, N., V. Trieu, Z. Yao, L. Louie, S. Ci, A. Yang, C. Tao, T. De, B. Beals, D. Dykes, P. Noker, R. Yao, E. Labao, M. Hawkins and P. Soon-Shiong (2006). Increased antitumor activity, intratumor paclitaxel concentrations, and endothelial cell transport of cremophor-free, albumin-bound paclitaxel, ABI-007, compared with cremophor-based paclitaxel. *Clin Cancer Res* **12**(4): 1317–1324.

Eckardt, J. R., E. Campbell, H. A. Burris, G. R. Weiss, G. I. Rodriguez, S. M. Fields, A. M. Thurman, N. W. Peacock, P. Cobb, M. L. Rothenberg and et al. (1994). A phase II trial of DaunoXome, liposome-encapsulated daunorubicin, in patients with metastatic adenocarcinoma of the colon. *Am J Clin Oncol* **17**(6): 498–501.

Foote, M. (2007). Using nanotechnology to improve the characteristics of antineoplastic drugs: Improved characteristics of nab-paclitaxel compared with solvent-based paclitaxel. *Biotechnol Annu Rev* **13**: 345–357.

Forssen, E. A., R. Male-Brune, J. P. Adler-Moore, M. J. Lee, P. G. Schmidt, T. B. Krasieva, S. Shimizu and B. J. Tromberg (1996). Fluorescence imaging studies for the disposition of daunorubicin liposomes (DaunoXome) within tumor tissue. *Cancer Res* **56**(9): 2066–2075.

Forssen, E. A. and Z. A. Tokes (1983). Improved therapeutic benefits of doxorubicin by entrapment in anionic liposomes. *Cancer Res* **43**(2): 546–550.

Forssen, E. A. and Z. A. Tokes (1981). Use of anionic liposomes for the reduction of chronic doxorubicin-induced cardiotoxicity. *Proc Natl Acad Sci U S A* **78**(3): 1873–1877.

Foss, F. M. (2001). Interleukin-2 fusion toxin: Targeted therapy for cutaneous T cell lymphoma. *Ann N Y Acad Sci* **941**: 166–176.

Frampton, J. E. (2010). Mifamurtide: A review of its use in the treatment of osteosarcoma. *Paediatr Drugs* **12**(3): 141–153.

Glantz, M. J., K. A. Jaeckle, M. C. Chamberlain, S. Phuphanich, L. Recht, L. J. Swinnen, B. Maria, S. LaFollette, G. B. Schumann, B. F. Cole and S. B. Howell (1999). A randomized controlled trial comparing intrathecal sustained-release cytarabine (DepoCyt) to intrathecal methotrexate in patients with neoplastic meningitis from solid tumors. *Clin Cancer Res* **5**(11): 3394–3402.

Gordon, E. M. and F. L. Hall (2010). Rexin-G, a targeted genetic medicine for cancer. *Expert Opin Biol Ther* **10**(5): 819–832.

Gordon, E. M. and F. L. Hall (2011). Critical stages in the development of the first targeted, injectable molecular-genetic medicine for cancer. In C. Kang (ed.) *Gene Therapy Applications*. InTech: 461–492.

Greish, K., J. Fang, T. Inutsuka, A. Nagamitsu and H. Maeda (2003). Macromolecular therapeutics: advantages and prospects with special emphasis on solid tumour targeting. *Clin Pharmacokinet* **42**(13): 1089–1105.

Guaglianone, P., K. Chan, E. DelaFlor-Weiss, R. Hanisch, S. Jeffers, D. Sharma and F. Muggia (1994). Phase I and pharmacologic study of liposomal daunorubicin (DaunoXome). *Invest New Drugs* **12**(2): 103–110.

Hershfield, M. (1993). Adenosine deaminase deficiency. In R. A. Pagon, M. P. Adam, T. D. Bird et al. *GeneReviews*. University of Washington, Seattle, WA.

Herzog, C., K. Hartmann, V. Kunzi, O. Kursteiner, R. Mischler, H. Lazar and R. Gluck (2009). Eleven years of Inflexal V-a virosomal adjuvanted influenza vaccine. *Vaccine* **27**(33): 4381–4387.

Hiemenz, J. W. and T. J. Walsh (1996). Lipid formulations of amphotericin B: Recent progress and future directions. *Clin Infect Dis* **22**(Suppl 2): S133–S144.

Ichikawa, K., Y. Takeuchi, S. Yonezawa, T. Hikita, K. Kurohane, Y. Namba and N. Oku (2004). Antiangiogenic photodynamic therapy (PDT) using Visudyne causes effective suppression of tumor growth. *Cancer Lett* **205**(1): 39–48.

Ishii, H., J. Furuse, M. Nagase, Y. Maru, M. Yoshino and T. Hayashi (2003). A phase I study of hepatic arterial infusion chemotherapy with zinostatin stimalamer alone for hepatocellular carcinoma. *Jpn J Clin Oncol* **33**(11): 570–573.

Janknegt, R., S. de Marie, I. A. Bakker-Woudenberg and D. J. Crommelin (1992). Liposomal and lipid formulations of amphotericin B. Clinical pharmacokinetics. *Clin Pharmacokinet* **23**(4): 279–291.

Juliano, R. L., S. Daoud, H. J. Krause and C. W. Grant (1987a). Membrane-to-membrane transfer of lipophilic drugs used against cancer or infectious disease. *Ann N Y Acad Sci* **507**: 89–103.

Juliano, R. L., C. W. Grant, K. R. Barber and M. A. Kalp (1987b). Mechanism of the selective toxicity of amphotericin B incorporated into liposomes. *Mol Pharmacol* **31**(1): 1–11.

Junghanns, J. U. and R. H. Muller (2008). Nanocrystal technology, drug delivery and clinical applications. *Int J Nanomed* **3**(3): 295–309.

Kim, S., D. J. Kim, M. A. Geyer and S. B. Howell (1987). Multivesicular liposomes containing 1-beta-D-arabinofuranosylcytosine for slow-release intrathecal therapy. *Cancer Res* **47**(15): 3935–3937.

Kreuter, J. (1994). *Colloidal Drug Delivery Systems*, CRC Press, Boca Raton, FL.

Maeda, H., K. Tsukigawa and J. Fang (2016). A retrospective 30 years after discovery of the enhanced permeability and retention effect of solid tumors: Next-generation chemotherapeutics and photodynamic therapy–problems, solutions, and prospects. *Microcirculation* **23**(3): 173–182.

Marty, J. J., R. C. Oppenheim and P. Speiser (1978). Nanoparticles–a new colloidal drug delivery system. *Pharm Acta Helv* **53**(1): 17–23.

Masuda, E. and H. Maeda (1995). Antitumor resistance induced by zinostatin stimalamer (ZSS), a polymer-conjugated neocarzinostatin (NCS) derivative. I. Meth A tumor eradication and tumor-neutralizing activity in mice pretreated with ZSS or NCS. *Cancer Immunol Immunother* **40**(5): 329–338.

Matsumura, Y. and H. Maeda (1986). A new concept for macromolecular therapeutics in cancer chemotherapy: Mechanism of tumoritropic accumulation of proteins and the antitumor agent smancs. *Cancer Res* **46**(12 Pt 1): 6387–6392.

Mischler, R. and I. C. Metcalfe (2002). Inflexal V a trivalent virosome subunit influenza vaccine: Production. *Vaccine* **20**(Suppl 5): B17–B23.

Northfelt, D. W., F. J. Martin, P. Working, P. A. Volberding, J. Russell, M. Newman, M. A. Amantea and L. D. Kaplan (1996). Doxorubicin encapsulated in liposomes containing surface-bound polyethylene glycol: Pharmacokinetics, tumor localization, and safety in patients with AIDS-related Kaposi's sarcoma. *J Clin Pharmacol* **36**(1): 55–63.

Oerlemans, C., W. Bult, M. Bos, G. Storm, J. F. Nijsen and W. E. Hennink (2010). Polymeric micelles in anticancer therapy: Targeting, imaging and triggered release. *Pharm Res* **27**(12): 2569–2589.

Pasero, C. and M. McCaffery (2005). Extended-release epidural morphine (DepoDur). *J Perianesth Nurs* **20**(5): 345–350.

Peng, Z. (2005). Current status of gendicine in China: Recombinant human Ad-p53 agent for treatment of cancers. *Hum Gene Ther* **16**(9): 1016–1027.

Rivera Gil, P., D. Huhn, L. L. del Mercato, D. Sasse and W. J. Parak (2010). Nanopharmacy: Inorganic nanoscale devices as vectors and active compounds. *Pharmacol Res* **62**(2): 115–125.

Saif, M. W. (2013). U.S. Food and drug administration approves Paclitaxel protein-bound particles (abraxane(R)) in combination with gemcitabine as first-line treatment of patients with metastatic pancreatic cancer. JOP **14**(6): 686–688.

Sartor, O. (2003). Eligard: Leuprolide acetate in a novel sustained-release delivery system. *Urology* **61**(2 Suppl 1): 25–31.

Schug, T. T., A. F. Johnson, D. M. Balshaw, S. Garantziotis, N. J. Walker, C. Weis, S. S. Nadadur and L. S. Birnbaum (2013). ONE Nano: NIEHS's strategic initiative on the health and safety effects of engineered nanomaterials. *Environ Health Perspect* **121**(4): 410–414.

Wang, R., Billone, P. S., Mullett, W. M. (2013). Nanomedicine in action: An overview of cancer Nanomedicine on the market and in Clinical Trials. *J Nanomater* **2013**: 1–12.

Weissig, V. and D. Guzman-Villanueva (2015). Nanopharmaceuticals (part 2): Products in the pipeline. *Int J Nanomed* **10**: 1245–1257.

Weissig, V., T. K. Pettinger and N. Murdock (2014). Nanopharmaceuticals (part 1): Products on the market. *Int J Nanomed* **9**: 4357–4373.

Whitesides, G. (2001). The one and future nanomachine. *Sci Am* **285**(3): 70–75.

2

Drug Nanocrystals

Mayank Singhal, Jouni Hirvonen, and Leena Peltonen
University of Helsinki

2.1 Introduction

Today more and more new active pharmaceutical ingredients (APIs) are poorly soluble due to more efficient screening methods: it has been approximated that 90% of all the new chemical entities are facing some kind of solubility issues, e.g., belonging to Biopharmaceutics Classification System (BCS) class II (70%) or class IV (20%) (Loftsson and Brewster 2010, Müller and Keck 2012).

Solubility properties of compounds can be manipulated in molecular or particular levels or by the formation of colloidal systems. Molecular-level approaches are utilization of prodrugs (Huttunen et al. 2011), salt formation (Serajuddin 2007), cosolvent systems (Seedher and Kanojia 2009), or cyclodextrins (Bilensoy and Hincal 2009). Particle-level techniques are, for example, metastable polymorphs (Censi and Di Martino 2015), co-crystals (Drozd et al. 2017), amorphous systems (Babu and Nangia, 2011) or particle size reduction (Peltonen and Hirvonen 2018). Examples of colloidal formulations are SEDDS/SMEDDS/SNEDDS, (micro/nano) emulsions or other lipid-based systems (Cerpnjak et al. 2013).

Nanocrystallization is a particulate-level approach to modify the intrinsic material properties, like solubility, and it is based on controlled reduction of drug particle size as compared to bulk or micronized drug material. When thinking about the structure of drug nanocrystals, they are solid drug nanoparticles surrounded by a stabilizer(s) layer (Liversidge and Cundy 1995, Wang et al. 2013, Peltonen and Hirvonen 2018); stabilizer(s) are needed to protect the inherently unstable nanoparticles against Ostwald ripening and/or aggregation both during the production period and in wholesale/pharmacy storage. Accordingly, the main role of stabilizing excipient(s) is to maintain constant particle size, but it should be kept in mind that many commonly used stabilizers may also, for example, enhance permeation or prolong *in vivo* the supersaturation period of the drug, which may lead higher bioavailability (Gao et al. 2012, Gao et al. 2014). The most important benefits of nanocrystals can be divided into three classes: (i) fast dissolution, (ii) higher solubility, and (iii) improved membrane adhesion. However, the most dominant effect reached with drug nanocrystals is higher dissolution rate due to the large increase in surface area per mass solid.

The research in pharmaceutical nanocrystals started widely in the beginning of the 1990s (Liversidge et al. 1992), and they are mostly utilized in formulations for poorly water-soluble drugs (Keck and Müller 2006, Valo et al. 2011). However, also controlled release applications have been studied (Kim et al. 2012, Valo et al. 2013). For drug nanocrystals, best candidates are BCS class II drugs, which are poorly soluble but well permeable (Liu et al. 2011). However, also class IV drugs may benefit on nanocrystallization, when the faster dissolution increases the concentration gradient between the intestine lumen and intestinal wall, hence improving the drug permeation (Gao et al. 2012). Gastrointestinal (GI) drug absorption involves release/dissolution of drug from formulation into luminal fluids followed by permeation across the GI epithelium. For BCS class II and IV drug, dissolution is known to be the rate-limiting step (Dressman and Reppas 2000), and via nanonization, this can be affected.

In drug developability criteria, BCS class II can be further divided into two subclasses: (i) class IIa with dissolution rate-limited brick-dust molecules and (ii) class IIb with solubility-limited grease ball molecules (Butler and Dressman 2010). Brick-dust compounds have poor aqueous solubility, but also solubility in lipids and organic solvents is low, and these compounds are good candidates for nanosizing (Borchard 2015). For grease ball molecules, lipid formulations are the first choice, because they normally have at least some lipid solubility (Bergström et al. 2016), but also nanocrystal formulations with these molecules do exist (Rydberg et al. 2016).

Accordingly, drug nanocrystals are good option for tailoring intrinsic material properties, like solubility, of APIs. In this chapter, first the most important and characteristic physical properties of drug nanocrystals are presented and the stabilizers and stabilization methods utilized in drug nanocrystals discussed. Then, different production approaches of nanocrystals as well as analytical techniques for nanocrystal characterization are reviewed. Finally, nanocrystal-based formulations and their drug delivery routes are presented followed by some future perspectives.

2.2 Characteristics of Drug Nanocrystals

As already mentioned before, the most important property of drug nanocrystals is the extremely high surface area per mass ratio. Increased surface area leads to (i) fast dissolution, (ii) high solubility, and (iii) improved membrane adhesion. Next, these phenomena are discussed more in detail.

2.2.1 Fast Dissolution

When the particle size is decreased, the surface area of the particles and interfacial area per mass is increased. To give an example, for spherical particles, if the particle size is reduced from 30 μm (typical particle size for bulk drug materials) to 300 nm (characteristic for drug nanocrystals), based on Noyes–Whitney equation (Eq 2.1), the dissolution rate is increased a hundredfold when taken into account only the effect of surface area change:

$$\frac{dC}{dt} = \frac{DS}{Vh}\left(C_s - C\right), \tag{2.1}$$

where C is the concentration at time t, D is the diffusion coefficient, S is the surface area, V is the dissolution volume, h is the diffusion layer thickness, and C_s is the saturation concentration. Indeed, faster dissolution may be challenging when making *in vitro* dissolution tests. Under sink conditions, the drug dissolution is often so fast that discriminating dissolution testing, for example, dissolution under non-sink conditions, is required for recognizing the critical differences in drug release profiles between different particle size fractions (Table 2.1) (Liu et al. 2011, Liu et al. 2013).

For drug nanocrystals, the higher dissolution rate is mainly reached via the increased surface area (S). However, the diffusion layer thickness (h) is also smaller in the case of small micron-sized particles and nanoparticles, and the apparent solubility values (C_s) of the nanosized particles are well above the thermodynamic solubility values; all these factors increase the dissolution rate of drug nanocrystals.

2.2.2 Higher Apparent Solubility

Higher apparent solubility is based on Ostwald–Freundlich equation (Eq. 2.2):

$$c_{\mathrm{NP}} = c_b \exp\left(\frac{2V_m\gamma}{RTr}\right), \tag{2.2}$$

where c_{NP} is the solubility of the nanoparticles with the radius r, c_{b} is the solubility of bulk material, V_m is the molar volume, γ is the interfacial tension, R is the gas constant, and T is the temperature. Though the theory was originally derived for liquid droplets in a gas phase, it was later accepted to be valid for solid particles in a liquid phase, too (Shchekin and Rusanov 2008).

TABLE 2.1 The Pair Comparisons of Dissolution Rates from Pharmacopoeia Paddle Method between Two Different Particle Size Fractions Using ANOVA Method with Tukey's Test

		Significance Level		
		Compared Particle Sizes		
Dissolution Conditions	Time/min	340–560 nm	560–1,300 nm	340–1,300 nm
Sink conditions	1	–	–	0.05
Ratio of drug used in the dissolution test to the saturation solubility of the drug = 1/4				
	2	0.05	–	0.01
	4	–	–	–
	6	–	–	–
Non-sink conditions	0.5	–	–	–
Ratio of drug used in the dissolution test to the saturation solubility of the drug = 1				
	1	0.001	–	0.001
	2	0.01	0.05	0.001
	4	0.01	0.01	0.001
	6	0.01	0.01	0.001
	15	–	0.05	0.01

Source: Modified from Liu et al. (2013).

Dissolution test was performed under either sink or non-sink conditions: the ratio of drug used in the dissolution test to the saturation solubility varied from 1/4 to 1. Nanosuspension particle size was an independent variable, and the dissolved amount was a dependent variable—meaning no significant difference.

The increased apparent solubility values reached with the nanosized particles were shown in a study by Sarnes et al. (2013). Concentration levels reached with different particle sizes were determined by flow through intrinsic dissolution test equipment and UV imaging technique. The measurements were performed from a constant flat surfaces in order to eliminate the surface area effect; for example, the higher solubility values were only caused by the higher apparent solubility values due to the nanosizing of the particles. In the intrinsic dissolution testings of nanocrystals with an approximated size of 580 nm, the apparent solubility value of indomethacin was tenfold higher than the corresponding value for bulk drug (Figure 2.1). In UV imaging, the surface concentration levels for drug nanocrystals were fivefold higher than the thermodynamic solubility value of the bulk drug.

2.2.3 Improved Mucoadhesion

If bioadhesion of drug nanocrystals to mucosa is taking place before dissolution, the retention time can be lengthened with the aid of nanosizing. Mucus penetration of solid drug nanocrystals has also been shown to be possible (Darville et al. 2015, Yu et al. 2016, Fu et al. 2017). Better adhesion on biological membranes is again one consequence of higher surface area per mass. This better adhesion means longer residence time *in vivo*, which can lead to improved bioavailability. But, it should be kept in mind that this effect is only reached if the particles have the time to adhere the membranes before drug dissolution. Often the dissolution with drug nanocrystals is so fast that no particle attachment into mucus layer is seen.

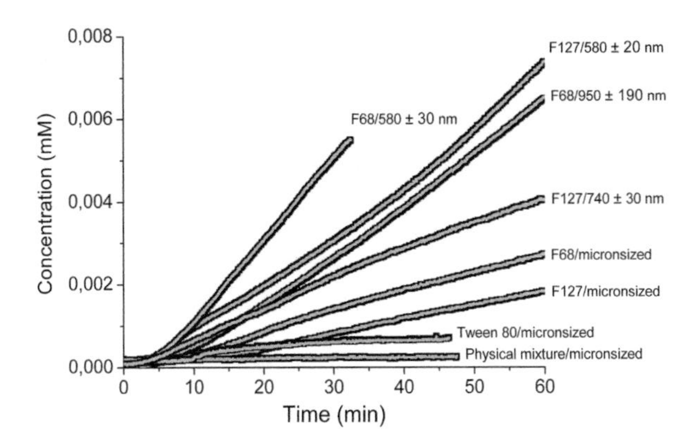

FIGURE 2.1 Intrinsic dissolution values for indomethacin from flat surfaces measured by flow through method. Drug nanocrystals were produced with two different poloxamers (F68 and F127) as a stabilizer, and the mean particle sizes are presented in the figure. As a reference sample, a physical mixture of stabilizer and bulk indomethacin (particle size tens of micrometers) were used. (Reprinted from European Journal of Pharmaceutical Sciences, Vol 50, A. Sarnes, J. Østergaard, S. Smedegaard Jensen, J. Aaltonen, J. Rantanen, J. Hirvonen, L. Peltonen, Dissolution study of nanocrystal powders of a poorly soluble drug by UV imaging and channel flow methods, 511–519, Copyright (2013), with permission from Elsevier.)

2.3 Stabilizers

Particle size diminishing results in the creation of new interfaces as well as positive Gibbs free energy change to take place. Accordingly, thermodynamically, the nanosuspensions are very unstable systems, and they tend to agglomerate or otherwise increase the particle size (Figure 2.2) (Wang et al. 2013). Poor solubility of the drug material and homogenous particle size minimize aggregation and Ostwald ripening (Möschwitzer et al. 2004), but most importantly, one or more stabilizing excipients are used to stabilize the newly formed drug nanocrystals against aggregation.

Stabilization can be based on steric and/or electrostatic impact, and often two stabilizers, typically a combination of one sterically stabilizing polymer and one electrostatically stabilizing ionic surfactant, are used for the improved stabilization (Van Eerdenbrugh et al. 2008, Ghosh et al. 2013, Krull et al. 2016). In the absence of efficient stabilizer(s), increased surface energy leads to Ostwald ripening, when mass transfer from the fine to coarse particles takes place, which increases particle size (Wu et al. 2011, Li et al. 2016). Aggregation or crystal growth during the production process or storage can affect dissolution rate and *in vivo* performance of the system due to the uncontrolled formation of larger particles with a reduced specific surface area.

2.3.1 Selection of Stabilizer

Though intensively studied, no systematic empirical or theoretical guidelines for stabilizer selection and optimization exist, and the knowledge about the relationship between the stabilizer efficacy and nanosuspension stability is also limited (Van Eerdenbrugh et al. 2008, Peltonen and Hirvonen 2010, Yue et al. 2013, Liu et al. 2015). Not a single universal stabilizer exists, and stabilizer selection and its amount for each drug formulation need to be studied empirically, which is the first step in Quality by Design (QbD) approach (He et al. 2015, Bhatia et al. 2016, Tuomela et al. 2016, Siewert et al. 2018). Polymers, such as cellulose derivatives (hydroxypropyl methylcellulose (HPMC) (Tuomela et al. 2014), methyl cellulose (MC) (Pattnaik et al. 2015)), poloxamers (block copolymers of poly(ethylene oxide)-poly(propylene oxide) (PEO-PPO-PEO)) (Liu et al. 2015), polyvinyl pyrrolidone (PVP) (Pattnaik et al. 2015), polyvinyl alcohol (PVA) (Lestari et al. 2014), D-α-tocopherol polyethylene glycol 1,000 succinate (TPGS 1000) (Gao et al. 2014), polyethylene glycol (PEG) (Liu et al. 2011), or surfactants, such as sodium dodecyl sulfate (SDS) (Lestari et al. 2014) and polysorbates (Liu et al. 2011) have been widely utilized as stabilizers. Also some proteins, like hydrophobins (Valo et al. 2011), soybean protein isolate, whey protein isolate, and β-lactoglobulin (He et al. 2013), and Janus dendrimers (Selin et al. 2018) have been tested as stabilizers in nanosuspension formulations.

When selecting the stabilizer(s) for nanocrystallization, final formulation should be taken into account. Ionic

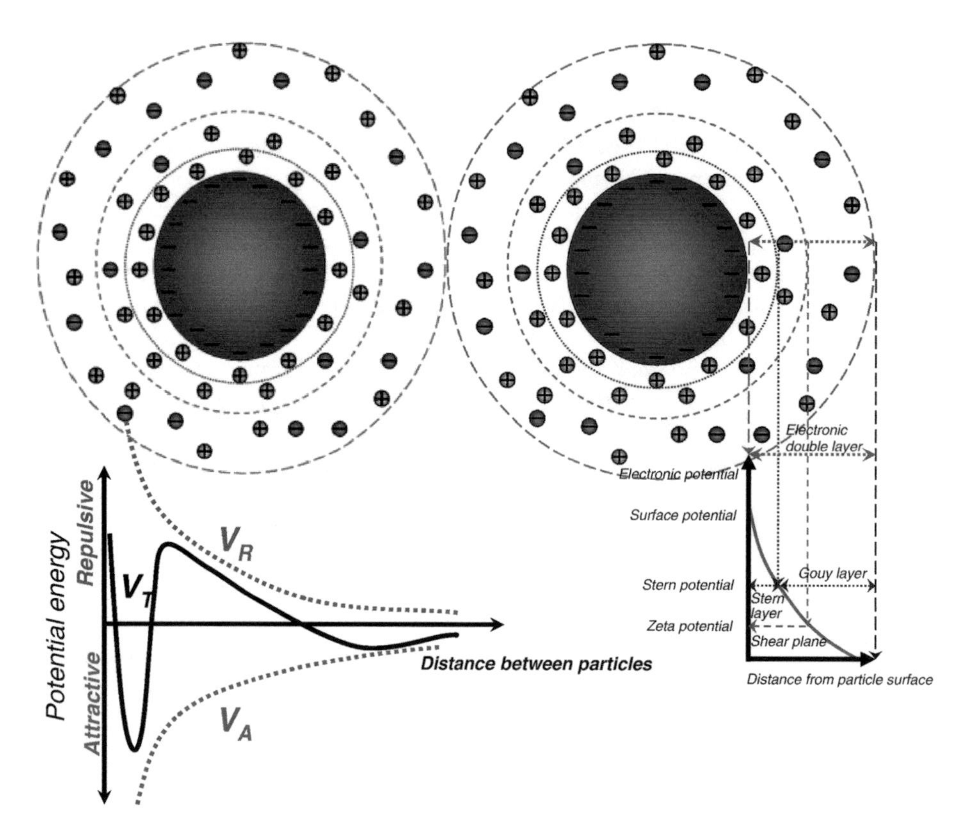

FIGURE 2.2 Schematic illustration of attractive and repulsive forces as of a function of distance between two separated particles based on classical DLVO theory. At small and large distances, attractive forces are dominating (primary and secondary minimum in potential energy). At intermediate distances, repulsive forces are prevailing being responsible for net repulsion between the dispersed particles. With the aid of good stabilizer(s), drug nanocrystals are kept at intermediate distances, and hence, aggregation is prevented. (Reprinted from Advanced Drug Delivery Reviews, Vol 63, L. Wu, J. Zhang, W. Watanabe, Physical and chemical stability of drug nanoparticles, 456–469, Copyright (2011), with permission from Elsevier.)

stabilizers like SDS are effective in suspension formulations, but not effective in dry nanocrystalline powder. Therefore, steric stabilizers, such as HPMC, PVP, and nonionic surfactants, together with poloxamers and polysorbates, are popular stabilizers in oral formulations (Tuomela et al. 2015, Li et al. 2018, Wei et al. 2018).

2.3.2 Stabilization Mechanisms

Ionic polymers/surfactants stabilize suspensions via electrostatic repulsion. With long-chain polymers and nonionic surfactants, the stabilization effect is based on steric repulsion (Figure 2.3). Stabilization can be based on both the approaches, and often two stabilizers, one steric and one electrostatic, are combined. In the gut, electrolysis may alter the stabilizing effect of ionic stabilizers, and there the steric stabilization can be a better choice (Müller et al. 2006, Mu et al. 2016).

Stabilization effect is based on fast absorption of the stabilizer on the solid drug particle surface. After absorption, hydrogen bonding and electrostatic interactions may take place, which further strengthen the stabilizer absorption. Efficient stabilization requires fast and strong absorption and full surface coverage of the stabilizer (Ghosh et al. 2012, Liu et al. 2015). Besides, in steric stabilization steric hindrance is ensured by long chains on top of the particles, which are forming dynamically rough surfaces due to the continuous thermal motion (Liu et al. 2015). In electrostatic stabilization, high enough surface charge is required in order for the particles to repel against each other. Recently, it was shown by NMR studies that in a nanosuspension, the crystalline drug core particles were surrounded by a semisolid phase of the drug and stabilizers (Kojima et al. 2018). The semisolid phase was shown to be in equilibrium with the solution phase in suspension state, and it was responsible for the stabilization of the nanoparticles by steric hindrance and electrostatic repulsion.

2.3.3 Impact of Stabilizers on Bioavailability

Though stabilizers are excipients in drug formulations, they may have some activity on biological systems. Permeation enhancing efficacy can be reached via opening of the tight junctions, creating more leaky cell layers or taking part on the active transport systems. For example, polysorbates and poloxamers have been shown to affect the drug transport activity and hence to act as permeation enhancers (Cornaire et al. 2004). Similarly, surfactants, which are often utilized as stabilizers, are known to affect the cell layer structures

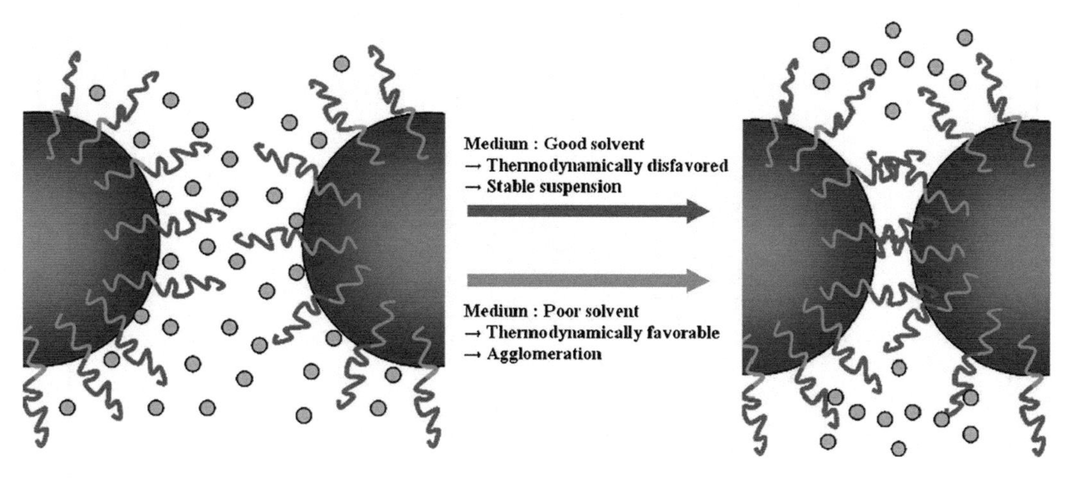

FIGURE 2.3 Schematic illustration of steric stabilization mechanisms. A positive Gibbs free energy means stable nanosuspension while negative Gibbs free energy leads particle aggregation. If the solvent is favorable for the stabilizing moiety, adsorbed stabilizing layer hinders the close contact of the particles leading to stable nanosuspension. In poor solvent, adsorbed layers can interpenetrate leading particle aggregation and unstable nanosuspension. (Reprinted from Advanced Drug Delivery Reviews, Vol 63, L. Wu, J. Zhang, W. Watanabe, Physical and chemical stability of drug nanoparticles, 456–469, Copyright (2011), with permission from Elsevier.)

due to their bio-mimicking molecular structures (Rege et al. 2002). On the other hand, celluloses can be mucoadhesive materials, which can increase the residence time *in vivo* and lead to higher bioavailability (Hansen et al. 2015).

In order to improve bioavailability with the aid of suitable transport activity, permeation enhancing effect, or mucoadhesivity of the stabilizer, the challenge for drug formulators is that the drug and the excipient should be in the absorption area simultaneously. For example, after endocytosis of TPGS-coated paclitaxel nanocrystals, the therapeutic efficiency was increased, due to the simultaneous presence of the drug and P-gp inhibitor (TPGS) inside the cells (Gao et al. 2014). However, when paclitaxel was given in a solution with TPGS, P-gp inhibition was not seen (Varmal and Panchagnula 2005). Accordingly, when the dissolved drug is in the cytoplasm, P-gp can pump the dissolved drug molecules out from the cell, if P-gp inhibitors are not available inside the cells simultaneously; this was the case with solution formulation.

2.4 Production of Drug Nanocrystals

Drug nanocrystals can be produced using either of the following two approaches: (i) bottom-up approach, where nanocrystals are synthesized starting from molecular level, or (ii) top-down approach, where larger particles are fractured into nanosized material. These processes are also known as nanonization. Both approaches have their own advantages and disadvantages.

Combinations of the two approaches have also been employed in order to overcome the shortcomings of each technique (Table 2.2) (Soliman et al. 2017, Zong et al. 2017). The suitability of a nanonization technique varies from molecule to molecule based on their physicochemical properties and also on the initial particle size (Gadadare et al. 2015, Liu et al. 2018). In most of the techniques, the nanosized drug is collected as a nanosuspension, and stabilizer(s) constitute an essential part of the formulation (Müller et al. 2011). The function of a stabilizer is partly process related; therefore, the differences between the production methods should be well understood (Verma et al. 2009, Bilgili and Afolabi 2012, Bi et al. 2015, Peltonen et al. 2015).

2.4.1 Top-Down Techniques

Top-down techniques, mainly media milling and high-pressure homogenization (HPH), are widely exploited by the pharmaceutical industry due to the ease in scalability, better control over particle size and wide acceptance by regulatory agencies (Möschwitzer 2013). These methodologies have led to the development of many marketed products, examples of which are listed in Table 2.3. Though these processes are more industrially accepted, they have their own limitations, such as long processing time to obtain particles smaller than 150–300 nm. Long processing time means high energy consumption and contamination due to wearing and tearing of the equipment (Juhnke et al. 2012, Li et al. 2015). Besides this, change in the solid state and chemical degradation may also occur during the processes (Sharma et al. 2009, Garad et al. 2010, Wu et al. 2011). However, chances to produce unstable amorphous form are low because of the presence of water, which enhances molecular mobility and reduces the glass transition temperature, hence stabilizing the crystalline product (Sharma et al. 2009).

Media Milling Technique

In the milling process, milling media, drug, and dispersion medium containing stabilizer(s) are filled in a milling chamber (Merisko-Liversidge et al. 2003, Peltonen and Hirvonen 2010, Bilgili and Afolabi 2012, Ghosh et al. 2013, Tuomela et al. 2015, Peltonen 2018b). The milling media

TABLE 2.2 An Overview of Top-Down and Bottom-Up Technologies for Drug Nanocrystal Production

Technology	Strength	Limitations	Critical Process Parameters	Remarks
Top-Down Techniques				
Media milling	Scalable, batch-to batch uniformity, stable formulation, better control of particle size, accepted by regulatory agencies	High-energy process, in some cases long processing time, possibility of contamination due to wear and tear, changes in solid state	Drug feed volume, viscosity of dispersion, loading percentage, original drug particle size, size, amount and density of milling beads, milling speed, milling duration, temperature, type, and amount of stabilizer(s)	Amount of drug in the milling chamber can vary from 2 to 30%, while the percentage of beads can be higher from 10 to 50% w/v of the slurry, preferred size of beads is <1 mm, duration of the milling process depends on the milling speed, higher the speed lesser the processing time, vice versa
High-pressure homogenization	Scalable, less prone to generating impurities due to wearing, variant techniques allow processing of thermolabile and water-sensitive drugs	Sometimes preprocessed material is required, high-energy process, chances of clogging are high, relatively complicate process as compared to media milling	Number of cycles, and homogenization pressure, geometry of the equipment, initial particle size and percentage of solid content, type and combination of surfactants, hardness of drugs	High number of passes may be required for hard drugs, which is not very production friendly using jet-stream homogenization, applied pressure can be increased gradually in order to avoid clogging
Bottom-Up Techniques				
Solvent–antisolvent precipitation	Relatively simple, low-cost equipment, low energy required, suitable for thermolabile drugs	Difficulty in scaling up, uncontrolled particle growth, change in polymorphic form, need of organic solvents making it regulatory unacceptable, laborious solvent removal, sometimes difficult to find antisolvent–solvent combination	Mixing speed, solvent and antisolvent properties, solvent-to-antisolvent ratio, mixing efficiency, medium viscosity and temperature, stabilizer, degree of supersaturation	The type of the antisolvent used in the precipitation process not only controls the particle size of the precipitated drug but also its physical properties, such as crystallinity and polymorphism, faster and uniform mixing is essential to get smaller particles with a narrower size distribution
Supercritical fluid (SCF) techniques	Rapid one-step processing, organic solvent free, easy to scale up, flexible, size, and morphology of the particles can be controlled. CO_2 as SCF nontoxic, inexpensive, innocuous, nonreactive, noninflammable, environmental friendly	Mostly applicable to drugs exhibiting fair solubility in selected SCFs, possible drug incompatibility with the SCF, elevated pressure required, high maintenance cost and requirement of the accessories equipment	Nozzle geometry, pre- and post-expansion pressure and temperature, drug concentration, degree of mixing, rate of antisolvent addition in supercritical antisolvent (SAS) and rapid expansion of a supercritical solution into aqueous solution (RESSAS) techniques	Properties of the SCF can be precisely adjusted, near to the critical point a slight change in temperature and pressure can result in changes in the SCF density
Liquid atomization technique	Fine and free-flowing powder is obtained, short processing times, low residual organic solvent, scalable, simple, cost-effective	Low yield, viscous materials cannot be atomized by pressure nozzles, difficulty in atomizing spray of ~ <100 nm	Nozzle geometry, atomizing pressure, drug concentration, drying process: temperature, air flow rate, nature of the solvent, air-to-feed ratio	Works on the principle of one-droplet-to-one-particle, air drying temperature can be reduced by increasing the air flow rate, fine nozzle aperture generates higher kinetic forces that gives finer and more wide-reaching spray

Source: Keck and Müller (2006), Peltonen et al. (2010), Girotra et al. (2013), Sinha et al. (2013).

TABLE 2.3 Examples of US FDA-Approved Drug Nanocrystal Products

Trade Name	Pharma Company	Technology	Administration Route	Year of FDA Approval
Gris-PEG® (Griseofulvin)	Valeant Pharmaceuticals International, Inc.	Bottom-up, coprecipitation	Oral (tablet)	1975
Naprelan®(Naproxen sodium)	Wyeth Pharmaceuticals	Top-down, media milling	Oral (tablet)	1996
Azopt® (brinzolamide)	Alcon Laboratories, Inc.	Top-down, media milling	Ocular (suspension)	1998
Verelan®PM (Verapamil)	Schwarz Pharma	Top-down, media milling	Oral (capsule)	1998
Rapamune® (Sirolimus)	Wyeth Pharmaceuticals	Top-down, media milling	Oral (tablet)	2000
Avinza® (Morphine sulfate)	King Pharmaceuticals, Inc.	Top-down, media milling	Oral (capsule)	2002 (discontinued)
Ritalin® LA (Methylphenidate HCl)	Novartis	Top-down, media milling	Oral (capsule)	2002
Zanaflex® (Tizanidine HCl)	Acorda Therapeutics Inc.	Top-down, media milling	Oral (capsule)	2002
Emend® (Aprepitant)	Merck & Co., Inc.	Top-down, media milling	Oral (capsule, suspension)	2003 (capsule.), 2015 (suspension)
Tricor®(Fenofibrate)	AbbVie Inc.	Top-down, media milling	Oral (tablet)	2004
Focalin® XR (Dexmethyl-phenidate HCl)	Novartis	Top-down, media milling	Oral (capsule)	2005
Megace ES® (Megestrol acetate)	Par Pharmaceutical Companies Inc.	Top-down, media milling	Oral (suspension)	2005
Triglide® (Fenofibrate)	Sciele Pharma Inc.	Bottom-up, jet-stream homogenization	Oral (tablet)	2005
Invega Sustenna® (paliperidone palmitate)	Janssen Pharmaceuticals, Inc.	Top-down, high-pressure homogenization	Parenteral (im injection)	2009
Ilevro® (Nepafenac)	Alcon Laboratories, Inc.	Top-down, media milling	Ocular (suspension)	2012
Ryanodex®(dantrolene sodium)	Eagle Pharmaceuticals, Inc.	Not reported	Parenteral (iv injection)	2014

Source: Modified from Bobo et al. (2016), Chen et al. (2017), Malamatari et al. (2018).

are milling pearls, typically made from ceramics, glass, or stainless steel, or they can be coated by highly cross-linked polystyrene resin. Rotation of the milling beads at high speed generates enough energy and shear forces resulting in high impaction of the milling medium with the drug and disintegration of microparticulate drug material into nano-sized particles (Knieke et al. 2009, Breitung-Faes and Kwade 2013). During the milling, due to the rotational motion of drug particles and beads, the likelihood of settling, flocculating, re-agglomerating, or crystal growth decreases.

In nanomilling, the process optimization leads to batch-to-batch uniformity and high quality of the product. The scale-up of the nanonization is also straightforward with the media milling, where the milling equipment suitable for laboratory and commercial purposes is available (Takatsuka et al. 2009, Van Eerdenbrugh et al. 2009, Singare et al. 2010, Srivalli and Mishra 2016, Yuminoki et al. 2016).

High-Pressure Homogenization

HPH consists of diminution of drug particles by shear, collision, and cavitation forces (Figure 2.4) (Kaialy and Al Shafiee 2016). It can be separated into two different approaches: (i) jet-stream (microfluidizer) homogenization (trademarked as insoluble drug delivery particle (IDD-P^{TM} technology)) (Haynes 1992), and (ii) piston-gap homogenization either in water also known as Dissocubes® technology (Müller et al. 1995) or alternatively in water-reduced/non-aqueous media known as Nanopure® technology (Müller et al. 2001).

Microfluidizer technology can generate nanosized particles by direct collision of two fluid streams in Y- or Z-shaped chamber under very high pressure (~1700 bars) (Alam et al. 2013). Product available in the market based on microfluidizer IDD-P technology is Triglide® oral tablet that contains fenofibrate in nanocrystals form (Shegokar and Müller 2010).

The piston-gap homogenization involves homogenization of drugs in either water or medium without or with reduced water, e.g., a mixture of water with PEG. This requires several homogenization cycles, increasing the pressure at each cycle, before reaching to the target particle size. Since the material is passed through a very thin gap (nearly ~25 μm), drug material should have appropriate initial particle size. If the initial particle size is not small enough, preprocesses, like milling, can be carried out in order to minimize the risk of clogging the equipment (Al-Kassas et al. 2017).

2.4.2 Bottom-Up Techniques

Bottom-up techniques are broadly referred to as precipitation/crystallization techniques since the drug is precipitated to nano-scale from a supersaturated solution. In comparison with top-down techniques, bottom-up techniques are low-energy-driven processes and require simple setup, which makes them relatively less expensive, due to which these have been popular at laboratory scale (Sinha et al. 2013). Above all, being able to operate at low temperature, these have been particularly suitable for thermolabile drugs. Unfortunately, difficulty in controlling the particle

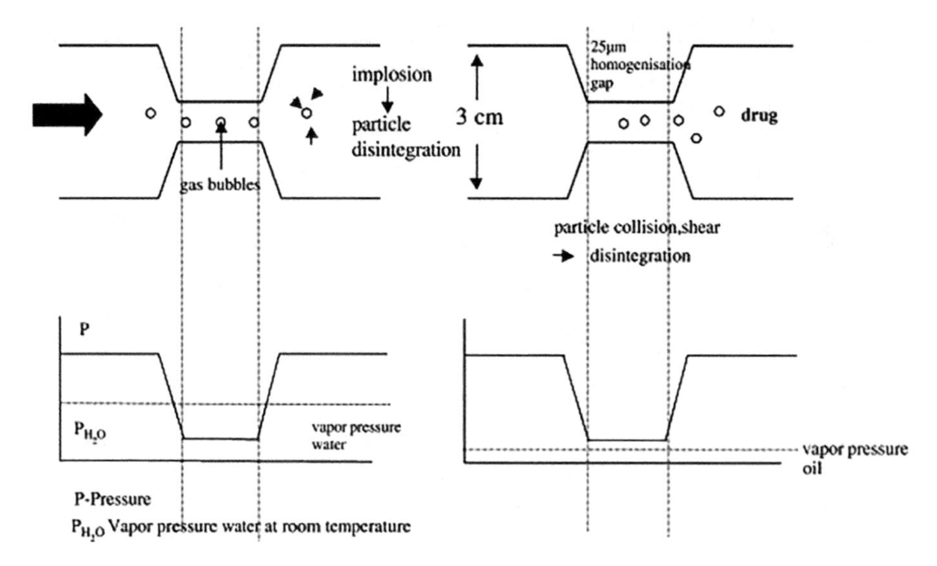

FIGURE 2.4 Schematic Figure of HPH process in (i) water (left) and (ii) water mixtures or water-free media (right). The level of the static pressure inside the homogenizer is shown in the lower figures. (Reprinted from European Journal of Pharmaceutics and Biopharmaceutics, Vol 62, C.M. Keck and R.H. Müller, Drug nanocrystals of poorly soluble drugs produced by high pressure homogenisation, 3–16, Copyright (2006), with permission from Elsevier.)

size growth makes this technique problematic for scaling up (Sinha et al. 2013). Other challenges are finding a suitable solvent/antisolvent combination for each drug and demand for the removal of an organic solvent.

Precipitation techniques can be categorized as (i) solvent–antisolvent precipitation techniques, (ii) supercritical fluid-based precipitation techniques, and (iii) liquid atomization-based techniques. The techniques are discussed more in detail in the following chapters.

Solvent–Antisolvent Precipitation Technique

This technique involves dissolution of a drug in solvent (usually organic solvent where the drug has generous solubility) followed by controlled addition of a second solvent (usually water). The second solvent acts as an antisolvent for the drug, but it should be miscible with the first solvent. Mixing of the drug solution with an antisolvent results in a supersaturated solution due to the drowning-out effect of the antisolvent, which reduces drug solubility and induces precipitation (Kaialy and Al Shafiee 2016). Mixing of the two phases can be done either by using simple static mixer or by using modified methods, such as ultrasonic waves (sonoprecipitation), high gravity reactive precipitation method, or evaporative precipitation into aqueous solution (Sinha et al. 2013). Modified techniques enable production of smaller particle sizes with narrower particle size distribution as a result of either faster mixing affecting the nucleation stage or by preventing particle growth.

Supercritical Fluid-Based Precipitation Techniques

Any material at a pressure and temperature above its critical point is a supercritical fluid (SCF). SCFs have high diffusion rates because of their low density and viscosity, which enables rapid mixing for fast precipitation (Girotra et al. 2013). Carbon dioxide is the preferred SCF because it can be easily transformed into the supercritical state as it has near ambient critical temperature (ca. 31°C) and a fairly low critical pressure (73.8 bar). Besides, it is easily available, inexpensive, non-flammable, and nontoxic making it more suitable for commercial use. The solvation ability of SCF can be controlled by manipulating the processing pressure and/or temperature, which affects the precipitation and generation of nanocrystals (Al-Kassas et al. 2017).

Among several SCF-based techniques, rapid expansion of supercritical solution (RESS) is a common technique where the desired drug is dissolved in a SCF with or without a cosolvent that is then expanded through a nozzle into a low-pressure chamber. This allows the SCF to transform into a gas in the expansion chamber due to the reduced pressure allowing the drug to precipitate (Hosseinpour et al. 2015). A modified form of RESS known as rapid expansion of a supercritical solution into aqueous solution (RESSAS) has also been developed: in the technique the expansion nozzle is placed inside a solvent phase instead of a low-pressure air or gas phase (He et al. 2011).

The utmost criteria for nanocrystals production using RESS or RESSAS are that the drug must have good solubility in a given SCF. SAS technique has been developed for molecules which are insoluble to SCFs. SAS includes the dissolution of drug in SCF miscible organic solvent, followed by drug nanoprecipitation by mixing drug solution with SCF, which acts as an antisolvent for the drug.

Liquid Atomization-Based Techniques

These techniques are based on solvent removal. In principle, it involves the spraying of a liquid or a dispersion feed into a hot drying medium to obtain a powdered solid phase. The technique involves two main steps: atomization and drying.

To get nanocrystals using this technique, it is important to achieve atomization at submicron scale (Peltonen et al. 2010). This can be achieved using effective liquid atomization methods, such as in Nano Spray Dryer from Büchi Labortechnik AG (Switzerland) (Baba and Nishida 2013).

In spray-freezing into liquid technique, drug solution is sprayed onto a cryogenic liquid, which results in immediate freezing of sprayed particles, which are then collected and lyophilized (Hu et al. 2004). The aerosol flow reactor method (AFRM) is another single-step process similar to nanospray-drying that can generate nanosized product with narrow particle size distributions (Eerikäinen et al. 2003). In AFRM, nanocrystals are formed with rapid solvent evaporation when the drug dissolved in a volatile biocompatible solvent is atomized using a collision-type air jet atomizer into a heated laminar flow tube containing a gas carrier.

NanoCrySP, a relatively new technique, generates nanocrystalline drug particles dispersed in the solidified matrix of small molecule excipients (Bansal et al. 2013, Bhatt et al. 2015, Shete et al. 2015, Shete and Bansal 2016). Excipients and process conditions are selected so as to induce crystallization and encourage nucleation. Briefly, a homogeneous solution of a drug with a crystallization inducing agent is spray-dried to obtain particles from 2 to 50 µm, where each particle comprises spatially distributed drug nanocrystals (Bansal et al. 2013).

2.4.3 Combination Techniques

If the specified critical quality attributes (CQAs) of the end product are unachievable using a single technique, combination of techniques can be utilized. This way, processing times can be significantly reduced, and often even smaller particles can be generated by coupling precipitation step with a high-energy process (Salazar et al. 2014, Soliman et al. 2017). Combination techniques often include (a) preprocess step (per say pre-milling or one of the bottom-up approaches) and (b) high-energy top-down process (such as pearl milling or HPH) (Zong et al. 2017). Process-related problems, such as clogging of HPH or chances of contamination during milling, can be completely evaded.

Combination of precipitation and HPH has been patented as Nanoedge® by Baxter International Inc. (Kipp et al. 2001). In this process, the drug is first precipitated in unstable fragile form by solvent–antisolvent precipitation technique. Then, this unstable form is transformed to a stable crystalline form in the presence of high energy generated during HPH processing. It was reported that the Nanoedge delivered more stable nanosuspension formulations than using either process alone (Zhao et al. 2010).

H69 technology is a variant of Nanoedge, where the cavitation and particle collisions take place in the high-energy zone of a homogenizer, within no more than 2 s after the particle formation (Müller and Möschwitzer 2007). Nanoedge and H69 techniques share the same disadvantages of the need of removing organic solvent residues. Two other related techniques, H96 and H42, are also combination techniques, but the pretreatment in H96 involves freeze-drying to obtain brittle particles and the pretreatment in H42 is spray-drying precipitation; both are followed by the HPH (Salazar et al. 2014).

Finally, the combination techniques are never the first choice because of the higher number of complicated processes involved and, consequently, increased overall costs. Hence, these are only recommended if desired CQAs cannot be reached using single technique only or if the raw material is, for example, so coarse that a pretreatment is required.

2.5 Characterization of Drug Nanocrystals

The interrelated characteristics such as those listed in Table 2.4 should be considered while evaluating the nanocrystal formulations (Peltonen and Hirvonen 2008, Peltonen 2018a). The critical properties of drug nanocrystals are described in detail in the following sections.

2.5.1 Particle Size and Particle Size Distribution

Size and size distribution are very critical characteristics of nanoparticle formulations as they directly affect the other properties, for example, saturation solubility, dissolution, physical stability, and even clinical efficacy. Also, the narrower the particle size distribution is, the more stable the product will be. As the particle size is decreased, the surface energy tends to increase, which supports aggregation. Photon correlation spectroscopy (PCS), static light scattering techniques, and microscopy are the most frequently used for particle size measurements of nanomaterials. PCS is commonly used to measure mean particle size and size distribution (polydispersity index) as it yields accurate results in fast and easy manner. However, if the sample is very heterogeneous in size, the reliability of PCS measurements can be low.

For intravenous delivery, it is essential to have all particles <5 µm since the diameter of the smallest blood capillary is ~ 5 µm. Having said that, even a small proportion of particles >5 µm may cause capillary blockade. The contents of microparticles in nanosuspensions can be strictly controlled by Coulter counter analysis since it gives an absolute value of particle size (Gao et al. 2008).

Difficulties in particle size analysis arise not only from the instrument but also from the sample to be analyzed. To achieve correct and reproducible results, it is of utmost importance for the sample to remain stable during the analysis. Common problems that exist during sample analysis are due to the contamination, crystal growth due to the Ostwald ripening, partial/complete dissolution, too concentrated/diluted sample affecting the light scattering, aggregation, settling of particles at the bottom of analytical vessel, and adsorption of the nanocrystals on the walls of sample cell.

TABLE 2.4 Commonly Used and More Recent Methods for Characterizing Nanocrystals

Characteristics	Method and Principle	Principle	Information	Data Type	Sample Requirements	Considerations	References
Size and morphology	Dynamic light scattering also called photon correlation spectroscopy	Fluctuation of Rayleigh scattering of light associated with Brownian motion of nanoparticles	Particle size, particle size distribution	Particle size distribution (polydispersity index), size (Z-average)	Suspension with suitable concentration, nondestructive	Suitable only for particles in nanometer size range, only for considerably monodisperse samples, viscosity of suspension and temperature affect results	Lu et al. (2014)
	Scanning electron microscopy (SEM)	Backscattering of electrons	Topographical information about particles	Scanning electron micrograph, particle morphology, size	Dry sample mounted on stage condition setup (vacuum)	Destructive sample preparation, amounts in microgram required for analysis, laborious for particle size analysis	Sarnes et al. (2014)
	Transmission electron microscopy (TEM)	Transmission of electrons	Density information	Transmission electron micrograph, morphology of cross sections, stabilizer nanocrystal interaction	Embedded cross-sectional preparation	Destructive sample preparation, amounts in microgram required for analysis, laborious for particle size analysis	Laaksonen et al. (2011)
Surface properties	Zeta-potential	Dynamic electrophoretic mobility under electric field	Surface charge (zeta-potential)	Zeta-potential, quantitative	Suspension with suitable concentration	Liquid environment affects the result	Van Eerdenbrugh et al. (2009)
	Surface plasmon resonance (SPR)	Changes in refractive index in the vicinity of a planar sensor surface	Surface adsorption	Spectrum, interaction between stabilizer drug crystals, qualitative and quantitative	Substrate on planar surface sensor required for indirect measurement of nanocrystals	Careful sample preparation required	Liu et al. (2015)
Solid state form	X-ray powder diffraction (XRPD)	Diffraction of X-rays from lattice planes	Polymorphic forms give unique diffraction peaks while amorphous form gives no peaks	Diffractogram, qualitative (type of polymorphic form) and quantitative (degree of crystallinity)	Powder, paste or slurry form, several sample presentation setups possible, amount required depends on setup	Nano materials give broad peaks due to decrease in crystal lattice size, anisotropic particle shape leads to preferred orientation effects (change in relative intensities of diffraction peaks)	Valo et al. (2010)
	Differential scanning calorimetry (DSC)	Determine the amount of energy absorbed (endothermic) or released (exothermic) by a sample as a function of time	Melting temperature, heat of fusion, glass transition temperature, crystalline phase transition temperature and energy	Thermogram, qualitative and quantitative	Powder form, few milligrams, destructive	Results are different with open and closed (hermetically sealed) pans	Valo et al. (2011)

(Continued)

TABLE 2.4 (*Continued*) Commonly Used and More Recent Methods for Characterizing Nanocrystals

Characteristics	Method and Principle	Principle	Information	Data Type	Sample Requirements	Considerations	References
	Infrared (IR) spectroscopy (mid-IR spectroscopy)	Change in dipole moment during molecular vibrations	Molecular interaction, chemical degradation, polymorphic form, crystallinity	Spectrum, qualitative and quantitative, suitable for multivariate analysis	Powder or tablet form, depends on sampling setup, few milligrams, wet samples are usually problematic, nondestructive	Sample preparation/measurement can involve pressure which can induce solid state transformations	Valo et al. (2013), Darville et al. (2014)
	Raman spectroscopy	Change in polarizability during molecular vibrations	Polymorphic form, crystallinity	Spectrum, qualitative and quantitative, suitable for multivariate analysis	Powder or suspension, few milligrams, fluorescent samples are problematic, Nondestructive	Sample heating can be problematic, samples can be in aqueous medium	Strachan et al. (2007), Darville et al. (2014)
Drug delivery	Dissolution testing dissolved drug analyzed over time, usually using UV spectroscopy or high-performance liquid chromatography (HPLC) coupled with a detector	–	Dissolution profile	Solution concentration vs time	Filtration or centrifugation is performed to obtain clear solution, protect sample from light, sampling should be done from the same place	Separating nanocrystals from dissolution medium can be problematic	Liu et al. (2013)
	Fluorescence microscopy	Fluorescence by endogenous or added fluorophores	Localization of nanocrystals in relation to cells and tissues	Fluorescence imaging	Non-fluorescent nanocrystals require fluorophore to physically entrapped into nanocrystals	Entrapment and leakage of fluorophore can be difficult or problematic	Valo et al. (2010)
	Nonlinear Raman microscopy	Change in polarizability during molecular vibrations	Label free localization of particles	Intensity of coherent anti-Stokes Raman scattering (CARS) shift (narrow band) or spectrum, (multiplex or broad band), most commonly qualitative, 2D or 3D images	Colored and two-photon fluorescent samples can interfere with signal can be coupled with other nonlinear phenomena such as second harmonic generation or two-photon electronic fluorescence	Label free, optimal lateral spatial resolution approximately 300–400 nm	Strachan et al. (2011), Darville et al (2014)

Source: Reprinted from Peltonen and Strachan (2015) by Creative Commons Attribution 4.0 International Public Licence.

2.5.2 Shape and Morphology

Besides the particle size information, transmission electron microscope (TEM), scanning electron microscope (SEM), and atomic force microscope (AFM) can be used to study the shape and morphology of drug nanoparticles (Figure 2.5). A wet formulation is required for TEM analysis, while for SEM and AFM, the sample should be in powder/dry state. Due to the potential crystal growth in the suspension state, it is preferred to obtain nanocrystals in the dry form. However, the process of drying can lead to agglomeration. Therefore, often the SEM is utilized to monitor changes in particle structures before and after the drying procedure (Gao et al. 2008).

2.5.3 Surface Properties

Most colloidal dispersions in aqueous media display an electric charge, which could arise due to the ionization of surface groups, differential loss of ions from the crystal lattice, or the surface adsorption of charged species. Therefore, charge measurements are greatly influenced by the pH, conductivity, and concentration of the formulation components. Net charge at the particle surface affects ions distribution in the surrounding interfacial region, resulting in increased concentration of counter ions close to the surface, which is known as the electric double layer around each particle.

Zeta-potential indicates how much charge is present on the particle surface by measuring particle's electrophoretic mobility in an electric field. Zeta-potential gives a clear indication of the potential stability of the nanosuspension formulation, if the suspension is stabilized only by electric forces. If particles in the suspension have zeta-potential values more positive than +30 mV or more negative than −30 mV, then they tend to repel each other, and thus, the formulation is considered to be stable one. However, if the particles have zeta-potential values close to zero, the particles may come closer and flocculate, if no steric stabilizer is present. Therefore, careful selection of the stabilizer is very important in the development of nanocrystal-based formulations (Liu et al. 2011).

2.5.4 Solid State Form

Determining solid state properties pre- and post-nanonization is highly significant, as a change in solid state, such as the degree of crystallinity, polymorphic form, or existence of solvate forms, can affect the stability, apparent solubility, as well as dissolution rate. Metastable crystalline and amorphous forms of drugs have better dissolution properties than the thermodynamically most stable crystalline form. However, development of formulations with such less stable crystalline forms or amorphous material needs extra care since uncontrolled shifting of the metastable polymorphic form or amorphous form to the stable crystal form can take place during product storage, affecting the drug bioavailability as a consequence (Surwase et al 2013).

Different nanocrystal processing conditions and environmental factors can affect the drug's solid state form. For example, hydrate forms are more stable but less soluble in aqueous media, and therefore, possibility for conversion of a potential hydrate forming drug should be thoroughly studied during the stability studies. Also most of the drugs are known to exist in more than one crystal form, e.g., having different polymorphs with various solubility properties. Techniques such as differential scanning calorimetry (DSC), XRPD, and vibrational spectroscopies (infrared (IR) and Raman) are used most commonly to determine potential transformations in the solid state (Table 2.3).

Bottom-up precipitation techniques may produce nanocrystals with partial amorphousness, with unfavorable effects on the nanocrystals stability (Ali et al. 2011). Liquid atomization-based techniques are specifically prone to generating partial or fully amorphous drug products, but

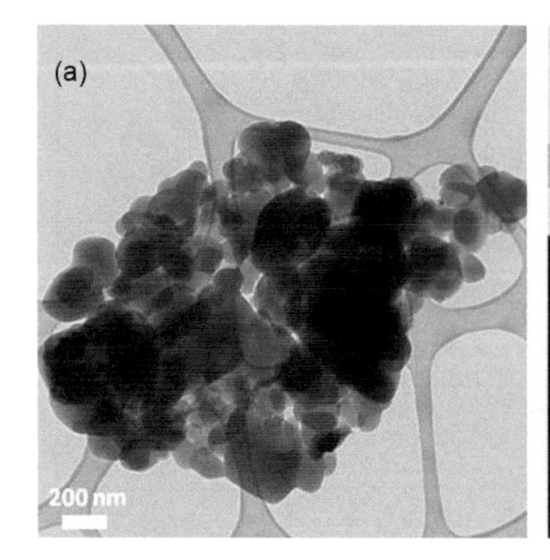

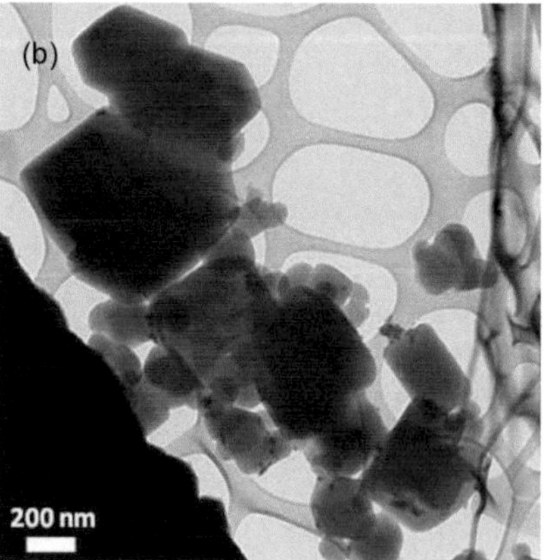

FIGURE 2.5 TEM images of budesonide nanosuspensions stabilized with (a) asolectin from soybean and (b) poloxamer. (Reprinted from International Journal of Pharmaceutics, Vol 441, J. Raula, A. Rahikkala, T. Halkola, J. Pessi, L. Peltonen, J. Hirvonen, K. Järvinen, T. Laaksonen, E.I. Kauppinen, Coated particle assemblies for the concomitant pulmonary administration of budesonide and salbutamol sulphate, 248–254, Copyright (2013), with permission from Elsevier.)

annealing can be used to achieve full crystallinity (Wang et al. 2012b). The high shear, collision, and cavitation stresses associated with HPH and shear stress with wet ball milling can also induce polymorphic changes. In milling or homogenization, water is used mainly as the dispersion solvent, which raises molecular mobility and thus reduces the tendency for drug amorphization.

2.5.5 Solubility and Dissolution

Solubility of a drug in its most stable crystalline form in a medium at certain temperature and pressure is its thermodynamic solubility, which is also referred to as the solubility at saturated state. Metastable polymorphic forms, higher energy amorphous forms, or nanosized drug particles temporarily show solubility values higher than the thermodynamic solubility, which is described as kinetic or apparent solubility. Faster dissolution of nanosized drug crystals leads to the high apparent solubility reaching supersaturated state, which is also known as the "spring effect" (Surwase et al. 2013).

Higher apparent solubility reached with drug nanocrystals was demonstrated in a study conducted by Ige et al. (2013), when the solubility of fenofibrate after size reduction from 80 μm to 460 nm was increased by ten folds in 0.5% SDS solution. Gahoi et al. investigated the effect of nanomilling on the dissolution of lumefantrine and found almost 100% dissolution from nanopowder formulation in comparison with only 50% release from unmilled material in 0.1N HCl dissolution media also containing 0.25% benzalkonium chloride (Gahoi et al. 2012). Similarly, Sarnes and colleagues found an increase in intrinsic dissolution rates of indomethacin after size reduction, which was 0.05 $\mu g/min/mm^2$ for the bulk drug and 0.50 $\mu g/min/mm^2$ for the nanoformulation in the presence of poloxamer F68 as stabilizer (Figure 2.1) (Sarnes et al. 2013). In the same study, the concentration levels next to the nanoparticle surfaces exceeded five times the thermodynamic saturation solubility concentration of the drug.

Separation of nanocrystals during dissolution and solubility testings requires great attention before the drug concentration determinations. Ultracentrifugation and filtration (Liu et al. 2013) or their combination (Valo et al. 2013) are generally used to separate nanocrystals during the sample preparation; however, they both have their own limitations. During filtration, the dissolved drug may get adsorbed on the surface of filter membrane and small particles can pass through it. Drug adsorption is also common with the centrifugation, where the adsorption happens between the drug and the centrifugation tube. In addition, centrifugation can further lead to drug dissolution due to the long run time and increased temperature (Liu et al. 2013). Careful selection of filters can avoid issues related to filtration, which makes the filtration often a preferred choice (Liu et al. 2013).

Another problem in reliable concentration determinations with nanoparticle systems can be the presence of undissolved nanoparticles. Absence of undissolved particles from the sample can be confirmed by testing the absorbance at two different UV wavelengths in a conventional UV spectrophotometer: one at drug's analytical wavelength (λ_{max} to quantify dissolved amount) and one at a wavelength, where there is no UV absorption by the dissolved drug (to determine the possible presence of undissolved particles) (Sarnes et al. 2013).

Recrystallization of the drug from a supersaturated solution to achieve thermodynamic solubility is not unknown. Furthermore, this is more challenging for drugs showing pH-dependent solubility, because variation in pH and ionic strength in the GI tract may ultimately affect the solubility and, thus, the tendency for crystallization (Figure 2.6). Efforts have been made to maintain the supersaturated state and hinder the precipitation (parachute effect) by adding some polymers to improve the bioavailability *in vivo*. Polymers that have been found effective in maintaining the supersaturated state are, for example, HPMC (Surwase et al. 2015), methacrylate copolymers (Jung et al. 1999), PVP (Abu-Diak et al. 2011), and HPMC acetate succinate (HPMC-AS) (Surwase et al. 2015).

2.6 Nanocrystal-Based Formulations

Physicochemical properties of drug particles affect both the drug substance and drug product. Therefore, many CQAs identified for a drug substance can be related to the quality target profile of the drug product. In recent years, nanocrystal-based formulations have shown great potential in drug delivery due to ease in scale-up and producing almost excipient-free end product (Kaialy and Al Shafiee 2016, Peltonen and Hirvonen 2018). Earlier, several potential molecules showing high *in vitro* potential used to get rejected on the basis of being poorly water soluble since it is difficult to translate such molecules into efficient drug delivery formulations *in vivo*. Nanonization has enabled formulators to (re)reconsider such molecules and take them further, which is demonstrated by the presence of increasing number of commercially available products based on nanocrystals (Table 2.3, Figures 2.7 and 2.8).

First nanosuspension formulation, Megace ES® (625 mg/5 mL daily dose), was approved for the delivery of megestrol acetate. In comparison with earlier formulations, nanosuspension formulation of megestrol acetate was less viscous and the administration volume was reduced to only 1/4. Reduction in the fed and fasted state variability was also noticed. This example shows that with the use of optimum excipients, particularly drugs with high doses can be prepared that will remain stable throughout the whole shelf life.

By virtues of ease of administration and physical stability, oral solid dosage forms of nanocrystals are usually preferred for commercialization (Figure 2.8). In oral drug delivery, nanocrystals contribute to improved drug absorption by

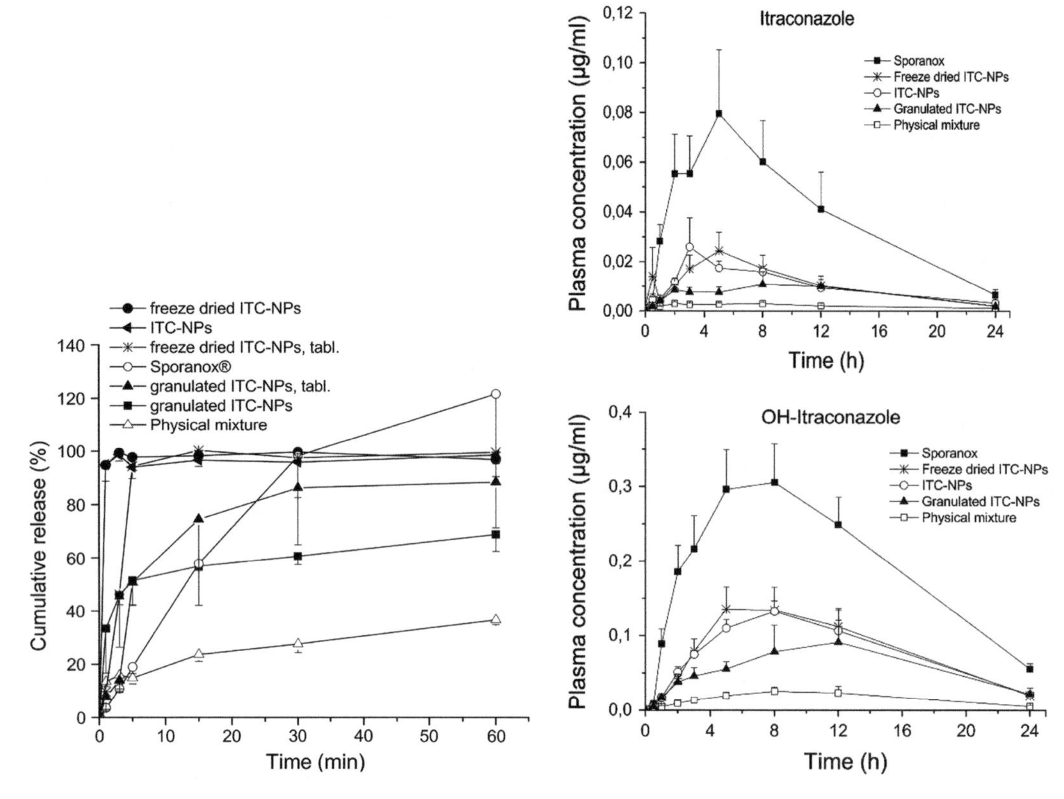

FIGURE 2.6 *In vitro* faster dissolution of the itraconazole nanocrystal formulations (ITC-NPs) as compared to Sporanox® granules (left) was not turned out to improved bioavailability: fast dissolved itraconazole from nanocrystals was precipitated in the intestine due to 250 times lower solubility in intestine as compared to the acidic stomach (right). (Reprinted from Journal of Controlled Release, Vol 180, A. Sarnes, M. Kovalainen, M.R. Häkkinen, T. Laaksonen, J. Laru, J. Kiesvaara, J. Ilkka, O. Oksala, S. Rönkkö, K. Järvinen, J. Hirvonen, L. Peltonen, Nanocrystal-based per-oral itraconazole delivery: superior in vitro dissolution enhancement versus Sporanox® is not realized in in vivo drug absorption, 109–116, Copyright (2014), with permission from Elsevier.)

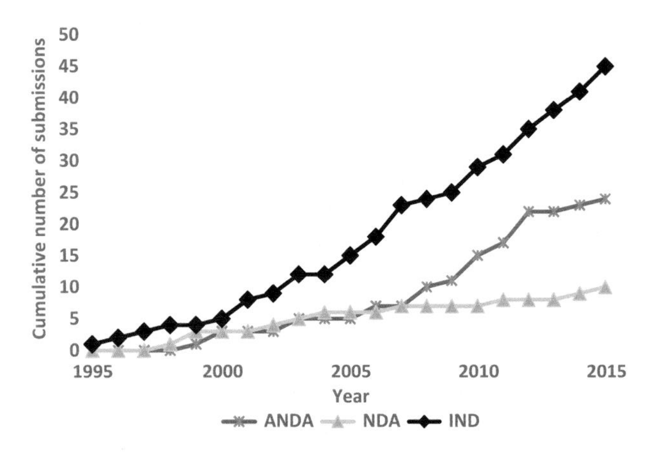

FIGURE 2.7 Cumulative number of abbreviated new drug applications (ANDAs), new drug applications (NDAs), and investigational new drug (IND) submissions for US FDA's Center for Drug Evaluation and Research containing drug nanocrystals between the years 1995 and 2015 (Chen et al. 2017).

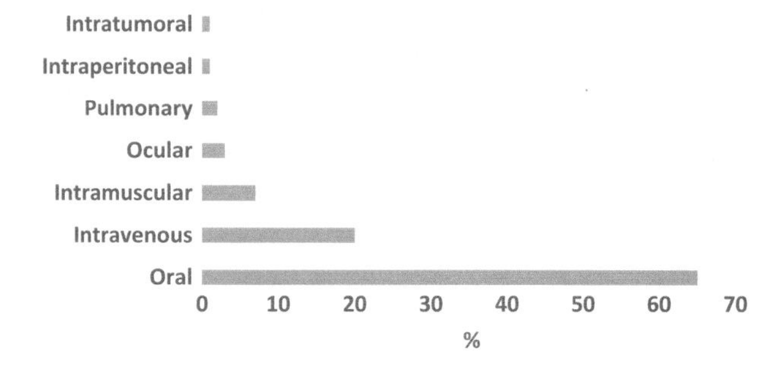

FIGURE 2.8 Different administration routes in percentages for all the FDA-submitted nanocrystal-based products during the years 1973 and 2015 (Chen et al. 2017).

improving solubility and dissolution, and by improving adherence to the GI lining. These two factors provide a high drug concentration gradient between the GI tract and first-pass blood circulation. However, the impact of GI factors, such as variation of pH and peristalsis, is difficult to predict on the performance of nanocrystal formulation.

Nanonization is just one way to manipulate the material properties of the drugs, and in order to make products for the patients, the final formulation is required. In final formulations, drug nanocrystals can be used in semisolid form without drying, like in ocular suspensions or topical ointments (Shegokar and Müller 2010, Tuomela et al. 2014). However, often nanosuspensions need to be dried, for example, for tableting, capsulation, or sterile iv formulations (Malamatari et al. 2016).

2.6.1 Drying of Drug Nanosuspensions

Tablets and capsules are the most readily available choices for product development for oral delivery, and they require solidification of the nanomaterial after nanonization. The primary nanodispersed system of drug particles after nanonization can be dried and then formulated into tablets, capsules, or pellets for oral delivery or simply as dry powders for various delivery routes (Malamatari et al. 2016). There are two strategies to obtain dry nanobased products: (i) spray-drying (Kumar et al. 2014b, Kumar et al. 2015), lyophilization (Tuomela et al. 2015), or oven drying (Cunha-Filho et al. 2008) to generate nanoparticles; and (ii) granulation and coating methods where the combination of spraying, drying, and shaping works together. Here, the nanosuspensions can be used as a granulating liquid in the granulation process or as layering dispersion in a fluidized bed process (Rabbani and Seville 2005, Figueroa and Bose 2013, Tan et al. 2017). Spray- and freeze-drying techniques are usually selected for preparing sterile (iv) formulations, whereas fluidized bed granulation, coating, and pelletization are selected for solid orals for being relatively more straightforward processes.

Fast re-dispersion and dissolution of the dried nanoproduct in the dispersion medium is an important parameter to evaluate the drying process. The critical process parameters, CPPs, of drying process, drug properties (hygroscopicity, melting point, etc.), the choice of dispersion media (polarity, boiling point, etc.), and nature of stabilizer can affect the re-dispersibility (Loh et al. 2011).

Freeze-drying process may induce stress on the nanoparticles; therefore. excipients such as cryo- and lyoprotectants are usually added to protect the formulation from freezing stress and drying stress, respectively (Fonte et al. 2016). In freeze-drying process, nature of cryo- and lyoprotectants (Kumar et al. 2014a), fast freezing, and high cryoprotectant concentration can be beneficial for re-dispersibility (Lee et al. 2009). The commonly used excipients include water-soluble sugars, such as mannitol, lactose, sucrose, trehalose, and microcrystalline cellulose (Kumar et al. 2014a) to protect the nanocrystals from aggregation. The presence of polymers used to stabilize nanosuspension can also sometimes exhibit good re-dispersibility after drying.

Lyophilization and spray-drying result in ultra-fine powders that need additional processing to ameliorate flow properties before converting the powder into a final capsule or tablet dosage form. Granulation of nanoparticles to obtain dense micron-sized granules with improved flow properties has also been studied, which is easier, more straightforward, and economical as it does not require maintaining long freezing conditions (Basa et al. 2008, Pieterjan et al. 2011). However, freeze-drying may prove to be advantageous in providing faster dissolution in comparison with the granulated product, since granulation may exhibit lag time in drug release (Sarnes et al. 2014). Bose and colleagues investigated spray granulation as a processing technique to convert a nanosuspension of a poorly soluble drug into a solid dosage form, where formulation parameters such as the level of stabilizer, granulation substrate, and drug loading in the granulation were evaluated (Bose et al. 2012). At drug loading of 10%, dissolution profile was comparable to the nanosuspension. However, at 20% of drug loading, slightly slower release was noticed due to the formation of agglomerates. *In vivo* PK studies in beagle dogs also showed an increase in AUC and $C_{\max}$ in comparison with coarse drug formulation.

Spray granulation process is typically governed by the physicochemical properties of the binder solution (primarily viscosity, density, and surface tension), process parameters (air velocity, binder atomization pressure, droplet size, and binder feed rate), and also the properties of the granulation substrate (Iveson et al. 2001). Nanosuspension dispersion medium with polymeric stabilizer can itself act as a binder solution. Binder solution atomized at high pressure produces very fine droplets, which could distribute uniformly on the powder substrate leading to minimal localized wetting. Further, low feed rate, high inlet air velocity, and optimum temperature reduce the availability of binder to form liquid bridges between the particles, thus giving friable granules (Bose et al. 2012, Figueroa and Bose 2013). With an aim of obtaining fast dissolution profile after drying, granules with poor strength are desirable, for which no additional binder is needed than the excipients already present in the nanosuspension. However, such granules tend to generate more fines, and thus, broad particle size distribution of granules could be achieved.

In tableting, the final amount of drug nanocrystals in the formulation affects the product performance. Tuomela and group studied different amounts of nanocrystals content in the tablet composition, when manufactured either by direct compression or by wet granulation (Tuomela et al. 2015). Nanocrystals of indomethacin and itraconazole produced by top-down wet milling approach were freeze-dried; part of the freeze-dried nanoparticles were used for direct compression, and the rest were granulated into microgranules. The disintegration time of tablets were found to be in decreasing order of amounts of nanocrystals present in the tablets. Less amounts of nanocrystals made the tablets more porous and less dense, which resulted in faster medium penetration. Therefore, by changing the percentage of nanocrystals in the formulation, drug release rates can be controlled.

The importance of understanding GI factors, such as variation of pH and peristalsis, and their impact on the performance of nanocrystal formulations were shown in another study, where *in vivo* pharmacokinetic studies of freeze-dried and granulated itraconazole nanocrystals were compared with a marketed formulation (Sporanox®, Janssen Pharmaceuticals, Inc.) (Sarnes et al. 2014). Though *in vitro* the nanoformulations demonstrated efficient dissolution (sink conditions), Sporanox outperformed all the study formulations of itraconazole *in vivo* (Figure 2.6). This was due to the rapid transit of the dissolved drug to small intestine, where itraconazole has low solubility leading to recrystallization. Accordingly, the extended parachute effect was not seen in the study. But, when the itraconazole nanocrystals were impeded into a cellulose matrix, bioavailability of nanocrystal formulations as compared to marketed Sporanox® granules was 1.2–1.3 times higher (Figure 2.9) (Valo et al. 2011). In this case, higher bioavailability reached with drug nanocrystals was probably due to the increased residence time of the formulation in the stomach, optimum dissolution area.

2.6.2 Semisolid Nanocrystal Formulations

In many drug delivery routes, nanosuspensions can be used without drying in semisolid form for drug delivery purposes. For example, aqueous nanosuspensions can be nebulized using mechanical or ultrasonic nebulizers for pulmonary drug delivery. Müller and Jacobs (2002) delivered budesonide nanoparticles with a nebulizer and the pharmacokinetic parameters showed comparable AUC, and higher $C_{\max}$ and lower $T_{\max}$ values as that of the control microparticulate dosage form. The small size of the nanoparticles presumably led to a more uniform distribution of budesonide in the lungs. The nanocrystals also increased the adhesiveness and prolonged the pulmonary residence time, in a similar fashion to oral drug delivery examples above.

In ocular drug delivery, drug suspensions have already been used previously, but the main benefits of nanosuspensions are lower eye irritation due to the smaller particle size. Ocular delivery of an antiglaucoma drug forskolin as a nanosuspension provided a more rapid or sustained release of the drug, depending on the drug formulation properties. According to Araújo et al. (2009), the poor bioavailability of drugs from ocular dosage forms is mainly due to precorneal loss factors (e.g., tear dynamics, nonproductive absorption, transient residence time in the cul-de-sac, and relative impermeability of the corneal epithelial membrane). Thus, there is a true need for effective topical ocular drug formulations capable of promoting drug penetration and maintaining therapeutic levels with a reasonable frequency of application and, at the same time, reducing the unwanted side effects.

As reported by Shegokar and Müller, for delivery to the skin nanocrystals were first exploited in the field of cosmetics to deliver natural antioxidants such as rutin and hesperidin and later expanded for drug delivery purposes (Shegokar and Müller 2010, Pireddu et al. 2016). Nanocrystals have better deposition/adhesion on the skin surface (in particular follicles), providing high concentration gradient because of higher kinetic saturation solubility, thus better partitioning contributing to increased dermal delivery (Vidlářová et al. 2016). Depending on the shape, nanocrystals have also demonstrated skin penetration (Larese Filon et al. 2015). Nanocrystals are formulated as gels and lotions for topical delivery where nanocrystals are retained on the surface for sufficient time period to slowly release the active constituent (Shaal et al. 2010, Mitri et al. 2011).

In parenteral delivery, reduced particle size enables intravenous administration of poorly soluble drugs without obstructing the blood capillaries. The parenteral route (iv, im, sc) of administration provides a quick onset of drug action and reduces the amount of the drug in the parenteral dosage form. Iv administration avoids the first-pass metabolism and potential drug irritation/degradation in the GI tract. Wang et al. (2012a) formulated puerarin nanocrystals and evaluated the pharmacokinetic parameters of the drug after intravenous delivery. Puerarin is used to treat coronary heart disease and presents a good candidate for iv delivery as a poorly water-soluble compound with adverse drug reactions in solution with cosolvents. Drug delivery and absorption results of puerarin in beagle dogs revealed that the puerarin nanocrystals significantly reduced $C_{\max}$ and clearance values, and resulted in a significantly greater MRT (mean residence time), clearance and

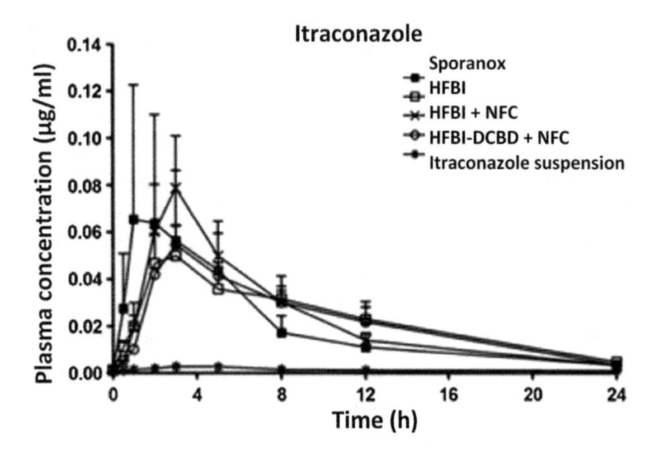

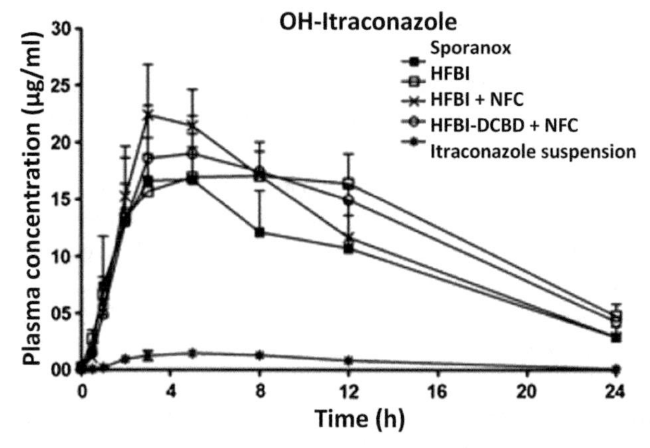

FIGURE 2.9 Plasma concentrations of itraconazole and OH-itraconazole as a function of time after oral administration in rats of three different hydrophobin-stabilized drug nanocrystal formulations (HFBI, HFBI + NFC, and HFBI-DCBD + NFC), Sporanox® granules and itraconazole microsuspension. (Reprinted from Journal of Controlled Release, Vol 156, H. Valo, M. Kovalainen, P. Laaksonen, M. Häkkinen, S. Auriola, L. Peltonen, M. Linder, K. Järvinen, J. Hirvonen, T. Laaksonen, Immobilization of protein-coated drug nanoparticles in nanofibrillar cellulose matrices – enhanced stability and release, 390–397, Copyright (2011), with permission from Elsevier.)

elimination half-life values compared to the puerarin solution, indicating higher drug efficacy and also drug safety by the nanocrystals.

Nanoparticles may also have potential as novel intravascular probes for both diagnostic (e.g., imaging) and therapeutic purposes (e.g., drug delivery). Critical factors for such nanoparticle delivery include the ability to target specific tissues and cell types, for example, the reticuloendothelial system (Åkerman et al. 2002). Targeting to macrophages may be based on the size and surface properties of the nanoformulation and changing of the stabilizer can easily alter the *in vivo* behavior (Lakshmi and Kumar 2010). The drug nanocrystals will be uptaken by the mononuclear phagocytic system to enable anti-mycobacterial, anti-fungal, or anti-leishmanial activity, as the infectious pathogens are often persisting intracellularly.

2.7 Conclusions and Future Perspectives

Drug nanocrystals have been widely studied for improving dissolution properties of poorly soluble drug materials, especially since the beginning of the 1990s, when poor aqueous solubility and dissolution properties restricted and prevented the development and commercialization of numerous new chemical entities. Nanocrystals are formed from 100% drug, which is covered by stabilizer(s) layer(s). The main benefit of nanocrystals compared to other drug nanosystems is their relative simplicity: a small enough particle size and few formulation components. These properties are also the main reasons for the early commercial successes of nanocrystalline products (Table 2.3, Figure 2.7). Thus, the nanocrystalline products have included product and formulation improvements to existing drugs, and later on also new chemical entities have increasingly been taken under research, development, and marketing.

The first nanocrystal products on the market were formulated as tablets, but concomitantly, nanocrystal formulations have been studied and/or commercialized also via ocular, parenteral, dermal, pulmonary, and buccal drug delivery routes (see the previous chapter); both dry nanocrystals and nanocrystals in suspension have been utilized. Overall, dissolution enhancing drug nanocrystals have been a big success story in drug development. This line of nanocrystalline research and new products on the market can be estimated to continue successfully also in the future.

Recently, nanoformulations including nanocrystals have been discussed frequently in the context of theranostics. Theranostics approach combines the utilization of nanocrystals towards medical diagnostics with subsequent drug treatment. Nanoparticles and nanocrystals exhibit significant potential in this area because many different functionalities can be combined in one particle/crystal (Kovalenko et al. 2015). Theranostic nanocrystals comprise a crystalline (drug containing) core with an optional organic coating by ligand(s), such as adsorbed protein(s). The core nanocrystals act as passive carriers, while the linked molecules on the particles surfaces might enable multifunctionality. These (multi)functional molecules may be ligands for specific targeting, for reduced interactions with the immune system, or for contrast agents in imaging purposes. Furthermore, perhaps one could add some functionalities also in the core of nanocrystal particles, such as fluorescence, superparamagnetism, or plasmon resonance (Kovalenko et al., 2015).

In the future, thus, building up and using of the nanocrystals might be suitable for combining drug containing fluorescent, magnetic, or radioactive cores with fluorescent, magnetic, or radioactive shells that could be detected with fluorescence microscopy, magnetic resonance imaging (MRI), or single-photon emission computed tomography. These examples are based on particles that report data for diagnosis to a physician (e.g., images of tissue, local analyte concentrations) and can, also, initiate the drug treatment, for example, in tumor therapy. However, as of today, the clinical use of theranostic nanocrystals is very little, if non-existent. There might be potential problems like cytotoxic effects of the nanocrystals and their core and surface contents. The main hurdle remaining is, however, the poorly controlled targeting of the nanocrystals—a problem intensified by the fast dissolution of the nanocrystalline core particles. Nanocrystals might enable, for example, passive tumor targeting governed by the small size and the enhanced permeation and retention (EPR) effect. However, passive targeting is oftentimes not sufficient and may involve significant side effects, such as nanocrystals clearance into the immune system or accumulation into the liver and spleen. Active targeting of nanocrystals by ligand specific reactions would, perhaps, increase the applicability of nanocrystalline formulations to new levels in the future.

References

Abu-Diak, O. A., Jones, D. S. and Andrews, G. P. 2011. An investigation into the dissolution properties of celecoxib melt extrudates: Understanding the role of polymer type and concentration in stabilizing supersaturated drug concentrations. *Mol Pharm* 8: 1362–71.

Alam, M. A., Al-Jenoobi, F. I. and Al-Mohizea, A. M. 2013. Commercially bioavailable proprietary technologies and their marketed products. *Drug Discov Today* 18: 936–49.

Ali, H. S., York, P., Ali, A. M. et al. 2011. Hydrocortisone nanosuspensions for ophthalmic delivery: A comparative study between microfluidic nanoprecipitation and wet milling. *J Control Release* 149: 175–81.

Al-Kassas, R., Bansal, M. and Shaw, J. 2017. Nanosizing techniques for improving bioavailability of drugs. *J Control Release* 260: 202–12.

Åkerman, M. E., Chan, W. C. W., Laakkonen, P. et al. 2002. Nanocrystal targeting in vivo. *PNAS* 99: 12617–21.

Araújo, J., Gonzales, E., Egea, M. A. et al. 2009. Nanomedicines for ocular NSAIDs: Safety on drug delivery. *Nanomedicine* 5: 394–401.

Baba, K. and Nishida, K. 2013. Steroid nanocrystals prepared using the nano spray dryer B-90. *Pharmaceutics* 5: 107–14.

Babu, N. J. and Nangia, A. 2011. Solubility advantage of amorphous drugs and pharmaceutical cocrystals. *Cryst Growth Des* 11: 2662–79.

Bansal, A. K., Dantuluri, A. K. R., Shete, G. et al. 2013. Nanocrystalline solid dispersion compositions and process of preparation thereof. WO Patent application WO2013132457A2.

Basa, S., Muniyappan, T., Karatgi, P. et al. 2008. Production and in vitro characterization of solid dosage form incorporating drug nanoparticles. *Drug Dev Ind Pharm* 34: 1209–18.

Bergström, C. A. S., Charman, W. N., Porter, C. J. H. 2016. Computational prediction of formulation strategies for beyond-rule-of-5 compounds. *Adv Drug Deliv Rev* 101: 6–21.

Bhatia, H., Read, E., Agarabi, C. et al. 2016. A design space exploration for control of critical quality attributes of mAb. *Int J Pharm* 512: 242–52.

Bhatt, V., Shete, G. and Bansal, A. K. 2015. Mechanism of generation of drug nanocrystals in celecoxib: mannitol nanocrystalline solid dispersion. *Int J Pharm* 495: 132–39.

Bi, Y., Liu, J., Wang, J. et al. 2015. Particle size control and the interactions between drug and stabilizers in an amorphous nanosuspension system. *J Drug Deliv Sci Tech* 29: 167–72.

Bilensoy, E. and Hincal, A. T. 2009. Recent advances and future directions in amphiphilic cyclodextrin nanoparticles. *Expert Opin Drug Deliv* 6: 1161–73.

Bilgili, E. and Afolabi, A. 2012. A combined microhydrodynamics-polymer adsorption analysis for elucidation of the roles of stabilizers in wet stirred media milling. *Int J Pharm* 439: 193–206.

Bobo, D., Robinson, K. J., Islam, J. et al. 2016. Nanoparticle-based medicines: A review of FDA-approved materials and clinical trials to date. *Pharm Res* 33: 2373–87.

Borchard, G. 2015. Drug nanocrystals. In *Non-biological Complex Drugs*, ed. D. Crommelin and J. de Vlieger, 171–189. Cham: Springer.

Bose, S., Schenck, D., Ghosh, I. et al. 2012. Application of spray granulation for conversion of a nanosuspension into a dry powder form. *Eur J Pharm Sci* 47: 35–43.

Breitung-Faes, S. and Kwade, A. 2013. Prediction of energy effective grinding conditions. *Miner Eng* 43–44: 36–43.

Butler, J. M. and Dressman, J. B. 2010. The developability classification system: Application of biopharmaceutics concepts to formulation development. *J Pharm Sci* 99: 4940–54.

Censi, R. and Di Martino P. 2015. Polymorph impact on the bioavailability and stability of poorly soluble drugs. Molecules 20: 18759–76.

Cerpnjak, K., Zvonar, A., Gasperlin, M. et al. 2013. Lipid-based systems as a promising approach for enhancing the bioavailability of poorly water-soluble drugs. *Acta Pharm* 63: 427–45.

Chen, M.-L., John, M., Lee, S. L. et al. 2017. Development considerations for nanocrystal drug products. *AAPS J* 19: 642–51.

Cornaire, G., Woodley, J., Hermann, P. et al. 2004. Impact of excipients on the absorption of P-glycoprotein substrates in vitro and in vivo. *Int J Pharm* 278: 119–31.

Cunha-Filho, M. S. S., Martínez-Pacheco, R. and Landín, M. 2008. Dissolution rate enhancement of the novel antitumoral β-lapachone by solvent change precipitation of microparticles. *Eur J Pharm Biopharm* 69: 871–77.

Darville, N., Saarinen, J., Isomäki, A. et al. 2015. Multimodal non-linear optical imaging for the investigation of drug nano-/microcrystal-cell interactions. *Eur J Pharm Biopharm* 96: 338–48.

Darville, N., Van Heerden, M., Vynckier, A. et al. 2014. Intramuscular administration of paliperidone palmitate extended-release injectable microsuspension induces a subclinical inflammatory reaction modulating the pharmacokinetics in rats. *J Pharm Sci* 103: 2072–87.

Dressman, J. B. and Reppas, C. 2000. In vitro-in vivo correlations for lipophilic, poorly water-soluble drugs. *Eur J Pharm Sci* 11(Suppl 2): S73–80.

Drozd, K. V., Manin, A. N., Churako, A. V. et al. 2017. Novel drug-drug cocrystals of carbamazepine with para-aminosalicylic acid: Screening, crystal structures and comparative study of carbamazepine cocrystal formation thermodynamics. *Cryst Eng Comm* 19: 4273–86.

Eerikäinen, H., Watanabe, W., Kauppinen, E. I. et al. 2003. Aerosol flow reactor method for synthesis of drug nanoparticles. *Eur J Pharm Biopharm* 55: 357–60.

Figueroa, C. E. and Bose, S. 2013. Spray granulation: importance of process parameters on in vitro and in vivo behavior of dried nanosuspensions. *Eur J Pharm Biopharm* 85: 1046–55.

Fonte, P., Reis, S. and Sarmento, B. 2016. Facts and evidences on the lyophilization of polymeric nanoparticles for drug delivery. *J Control Release* 225: 75–86.

Fu, Q., Ma, M., Li, M. et al. 2017. Improvement of oral bioavailability for nisoldipine using nanocrystals. *Powder Technol* 305: 757–63.

Gadadare, R., Mandpe, L. and Pokharkar, V. 2015. Ultra rapidly dissolving repaglinide nanosized crystals prepared via bottom-up and top-down approach: Influence of food on pharmacokinetics behavior. *AAPS PharmSciTech* 16: 787–99.

Gahoi, S., Jain, G. K., Tripathi, R. et al. 2012. Enhanced antimalarial activity of lumefantrine nanopowder prepared by wet-milling DYNO MILL technique. *Colloids Surf B Biointerfaces* 95: 16–22.

Gao, L., Zhang, D. and Chen, M. 2008. Drug nanocrystals for the formulation of poorly soluble drugs and its application as a potential drug delivery system. *J Nanopart. Res* 10: 845–62.

Gao, L., Liu, G., Ma, J. et al. 2012. Drug nanocrystals: In vivo performances. *J Control Release* 160: 418–30.

Gao, L., Liu, G., Ma, J. et al. 2014. Paclitaxel nanosuspension coated with P-gp inhibitory surfactants: II ability to reverse the drug-resistance of H460 human lung cancer cells. *Colloids Surf B Biointerfaces* 117: 122–27.

Garad, S., Wang, J., Joshi, Y. et al. 2010. Preclinical development for suspensions. In *Pharmaceutical Suspensions: From Formulation Development to Manufacturing*, ed. A. K. Kulshreshtha, O. N. Singh, O. N. and G. M. Wall 127–176. New York: Springer.

Ghosh, I., Schenck, D., Bose, S. et al. 2013. Identification of critical process parameters and its interplay with nanosuspension formulation prepared by top down media milling technology—A QbD perspective. *Pharm Dev Technol* 18: 719–29.

Ghosh, I., Schenck, D., Bose, S. et al. 2012. Optimization of formulation and process parameters for the production of nanosuspension by wetmedia milling technique: effect of vitamin E TPGS and nanocrystal particle size on oral absorption. *Eur J Pharm Sci*, 47: 718–28.

Girotra, P., Singh, S. K. and Nagpal, K. 2013. Supercritical fluid technology: A promising approach in pharmaceutical research. *Pharm Dev Technol* 18: 22–38.

Hansen, K., Kim, G., Desai, K. G. et al. 2015. Feasibility investigation of cellulose polymers for mucoadhesive nasal drug delivery applications. *Mol Pharm* 12: 2732–41.

Haynes, D. H. 1992. Phospholipid-coated microcrystals: Injectable formulations of water insoluble drugs. US Patent 5,091,188.

He, W., Lu, Y., Qi, J. et al. 2013. Formulating food protein-stabilized indomethacin nanosuspensions into pellets by fluid-bed coating technology: Physical characterization, redispersity, and dissolution. *Int J Nanomed.* 8: 3119–28.

He, S., Yang, H., Zhang, R. et al. 2015. Preparation and in vitro-in vivo evaluation of teniposide nanosuspensions. *Int J Pharm* 478: 131–37.

He, S., Zhou, B., Zhang, S. et al. 2011. Preparation of nanoparticles of magnolia bark extract by rapid expansion from supercritical solution into aqueous solutions. *J Microencapsul* 28: 183–89.

Hosseinpour, M., Vatanara, A. and Zarghami, R. 2015. Formation and characterization of beclomethasone dipropionate nanoparticles using rapid expansion of supercritical solution. *Adv Pharm Bull* 5: 343–49.

Hu, J., Johnston, K. P. and Williams 3rd, R. O., 2004. Rapid dissolving high potency danazol powders produced by spray freezing into liquid process. *Int J Pharm* 271: 145–54.

Huttunen, K. M., Raunio, H. and Rautio, J. 2011. Prodrugs – from serendipity to rational design. *Pharmacol Rev* 63: 750–71.

Ige, P. P., Baria, R. K. and Gattani, S. G. 2013. Fabrication of fenofibrate nanocrystals by probe sonication method for enhancement of dissolution rate and oral bioavailability. *Colloids Surf B Biointerfaces* 108: 366–73.

Iveson, S. M., Wauters, P. A. L., Forrest, S. et al. 2001. Growth regime map for liquid-bound granules: further development and experimental validation. *Powder Technol* 117: 83–97.

Juhnke, M., Märtin, D. and John, E. 2012. Generation of wear during the production of drug nanosuspensions by wet media milling. *Eur J Pharm Biopharm* 81: 214–22.

Jung, J.-Y., Yoo, S. D., Lee, S.-H. et al. 1999. Enhanced solubility and dissolution rate of itraconazole by a solid dispersion technique. *Int J Pharm* 187: 209–18.

Kaialy, W. and Al Shafiee, M. 2016. Recent advances in the engineering of nanosized active pharmaceutical ingredients: Promises and challenges. *Adv Colloid Interface Sci* 228: 71–91.

Keck, C. M. and Müller, R. H. 2006. Drug nanocrystals of poorly soluble drugs produced by high pressure homogenisation. *Eur J Pharm Biopharm* 62: 3–16.

Kim, S., Solari, H., Weiden, P.J. et al. 2012. Paliperidone palmitate injection for the acute and maintenance treatment of schizophrenia in adults. *Patient Prefer Adherence* 6: 533–45.

Kipp, J. E., Wong, J. C. T., Doty, M. J. et al. 2001. Microprecipitation method for preparing submicron suspensions. US Patent 6,607,784.

Knieke, C., Sommer, M., Peukert, W. 2009. Identifying the apparent and true grinding limit. *Powder Technol* 195: 25–30.

Kojima, T., Karashima, M., Yamamoto, K. et al. 2018. Combination of NMR methods to reveal the interfacial structure of a pharmaceutical nanocrystal and nanococrystal in the suspended state. *Mol Pharm* 15: 3901–08.

Kovalenko, M. V., Manna, L., Cabot, A. et al. 2015. Prospects of nanoscience with nanocrystals. *ACS Nano* 9: 1012–57.

Krull, S. M., Patel, H. V., Li, M. et al. 2016. Critical material attributes (CMAs) of strip films loaded with poorly water-soluble drug nanoparticles: I. Impact of plasticizer on film properties and dissolution. *Eur J Pharm Sci* 92: 146–55.

Kumar, S., Gokhale, R. and Burgess, D. J. 2014a. Sugars as bulking agents to prevent nano-crystal aggregation during spray or freeze-drying. *Int J Pharm* 471: 303–11.

Kumar, S., Jog, R., Shen, J. et al. 2015. In vitro and in vivo performance of different sized spray-dried crystalline itraconazole. *J Pharm Sci* 104: 3018–28.

Kumar, S., Shen, J. and Burgess, D. J. 2014b. Nano-amorphous spray dried powder to improve oral bioavailability of itraconazole. *J Control Release* 192: 95–102.

Laaksonen, T., Liu, P., Rahikkala., A. et al. 2011. Intact nanoparticulate indomethacin in fast-dissolving carrier particles by combined wet milling and aerosol flow reactor methods. *Pharm Res* 28: 2403–11.

Lakshmi, P. and Kumar, G.A. 2010. Nano-suspension technology – a review. *Int J Pharm Pharm Sci* 2(Suppl 4): 35–40.

Larese Filon, F., Mauro, M., Adami, G. et al. 2015. Nanoparticles skin absorption: New aspects for a safety profile evaluation. *Regul Toxicol Pharmacol* 72: 310–22.

Lee, M. K., Kim, M. Y., Kim, S. et al. 2009. Cryoprotectants for freeze drying of drug nano-suspensions: Effect of freezing rate. *J Pharm Sci* 98: 4808–17.

Lestari, M. L. A. D., Müller, R. H. and Möschwitzer, J. P. 2014. Systematic screening of different surface modifiers for the production of physically stable nanosuspensions. *J Pharm Sci* 104: 1128–40.

Li, M., Alvarez, P., Orbe, P. et al. 2018. Multi-faceted characterization of wet-milled griseofulvin nanosuspensions for elucidation of aggregation state and stabilization mechanisms. *AAPS PharmSciTech* 19: 1789–1801.

Li, M., Azad, M., Dave, R. et al. 2016. Nanomilling of drugs for bioavailability enhancement: A holistic formulation-process perspective. *Pharmaceutics* 8: 17.

Li, M., Yaragudi, N., Afolabi, A. et al. 2015. Sub-100 nm drug particle suspensions prepared via wet milling with low bead contamination through novel process intensification. *Chem Eng Sci* 130: 207–20.

Liu, P., De Wulf, O., Laru, J. et al. 2013. Dissolution studies of poorly soluble drug nanosuspensions in non-sink conditions. *AAPS PharmSciTech* 14: 748–56.

Liu, P., Rong, X., Laru, J. et al. 2011. Nanosuspensions of poorly soluble drugs: Preparation and development by wet milling. *Int J Pharm* 411: 215–22.

Liu, P., Viitala, T., Kartal-Hodzig, A. et al. 2015. Interaction studies between indomethacin nanocrystals and PEO/PPO copolymer stabilizers. *Pharm Res* 32: 628–39.

Liu, T., Müller, R. H. and Möschwitzer, J. P. 2018. Production of drug nanosuspensions: Effect of drug physical properties on nanosizing efficiency. *Drug Dev Ind Pharm* 44: 233–42.

Liversidge, G. G., Cundy, K. C., Bishop, J. F. et al. 1992. Surface modified drug nanoparticles. US Patent 5,145,684.

Liversidge, G. G. and Cundy, K. C. 1995. Particle size reduction for improvement of oral bioavailability of hydrophobic drugs: I. Absolute oral bioavailability of nanocrystalline danazol in beagle dogs. *Int J Pharm* 125: 91–97.

Loftsson, T. and Brewster, M. E. 2010. Pharmaceutical applications of cyclodextrins: Basic science and product development. *J Pharm Pharmacol* 62: 1607–21.

Loh, Z. H., Er, D. Z., Chan, L. W. et al. 2011. Spray granulation for drug formulation. *Expert Opin Drug Deliv* 8: 1645–61.

Lu, Y., Wang, Z. H., Li, T. et al. 2014. Development and evaluation of transferrin-stabilized paclitaxel nanocrystal formulation. *J Control Release* 176: 76–85.

Malamatari, M., Somavarapu, S., Taylor, K. M. et al. 2016. Solidification of nanosuspensions for the production of solid oral dosage forms and inhalable dry powders. *Expert Opin Drug Deliv* 13: 435–50.

Malamatari, M., Taylor, K. M. G., Malamataris, S. et al. 2018. Pharmaceutical nanocrystals: Production by wet milling and applications. *Drug Discov Today* 23: 534–47.

Merisko-Liversidge, E., Liversidge, G. G. and Cooper, E. R. 2003. Nanosizing: A formulation approach for poorly-water-soluble compounds. *Eur J Pharm Sci* 18: 113–20.

Mitri, K., Shegokar, R., Gohla, S. et al. 2011. Lutein nanocrystals as antioxidant formulation for oral and dermal delivery. *Int J Pharm* 420: 141–46.

Mu, S., Li, M., Guo, M. et al. 2016. Spironolactone nanocrystals for oral administration: Different pharmacokinetic performances induced by stabilizers. *Colloids Surf B Biointerfaces* 147: 73–80.

Müller, R. H., Becker, R., Kruss, B. et al. 1995. Pharmaceutical nanosuspensions for medicament administration as systems with increased saturation solubility and rate of solution. US Patent 5,858,410.

Müller, R. H., Gohla, S. and Keck, C. M. 2011. State of the art of nanocrystals – special features, production, nanotoxicology aspects and intracellular delivery. *Eur J Pharm Biopharm* 78: 1–9.

Müller, R. H., Jacobs, C., 2002. Production and characterization of budesonide nanosuspension for pulmonary administration. *Pharm Res* 19: 189–94.

Müller, R. H. and Keck, C. M. 2012. Twenty years of drug nanocrystals: Where are we, and where do we go? *Eur J Pharm Biopharm* 80: 1–3.

Müller, R.H., Krause, K. and Mäder, K. 2001. Method for controlled production of ultrafine microparticles and nanoparticles. WO Patent application WO2001003670A1.

Müller, R.H. and Möschwitzer, J. 2007. Method and device for producing very fine particles and coating such particles. US Patent 9,168,498.

Müller, R.-H., Möschwitzer, J. and Bushrab, F.-N. 2006. Manufacturing of nanoparticles by milling and homogenization techniques. In *Nanoparticle Technology for Drug Delivery*, eds. R. B. Gupta and U. B. Kompella, 21–51. Boca Raton, FL: CRC Press.

Möschwitzer, J., Achleitner, G., Pomper, H. et al. 2004. Development of an intravenously injectable chemically stable aqueous omeprazole formulation using nanosuspension technology. *Eur J Pharm Biopharm* 58: 615–19.

Möschwitzer, J. P. 2013. Drug nanocrystals in the commercial pharmaceutical development process. *Int J Pharm* 453: 142–56.

Pattnaik, S., Swain, K., Rao, J. V. et al. 2015. Aceclofenac nanocrystals for improved dissolution: influence of polymeric stabilizers. *RSC Adv* 5: 91960–65.

Peltonen, L. 2018a. Practical guidelines for the characterization and quality control of pure drug nanoparticles and nano-cocrystals in the pharmaceutical industry. *Adv Drug Deliv Rev* 131: 101–15.

Peltonen, L. 2018b. Design space and QbD approach for production of drug nanocrystals by wet media milling techniques. *Pharmaceutics* 10: 104.

Peltonen, L. and Hirvonen, J. 2008. Physicochemical characterization of nano- and microparticles. *Curr Nanosci* 4: 101–7.

Peltonen, L. and Hirvonen, J. 2010. Pharmaceutical nanocrystals by nanomilling: Critical process parameters, particle fracturing and stabilization methods. *J Pharm Pharmacol* 62: 1569–79.

Peltonen, L. and Hirvonen, J. 2018. Drug nanocrystals – versatile option for formulation of poorly soluble materials. *Int J Pharm* 537: 73–83.

Peltonen, L. and Strachan, C. 2015. Understanding critical quality attributes for nanocrystals from preparation to delivery. *Molecules* 20: 19851.

Peltonen, L., Tuomela, A. and Hirvonen, J. 2015. Polymeric stabilizers for drug nanocrystals. In *Handbook of Polymers for Pharmaceutical Technologies*, eds. V. K. Thakur and M. K. Thakur, 67–87. Salem, MA: Scrivener Publishing LLC.

Peltonen, L., Valo, H., Kolakovic, R. et al. 2010. Electrospraying, spray drying and related techniques for production and formulation of drug nanoparticles. *Expert Opin Drug Deliv* 7: 705–19.

Pieterjan, K., Michaël, A. and Guy, V. D. M. 2011. Bead layering as a process to stabilize nanosuspensions: Influence of drug hydrophobicity on nanocrystal reagglomeration following in-vitro release from sugar beads. *J Pharm Pharmacol* 63: 1446–53.

Pireddu, R., Caddeo, C., Valenti, D. et al. 2016. Diclofenac acid nanocrystals as an effective strategy to reduce in vivo skin inflammation by improving dermal drug bioavailability. Colloids Surf B Biointerfaces 143: 64–70.

Rabbani, N. R. and Seville, P. C. 2005. The influence of formulation components on the aerosolisation properties of spray-dried powders. *J Control Release* 110: 130–40.

Raula, J., Rahikkala, A., Halkola, T. et al. 2013. Coated particle assemblies for the concomitant pulmonary administration of budesonide and salbutamol sulphate. *Int J Pharm* 441: 248–54.

Rege, B. D., Kao, J. P., Polli, J. E. 2002. Effects of nonionic surfactants on membrane transporters in Caco-2 cell monolayers. *Eur J Pharm Sci* 16: 237–46.

Rydberg, H. A., Yanez Arteta, M., Berg, S. et al. 2016. Probing adsorption of DSPE-PEG2000 and DSPE-PEG5000 to the surface of felodipine and griseofulvin nanocrystals. *Int J Pharm* 510: 232–39.

Salazar, J., Müller, R. H. and Möschwitzer, J. P. 2014. Combinative particle size reduction technologies for the production of drug nanocrystals. *J Pharm (Cairo)* 2014: 265754.

Sarnes, A., Kovalainen, M., Häkkinen, M. R. et al. 2014. Nanocrystal-based per-oral itraconazole delivery: Superior in vitro dissolution enhancement versus Sporanox(R) is not realized in in vivo drug absorption. *J Control Release* 180: 109–16.

Sarnes, A., Østergaard, J., Jensen, S. S. et al. 2013. Dissolution study of nanocrystal powders of a poorly soluble drug by UV imaging and channel flow methods. *Eur J Pharm Sci* 50: 511–19.

Seedher, N. and Kanojia, M. 2009. Co-solvent solubilization of some poorly-soluble antidiabetic drugs. *Pharm Dev Technol* 14: 185–92.

Selin, M., Nummelin, S., Deleu, J. et al. 2018. High-generation amphiphilic Janus-dendrimers as stabilizing agents for drug suspensions. *Biomacromolecules* 19, 3983–93.

Serajuddin, A.T. 2007. Salt formation to improve drug solubility. *Adv Drug Deliv Rev* 59: 603–16.

Shaal, L. A., Müller, R. H., Keck, C. M. 2010. Preserving hesperetin nanosuspensions for dermal application. *Pharmazie* 65: 86–92.

Sharma, P., Denny, W. A. and Garg, S. 2009. Effect of wet milling process on the solid state of indomethacin and simvastatin. *Int J Pharm* 380: 40–48.

Shchekin, A. K., Rusanov, A.I. 2008. Generalization of the Gibbs-Kelvin-Köhler and Ostwald -Freundlich equations for a liquid film on a soluble nanoparticle. *J Chem Phys* 129: 154116.

Shegokar, R. and Müller, R. H. 2010. Nanocrystals: Industrially feasible multifunctional formulation technology for poorly soluble actives. *Int J Pharm* 399: 129–39.

Shete, G. and Bansal, A. K. 2016. NanoCrySP technology for generation of drug nanocrystals: Translational aspects and business potential. *Drug Deliv Transl Res* 6: 392–98.

Shete, G., Pawar, Y. B., Thanki, K. et al. 2015. Oral bioavailability and pharmacodynamic activity of hesperetin nanocrystals generated using a novel bottom-up technology. *Mol Pharm* 12: 1158–70.

Siewert, C., Moog, R., Alex, R. et al. 2018. Process and scaling parameters for wet media milling in early phase drug development: A knowledge based approach. *Eur J Pharm Sci* 115: 126–31.

Singare, D. S., Marella, S., Gowthamrajan, K. et al. 2010. Optimization of formulation and process variable of nanosuspension: An industrial perspective. *Int J Pharm* 402: 213–20.

Sinha, B., Müller, R. H. and Möschwitzer, J. P. 2013. Bottom-up approaches for preparing drug nanocrystals: Formulations and factors affecting particle size. *Int J Pharm* 453: 126–41.

Soliman, K. A., Ibrahim, H. K. and Ghorab, M. M. 2017. Effects of different combinations of nanocrystallization technologies on avanafil nanoparticles: In vitro, in vivo and stability evaluation. *Int J Pharm* 517: 148–56.

Srivalli, K. M. R. and Mishra, B. 2016. Drug nanocrystals: a way toward scale-up. *Saudi Pharm J* 24: 386–404.

Strachan, C. J., Rades, T., Gordon, K. C. et al. 2007. Raman spectroscopy for quantitative analysis of pharmaceutical solids. *J Pharm Pharmacol* 59: 179–92.

Strachan, C. J., Windbergs, M. and Offerhaus, H. L. 2011. Pharmaceutical applications of non-linear imaging. *Int J Pharm* 417: 163–72.

Surwase, S. A., Boetker, J. P., Saville, D. et al. 2013. Indomethacin: new polymorphs of an old drug. *Mol Pharm.* 10: 4472–80.

Surwase, S. A., Itkonen, L., Aaltonen, J. et al. 2015. Polymer incorporation method affects the physical stability of amorphous indomethacin in aqueous suspension. *Eur J Pharm Biopharm* 96: 32–43.

Takatsuka, T., Endo, T., Jianguo, Y. et al. 2009. Nanosizing of poorly water soluble compounds using rotation/revolution mixer. *Chem Pharm Bull* 57: 1061–67.

Tan, E. H., Parmentier, J., Low, A. et al. 2017. Downstream drug product processing of itraconazole nanosuspension: Factors influencing tablet material properties and dissolution of compacted nanosuspension-layered sugar beads. *Int J Pharm* 532: 131–38.

Tuomela, A., Hirvonen, J., Peltonen, L. 2016. Stabilizing agents for drug nanocrystals: Effect on bioavailability. *Pharmaceutics* 8: 16.

Tuomela, A., Laaksonen, T., Laru, J. et al. 2015. Solid formulations by a nanocrystal approach: Critical process parameters regarding scale-ability of nanocrystals for tableting applications. *Int J Pharm* 485: 77–86.

Tuomela, A., Liu, P., Puranen, J. et al. 2014. Brinzolamide nanocrystal formulations for ophthalmic delivery: Reduction of elevated intraocular pressure in vivo. *Int J Pharm* 467: 34–41.

Valo, H., Arola, S., Laaksonen, P. et al. 2013. Drug release from nanoparticles embedded in four different nanofibrillar cellulose aerogels. *Eur J Pharm* Sci 50: 69–77.

Valo, H., Kovalainen, M., Laaksonen, P. et al. 2011. Immobilization of protein-coated drug nanoparticles in nanofibrillar cellulose matrices - enhanced stability and release. *J Control Release* 156: 390–97.

Valo, H. K., Laaksonen, P. H., Peltonen, L. J. et al. 2010. Multifunctional hydrophobin: Toward functional coatings for drug nanoparticles. *ACS Nano* 4: 1750–58.

Van Eerdenbrugh, B., Van Den Mooter, G., Augustijns, P. 2008. Top-down production of drug nanocrystals: Nanosuspension stabilization, miniaturization and transformation into solid products. *Int J Pharm* 364: 64–75.

Van Eerdenbrugh, B., Vermaat, J., Martens, J. A. et al. 2009. A screening study of surface stabilization during the production of drug nanocrystals. *J Pharm Sci* 98: 2091–103.

Wang, Y., Ma, Y., Du, Y. et al. 2012a. Formulation and pharmacokinetics evaluation of puerarin nanocrystals for intravenous delivery. *J Nanosci Nanotechnol* 12: 6176–84.

Wang, M., Rutledge, G. C., Myerson, A. S. et al. 2012b. Production and characterization of carbamazepine nanocrystals by electrospraying for continuous pharmaceutical manufacturing. *J Pharm Sci* 101: 1178–88.

Wang, Y., Zheng, Y., Zhang, L. et al. 2013. Stability of nanosuspensions in drug delivery. *J Control Release* 172: 1126–41.

Varmal, M. V. and Panchagnula, R. 2005. Enhanced oral paclitaxel absorption with vitamin E-TPGS: Effect on solubility and permeability in vitro, in situ and in vivo. *Eur J Pharm Sci* 25: 445–53.

Wei, Q., Keck, C. M. and Müller, R. H. 2018. Solidification of hesperidin nanosuspension by spray drying optimized by design of experiment (DoE). *Drug Dev Ind Pharm* 44: 1–12.

Verma, S., Gokhale, R. and Burgess, D. J. 2009. A comparative study of top-down and bottom-up approaches for the preparation of micro/nanosuspensions. *Int J Pharm* 380: 216–22.

Vidlářová, L., Romero, G. B., Hanus, J. et al. 2016. Nanocrystals for dermal penetration enhancement – Effect of concentration and underlying mechanisms using curcumin as model. *Eur J Pharm Biopharm* 104: 216–25.

Wu, L., Zhang, J. and Watanabe, W. 2011. Physical and chemical stability of drug nanoparticles. *Adv Drug Deliv Rev* 63: 456–69.

Yu, T., Choi, W.-J., Anonuevo, A. et al. 2016. Mucus-penetrating nanosuspensions for enhanced delivery of poorly soluble drugs to mucosal surfaces. *Adv Healthc Mater* 5: 2745–50.

Yue, P.-F., Li, Y., Wan, W. et al. 2013. Study on formability of solid nanosuspensions during nanodispersion and solidification: I. Novel role of stabilizer/drug property. *Int J Pharm* 454: 269–77.

Yuminoki, K., Tachibana, S., Nishimura, Y. et al. 2016. Scaling up nano-milling of poorly water soluble compounds using a rotation/revolution pulverizer. *Pharmazie* 71: 56–64.

Zhao, Y. X., Hua, H. Y., Chang, M. et al. 2010. Preparation and cytotoxic activity of hydroxycamptothecin nanosuspensions. *Int J Pharm* 392: 64–71.

Zong, L., Li, X., Wang, H. et al. 2017. Formulation and characterization of biocompatible and stable I.V. Itraconazole nanosuspensions stabilized by a new stabilizer polyethylene glycol-poly(β-Benzyl-l-aspartate) (PEG-PBLA). *Int J Pharm* 531: 108–17.

3

Porous Inorganic Nanoparticles for Drug Delivery

Gonzalo Villaverde and
Alejandro Baeza
Universidad Politécnica de Madrid

3.1 Introduction

The use of nanometric particles as drug carriers has emerged as one of the most promising tools for the controlled delivery of therapeutic agents in the 21st century.[1] Nanocarriers are capable to transport drugs with very different nature (hydrophilic or hydrophobic, small molecules or macromolecules as proteins, enzymes and even genes) improving their pharmacokinetic profile and protecting the housed drugs against external or internal insults during the manufacturing and storage process, and also through their journey within the patient.[2] The nanometric size of these objects, similar to viruses, turns then into a suitable instrument which allows the intimate interaction with the cells. Their careful design can achieve a precise control in the drug release in terms of time and place in which the drug release takes place.[3] This property is extremely important in the case of the delivery of highly toxic therapeutic agents such as chemotherapeutic drugs in oncology because the fact to improve the selectivity of the therapy is a goal of paramount importance in order to enhance the efficacy and eliminate the dramatic side effects of the current treatments.[4] For this reason, one of the main applications of nanoparticles for drug delivery has been in oncology. The origin of the use of nanocarriers in antitumoral uses dates from 1986, when Maeda and Matsumura reported the passive accumulation in tumoral masses of macromolecules higher than 40 KDa of molecular weight injected in the blood stream.[5] This effect is originated simply by the nanometric size of the macromolecules and was also observed with nanoparticles, being the result of the irregular blood vessel architecture existing inside solid tumors. These tumoral blood vessels are highly tortuous and present fenestrations and pores of up to a few hundreds of nanometers, whereas the interendothelial junctions present in the healthy blood vessels are only a few of nanometers. Therefore, when the nanocarriers reach the neoplastic tissues, they are able to pass throughout these pores being accumulated mainly there, while they cannot pass the healthy epithelia.[6] Additionally, solid tumors lack of an effective lymphatic drainage system and, as a consequence of this, the nanoparticles which reached the tumoral zone are not capable to leave the diseased tissue resulting accumulated there during long periods of time.[7] Both properties, high permeability of tumoral blood vessels and inefficient drainage system, explain the accumulation of nanometric objects inside solid tumors and receive the name of *enhanced permeation and retention* (EPR) effect (Figure 3.1).

In addition to the passive nanocarrier accumulation in neoplastic tissues, the surface of the nanosystems can be decorated with molecules able to interact with specific membrane receptors of tumoral cells, and therefore, it is possible to achieve the selective internalization of the nanoparticles within the malignant cells leaving the surrounding healthy cells unaltered.[8] High permeability of blood vessels is also found in tissues affected by inflammation-mediated processes such as atherosclerosis or bacterial infection, and thus, nanoparticle-based therapies for these pathologies have also received increasing attention.[9,10] In order to exploit this effect, a myriad of different types of nanoparticles have been reported both inorganic, such as metallic, ceramic and based on carbon allotropes, and organic nanocarriers as liposomes, polymersomes, micelles or dense polymeric nanocarriers.[8,11] Among these different nanosystems, porous inorganic nanoparticles emerged as promising systems due to their unique properties

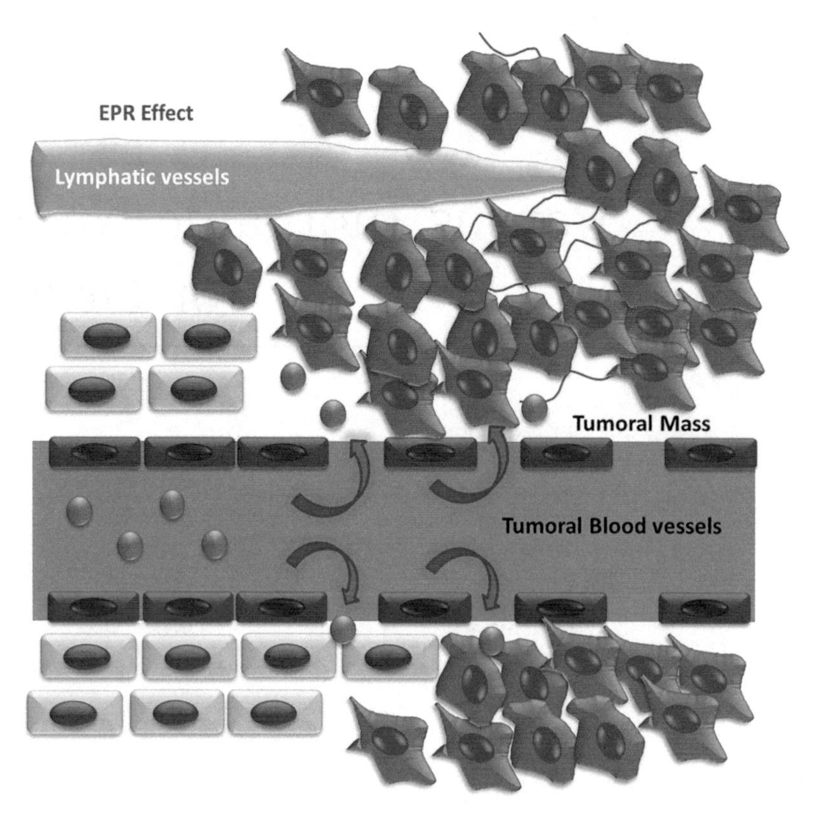

FIGURE 3.1 EPR effect.

such as high loading capacity, chemical and mechanical resistance, tolerance to organic solvents which is useful in order to access to a wide number of manufacturing processes, low toxicity and immunogenicity, and ease to scale up their production.[12] Moreover, the pores present in these systems can be tuned in order to house very different molecules from pores with diameter comprised between 1 and 3 nm, suitable for the transportation of small drugs to large pores of some tens of nanometers, which can hold proteins or even genes. In this chapter, the advances in the development of porous inorganic nanoparticles, as porous silicon, mesoporous silica and metal-organic nanoparticles, carried out in the last years will be described. The intention of this chapter is not to provide a comprehensive review of these systems but to offer a panoramic view about the current state of the art, presenting illustrative examples of each type of material which point out their advantages and limitations in order to present a clear picture about the potentiality of these systems in the clinical field.

3.2 Porous Silicon Nanoparticles (PSNs)

Silicon is the most abundant element in the earth crust and is part of the composition of many minerals and rocks, mainly in the form of oxides. Pure silicon is not present in the nature and must be artificially produced. Porous silicon was discovered by serendipity in the Bell Laboratories in the middle of the last century, when the researchers of the company were trying to develop an electrochemical method for produce silicon wafers for electronic applications.[13] After its discovery, the interest in this material was scarce until its highly porous structure aroused the attention of the scientists in order to produce photoluminescence materials by quantum confinement effects.[14] The present of these pores in this material makes porous silicon into an excellent system for drug delivery applications in nanoparticulate form (PSN) in which the drugs can be housed within the channels. PSNs are commonly produced following top-bottom approaches such as chemical or laser-induced etching or milling.[15] PSNs present high loading capacity because their external surface is comprised of between 200 and 500 m^2/g and the pore diameter can be tuned from 2 to 50 nm, allowing the housing of many different types of therapeutic compounds. Moreover, porous silicon is a harmless and non-immunogenic material, and is degradable in physiological conditions producing non-toxic compounds, as silicic acid. The tunable pore structure can be used for the transportation of many different types of molecules which can be loaded within the silicon channels simply by immersion in a concentrated solution of the desired drug or grafting the drugs on the silicon surface through covalent bonds. In the first approach, the drug departure can be adjusted playing with the electrostatic or hydrophobic interactions of the drug with the silicon surface, which usually present negative surface charge in the oxidized form, but it can be modified through chemical modifications.[16] Thus, thermally carbonized PSNs which suffer a thermic treatment at more than 600°C exhibit hydrophobic surface being suitable

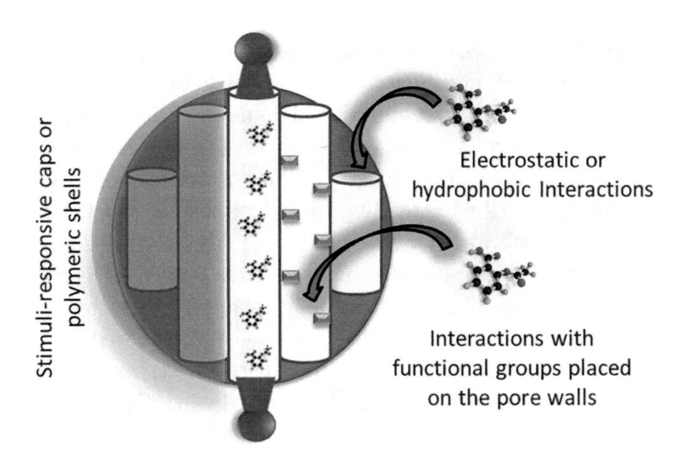

FIGURE 3.2 Loading strategies for porous silicon nanoparticles.

for the transportation of hydrophobic drugs. The silicon surface can be functionalized with many different chemical groups by hydrosilylation and silanization, in order to prepare suitable surfaces for enhancing the drug adsorption process or for providing functional groups suitable by the covalent grafting of the therapeutic compounds on PSN surface.[17] Other chemical methods have also been reported for the surface modification of PSN in order to load drugs by different strategies (Figure 3.2).[18]

Santos et al. have described the use of catechol-metal complexation in order to retain a potent cytotoxic compound, doxorubicin (Dox) in PSNs.[19] In this work, the PSNs were functionalized with undecylenic acid in order to introduce carboxylic groups on the surface, and then, these carboxylic groups were employed for the covalent grafting of an aminopropyl derivative of quercetin using the well-known carbodiimide chemistry. Fe^{3+} acted as the glue between the quercetin moiety and the Dox which are complexed to the metallic ion. One of the main advantages of this strategy is that Dox is released following a pH-sensitive manner yielding higher drug departure when the pH is mild acidic (pH 5). Styrene molecules have also been grafted on the surface of PSNs in order to load Dox by π–π stacking interactions between the aromatic rings of Dox and the ones attached on the surface.[20] Sailor et al. have reported a novel method for trapping siRNA inside the PSN pores based on the precipitation of Ca_2SiO_4 on the pore surface.[21] In this case, the surface of PSN is partially dissolved in the presence of Ca^{2+}, and the RNA yielding the formation of calcium silicate that seals the pore entrance trapping the RNA inside the silicon matrix. The formation of this insoluble specie retards the dissolution of the silicon and therefore slows the RNA release. Additionally, the calcium silicate surface was decorated with two peptide sequences, one of them derived from rabies virus glycoprotein (RVG) for providing selectivity towards neurons, and the other one a cell-penetrating peptide, which induces a rapid endosomal scape of the nanoparticle. This last property is especially important in the case of the transportation of highly sensitive therapeutic agents as proteins or oligonucleotides, which can be degraded by the acid environment and the presence of digestive enzymes characteristics of these organelles. These particles were able to knock down the expression of peptidylprolyl isomerase B in neuroblastoma cells through the selective delivery of one specific siRNA. The adsorption of polycationic polymers as polyethyleneimine (PEI) on the PSN surface was also employed as capping moiety which retains siRNA within the pores.[22] This system was able to block the expression of multidrug resistance protein 1 in glioblastoma cells, which is a protein on charge of the removal of chemotherapeutic agents from the cytosol. In the case of covalent grafting of therapeutic compounds, the departure of these agents takes place when the silicon is degraded, or when the functional groups which maintain the drug attached on the surface are broken in the presence of certain conditions providing to the system a stimuli-responsive behavior.[23] The possibility to release the therapeutic drugs on demand only when the nanocarrier reaches the diseased zone is one aspect of paramount importance in the clinical field, and therefore, the development of stimuli-responsive materials has received a huge interest in the recent years. The controlled drug release in one specific location achieves a high concentration of therapeutic compounds in this area enhancing their effect and reduces dramatically the side effects caused by these drugs, because the release has taken place locally instead of systemically. Lehto et al. have attached a thermosensitive polymer based on a poly-*N*-isopropylacrylamide (PNIPAM) derivative which exhibits transition temperature above the physiological conditions in order to control the departure process of the drugs trapped within the pore channels.[24] This system was capable to release Dox inside different tumoral cells in response to infrared and radiofrequency irradiation thanks to the capacity of PSN to generate heat in response to these stimuli[25] and the temperature-responsive behavior of PNIPAM which suffers a strong contraction when the temperatures reaches its transition temperature. Indocyanine-based dye (IR820) has electrostatically adsorbed on the pores of PSN in order to provide light-triggered drug release behavior.[26] The presence of this dye inside the pores allows the detection of the particles inside tumoral cells and the on/off release of Dox by the application of near-infrared radiation (NIR) providing a theranostic nanoplatform. Shahbazi et al. have recently reported the development of a nanoplatform based on PSN able to release multiple therapeutic agents with different chemical nature (hydrophilic and hydrophobic) inside the tumoral cells in a selective manner, escaping rapidly from the endosomes.[27] This nanoplatform was synthesized attaching DNA intercalators (9-aminoacridine) on the surface of PSN, which were previously loaded with the therapeutic agents, using redox-sensitive crosslinkers. Then, the system was sealed by the addition of DNA strands which are complexed by the acridine. Finally, the system was coated with polyethyleneimine (PEI)-poly (methyl vinyl ether-alt-maleic acid) (PMVE-MA) which provides cell-penetrating and endosomal escape abilities. This novel nanoplatform was able to be engulfed by

MDA-MB-231 breast cancer cells and, once there, to reach the cytosolic compartment where the drugs were released causing the destruction of the tumoral cells. One important property of this system is the possibility to employ therapeutic DNA strands as capping systems in order to enhance even more the therapeutic effect. The surface decoration with different groups not only allows the control of the drug release process, but it can be used for providing other important properties as particle penetration within the tumoral zone. Thus, Yong et al. have reported that the surface functionalization of PSN with undecylenic acid achieved the rapid particle excretion from the cells. This fact allows the propagation of the nanocarriers along the tumoral tissue, which enhanced their therapeutic efficacy because a more homogeneous distribution of these agents was observed.[28] Additionally, imaging agents can be placed on the surface in order to obtain valuable information in real time of the disease progression.[29] PSNs have been also employed for the treatment of different pathologies rather than cancer. Thus, Voelcker et al. have employed PSN loaded with rapamycin in order to induce a maturation-resistant phenotype in dendritic cells, which were able to suppress the allogenic T-cell proliferation, an important point in immunosuppressive treatments.[30] In other work by the same research group, the authors employed PSN for delivering Flightless I (Flii) neutralizing antibodies in order to suppress the inhibitor role of Flii in the wound-healing process, which could be very valuable for the treatment of diabetes-associated ulcers.[31] Santos et al. have reported the use of PSN coated with chitosan functionalized either with L-cysteine or oligoarginine R9 cell-penetrating peptide for the oral delivery of insulin.[32] The presence of these peptides induces a considerable increase in the permeability of the particles in the intestinal cells resulting in 1.86- and 2.03-fold increment of available insulin. In other recent work, PSNs loaded with a therapeutic drug for the treatment of acute liver failure were co-encapsulated with gold nanoparticles inside an acetalated dextran organic matrix using a microfluidic method.[33] The presence of gold nanoparticles allowed to identify the pathological alterations caused by the disease progression by computer tomography. The PSN nanoparticles acted as depots for the delivery of therapeutic compounds into the diseased tissue.

3.3 Mesoporous Silica Nanoparticles (MSNs)

Among the different types of porous inorganic nanoparticles, MSNs are probably one of the most important ones. MSNs have received an increasing amount of research in drug delivery since the discovery of this type of material first reported by Kuroda and co-workers[34] and later by scientists of the Mobil corporation.[35] The reasons of this high attention lie on the unique properties of mesoporous silica such as high loading capacity, as a consequence of its high external surface (around 1,000 m^2/g) chemical and mechanical robustness, excellent biocompatibility, low immunogenicity, easy and low-cost production, tunable pore diameter and the high possibilities of functionalization exploiting the rich chemistry of the silanol groups present on its surface.[36] Different families of MSN have been described attending to architecture of the mesoporous channels as SBA-15, MCM-48 or MCM-41, to name just a few of them. Among these systems, MCM-41 is the most studied system for drug delivery applications by far because this material presents a honeycomb ordered pore structure in which the pores are distributed in parallel without any interconnections between them.[37] Thus, one of the main advantages of this pore configuration is that the drugs loaded within the silica channels can be retained there simply placing caps on the pore entrances in such a way that the drug departure takes place only when the caps are removed by the presence of certain stimuli.[38] The synthesis of MCM-41 is based on the soft-matter templating method in basic aqueous conditions in which small cationic surfactants such as hexadecyltrimethylammonium bromide (CTAB) are employed as structure-directing agent being the responsible of the ordered pore formation.[37] The pore diameter can be tuned from two to a few tens of nanometers by modifications of the length of the alkyl chain of the surfactant or by the addition of swelling agents as mesitylene, which produce the expansion of the surfactants micelles in the aqueous phase. This versatility allows loading a great number of different therapeutic agents, from small molecules, as is the case of the usual chemotherapeutic agents, antibiotics or cardiovascular drugs to large macromolecules as proteins, enzymes or oligonucleotide sequences. Employing the well-known silanol chemistry and the enriched number of silanol groups present on the MSN, it is possible to introduce many different functional groups both on the external surface of the particles and in the inner walls of the pores.[39] Thus, folic acid has been anchored on the surface of MSN loaded with camptothecin (CPT) in order to destroy the pancreatic tumoral cells by a selective manner because these malignant cells overexpress folate receptors, which allow them to capture this important vitamin required for sustaining their accelerated growth.[40] These targeted nanoparticles were injected in pancreatic xenograft mice model being able to reach the tumoral zone and producing remarkable tumoral growth inhibition in comparison with the non-targeted particles. Tumoral cells required higher amounts of other nutrients such as sugar or iron, among others, for sustaining their accelerated metabolism and growing rate. Therefore, mannose has been grafted on the MSN which transports photosensitizers usually employed in photodynamic therapy.[41] Zink et al. have reported the MSN surface decoration with transferrin, the protein on charge of the transportation of iron in the organism, for the selective transportation of highly lipophilic drugs inside the pore channels.[42] These particles were able to destroy pancreatic and breast tumoral cells (PANC-1 and BT-549, respectively) producing only a slight effect in human foreskin fibroblast cells, used as control.

In this work, cyclic-RGD peptide was also employed as targeting agent due to the capacity of this moiety to bind α,β-integrins receptors also overexpressed by many tumoral cells, as the metastatic breast cancer cell line MDA-MB 435. Other targeting moieties which have been widely employed for enhancing the selectivity against tumoral cells are antibodies (anti-CD44[43] or anti-HER2[44], to name only a few of them) and aptamers, which are oligonucleotide sequences that selectively bind to specific cell membrane receptors of the malignant cells.[45–47] The production or isolation of these macromolecules is usually a time-consuming process, and therefore, these targeting moieties are commonly expensive and only available in low amounts. Moreover, antibodies can elicit immune responses which can compromise the safe application of these targeted nanodevices. As alternative to antibodies, short peptide fragments, which exhibit similar binding properties as the complete antibodies but lesser immune responses and much lower fabrication costs, can be used in order to achieve selective targeting.[48] It is also possible to employ synthetic analogs of natural molecules which bind to these tumoral receptors in order to enhance even more the affinity of the ligand by the overexpressed receptor. Villaverde et al. have reported the attachment of *meta*-aminobenzylguanidine (MABG) on the surface of MSN as synthetic analog of norepinephrine for the selective recognition of neuroblastoma cells.[49] MABG presents strong affinity by the norepinephrine transporter, which is a transmembrane protein on charge of the recognition of norepinephrine and is usually overexpressed in more than 90% of the neuroblastoma cells. Thus, the decoration with these molecules produces a remarkable enhancement of the MSN uptake by the malignant cells allowing the use of these particles for the detection and treatment of this disease. The use of synthetic small molecules for targeting has opened a very interesting way for providing target abilities to nanoparticles, because these molecules can be easily produced in the lab allowing the introduction of large number of structural modifications in a similar way than the structure–activity relationship (SAR) studies frequently employed by the pharmaceutical industry. Therefore, it is possible to explore many different molecules in one single nanoplatform in order to find the system which achieve the best performance.[50]

Targeting skills are important for the selective delivery of therapeutic agents into the diseased place, especially when the transported drugs are toxic or can produce severe side effects if the delivery has been taken place in an incorrect location. Additionally, other important property which should have an efficient nanocarrier is the ability to retain the cargo until it reaches its final location. As was mentioned above, MSNs with MCM-41 structure are probably one of the most employed systems which can exhibit “zero-premature release” of the transported cargo, because the pore outlets can be sealed with reversible caps or detachable layers which can be removed by the application to certain stimuli in order to control the drug departure process at will (Figure 3.3).[51]

The combination of these two properties, targeting abilities and stimuli-responsive behavior, allows a fine control of the drug release process in a temporary and spatial way, yielding the drug departure only in the desired place and with the exact dosage. The stimulus which triggers the drug release can be externally applied as is the case of light, magnetic fields or ultrasounds (US) or can be an internal stimulus characteristic of the own pathological process such as the overproduction of certain enzymes, presence of acidic environments or different redox conditions, among others. In order to provide a clear vision of the current state of the art of stimuli-responsive MSN, a brief description of a few examples of each type of system, triggered by external or internal stimuli, will be provided in order to offer an idea about the great potential of these nanodevices.

3.3.1 Externally Triggered MSN

As it has been mentioned above, the drug departure can be triggered by the application of certain external stimulus such as light, magnetic fields or ultrasounds. This strategy achieves a precise control in the drug departure process being capable to administrate the transported therapeutic compounds in precise dosages because the trigger

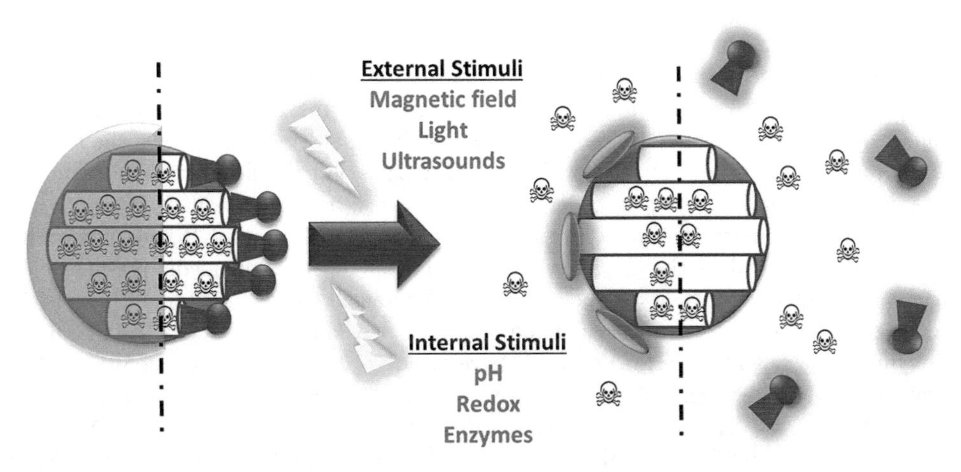

FIGURE 3.3 External and internal stimuli which trigger the drug departure process in MSN.

stimulus is completely exogenous to the patient and can be applied by the clinicians on demand. However, its main drawback is the necessity to employ complex infrastructures in many cases and specialized technical personnel. Light has been widely employed for this purpose because this stimulus presents significant advantages such as the good biocompatibility and the capacity to focus the light beam in really small areas, which is important for the precise treatment of tumoral lesions in vital organs. The main liability of light is its poor penetration in living tissues, except in the case of NIR in which this radiation is able to penetrate a few millimeters.[52] Despite this important drawback, light can be used for the treatment of accessible tumoral lesions in eyes, digestive tracts and lung, among others, or can be delivered to the diseased tissue employing optic fibers. Therefore, different light sources have been widely employed for triggering the drug release in light-sensitive MSN, from UV[53] to NIR.[54] These systems usually employ caps which are attached on the pore outlets through light-cleavable bonds[55] or light-sensitive shells which are detached by the exposition to radiations.[56] These sensitive moieties are mainly sensitive to UV or visible light instead of the more penetrating radiation NIR, which restrict their use to exposed locations. One interesting approach to overcome this limitation is to employ upconverting nanoparticles (UCNPs) in combination with MSN decorated with these sensitive moieties.[57] UCNPs are nanoparticles based on earth-rare elements which are capable to transform NIR into UV or visible light.[58] Thus, these nanosystems can take the advantage of the higher penetration capacity of NIR and the large number of UV- or VIS-sensitive bonds for triggering the drug release.[59] Magnetic field is also a biocompatible stimulus, and it is more penetrating than light, and therefore, it has also been used as trigger stimuli in MSN through the introduction of superparamagnetic cores inside the silica matrix or on the external surface.[60] The usual strategy in this case is to place temperature-responsive gates on the MSN surface in such a way that these gates open or close the pores in response of the heat produced by these magnetic cores when they are exposed to alternating magnetic fields (AMFs). Thus, MSNs which contain superparamagnetic iron oxide nanoparticles (SPIONs) inside have been decorated with DNA double strands which act as reversible caps because the double helix suffers dehybridization when the temperature exceeds certain value, which can be precisely tuned by modifications of the base pair composition.[61] Other alternative is to coat magnetic MSN with thermosensitive polymers as PNIPAM in order to control the drug departure by the exposition or not to AMF.[62] Guisasola et al. have controlled the drug release placing on the surface of magnetic MSN a thermosensitive polymer formed by a mixture of *N*-isopropylacrylamide (NIPAM), *N*-(hydroxymethyl)acrylamide (NHMA) and *N*, *N'*-methylenebis(acrylamide) (MBA) which exhibit a transition temperature of 42–43°C.[63] The drug departure occurred as consequence of the conformational change of the polymer shell at temperatures above this value. Interestingly, this system was able to release a drug model (fluorescein) in response to AMF, even if the macroscopic temperature was maintained at physiological conditions (37°C) placing the sample in a thermostatic chamber, which pointed out that this release was triggered by the local temperature increase in the close vicinity of the SPIONs trapped within the silica matrix (*hot spot effect*). Finally, the same research group tested the capacity of these nanoparticles to release a potent cytotoxic agent (Dox) inside a tumoral lesion in mice models showing an excellent capacity to destroy the tumoral cells when AMF was applied as consequence of the triggered Dox release and also by the synergic effect caused by the cytotoxic drug and the local temperature increase.[64] Finally, ultrasounds have also been employed for the controlled release of cytotoxic compounds due to the excellent penetration capacity of these mechanical waves in living tissues and the possibility to focus these waves in specific points of the body through the use of focused ultrasounds. Thus, Paris et al. have coated the external surface of MSN with a thermosensitive polymer which contains a hydrophobic monomer with an acetal group sensitive to US.[65] This thermosensitive polymer is collapsed at physiological temperature because its transition temperature is higher which maintains the pore closed. When US is applied, the acetal group of the hydrophobic monomer is broken making more hydrophilic the polymer coat.Therefore, its transition temperature drops to values lower than 37°C which allows the drug departure.

3.3.2 Internally Triggered MSN

Internal stimuli as trigger of the drug release present as advantage that does not require the use of sophisticated equipment, as in the case of externally triggered systems. However, as main liability, the control of the release process is lesser precise than that in the case of the external ones. One of the most employed internal stimuli for drug delivery applications is the pH. In the organism, there are natural pH gradients which can be exploited for triggering the drug departure, as, for example, the strong acidic condition presents in the stomach or the mild-basic environment in the colon. Lee et al. have attached trimethylammonium (TA) groups on the pore walls using pH-sensitive hydrazone bonds.[66] The positive charge provided by these groups allows the loading of high amounts of anti-inflammatory drugs used in the treatment of intestinal pathologies as irritable bowel syndrome or Crohn's disease which are strongly retained inside the silica matrix by electrostatic interactions. When these particles were orally administered, the acidic media present in the stomach broke the hydrazone bonds releasing the positive charges from the surface, and then, the anti-inflammatory drugs were released in the mild-basic conditions of the colon. The same pH gradient has been employed for the controlled delivery of prednisolone using MSN coated with succinylated ε-polylysine.[67] This polymer coat maintained the drugs trapped within the silica matrix when the particle passed through the acidic environment of the stomach and small intestine, but released them when the system reached the colon, as consequence of the repulsive forces between the negatively charged polymeric chains which produced the swelling of the polymeric shell. Tumoral masses also exhibit pH gradient which can be exploited being more acidic than healthy tissues due to their accelerated glycolysis in anaerobic conditions.[68] Therefore, MSNs have been coated with pH-sensitive polymers[69] or caps[70] in order to deliver chemotherapeutic agents selectively to the tumoral tissues. Finally, the acidic environment of the endosomes and lysosomes can be exploited for the controlled release of therapeutic agents inside these organelles.[71] Other gradient which has been employed for triggering the drug departure inside the tumoral cells is the higher amount of reductive species present in the cellular cytosol in comparison with the extracellular environment. The common strategy employed for exploiting this characteristic is the attachment of pore blockers through linkers which contains a –S-S- bond cleavable by the presence of glutathione inside the cell.[72,73] Moreover, polymeric coatings can also be attached on the particle surface by disulfide bonds,[74] or the own polymer shell can be formed by monomers which contains these redox-sensitive bonds in such a way that the presence of intracellular reductive species produces the disassembly of the polymeric layer.[75] Even it is also possible to employ silanes which contains a disulfide bond, as bis(triethoxysilyl-propyl)disulfide, in order to synthesize nanoparticles which are breakable in the intracellular environment.[76] Finally, many different types of pathologies course with the overproduction of certain enzymes which are present in lower amount in healthy tissues generating a gradient which can be used for controlled drug delivery applications. One of these pathologies is cancer, in which the malignant cells produce high amounts of proteolytic enzymes as metalloproteinases or collagenase, in order to break the boundaries of the host tissue and colonize the surrounding zones. Bein et al. have described the use of avidin as pore blocker which was bound on the pore entrance by a specific peptide sequence sensitive to metalloproteinase-9 that avoided the premature release of cisplatin trapped inside until the system reached the tumoral tissue, where this enzyme was present.[77] Manna et al. have grafted bilirubin on the surface of MSN in order to seal the system using bovine serum albumin (BSA) thanks to the great affinity of this molecule for the protein.[78] This system has been tested using HCT-116 human colon cancer cells being able to release cytotoxic compounds only when proteases were present in the media. This system presents the advantage that bilirubin exhibits antitumoral effect by itself because it induces oxidative stress and activates tumor suppressor proteins, such as phosphatase and tensin homologue (PTEN). NAD(P)H:quinone oxidoreductase isozyme 1 (NQO1) is overexpressed in different types of cancer as pancreas, colon, breast or lung, among others. Wu et al. have anchored rotaxanes on MSN surface as pore blockers employing a benzoquinone-based stalk in order to retain the rotaxane moiety. The existence of NQO1 induces the transformation of the benzoquinone into hydroquinone resulting in the removal of the stalk and therefore in the pore uncapping.[79] The presence of other enzymes, such as pepsin[80] or hyaluronidase,[81] among others, or even proteins such as thrombin has been employed as triggering mechanism providing a great arsenal of stimuli-responsive materials for drug delivery applications.[82]

3.4 Metal-Organic Frameworks in Nanomedicine

Although metal-organic frameworks (MOFs) structures were discovered in the late 1980s, they have been recently rediscovered for the applicability of their tunable properties such as sensors, nanocarriers of drug delivery and theranostic, inside the nanomedicine field.[83,84] MOFs are porous and versatile hybrid materials based on units of a central metal atom decorated with an organic, at least, bidentate ligand as linker for forming crystal nets in one, two and even three dimensions.

High loading capacity, easy tunable composition, size and porosity, possibility to behave as vehicle of drug or image agent at nanoscale, possibility to belong part of the treatment or diagnosis by active structure compounds and tunable biodegradability, among others are the features that make MOFs a real alternative for nanomedicine applications.[85]

There are three classical approximations for the synthesis of these materials, and the election depends on the nature of the own MOF.[86] Solvo-thermal synthesis usually is applied for Fe Ca or Mn materials with ligands such as terephthalic acid or zoledronate derivates. Temperature-assisted microwave or sono-chemical techniques are very common in order to support this kind of reactions.

On the other hand, reverse phase micro-emulsions are commonly used for Gd or Zn MOF generation. The method is based on the stabilization of water drops in organic solvents. This strategy is usually carried out under mild conditions. This fact makes the possibility to include biomolecules in the own MOF structures, sometimes unstable in harder conditions. Further, surfactant can also be used as templates for incorporating in the structure or for directing the pore formation.

One of the great advances in the MOF chemistry which had notable impact on their application mostly on catalysis but also on nanomedicine field is the absolute control over the their structure by the synthesis process and starting materials.

Reticular synthesis is the main strategy assumed in order to afford materials with same topology but different porous length and shapes, from meso- to microporosity cavities depending on the substrate to be loaded. In this way, MOF chemistry has introduced a new concept such as iso-reticular expansion (IEE). The IEE is the synthesis of family or expanded versions of a determined MOF, varying the pore size, as well as the specific surface, and keeping the rest of the properties intact looking forward to found the best candidates for loading the cargoes at issue (Figure 3.4).

In order to give an idea of MOF versatility, in theory, there are infinite combinations of linkers and metallic centers that are able to build these interesting structures. Further, there are infinite combinations of them (metal centers and linkers), and each combination may confer different properties to the material. Thus, the application of MOFs range from gas storage, separation and purification, heterogeneous catalysis, biomedicine, nanotechnology and specific nanomedicine, among others.

There are few main types of ligands used for synthetizing MOFs and in particular Nanoparticle Metal-Organic Framework (NMOF) for biological use: carboxylate, phosphonate, imidazolates, amines, phenols and sulfonate of different length and strength. Benzyl dicarboxylate ligands (BDCs) usually generate 1D linear structures which applied to each other forming 3D structures with easy control of the pore size by elongating the aromatic core. Further, usually nonlinear binaphthyl phosphonate ligands dimers also are used for forming chiral structures.

In order to build 3D structures in a direct manner, multidentate ligands are commonly used; trimers or tetra-dentate ligands from carboxylic acids or imidazole type are the most frequented and generate more stable structures through π–π staking.

On the other hand, metal centers such as Zn, Cu, Ca or Mg are mostly chosen, not only for modeling the topology of the final porous material but also for their low toxicity and biocompatibility. Further, Gd and Fe for theranostic purposes is the typical metallic centers applied for forming these crystal structures.[86]

For the efficient application of MOF in nanomedicine is necessary to control, not only the composition and structure, but also the macromolecular size of the material. In biomedicine, the main applications of these materials can be separated by administration via cutaneous, oral or parental.

Nanotechnology is mostly used for parental treatment as drug nanocarriers, image or theranostic. In order to transfer the properties of these bulk materials to the nanometer scale, two main methods have been developed: the use of crystal grow inhibitors that are added in the final step of the synthesis after the first nucleation, and the second one, the use of surfactant combination space agents forming micelles which limit the space of the reaction to inside of them. Both methods allow the limited growing of the crystal at nanometric size. The resulting system NMOF behaves as nanocarrier such as the PSN or MSN previously described.

The biocompatibility is one of the aspects where the composition of the MOF plays a highlighted role. The possible leaching process during the contact and circulation time with the body fluids from the ligand, as well as the metal center, including the possible cytotoxic effects caused from the material after its decomposition, could be controlled by using biocompatible starting materials. On the other side, the toxic effects of the own system when it is put in contact with the patient body, such as inflammation, autoimmune responses or other sides effects, could be handled by tuning their surface with biocompatible

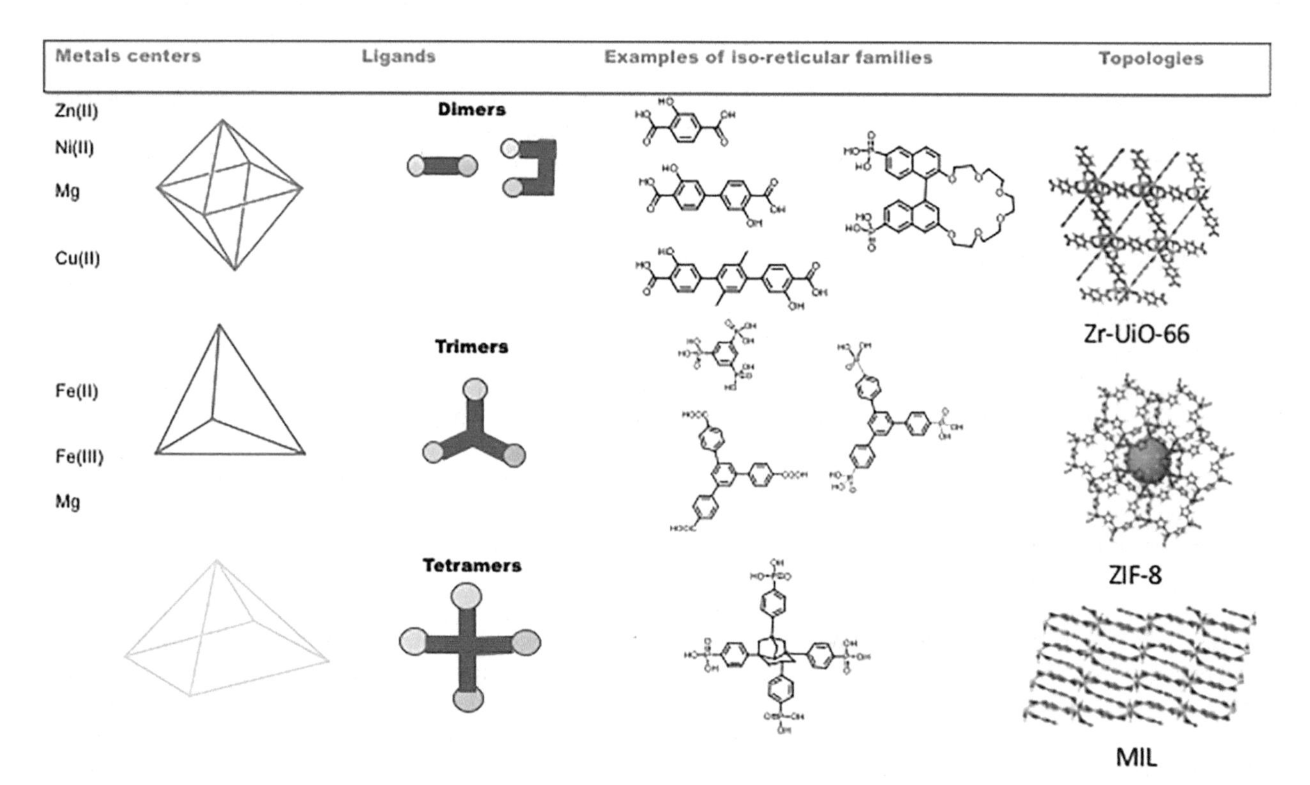

FIGURE 3.4 Correlation between metal, linker election and resulting structure of biocompatible MOFs.

polymers and polyethylene glycol, among others, with methodology widely used in many other materials for biological applications.

Indeed, NMOF presents several pros in terms of their capacity to play as nano-agent for diagnosis, treatment or theranostic. However, the way to introduce them in the nanomedicine is based on the same principles. Thus, EFR or passive targeting is commonly also accepted for NMOF as PSN or MSN or from different nature for cancer treatment. The easy functionalization of their surfaces makes also possible the inclusion of active targeting moieties or stimulus-responsive taping properties. Further, these materials could also be applied to others diseases such as arterial hypertension, anti-inflammatory treatment or antibacterial properties, to name only a few of them.

3.4.1 NMOF in Cancer Therapy, Image and Theranostics

Herein, we describe few didactic and highlighted examples for describing the main utilities of these materials as nanovehicles for oncology.

- Encapsulation and transportation of small molecules

Since the work reported by Horcajada et al. which described for the very first time the drug load capacity of MOFs in 2006 with ibuprofen in MIL-101 and MIL-100, MOF have attracted more and more attention in drug delivery.[84]

These novel MOFs made of trimers of octahedral Cr^{2+} conjugated with tri (MIL-100)- and di (MIL-101)-benzene carboxylic acids as linkers showed comparable performance to typical MCM-41 in terms of kinetic delivery but also more loading capacity and prolonged sustain delivery, probably due to higher electrostatic interactions with the metal center and the ibuprofen but also by the notably improvement of the surface area versus MCM-41. With this first example in mind, the use of these types of materials as drug carriers has been widely studied.

Lately, Cr (II) MIL-101 probed in bulk phase for ibuprofen controlled release was transformed onto a nanoparticle for its application in cancer treatment with cisplatin as cargo.[87] This work represents a great example of the potential of the NMOFs because the real value of MOF structures is their possibility of varying their properties. In this case, the authors transferred the simple initial M-101 bulk material properties to a smart targeted nanomaterial based on MOF for its application on nanomedicine for cancer treatment (Figure 3.5).

Cis-platin and its analogs are one of the most used drugs in chemotherapy. However, its efficacy could be improved by transporting it to the diseased tissue by raising its concentration in the problematic location even selectively inside the tumoral cells versus healthy ones.

In order to apply bulk MOF technology of the original M-101 to nanomedicine, the material was tuned in the macroscopic size by forming nanoparticles. On the other hand, composition of the NMOF was transformed through changing the metal center to Fe (III) making the structure more biocompatible and susceptible for magnetic resonance imaging (MRI) but keeping the topology intact. On the other hand, linkers were also changed in a controlled proportion by adding amine functionalization in their structure, allowing the functionalization of different cargoes to be delivered by the degradation of the NMOF itself, herein cisplatin derivate. Further, postsynthetic coating of silica layer made possible improvements in terms of stability of the system prolonging the time of the release of the cargoes,

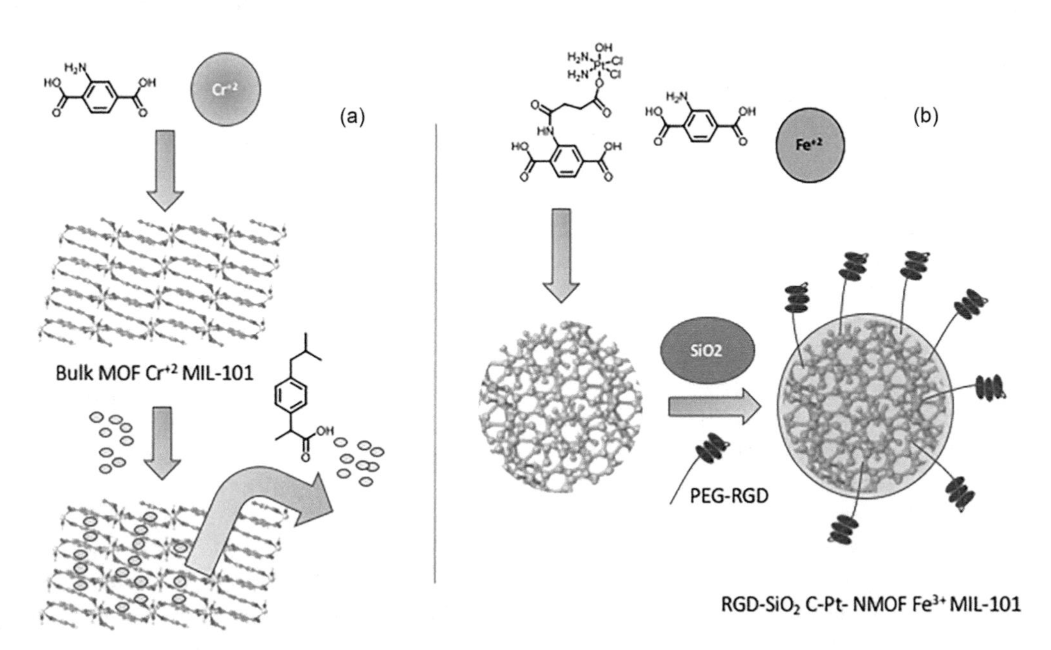

FIGURE 3.5 (a) First MOF Cr-MIL-101, with drug delivery functionality described for ibuprofen release. (b) Biocompatible Fe MIL-101 Smart NMOF, tuned and modified for targeted cisplatin delivery.

and finally, RGD-type targeting moieties were also grafted to the silica surface for vectorization purposes to typical α,β-integrin receptor present in various cancer cell lines. All these modifications provoked the generation a tuned hybrid NMOF material with promising performance in drug delivery.

Other small molecule that attracts a lot of attention for its selective delivery in nanomedicine is doxorubicin (Dox). Dox is probably the cytotoxic drug most common in the chemotherapy; however, due to its very poorly soluble nature, it is therefore difficult to find manner to transport into the body to diseased tissues and malignant cells without elevating the doses.

One of the latest works for Dox loading in MOF materials by Vasconcelos et al.[88] Dox was loaded in lamellar ZnBDC-MOF. The material showed great loading capacities that come from the high porosity but also from the high interaction between the aromatic structure of the drug and the aromatic ligand (1,4-benzene-dicarboxylate, BDC) through π–π interaction. This phenomenon helps not only to load the drug but also to retain it into the structure. This interaction within the aromatic rings (ligand-substrate) also confers a pH stimulus-responsive triggered delivery system to the final material by decomposition of the crystal structure. This idea was transferred to a other smart nanomaterial in the work recently presented by Chen et al.[89] In this recent paper, ^{89}Zr-UiO-66/Py–PGA-PEG-F3 has been synthetized by crystallization of aromatic ligands (BDC and benzoic acid) with ^{89}Zr-forming nanoparticles with 3D UiO-66 structure also for theranostic purposes. Theranostic applications are widely used in the NMOF systems, MRI with the inclusion of Fe or positron emission tomography (PET) with the use of ^{89}Zr as metal centers in the structure are the most common techniques. Further, the transporter was functionalized with PEG-peptide moiety in its surface for targeting purposes Herein, we have the typical example of multi-tasking material, since it has high loading capacity typical for NMOF, pH stimulus release, and high selectivity in its distribution in the tumor mass and finally able to give PET image for diagnosis being a promising smart multifunctional theranostic based on NMOF material.

Other chemotherapeutic agents such as methotrexate or 5-Fluorouracil have also been widely studied as cargoes in NMOF following the same philosophy that shown for Dox materials, the pH triggered release based on the aromatic interaction between drugs and MOF structure. The reason is that NMOF structures such as ZIF-8 (zeolitic imidazolate framework) with Zn as metal center and imidazolate as aromatic linker present similar characteristics than others presented previously.[90,91] However, other strategy recently reported by Huxford et al.[92] described the incorporation of methotrexate as building block into the NMOF structure. Methotrexate has planar aromatic core and two acid groups in its structure that allow the active molecule to emulate a MOF linker by coordination with the metal center of the primary structure. This strategy allows the improvement in the loading capacity versus other systems based on the π–π interaction. With the desired drug belonging to the structure as linker, the material capacity of load rises to 79% in weight of cargo; however, its structure loses crystallinity and becomes amorphous. Leaching problems probably occur. This fact may provokes the non-controlled delivery of the drug. In the paper, the authors tried to abolish this leaching by incorporating a lipid bilayer protection. The lipid surface was also tuned with targeting moiety for active vectorization to diseased cells. This is another example of cooperation between different types of materials for a common objective of stabilization and controlled targeted release, creating a new hybrid NMOF/lipid system for nanomedicine applications.

As has been showed in the above examples, the substrate usually presents aromatic moieties in their structure or functional groups able to be coordination points that helps in its loading and retention efficiency for a controlled release processes. However, in other cases, the situation is more complicate. For example, small-molecule Busulfan (Bu) is an antitumoral drug commonly used in multidrug treatments in pediatric leukemia. Its structure is based on a simple alkane chain with sulfonate groups in each extreme. This drug presents high hepatic toxicity and also poor stability in water, and the systemic injection must be in co-solution with toxic organic solvents making it still more harmful. On the other hand, liposomes and other organic-based nanotransporters presented low loading capacities because of their low interaction with the matrix. Horcajada et al.[93] recently studied the use of NMOF for solving this issue. For this aim, an iso-reticular family of Fe (III) MIL-type MOF was studied, resulting the MIL-100, the more efficient in loading capacity with a 25 %wt. The new NMOF was functionalized with PEG for improving their stability, and in the end, the encapsulation allowed the improvement of the formulation stability of the active principle, the main goal in this case to minimize the doses and reduce side effects.

- Encapsulation and transportation of big molecules

One of the most famous examples of the NMOF technology used for drug delivery is to tune the size of the pores in order to transport small but also big macromolecular substances such as DNA or RNA fragments for transfection purposes, proteins and enzymes, among others.

The transportation of DNA or RNA strains for transfection purposes is one of the methodologies for gene therapy used in the last years for supporting the chemotherapy. The transfection process must to be intracellular. However, oligonucleotide chains cannot usually pass through the cell membrane. Thus, vectorization agents are needed in order to allow the uptake and the endosomal scape for deliver the payload in the cytosol.

Several NMOF materials have been tested as nanotransporters for siRNAs for gene silencing;[94,95] however, several issues such as strong affinity for the nano-transporter by the nucleotide strain as well as low loading capacity have been found. Recently, Peng et al.[96] reported a great capacity of NMOFs for increasing the concentration of gene

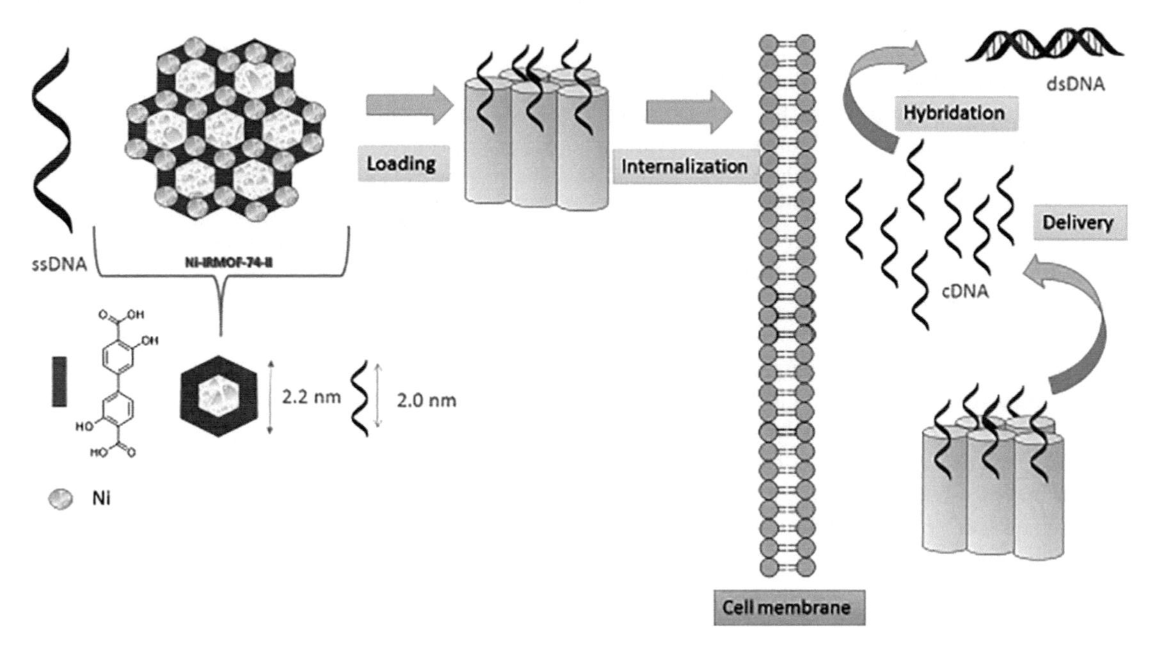

FIGURE 3.6 Ni-IRMOF-II specifically designed for loading and intracellular controlled releasing of ssDNA.

silencing agents in the cytosol of cancer cells by the transportation of the genetic material by Ni-IRMOF-74 family. The nano-transporter used was a Ni MOF synthetized using Ni^{2+} as metal center and bi-functional derivate of salicylic acid as organic linkers. Iso-reticular family of this MOF with hexagonal topology in 1D pores. The material with larger linkers exhibited mesoporous typical N_2 isotherm with great capacity (Figure 3.6).

Further, nanocrystals of these materials of barley 100 nm were synthetized and loaded with a single-stranded DNA (ssDNA). The ssDNA chains with 2.0 nm of size are accommodated in the 2.2 nm pores of Ni-IRMOF-74-II. This interaction is possible because its encapsulation has been practically made to measure by choosing the best candidate within the family for the substrate. In this way, the release of the ssDNA is triggered only by the presence of its complementary strain. This reflects the high selectivity for delivering the gene material in desired spot, inculcated by the NMOF smart system.

On other hand, enzyme replacement therapies are lately becoming in a supportive treatment for genetic manipulations and are being handled in the nanomedicine for oncology applications.

For this aim, the selective enzyme cytosol delivery and high activity performance by enzyme are necessary to afford efficient treatments. Biocatalysts is one of the most used applications of MOFs,[97] and it is known that MOFs are able to anchor enzymes or proteins keeping their properties, structure and activity even rising the half-life in multiple ex vivo situations. Thus, the encapsulation of the proteins or enzymes in MOF structure is widely known, and there multiple methods to do it.

On the other hand, protein due its proportions and nature hardly penetrates to the cellular cytosol. Further, once there they have short half-lives if the protein is not endogenous from the cell in question. MOF enzyme transportation with in-cell activity may be one of the future approximations for using enzymatic or proteins. Lian et al.[98] proposes the concept MOF enzyme nanofactories. This concept is based on a MOF nano-cages based on aluminum-4,4′,4″-*s*-triazine-2,4,6-triyl-tribenzoic (TATB) acid ligands, with ultra-high enzyme encapsulation capacity. This NMOF is able to transport and maintain the activity inside the lysosomes of the living cells.

3.4.2 Other Applications of NMOF

Due their high capacity of tunable pore size and versatile synthetic and functionalization methodology, NMOF has become one of the most studied platforms in the nanomedicine field. NMOF is useful not only for cancer but also for other different diseases by transporting several other active principles for the treatment of inflammatory or autoimmune disseases[99] such as pulmonary atrial hypertension (PAH)[100], among others.

Further, the encapsulation of biomolecules under mild conditions is one of the best characteristics of these materials for bio-application. Thus, nowadays, MOF bio-composite is an emerging study line in the last years. Further, the concept can be extended to viruses for vaccine formulations and even cells encapsulation for regenerative cell therapy, meaning MOF coating or biomineralization.[101]

3.5 Conclusions

Along this chapter, some of the most important types of porous inorganic nanomaterials have been presented. Their use for the treatment of complex pathologies, such as cancer, has provided to the clinicians novel weapons which can help for the complete eradication of unmet diseases. The excellent cargo capacity of these materials, in combination with their robustness and, in general, highly versatile nature, present important advantages in comparison with other types of nanocarriers. However, there are still some issues which should be addressed in order to improve their performance. For example, toxic metals which usually compose MOFs should be replaced by more biocompatible ions. Related to this issue, despite the fact that silicon-based materials (PSN and MSNs) are, generally speaking, biocompatible materials, their lack of toxicity should be studied at long times in order to apply these materials for the treatment of chronic diseases which requires their administration during long periods of time. This is particularly required in the case of hybrid systems which are not only composed by harmless silica derivatives but also by organo/inorganic coatings or gatekeepers, in which the biocompatibility of their components has not been studied in detail. In this chapter, only a few examples of their application have been described with the aim to show their possibilities. The future of these types of materials in the clinical field is plenty of possibilities. More work is needed in order to find new uses, but, in any case, in the coming years, a new generation of porous inorganic nanocarriers will arrive to the market offering novel ways to improve the quality of life of the patients and the society in general.

References

1. Tibbitt, M. W.; Dahlman, J. E.; Langer, R. Emerging frontiers in drug delivery. *J. Am. Chem. Soc.* **2016**, *138*, 704–717.
2. Dawidczyk, C. M.; Kim, C.; Park, J. H.; Russell, L. M.; Lee, K. H.; Pomper, M. G.; Searson, P. C. State-of-the-art in design rules for drug delivery platforms: Lessons learned from FDA-approved nanomedicines. *J. Control. Release* **2014**, *187*, 133–144.
3. Nguyen, K. T.; Zhao, Y. Engineered hybrid nanoparticles for on-demand diagnostics and therapeutics. *Acc. Chem. Res.* **2015**, *48*, 3016–3025.
4. Mitra, A. K.; Agrahari, V.; Mandal, A.; Cholkar, K.; Natarajan, C.; Shah, S.; Joseph, M.; Trinh, H. M.; Vaishya, R.; Yang, X.; et al. Novel delivery approaches for cancer therapeutics. *J. Control. Release* **2015**, *219*, 248–268.
5. Matsumura, Y.; Maeda, H. A new concept for macromolecular therapeutics in cancer-chemotherapy—mechanism of tumoritropic accumulation of proteins and the antitumor agent smancs. *Cancer Res.* **1986**, *46*, 6387–6392.
6. Jain, R. K.; Stylianopoulos, T. Delivering nanomedicine to solid tumors. *Nat. Rev. Clin. Oncol.* **2010**, *7*, 653–664.
7. Fang, J.; Nakamura, H.; Maeda, H. The EPR effect: Unique features of tumor blood vessels for drug delivery, factors involved, and limitations and augmentation of the effect. *Adv. Drug Deliv. Rev.* **2011**, *63*, 136–151.
8. Ediriwickrema, A.; Saltzman, W. M. Nanotherapy for cancer: Targeting and multifunctionality in the future of cancer therapies. *ACS Biomater. Sci. Eng.* **2015**, *1*, 64–78.
9. Godin, B.; Sakamoto, J. H.; Serda, R. E.; Grattoni, A.; Bouamrani, A.; Ferrari, M. Emerging applications of nanomedicine for the diagnosis and treatment of cardiovascular diseases. *Trends Pharmacol. Sci.* **2010**, *31*, 199–205.
10. Zazo, H.; Colino, C. I.; Lanao, J. M. Current applications of nanoparticles in infectious diseases. *J. Control. Release* **2016**, *224*, 86–102.
11. Allen, T. M.; Cullis, P. R. Liposomal drug delivery systems: From concept to clinical applications. *Adv. Drug Deliv. Rev.* **2013**, *65*, 36–48.
12. Baeza, A.; Ruiz-Molina, D.; Vallet-Regí, M. Recent advances in porous nanoparticles for drug delivery in antitumoral applications: Inorganic nanoparticles and nanoscale metal-organic frameworks. *Expert Opin. Drug Deliv.* **2017**, *14*, 783–796.
13. Uhlir, A. Electrolytic shaping of germanium and silicon. *Bell Syst. Tech. J.* **1956**, *35*, 333–347.
14. Canham, L. T. Silicon quantum wire array fabrication by electrochemical and chemical Ddssolution of wafers. *Appl. Phys. Lett.* **1990**, *57*, 1046–1048.
15. Santos, H. A.; Mäkilä, E.; Airaksinen, A. J.; Bimbo, L. M.; Hirvonen, J. Porous silicon nanoparticles for nanomedicine: Preparation and biomedical applications. *Nanomedicine* **2014**, *9*, 535–554.
16. Santos, H. A.; Hirvonen, J. Nanostructured porous silicon materials: Potential candidates for improving drug delivery. *Nanomedicine* **2012**, *7*, 1281–1284.
17. Buriak, J. M.; Stewart, M. P.; Geders, T. W.; Allen, M. J.; Choi, H. C.; Smith, J.; Raftery, D.; Canham, L. T. Lewis acid mediated hydrosilylation on porous silicon surfaces. *J. Am. Chem. Soc.* **1999**, *121*, 11491–11502.
18. Li, W.; Liu, Z.; Fontana, F.; Ding, Y.; Liu, D.; Hirvonen, J. T.; Santos, H. A. Tailoring porous silicon for biomedical applications: From drug delivery to cancer immunotherapy. *Adv. Mater.* **2018**, *1703740.*
19. Liu, Z.; Balasubramanian, V.; Bhat, C.; Vahermo, M.; Mäkilä, E.; Kemell, M.; Fontana, F.; Janoniene, A.; Petrikaite, V.; Salonen, J.; et al. Quercetin-based modified porous silicon nanoparticles for enhanced inhibition of doxorubicin-resistant cancer cells. *Adv. Healthc. Mater.* **2017**, *6*, 1601009.
20. Xia, B.; Wang, B.; Zhang, W.; Shi, J. High loading of doxorubicin into styrene-terminated porous silicon nanoparticles via π-stacking for cancer treatments in vitro. *RSC Adv.* **2015**, *5*, 44660–44665.
21. Kang, J.; Joo, J.; Kwon, E. J.; Skalak, M.; Hussain, S.; She, Z.-G.; Ruoslahti, E.; Bhatia, S. N.;

Sailor, M. J. Self-sealing porous silicon-calcium silicate core-shell nanoparticles for targeted siRNA delivery to the injured brain. *Adv. Mater.* **2016**, *28*, 7962–7969.
22. Tong, W. Y.; Alnakhli, M.; Bhardwaj, R.; Apostolou, S.; Sinha, S.; Fraser, C.; Kuchel, T.; Kuss, B.; Voelcker, N. H. Delivery of siRNA in vitro and in vivo using PEI-capped porous silicon nanoparticles to silence MRP1 and inhibit proliferation in glioblastoma. *J. Nanobiotechnol.* **2018**, *16*, 38.
23. Anglin, E.; Cheng, L.; Freeman, W.; Sailor, M. Porous silicon in drug delivery devices and materials. *Adv. Drug Deliv. Rev.* **2008**, *60*, 1266–1277.
24. Tamarov, K.; Xu, W.; Osminkina, L.; Zinovyev, S.; Soininen, P.; Kudryavtsev, A.; Gongalsky, M.; Gaydarova, A.; Närvänen, A.; Timoshenko, V.; et al. Temperature responsive porous silicon nanoparticles for cancer therapy – spatiotemporal triggering through infrared and radiofrequency electromagnetic heating. *J. Control. Release* **2016**, *241*, 220–228.
25. Timoshenko, V. Y.; Dittrich, T.; Lysenko, V.; Lisachenko, M. G.; Koch, F. Free charge carriers in mesoporous silicon. *Phys. Rev. B* **2001**, *64*, 85314.
26. Xia, B.; Wang, B.; Chen, Z.; Zhang, Q.; Shi, J. Near-infrared light-triggered intracellular delivery of anticancer drugs using porous silicon nanoparticles conjugated with IR820 dyes. *Adv. Mater. Interfaces* **2016**, *3*, 1500715.
27. Shahbazi, M.-A.; Almeida, P. V.; Correia, A.; Herranz-Blanco, B.; Shrestha, N.; Mäkilä, E.; Salonen, J.; Hirvonen, J.; Santos, H. A. Intracellular responsive dual delivery by endosomolytic polyplexes carrying DNA anchored porous silicon nanoparticles. *J. Control. Release* **2017**, *249*, 111–122.
28. Yong, T.; Hu, J.; Zhang, X.; Li, F.; Yang, H.; Gan, L.; Yang, X. Domino-like intercellular delivery of undecylenic acid-conjugated porous silicon nanoparticles for deep tumor penetration. *ACS Appl. Mater. Interfaces* **2016**, *8*, 27611–27621.
29. Wang, C.-F.; Sarparanta, M. P.; Mäkilä, E. M.; Laakkonen, P. M.; Salonen, J. J.; Hirvonen, J. T.; Airaksinen, A. J.; Santos, H. A.; Hyvönen, M. L. K.; Laakkonen, P. M.; et al. Multifunctional porous silicon nanoparticles for cancer theranostics. *Biomaterials* **2015**, *48*, 108–118.
30. Stead, S. O.; McInnes, S. J. P.; Kireta, S.; Rose, P. D.; Jesudason, S.; Rojas-Canales, D.; Warther, D.; Cunin, F.; Durand, J. O.; Drogemuller, C. J.; et al. Manipulating human dendritic cell phenotype and function with targeted porous silicon nanoparticles. *Biomaterials* **2018**, *155*, 92–102.
31. Turner, C. T.; McInnes, S. J. P.; Melville, E.; Cowin, A. J.; Voelcker, N. H. Delivery of flightless I neutralizing antibody from porous silicon nanoparticles improves wound healing in diabetic mice. *Adv. Healthc. Mater.* **2017**, *6*, 1600707.
32. Shrestha, N.; Araújo, F.; Shahbazi, M.-A.; Mäkilä, E.; Gomes, M. J.; Herranz-Blanco, B.; Lindgren, R.; Granroth, S.; Kukk, E.; Salonen, J.; et al. Thiolation and cell-penetrating peptide surface functionalization of porous silicon nanoparticles for oral delivery of insulin. *Adv. Funct. Mater.* **2016**, *26*, 3405–3416.
33. Liu, Z.; Li, Y.; Li, W.; Xiao, C.; Liu, D.; Dong, C.; Zhang, M.; Mäkilä, E.; Kemell, M.; Salonen, J.; et al. Multifunctional nanohybrid based on porous silicon nanoparticles, gold nanoparticles, and acetalated dextran for liver regeneration and acute liver failure theranostics. *Adv. Mater.* **2017**, *30*, 1703393.
34. Yanagisawa, T.; Shimizu, T.; Kuroda, K.; Kato, C. The preparation of alkyltriinethylaininonium–kaneinite complexes and their conversion to microporous materials. *Bull. Chem. Soc. Jpn.* **1990**, *63*, 988–992.
35. Kresge, C. T.; Leonowicz, M. E.; Roth, W. J.; Vartuli, J. C.; Beck, J. S. Ordered mesoporous molecular sieves synthesized by a liquid-crystal template mechanism. *Nature* **1992**, *359*, 710–712.
36. Vallet-Regí, M.; Balas, F.; Arcos, D. Mesoporous materials for drug delivery. *Angew. Chemie Int. Ed.* **2007**, *46*, 7548–7558.
37. Hoffmann, F.; Cornelius, M.; Morell, J.; Fröba, M. Silica-based mesoporous organic–inorganic hybrid materials. *Angew. Chemie Int. Ed.* **2006**, *45*, 3216–3251.
38. Cotí, K. K.; Belowich, M. E.; Liong, M.; Ambrogio, M. W.; Lau, Y. A.; Khatib, H. A.; Zink, J. I.; Khashab, N. M.; Stoddart, J. F. Mechanised nanoparticles for drug delivery. *Nanoscale* **2009**, *1*, 16–39.
39. Huh, S.; Wiench, J. W.; Yoo, J. C.; Pruski, M.; Lin, V. S. Y. Organic functionalization and morphology control of mesoporous silicas via a co-condensation synthesis method. *Chem. Mater.* **2003**, *15*, 4247–4256.
40. Lu, J.; Li, Z.; Zink, J. I.; Tamanoi, F. In vivo tumor suppression efficacy of mesoporous silica nanoparticles-based drug-delivery system: Enhanced efficacy by folate modification. *Nanomed. Nanotechnol., Biol. Med.* **2012**, *8*, 212–220.
41. Gary-Bobo, M.; Mir, Y.; Rouxel, C.; Brevet, D.; Basile, I.; Maynadier, M.; Vaillant, O.; Mongin, O.; Blanchard-Desce, M.; Morere, A.; et al. Mannose-functionalized mesoporous silica nanoparticles for efficient two-photon photodynamic therapy of solid tumors. *Angew. Chemie-Int. Ed.* **2011**, *50*, 11425–11429.
42. Ferris, D. P.; Lu, J.; Gothard, C.; Yanes, R.; Thomas, C. R.; Olsen, J.-C.; Stoddart, J. F.; Tamanoi, F.; Zink, J. I. Synthesis of biomolecule-modified mesoporous silica nanoparticles for targeted hydrophobic drug delivery to cancer cells. *Small* **2011**, *7*, 1816–1826.

43. Wang, X.; Liu, Y.; Wang, S.; Shi, D.; Zhou, X.; Wang, C.; Wu, J.; Zeng, Z.; Li, Y.; Sun, J.; et al. CD44-engineered mesoporous silica nanoparticles for overcoming multidrug resistance in breast cancer. *Appl. Surf. Sci.* **2015**, *332*, 308–317.
44. Ngamcherdtrakul, W.; Morry, J.; Gu, S.; Castro, D. J.; Goodyear, S. M.; Sangvanich, T.; Reda, M. M.; Lee, R.; Mihelic, S. a.; Beckman, B. L.; et al. Cationic polymer modified mesoporous silica nanoparticles for targeted siRNA delivery to HER2 + breast cancer. *Adv. Funct. Mater.* **2015**, *25*, 2646–2659.
45. Gao, L.; Cui, Y.; He, Q.; Yang, Y.; Fei, J.; Li, J. Selective recognition of Co-assembled thrombin aptamer and docetaxel on mesoporous silica nanoparticles against tumor cell proliferation. *Chem. A Eur. J.* **2011**, *17*, 13170–13174.
46. Wang, K.; Yao, H.; Meng, Y.; Wang, Y.; Yan, X.; Huang, R. Specific aptamer-conjugated mesoporous silica–carbon nanoparticles for HER2-targeted chemo-photothermal combined therapy. *Acta Biomater.* **2015**, *16*, 196–205.
47. Tang, Y.; Hu, H.; Zhang, M. G.; Song, J.; Nie, L.; Wang, S.; Niu, G.; Huang, P.; Lu, G.; Chen, X. An aptamer-targeting photoresponsive drug delivery system using "off–on" graphene oxide wrapped mesoporous silica nanoparticles. *Nanoscale* **2015**, *7*, 6304–6310.
48. Cheng, K.; El-Boubbou, K.; Landry, C. C. Binding of HIV-1 gp120 glycoprotein to silica nanoparticles modified with CD4 glycoprotein and CD4 peptide fragments. *ACS Appl. Mater. Interfaces* **2012**, *4*, 235–243.
49. Villaverde, G.; Baeza, A.; Melen, G. J.; Alfranca, A.; Ramirez, M.; Vallet-Regí, M.; Marchesan, S.; Prato, M.; Knežević, N. Ž.; Durand, J.-O.; et al. A new targeting agent for the selective drug delivery of nanocarriers for treating neuroblastoma. *J. Mater. Chem. B* **2015**, *3*, 4831–4842.
50. Weissleder, R.; Kelly, K.; Sun, E. Y.; Shtatland, T.; Josephson, L. Cell-specific targeting of nanoparticles by multivalent attachment of small molecules. *Nat. Biotechnol.* **2005**, *23*, 1418–1423.
51. Baeza, A.; Colilla, M.; Vallet-Regí, M. Advances in mesoporous silica nanoparticles for targeted stimuli-responsive drug delivery. *Expert Opin. Drug Deliv.* **2015**, *12*, 319–337.
52. Stolik, S.; Delgado, J. A.; Pérez, A.; Anasagasti, L. Measurement of the penetration depths of red and near infrared light in human "ex Vivo" tissues. *J. Photochem. Photobiol. B Biol.* **2000**, *57*, 90–93.
53. Mal, N. K.; Fujiwara, M.; Tanaka, Y. Photocontrolled reversible release of guest molecules from coumarin-modified mesoporous silica. *Nature* **2003**, *421*, 350–353.
54. Li, H.; Tan, L.-L.; Jia, P.; Li, Q.-L.; Sun, Y.-L.; Zhang, J.; Ning, Y.-Q.; Yu, J.; Yang, Y.-W. Near-unfrared light-responsive supramolecular nanovalve based on mesoporous silica-coated gold nanorods. *Chem. Sci.* **2014**, *5*, 2804.
55. Martínez-Carmona, M.; Baeza, A.; Rodriguez-Milla, M. A.; García-Castro, J.; Vallet-Regí, M. Mesoporous silica nanoparticles grafted with a light-responsive protein shell for highly cytotoxic antitumoral therapy. *J. Mater. Chem. B* **2015**, *3*, 5746–5752.
56. Yang, S.; Li, N.; Chen, D.; Qi, X.; Xu, Y.; Xu, Y.; Xu, Q.; Li, H.; Lu, J. Visible-light degradable polymer coated hollow mesoporous silica nanoparticles for controlled drug release and cell imaging. *J. Mater. Chem. B* **2013**, *1*, 4628.
57. Gnanasammandhan, M. K.; Idris, N. M.; Bansal, A.; Huang, K.; Zhang, Y. Near-IR photoactivation using mesoporous silica–coated NaYF4:Yb,Er/Tm upconversion nanoparticles. *Nat. Protoc.* **2016**, *11*, 688–713.
58. Wang, F.; Liu, X. Recent advances in the chemistry of lanthanide-doped upconversion nanocrystals. *Chem. Soc. Rev.* **2009**, *38*, 976.
59. Wu, S.; Butt, H.-J. Near-infrared-sensitive materials based on upconverting nanoparticles. *Adv. Mater.* **2016**, *28*, 1208–1226.
60. Baeza, A.; Arcos, D.; Vallet-Regí, M. Thermoseeds for interstitial magnetic hyperthermia: From bioceramics to nanoparticles. *J. Phys. Condens. Matter* **2013**, *25*, 484003.
61. Ruiz-Hernandez, E.; Baeza, A.; Vallet-Regi, M. Smart drug delivery through DNA/magnetic nanoparticle gates. *ACS Nano* **2011**, *5*, 1259–1266.
62. Baeza, A.; Guisasola, E.; Ruiz-Hernández, E.; Vallet-Regí, M. Magnetically triggered multidrug release by hybrid mesoporous silica nanoparticles. *Chem. Mater.* **2012**, *24*, 517–524.
63. Guisasola, E.; Baeza, A.; Talelli, M.; Arcos, D.; Moros, M.; de la Fuente, J. M.; Vallet-Regí, M. Magnetic-responsive release controlled by hot spot effect. *Langmuir* **2015**, *31*, 12777–12782.
64. Guisasola, E.; Asín, L.; Beola, L.; de la Fuente, J. M.; Baeza, A.; Vallet-Regí, M. Beyond traditional hyperthermia: In vivo cancer treatment with magnetic-responsive mesoporous silica nanocarriers. *ACS Appl. Mater. Interfaces* **2018**, *10*, 12518–12525.
65. Paris, J. L.; Cabañas, M. V.; Manzano, M.; Vallet-Regí, M. Polymer-grafted mesoporous silica nanoparticles as ultrasound-responsive drug carriers. *ACS Nano* **2015**, *9*, 11023–11033.
66. Cheng, S.-H.; Liao, W.-N.; Chen, L.-M.; Lee, C.-H. pH-controllable release using functionalized mesoporous silica nanoparticles as an oral drug delivery system. *J. Mater. Chem.* **2011**, *21*, 7130.
67. Nguyen, C. T. H.; Webb, R. I.; Lambert, L. K.; Strounina, E.; Lee, E. C.; Parat, M.-O.; McGuckin, M. A.; Popat, A.; Cabot, P. J.; Ross, B. P. Bifunctional succinylated ε-polylysine-coated mesoporous

silica nanoparticles for pH-responsive and intracellular drug delivery targeting the colon. *ACS Appl. Mater. Interfaces* **2017**, *9*, 9470–9483.

68. Du, J.; Lane, L. A.; Nie, S. Stimuli-responsive nanoparticles for targeting the tumor microenvironment. *J. Control. Release* **2015**, *219*, 205–214.
69. Niedermayer, S.; Weiss, V.; Herrmann, A.; Schmidt, A.; Datz, S.; Müller, K.; Wagner, E.; Bein, T.; Bräuchle, C. Multifunctional polymer-capped mesoporous silica nanoparticles for pH-responsive targeted drug delivery. *Nanoscale* **2015**, *7*, 7953–7964.
70. Chen, Y.; Ai, K.; Liu, J.; Sun, G.; Yin, Q.; Lu, L. Multifunctional envelope-type mesoporous silica nanoparticles for pH-responsive drug delivery and magnetic resonance imaging. *Biomaterials* **2015**, *60*, 111–120.
71. Lin, Z.; Li, J.; He, H.; Kuang, H.; Chen, X.; Xie, Z.; Jing, X.; Huang, Y. Acetalated-dextran as valves of mesoporous silica particles for pH responsive intracellular drug delivery. *RSC Adv.* **2015**, *5*, 9546–9555.
72. Zhang, B.; Luo, Z.; Liu, J.; Ding, X.; Li, J.; Cai, K. Cytochrome c end-capped mesoporous silica nanoparticles as redox-responsive drug delivery vehicles for liver tumor-targeted triplex therapy in vitro and in vivo. *J. Control. Release* **2014**, *192*, 192–201.
73. Kim, H.; Kim, S.; Park, C.; Lee, H.; Park, H. J.; Kim, C. Glutathione-induced intracellular release of guests from mesoporous silica nanocontainers with cyclodextrin gatekeepers. *Adv. Mater.* **2010**, *22*, 4280–4283.
74. Luo, Z.; Cai, K.; Hu, Y.; Zhao, L.; Liu, P.; Duan, L.; Yang, W. Mesoporous silica nanoparticles end-capped with collagen: Redox-responsive nanoreservoirs for targeted drug delivery. *Angew. Chemie Int. Ed.* **2011**, *50*, 640–643.
75. Sun, L.; Liu, Y.-J.; Yang, Z.-Z.; Qi, X.-R. Tumor specific delivery with redox-triggered mesoporous silica nanoparticles inducing neovascularization suppression and vascular normalization. *RSC Adv.* **2015**, *5*, 55566–55578.
76. Maggini, L.; Cabrera, I.; Ruiz-Carretero, A.; Prasetyanto, E. A.; Robinet, E.; De Cola, L. Breakable mesoporous silica nanoparticles for targeted drug delivery. *Nanoscale* **2016**, *8*, 7240–7247.
77. van Rijt, S. H.; Bölükbas, D. A.; Argyo, C.; Datz, S.; Lindner, M.; Eickelberg, O.; Königshoff, M.; Bein, T.; Meiners, S. Protease-mediated release of chemotherapeutics from mesoporous silica nanoparticles to ex vivo human and mouse lung tumors. *ACS Nano* **2015**, *9*, 2377–2389.
78. Srivastava, P.; Hira, S. K.; Srivastava, D. N.; Gupta, U.; Sen, P.; Singh, R. A.; Manna, P. P. Protease-responsive targeted delivery of doxorubicin from bilirubin-BSA-capped mesoporous silica nanoparticles against colon cancer. *ACS Biomater. Sci. Eng.* **2017**, *3*, 3376–3385.
79. Gayam, S. R.; Venkatesan, P.; Sung, Y.-M.; Sung, S.-Y.; Hu, S.-H.; Hsu, H.-Y.; Wu, S.-P. An NAD(P)H:quinone oxidoreductase 1 (NQO1) enzyme responsive nanocarrier based on mesoporous silica nanoparticles for tumor targeted drug delivery in vitro and in vivo. *Nanoscale* **2016**, *8*, 12307–12317.
80. Hu, C.; Huang, P.; Zheng, Z.; Yang, Z.; Wang, X. A facile strategy to prepare an enzyme-responsive mussel mimetic coating for drug delivery based on mesoporous silica nanoparticles. *Langmuir* **2017**, *33*, 5511–5518.
81. Jiang, H.; Shi, X.; Yu, X.; He, X.; An, Y.; Lu, H. Hyaluronidase enzyme-responsive targeted nanoparticles for effective delivery of 5-fluorouracil in colon cancer. *Pharm. Res.* **2018**, *35*, 73.
82. Bhat, R.; Ribes, À.; Mas, N.; Aznar, E.; Sancenón, F.; Marcos, M. D.; Murguía, J. R.; Venkataraman, A.; Martínez-Máñez, R. Thrombin-responsive gated silica mesoporous nanoparticles as coagulation regulators. *Langmuir* **2016**, *32*, 1195–1200.
83. Horcajada, P.; Gref, R.; Baati, T.; Allan, P. K.; Maurin, G.; Couvreur, P.; Férey, G.; Morris, R. E.; Serre, C. Metal-organic frameworks in biomedicine. *Chem. Rev.* **2012**, *112*, 1232–1268.
84. Horcajada, P.; Serre, C.; Vallet-Regí, M.; Sebban, M.; Taulelle, F.; Férey, G. Metal-organic frameworks as efficient materials for drug delivery. *Angew. Chemie Int. Ed.* **2006**, *45*, 5974–5978.
85. He, C.; Liu, D.; Lin, W. Nanomedicine applications of hybrid nanomaterials built from metal-ligand coordination bonds: Nanoscale metal-organic frameworks and nanoscale coordination polymers. *Chem. Rev.* **2015**, *115*, 11079–11108.
86. Liu, R.; Yu, T.; Shi, Z.; Wang, Z. The preparation of metal organic frameworks and their biomedical application. *Int. J. Nanomed.* **2016**, *11*, 1187.
87. Taylor-Pashow, K. M. L.; Rocca, J. Della; Xie, Z.; Tran, S. Post-synthetic modifications of iron-carboxylate nanoscale metal-organic frameworks for imaging and drug delivery. *J. Am. Chem. Soc.* **2010**, *131*, 14261–14263.
88. Vasconcelos, I. B.; Wanderley, K. A.; Rodrigues, N. M.; da Costa, N. B.; Freire, R. O.; Junior, S. A. Host-guest interaction of ZnBDC-MOF + Doxorubicin: A theoretical and experimental study. *J. Mol. Struct.* **2017**, *1131*, 36–42.
89. Chen, D.; Yang, D.; Dougherty, C. A.; Lu, W.; Wu, H.; He, X.; Cai, T.; Van Dort, M. E.; Ross, B. D.; Hong, H. In vivo targeting and positron emission tomography imaging of tumor with intrinsically radioactive metal-organic frameworks nanomaterials. *ACS Nano* **2017**, *11*, 4315–4327.

90. Sun, C.-Y.; Qin, C.; Wang, X.-L.; Yang, G.-S.; Shao, K.-Z.; Lan, Y.-Q.; Su, Z.-M.; Huang, P.; Wang, C.-G.; Wang, E.-B. Zeolitic imidazolate framework-8 as efficient pH-sensitive drug delivery vehicle. *Dalt. Trans.* **2012**, *41*, 6906.
91. Adhikari, C.; Das, A.; Chakraborty, A. Zeolitic imidazole framework (ZIF) nanospheres for easy encapsulation and controlled release of an anticancer drug doxorubicin under different external stimuli: A way toward smart drug delivery system. *Mol. Pharm.* **2015**, *12*, 3158–3166.
92. Huxford, R. C; deKrafft, K. E; Boyle, W. S.; Liu, D.; Lin, W. Lipid-coated nanoscale coordination polymers for targeted delivery of antifolates to cancer cells. *Chem. Sci.* **2015**, *25*, 713–724.
93. Horcajada, P.; Chalati, T.; Serre, C.; Gillet, B.; Sebrie, C.; Baati, T.; Eubank, J. F.; Heurtaux, D.; Clayette, P.; Kreuz, C.; et al. Porous metal-organic-framework nanoscale carriers as a potential platform for drug delivery and imaging. *Nat. Mater.* **2010**, *9*, 172–178.
94. Wu, Y.; Han, J.; Xue, P.; Xu, R.; Kang, Y. Nano metal–organic framework (NMOF)-based strategies for multiplexed microRNA detection in solution and living cancer cells. *Nanoscale* **2015**, *7*, 1753–1759.
95. He, C.; Lu, K.; Liu, D.; Lin, W. Nanoscale metal-organic frameworks for the co-delivery of cisplatin and pooled siRNAs to enhance therapeutic efficacy in drug-resistant ovarian cancer cells. *J. Am. Chem. Soc.* **2014**, *136*, 5181–5184.
96. Peng, S.; Bie, B.; Sun, Y.; Liu, M.; Cong, H.; Zhou, W.; Xia, Y.; Tang, H.; Deng, H.; Zhou, X. Metal-organic frameworks for precise inclusion of single-stranded DNA and transfection in immune cells. *Nat. Commun.* **2018**, *9*, 1293.
97. Lian, X.; Fang, Y.; Joseph, E.; Wang, Q.; Li, J.; Banerjee, S.; Lollar, C.; Wang, X.; Zhou, H.-C. Enzyme–MOF (Metal–organic Framework) composites. *Chem. Soc. Rev.* **2017**, *46*, 3386–3401.
98. Lian, X.; Erazo-Oliveras, A.; Pellois, J. P.; Zhou, H. C. High efficiency and long-term intracellular activity of an enzymatic nanofactory based on metal-organic frameworks. *Nat. Commun.* **2017**, *8*, 1–10.
99. Haydar, A.; Abid, H. R.; Sunderland, B.; Wang, S. Metal-organic-frameworks-as-a-drug-delivery-system-for-flurbiprofen. *Drug. Des. Devel. Ther.* **2017**, *11*, 2685–2695.
100. Mohamed, N. A.; Davies, R. P.; Lickiss, P. D.; Ahmetaj-Shala, B.; Reed, D. M.; Gashaw, H. H.; Saleem, H.; Freeman, G. R.; George, P. M.; Wort, S. J.; et al. Chemical and biological assessment of metal organic frameworks (MOFs) in pulmonary cells and in an acute in vivo model: Relevance to pulmonary arterial hypertension Therapy. *Pulm. Circ.* **2017**, *7*, 643–653.
101. Riccò, R.; Liang, W.; Li, S.; Gassensmith, J. J.; Caruso, F.; Doonan, C.; Falcaro, P. Metal-organic frameworks for cell and virus biology: A perspective. *ACS Nano* **2018**, *12*, 13–23.

4

Smart Nanoparticles in Drug/Gene Delivery

Mahdi Karimi
Iran University of Medical Sciences

Michael R. Hamblin
Harvard Medical School

4.1 Introduction

In recent decades, nanotechnology has emerged as a highly innovative field showing great potential in various areas of science and technology. Nanotechnology is now exerting a major influence on pure science (e.g., chemistry and physics), materials science, energy science, biotechnology, biomedicine, and pharmaceutics. Due to the widespread and increasing burden of serious diseases, such as drug-resistant infections, cancer, Alzheimer's disease, diabetes, hepatitis, cardiovascular disease, and systemic inflammatory disorders, better therapies are urgently required with a focus on targeting to the diseased site and individualized therapy for the specific patient. Furthermore, the combination of diagnostics and imaging with therapy has become of interest and is now called "theranostics". There is an increasing need for well-controlled clinical trials in nanomedicine that has already resulted in successes, and more nanoparticles (NP) are expecting to receive approval by the US Food and Drug Administration (FDA) [1–6].

Micro/nanosystems have been applied for drug delivery using various materials and approaches such as nanostructured particles and surfaces, and diffusion-controlled delivery systems. Other new applications in biosensing and implantable devices such as drug-eluting/bioresorbable stents can be improved by nanotechnology [7–10]. The administration of different nano/microparticle-based drug/gene delivery systems (DGDS) has been suggested for targeted delivery of therapeutic agents to the specific disease site inside the body, with important advantages such as reduced toxicity and less damage to normal tissues and cells, enhanced drug solubility, better treatment of diseases, minimal/controllable side effects of drugs [11]. Furthermore, macromolecules are increasingly being used as therapeutic agents, and their targeted delivery is also an important challenge [12]. The delivery of such macromolecules should be both time-controlled and site-specific [13]. Among DGDS, smart targeting/delivery approaches are of great value, and in this area, an important group of such smart systems is classified as "stimuli-responsive systems". Therefore, design of intelligent systems with controllable and accurate feedback to stimulation has been extensively investigated. Recently, newly developed smart nano/microparticles have illustrated great potential in targeted delivery of drugs/genes [14]. In these smart systems, a wide variety of either external or internal stimuli or triggers can be used to kick-start the delivery and the release of the therapeutic agents in a controlled manner [15]. This needs a high sensitivity of the particular NP to the different stimuli, producing physicochemical alterations in the structure [16,17].

The human body is a complex collection of many different kinds of cells each with its own environment characterized by specific physical and chemical parameters. Each cell is surrounded by special receptors on its surface and contains specific enzymes as well as other molecules to change its redox-potential and pH from that of normal body fluids [18,19]. The various features of different cells and tissues can be taken advantage of to enable specific targeting and delivery of drugs. Several smart-nanocarriers have been reported that can respond to sequences of DNA, have antibody-mediated specificity, or are sensitive to different enzyme reactions. In addition, every individual enzyme possesses its own optimum pH, co-enzymes, temperature, and site of action so that an enzyme-responsive nanocarrier could be fabricated with a multi-responsive action. The presence of specific enzymes on the cell surface of individual cells can specifically determine the uptake of the drug into that cell, while different enzymes present within the cell could control intracellular drug release. In the environment of tumors, there are alterations in pH, Adenosine Triphosphate (ATP) content, reactive oxygen species (ROS, e.g., H_2O_2), glucose, and glutathione concentrations. ROS and reactive nitrogen species (RNS) both exist in many biological microenvironments. For example, high levels of ROS (known to be involved signaling pathways) are found in cancer cells suggesting an oxidative environment which could be a promising target for anticancer drug delivery [20–23]. The highly acidic environment of tumor tissue or the overexpression of a particular enzyme in cancer cells can guide the laboratory modification of carrier vehicles so that drug release can be triggered by various stimuli including internal or external triggers [18,24,25].

Different external physical stimuli can take the form of changes in magnetic and electric fields, light irradiation, application of ultrasound and heating to raise the temperature, and the use of mechanical force. Internally triggered delivery systems can be controlled via taking advantage of the pH value of a specific biological site, responding to redox-potentials of different biological sites, as well as responding to the activity of specific biomolecules such as enzymes, adenosine-5′-triphosphate (ATP), glucose, glutathione, and ROS. [16,17,23,26]. Figure 4.1 shows a schematic depiction of the various external and internal stimuli that can be utilized in smart DGDS. Furthermore, an innovative advance in this area is to combine two or more of the stimuli, e.g., dual or even triple stimuli-responsive release [17,27–29].

In some cases, the use of smart DGDS can eliminate the risks and drawbacks of using other different carrier systems, such as using viral vectors in clinical gene therapy [30]. Although NP-based nanocarriers generally show only low cytotoxicity towards normal cells and in biological environments [31], the various effects NP could have on biological environments have led to the new field called "nanotoxicology", and these must be considered in the design of new nanocarriers. These toxicity issues have been one of the main concerns in recent literature [32] and have worried the general public, and efforts have been taken to define and, if necessary, reduce this toxicity [33,34]. In addition, interactions of NP with biological molecules and materials, and the occurrence of phenomena such as the coating with proteins known as a "corona" and the cell-type-specific effect known as "cell vision" can substantially affect the biological fate of NP, their targeting ability [35–37], and their cytotoxicity [38]. Smart NPs have demonstrated noticeable therapeutic potential particularly in cancer therapy by being designed to be triggered in tumor sites [39]. Smart NP can respond to a variety of tumor specific stimuli [40] and can dramatically improve the cytotoxicity of anticancer drugs to malignant cells while reducing toxicity towards normal cells [41], and large-scale molecular simulations and systems biology approaches can be used to model these effects [42] (Figure 4.2).

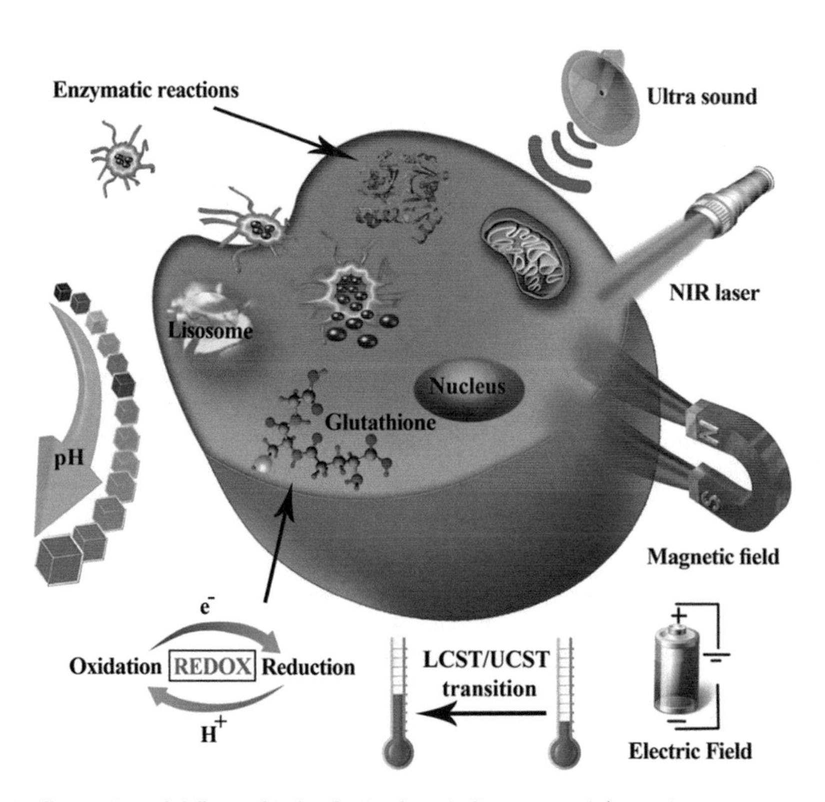

FIGURE 4.1 Schematic illustration of different kinds of stimuli including external (e.g., electric, magnetic fields, light irradiation, and ultrasound) and internal (e.g., pH alterations, enzymatic activity, and redox-potential) that can act as triggers for design of smart stimuli-responsive targeted DGDS.

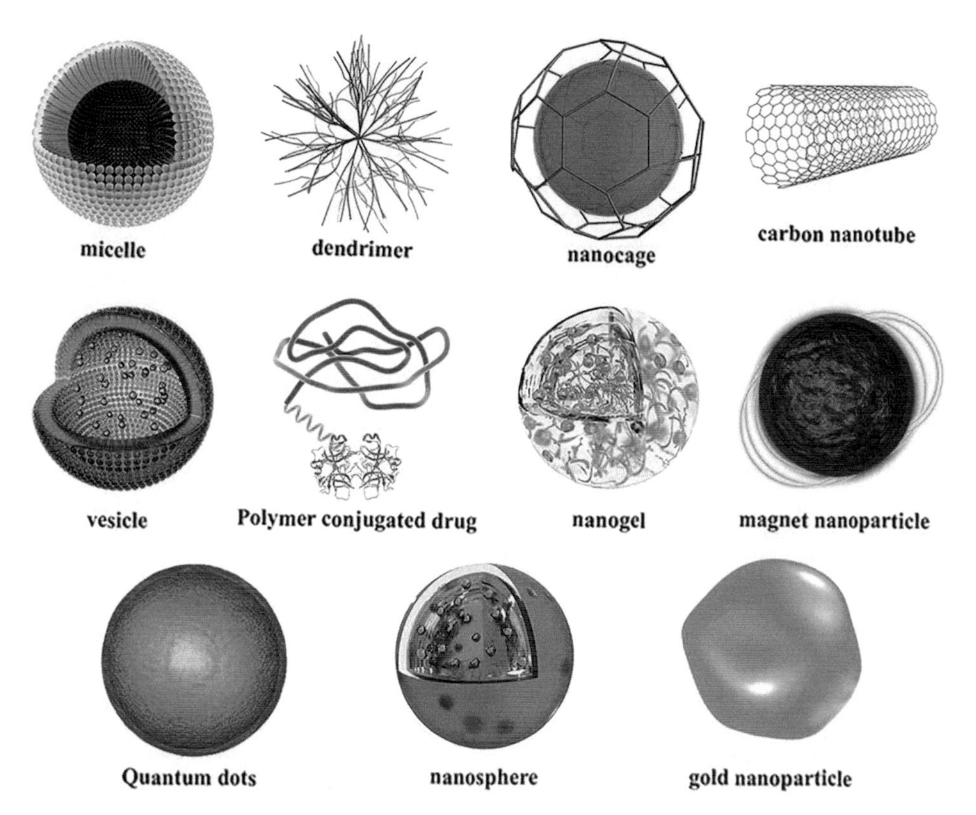

FIGURE 4.2 Different types of NP used for smart micro/nanocarriers including micelles, dendrimers, nanocages, carbon nanotubes (CNT), polymeric conjugates, nanogels, magnetic NP, quantum dots (QDs), nanospheres, and gold (Au) NP.

4.2 External Physical Stimuli-Responsive MNPs

4.2.1 Thermo-Responsive MNPs

Inflamed or pathological sites and tumors can be characterized by temperatures higher than the basal levels of the organism (37°C). Increased temperature can act as stimulus for thermo-responsive MNP to make stimuli-responsive drug delivery system (DDS) capable of being triggered by both internally and externally applied changes in temperature [43]. Thermo-responsive DDSs have a fast response to thermal changes and can be used as injectable fluids [44–48].

In thermo-responsive polymers such as hydrogels, the nanoparticles respond to temperature changes by a phase transition which causes the release of the encapsulated cargo [49]. Alteration of their solubility behavior after a change in temperature and the subsequent phase transition affects their volume. During this phase transition, volumetric shrinkage and water squeezing occur. This volumetric change is reversible which is called "swelling-shrinkage" behavior. When the temperature changes to a value above or below a critical temperature, the collapse and transition of the nanoparticles to a shrunken and gelated structure occurs [50–52]. The lower critical solution temperature (LCST) for polymeric particles is where solubility of the polymers is higher below the LCST. This is due to hydrogen bonds forming between water and functional groups of polymer particles; thus, the swelling of the polymer occurs [53,54]. Other polymeric particles possess an upper critical solution temperature (UCST). Here, the swelling of the polymer occurs when the temperature increases above the UCST [53–55]. Polymers with LCST are called negative thermo-responsive particles and those with an UCST called positive thermo-responsive particles. By tailoring the LCST/UCST, respectively, controlled delivery and release of drugs can be obtained. The LCST/UCST transition temperature of the polymeric NPs should be near physiological temperature conditions. Transition temperatures of these hydrogels have been reported to range from low (e.g., 15°C) [56] to high temperatures (e.g., 60°C) [57].

Different classes of nanoparticles have been reported to have thermo-responsive capabilities, including polymeric micelles [58,59], core–shell particles [60,61], hydrogel polymers [55,62,63], and layer-by-layer (LBL) assembled nanocapsules [64]. Moreover, several naturally occurring materials (e.g., chitosan and hyaluronic acid (HA)) have been utilized for drug encapsulation and release in thermo-responsive nanoparticles [65,66].

Hydrogels belong to one of the most important categories of thermo-responsive nanoparticles, for example, hydrogels composed of poly-(*N*-isopropylacrylamide-*co*-acrylic acid), PNIPAAm, and its derivatives [67] and pluronic copolymers [63], cellulose derivatives [68], PUA [69], and poly lactic-co-glycolic acid (PLGA) [70]. Hydrogels can be prepared by copolymerization and grafting of PNIPAAm with other monomers such as PEO [71] and PEG [72].

Thermo-responsive DGDS potentially can be used for the development of anticancer drugs based on nanocarriers. Nakayama et al. prepared synthetic polymer micelles sensitive to temperature changes and analyzed the anticancer capability of these doxorubicin (DOX)-loaded nanocarriers against human breast cancer cells (MCF-7). The results showed that the DOX-loaded thermo-responsive micelles underwent release induced by a temperature changes above the LCST. Above the LCST of the micelles, the anticancer drug, DOX, was localized in the intracellular compartments by triggered phase transition of the outer shell and the interaction of the micelles with the cells, so the cytotoxicity towards cancer cells improved. In contrast, below the LCST, the micelles interacted only minimally with the cells, and the amount of the DOX delivered inside the cells significantly decreased [73]. In addition, in thermo-responsive DDS, the cancer cells are directly eradicated by the external thermal stimulation, and as known in the literature, cancer cells are more vulnerable to heat compared with normal cells [74]. This hyperthermia effect can be considered in design of thermo-responsive DDS, so their destructive effect could be substantially enhanced [75].

In a study, DOX release from a liposomal carrier occurred at 42°C, causing toxicity for other healthy tissues. To eliminate this drawback, and in order to render the liposomes thermo-responsive and reduce toxicity, *N*-(2-hydroxypropyl) methacrylamide mono/dilactate was conjugated to the liposome surface using different molecular weight polymers, and high-intensity focused ultrasound (HIFU) was applied as a thermal generator. DOX release was obtained in 10 min at 42°C using this hyperthermia technique. Results also showed that by increasing the polymer molecular weight, the temperature of DOX release was reduced [76].

Temperature-sensitive nano/microparticles have been shown to be promising candidates for efficient delivery of plasmid DNA, siRNA, and oligodeoxynucleotides (ODN) [77–81]. In smart gene delivery systems, thermo-responsive polymer particles have shown improved results due to their stimuli-responsive characteristics [82]. For example, in polymers with a LCST property, the encapsulated agents are released by the temperature-induced phase transition. Furthermore, these polymer nanoparticles potentially can form strong complexes with nucleic acids followed by their transport to the target cells at temperatures below the LCST. Subsequently, the cargo is released due to the loosening of bonding between the complexes by increasing the temperature above the LCST. This procedure results in improved gene transfection efficiency [83–85]. In thermo-responsive gene delivery systems, diverse advantages have been obtained including effective gene expression, reduced cytotoxicity, maintained bioactivity, good bonding of the nucleic acid with the nanocarrier, enhanced transgene expression, better cell viability, and improved capability of nanocarriers for condensing nucleic acids. [86–90] (Figure 4.3).

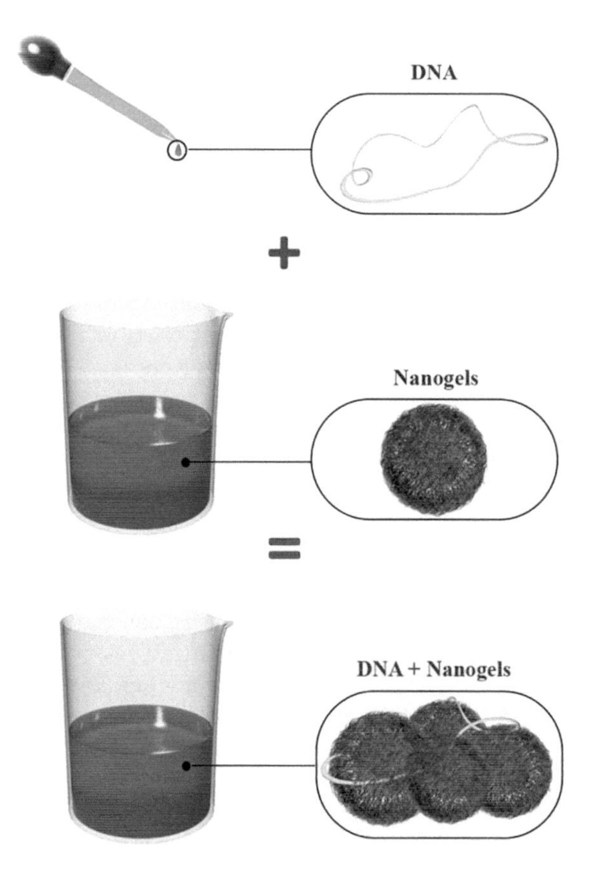

FIGURE 4.3 Complex formation between pH and temperature-sensitive nanogels with DNA.

4.2.2 Magnetic-Responsive MNPs

Magnetism is an external non-invasive method of activation that can be controlled in a temporal and spatial manner. Magnetic fields rarely interact with the patient's body in comparison with other external stimuli, e.g., pH, ultrasound, or light [91–93]. The first use of a magnetic field as an efficient trigger to release drugs dates back to 1960 [94]. Twenty years later at Northwestern University, Widder et al. [95] initially proposed the concept of using magnetic nanoparticles (MNP) for delivery of drugs that had been bound to carbon mixed with iron. Their results showed that magnetic targeted carriers (MTCs) ($\geq$2 μm) could absorb and release such drugs as DOX. Their results demonstrated that MTC mixed with a drug could be easily delivered to a liver tumor if they were passed through a catheter into a hepatic artery branch upstream of the tumor. Figure 4.4 schematically illustrates this mechanism of drug release under magnetic stimulation at a specific site. This method has one limitation, in that they needed to use elemental iron magnets with a greater penetration depth of the applied field instead of the more usual ferrous oxide magnets.

When alternating magnetic fields are applied to magnetic materials, heat is produced and can be used as a source of hyperthermia in tissue as well as for drug release. The incorporation of micro- and nanoscale metal particles into hydrogel matrices has attracted the attention of a great number of researchers, so that in last decades, the development of unique materials with novel properties has allowed controlled drug delivery [96]. For example, a nanoscale hybrid system has been developed for drug delivery. Giani et al. [97] cross-linked a biodegradable polymer, carboxymethyl cellulose (CMC), to $CoFe_2O_4$ NPs to synthesize a hybrid hydrogel and used it to prepare magnetic-responsive NP. The functionalized MNP with (3-aminopropyl)-trimethoxysilane (APTMS) displayed NH_2 groups on their surface. Consequently, controlled drug release from the matrix under influence of magnetic fields was achieved.

Several unique properties of lipid-polymer hybrid nanoparticles make them promising materials to develop a DDS. Three of these properties, enhanced drug release kinetics, surface functionalization procedures, and a simple synthesis methods, are most important [98]. In one study, Kong et al. [99] developed a system to release camptothecin (CPT) based on a lipid-polymer hybrid NP containing Fe_3O_4 magnetic beads. In another part of this study, the treatment of MT2 breast cancer in mice was reported by applying external magnetic fields.

Recently, it was shown that the effects of heat on cancer cells were more cytotoxic than on the surrounding normal cells [100]. This had been studied by Gilchrist et al. [101] in 1957. Cancer cells were locally heated using MNP with a 1.2 MHz magnetic field. Magnetic-induced hyperthermia in animals that had been directly injected with MNP could produce tumor regression by applying magnetic fields to the solid tumors [102].

MNPs have also been utilized in order to produce a smart gene delivery system. This approach for triggered gene delivery is called magnetofection. In 2009, Mah et al. [103] used a magnetofection method, linking MNP with viral vectors, and reported that the biocompatibility of photoluminescent nanoparticles or MNP could be increased by applying a nanoscale coating with a silicon layer in a stable manner.

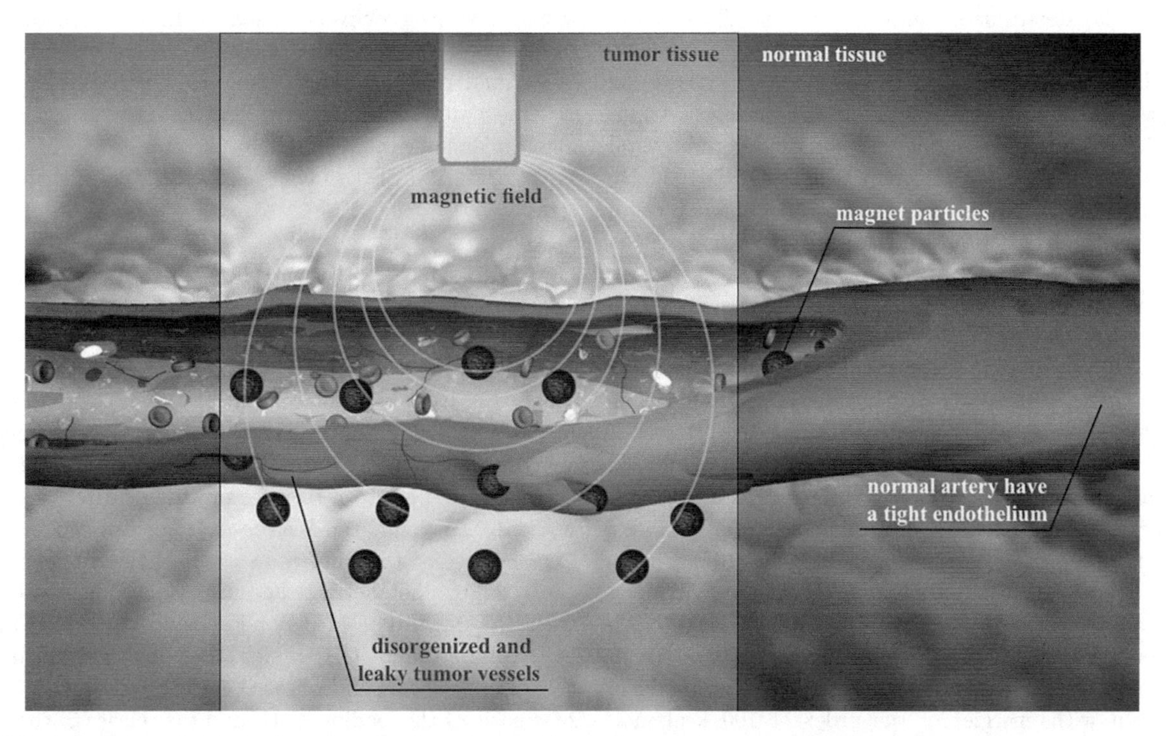

FIGURE 4.4 Schematic illustration of mechanism of magnetic sensitive carriers and triggered drug release.

A smart system for controlled drug/gene delivery was developed by Ruiz-Hernandez et al. [104], in which the "caps" formed from iron oxide nanoparticles (IONP) could entrap drugs or genes. In this system, the complementary DNA strand was linked to the mesoporous silica membrane, and the caps were conjugated with the strands of DNA. The mechanism of this stimuli-responsive delivery works as an "on-off" switch, in which the temperature of the nanostructure was increased by exposure to a magnetic field, as a consequence, the DNA dehybridized, and the encapsulated genes were released from the matrix. However, in the "off" mode, no genes were released from the structure owing to the double-helix DNA binding that maintained the closed conformation of the caps.

4.2.3 Light-Responsive MNPs

The use of light for therapeutic purposes has a history of several thousands of years. Furthermore, in recent decades, nanoparticles that can be activated by light to mediate photodynamic therapy have been suggested for therapeutic purposes. According to the physics of photons interacting with the electronic structure of the material, the UV or visible photons are absorbed corresponding to the energy band gap of the molecular orbitals of the chromophore. The specific band gap of a material allows tuning of the wavelength and intensity of the light. Considering the overall non-invasiveness of light, and the fact it is a clean and efficient stimulus without need for any physical contact, this approach can be used to prepare light-responsive nanocarriers for smart DGDS [105].

Different photo-responsive materials, polyelectrolyte multilayer membrane, polymer grafted porous membrane, polymer gels, dendrimers, and nanoporous silica materials have been tested. The most important mechanism responsible for photo-triggered functions is based on photo-switchable molecules including azobenzenes, triphenylmethanes, spiropyrans, polypeptides, and liquid crystal materials which have all been explored as DDS [106]. By modifying the light parameters such as beam diameter and wavelength, the irradiation exposure for a specific tissue is controllable [107].

Near-infrared (NIR) light is the most common wavelength to be used because it can deeply penetrate into living tissue [108],due to its lower absorption and scattering. Moreover, NIR irradiation causes less damage to tissue than visible light [109]. However, NIR lasers with high intensities (of the order of several W/cm^2) can produce overheating and photo-damage to biological tissue [110].

In the wavelength range of visible light, only a few supramolecular assemblies have been designed [108]. Consequently, irradiation using NIR or higher energy UV wavelengths are preferred. UV light (100–400nm) has much higher energy per photon than visible light, which can induce ionization and cleavage of strong covalent chemical bonds even with energies of the order of 100 kcal/mol. This wavelength can have deleterious effects on living tissues especially with far-UV range (wavelengths under 200 nm). Also in this spectrum, the photoactive molecules can be readily destroyed. UV irradiation has more cytotoxic effects and also, due to its absorption by endogenous chromophores (e.g., lipid, water, and deoxy/oxy-hemoglobin), is not capable of deep penetration in tissue. Therefore, the UV spectrum is not generally suitable for therapeutic applications. However, the visible and near-infrared wavelengths have safer characteristics for utilization in medicine and smart DGDS.

Diverse light-responsive DDSs fabricated from mesoporous silica nanoparticles (MSNs) have been shown to have advantages such as tumor-selective accumulation by the enhanced permeability and retention effect [111], an extended surface area, isostructural mesoporosity, biocompatibility [112,113], easy surface modification, and accurate controlled release [114]. Furthermore, MSN-based stimuli-responsive DGDS can be designed using gatekeepers sensitive to photon irradiation [114]. Carbon dots (CD) are another novel material with strong fluorescent properties [115]. Metal nanoparticles such as gold nanoparticles can be utilized for localized heat generation upon photo-thermal induction. Polymeric nanoparticles such as hydrogels are another important group of light-responsive nanomaterials which can be triggered via reversible hydration–dehydration transition mechanisms activated by the absorption of externally delivered light. The phase transition occurs as a result of a temperature increase induced via the conversion of light irradiation to thermal energy [116].

Recently, it has been shown that light irradiation can lead to the occurrence of electrostatic interactions. This phenomenon provided the ability to design photo-responsive drug delivery systems. Li et al. showed the potential of nanoscale-colloidosome capsules for encapsulation and release of therapeutic cargos [117]. The colloidosome nanocarriers contained a bridged nitrophenylene-alkoxysilane derivative which, when irradiated with 365 nm laser for 10 min, led to charge reversal and efficient disassembly.

Another study described self-assembled monolayers (SAM) in which both controlled drug loading and drug release was obtained in a single system due to surface-charge inversion occurring after photoreactions induced by different wavelengths of visible light [118]. The photosensitizers used were biocompatible in biological environments with low cytotoxicity. The photoreactions relied on a substituted 4-picolinium (NAP) ester based on thiolene chemistry and electron transfer processes. After irradiation with 515 nm light, the NAP ester-tagged cargo molecules were bonded to the thiolated surface of the SAM substrate catalyzed by photoactivated eosin Y. Subsequent irradiation with 452 nm produced a bond scission reaction catalyzed by ($[Ru(2,2'\text{-bipy})_3]Cl_2$) which then inverted the surface charge of the SAM and released the cargo. Figure 4.5 shows schematic of the loading (I–II) and the release of the cargos (II–III) from the substrate.

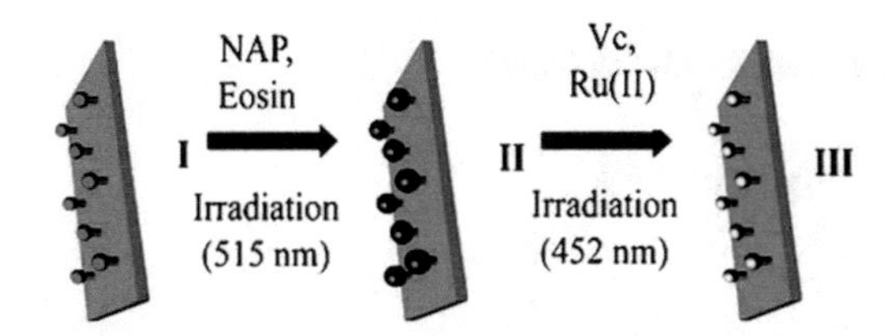

FIGURE 4.5 Schematic of the cargo loading and release via different photo-triggering steps [118].

Light-responsive nanoparticles can be constructed from different materials, but they all have in common a chromophore or photosensitizer in their architecture to absorb the light. This chromophore is the key part of these light-responsive nanoparticles and triggers the cargo-release process [119] by photo-isomerization, photo-induced cleavage, photo-induced bonding, photo-oxidation, or reversible photo-cross-linking [120].

In MSN, drugs can be loaded into the mesopores, and then, the pores are capped by various photosensitive capping agents. Payload molecules are only released from light-responsive MSN nanocarriers when the correct wavelength of light is absorbed. In one study, sulforhodamine 101 (Sr101) as a model cargo was loaded into the pores of mercaptopropyl-functionalized MSN. (Ru(bpy)(2)(PPh(3)) catalyst was utilized as a capping agent to seal the Sr101 molecules inside the mesopores. The Ru compound formed a coordination bond with the mercaptopropyl groups on the silica surface, and light irradiation was able to release the capping moieties and hence the Sr101 dye [108] (Figure 4.6).

4.2.4 Ultrasound-Responsive MNPs

Ultrasound (US) plays a significant role in many medical applications ranging from high-frequency US for medical imaging to low-frequency US for therapeutic applications including the removal of tumors. US-sensitive nanoparticles (usually containing gas bubbles) can enhance the diagnostic and therapeutic applications of US, functioning as contrast agents and as delivery vehicles for therapeutic agents including, drug, genes, and hormones [121–123]. These rely on a difference in speed of sound between the gas core and surrounding tissue [124]. Another advantage is their compressibility and tendency to oscillate in response to different US pulses in a size-dependent manner [125–127]. US has been widely used to enhance the permeability of different biological barriers [128–135].

Microbubbles (MB) have been widely utilized as US-responsive drug carriers and can also be used to enhance ultrasound imaging [136–138]. There are different forms of MB, depending on the gas core, shell composition, and whether they are targeted or not. MB possess a gas-filled core surrounded by a shell of polymer, lipid, or protein [139]. The gas core affects the echogenicity [137] and can be nitrogen, perfluorocarbon, or air [140]. Palmitic acid, phospholipids, albumin, polymers, and lipids can serve as components of the MB shell [140] (Figure 4.7).

The mechanical elasticity of the MB is governed by the shell composition. The higher the elasticity of the material is, the better it can absorb the acoustic energy without bursting [141]. US increases the permeability of biological barriers (cell membranes, blood–brain barrier, etc.) by increasing the local temperature resulting in enhanced drug diffusion and also by producing cavitation bubbles [142]. Rapoport et al. [143] loaded 5 mg PTX into NP prepared from 20–50 mg poly(ethylene glycol)-poly(D-lactide)(PEG-PDLA) block copolymer, including perfluoro-15-crown-5-ether(PFCE) for the treatment of pancreatic cancer with a focused US beam. They used a polymer with a molecular weight of 2000 Da, and the NP had a diameter about 200–300 nm. Their results showed that US delivery enhanced tumor regression.

Yune et al. [144] used US-mediated collapse of MB to deliver siRNA to enhance therapy against yolk sac carcinoma in vitro. They suggest that owing to the negative charge of DNA, cationic lipid MB would be a suitable

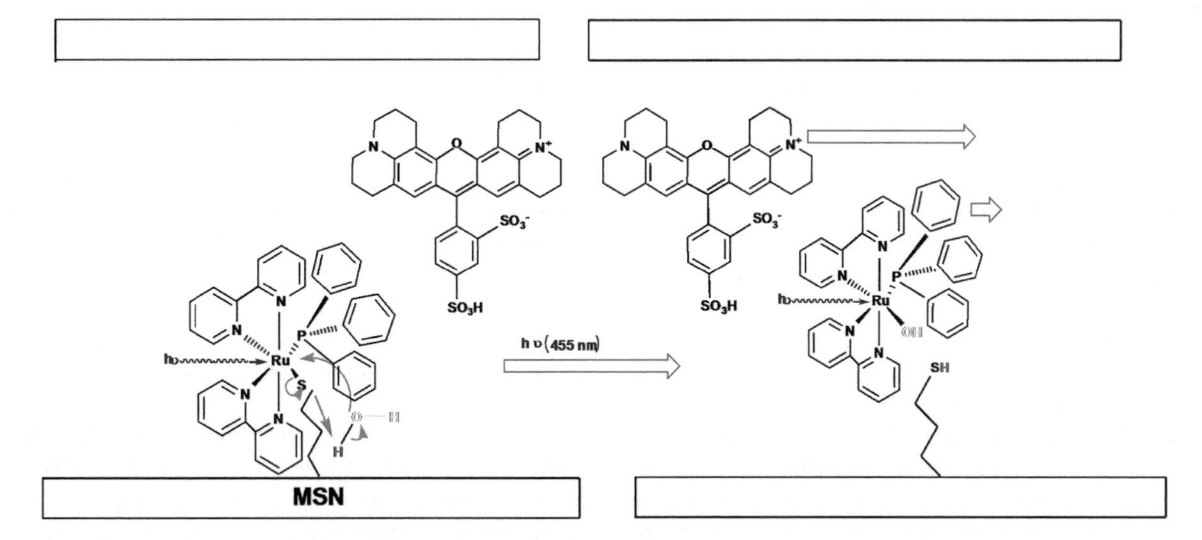

FIGURE 4.6 Photo-transformation of the capping agent (Ru(bpy)2(pph3)) results in cargo (Sr101) release from MSN.

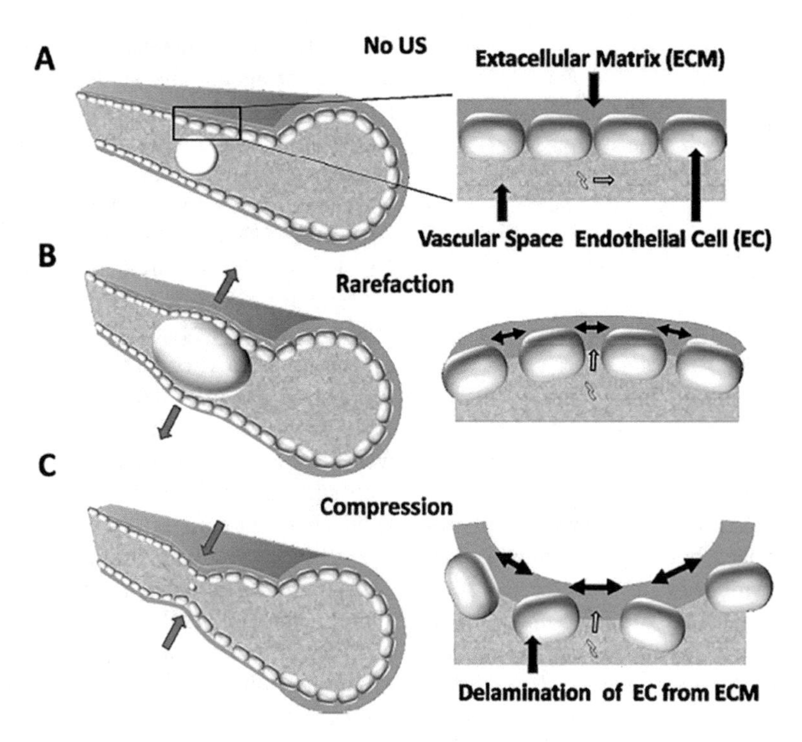

FIGURE 4.7 Schematic interaction of US and MB for drug delivery across endothelial barriers.

candidate for gene delivery. Unger et al. [142] showed that US could increase gene expression using cationic liposomes (dipalmitoylphosphocholine) loaded into HeLa, NIH-3T3, and C1271 cells. They found that the application of US (0.5 W/cm^2) for only 30 s significantly improved gene expression. Application of 2 MHz pulsed Doppler US for 60s enhanced transfection of lipoplexes containing EGFP (enhanced GFP) DNA into rodent (9L) and canine (J3T) glioma cells [145]. Lawrie et al. [146] reported that US in the presence of MB-contrast agents enhanced vascular gene delivery (naked DNA) by 3,000-fold to combat restenosis after angioplasty.

4.2.5 Electrical-Responsive MNPs

An electrical field is an external stimulus that can activate electro-sensitive materials in applications such as, artificial muscle actuators, sound dampening, controlled drug delivery, and energy transduction. Electrically controlled delivery systems on a macro-scale are used in such techniques as infusion pumps, iontophoresis, and electroporation [147]. The conversion of electric energy into mechanical energy, the ability to wield precise control over the duration of electrical pulses, the wide range of magnitude of current that can be used, and the common availability of equipment are mentioned as the most important strengths of electrical field-responsive materials [148].

Smart delivery nanosystems exploiting an electrical stimulus have been generated from polyelectrolytes, which possess a high concentration of ionizable groups along the backbone chain. This property allows these polymers to respond to pH changes, as well as electrical stimuli [148,149]. In other words, the electric pulses can change the pH value which causes the disruption of hydrogen bonding between the polymer chains, leads to bending, deformation, or degradation of the polymer chain. Under the influence of the shrinking or swelling of polymers, the rate of drug release can be controlled [149,150].

The changes that occurred in hydrogels during the exposure to electrical stimuli (e.g., shrinking, swelling, or bending) depend on factors such as osmotic pressure, applied voltage, differences in the shape and thickness of the gel, and the position of the gel relative to the electrodes. Although there are many advantages in this type of DDS, the use of electric current encounters a problem that in order to cause drug release, it may stimulate the nerve endings in the surrounding tissue, thus causing unacceptable pain [151,152].

Partially hydrolyzed polyacrylamide (PAM) hydrogels were used in order to develop an electro-sensitive DDS. When the electric potential was applied to the structure, H^+ ions migrated towards the cathode, which resulted in a loss of water at the anodic region. Furthermore, a uniaxial stress along the gel axis was created mainly due to the presence of an electrostatic attraction between the negatively charged acrylic acid groups and the anode surface. This stress gradient contributed to the anisotropic gel shrinking on the anodic side, and as a consequence, drug was released [153–155].

Miller et al. [156] presented a new approach to electrically controlled drug delivery, using polymers which bind and then release bioactive molecules in response to electrical stimuli. They reported that these polymers have two redox states, only one of which is appropriate for ion binding.

In this system, the drug ions were bound in one redox state and released from the other redox state.

In most cases of electrical field-sensitive polymers, these synthetic compositions, vinyl alcohol, allyl amine, vinylacrylic acid, acrylonitrile, and methacrylic acid, have been used. However, in recent studies, other naturally occurring polymeric gels (e.g., HA, chitosan, and alginate) with unique properties have been found to show electrical field-responsivity [157,158]. Neutral polymers which respond to electrical pulses need to contain a polarizable component.

Recently, a DDS was developed based on cross-linked poly(dimethylsiloxane)-electro-sensitive colloidal SiO_2 particles, and they observed a rapid electrically-stimulated bending of the gel in silicon oil. In another study, Zhao et al. [159] used the polymer poly (2-acrylamido-2-methylpropane sulfonic acid-*co*-*n*-butylmethacrylate) as a controlled delivery system for edrophonium hydrochloride and hydrocortisone. Their results showed that the drug release rate strongly depended on the intensity of electrical stimulation in distilled water. Furthermore, the release of a positively charged drug depended on ion exchange between positively charged solute and the hydrogen ions produced in the electrolysis of water.

4.2.6 Mechanical-Responsive MNPs

Mechanically activated delivery systems harness existing physiological and/or externally applied forces to provide spatiotemporal control over the release of active cargos [160]. Current strategies to deliver therapeutic proteins and drugs use three types of mechanical stimuli: compression, tension, and shear stress. Based on the intended application, each stimulus requires selection of a specific material, in terms of substrate composition and size (e.g., macrostructured materials and nanomaterials), for optimal in vitro and in vivo performance. For example, compressive systems typically utilize hydrogels or elastomeric substrates that respond to and withstand cyclic compressive loading, whereas tension-responsive systems use composites to compartmentalize payloads. Finally, shear-activated systems are based on nanoassemblies or microaggregates that respond to physiological or externally applied shear stresses.

Mechanically activated systems are triggered by mechanical forces in the body that either occur physiologically or are exerted on the body by external devices. The forces exerted in both of these approaches can vary over a wide range of magnitude. Generally, an unopposed force exerted on an object accelerates its motion. The distribution of the force on the object is described as the mechanical stress, which can result in deformation. Microscopic cellular forces [49–54] are present and coordinate into macroscopic forces for processes such as wound repair and inflammation. Further coordination results in the exertion of even greater forces by various systems, such as the musculoskeletal [55,56], cardiovascular [57–59], and respiratory systems [60,61]. Alternatively, external triggers can be applied by medical devices such as stents [62–65] and catheters [66,67] that mechanically open blocked or narrowed structures, or are used to deliver therapeutics. Therefore, drug and protein delivery systems that respond to mechanical forces can provide on-demand release. Designing such mechanoresponsive systems that account for the dynamic nature of the human body will add another weapon to the arsenal of smart delivery systems.

Smart DDS could also be based on mechanical stimulation of an implant [161,162]. Mechanical forces are signals that could easily stimulate well-designed polymer matrices, which could, for instance, trigger release of a growth factor in mechanically stressed environments.

Lee et al. [163] reported a DDS in which alginate hydrogels could respond to mechanical stress by triggering the release of vascular endothelial growth factor (VEGF). In this approach which can be compared to "squeezing the drug out of a sponge", molecules that have been encapsulated within the polymer shell are released during the application of mechanical compressive forces. However, when the strain is removed, the hydrogel returns to its initial volume. In this system, a reversible bond between the protein growth factor and polymeric matrix allows the system to respond to repeated deformation stimuli.

Commonly used materials for compression are elastomeric polymers, which are viscoelastic; that is, they have both viscous (resistance to flow) and elastic properties (tendency to return to its original shape after removal of stress). Examples of elastomers include rubbers, silicones, and hydrogels such as alginate, chitosan, collagen, HA, poly(hydroxyethylmethacrylate), polyacrylamides, poly(ethylene glycol), poly(vinyl alcohol), and poly(*N*-isopropyl acrylamide). Due to their biocompatibility and aqueous loading environment, elastomeric hydrogels are widely used for tissue engineering [164], diagnostic assays and imaging [165], and drug and protein delivery [166].

Elastomeric substrates possess the structural integrity required for compressive release. Wang reported [167] a two-compartment silicone implant that could release insulin to reduce hyperglycemia in an in vivo diabetic rat model. The first compartment allowed the influx of serous fluid to solubilize the insulin powder contained in the second compartment. Compression then drove the efflux of the insulin-dissolved serum. Efficient insulin delivery was achieved by compression (2 s followed by 1 min massage) once a day giving significant reduction in blood sugar levels

Yang et al. described [168] cyclical compressive release of bovine serum albumin (BSA) from a porous matrix. In their study, BSA-loaded microspheres were incorporated within a block copolymer poly(ethylene glycol)-*b*-poly(L-lactide) (PELA) scaffold. Compression of the scaffolds (1 Hz) for 3 h each day (4–5% compressive strain) for 30 days accelerated BSA release compared to the scaffold under static conditions. Half of the total BSA concentration was released after 4 days.

4.3 Internal Chemical/Biological Stimuli-Responsive MNPs

4.3.1 pH-Responsive MNPs

pH-responsive nanostructures have been used in different technologies such as pH-sensors [169], theranostic applications [170,171], and drug/gene delivery systems. In addition pH-responsive smart nano/microparticles can be utilized in controllable switches, controlled-release surfaces, controllable wettability, and specific recognition of cells [172].

There are distinct pH differences between various tissues and cellular compartments of the human body, and also pH gradients have been shown to exist in the case of both healthy tissues and pathological conditions. For example, there are pH gradients in the sub-cellular environment, with pH differences between the lysosomes (pH 4.5–5), endosomes (pH 5.5–6), cytosol (pH 7.4), Golgi apparatus (pH 6.4), etc. [173] The sharp pH changes seen along the length of the gastrointestinal tract is an example of these pH variations. Furthermore, the growth of microorganisms can change the pH of their immediate environment either directly or indirectly by inducing the release of host defense enzymes. Healing and non-healing wounds can also create acidic and alkaline milieus, respectively [174].

Smart DGDS can release their loaded drugs or genes in response to a pH change between normal physiological pH (i.e., 7.4) and the lower pH values found in cancer. The pH profiles in cancerous tissues differ from those in normal tissues; due to the rapid proliferation of cancerous cells, their vascular system is often inefficient and they do not receive adequate supplies of oxygen and nutrients. Under anaerobic conditions, cancerous tissues carry out glycolysis (Warburg effect) rather than oxidative phosphorylation and therefore produce lactic acid which makes their environment more acidic, leading to a extracellular pH value that is lower than normal blood pH (i.e., 7.4) [175,176]. In fact, lactic acid and carbonic acid are the two main sources of low pH in the tumor mass [177]. According to the Warburg effect, the preference for glucose fermentation persists even in the presence of adequate oxygen, and it is believed by some experts to be a cause of cancer rather than the result [178,179]. Furthermore, the pH of cancer cells varies depending upon cancer type, size, and its location [180]. Sometimes, the environmental pH increases due to necrosis occurring in the large tumor mass [180]. Moreover, in addition to the lower extracellular pH of tumor tissues, in the intracellular environment, the (adenosine triphosphate) ATP-driven H^+ pump, which is located in the endosomal membrane, pulls H^+ ions into the endosomal lumen from the cytosol and further reduces the pH [181].

This lower pH can cause several problems in cancer treatment. For example, the low pH reduces the activity of many common anticancer drugs, such as DOX [182].

The pH-sensitive nanomaterials used in DGDS can be classified into organic, inorganic, or hybrid materials according to their constituents [183]. Various nanomaterials have been evaluated as pH-sensitive nanocarriers. A summary of the various types of pH-responsive nanocarriers including polymeric micelles, liposomes, nanogels, core–shell nanoparticles, polymer–drug conjugates, and inorganic nanoparticles as well as their effects on drug/gene delivery systems is shown in Table 4.1.

The drug release in pH-responsive nanoparticles occurs due to two different mechanisms which are dissolution of the nanocarrier [215], swelling of the polymeric nanocarrier [216,217], or both dissolution and swelling mechanisms [218] at specific pH values due to acid-swellable groups [219]. In pH-responsive polymeric particle-based drug delivery systems, the drug is conjugated to polymeric particles through pH-responsive spacers. These spacers can be degraded at low pH of tumors or endosomes/lysosome [220]. The pH responsiveness depends on the hydrophobic carbon-chain length or monomer type, as well as the proportions of the monomers [221,222]. In liposomes, internalization of the modified liposomes into cells occurs through endocytosis into endosomes and subsequent fusion with lysosomes. Then, the cargo is delivered to the cytoplasm [194]. The pH sensitivity of micelles is based on protonation of moieties such as carboxylic groups or titratable amines attached to the surface of the polymeric particle. The form of these particles changes as the pH change occurs in the environment [223]. In dendrimers, one important pH-responsive method is the cleavage of particle occurs by using hydrophobic groups sensitive to acids, after which drugs can be released [224]. pH-Responsive lipid-based delivery systems utilize nanocarriers including liposomes, acid-labile zwitterionic peptide lipid derivatives, etc. [225].

Ionizable chemical moieties such as amines, phosphoric acids, and carboxylic acid accept or donate protons in response to pH alteration. During a pH change, the weak acid serves as a proton donor, and its conjugated base is a proton acceptor. Carboxylic acids are weak acids which are used in anionic polymers including poly(acrylic acid) (PAA), poly(butylacrylic acid) (PBAA), poly(propylacrylic acid) (PPAA), poly(ethylacrylicacid) (PEAA), and poly(methylacrylic acid) (PMAA) [226–229].

Cationic polymers acting based on the proton-sponge effect have displayed enhanced delivery efficiency for therapeutics such as nucleic acids, drugs [230], and proteins [231,232]. Poly(L-lysine) (PLL) (widely used for DNA delivery) [233] and polyethylenimine (PEI) (in both linear and branched forms) are cationic polymers showing high efficiencies for delivery of oligonucleotides [234], plasmid DNA (pDNA) [235,236], and RNA such as siRNA [237,238]. Furthermore, pH-responsive nanocarriers have been shown to cause facilitated endosomal membrane rupture through proton-sponge effect as well as other endosome destabilization mechanisms such as surface-charge reversion of nanocarrier and deshielding (i.e., active ligand exposure at the outer surface of nanocarrier) at extracellular pH of tumor milieu [239].

TABLE 4.1 Examples of Different Kinds of pH-Responsive Nanocarriers

Nanocarrier Type	Particle Formation Method	Characteristics of the Particle	pH-Dependent Therapeutic Effect or Outcome	Ref.
Polymeric micelles	• Self-assembled spherical supramolecular aggregates formed by amphiphilic block copolymers	• Capable of solubilizing insoluble drug molecules, enhanced solubility, reduced toxicity, and the improved permeability and retention (EPR) effect for passive targeting • The weakly water-soluble drug molecules can be encapsulated in the core of micellar structures	• Prevention of drug release at physiological pH but facilitated drug release at endosomal pH (pH 5.0) • Tumor-infiltrating permeability, effective antitumor activity, and low toxicity • Increased serum stability of the nanocarriers in the blood • Elimination of premature drug release	[184–188]
Liposomes	Self-assembled spherical vesicles made up of lipid bilayer structures	• Can be used as carriers of both hydrophobic and hydrophilic molecules, variation in size, localization to diseased tissues, via surface charge, and functionality • Better compatibility than micelles • Limitations such as low encapsulation effectiveness, fast release rate of drug, low storage stability, and lack of tunable triggers for drug release	• Improved targeting and drug release by surface modification and utilization of LBL liposome nanoparticle • Enhanced gene targeting in the liver followed by reduction in plasma cholesterol • Multi-drug pH-sensitive delivery capability with controlled and different drug release rates and improved biodistribution profile • Antigen delivery • Vaccine delivery • Increased cellular uptake of DOX-loaded liposomal carrier at pH 6.4	[186,189–196]
Polymeric nanogels	Highly porous three-dimensional network by self-assembly or covalent bonding of cross-linked hydrophilic polymer chains	• Higher drug loading capacity than micelles and liposomes except for hydrophobic drugs, efficient targeted delivery	• Biocompatible pH-responsive nanogels with capability for loading hydrophobic drugs • Dual pH/temperature-responsive, hyperthermia therapy of cancer cell with increased therapeutic activity of the drugs	[197–200]
Polymer–drug conjugates	Polymeric chains	• Drugs covalently conjugated to pH-sensitive polymeric chains, high blood circulation time	• Increased water solubility of drugs and tumor targeting • High sensitivity with a strong radio-sensitizing effect • Loading capability for poorly soluble drugs	[201–203]
Core–shell nanoparticles	Polymeric colloidal particles with spherical, branched and core–shell forms	• Core–shell nanoparticles can be prepared via synthetic or natural polymers	• Core–shell drug carriers with antitumor ability, faster drug release at lower pH • Capability for protein delivery • Noticeable hypoglycemic effects and enhanced insulin-relative bioavailability • High-sensitive dual pH/temperature responsiveness, sharp tunable phase transitions of dual temperature, and pH-responses	[204–208]
Inorganic nanoparticles	Ceramic (e.g., MSNs) or metal nanoparticles (e.g., gold) with functional pH-sensitive polymeric moieties	• Controllable surface functionalization with good encapsulation efficiency for drugs/genes	• Increased cytotoxicity through enhanced cellular uptake mechanisms • In MSN, blocked pores at neutral pH and drug release in acidic condition, low cytotoxicity, and improved biocompatibility • High drug release and targeting capability • High drug loading capability	[204,209–214]

Chitosan is considered as a natural, cationic aminopolysaccharide derived from partial deacetylation of chitin and is a copolymer of *N*-acetylglucosamine and glucosamine linked by a 1–4 glycosidic linker. Sung et al. reviewed advances in chitosan-based nanoparticles-based macromolecular delivery [240]. Furthermore, chitosan is nontoxic and biodegradable, and is widely applied in drug delivery systems for controlled release of proteins (e.g., insulin) [241] and controlled oral delivery as well as entrapment of enzymes [242]. The significant properties of chitosan and its derivatives, including adhesion to mucosal surfaces and the capability of penetrating between tight junctions in epithelial cells, allow them to serve as favorable candidates for oral delivery of therapeutics [29,243]. Diverse chitosan-based nano-assembled nanoparticles are illustrated in Figure 4.8. In one study, a pH-triggered polyelectrolyte "complex sandwich" microparticle was fabricated from alginate/oligochitosan/Eudragit(®) L100-55. The results showed the pH-sensitive release of drugs in conditions of which simulated the intestinal environment (pH 6.8) [244].

The self-assembly behavior of chitosan is enabled by the presence of functional moieties including $-OH$ and $-NH_2$. Progress in the synthesis of self-assembled chitosan-based nanoparticles such as nanoshells, nanocomplexes, micelles as systems for delivery of drugs, genes, and small molecules has been reviewed [245]. Nogueira et al. developed methotrexate-loaded chitosan nanoparticles produced from a modified ionotropic complexation process which led to apoptotic effects in tumor cells [246]. In addition, nucleic acids can be delivered via chitosan-based nanoparticles such as polyaspartamide (PASPAM) [247], poly(malic acid) [248,249], and poly(histidine) [187]. In a study by Cheng et al., a pH-triggered and non-viral gene delivery system was prepared. These nanoparticles showed low toxicity to HepG2 and KB cells, and despite the fact that gene transfection and expression could be enhanced in KB cells, free folic acid (FA) significantly inhibited this process [250].

MSNs have shown great potential in stimuli-responsive drug/gene delivery systems especially pH-sensitive nanocarriers. Wen et al. [251] reported the synthesis of pH-responsive composite microspheres via the distillation precipitation polymerization process. This nanocarrier was composed of a core of Fe_3O_4nanoparticle, a triple layer of MSN and coated with cross-linked poly(methacrylic acid) (PMAA) as a shell. The results indicated that the cumulative drug release rate from the microspheres was significantly higher below its pKa value than above the pKa. Furthermore, pH-sensitive supramolecular nanovalves that capped MSN pores have been utilized in smart delivery systems [252–254]. Nguyen et al. [253] showed that the pores of coumarin 460-loaded mesoporous silica MSM-41 could be capped with dibenzo[24]crown-8 (DB24C8) nanovalves.

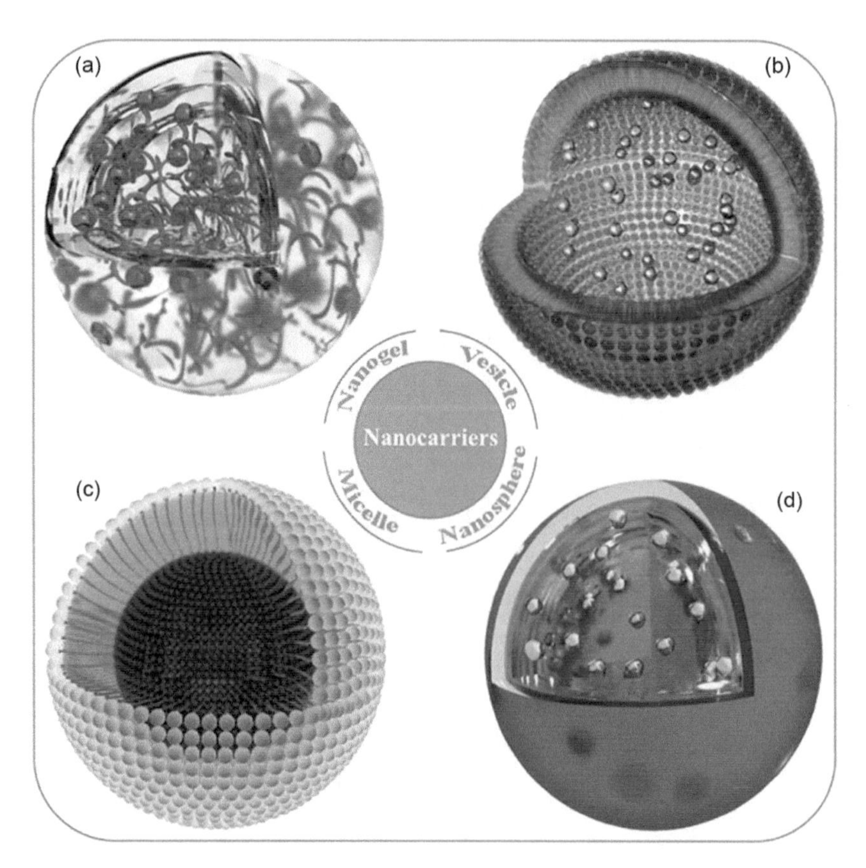

FIGURE 4.8 Different chitosan-based nanoassemblies used in biomedicine: (a) nanogel, (b) nanovesicle, (c) micellar structure, (d) nanosphere.

4.3.2 Redox-Responsive MNPs

In recent decades, research on redox-responsive systems has increased considerably [255,256]. The advantages of redox are the fact that glutathione disulfide/reduced glutathione (GSSG/GSH) is a very important redox couple in mammalian cells and can easily be utilized in reduction-responsive DDS [28,257–262]. GSH is a tripeptide where the normal intracellular concentration falls within a wide range of 2–10 mM. In contrast, the GSH concentration in extracellular fluid is only about 2–20 μM [263]. Moreover, elevated GSH is fairly specific for tumors as compared to normal tissues [264]. However, the exact ratio of GSH does vary between different types of tumor, and it can be elevated as much as tenfold, but can also be reduced below 1.0. Different GSH-responsive nanocarriers (e.g., nanogels, polymersomes, capsules, and micelles) have been reported.

Nanogels, (nanoparticles composed of a cross-linked hydrophilic polymer network [265]), are biocompatible three-dimensional materials that can be used for the encapsulation of cargos including anticancer drugs, plasmid DNA, and imaging probes [266]. Nanogels have a high water content and their size ranges from tens of nanometers to submicrons [266,267]. Matyjaszewski and co-workers [268] synthesized a reduction-sensitive nanogel, in which they used the disulfide-thiol exchange reaction and the process of inverse mini-emulsion atom transfer radical polymerization (ATRP). These nanogels could be loaded with a wide range of water-soluble biomolecules, ranging from anticancer drugs to proteins, to carbohydrates, to nucleic acids [269,270]. For example, nanogels were loaded with DOX with 50%–70% efficiency, and the disulfide cross-linked nanovehicles were nontoxic, but after the addition of 20 wt.% GSH, they significantly inhibited HeLa cell growth. Caruso et al. recently reported the preparation of reduction-sensitive DOX-loaded PEGylated nanoporous polymer-spheres (NPSPEG-DOX). First, they loaded and immobilized alkyne- or azide-functionalized PEG into a mesoporous template (MSN) via click chemistry; second, they carried out cross-linking of PEG and covalent binding of DOX through biocompatible linkers containing disulfide bonds; and finally, they dissolved away the MSN templates [271]. Under reducing conditions (5 mM GSH), the nanospheres disassembled to release DOX. Biocompatible and degradable nanogels have also been prepared by covalent binding of thiol-functionalized star-shaped poly(ethyleneoxide-*co*-propylene oxide) with linear polyglycidol in an inverse mini-emulsion, via the formation of disulfide bonds [272]. These nanogels were degraded after 6-h incubation in 10 mM GSH.

Hubell et al. [273–275] synthesized polymersomes that could be triggered by either oxidation [274–276] or reduction [273]. In the case of reduction triggering, the PPS block contained hydrophobic thioether moieties that stabilized the vesicular bilayer. However, for oxidation-responsive triggering, the polymersomes contained a triblock copolymer connected to two hydrophilic blocks of PEG surrounding a hydrophobic block of poly(propylene sulfide) (PPS) and PEG-PPS-PEG. The functional groups were oxidized by hydrogen peroxide producing hydrophilic sulfoxide and sulfone groups. Loss of the hydrophobic character destabilized the vesicle by rupturing the lamellar bilayer [274–276].

A new redox-responsive nanocapsule was developed by Zelikin et al. [277,278] based on LBL assembly using poly(vinylpyrrolidone) (PVPON) and thiolated poly(methacrylic acid) (PMASH) loaded onto silica gel. In this system, the silica-based core was disrupted by the reduction of disulfide bonds to free thiol groups, leading to the loss of PVPON and release of the encapsulated drug. In oxidizing conditions, the disulfide bonds and the capsules remained stable.

Kim et al. [279] recently reported a new template-free method to synthesize reduction-responsive polymer nanocapsules based on the self-assembly of amphiphilic cucurbit[6]uril (CB[6]) followed by shell-cross-linking with a disulfide-containing cross-linker. The average diameter of the resulting capsules was approximately 70 nm, and they had a hollow interior, surrounded by an approximately 2.0 nm thick shell. The nanocapsules aggregated and collapsed after a 30 min treatment with 100mM DTT. In vitro release studies demonstrated that encapsulated carboxyfluorescein (CF) was rapidly released. The galactose-coated capsules were taken up by HepG2 cells and showed burst release kinetics of CF inside the cells.

Zhang et al. [280] reported the preparation of reduction-sensitive hollow polyelectrolyte nanocapsules made from dextran sulfate and chitosan-linked cysteamine by LBL adsorption on cyclodextrin-coated nanospheres of silica that were then thiolcross-linked to allow the silica core to be removed.

De-cross-linking and disassembly or full destabilization of redox-sensitive micelles may be caused by GSH-mediated reduction of disulfide bonds [281–284]. Takeoka et al. [285] showed that a hydrophobic model drug released from these micelles could be controlled by selective electrochemical oxidation of the ferrocenyl-alkyl moiety, with zero-order kinetics. Wang et al. [286,287] synthesized a reduction-sensitive shell-detachable micelle. Their results demonstrated that a disulfide-linked copolymer of PCL and the hydrophilic poly(ethyl ethylene phosphate)(PEEP) (PCL-SS-PEEP) could release DOX in the presence of GSH. Recently, there was described a bioreducible amphiphilic triblock-polymersome with a diameter of 256 nm. At pH<7.4, the drug release increased in the presence of 10 mm GSH [259]. Yoo and Park [288] reported GSH-triggered drug release from camptothecin (CPT)-loaded PEG-SS-poly(g-benzyl L-glutamate) (PEG-SS-PBLG) micelles, resulting in higher toxicity towards SCC7 cancer cells, compared with CPT loaded into PEG-*b*-PBLG micelles (reduction-non-sensitive control). Yuan et al. determined that the intracellular GSH concentration could affect gold nanoparticles that were loaded with a micro-RNA (miR-122) that targeted the Bcl-W pathway and resulted in apoptosis of liver cancer cells (Hep G2). Conjugation of FA additionally helped to target the cancer cells [289]. In another study, the authors tested the attachment of CPT to a MSN containing disulfide bridges formed from a mercapto-functionalized silica hybrid that could release drug to kill HeLa cells under the influence of GSH [290].

The use of disulfide cross-links reversible under intracellular conditions is a good strategy to solve the stability/drug release dilemma of micelles. Using divalent metal cations (Ca^{2+}) as a template and cross-linking the ionic cores with cystamine, Bronich et al. [291] prepared poly(ethylene oxide)-*b*-poly(methacrylic acid) (PEO-*b*-PMAc) micelles that showed significant acceleration of DOX release in the presence of GSH or cysteine in the medium; 75% of DOX was released in 1 h in response to 10 mM GSH. Stenzel et al. [292] produced stable nucleoside-containing block copolymer micelles by sequential reversible addition-fragmentation chain transfer (RAFT) copolymerization of polyethylene glycol methyl ether methacrylate, 5-*O*-methacryloyl-uridine and bis (2-methacryloyloxyethyl)-disulfide (DSDMA, a bioreducible cross-linker). In the presence of 0.65 mM DTT, the CCL micelles hydrolyzed in less than 60 min into free block copolymers. As expected, CCL micelles showed a rather slow release of riboflavin. By contrast, the addition of 0.65 mM DTT led to fast drug release, with a pattern akin to that of the non-cross-linked control (about 60–70% release in 7 h). Liu et al. [293] also used RAFT polymerization to prepare two types of degradable thermo-responsive CCL micelles.

4.3.3 Enzyme-Responsive MNPs

Enzymes play a major role as components of the bio-nanotechnology toolkit. The concept of enzyme action was described more than 100 years ago by Paul Ehrlich. Their unique properties, including an unparalleled biorecognition capacity, and their exquisite catalytic functions lead to very selective activity against specific substrates [294]. Since enzymatic activity can be associated with a particular tissue, or the enzyme is found at higher concentrations at the target site, an enzyme-responsive nanomaterial can be programmed to deliver and release drugs. Moreover, the detection of enzyme activity can be an extremely useful tool in diagnostics. Designing a DDS to take advantage of enzyme activity can be based on either a physical or chemical mechanism. In the chemical mode of action, it is possible to program the nanomaterial to release its cargo by enzymatic degradation of the polymeric shell, for example, using amphiphile-based nanoparticles [295]. In the physical mode, enzyme-responsive nanoparticles can be designed, so their macro-scale structure or conformation is altered by the enzyme action, thus releasing the cargo. For example, the surface can be modified with molecules that generate a change in the physical properties of the nanoparticle itself upon enzymatic transformation. However, there are some limitations of enzyme-responsive DDS as follows. First, enzymes generally work under specific physiological conditions. Compared to biomolecules, polymers are not generally enzyme-sensitive. Enzyme-responsive DDSs need to be stable in complex biological environments until they encounter their target enzyme.

The products of the reaction catalyzed by an immobilized enzyme contained in a smart hydrogel could themselves cause the gel phase transition to occur. It would then be possible to transform a chemical signal such as the presence of the substrate, into an environmental signal such as a pH change [296] or into a mechanical signal, such as shrinking or swelling of the smart gel. ERMs can be categorized into two different groups: hydrogels and nanoparticles (Figure 4.9).

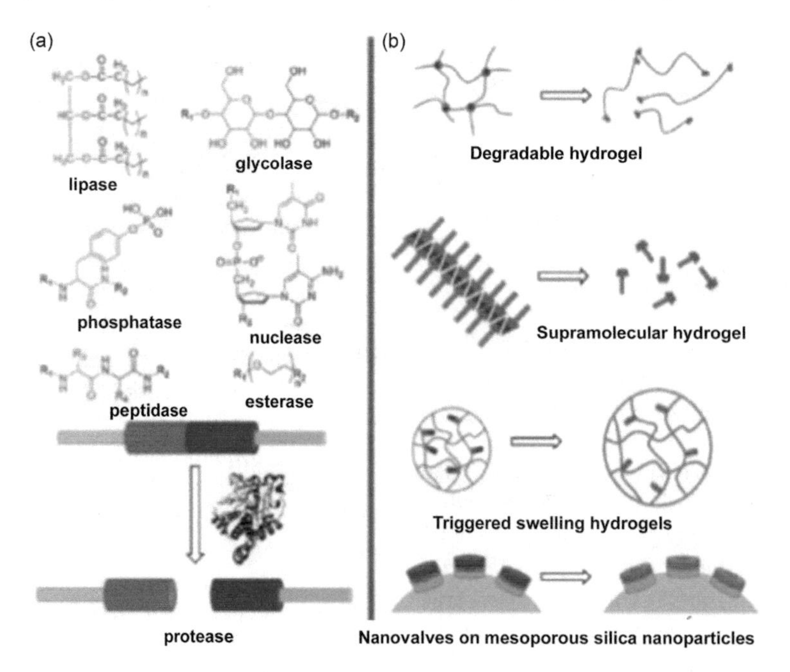

FIGURE 4.9 Examples of molecules that can be used as the substrate for different enzymes (a), schematic description of enzyme-responsive materials (b).

Enzymatic cleavage of different biological molecules changes the physiochemical properties of the nanomaterials [297].

The main substrates which are used in the preparation of enzyme-responsive hydrogels are naturally occurring polymers such as polypeptides and polysaccharides, and synthetic polymers such as poly(*N*-isopropylacrylamide) (PNIPAAm) and PEG/poly(ethylene oxide) (PEO) [298,299]. Synthetic polymers require the attachment of enzyme-sensitive moieties such as short peptide sequences. Polymer hydrogels can be prepared so that they are susceptible to degradation by enzymes such as elastase and metalloproteinases. Using enzymes such as trypsin, thermolysin, and elastase, the degree of swelling in polymers can be altered, and the drug is released accordingly [300–303].

Su et al. [304] synthesized a new enzyme-responsive hydrogel using an *N*-hydroxyamide–heparin conjugate and glucose oxidase (GOx) to catalyze radical polymerization. A different enzyme, heparanase, which is overexpressed in cancers and sites of inflammation served to cleave Hep(DOX)SN gel and release the encapsulated drug cargo, DOX.

Polymer hydrogels can be directly degraded by enzymes such as elastase and metalloproteinases [305–308]. Enzymes can not only destroy the polymer hydrogel structure but also control the morphology of the hydrogel particles as well. Using enzymes such as trypsin, thermolysin, and elastase the degree of swelling in polymers can be altered leading to drug release [300–303]. Klinger et al. designed an enzyme-degradable hydrogel by cross-linking poly(PAAm) with dextran-methacrylate (Dex-MA) that was partially biodegradable by enzymatic cleavage of the methacryl-functionalized polysaccharide chains [309]. 1,4-Hydroxymandelic acid was used as a framework to attach an enzyme-cleavable group such as phenylacetic acid which was cleaved by bacterial penicillin G amidase. A TAT peptide acted as a protein-transduction domain, and nalidixic acid was the antibacterial cargo. The enzyme-activated system exhibited a minimum inhibitory concentration (MIC) value against *Escherichia coli* about 70 times lower (1.9 μM) than that of free nalidixic acid (138 μM) [310–313].

Gu and Tang devised a protein nanocapsule interwoven with an enzyme-degradable polymeric network [314–317]. Proteins were functionally encapsulated using a cocoon-like polymeric nanocapsule formed by interfacial polymerization. The nanocapsule was cross-linked by peptides that could be proteolyzed, thus releasing the protein cargo. The protease-mediated degradation process could be controlled in a spatiotemporal fashion through modification of the peptide cross-linker with photolabile moieties. They demonstrated the utility of this approach through the cytoplasmic delivery of the apoptosis inducing caspase-3 to cancer cells.

Glucose is a major target in approaches to treat diabetes mellitus causing kidney failure, blindness, heart attacks, and death [318]. In recent years, DDSs employing an implantable polymer matrix that can function as self-regulated insulin delivery systems through molecularly programmed release have attracted much attention [319–323]. In these DDS, hydrogels that are sensitive to glucose are programmed to release insulin in a controlled manner [319–323]. The diffusion of increased glucose into the hydrogel and the conversion of glucose into gluconic acid catalyzed by glucose oxidase is the main mechanism. This reaction is usually accompanied by pH changes so the polymeric hydrogel swells, and the diffusion and release of insulin is triggered [324–328].

Trypsin can degrade protamines and chondroitin sulfate to release the encapsulated cargo [329]. Phospholipase activity is increased in infections and in pancreatic and prostate cancer [330]. Elastase is particularly increased in inflammation, and can digest the collagen and elastin components of nanoparticles to release drugs [331]. The carrier nanomaterial can be digested by the hydrolase, whenever the concentration of both species is high enough [295].

Proteases are ideal enzymes for designing smart DDS, because of their highly controlled reactions and their involvement in normal biological processes, such as DNA replication, transcription, wound repair, blood coagulation, and stem cell mobilization [332]. Proteases are produced both by normal cells and by tumor cells, and they can cleave proteins, turning them into smaller peptides or amino acids [333,334]. In a study by Vicent et al. [335], they showed that a specific peptide sequence could efficiently control the release of aminoglutethimide (AGM) and DOX when used as a linker to *N*-(2-hydroxypropyl) methacrylamide (HPMA) nanoparticles.

The expression of matrix metalloproteinases (MMP) is often specific to the tumor microenvironment and/or inside cancer cells [336,337]. Jiang et al. developed a peptide-modified nanoparticle DDS triggered by MMP-2 and MMP-9 that could be used to deliver the Cy5 fluorophore, gadolinium, or both species at once [338,339].

Radhakrishnan et al. [329] demonstrated the effect of either trypsin or hyaluronidase enzymes, which are over-expressed under pathological conditions, to disintegrate capsules fabricated from protamine and chondroitinsulfate. When these nanocapsules were exposed to pH 7.4, the cross-linking was maintained, while the drug molecules were rapidly released in the presence of either one of the triggering enzymes [340]. Wang et al. synthesized a four-component nanocomposite, from trypsin-immobilized polyaniline-coated Fe_3O_4/carbon nanotubes as a composite, for highly efficient protein digestion. This biocomposite offered considerable promise for protein analysis due to its high magnetic responsivity and excellent dispersibility [341].

4.4 Nanotoxicology and Future Perspectives

The extremely small size of nanomaterials also means that they much more readily gain entry into cells than larger sized particles. How these nanoparticles behave inside the body is one concern of the new science of nanotoxicology. The large surface area-to-volume ratio of NP imparts unique biological properties compared to the same material in bulk form. The biological behavior of NP is a function of their physiochemical properties such as the shape and size of particles, their composition, stability, chemistry, and the surface reactivity between the NP with the surrounding tissues and cells. Although the exact mechanism of NP toxicity is as yet unknown, it seems that oxidative stress and activation of pro-inflammatory genes play significant role in nanotoxicity [342]. A large burden of NP could overload phagocytes, cells that ingest and destroy foreign matter, thereby triggering stress reactions that lead to inflammation and weaken the host defense against other pathogens. In addition to questions about what happens if non-degradable or slowly biodegradable nanoparticles accumulate over time in bodily organs, another concern is their potential interaction or interference with normal biological processes occurring inside the body. Because of their large surface area, nanoparticles can immediately be adsorbed onto the surface of some of the macromolecules they encounter. This binding could affect the regulatory mechanisms of enzymes and other proteins.

Fluorescent QDs play an important role as promising NP for biomedical/biological applications including bio-imaging and therapeutic agent delivery. In respect of nanotoxicology evaluations of QD, complex and inconsistent results have been obtained and more assessments are needed concerning the toxicology and pharmacokinetics of QD. In this regard, the toxicity of QD nanomaterials can be attributed to materials in the core nanocrystal, as well as the specific nanosize of the materials. The most worrying potentially toxic element of the semiconductor nanocrystal is cadmium [343], although selenium and tellurium have also engendered some concern. The degree to which the core nanocrystal is encapsulated with capping materials, the composition of the shell, and the coating, and also the role of functional groups on the QD, and various models such as animals will have an effect on the overall nanotoxicity. Furthermore, diverse-related parameters such as size, dose, and exposure of QD have to be studied in in vivo and in vitro evaluations [344].

The toxicity of carbon-based nanomaterials such as carbon nanotubes (CNTs) and graphene has also been considered a critical issue. CNTs have a rather similar size and shape to asbestos fibers that are well known to have major toxicity issues including carcinogenicity and the propensity to cause serious lung damage. The biological effect and toxicological response of CNT and other carbon nanomaterials in vivo and in vitro conditions depend on their bio-physico-chemical properties including shape, dose, surface chemistry, purity, and exposure route, cellular uptake, and particle kinetics. The issues such as the ROS-mediated toxicity of CNT have been suggested to be reduced through surface modification. Previous studies have suggested that inconsistent toxicity results ranging from no toxicity to significant toxicity could be attributed to the use of different biological models for testing, with no standardization of experimental conditions and nanomaterial type utilized. Therefore, a broad of evaluation of toxicity issues using both in vitro and in vivo studies for risk assessment should be conducted for carbon-based nanomaterials [345].

Graphene is also emerging as one of the most promising materials, not only for industrial applications but also for drug delivery, gene delivery, and tissue engineering. Although it has been widely reported that functionalized graphene-based nanomaterials are less toxic than the non-functionalized counterparts, the biocompatibility and toxicity of graphene and its derivatives are not yet completely understood [346]. The concentration and dose of graphene, and whether a tumor is present or not are the main factors that influence the toxicity, and further studies are needed to understand the toxicity of graphene and its derivatives.

In recent years, gold nanoparticles (AuNP) have been synthesized for a wide range of medical applications, especially drug/gene delivery, imaging, and anticancer therapeutics. Although several studies have been conducted on the possible toxicity of AuNP, there are many unknown factors that have to be considered in order to standardize the biocompatibility toxicological trials of AuNP.

Even though the toxicity of such insoluble nanoparticles as AuNP could be influenced by many factors, the effect of surface-bound ligands is the most important factor. The variable curvature in the surface of AuNP can lead to different grafting density (GD) of ligands that may be distinguished from remained contaminants of chemical synthesis and the effects of NPs size. The authors strongly recommend that future studies should focus on the determination of AuNP-mediated toxicity independent from the size effects of AuNP and compare to surfactant-free and ligand-free AuNP standards. Another equally important parameter is the particle size distribution which enables the calculation of the effective particle surface area dose. Last but not least, the nanoparticle dose (which is usually reported as mass per volume) is a key factor in all toxicological studies [347].

In recent years, polymeric nanoparticles such as poly(alkyl-cyanoacrylate) (PACA) and poly(lactic acid), and/or poly(glycolic acid) nanoparticles have revolutionized many fields of medicine, in applications such as gene therapy, insulin delivery and release, artificial hemoglobin, and vehicles for transporting other drugs in the bloodstream. Unfortunately, despite many studies on the efficacy of polymeric NP, the crucial issue of toxicity has not so far been addressed sufficiently. As mentioned in previous chapters of this ebook, most of the compounds utilized in smart DDS are biodegradable polymers that can be triggered by different stimuli to release drugs. Furthermore, polymeric materials are widely used for coating of other NP in order to prevent agglomeration. Minimizing the side effects of polymeric NP, reducing their unintentional inhalation, and avoiding any potential neurotoxicity of polymeric NP in the nervous system are the goals in the development of new DDS which are not only effective but also shown to be safe [348].

Solid lipid nanoparticles (SLNs) and nanostructured lipid carriers (NLCs) have been recently developed, and are considered to be well-tolerated carriers with low cytotoxicity, although questions of oxidative stress and hemocompatibility still remain concerns. The susceptibility of "normal cell lines" and "cancer cell lines" towards SLN/NLC should be further investigated. Although the delivery of chemotherapeutic drugs by SLNs/NLCs have been frequently reported, the toxicity of non-loaded SLN/NLC has not been much studied, and some contradictory results have been obtained. Furthermore, there are still questions about the effects of the administration route focusing on the oral route (due to its higher nanoparticle uptake efficiency) and also concerns about the systemic toxicity to model organisms, and the biodistribution, pharmacokinetics, and tolerability of SLN/NLC have been raised [349].

In order for smart nanoparticles to gain widespread acceptance as DGDS, it will be necessary to reconcile the above concerns about potential toxicity, with the extra benefit they can confer for drug and gene delivery as has been discussed in the previous sections. The biocompatibility of the designed vehicles, their cost-effectiveness, efficiency, and convenience for patient and physician administration have to be taken into consideration while designing a stable and efficient smart DDS. Therefore, far-reaching cooperation between engineers (e.g., materials scientists and polymer scientists), biologists, geneticists, and medical doctors will be needed to pave the way to success in this research field.

Although significant advancements in the performances of DDSs have occurred, the traditional approaches of drug administration (e.g., subcutaneous or intravenous injection, and pills) are still the prevailing method for drug administration. As far as the future scope of stimulus-responsive DDS is concerned, there may be other factors that need to be considered; two of which the synthesis of carriers with the ability for controlled release of cargo, and the efficient delivery of drugs to a targeted site with a predetermined rate over a specified time are the main ones. In a summary of the previous sections, the field of smart DDS has to focus on the following challenges: optimum performance, maximum efficiency, convenience for patients and physicians, lack of general toxicity to humans, animals and the environment, effective local delivery of drugs with minimum side effects, simple fabrication and application in real-life, keeping up with new and more advanced techniques that regularly appear.

If these requirements are met, stimuli-responsive smart DGDS could become a powerful weapon to overcome a wide range of diseases in the near future.

Acknowledgments

Research in the Hamblin laboratory is supported by US NIH grants R01AI050875 and R21AI121700.

References

1. Gaheen, S., G. W. Hinkal, S. A. Morris, M. Lijowski, M. Heiskanen and J. D. Klemm. (2013). caNanoLab: Data sharing to expedite the use of nanotechnology in biomedicine. *Computational Science & Discovery*. **6**, 014010.
2. Weintraub, K. (2013). Biomedicine: The new gold standard. *Nature*. **495**, S14–S16.
3. Mirkin, C. A., T. J. Meade, S. H. Petrosko and A. H. Stegh. (2015). *Nanotechnology-Based Precision Tools for the Detection and Treatment of Cancer*, Springer, Heidelberg.
4. Gallo, J. and N. J. Long. (2014). in Nanoparticulate MRI Contrast Agents.
5. Matoba, T. and K. Egashira. (2014). Nanoparticle-mediated drug delivery system for cardiovascular disease. *International Heart Journal*. **55**, 281–286.
6. Tiwari, P. (2015). Recent trends in therapeutic approaches for diabetes management: A comprehensive update. *Journal of Diabetes Research*. **501**, 340838.
7. LaVan, D. A., T. McGuire and R. Langer. (2003). Small-scale systems for in vivo drug delivery. *Nature Biotechnology* **21**, 1184–1191.
8. Son, D., J. Lee, D. J. Lee, R. Ghaffari, S. Yun, S. J. Kim, J. E. Lee, H. R. Cho, S. Yoon, S. Yang, S. Lee, S. Qiao, D. Ling, S. Shin, J.-K. Song, J. Kim, T. Kim, H. Lee, J. Kim, M. Soh, N. Lee, C. S. Hwang, S. Nam, N. Lu, T. Hyeon, S. H. Choi and D.-H. Kim. (2015). Bioresorbable electronic stent integrated with therapeutic nanoparticles for endovascular diseases. *ACS Nano*. **9**, 5937–5946.
9. Takahashi, H., D. Letourneur and D. W. Grainger. (2007). Delivery of large biopharmaceuticals from cardiovascular stents: A review. *Biomacromolecules*. **8**, 3281–3293.
10. Ruedas-Rama, M. J., J. D. Walters, A. Orte and E. A. Hall. (2012). Fluorescent nanoparticles for intracellular sensing: A review. *Analytica Chimica Acta*. **751**, 1–23.
11. Torchilin, V. (2011). Tumor delivery of macromolecular drugs based on the EPR effect. *Advanced Drug Delivery Reviews*. **63**, 131–135.
12. Berg, K., P. K. Selbo, L. Prasmickaite, T. E. Tjelle, K. Sandvig, J. Moan, G. Gaudernack, Ø. Fodstad, S. Kjølsrud and H. Anholt. (1999). Photochemical internalization a novel technology for delivery of macromolecules into cytosol. *Cancer Research*. **59**, 1180–1183.
13. Son, S., E. Shin and B.-S. Kim. (2014). Light-Responsive Micelles of Spiropyran Initiated Hyperbranched Polyglycerol for Smart Drug Delivery. *Biomacromolecules*. **15**, 628–634
14. Motornov, M., Y. Roiter, I. Tokarev and S. Minko. (2010). Stimuli-responsive nanoparticles, nanogels and capsules for integrated multifunctional

intelligent systems. *Progress in Polymer Science*. **35**, 174–211.
15. Tirelli, N. (2006). (Bio) Responsive nanoparticles. *Current Opinion in Colloid & Interface Science*. **11**, 210–216.
16. Mura, S., J. Nicolas and P. Couvreur. (2013). Stimuli-responsive nanocarriers for drug delivery. *Nature Materials*. **12**, 991–1003.
17. Cheng, R., F. Meng, C. Deng, H.-A. Klok and Z. Zhong. (2013). Dual and multi-stimuli responsive polymeric nanoparticles for programmed site-specific drug delivery. *Biomaterials*. **34**, 3647–3657.
18. Wu, L., Y. Zou, C. Deng, R. Cheng, F. Meng and Z. Zhong. (2013). Intracellular release of doxorubicin from core-crosslinked polypeptide micelles triggered by both pH and reduction conditions. *Biomaterials*. **34**, 5262–5272.
19. Cheng, R., F. Meng, S. Ma, H. Xu, H. Liu, X. Jing and Z. Zhong. (2011). Reduction and temperature dual-responsive crosslinked polymersomes for targeted intracellular protein delivery. *Journal of Materials Chemistry*. **21**, 19013–19020.
20. Wu, H., H. Shi, H. Zhang, X. Wang, Y. Yang, C. Yu, C. Hao, J. Du, H. Hu and S. Yang. (2014). Prostate stem cell antigen antibody-conjugated multiwalled carbon nanotubes for targeted ultrasound imaging and drug delivery. *Biomaterials*. **35**, 5369–5380.
21. Yu, J., Y. Zhang, Y. Ye, R. DiSanto, W. Sun, D. Ranson, F. S. Ligler, J. B. Buse and Z. Gu. (2015). Microneedle-array patches loaded with hypoxia-sensitive vesicles provide fast glucose-responsive insulin delivery. *Proceedings of the National Academy of Sciences*. **112**, 8260–8265.
22. Joshi-Barr, S., C. de Gracia Lux, E. Mahmoud and A. Almutairi. (2014). Exploiting oxidative microenvironments in the body as triggers for drug delivery systems. *Antioxidants & Redox Signaling*. **21**, 730–754.
23. Zhu, C.-L., X.-W. Wang, Z.-Z. Lin, Z.-H. Xie and X.-R. Wang. (2014). Cell microenvironment stimuli-responsive controlled-release delivery systems based on mesoporous silica nanoparticles. *Journal of Food and Drug Analysis*. **22**, 18–28.
24. Alvarez-Lorenzo, C., L. Bromberg and A. Concheiro. (2009). Light-sensitive intelligent drug delivery systems. *Photochemistry and Photobiology*. **85**, 848–860.
25. Chan, A., R. P. Orme, R. A. Fricker and P. Roach. (2013). Remote and local control of stimuli responsive materials for therapeutic applications. *Advanced Drug Delivery Reviews*. **65**, 497–514.
26. You, J.-O., D. Almeda, J. George and D. T. Auguste. (2010). Bioresponsive matrices in drug delivery. *Journal of Biological Engineering*. **4**, 1–12.
27. Huang, X., X. Jiang, Q. Yang, Y. Chu, G. Zhang, B. Yang and R. Zhuo. (2013). Triple-stimuli (pH/thermo/reduction) sensitive copolymers for intracellular drug delivery. *Journal of Materials Chemistry B*. **1**, 1860–1868.
28. Han, D., X. Tong and Y. Zhao. (2012). Block copolymer micelles with a dual-stimuli-responsive core for fast or slow degradation. *Langmuir*. **28**, 2327–2331.
29. Dong, J., Y. Wang, J. Zhang, X. Zhan, S. Zhu, H. Yang and G. Wang. (2013). Multiple stimuli-responsive polymeric micelles for controlled release. *Soft Matter*. **9**, 370–373.
30. Nishiyama, N., A. Iriyama, W.-D. Jang, K. Miyata, K. Itaka, Y. Inoue, H. Takahashi, Y. Yanagi, Y. Tamaki and H. Koyama. (2005). Light-induced gene transfer from packaged DNA enveloped in a dendrimeric photosensitizer. *Nature Materials*. **4**, 934–941.
31. Chang, Y. T., P. Y. Liao, H. S. Sheu, Y. J. Tseng, F. Y. Cheng and C. S. Yeh. (2012). Near-infrared light-responsive intracellular drug and siRNA release using Au nanoensembles with oligonucleotide-capped silica shell. *Advanced Materials*. **24**, 3309–3314.
32. Shah, V., O. Taratula, O. B. Garbuzenko, M. L. Patil, R. Savla, M. Zhang and T. Minko. (2013). Genotoxicity of different nanocarriers: Possible modifications for the delivery of nucleic acids. *Current Drug Discovery Technologies*. **10**, 8.
33. Luo, M., C. Shen, B. N. Feltis, L. L. Martin, A. E. Hughes, P. F. Wright and T. W. Turney. (2014). Reducing ZnO nanoparticle cytotoxicity by surface modification. *Nanoscale*. **6**, 5791–5798.
34. Hu, X., Y. Wang and B. Peng. (2014). Chitosan-capped mesoporous silica nanoparticles as pH-responsive nanocarriers for controlled drug release. *Chemistry Asian Journal* **9**, 319–327.
35. Mahmoudi, M., S. N. Saeedi-Eslami, M. A. Shokrgozar, K. Azadmanesh, M. Hassanlou, H. R. Kalhor, C. Burtea, B. Rothen-Rutishauser, S. Laurent and S. Sheibani. (2012). Cell "vision": Complementary factor of protein corona in nanotoxicology. *Nanoscale*. **4**, 5461–5468.
36. Mahmoudi, M., S. E. Lohse, C. J. Murphy, A. Fathizadeh, A. Montazeri and K. S. Suslick. (2013). Variation of protein corona composition of gold nanoparticles following plasmonic heating. *Nano Letters*. **14**, 6–12.
37. Mirshafiee, V., M. Mahmoudi, K. Lou, J. Cheng and M. L. Kraft. (2013). Protein corona significantly reduces active targeting yield. *Chemical Communications*. **49**, 2557–2559.
38. Mortensen, N. P., G. B. Hurst, W. Wang, C. M. Foster, P. D. Nallathamby and S. T. Retterer. (2013). Dynamic development of the protein corona on silica nanoparticles: Composition and role in toxicity. *Nanoscale*. **5**, 6372–6380.
39. Karimi, M., N. Solati, A. Ghasemi, M. A. Estiar, M. Hashemkhani, P. Kiani, E. Mohamed, A. Saeidi, M.

Taheri and P. Avci. (2015). Carbon nanotubes part II: A remarkable carrier for drug and gene delivery. *Expert Opinion on Drug Delivery.* **12**, 1–17.

40. Fang, Z., L.-Y. Wan, L.-Y. Chu, Y.-Q. Zhang and J.-F. Wu. (2015). 'Smart'nanoparticles as drug delivery systems for applications in tumor therapy. *Expert Opinion on Drug Delivery.* **12**, 1–11.
41. Zhao, Z., D. Huang, Z. Yin, X. Chi, X. Wang and J. Gao. (2012). Magnetite nanoparticles as smart carriers to manipulate the cytotoxicity of anticancer drugs: Magnetic control and pH-responsive release. *Journal of Materials Chemistry.* **22**, 15717–15725.
42. Jimenez-Cruz, C. A., S. G. Kang and R. Zhou. (2014). Large scale molecular simulations of nanotoxicity. *Wiley Interdisciplinary Reviews: Systems Biology and Medicine.* **6**, 329–343.
43. Karimi, M., P. Sahandi Zangabad, A. Ghasemi, M. Amiri, M. Bahrami, H. Malekzad, H. Ghahramanzadeh Asl, Z. Mahdieh, M. Bozorgomid, A. Ghasemi, M. R. Rahmani Taji Boyuk and M. R. Hamblin. (2016). Temperature-responsive smart nanocarriers for delivery of therapeutic agents: applications and recent advances. *ACS Appl Mater Interfaces.* **8**, 21107–21133.
44. Ruel-Gariépy, E. and J.-C. Leroux. (2004). In situ-forming hydrogels—review of temperature-sensitive systems. *European Journal of Pharmaceutics and Biopharmaceutics.* **58**, 409–426.
45. Bölgen, N., M. R. Aguilar, M. d. M. Fernández, S. Gonzalo-Flores, S. Villar-Rodil, J. San Román and E. Piskin. (2015). Thermoresponsive biodegradable HEMA-Lactate-Dextran-co-NIPA cryogels for controlled release of simvastatin. *Artificial Cells, Nanomedicine, and Biotechnology.* **43**, 40–49.
46. Ulasan, M., E. Yavuz, E. U. Bagriacik, Y. Cengeloglu and M. S. Yavuz. (2015). Biocompatible thermoresponsive PEGMA nanoparticles crosslinked with cleavable disulfide-based crosslinker for dual drug release. *Journal of Biomedical Materials Research Part A.* **103**, 243–251.
47. Yuan, Q., W. A. Yeudall and H. Yang. (2014). Thermoresponsive dendritic facial amphiphiles for gene delivery. *Nanomedicine and Nanobiology.* **1**, 64–69.
48. Cardoso, A. M., M. T. Calejo, C. M. Morais, A. L. Cardoso, R. Cruz, K. Zhu, M. C. Pedroso de Lima, A. l. S. Jurado and B. Nyström. (2014). Application of thermoresponsive PNIPAAM-b-PAMPTMA diblock copolymers in siRNA delivery. *Molecular Pharmaceutics.* **11**, 819–827.
49. Li, Y., F. Wang, T. Sun, J. Du, X. Yang and J. Wang. (2014). Surface-modulated and thermoresponsive polyphosphoester nanoparticles for enhanced intracellular drug delivery. *Science China Chemistry.* **57**, 579–585.
50. Soni, G. and K. S. Yadav. (2013). High encapsulation efficiency of poloxamer-based injectable thermoresponsive hydrogels of etoposide. *Pharmaceutical Development and Technology.* **19**, 651–661.
51. Cheng, X., Y. Jin, T. Sun, R. Qi, B. Fan and H. Li. (2015). Oxidation-and thermo-responsive poly (N-isopropylacrylamide-co-2-hydroxyethyl acrylate) hydrogels cross-linked via diselenides for controlled drug delivery. *RSC Advances.* **5**, 4162–4170.
52. Yang, J., R. van Lith, K. Baler, R. A. Hoshi and G. A. Ameer. (2014). A thermoresponsive biodegradable polymer with intrinsic antioxidant properties. *Biomacromolecules.* **15**, 3942–3952.
53. Indiana, I. (2001). Triggering in drug delivery systems. *Advanced Drug Delivery Reviews.* **53**, 245.
54. Ward, M. A. and T. K. Georgiou. (2011). Thermoresponsive polymers for biomedical applications. *Polymers.* **3**, 1215–1242.
55. Schmaljohann, D. (2006). Thermo-and pH-responsive polymers in drug delivery. *Advanced Drug Delivery Reviews.* **58**, 1655–1670.
56. Yao, Y., H. Shen, G. Zhang, J. Yang and X. Jin. (2014). Synthesis of poly (N-isopropylacrylamide)-co-poly (phenylboronate ester) acrylate and study on their glucose-responsive behavior. *Journal of Colloid and Interface Science.* **431**, 216–222.
57. Ha, W., J. Yu, X.-y. Song, J. Chen and Y.-p. Shi. (2014). Tunable temperature-responsive supramolecular hydrogels formed by prodrugs as a codelivery system. *ACS Applied Materials & Interfaces.* **6**, 10623–10630.
58. Yeh, J.-C., Y.-T. Hsu, C.-M. Su, M.-C. Wang, T.-H. Lee and S.-L. Lou. (2014). Preparation and characterization of biocompatible and thermoresponsive micelles based on poly (N-isopropylacrylamide-co-N, N-dimethylacrylamide) grafted on polysuccinimide for drug delivery. *Journal of Biomaterials Applications.* doi: 10.1177/0885328214533736.
59. Sun, F., Y. Wang, Y. Wei, G. Cheng and G. Ma. (2014). Thermo-triggered drug delivery from polymeric micelles of poly (N-isopropylacrylamide-co-acrylamide)-b-poly (n-butyl methacrylate) for tumor targeting. *Journal of Bioactive and Compatible Polymers: Biomedical Applications.* doi:10.1177/0883911514535288.
60. Kashyap, S. and M. Jayakannan. (2014). Thermo-responsive and shape transformable amphiphilic scaffolds for loading and delivering anticancer drugs. *Journal of Materials Chemistry B.* **2**, 4142–4152.
61. Picos-Corrales, L. A., A. Licea-Claveríe and K.-F. Arndt. (2014). Bisensitive core–shell nanohydrogels by e-Beam irradiation of micelles. *Reactive and Functional Polymers.* **75**, 31–40.
62. Chen, Y.-Y., H.-C. Wu, J.-S. Sun, G.-C. Dong and T.-W. Wang. (2013). Injectable and thermoresponsive self-assembled nanocomposite hydrogel for

long-term anticancer drug delivery. *Langmuir*. **29**, 3721–3729.

63. Elluru, M., H. Ma, M. Hadjiargyrou, B. S. Hsiao and B. Chu. (2013). Synthesis and characterization of biocompatible hydrogel using Pluronics-based block copolymers. *Polymer*. **54**, 2088–2095.
64. Zhou, J., M. V. Pishko and J. L. Lutkenhaus. (2014). Thermoresponsive layer-by-layer assemblies for nanoparticle-based drug delivery. *Langmuir*. **30**, 5903–5910.
65. Li, Z., S. Cho, I. C. Kwon, M. M. Janát-Amsbury and K. M. Huh. (2013). Preparation and characterization of glycol chitin as a new thermogelling polymer for biomedical applications. *Carbohydrate Polymers*. **92**, 2267–2275.
66. Muzzarelli, R. A., F. Greco, A. Busilacchi, V. Sollazzo and A. Gigante. (2012). Chitosan, hyaluronan and chondroitin sulfate in tissue engineering for cartilage regeneration: A review. *Carbohydrate Polymers*. **89**, 723–739.
67. Beija, M., J.-D. Marty and M. Destarac. (2011). Thermoresponsive poly (N-vinyl caprolactam)-coated gold nanoparticles: Sharp reversible response and easy tunability. *Chemical Communications*. **47**, 2826–2828.
68. Mayol, L., D. De Stefano, F. De Falco, R. Carnuccio, M. C. Maiuri and G. De Rosa. (2014). Effect of hyaluronic acid on the thermogelation and biocompatibility of its blends with methyl cellulose. *Carbohydrate polymers*. **112**, 480–485.
69. Manokruang, K., J. S. Lym and D. S. Lee. (2014). Injectable hydrogels based on poly (amino urethane) conjugated bovine serum albumin. *Materials Letters*. **124**, 105–109.
70. Zhao, J., B. Guo and P. X. Ma. (2014). Injectable alginate microsphere/PLGA–PEG–PLGA composite hydrogels for sustained drug release. *RSC Advances*. **4**, 17736–17742.
71. Zheng, Y. and S. Zheng. (2012). Poly (ethylene oxide)-grafted poly (N-isopropylacrylamide) networks: Preparation, characterization and rapid deswelling and reswelling behavior of hydrogels. *Reactive and Functional Polymers*. **72**, 176–184.
72. Guo, P., H. Li, W. Ren, J. Zhu, F. Xiao, S. Xu and J. Wang. (2015). Unusual thermo-responsive behaviors of poly(NIPAM-co-AM)/PEG/PTA composite hydrogels. *Materials Letters*. **143**, 24–26.
73. Nakayama, M., J. Chung, T. Miyazaki, M. Yokoyama, K. Sakai and T. Okano. (2007). Thermal modulation of intracellular drug distribution using thermoresponsive polymeric micelles. *Reactive and Functional Polymers*. **67**, 1398–1407.
74. Wust, P., B. Hildebrandt, G. Sreenivasa, B. Rau, J. Gellermann, H. Riess, R. Felix and P. Schlag. (2002). Hyperthermia in combined treatment of cancer. *The Lancet Oncology*. **3**, 487–497.
75. Meyer, D. E., B. Shin, G. Kong, M. Dewhirst and A. Chilkoti. (2001). Drug targeting using thermally responsive polymers and local hyperthermia. *Journal of Controlled Release*. **74**, 213–224.
76. Van Elk, M., R. Deckers, C. Oerlemans, Y. Shi, G. Storm, T. Vermonden and W. E. Hennink. (2014). Triggered release of doxorubicin from temperature-sensitive poly (N-(2-hydroxypropyl)-methacrylamide mono/dilactate) grafted liposomes. *Biomacromolecules*. **15**, 1002–1009.
77. Lee, S. H., S. H. Choi, S. H. Kim and T. G. Park. (2008). Thermally sensitive cationic polymer nanocapsules for specific cytosolic delivery and efficient gene silencing of siRNA: Swelling induced physical disruption of endosome by cold shock. *Journal of Controlled Release*. **125**, 25–32.
78. Movahedi, F., R. G. Hu, D. L. Becker and C. Xu. (2015). Stimuli-responsive liposomes for the delivery of nucleic acid therapeutics. *Nanomedicine: Nanotechnology, Biology and Medicine*. **11**, 1575–1584.
79. Feng, G., H. Chen, J. Li, Q. Huang, M. J. Gupte, H. Liu, Y. Song and Z. Ge. (2015). Gene therapy for nucleus pulposus regeneration by heme oxygenase-1 plasmid DNA carried by mixed polyplex micelles with thermo-responsive heterogeneous coronas. *Biomaterials*. **52**, 1–13.
80. Tamaddon, A. M., F. H. Shirazi and H. R. Moghimi. (2007). Modeling cytoplasmic release of encapsulated oligonucleotides from cationic liposomes. *International Journal of Pharmaceutics*. **336**, 174–182.
81. Peng, Z., C. Wang, E. Fang, X. Lu, G. Wang and Q. Tong. (2014). Co-delivery of doxorubicin and SATB1 shRNA by thermosensitive magnetic cationic liposomes for gastric cancer therapy. *PLoS One*. **9**, e92924.
82. Calejo, M.T., A. M. S. Cardoso, A. Kjøniksen, K. Zhu, C. M. Moraisc, S. Sande, A. Cardoso, M. C. P. de Lima, A. Jurado and B. Nyström. (2013). Temperature-responsive cationic block copolymers as nanocarriers for gene delivery. *International Journal of Pharmaceutics* **448**, 105–114.
83. Saravanakumar, G. and W. J. Kim. (2014). Stimuli-responsive polymeric nanocarriers as promising drug and gene delivery systems. *Fundamental Biomedical Technologies* **7**, 55–91.
84. Chang, Y., W. Yandi, W.-Y. Chen, Y.-J. Shih, C.-C. Yang, Y.Chang, Q.-D. Ling and A.Higuchi. (2010). Tunable bioadhesive copolymer hydrogels of thermoresponsive poly(N-isopropyl acrylamide) containing zwitterionic polysulfobetaine. *Biomacromolecules*. **11**, 1101–1110.
85. Caldorera-Moore, M. E., W. B. Liechty and N. A. Peppas. (2011). Responsive theranostic systems: Integration of diagnostic imaging agents and

responsive controlled release drug delivery carriers. *Accounts of Chemical Research.* **44**, 1061–1070.
86. (Türkoğlu) Laçin, N., G. (Guven) Utkan, T. Kutsal and E. Pişkin. (2012). A thermo-sensitive NIPA-based co-polymer and monosize polycationic nanoparticle for non-viral gene transfer to smooth muscle cells *Biomaterials Science.* **23**, 577–592.
87. Zhang, R., Y. Wang, F.-S. Du, Y.-L. Wang, Y.-X. Tan, S.-P. Ji and Z.-C. Li. (2011). Thermoresponsive gene carriers based on polyethylenimine-graft-Poly[oligo(ethyleneglycol) methacrylate]. *Macromolecular Bioscience* **11**, 1393–1406.
88. Gandhi, A., A. Paul, S. Oommen Sen and K. K. Sen. (2014). Studies on thermoresponsive polymers: Phase behaviour, drug delivery and biomedical applications. *Pharmaceutical Sciences.* **10**, 99–107.
89. Iwai, R., S. Kusakabe, Y. Nemoto and Y. Nakayama. (2012). Deposition gene transfection using bioconjugates of DNA and thermoresponsive cationic homopolymer. *Bioconjugate Chemistry.* **23**, 751–757.
90. Ivanova, E. D., N. I. Ivanova, M. D. Apostolova, S. C. Turmanova and I. V. Dimitrov. (2013). Polymer gene delivery vectors encapsulated in thermally sensitive bioreducible shell. *Bioorganic Medicinal Chemistry Letters.* **23**, 4080–4084.
91. Thévenot, J., H. Oliveira, O. Sandre and S. Lecommandoux. (2013). Magnetic responsive polymer composite materials. *Chemical Society Reviews.* **42**, 7099–7116.
92. Whitesides, G. M. (2003). The 'right' size in nanobiotechnology. *Nature Biotechnology.* **21**, 1161–1165.
93. Shubayev, V. I., T. R. Pisanic II and S. Jin. (2009). Magnetic nanoparticles for theragnostics. *Advanced Drug Delivery Reviews.* **61**, 467–477.
94. Freeman, M., A. Arrott and J. Watson. (1960). Magnetism in medicine. *Journal of Applied Physics.* **31**, S404–S405.
95. Widder, K., G. Flouret and A. Senyei. (1979). Magnetic microspheres: Synthesis of a novel parenteral drug carrier. *Journal of Pharmaceutical Sciences.* **68**, 79–82.
96. Schexnailder, P. and G. Schmidt. (2009). Nanocomposite polymer hydrogels. *Colloid and Polymer Science.* **287**, 1–11.
97. Giani, G., S. Fedi and R. Barbucci. (2012). Hybrid magnetic hydrogel: A potential system for controlled drug delivery by means of alternating magnetic fields. *Polymers.* **4**, 1157–1169.
98. Fang, R. H., S. Aryal, C.-M. J. Hu and L. Zhang. (2010). Quick synthesis of lipid−polymer hybrid nanoparticles with low polydispersity using a single-step sonication method. *Langmuir.* **26**, 16958–16962.
99. Deok Kong, S., M. Sartor, C.-M. Jack Hu, W. Zhang, L. Zhang and S. Jin. (2013). Magnetic field activated lipid–polymer hybrid nanoparticles for stimuli-responsive drug release. *Acta Biomaterialia.* **9**, 5447–5452.
100. Jordan, A., P. Wust, H. Fähling, W. John, A. Hinz and R. Felix. (2009). Inductive heating of ferrimagnetic particles and magnetic fluids: Physical evaluation of their potential for hyperthermia. *International Journal of Hyperthermia.* **25**, 499–511.
101. Gilchrist, R., R. Medal, W. D. Shorey, R. C. Hanselman, J. C. Parrott and C. B. Taylor. (1957). Selective inductive heating of lymph nodes. *Annals of Surgery.* **146**, 596.
102. Ito, A., K. Tanaka, H. Honda, S. Abe, H. Yamaguchi and T. Kobayashi. (2003). Complete regression of mouse mammary carcinoma with a size greater than 15 mm by frequent repeated hyperthermia using magnetite nanoparticles. *Journal of Bioscience and Bioengineering.* **96**, 364–369.
103. Mah, C., T. J. Fraites, I. Zolotukhin, S. Song, T. R. Flotte, J. Dobson, C. Batich and B. J. Byrne. (2002). Improved method of recombinant AAV2 delivery for systemic targeted gene therapy. *Molecular Therapy.* **6**, 106–112.
104. Ruiz-Hernandez, E., A. Baeza and M. A. Vallet-Regí. (2011). Smart drug delivery through DNA/magnetic nanoparticle gates. *ACS Nano.* **5**, 1259–1266.
105. Karimi, M., P. Sahandi Zangabad, S. Baghaee-Ravari, M. Ghazadeh, H. Mirshekari and M. R. Hamblin. (2017). Smart nanostructures for cargo delivery: Uncaging and activating by light. *Journal of the American Chemical Society.* **139**, 4584–4610.
106. Nicoletta, F. P., D. Cupelli, P. Formoso, G. De Filpo, V. Colella and A. Gugliuzza. (2012). Light responsive polymer membranes: A review. *Membranes.* **2**, 134–197.
107. Alatorre-Meda, M. (2013). UV and near-IR triggered release from polymeric micelles and nanoparticles. *Smart Materials for Drug Delivery.* **1**, 304.
108. Knežević, N. Ž., B. G. Trewyn and V. S.-Y. Lin. (2011). Functionalized mesoporous silica nanoparticle-based visible light responsive controlled release delivery system. *Chemical Communications* **47**, 2817–2819.
109. Yi, Q. and G. B. Sukhorukov. (2013). UV light stimulated encapsulation and release by polyelectrolyte microcapsules. *Advances in Colloid and Interface Science.* **207**, 280–289
110. Xie, X., N. Gao, R. Deng, Q. Sun, Q.-H. Xu and X. Liu. (2013). Mechanistic investigation of photon upconversion in Nd3+-sensitized core–shell nanoparticles. *Journal of the American Chemical Society.* **135**, 12608–12611.
111. Meng, H., M. Xue, T. Xia, Z. Ji, D. Y. Tarn, J. I. Zink and A. E. Nel. (2011). Use of size and a copolymer design feature to improve the biodistribution and the enhanced permeability and retention effect of

doxorubicin-loaded mesoporous silica nanoparticles in a murine xenograft tumor model. *Acs Nano.* **5**, 4131–4144.

112. Lu, J., M. Liong, Z. Li, J. I. Zink and F. Tamanoi. (2010). Biocompatibility, biodistribution, and drug-delivery efficiency of mesoporous silica nanoparticles for cancer therapy in animals. *Small.* **6**, 1794–1805.
113. Huang, X., X. Teng, D. Chen, F. Tang and J. He. (2010). The effect of the shape of mesoporous silica nanoparticles on cellular uptake and cell function. *Biomaterials.* **31**, 438–448.
114. Luo, G., W. Chen, H. Jia, Y. Sun, H. Cheng, R. Zhuo and X. Zhang. An indicator-guided photo-controlled drug delivery system based on mesoporous silica/gold nanocomposites. *Nano Research.* **8**, 1–13.
115. Karthik, S., B. Saha, S. K. Ghosh and N. P. Singh. (2013). Photoresponsive quinoline tethered fluorescent carbon dots for regulated anticancer drug delivery. *Chemical Communications.* **49**, 10471–10473.
116. Lo, C.-W., D. Zhu and H. Jiang. (2011). An infrared-light responsive graphene-oxide incorporated poly (N-isopropylacrylamide) hydrogel nanocomposite. *Soft Matter.* **7**, 5604–5609.
117. Li, S., B. A. Moosa, J. G. Croissant and N. M. Khashab. (2015). Electrostatic assembly/disassembly of nanoscaled colloidosomes for light-triggered cargo release. *Angewandte Chemie.* **127**, 6908–6912.
118. Yu, Y., X. Kang, X. Yang, L. Yuan, W. Feng and S. Cui. (2013). Surface charge inversion of self-assembled monolayers by visible light irradiation: Cargo loading and release by photoreactions. *Chemical Communications.* **49**, 3431–3433.
119. Alvarez-Lorenzo, C., S. Deshmukh, L. Bromberg, T. A. Hatton, I. Sández-Macho and A. Concheiro. (2007). Temperature-and light-responsive blends of pluronic F127 and poly (N, N-dimethylacrylamide-co-methacryloyloxyazobenzene). *Langmuir.* **23**, 11475–11481.
120. Jiang, J., B. Qi, M. Lepage and Y. Zhao. (2007). Polymer micelles stabilization on demand through reversible photo-cross-linking. *Macromolecules.* **40**, 790–792.
121. Hill, C. R., J. C. Bamber and G. Haar. (2004). *Physical Principles of Medical Ultrasonics*, Wiley Online Library.
122. Schroeder, A. (2013). Using ultrasound to formulate nanotherapeutics. *Chimica Oggi-Chemistry Today.* **31**, 6.
123. Di, J., J. Price, X. Gu, X. Jiang, Y. Jing and Z. Gu. (2013). Ultrasound-triggered regulation of blood glucose levels using injectable nano-network. *Advanced Healthcare Materials.* doi:10.1002/adhm.201300490.
124. Morgan, K. E., J. S. Allen, P. A. Dayton, J. E. Chomas, A. Klibaov and K. W. Ferrara. (2000). Experimental and theoretical evaluation of microbubble behavior: Effect of transmitted phase and bubble size. *IEEE Transactions on Ultrasonics, Ferroelectrics and Frequency Control.* **47**, 1494–1509.
125. Van Der Meer, S., M. Versluis, D. Lohse, C. Chin, A. Bouakaz and N. De Jong. (2004) Ultrasonics Symposium, 2004 IEEE, pp. 343–345.
126. Goertz, D. E., N. de Jong and A. F. van der Steen. (2007). Attenuation and size distribution measurements of definity™ and manipulated definity™ populations. *Ultrasound in Medicine & Biology.* **33**, 1376–1388.
127. Shi, W. T. and F. Forsberg. (2000). Ultrasonic characterization of the nonlinear properties of contrast microbubbles. *Ultrasound in Medicine & Biology.* **26**, 93–104.
128. Moonen, C. and I. Lentacker. (2014). Ultrasound assisted drug delivery. *Advanced Drug Delivery Reviews.* **72**, 1.
129. Khaibullina, A., B.-S. Jang, H. Sun, N. Le, S. Yu, V. Frenkel, J. A. Carrasquillo, I. Pastan, K. C. Li and C. H. Paik. (2008). Pulsed high-intensity focused ultrasound enhances uptake of radiolabeled monoclonal antibody to human epidermoid tumor in nude mice. *Journal of Nuclear Medicine.* **49**, 295–302.
130. Sundaram, J., B. R. Mellein and S. Mitragotri. (2003). An experimental and theoretical analysis of ultrasound-induced permeabilization of cell membranes. *Biophysical Journal.* **84**, 3087–3101.
131. Stone, M. J., V. Frenkel, S. Dromi, P. Thomas, R. P. Lewis, K. C. Li, M. Horne III and B. J. Wood. (2007). Pulsed-high intensity focused ultrasound enhanced tPA mediated thrombolysis in a novel *in vivo* clot model, a pilot study. *Thrombosis Research.* **121**, 193–202.
132. Karshafian, R., P. Bevan, P. Burns, S. Samac and M. Banerjee (2005). *Ultrasonics Symposium, 2005 IEEE*, pp. 13–16.
133. Frenkel, V., A. Etherington, M. Greene, J. Quijano, J. Xie, F. Hunter, S. Dromi and K. C. Li. (2006). Delivery of liposomal doxorubicin (Doxil) in a breast cancer tumor model: Investigation of potential enhancement by pulsed-high intensity focused ultrasound exposure. *Academic Radiology.* **13**, 469–479.
134. Duvshani-Eshet, M., L. Baruch, E. Kesselman, E. Shimoni and M. Machluf. (2005). Therapeutic ultrasound-mediated DNA to cell and nucleus: Bioeffects revealed by confocal and atomic force microscopy. *Gene Therapy.* **13**, 163–172.
135. Dittmar, K. M., J. Xie, F. Hunter, C. Trimble, M. Bur, V. Frenkel and K. C. Li. (2005). Pulsed high-intensity focused ultrasound enhances systemic administration of naked DNA in squamous cell

carcinoma model: Initial experience 1. *Radiology.* **235**, 541–546.
136. Lentacker, I., S. C. De Smedt and N. N. Sanders. (2009). Drug loaded microbubble design for ultrasound triggered delivery. *Soft Matter.* **5**, 2161–2170.
137. Qin, S., C. F. Caskey and K. W. Ferrara. (2009). Ultrasound contrast microbubbles in imaging and therapy: Physical principles and engineering. *Physics in Medicine and Biology.* **54**, R27.
138. Sirsi, S. R. and M. A. Borden. (2012). Advances in ultrasound mediated gene therapy using microbubble contrast agents. *Theranostics.* **2**, 1208.
139. Sirsi, S. and M. Borden. (2009). Microbubble compositions, properties and biomedical applications. *Bubble Science, Engineering & Technology.* **1**, 3–17.
140. Lindner, J. R. (2004). Microbubbles in medical imaging: Current applications and future directions. *Nature Reviews Drug Discovery.* **3**, 527–533.
141. McCulloch, M., C. Gresser, S. Moos, J. Odabashian, S. Jasper, J. Bednarz, P. Burgess, D. Carney, V. Moore and E. Sisk. (2000). Ultrasound contrast physics: A series on contrast echocardiography, article 3. *Journal of the American Society of Echocardiography.* **13**, 959–967.
142. Unger, E. C., T. P. McCreery and R. H. Sweitzer. (1997). Ultrasound enhances gene expression of liposomal transfection. *Investigative Radiology.* **32**, 723–727.
143. Rapoport, N., A. Payne, C. Dillon, J. Shea, C. Scaife and R. Gupta. (2013). Focused ultrasound-mediated drug delivery to pancreatic cancer in a mouse model. *Journal of Therapeutic Ultrasound.* **1**, 11.
144. He, Y., Y. Bi, Y. Hua, D. Liu, S. Wen, Q. Wang, M. Li, J. Zhu, T. Lin and D. He. (2011). Ultrasound microbubble-mediated delivery of the siRNAs targeting MDR1 reduces drug resistance of yolk sac carcinoma L2 cells. *J Exp Clin Cancer Res.* **30**, 104.
145. Unger, E. C., T. Porter, W. Culp, R. Labell, T. Matsunaga and R. Zutshi. (2004). Therapeutic applications of lipid-coated microbubbles. *Advanced Drug Delivery Reviews.* **56**, 1291–1314.
146. Lawrie, A., A. F. Brisken, S. E. Francis, D. I. Tayler, J. Chamberlain, D. C. Crossman, D. C. Cumberland and C. M. Newman. (1999). Ultrasound enhances reporter gene expression after transfection of vascular cells in vitro. *Circulation.* **99**, 2617–2620.
147. Aoki, T., M. Muramatsu, A. Nishina, K. Sanui and N. Ogata. (2004). Thermosensitivity of optically active hydrogels constructed with N-(L)-(1-hydroxymethyl) propylmethacrylamide. *Macromolecular Bioscience.* **4**, 943–949.
148. Anal, A. K. (2007). Stimuli-induced pulsatile or triggered release delivery systems for bioactive compounds. *Recent Patents on Endocrine, Metabolic & Immune Drug Discovery.* **1**, 83–90.
149. Qiu, Y. and K. Park. (2012). Environment-sensitive hydrogels for drug delivery. *Advanced Drug Delivery Reviews.* **64**, 49–60.
150. Shiga, T. (1997). Deformation and viscoelastic behavior of polymer gels in electric fields.
151. Tanaka, T., I. Nishio, S. Sun and S. Ueno-Nishio. (1982). Collapse of gels in an electric field. *Electroactivity in Polymeric Materials.* **218**(4571), 467–469.
152. Murdan, S. (2003). Electro-responsive drug delivery from hydrogels. *Journal of Controlled Release.* **92**, 1–17.
153. Gong, J., T. Nitta and Y. Osada. (1994). Electrokinetic modeling of the contractile phenomena of polyelectrolyte gels. One-dimensional capillary model. *The Journal of Physical Chemistry.* **98**, 9583–9587.
154. Tanaka, T., I. Nishio, S.-T. Sun and S. Ueno-Nishio. (1982). Collapse of gels in an electric field. *Science.* **218**, 467–469.
155. Kwon, I. C., Y. H. Bae, T. Okano and S. W. Kim. (1991). Drug release from electric current sensitive polymers. *Journal of controlled release.* **17**, 149–156.
156. Miller, L. L., G. A. Smith, A.-C. Chang and Q.-X. Zhou. (1987). Electrochemically controlled release. *Journal of Controlled Release.* **6**, 293–296.
157. D'Emanuele, A. and J. Stainforth. (1989). *Proc Int Symp Controlled Release Bioact Mater*, pp. 45–46.
158. D'Emanuele, A., J. Stainforth and R. Maraden (1988). *Proc Int Symp Controlled Release Bioact Mater*, pp. 76–77.
159. Zhao, X., J. Kim, C. A. Cezar, N. Huebsch, K. Lee, K. Bouhadir and D. J. Mooney. (2011). Active scaffolds for on-demand drug and cell delivery. *Proceedings of the National Academy of Sciences.* **108**, 67–72.
160. Wang, J., J. A. Kaplan, Y. L. Colson and M. W. Grinstaff. (2017). Mechanoresponsive materials for drug delivery: Harnessing forces for controlled release. *Advanced Drug Delivery Reviews.* **108**, 68–82.
161. Sershen, S. and J. West. (2002). Implantable, polymeric systems for modulated drug delivery. *Advanced Drug Delivery Reviews.* **54**, 1225–1235.
162. Kumar, G. A., A. Bhat, A. P. Lakshmi and K. Reddy. (2010). An overview of stimuli-induced pulsatile drug delivery systems. *International Journal of Pharm Tech Research.* **2**, 3658–2375.
163. Lee, K. Y., M. C. Peters, K. W. Anderson and D. J. Mooney. (2000). Controlled growth factor release from synthetic extracellular matrices. *Nature.* **408**, 998–1000.
164. Park, H., S. W. Kang, B. S. Kim, D. J. Mooney and K. Y. Lee. (2009). Shear-reversibly crosslinked alginate hydrogels for tissue engineering. *Macromolecular Bioscience.* **9**, 895–901.

165. Xiao, L., J. Zhu, D. J. Londono, D. J. Pochan and X. Jia. (2012). Mechano-responsive hydrogels crosslinked by block copolymer micelles. *Soft Matter.* **8**, 10233–10237.
166. Annabi, N., A. Tamayol, J. A. Uquillas, M. Akbari, L. E. Bertassoni, C. Cha, G. Camci-Unal, M. R. Dokmeci, N. A. Peppas and A. Khademhosseini. (2014). 25th anniversary article: Rational design and applications of hydrogels in regenerative medicine. *Advanced Materials.* **26**, 85–123.
167. Wang, P. Y. (1989). Implantable reservoir for supplemental insulin delivery on demand by external compression. *Biomaterials.* **10**, 197–201.
168. Yang, Y., G. Tang, H. Zhang, Y. Zhao, X. Yuan, Y. Fan and M. Wang. (2011). Controlled release of BSA by microsphere-incorporated PLGA scaffolds under cyclic loading. *Materials Science and Engineering: C.* **31**, 350–356.
169. Joshi, G. K., M. A. Johnson and R. Sardar. (2014). Novel pH-responsive nanoplasmonic sensor: Controlling polymer structural change to modulate localized surface plasmon resonance response. *RSC Advances.* **4**, 15807–15815.
170. Yu, B., X. Li, W. Zheng, Y. Feng, Y.-S. Wong and T. Chen. (2014). pH-responsive cancer-targeted selenium nanoparticles: A transformable drug carrier with enhanced theranostic effects. *Journal of Materials Chemistry B.* **2**, 5409–5418.
171. He, L., T. Wang, J. An, X. Li, L. Zhang, L. Li, G. Li, X. Wu, Z. Su and C. Wang. (2014). Carbon nanodots@ zeolitic imidazolate framework-8 nanoparticles for simultaneous pH-responsive drug delivery and fluorescence imaging. *CrystEngComm.* **16**, 3259–3263.
172. Rasouli, S., S. Davaran, F. Rasouli, M. Mahkam and R. Salehi. (2014). Synthesis, characterization and pH-controllable methotrexate release from biocompatible polymer/silica nanocomposite for anticancer drug delivery. *Drug Delivery.* **21**, 155–163.
173. Alvarez-Lorenzo, C. and A. Concheiro. (2014). Smart drug delivery systems: From fundamentals to the clinic. *Chemical Communications.* **50**, 7743–7765.
174. Alvarez-Lorenzo, C. and A. Concheiro. (2013). From drug dosage forms to intelligent drug-delivery systems: A change of paradigm. *Smart Materials for Drug Delivery.* **1**, 1.
175. Wike-Hooley, J., J. Haveman and H. Reinhold. (1984). The relevance of tumour pH to the treatment of malignant disease. *Radiotherapy and Oncology.* **2**, 343–366.
176. Vaupel, P., F. Kallinowski and P. Okunieff. (1989). Blood flow, oxygen and nutrient supply, and metabolic microenvironment of human tumors: A review. *Cancer Research.* **49**, 6449–6465.
177. Tian, L. and Y. H. Bae. (2012). Cancer nanomedicines targeting tumor extracellular pH. *Colloids and Surfaces B: Biointerfaces.* **99**, 116–126.
178. Hsu, P. P. and D. M. Sabatini. (2008). Cancer cell metabolism: Warburg and beyond. *Cell.* **134**, 703–707.
179. Vander Heiden, M. G., L. C. Cantley and C. B. Thompson. (2009). Understanding the Warburg effect: The metabolic requirements of cell proliferation. *Science.* **324**, 1029–1033.
180. Volk, T., E. Jähde, H. Fortmeyer, K. Glüsenkamp and M. Rajewsky. (1993). pH in human tumour xenografts: Effect of intravenous administration of glucose. *British Journal of Cancer.* **68**, 492.
181. Mellman, I., R. Fuchs and A. Helenius. (1986). Acidification of the endocytic and exocytic pathways. *Annual Review of Biochemistry.* **55**, 663–700.
182. Bae, Y., N. Nishiyama, S. Fukushima, H. Koyama, M. Yasuhiro and K. Kataoka. (2005). Preparation and biological characterization of polymeric micelle drug carriers with intracellular pH-triggered drug release property: Tumor permeability, controlled subcellular drug distribution, and enhanced in vivo antitumor efficacy. *Bioconjugate Chemistry.* **16**, 122–130.
183. Liu, J., Y. Huang, A. Kumar, A. Tan, S. Jin, A. Mozhi and X.-J. Liang. (2013). Ph-sensitive nano-systems for drug delivery in cancer therapy. *Biotechnology Advances.* **32**, 693–710.
184. Lv, Y., H. Huang, B. Yang, H. Liu, Y. Li and J. Wang. (2014). A robust pH-sensitive drug carrier: Aqueous micelles mineralized by calcium phosphate based on chitosan. *Carbohydrate Polymers.* **111**, 101–107.
185. Quader, S., H. Cabral, Y. Mochida, T. Ishii, X. Liu, K. Toh, H. Kinoh, Y. Miura, N. Nishiyama and K. Kataoka. (2014). Selective intracellular delivery of proteasome inhibitors through pH-sensitive polymeric micelles directed to efficient antitumor therapy. *Journal of Controlled Release.* **188**, 67–77.
186. Kamimura, M. and Y. Nagasaki. (2013). pH-sensitive polymeric micelles for enhanced intracellular anticancer drug delivery. *Journal of Photopolymer Science and Technology.* **26**, 161–164.
187. Wu, H., L. Zhu and V. P. Torchilin. (2013). pH-sensitive poly (histidine)-PEG/DSPE-PEG copolymer micelles for cytosolic drug delivery. *Biomaterials.* **34**, 1213–1222.
188. Kataoka, K., A. Harada and Y. Nagasaki. (2001). Block copolymer micelles for drug delivery: Design, characterization and biological significance. *Advanced Drug Delivery Reviews.* **47**, 113–131.
189. Hatakeyama, H., M. Murata, Y. Sato, M. Takahashi, N. Minakawa, A. Matsuda and H. Harashima. (2014). The systemic administration of an anti-miRNA oligonucleotide encapsulated pH-sensitive liposome results in reduced level of hepatic microRNA-122 in mice. *Journal of Controlled Release.* **173**, 43–50.
190. Ramasamy, T., Z. S. Haidar, T. H. Tran, J. Y. Choi, J.-H. Jeong, B. S. Shin, H.-G. Choi, C. S. Yong

and J. O. Kim. (2014). Layer-by-layer assembly of liposomal nanoparticles with PEGylated polyelectrolytes enhances systemic delivery of multiple anticancer drugs. *Acta Biomaterialia*. **10**, 5116–5127.

191. Yoshizaki, Y., E. Yuba, N. Sakaguchi, K. Koiwai, A. Harada and K. Kono. (2014). Potentiation of pH-sensitive polymer-modified liposomes with cationic lipid inclusion as antigen delivery carriers for cancer immunotherapy. *Biomaterials*. **35**, 8186–8196.
192. Watarai, S., T. Iwase, T. Tajima, E. Yuba and K. Kono. (2013). Efficiency of pH-sensitive fusogenic polymer-modified liposomes as a vaccine carrier. *The Scientific World Journal*. **2013**, 903234.
193. Xu, H., W. Zhang, Y. Li, F. Y. Fei, P. P. Yin, X. Yu, M. N. Hu, Y. S. Fu, C. Wang and D. J. Shang. (2014). The bifunctional liposomes constructed by poly (2-ethyl-oxazoline)-cholesteryl methyl carbonate: An effectual approach to enhance liposomal circulation time, pH-sensitivity and endosomal escape. *Pharmaceutical Research*. **31**, 3038–3050.
194. Torchilin, V. P. (2005). Recent advances with liposomes as pharmaceutical carriers. *Nature Reviews Drug Discovery*. **4**, 145–160.
195. Torchilin, V. P. (2007). Micellar nanocarriers: Pharmaceutical perspectives. *Pharmaceutical Research*. **24**, 1–16.
196. Sihorkar, V. and S. Vyas. (2001). Potential of polysaccharide anchored liposomes in drug delivery, targeting and immunization. *Journal of Pharmacy and Pharmaceutical Sciences*. **4**, 138–158.
197. Raemdonck, K., J. Demeester and S. De Smedt. (2009). Advanced nanogel engineering for drug delivery. *Soft Matter*. **5**, 707–715.
198. Soni, G. and K. S. Yadav. (2014). Nanogels as potential nanomedicine carrier for treatment of cancer: A mini review of the state of the art. *Saudi Pharmaceutical Journal*. **24**, 133–139.
199. Abandansari, H. S., M. R. Nabid, S. J. T. Rezaei and H. Niknejad. (2014). pH-sensitive nanogels based on Boltorn® H40 and poly (vinylpyridine) using mini-emulsion polymerization for delivery of hydrophobic anticancer drugs. *Polymer*. **55**, 3579–3590.
200. Chacko, R. T., J. Ventura, J. Zhuang and S. Thayumanavan. (2012). Polymer nanogels: A versatile nanoscopic drug delivery platform. *Advanced Drug Delivery Reviews*. **64**, 836–851.
201. Du, C., D. Deng, L. Shan, S. Wan, J. Cao, J. Tian, S. Achilefu and Y. Gu. (2013). A pH-sensitive doxorubicin prodrug based on folate-conjugated BSA for tumor-targeted drug delivery. *Biomaterials*. **34**, 3087–3097.
202. Lv, S., Z. Tang, D. Zhang, W. Song, M. Li, J. Lin, H. Liu and X. Chen. (2014). Well-defined polymer-drug conjugate engineered with redox and pH-sensitive release mechanism for efficient delivery of paclitaxel. *Journal of Controlled Release*. **194**, 220–227.
203. Rigogliuso, S., M. A. Sabatino, G. Adamo, N. Grimaldi, C. Dispenza and G. Ghersi. (2012). Polymeric nanogels: Nanocarriers for drug delivery application. *Chemical Engineering*. **27**, 247–252.
204. Xu, D., F. Wu, Y. Chen, L. Wei and W. Yuan. (2013). pH-sensitive degradable nanoparticles for highly efficient intracellular delivery of exogenous protein. *International Journal of Nanomedicine*. **8**, 3405.
205. Soppimath, K. S., T. M. Aminabhavi, A. R. Kulkarni and W. E. Rudzinski. (2001). Biodegradable polymeric nanoparticles as drug delivery devices. *Journal of Controlled Release*. **70**, 1–20.
206. Liu, R., D. Li, B. He, X. Xu, M. Sheng, Y. Lai, G. Wang and Z. Gu. (2011). Anti-tumor drug delivery of pH-sensitive poly (ethylene glycol)-poly (L-histidine-)-poly (L-lactide) nanoparticles. *Journal of Controlled Release*. **152**, 49–56.
207. Ma, L., M. Liu, H. Liu, J. Chen and D. Cui. (2010). In vitro cytotoxicity and drug release properties of pH-and temperature-sensitive core–shell hydrogel microspheres. *International Journal of Pharmaceutics*. **385**, 86–91.
208. Mukhopadhyay, P., S. Chakraborty, S. Bhattacharya, R. Mishra and P. Kundu. (2015). pH-sensitive chitosan/alginate core-shell nanoparticles for efficient and safe oral insulin delivery. *International Journal of Biological Macromolecules*. **72**, 640–648.
209. Yuan, L., Q. Tang, D. Yang, J. Z. Zhang, F. Zhang and J. Hu. (2011). Preparation of pH-responsive mesoporous silica nanoparticles and their application in controlled drug delivery. *The Journal of Physical Chemistry C*. **115**, 9926–9932.
210. Zheng, Q., T. Lin, H. Wu, L. Guo, P. Ye, Y. Hao, Q. Guo, J. Jiang, F. Fu and G. Chen. (2014). Mussel-inspired polydopamine coated mesoporous silica nanoparticles as pH-sensitive nanocarriers for controlled release. *International Journal of Pharmaceutics*. **463**, 22–26.
211. Yang, S., D. Chen, N. Li, X. Mei, X. Qi, H. Li, Q. Xu and J. Lu. (2012). A facile preparation of targetable pH-sensitive polymeric nanocarriers with encapsulated magnetic nanoparticles for controlled drug release. *Journal of Materials Chemistry*. **22**, 25354–25361.
212. Kim, S. and C. B. Park. (2010). Mussel-inspired transformation of $CaCO_3$ to bone minerals. *Biomaterials*. **31**, 6628–6634.
213. Wei, W., G.-H. Ma, G. Hu, D. Yu, T. Mcleish, Z.-G. Su and Z.-Y. Shen. (2008). Preparation of hierarchical hollow $CaCO_3$ particles and the application as anticancer drug carrier. *Journal of the American Chemical Society*. **130**, 15808–15810.
214. Deng, Z., Z. Zhen, X. Hu, S. Wu, Z. Xu and P. K. Chu. (2011). Hollow chitosan–silica nanospheres

as pH-sensitive targeted delivery carriers in breast cancer therapy. *Biomaterials.* **32**, 4976–4986.

215. Sonaje, K., K. J. Lin, J. J. Wang, F. L. Mi, C. T. Chen, J. H. Juang and H. W. Sung. (2010). Self-assembled pH-sensitive nanoparticles: A platform for oral delivery of protein drugs. *Advanced Functional Materials.* **20**, 3695–3700.
216. Boppana, R., G. K. Mohan, U. Nayak, S. Mutalik, B. Sa and R. V. Kulkarni. (2015). Novel pH-sensitive IPNs of polyacrylamide-g-gum ghatti and sodium alginate for gastro-protective drug delivery. *International Journal of Biological Macromolecules.* **75**, 133–143.
217. Rao, K. M., K. K. Rao, G. Ramanjaneyulu and C.-S. Ha. (2015). Curcumin encapsulated pH sensitive gelatin based interpenetrating polymeric network nanogels for anti cancer drug delivery. *International Journal of Pharmaceutics.* **478**, 788–795.
218. Zhang, H., H. Wu, L. Fan, F. Li, C. h. Gu and M. Jia. (2009). Preparation and characteristics of pH-sensitive derivated dextran hydrogel nanoparticles. *Polymer Composites.* **30**, 1243–1250.
219. Yoo, J.-Y., S.-Y. Kim, J.-A. Hwang, S.-H. Hong, A. Shin, I. J. Choi and Y.-S. Lee. (2012). Association study between folate pathway gene single nucleotide polymorphisms and gastric cancer in Koreans. *Genomics & Informatics.* **10**, 184–193.
220. Kamada, H., Y. Tsutsumi, Y. Yoshioka, Y. Yamamoto, H. Kodaira, S.-I. Tsunoda, T. Okamoto, Y. Mukai, H. Shibata and S. Nakagawa. (2004). Design of a pH-sensitive polymeric carrier for drug release and its application in cancer therapy. *Clinical Cancer Research.* **10**, 2545–2550.
221. Na, K., E. S. Lee and Y. H. Bae. (2003). Adriamycin loaded pullulan acetate/sulfonamide conjugate nanoparticles responding to tumor pH: pH-dependent cell interaction, internalization and cytotoxicity in vitro. *Journal of Controlled Release.* **87**, 3–13.
222. Stayton, P., M. El-Sayed, N. Murthy, V. Bulmus, C. Lackey, C. Cheung and A. Hoffman. (2005). 'Smart'delivery systems for biomolecular therapeutics. *Orthodontics & Craniofacial Research.* **8**, 219–225.
223. Ge, Y., S. Li, S. Wang and R. Moore. (2014) *Nanomedicine: Principles and Perspectives*, Springer, New York.
224. Gillies, E. R., T. B. Jonsson and J. M. Fréchet. (2004). Stimuli-responsive supramolecular assemblies of linear-dendritic copolymers. *Journal of the American Chemical Society.* **126**, 11936–11943.
225. Ganta, S., M. Talekar, A. Singh, T. P. Coleman and M. M. Amiji. (2014). Nanoemulsions in translational research—opportunities and challenges in targeted cancer therapy. *AAPS PharmSciTech.* **15**, 694–708.
226. Jones, R., C. Cheung, F. Black, J. Zia, P. Stayton, A. Hoffman and M. Wilson. (2003). Poly (2-alkylacrylic acid) polymers deliver molecules to the cytosol by pH-sensitive disruption of endosomal vesicles. *Biochemical Journal.* **372**, 65–75.
227. Thomas, J. L., S. W. Barton and D. A. Tirrell. (1994). Membrane solubilization by a hydrophobic polyelectrolyte: Surface activity and membrane binding. *Biophysical Journal.* **67**, 1101–1106.
228. Thomas, J. L. and D. A. Tirrell. (1992). Polyelectrolyte-sensitized phospholipid vesicles. *Accounts of Chemical Research.* **25**, 336–342.
229. Foster, S., C. L. Duvall, E. F. Crownover, A. S. Hoffman and P. S. Stayton. (2010). Intracellular delivery of a protein antigen with an endosomal-releasing polymer enhances CD8 T-cell production and prophylactic vaccine efficacy. *Bioconjugate Chemistry.* **21**, 2205–2212.
230. Pietersz, G. A., C.-K. Tang and V. Apostolopoulos. (2006). Structure and design of polycationic carriers for gene delivery. *Mini Reviews in Medicinal Chemistry.* **6**, 1285–1298.
231. Kitazoe, M., J. Futami, M. Nishikawa, H. Yamada and Y. Maeda. (2010). Polyethylenimine-cationized β-catenin protein transduction activates the Wnt canonical signaling pathway more effectively than cationic lipid-based transduction. *Biotechnology Journal.* **5**, 385–392.
232. Futami, J., M. Kitazoe, T. Maeda, E. Nukui, M. Sakaguchi, J. Kosaka, M. Miyazaki, M. Kosaka, H. Tada and M. Seno. (2005). Intracellular delivery of proteins into mammalian living cells by polyethylenimine-cationization. *Journal of Bioscience and Bioengineering.* **99**, 95–103.
233. Zauner, W., M. Ogris and E. Wagner. (1998). Polylysine-based transfection systems utilizing receptor-mediated delivery. *Advanced Drug Delivery Reviews.* **30**, 97–113.
234. Boussif, O., M. A. Zanta and J.-P. Behr. (1996). Optimized galenics improve in vitro gene transfer with cationic molecules up to 1000-fold. *Gene Therapy.* **3**, 1074–1080.
235. Kichler, A., C. Leborgne, E. Coeytaux and O. Danos. (2001). Polyethylenimine-mediated gene delivery: A mechanistic study. *The Journal of Gene Medicine.* **3**, 135–144.
236. Oh, Y., D. Suh, J. Kim, H. Choi, K. Shin and J. Ko. (2002). Polyethylenimine-mediated cellular uptake, nucleus trafficking and expression of cytokine plasmid DNA. *Gene Therapy.* **9**, 1627–1632.
237. Segura, T. and J. A. Hubbell. (2007). Synthesis and in vitro characterization of an ABC triblock copolymer for siRNA delivery. *Bioconjugate Chemistry.* **18**, 736–745.
238. Park, S.-C., J.-P. Nam, Y.-M. Kim, J.-H. Kim, J.-W. Nah and M.-K. Jang. (2013). Branched polyethylenimine-grafted-carboxymethyl chitosan copolymer enhances the delivery of pDNA or siRNA

in vitro and in vivo. *International Journal of Nanomedicine.* **8**, 3663.
239. Meng, F., Y. Zhong, R. Cheng, C. Deng and Z. Zhong. (2014). pH-sensitive polymeric nanoparticles for tumor-targeting doxorubicin delivery: Concept and recent advances. *Nanomedicine.* **9**, 487–499.
240. Chen, M.-C., F.-L. Mi, Z.-X. Liao, C.-W. Hsiao, K. Sonaje, M.-F. Chung, L.-W. Hsu and H.-W. Sung. (2013). Recent advances in chitosan-based nanoparticles for oral delivery of macromolecules. *Advanced Drug Delivery Reviews.* **65**, 865–879.
241. Bugamelli, F., M. A. Raggi, I. Orienti and V. Zecchi. (1998). Controlled insulin release from chitosan microparticles. *Archiv der Pharmazie.* **331**, 133–138.
242. Jayakumar, R., R. Reis and J. Mano. (2006). Phosphorous containing chitosan beads for controlled oral drug delivery. *Journal of Bioactive and Compatible Polymers.* **21**, 327–340.
243. Jana, S., N. Maji, A. K. Nayak, K. K. Sen and S. K. Basu. (2013). Development of chitosan-based nanoparticles through inter-polymeric complexation for oral drug delivery. *Carbohydrate Polymers.* **98**, 870–876.
244. Čalija, B., N. Cekić, S. Savić, R. Daniels, B. Marković and J. Milić. (2013). pH-sensitive microparticles for oral drug delivery based on alginate/oligochitosan/Eudragit® L100-55 "sandwich" polyelectrolyte complex. *Colloids and Surfaces B: Biointerfaces.* **110**, 395–402.
245. Yang, Y., S. Wang, Y. Wang, X. Wang, Q. Wang and M. Chen. (2014). Advances in self-assembled chitosan nanomaterials for drug delivery. *Biotechnology Advances.* **32**, 1301–1316.
246. Nogueira, D. R., L. Tavano, M. Mitjans, L. Pérez, M. R. Infante and M. P. Vinardell. (2013). *In vitro* antitumor activity of methotrexate via pH-sensitive chitosan nanoparticles. *Biomaterials.* **34**, 2758–2772.
247. Cavallaro, G., M. Licciardi, G. Amato, C. Sardo, G. Giammona, R. Farra, B. Dapas, M. Grassi and G. Grassi. (2014). Synthesis and characterization of polyaspartamide copolymers obtained by ATRP for nucleic acid delivery. *International Journal of Pharmaceutics.* **466**, 246–257.
248. Wang, J., C. Ni, Y. Zhang, M. Zhang, W. Li, B. Yao and L. Zhang. (2014). Preparation and pH controlled release of polyelectrolyte complex of poly (l-malic acid-co-d, l-lactic acid) and chitosan. *Colloids and Surfaces B: Biointerfaces.* **115**, 275–279.
249. Lanz-Landázuri, A., A. Martínez de Ilarduya, M. García-Alvarez and S. Muñoz-Guerra. (2014). Poly (β, L-malic acid)/Doxorubicin ionic complex: A pH-dependent delivery system. *Reactive and Functional Polymers.* **81**, 45–53.
250. Wang, M., H. Hu, Y. Sun, L. Qiu, J. Zhang, G. Guan, X. Zhao, M. Qiao, L. Cheng and L. Cheng. (2013). A pH-sensitive gene delivery system based on folic acid-PEG-chitosan–PAMAM-plasmid DNA complexes for cancer cell targeting. *Biomaterials.* **34**, 10120–10132.
251. Wen, H., J. Guo, B. Chang and W. Yang. (2013). pH-responsive composite microspheres based on magnetic mesoporous silica nanoparticle for drug delivery. *European Journal of Pharmaceutics and Biopharmaceutics.* **84**, 91–98.
252. Leung, K. C.-F., T. D. Nguyen, J. F. Stoddart and J. I. Zink. (2006). Supramolecular nanovalves controlled by proton abstraction and competitive binding. *Chemistry of Materials.* **18**, 5919–5928.
253. Nguyen, T. D., K. C.-F. Leung, M. Liong, C. D. Pentecost, J. F. Stoddart and J. I. Zink. (2006). Construction of a pH-driven supramolecular nanovalve. *Organic Letters.* **8**, 3363–3366.
254. Roik, N. and L. Belyakova. (2014). Chemical design of pH-sensitive nanovalves on the outer surface of mesoporous silicas for controlled storage and release of aromatic amino acid. *Journal of Solid State Chemistry.* **215**, 284–291.
255. Chuan, X., Q. Song, J. Lin, X. Chen, H. Zhang, W. Dai, B. He, X. Wang and Q. Zhang. (2014). Novel free-paclitaxel-loaded redox-responsive nanoparticles based on a disulfide-linked poly (ethylene glycol)-drug conjugate for intracellular drug delivery: Synthesis, characterization, and antitumor activity in vitro and in vivo. *Molecular Pharmaceutics.* **11**, 3656–3670.
256. Phillips, D. J. and M. I. Gibson. (2013). Redox-sensitive materials for drug delivery: Targeting the correct intracellular environment, tuning release rates, and appropriate predictive systems. *Antioxidants & Redox Signaling.* **21**, 786–803
257. Zhang, J., F. Yang, H. Shen and D. Wu. (2012). Controlled formation of microgels/nanogels from a disulfide-linked core/shell hyperbranched polymer. *ACS Macro Letters.* **1**, 1295–1299.
258. Yuan, W., H. Zou, W. Guo, T. Shen and J. Ren. (2013). Supramolecular micelles with dual temperature and redox responses for multi-controlled drug release. *Polymer Chemistry.* **4**, 2658–2661.
259. Thambi, T., V. Deepagan, H. Ko, D. S. Lee and J. H. Park. (2012). Bioreducible polymersomes for intracellular dual-drug delivery. *Journal of Materials Chemistry.* **22**, 22028–22036.
260. Yu, Z.-Q., J.-T. Sun, C.-Y. Pan and C.-Y. Hong. (2012). Bioreducible nanogels/microgels easily prepared via temperature induced self-assembly and self-crosslinking. *Chemical Communications.* **48**, 5623–5625.
261. Yan, Y., Y. Wang, J. K. Heath, E. C. Nice and F. Caruso. (2011). Cellular association and cargo release of redox-responsive polymer capsules

mediated by exofacial thiols. *Advanced Materials.* **23**, 3916–3921.

262. Aleksanian, S., B. Khorsand, R. Schmidt and J. K. Oh. (2012). Rapidly thiol-responsive degradable block copolymer nanocarriers with facile bioconjugation. *Polymer Chemistry.* **3**, 2138–2147.
263. Cheng, R., F. Feng, F. Meng, C. Deng, J. Feijen and Z. Zhong. (2011). Glutathione-responsive nano-vehicles as a promising platform for targeted intracellular drug and gene delivery. *Journal of Controlled Release.* **152**, 2–12.
264. Gamcsik, M. P., M. S. Kasibhatla, S. D. Teeter and O. M. Colvin. (2012). Glutathione levels in human tumors. *Biomarkers.* **17**, 671–691.
265. Sultana, F., M. Manirujjaman, M. A. Imran-Ul-Haque and S. Sharmin. (2013). An overview of nanogel drug delivery system. *Journal of Applied Pharmaceutical Science.* **3**, S95–S105.
266. Kabanov, A. V. and S. V. Vinogradov. (2009). Nanogels as pharmaceutical carriers: Finite networks of infinite capabilities. *Angewandte Chemie International Edition.* **48**, 5418–5429.
267. Oh, J. K., R. Drumright, D. J. Siegwart and K. Matyjaszewski. (2008). The development of microgels/nanogels for drug delivery applications. *Progress in Polymer Science.* **33**, 448–477.
268. Averick, S. E., E. Paredes, A. Irastorza, A. R. Shrivats, A. Srinivasan, D. J. Siegwart, A. J. Magenau, H. Y. Cho, E. Hsu, A. A. Averick, J. Kim, S. Liu, J. O. Hollinger, S. R. Das and K. Matyjaszewski. (2012). Preparation of cationic nanogels for nucleic acid delivery. *Biomacromolecules.* **13**, 3445–3449.
269. Oh, J. K., D. J. Siegwart, H.-I. Lee, G. Sherwood, L. Peteanu, J. O. Hollinger, K. Kataoka and K. Matyjaszewski. (2007). Biodegradable nanogels prepared by atom transfer radical polymerization as potential drug delivery carriers: Synthesis, biodegradation, in vitro release, and bioconjugation. *Journal of the American Chemical Society.* **129**, 5939–5945.
270. Oh, J. K., D. J. Siegwart and K. Matyjaszewski. (2007). Synthesis and biodegradation of nanogels as delivery carriers for carbohydrate drugs. *Biomacromolecules.* **8**, 3326–3331.
271. Yap, H. P., A. P. Johnston, G. K. Such, Y. Yan and F. Caruso. (2009). Click-Engineered, Bioresponsive, Drug-Loaded PEG Spheres. *Advanced Materials.* **21**, 4348–4352.
272. Groll, J., S. Singh, K. Albrecht and M. Moeller. (2009). Biocompatible and degradable nanogels via oxidation reactions of synthetic thiomers in inverse miniemulsion. *Journal of Polymer Science Part A: Polymer Chemistry.* **47**, 5543–5549.
273. Cerritelli, S., D. Velluto and J. A. Hubbell. (2007). PEG-SS-PPS: Reduction-sensitive disulfide block copolymer vesicles for intracellular drug delivery. *Biomacromolecules.* **8**, 1966–1972.
274. Napoli, A., M. Valentini, N. Tirelli, M. Müller and J. A. Hubbell. (2004). Oxidation-responsive polymeric vesicles. *Nature Materials.* **3**, 183–189.
275. Napoli, A., M. J. Boerakker, N. Tirelli, R. J. Nolte, N. A. Sommerdijk and J. A. Hubbell. (2004). Glucose-oxidase based self-destructing polymeric vesicles. *Langmuir.* **20**, 3487–3491.
276. Zhang, H., F. Xin, W. An, A. Hao, X. Wang, X. Zhao, Z. Liu and L. Sun. (2010). Oxidizing-responsive vesicles made from "tadpole-like supramolecular amphiphiles" based on inclusion complexes between driving molecules and β-cyclodextrin. *Colloids and Surfaces A: Physicochemical and Engineering Aspects.* **363**, 78–85.
277. Zelikin, A. N., Q. Li and F. Caruso. (2008). Disulfide-stabilized poly (methacrylic acid) capsules: Formation, cross-linking, and degradation behavior. *Chemistry of Materials.* **20**, 2655–2661.
278. Zelikin, A. N., J. F. Quinn and F. Caruso. (2006). Disulfide cross-linked polymer capsules: En route to biodeconstructible systems. *Biomacromolecules.* **7**, 27–30.
279. Kim, E., D. Kim, H. Jung, J. Lee, S. Paul, N. Selvapalam, Y. Yang, N. Lim, C. G. Park and K. Kim. (2010). Facile, template-free synthesis of stimuli-responsive polymer nanocapsules for targeted drug delivery. *Angewandte Chemie.* **122**, 4507–4510.
280. Shu, S., X. Zhang, Z. Wu, Z. Wang and C. Li. (2010). Gradient cross-linked biodegradable polyelectrolyte nanocapsules for intracellular protein drug delivery. *Biomaterials.* **31**, 6039–6049.
281. Li, Y., B. S. Lokitz, S. P. Armes and C. L. McCormick. (2006). Synthesis of reversible shell cross-linked micelles for controlled release of bioactive agents. *Macromolecules.* **39**, 2726–2728.
282. Kakizawa, Y., A. Harada and K. Kataoka. (2001). Glutathione-sensitive stabilization of block copolymer micelles composed of antisense DNA and thiolated poly (ethylene glycol)-b lock-poly (l-lysine): A potential carrier for systemic delivery of antisense DNA. *Biomacromolecules.* **2**, 491–497.
283. Kakizawa, Y., A. Harada and K. Kataoka. (1999). Environment-sensitive stabilization of core-shell structured polyion complex micelle by reversible cross-linking of the core through disulfide bond. *Journal of the American Chemical Society.* **121**, 11247–11248.
284. Ghosh, S., S. Basu and S. Thayumanavan. (2006). Simultaneous and reversible functionalization of copolymers for biological applications. *Macromolecules.* **39**, 5595–5597.
285. Takeoka, Y., T. Aoki, K. Sanui, N. Ogata, M. Yokoyama, T. Okano, Y. Sakurai and M. Watanabe. (1995). Electrochemical control of drug release from redox-active micelles. *Journal of Controlled Release.* **33**, 79–87.

286. Wang, Y.-C., F. Wang, T.-M. Sun and J. Wang. (2011). Redox-responsive nanoparticles from the single disulfide bond-bridged block copolymer as drug carriers for overcoming multidrug resistance in cancer cells. *Bioconjugate Chemistry*. **22**, 1939–1945.
287. Tang, L.-Y., Y.-C. Wang, Y. Li, J.-Z. Du and J. Wang. (2009). Shell-detachable micelles based on disulfide-linked block copolymer as potential carrier for intracellular drug delivery. *Bioconjugate Chemistry*. **20**, 1095–1099.
288. Thambi, T., H. Y. Yoon, K. Kim, I. C. Kwon, C. K. Yoo and J. H. Park. (2011). Bioreducible block copolymers based on poly (ethylene glycol) and poly (γ-benzyl L-glutamate) for intracellular delivery of camptothecin. *Bioconjugate Chemistry*. **22**, 1924–1931.
289. Yuan, Y., X. Zhang, X. Zeng, B. Liu, F. Hu and G. Zhang. (2014). Glutathione-mediated release of functional MiR-122 from gold nanoparticles for targeted induction of apoptosis in cancer treatment. *Journal of Nanoscience and Nanotechnology*. **14**, 5620–5627.
290. Botella, P., C. Muniesa, V. Vicente, K. Fabregat and A. Cabrera. (2014). Controlled Intracellular Release of Camptothecin by Glutathione-Driven Mechanism. TechConnect Briefs. *Nanotechnology*. **2**, 347–350.
291. Kim, J. O., G. Sahay, A. V. Kabanov and T. K. Bronich. (2010). Polymeric micelles with ionic cores containing biodegradable cross-links for delivery of chemotherapeutic agents. *Biomacromolecules*. **11**, 919–926.
292. Zhang, L., W. Liu, L. Lin, D. Chen and M. H. Stenzel. (2008). Degradable disulfide core-cross-linked micelles as a drug delivery system prepared from vinyl functionalized nucleosides via the RAFT process. *Biomacromolecules*. **9**, 3321–3331.
293. Zhang, J., X. Jiang, Y. Zhang, Y. Li and S. Liu. (2007). Facile fabrication of reversible core cross-linked micelles possessing thermosensitive swellability. *Macromolecules*. **40**, 9125–9132.
294. Hu, J., G. Zhang and S. Liu. (2012). Enzyme-responsive polymeric assemblies, nanoparticles and hydrogels. *Chemical Society Reviews*. **41**, 5933–5949.
295. de la Rica, R., D. Aili and M. M. Stevens. (2012). Enzyme-responsive nanoparticles for drug release and diagnostics. *Advanced Drug Delivery Reviews*. **64**, 967–978.
296. Hoshino, K., M. Taniguchi, H. Ueoka, M. Ohkuwa, C. Chida, S. Morohashi and T. Sasakura. (1996). Repeated utilization of β-glucosidase immobilized on a reversibly soluble-insoluble polymer for hydrolysis of phloridzin as a model reaction producing a water-insoluble product. *Journal of Fermentation and Bioengineering*. **82**, 253–258.
297. Hu, Q., P. S. Katti and Z. Gu. (2014). Enzyme-responsive nanomaterials for controlled drug delivery. *Nanoscale*. **6**, 12273–12286.
298. Sperinde, J. J. and L. G. Griffith. (1997). Synthesis and characterization of enzymatically-cross-linked poly (ethylene glycol) hydrogels. *Macromolecules*. **30**, 5255–5264.
299. Wichterle, O. and D. Lim. (1960). Hydrophilic gels for biological use. *Nature*. **185**, 117–118.
300. Aguilar, M. R. and J. Román. (2014) *Smart Polymers and Their Applications*, Elsevier, Cambridge.
301. Thornton, P. D., R. J. Mart, S. J. Webb and R. V. Ulijn. (2008). Enzyme-responsive hydrogel particles for the controlled release of proteins: Designing peptide actuators to match payload. *Soft Matter*. **4**, 821–827.
302. Patrick, A. G. and R. V. Ulijn. (2010). Hydrogels for the detection and management of protease levels. *Macromolecular Bioscience*. **10**, 1184–1193.
303. McDonald, T. O., H. Qu, B. R. Saunders and R. V. Ulijn. (2009). Branched peptide actuators for enzyme responsive hydrogel particles. *Soft Matter*. **5**, 1728–1734.
304. Su, T., Z. Tang, H. He, W. Li, X. Wang, C. Liao, Y. Sun and Q. Wang. (2014). Glucose oxidase triggers gelation of N-hydroxyimide–heparin conjugates to form enzyme-responsive hydrogels for cell-specific drug delivery. *Chemical Science*. **5**, 4204–4209.
305. van Dijk, M., C. F. van Nostrum, W. E. Hennink, D. T. Rijkers and R. M. Liskamp. (2010). Synthesis and characterization of enzymatically biodegradable PEG and peptide-based hydrogels prepared by click chemistry. *Biomacromolecules*. **11**, 1608–1614.
306. Yang, J., M. T. Jacobsen, H. Pan and J. Kopeček. (2010). Synthesis and characterization of enzymatically degradable PEG-based peptide-containing hydrogels. *Macromolecular Bioscience*. **10**, 445–454.
307. Lutolf, M. and J. Hubbell. (2003). Synthesis and physicochemical characterization of end-linked poly (ethylene glycol)-co-peptide hydrogels formed by Michael-type addition. *Biomacromolecules*. **4**, 713–722.
308. Kim, S. and K. E. Healy. (2003). Synthesis and characterization of injectable poly (N-isopropylacrylamide-co-acrylic acid) hydrogels with proteolytically degradable cross-links. *Biomacromolecules*. **4**, 1214–1223.
309. Klinger, D., E. M. Aschenbrenner, C. K. Weiss and K. Landfester. (2012). Enzymatically degradable nanogels by inverse miniemulsion copolymerization of acrylamide with dextran methacrylates as crosslinkers. *Polymer Chemistry*. **3**, 204–216.
310. Lee, M. R., K. H. Baek, H. J. Jin, Y. G. Jung and I. Shin. (2004). Targeted enzyme-responsive drug carriers: Studies on the delivery of a combination of

drugs. *Angewandte Chemie International Edition.* **43**, 1675–1678.

311. Gu, G., H. Xia, Q. Hu, Z. Liu, M. Jiang, T. Kang, D. Miao, Y. Tu, Z. Pang and Q. Song. (2013). PEG-co-PCL nanoparticles modified with MMP-2/9 activatable low molecular weight protamine for enhanced targeted glioblastoma therapy. *Biomaterials.* **34**, 196–208.
312. Petros, R. A. and J. M. DeSimone. (2010). Strategies in the design of nanoparticles for therapeutic applications. *Nature Reviews Drug Discovery.* **9**, 615–627.
313. Davis, M. E. (2008). Nanoparticle therapeutics: An emerging treatment modality for cancer. *Nature Reviews Drug Discovery.* **7**, 771–782.
314. Gu, Z. and Y. Tang. (2010). Enzyme-assisted photolithography for spatial functionalization of hydrogels. *Lab on a Chip.* **10**, 1946–1951.
315. Gu, Z., A. Biswas, K.-I. Joo, B. Hu, P. Wang and Y. Tang. (2010). Probing protease activity by single-fluorescent-protein nanocapsules. *Chemical Communications.* **46**, 6467–6469.
316. Gu, Z., M. Yan, B. Hu, K.-I. Joo, A. Biswas, Y. Huang, Y. Lu, P. Wang and Y. Tang. (2009). Protein nanocapsule weaved with enzymatically degradable polymeric network. *Nano Letters.* **9**, 4533–4538.
317. Chien, M.-P., A. S. Carlini, D. Hu, C. V. Barback, A. M. Rush, D. J. Hall, G. Orr and N. C. Gianneschi. (2013). Enzyme-directed assembly of nanoparticles in tumors monitored by in vivo whole animal imaging and ex vivo super-resolution fluorescence imaging. *Journal of the American Chemical Society.* **135**, 18710–18713.
318. Wang, J. (2008). Electrochemical glucose biosensors. *Chemical Reviews.* **108**, 814–825.
319. Jin, R., C. Hiemstra, Z. Zhong and J. Feijen. (2007). Enzyme-mediated fast in situ formation of hydrogels from dextran–tyramine conjugates. *Biomaterials.* **28**, 2791–2800.
320. Brown, L. R., E. R. Edelman, F. Fischel-Ghodsian and R. Langer. (1996). Characterization of glucose-mediated insulin release from implantable polymers. *Journal of Pharmaceutical Sciences.* **85**, 1341–1345.
321. Chandy, T. and C. P. Sharma. (1992). Glucose-responsive insulin release from poly (vinyl alcohol)-blended polyacrylamide membranes containing glucose oxidase. *Journal of Applied Polymer Science.* **46**, 1159–1167.
322. Kikuchi, A. and T. Okano. (2002). Pulsatile drug release control using hydrogels. *Advanced Drug Delivery Reviews.* **54**, 53–77.
323. Wu, W. and S. Zhou. (2013). Responsive materials for self-regulated insulin delivery. *Macromolecular Bioscience.* **13**, 1464–1477.
324. Imanishi, Y. and Y. Ito. (1995). Glucose-sensitive insulin-releasing molecular systems. *Pure and Applied Chemistry.* **67**, 2015–2022.
325. Kai, G. and Y. Min. (2001). AAc photografted porous polycabonate films and its controlled release system. *Journal of Controlled Release.* **71**, 221–225.
326. Tang, M., R. Zhang, A. Bowyer, R. Eisenthal and J. Hubble. (2003). A reversible hydrogel membrane for controlling the delivery of macromolecules. *Biotechnology and Bioengineering.* **82**, 47–53.
327. Podual, K., F. J. Doyle III and N. A. Peppas. (2000). Glucose-sensitivity of glucose oxidase-containing cationic copolymer hydrogels having poly (ethylene glycol) grafts. *Journal of Controlled Release.* **67**, 9–17.
328. Miyata, T., T. Uragami and K. Nakamae. (2002). Biomolecule-sensitive hydrogels. *Advanced Drug Delivery Reviews.* **54**, 79–98.
329. Radhakrishnan, K., J. Tripathy and A. M. Raichur. (2013). Dual enzyme responsive microcapsules simulating an "OR" logic gate for biologically triggered drug delivery applications. *Chemical Communications.* **49**, 5390–5392.
330. Wells, A. and J. R. Grandis. (2003). Phospholipase C-γ1 in tumor progression. *Clinical & Experimental Metastasis.* **20**, 285–290.
331. Ferreira, A. V. F. (2013). Incorporation of elastase inhibitor in silk fibroin nanoparticles for transdermal delivery.
332. López-Otín, C. and J. S. Bond. (2008). Proteases: Multifunctional enzymes in life and disease. *Journal of Biological Chemistry.* **283**, 30433–30437.
333. López-Otín, C. and T. Hunter. (2010). The regulatory crosstalk between kinases and proteases in cancer. *Nature Reviews Cancer.* **10**, 278–292.
334. Basel, M. T., T. B. Shrestha, D. L. Troyer and S. H. Bossmann. (2011). Protease-sensitive, polymer-caged liposomes: A method for making highly targeted liposomes using triggered release. *ACS Nano.* **5**, 2162–2175.
335. Vicent, M. J., F. Greco, R. I. Nicholson, A. Paul, P. C. Griffiths and R. Duncan. (2005). Polymer therapeutics designed for a combination therapy of hormone-dependent cancer. *Angewandte Chemie.* **117**, 4129–4134.
336. Yoon, S.-O., S.-J. Park, C.-H. Yun and A.-S. Chung. (2003). Roles of matrix metalloproteinases in tumor metastasis and angiogenesis. *Journal of Biochemistry and Molecular Biology.* **36**, 128–137.
337. Crawford, H. and L. Matrisian. (1993). Tumor and stromal expression of matrix metalloproteinases and their role in tumor progression. *Invasion & Metastasis.* **14**, 234–245.
338. Olson, E. S., T. Jiang, T. A. Aguilera, Q. T. Nguyen, L. G. Ellies, M. Scadeng and R. Y. Tsien. (2010). Activatable cell penetrating peptides linked to nanoparticles as dual probes for in vivo fluorescence and MR imaging of proteases. *Proceedings of the National Academy of Sciences.* **107**, 4311–4316.

339. Jiang, T., E. S. Olson, Q. T. Nguyen, M. Roy, P. A. Jennings and R. Y. Tsien. (2004). Tumor imaging by means of proteolytic activation of cell-penetrating peptides. *Proceedings of the National Academy of Sciences of the United States of America.* **101**, 17867–17872.
340. Radhakrishnan, K., J. Tripathy, D. P. Gnanadhas, D. Chakravortty and A. M. Raichur. (2014). Dual enzyme responsive and targeted nanocapsules for intracellular delivery of anticancer agents. *RSC Advances.* **4**, 45961–45968.
341. Xu, F., W.-H. Wang, Y.-J. Tan and M. L. Bruening. (2010). Facile trypsin immobilization in polymeric membranes for rapid, efficient protein digestion. *Analytical chemistry.* **82**, 10045–10051.
342. Krug, H. F. (2014). Nanosafety research—Are we on the right track? *Angewandte Chemie International Edition.* **53**, 12304–12319.
343. Wang, F., L. Shu, J. Wang, X. Pan, R. Huang, Y. Lin and X. Cai. (2013). Perspectives on the toxicology of cadmium-based quantum dots. *Current Drug Metabolism.* **14**, 847–856.
344. Ghaderi, S., B. Ramesh and A. M. Seifalian. (2011). Fluorescence nanoparticles "quantum dots" as drug delivery system and their toxicity: A review. *Journal of Drug Targeting.* **19**, 475–486.
345. Zhang, Y., D. Petibone, Y. Xu, M. Mahmood, A. Karmakar, D. Casciano, S. Ali and A. S. Biris. (2014). Toxicity and efficacy of carbon nanotubes and graphene: The utility of carbon-based nanoparticles in nanomedicine. *Drug Metabolism Reviews.* **46**, 232–246.
346. Tonelli, F. M., V. A. Goulart, K. N. Gomes, M. S. Ladeira, A. K. Santos, E. Lorençon, L. O. Ladeira and R. R. Resende. (2015). Graphene-based nanomaterials: Biological and medical applications and toxicity. *Nanomedicine.* **10**, 1–28.
347. Taylor, U., C. Rehbock, C. Streich, D. Rath and S. Barcikowski. (2014). Rational design of gold nanoparticle toxicology assays: A question of exposure scenario, dose and experimental setup. *Nanomedicine.* **9**, 1971–1989.
348. De Jong, W. H. and P. J. Borm. (2008). Drug delivery and nanoparticles: Applications and hazards. *International Journal of Nanomedicine.* **3**, 133.
349. Doktorovova, S., E. B. Souto and A. M. Silva. (2014). Nanotoxicology applied to solid lipid nanoparticles and nanostructured lipid carriers–a systematic review of in vitro data. *European Journal of Pharmaceutics and Biopharmaceutics.* **87**, 1–18.

5

Nanomaterials for the Delivery of Therapeutic Nucleic Acids

Michael Riley II and Wilfred Vermerris
University of Florida

5.1 Introduction

Nanomedicine is the application of nanotechnology in medicine for diagnostic or therapeutic purposes. Nanomedicine-based systems take advantage of the small size and tunable properties of nanomaterials that enable smart delivery, i.e. specific delivery to target cells or tissues, and detection. Therapeutic applications of nanomedicine with great promise include the treatment of cancer and genetic diseases based on the potential to deliver therapeutic nucleic acids (DNA or RNA) to specific targets (tissues or cells) within the patient's body.

In the case of cancer, the goal is to target tumor cells with nano-scale delivery vehicles carrying DNA or RNA that will interfere with tumor cell's metabolism and ultimately cause the tumor to die. In the case of gene therapy, the goal is to provide a patient who has a genetic disorder or age-induced deficiency with a functional copy of a defective gene or with a therapeutic nucleic acid that will correct the mis-expression of a gene. This can include the overexpression of a gene (greater number of transcripts), expression of a gene in a different tissue, or interfering with the correct processing of a transcript (Glover, Lipps, and Jans 2005).

The abovementioned approach overcomes limitations of alternative therapeutic options, such as the use of chemotherapy agents or the delivery of genes with the help of viral vectors. It is important to realize, however, that the use of nanomedicine for the delivery of therapeutic nucleic acids is still in its infancy. Most of the scientific studies reported to date are based on studies involving cell cultures, with only a limited number of studies having been conducted on animals.

This chapter will provide an overview of different nanomaterials (see Figure 5.1) with potential to deliver therapeutic nucleic acids, potential advantages over alternative treatments, and design criteria for nanomaterials necessary to ensure compatibility with the patient's body.

5.2 Viral versus Non-viral Vectors for Gene Therapy

In 2012, the European Medicines Agency (EMA) approved Glybera (alipogene tiparvovec) as the first gene therapy drug, designed to treat lipoprotein lipase deficiency (Ylä-Herttuala 2012). Glybera relies on an adeno-associated viral vector for its treatment mechanism. The first gene therapy drug in the United States was approved by the U.S. Food and Drug Administration (FDA) in 2017 and is known as Tisagenlecleucel, marketed under the brand name Kymriah, as a treatment for B-cell acute lymphoblastic leukemia. The basis for Kymriah is chimeric antigen receptor (CAR) T-cell therapy, in which T-cells express a chimeric receptor engineered (*ex vivo*) to graph specificity against B-lymphocyte antigen CD19 (Jin et al. 2016). The reprogramming of T-cell receptors into the CAR-modified T-cell receptor is accomplished with the use of viral vectors, such as retroviruses or lentiviruses.

While both of these gene therapy drugs rely on viral vectors, delivery of different types of nucleic acids for gene delivery can also be accomplished by using non-viral vectors.

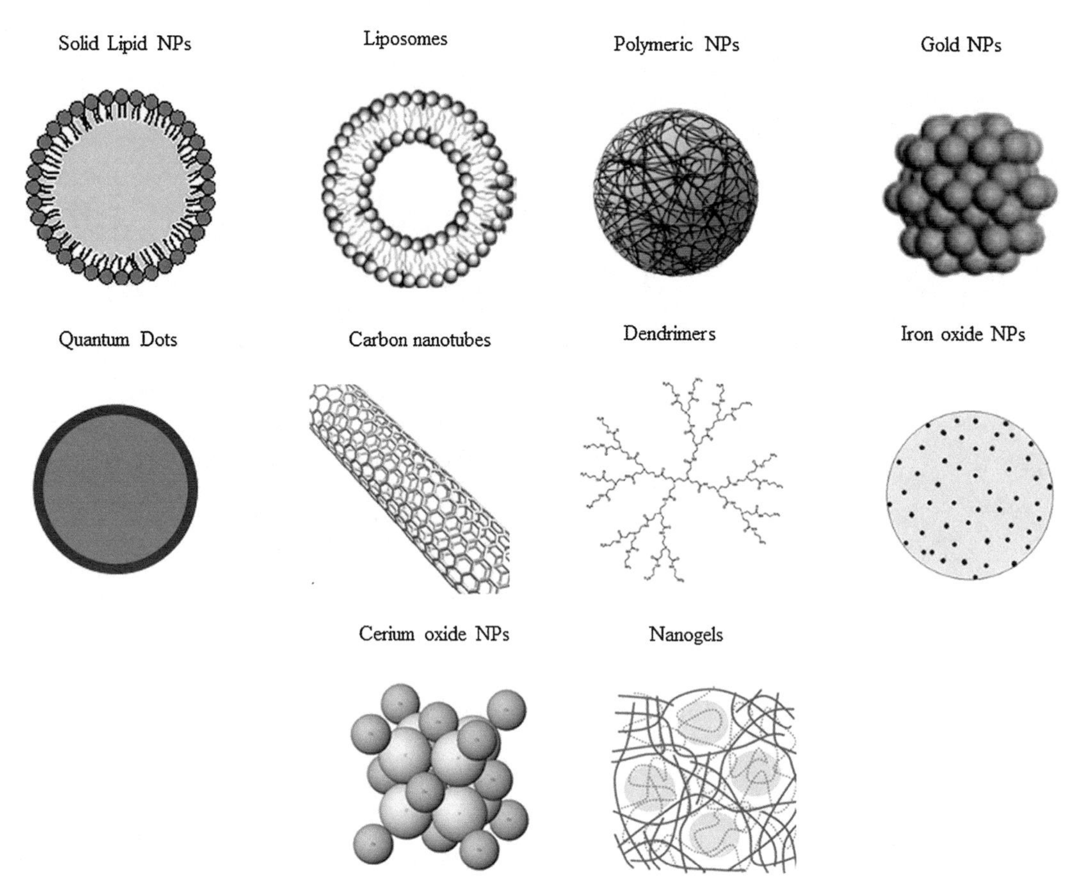

FIGURE 5.1 Different types of nanomaterials for biomedical use. Nanomaterials are defined as objects with dimensions 1–100 nm, which include nanogels, nanofibers, nanotubes, and nanoparticles (NPs). (Reproduced with permission from Re, Gregori, and Masserini 2012.)

Both types of vectors have positive and negative attributes, which will be discussed in the subsequent sections.

Viral vectors are designed by inactivating pathogenic viruses through removal of part of the viral genome, while retaining the genes necessary for replication. Viral vectors can be derived from retroviruses, adenovirus, adeno-associated viruses, and herpes simplex virus (Robbins and Ghivizzani 1998), each having advantages and drawbacks for use in gene therapy.

Retroviruses are single-stranded RNA (ssRNA) viruses from the family Retroviridae. The term "retro" in the name refers to the reliance of these viruses on the enzyme reverse transcriptase to reverse transcribe their RNA genome into DNA (Poiesz et al. 1980). This newly transcribed DNA is subsequently integrated into the host's genome via the action of the enzyme integrase. Upon integration, the so-called provirus relies on the host's replication, transcription, and translation machinery to reproduce. Most retroviral vectors used in gene therapy applications are derived from the genera Lentivirus and Gammaretrovirus, with murine leukemia virus (MLV), a gammaretrovirus, being the first and most commonly used (Robbins and Ghivizzani 1998; Cannon and Anderson 2003). The ability of retroviral vectors to integrate into the host's genome offers the advantage of stability and (temporary) escape from the host's immune system. A drawback of MLV is that it is unable to infect nondividing cells. Consequently, lentiviral vectors are used in gene therapy studies because of their ability to infect both nondividing and dividing cells (Miller, Adam, and Miller 1990; Naldini et al. 1996).

Adenoviruses are double-stranded linear DNA (dsDNA) viruses from the family Adenoviridae, with a genome of approximately 36 kilobasepairs (kb) (Graham and Prevec 1995). Unlike retroviruses, adenoviral vectors do not integrate into the host genome, limiting their long-term expression potential. This limitation, however, makes adenoviral vectors a prime candidate for vaccine development. Due to its large genome size, adenovirus-derived vectors can accommodate large DNA inserts, which increases their potential use in gene therapy applications involving entire genes (as opposed to short DNA or RNA fragments). One of the major drawbacks of adenoviral vectors is the immune response they trigger in patients receiving the recombinant virus. This could not only be detrimental to the patient's overall health but also reduce the effectiveness of the therapeutic DNA being delivered.

Adeno-associated vectors (AAVs) were designed to overcome the shortcomings of both retroviral and adenoviral vectors. AAV is a single-stranded DNA virus (ssDNA) from the family Parvoviridae, with a genome of about 4.7 kb.

One of the features that makes AAV an attractive vector choice is its reliable and stable integration into human chromosome 19. This is in contrast to retroviruses, which integrate randomly, potentially causing disruption of the gene in which they integrate. Another benefit is that AAV has limited pathogenicity compared to adenovirus and retroviruses. However, AAV has limited capacity, restricting its use in gene therapy to applications involving relatively short DNA sequences.

While the bulk of gene therapy research has focused on the use of viral vectors, non-viral vectors have started to emerge. They include polyplexes, liposomes, vectors based on inorganic materials, nanomaterials, and other physical means of enhancing delivery of therapeutic nucleic acids. One of the reasons non-viral vectors have piqued an interest is because of their low immunogenicity relative to viral vectors and the ease with which they can deliver short DNA or RNA molecules (see Section 5.3). In addition, their ease of synthesis and fabrication relative to viral vectors enables large-scale production. Among the drawbacks of non-viral vectors are the lower transfection rates, cytotoxicity, and potential to be eliminated by the host's immune system prior to delivery of their cargo. Active research on non-viral vectors has begun to improve these factors, making non-viral vectors a promising alternative for gene therapy. The remainder of this chapter will focus on the use of different classes of nanomaterials for the delivery of therapeutic nucleic acids.

5.3 Nucleic Acids Used in Non-viral Methods of Gene Therapy

As mentioned in the Introduction, gene therapy involves the use of therapeutic nucleic acids to modulate gene expression. While the term "gene therapy" encompasses "gene", this does not mean entire genes have to be delivered to target cells or tissues. Size constraints of different vectors may preclude this, while in other instances, it may simply be more effective or efficient to modulate gene expression through the use of short nucleic acids.

The different types of nucleic acids used in therapeutic gene delivery include linear or circular (plasmid) DNA, small interfering RNAs (siRNAs), micro RNAs (miRNAs), short hairpin RNAs (shRNAs), and antisense oligonucleotides (AONs).

5.3.1 siRNAs

siRNAs are 21–25 nucleotides in length and were first discovered to play a role in post-transcriptional gene silencing in plants as a defense against viruses (Hamilton and Baulcombe, 1999), and subsequently also shown to exist in mammalian cells (Elbashir et al. 2001). siRNAs are an integral part of RNA interference (RNAi), a regulatory process initially discovered in the nematode *Caenorhabditis elegans* by Nobel prize winners Fire and Mellow (Fire et al., 1998), whereby the presence of double-stranded RNA (dsRNA) molecules was found to induce gene silencing through a mechanism that has since been elucidated (Figure 5.2). RNAi has as a wide array of functions in eukaryotes, ranging from antiviral defense (Hamilton and Baulcombe 1999), post-translational modification (Fire et al. 1998), heterochromatic silencing, and epigenetic gene expression (Volpe 2002).

The siRNA-mediated RNAi pathway begins with dsRNA being processed and cleaved by the enzyme Dicer. The dsRNA can either have an exogenous origin (including delivery via nanoparticles (NPs)) or can be produced by an infecting pathogen (including viruses). The Dicer enzyme cleaves this dsRNA molecule into siRNAs of 21–25 nucleotides that are fully complementary to the target mRNA. After loading the ds siRNAs into the RNA-induced silencing complex (RISC) with the help of RISC-loading complexes (RLCs), one of the strands, dubbed the passenger strand, is degraded by the Argonaute 2 protein that is part of the RISC, while the remaining guide strand targets the RISC to the complementary mRNA. Once bound, the target mRNA is degraded (Figure 5.2).

Elbashir et al. (2001) mentioned the potential to use siRNA for therapeutic purposes, and considerable effort has been made in it since then. However, major challenges associated with the administration of siRNAs for therapeutic purposes are their various interactions with blood serum proteins, which could lead to their degradation before they reach their target site(s) (Haupenthal et al. 2006; Choung et al. 2006). Improvement of siRNA survival and the associated prolongation of effects on gene expression can be accomplished through chemical modification, encapsulation in a non-viral vector, or modification of the genome of a suitable virus for viral-mediated delivery, with each method having benefits and disadvantages.

5.3.2 miRNAs

MicroRNAs are 21–28 nucleotides in length. These small RNA molecules are involved in post-translational gene regulation. They were first identified in the nematode *C. elegans*, when the Ambros group noticed that a *lin-4* mutant was developmentally arrested at the first larval stage (Lee, Feinbaum, and Ambros 1993). *Lin-4* was found to encode short 22-nucleotide RNAs complementary to transcripts of another gene, *Lin-14*, and the lack of miRNAs in the mutant caused developmental arrest. Unlike siRNAs, miRNAs have an endogenous origin: they are encoded by the introns or exons of an organism's own genes (Rodriguez 2004), but their mode of action also involves the RNAi pathway.

The biogenesis of miRNA starts with transcription of genomic DNA encoding a primary pre-miRNA (pri-miRNA) by RNA polymerase II that is subsequently polyadenylated at the 3' end and capped at the 5' end (Figure 5.2). The ends of the pri-miRNA are cleaved by the enzyme Drosha, resulting in a pre-miRNA of 70–90 nucleotides that is exported from the nucleus to the cytoplasm, where Dicer

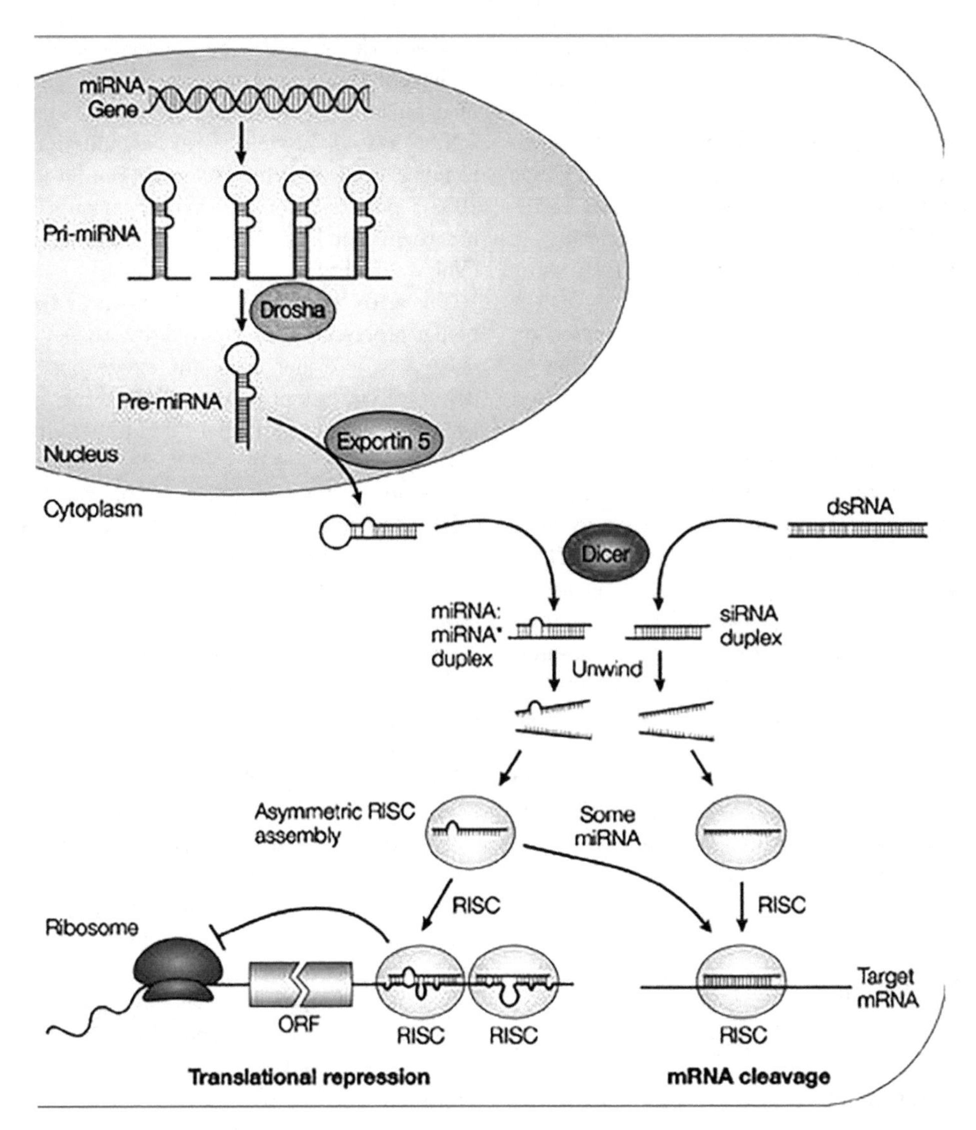

FIGURE 5.2 The current model for the biogenesis and post-transcriptional suppression of miRNAs and siRNAs. Exportin 5 facilitates transport of the pre-miRNA into the cytosol. (Reproduced with permission from He and Hannon 2004.)

cleaves it into mature miRNAs that are loaded in the RISC. Two differences with siRNAs (Section 5.3.1) are that the passenger strand is discarded, rather than degraded, and that miRNAs are only partially complementary to their target mRNAs and tend to target the 3' untranslated region (UTR) of the mRNA. Binding of the RISC complex to target RNAs can cause the translation of mRNA to be stalled or result in RNA cleavage or degradation (Zamore et al. 2000) (Figure 5.2).

MicroRNAs have the potential to bind and activate Toll-like receptors (TLRs), which are receptors involved in the recognition of pathogen-associated molecular patterns (PAMPs) during the immune response to foreign antigens (Fabbri et al. 2013). The induction of the innate immune response by miRNAs is an undesirable effect when using them for therapeutic purposes. One method to avoid this activation of the innate immune is to chemically modify the miRNAs, which can also contribute to their overall stability (Lam et al. 2015).

5.3.3 shRNAs

Short hairpin RNAs are, as the name implies, short RNA molecules with a stem-loop structure. They differ from exogenously supplied siRNAs in that they are synthesized in the cell nucleus following transcription of DNA, and they differ from miRNAs in that they have an exogenous origin. They can be delivered with the help of viral vectors or via transformation with plasmid DNA encoding an shRNA. Following transcription, they are processed analogous to pri-miRNAs and their effect also relies on the RNAi pathway (Figure 5.2) to modulate gene expression (Silva et al. 2005). The specific shRNA design and incorporation of shRNA into the RISC can result in two outcomes: mRNA cleavage or suppression of mRNA translation. A particular type of shRNA, known as bifunctional shRNA, allows both mRNA cleavage and suppression of mRNA translation. This is accomplished by designing a single shRNA containing two stem-loop structures, one with a perfect and another with an imperfect match to the target mRNA, to induce both cleavage-dependent and cleavage-independent RISC pathways (Rao et al. 2009). One of the challenges with utilizing shRNA is avoiding induction of the interferon response of the immune system (Bridge et al. 2003). In some instances, such as in viral infection or cancer, this immune activation is, however, beneficial. Therefore, the clinical team should determine on a case-by-case basis whether shRNA molecules need to have dual RNAi and immunostimulatory activity (Meng and Lu 2017).

5.3.4 Antisense Oligonucleotides

While siRNAs, miRNAs, and shRNAs all have their origins as dsRNA molecules and rely on the RISC to target complementary mRNA molecules with the goal of silencing expression, AONs can be provided as single-stranded molecules that directly target mRNA and can include DNA or DNA derivatives. An example of AONs used for therapeutic purposes is the drug nusinersen (Spinraza), which was approved in December 2016 and which is used to treat spinal muscular atrophy (SMA), a crippling disease caused by improper processing of a pre-mRNA transcribed from the *SMN1* gene encoding a survival motor neuron protein (Cartegni and Krainer 2002). Patients, preferably at a young age, receive an injection of the drug in their spine that can correct the processing error and restore normal function for several months.

Their single-stranded nature makes OANs more susceptible to cellular nucleases than siRNAs (Eckstein 2002). One way to avoid the degradation by cellular nucleases is to chemically modify the AONs, as in the case of nusinersen. These chemical modifications can also increase the affinity towards the target mRNA (Straarup et al. 2010; Thiel and Giangrande 2009; Chiang et al. 1991). While nusinersen can be effectively delivered into the spinal cord, owing to the expression pattern of the *SMN1* gene, AONs targeting different genes in different tissues may require delivery via nanomaterial-derived vectors.

In summary, each of the various types of short nucleotides discussed in the previous sections has different strengths and weaknesses. This means that it is necessary to determine the most effective treatment and delivery option for each application.

5.4 Mechanisms to Ensure Successful Delivery of Therapeutic Nucleic Acids to Their Target Cells

While viruses have the benefit of having evolved to target specific cells or tissue types to reproduce efficiently, with the assistance of their host's cellular machinery, targeted delivery of small nucleic acid species to specific cells or tissues in the body with the use of non-viral vectors faces several obstacles. First, non-viral vectors do not have the innate target specificity of viral vectors and have to be chemically modified to obtain target specificity. Additionally, non-viral NP vectors have to overcome several biological barriers, which include resistance to opsonization via protein absorption, avoiding the mononuclear phagocyte system (also known as the reticuloendothelial system (RES)), and avoiding cellular endosomal degradation (Nie 2010), as discussed in more detail in the next section.

Opsonization is the process by which a foreign molecule or antigen is coated with serum proteins, to make them more susceptible to recognition by immune cells, specifically macrophages that can take up these tagged molecules via phagocytosis. The best known opsonins are antibodies and complement proteins. These two groups of proteins are responsible for neutralizing toxins, pathogens, and other foreign particles and then facilitate macrophage phagocytosis (Gordon 2016). Other nonspecific proteins such as collectins, ficolins, and serum albumin proteins can also function in an opsonization-like manner. Consequently, non-viral vectors have to avoid these different classes of proteins to avoid being destroyed or removed before they reach their target site. Research has shown that hydrophobic particles tend to have a higher degree of opsonization and subsequent inactivation and removal compared to hydrophilic particles (Carrstensen, Müller, and Müller 1992). Grafting the shielding groups onto the NP is an effective way to minimize opsonization, and is most commonly accomplished with polyethylene glycol (PEG) and related compounds (Peracchia et al. 1999). Once the NPs have been shielded, they have a better chance at avoiding the RES. This system is part of the immune system and is comprised of monocytes and macrophages in the liver and spleen that take up particles tagged by opsonin proteins. It is important to realize that opsonin proteins facilitate the uptake by macrophages, but that the macrophages and the RES itself can still remove and clear NPs even when they are not modified with opsonin proteins. Hence, in addition to shielding the NPs, decreasing their size can help prevent clearance by the RES (Gaur et al. 2000).

NPs that are able to avoid opsonization and removal by the RES system are able to increase their circulation time. The entire goal of this targeting approach, known as passive targeting, is to simply increase the circulation time so that the NP has an increased chance to reach its target site before being cleared by the RES system. In contrast to the passive targeting system, there is the active targeting system. The active targeting system relies on the ability to conjugate a molecule onto the NP so that it is targeted to a specific tissue or cell type, this active targeting technique is usually coupled with the ability of NP to be up taken by the cell. For active targeting to be a success, the NP has to be chemically modified to allow a specific ligand to be conjugated onto the NP. Generally, most of the chemical modifications are either covalent or non-covalent with majority of them being covalent. Some of the most common covalent reactions utilize reactive carbonyl groups, amine groups, and sulfhydryl groups (Werengowska-Ciećwierz et al. 2015).

Once an NP has reached its target cell, the NP has to cross the cell membrane. The surface chemistry of an NP will determine the entry mechanism into the cell. Most NPs will enter the cell via endocytosis, a process by which vesicles derived from the cell membrane are internalized. Several different subtypes of endocytosis exist, including phagocytosis, clathrin-mediated endocytosis, caveolin-mediated endocytosis, clathrin/caveolae-independent endocytosis, and macropinocytosis (Oh and Park 2014). NPs taken up via endocytosis will be inside an endosome (the internalized membrane vesicle), preventing their ability to deliver the therapeutic nucleic acids. As a consequence, mechanisms to escape the endosome and rescue the NP and its cargo are of interest. Three approaches are being explored to accomplish this. The first method relies on disturbing the internal low pH of the endosome by using a polymer that absorbs the excess protons (H^+) responsible for the low endosomal pH. Additional protons, along with chloride counter ions, are pumped into the endosome to restore the pH, while at the same time, the osmotic potential increases. This eventually causes the endosome to rupture, releasing the cargo. This mechanism is referred to as the proton sponge effect, and the polymer polyethylenimine (PEI) can be used for this purpose. The second mechanism is based on the observation that enveloped viruses escape from the endosome by fusing their viral envelope with the endosomal membrane. Hence, particles assembled from lipids or other amphiphilic molecules are expected to be able to fuse with the endosomal membrane and deliver their cargo directly into the cytoplasm. The third mechanism relies on the action of the cargo interacting directly with the endosomal membrane to either destabilize the membrane or create pores in the membrane that allow the cargo to diffuse out of the membrane. This can be accomplished by the direct interaction of polymers or peptides with the membrane itself (Nakase, Kobayashi, and Futaki 2010; Tian and Ma 2012; Selby et al. 2017). Depending on the exact application, it may be possible to combine mechanisms to accomplish endosomal escape with a greater success rate.

5.5 Nanomaterials for Nucleic Acid Delivery

This section summarizes recent developments related to non-viral NP gene delivery systems and is organized based on the nature of the nanomaterial system: inorganic nanomaterials, carbon-based nanomaterials, protein- and peptide-based nanomaterials, lipid-based nanomaterials, and polymer-based nanomaterials.

5.5.1 Inorganic Nanomaterials

Inorganic nanomaterials derived from gold, silver, calcium phosphate, iron oxides, copper, selenium, cadmium, and zinc are being explored for gene therapy uses and were selected because of their attractive optical and/or electrical properties. Compared to nanomaterials based on polymers, inorganic nanomaterials tend to be easier to synthesize and functionalize. As will be discussed in further detail below, gold nanomaterials are widely used because of their surface chemistry properties, ease of functionalization, and low cytotoxicity (Qiu et al. 2010; Fratoddi et al. 2014). Magnetic nanomaterials are a subset of inorganic nanomaterials that have sparked an interest because of their utility in imaging, while semiconductor-based nanocrystals known as quantum dots (QDs) are of interest because of their attractive optical and electrical properties that are a function of their size (Murray et al., 2000).

Gold NPs, and to a lesser extent other inorganic NPs such a silver NPs, exhibit a phenomenon called localized surface plasmon resonance (LSPR) (Petryayeva and Krull 2011) in which the NPs absorb and/or scatter light at specific wavelengths. This LSPR property allows inorganic NPs to be used in photothermal therapy (Zhang, Wang, and Chen 2013; Hwang et al. 2014), fluorescent imaging (Sun and Jin 2014), radiation therapy (Liu et al. 2010), and easy surface functionalization (Qiu et al. 2010; Fratoddi et al. 2014). Recently, Deng et al. (2018) demonstrated how gold NPs can be used to treat acute myeloid leukemia. The group designed a nanomaterial complex consisting of gold NPs functionalized with a nuclear localization signal (NLS) that enables the particles and their cargo to enter the cell nucleus to deliver anticancer agents AS1411 (a modified short DNA molecule) and anti-221. The anti-221 targets miRNA-221, which is highly upregulated in cancer cells and stimulates tumor growth by targeting tumor suppressor gene *p27kip1*. In both cell cultures and mice with leukemia, the treatment with the gold NPs negatively affected cell growth, illustrating the potential of this method for treating this type of leukemia, pending the outcomes of clinical trials in human patients. Gold NPs can also be functionalized with peptides to increase transfection efficiency. For example, Peng et al. (2016) used antimicrobial peptides to coat gold NPs to transfect bone marrow-derived mesenchymal stem cells with plasmid DNA encoding vascular endothelial growth factor-165 (VEGF) with greater efficiency than naked gold NPs. Combining the ability to functionalize gold NPs with proteins and their LSPR feature enabled Wang et al. (2018) to increase transfection efficiency, by designing a laser-triggered release of gold NPs encapsulated in lipid membranes (see Figure 5.3). These gold NPs were designed to deliver and release plasmid DNA encoding the Cas9 endonuclease, which is widely used as part of the genome editing tool CRISPR (clustered regularly interspaced short palindromic repeats)/Cas9 (Jinek et al. 2012; Ran et al. 2013). Their objective was to release Cas9 DNA when triggered by a laser, to have it cut, and thereby inactivate the *Plk-1* gene in tumor cells.

QDs are semiconductor-based nanomaterials that are synthesized either via high-temperature colloidal synthesis, plasma synthesis, or a bottom-up approach using lithography or a thin-film deposition method known as molecular beam epitaxy (Valizadeh et al. 2012;

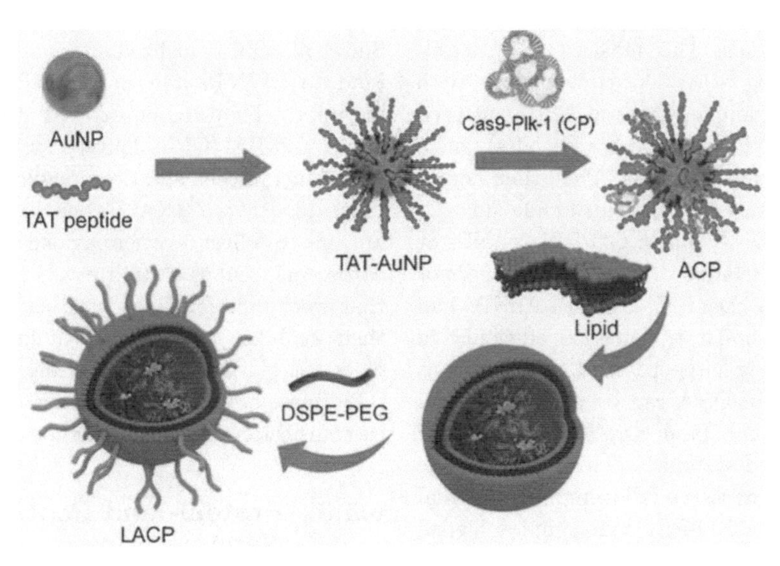

FIGURE 5.3 Synthesis process for lipid-encapsulated gold NPs with condensed plasmid DNA (LACP). AuNP = gold nanoparticle; TAT peptide facilitates nuclear import; Cas9-PLK1 (CP) = condensed plasmid DNA encoding Cas9 endonuclease and *Plk1* single guide RNA; ACP = encapsulated gold nanoparticles with condensed plasmid DNA; DSPE-PEG = 1,2-distearoyl-sn-glycero-3-phosphoethanolamine (DSPE)-conjugated PEG. (Reproduced with permission from Wang et al. 2018.)

Brichkin and Razumov 2016). The optical characteristics of QDs make them the ideal delivery vehicles for neuronal cells and enable them to be tracked without the need for additional labeling. Walters et al. (2012, 2015) demonstrated how QDs with a CdSe core, a ZnS shell, and a negatively charged compact molecular ligand coating (CL4) could be used to selectively target neurons rather than glia in the brain. This selective uptake was influenced by the extent of the negative charge on the QD coating. Getz et al. (2016) used this same type of CL4-coated CdSe/ZnS QDs, conjugated with a positively charged JB577 peptide, to deliver siRNA designed to target one of three sphingomyelinase genes active in the central nervous system. The QD-siRNA complex was shown to be able to escape the endosome and downregulate each of the sphingomyelinases in a dose-dependent manner.

Magnetic NPs are a class of NPs that have application in magnetic resonance imaging and can also be used for therapeutic uses whereby a magnet is used to target the NPs to the desired tissue. This makes magnetic NPs the driving force behind theranostics, a field of medicine focused on personalized medicine whereby the therapeutic agent is used to both diagnose (via imaging) and provide therapy simultaneously (Chen, Ehlerding, and Cai 2014). Well-studied magnetic NPs include iron oxides such as γ-Fe_2O_3 (maghemite) and α-Fe_2O_3 (hematite) (Lu, Salabas, and Schüth 2007). Once the size of the iron oxide gets small enough (<128 nm), the NP can undergo superparamagnetism (Lu, Salabas, and Schüth 2007). Superparamagnetism is a process that occurs when random, temperature-induced changes in the directionality of magnetization effectively cause loss of magnetization unless there is an external magnetic field that can induce magnetization. Superparamagnetic iron oxide nanoparticles (SPIONs) have high surface energy, which leads to inherent aggregation and hinders their potential application (Jeon et al. 2016). Surface chemistry modifications such as coatings with polymers of carbon-based materials can help reduce aggregation. Additionally, these surface modifications make them less susceptible to oxidation, which can otherwise reduce their magnetism (Hola et al. 2015). Utilizing a theranostics approach, Mahajan et al. (2016) designed a SPION complexed with siRNA against polo-like kinase-1 (siPLK1-StAv-SPIONs) to target pancreatic cancer. This dual-purpose NP was utilized to deliver siRNA and to observe the tumor response.

5.5.2 Carbon-Based Nanomaterials

Carbon-based nanomaterials are important because of their diverse chemical and physical properties, offering a range in mechanical and optical properties, and electrical conductivity. This diversity reflects the fact that there are multiple carbon allotropes, i.e. different configurations in which neighboring carbon atoms are bound to each other (Cha et al. 2013). The most common carbon-based nanomaterials used in gene therapy are graphene, in which carbon atoms are arranged in a sheet-like configuration consisting of a single layer of atoms, graphene oxide (an oxidized version of graphene), and single- and multi-walled carbon nanotubes (MWCNTs), which are tubular structures made of graphene sheets, the spherical C_{60} buckminsterfullerene (also known as Buckyballs), and nanodiamonds.

Graphene is an attractive nanomaterial because it has optical, thermal, and electrical properties that are similar to inorganic nanomaterials. Graphene oxide (GO), an oxidized form of grapheme, is highly soluble in aqueous environments, which prevents GO aggregation in comparison to other inorganic nanomaterials. This makes GO an excellent candidate for shielding other inorganic molecules to prevent aggregation. For example, Xu et al. (2013) utilized GO to encapsulate gold NPs and nanorods (NRs) via a novel electrostatic self-assembly process. They then conjugated PEI onto the GO surface using an amide linkage, which enabled binding of DNA. These GO-PEI-Au NPs or NRs reduced aggregation (relative to the use of Au NPs or NRs). In comparison to 25-kDa PEI, GO-PEI-AuNPs had a lower cytotoxicity and similar transfection efficiency in HeLa cells. GO can carry a large payload, which means a relatively large amount of DNA can be accommodated, enabling slow release over time. Paul et al. (2014) developed an injectable hydrogel-based system to deliver DNA to the heart to enable repair of scar tissue following a myocardial infarction. They prepared a GO-PEI nanocomplex loaded with a plasmid DNA encoding VEGF-165 and encapsulated this in a hydrogel synthesized from gelatin coupled with methacrylate, which gives the hydrogel suitable mechanical and degradation properties. This hydrogel nanocomplex was then injected in infarcted rat hearts and shown to reduce fibrosis, stimulate angiogenesis, and improve cardiac function in comparison to the hydrogel and the hydrogel/DNA controls.

Carbon nanotubes are cylindrical-shaped structures composed of single or multiple sheets of graphene, which were first described in 1991 (Iijima 1991, 2002; Iijima and Ichihashi 1993). Iijima discovered multi-walled carbon nanotubes (MWNT) first, and then later discovered single-wall carbon nanotubes (SWNT). Carbon nanotubes are usually functionalized to increase their solubility in aqueous environments. This can be accomplished via oxidation or coating with amphiphilic molecules. As an example illustrating the use of functionalized SWNTs for gene delivery, Wang et al. (2013) used a covalently functionalized SWNT/PEI/NGR/siRNA system (NGR is (Cys-Asn-Gly-Arg-Cys)-peptide) to deliver siRNA against transcripts encoding human telomerase reverse transcriptase (hTERT), an enzyme important in carcinogenesis. The tumor-targeting peptide NGR was used to increase transfection efficiency of the system into the prostate cancer cell line PC-3. Even though covalent functionalization can substantially alter their intrinsic properties, carbon nanotubes still possess the ability to absorb radiation. Wang et al. (2013) utilized a combination therapy in which siRNA was delivered to a tumor and combined this with phototherapy based on the ability of the nanotube to absorb NIR radiation at a wavelength of 808 nm. In contrast to covalent functionalization, as in the example above, non-covalent functionalization tends to have a smaller effect on the intrinsic physical properties of the carbon nanotubes. Behnam et al. (2013) functionalized SWNTs with non-covalently bound hydrophobic moieties to which PEI was subsequently covalently bound, thus enabling association with DNA. The resulting complexes were used to transfect neuroblastoma cells and mice and shown to have increased transfection rate as well as decreased cytotoxicity compared to DNA delivery mediated by PEI alone. Siu et al. (2014) utilized succinated-PEI to non-covalently bind to SWNTs and deliver siRNA to treat melanoma in mice. They demonstrated that this water-soluble, succinated-PEI/CNT delivery system could deliver siRNA in melanoma cells and downregulate the gene encoding the protein B-Raf via RNAi. A benefit of MWCNTs is that they are able to deliver a larger payload compared to single-walled carbon nanotubes. Munk et al. (2017) utilized MWCNTs that were functionalized with carboxyl groups to improve their solubility and biocompatibility. The group successfully delivered plasmid DNA encoding green fluorescent protein (GFP) into bovine fibroblast cells, which tend to be resistant to transfection via other methods.

5.5.3 Protein- and Peptide-Based Nanomaterials

Proteins and peptide-based nanomaterials are highly sought-after because of their high biocompatibility and their ease of biodegradability. Proteins being biological molecules are less cytotoxic compared to carbon nanotubes and inorganic nanomaterials. Additionally, proteins can interact with both solvent and their cargo because of their amphiphilic nature (polar and apolar characteristics in the same molecule). There is a wide array of different types of proteins that can be utilized for gene delivery. One of the most common proteins used for gene delivery is gelatin (Jahanshahi and Babaei 2008). Gelatin is derived from the hydrolysis of collagen and, depending upon the isolation method, gelatin can be classified as type A or type B. The choice of using type A or type B depends upon the purpose of the gelatin because of the difference in isoelectric point between the two types (Kumar 2005). Gelatin B has an isoelectric point between 4.8 and 5.2, which causes the gelatin to have a negative charge at physiological pH. When gelatin B is internalized inside an endosome, gelatin becomes positively charged, which causes the nanocomplex to release its cargo or therapeutic agent. Morán et al. (2015) utilized gelatin B complexed with protamine sulfate to deliver DNA. This gelatin complex offers controlled release because the strength of the gelatin as well as the initial concentration of the DNA determines the rate of the DNA release.

Another class of peptide-based NPs are the elastin-like peptides (ELPs). ELPs are artificial peptides with a repeating sequences of polypeptides $(\text{Val-Pro-Gly-X-Gly})_n$ with X designating any amino acid (Lohcharoenkal et al. 2014). Depending on the choice of amino acid, "X" ELPs can have a tunable thermal transition temperature (T_t). Below their T_t, ELPs are soluble, whereas above their T_t, ELPs form a viscous fluid called coacervate (Urry 1997). Dash et al. (2015) designed a dual ELP injectable system composed of an ELP gel-based scaffold and ELP hollow spheres, to deliver two different therapeutic gene doses. This system allowed them to deliver plasmids encoding nitric oxide synthase and interleukin-10 to human umbilical vein cells (HUVECs) to induce angiogenesis and reduce inflammation to treat critical limb ischemia, which manifests itself as obstruction of arteries and which can ultimately lead to loss of limbs. Monfort and Koria (2017) constructed heterogeneous NPs composed of two chimeric ELP fusion proteins with low-density lipoprotein receptor repeat 3 (LDLR3) and keratinocyte growth factor (KGF), to selectively deliver a lentiviral vector encoding GFP to cells expressing high levels of KGF. This novel NP approach provides target specificity, whereas the lentivirus by itself infects both dividing and nondividing cells, creating potential off-target effects, increased immunogenicity, and toxicity.

5.5.4 Lipid-Based Nanomaterials

Lipids include various derivatives of fats, waxes, oil, cholesterol, phospholipids, and the fat-soluble vitamins (A, D, E, K). Lipids are defined by their amphiphilic nature and easily form vesicles (liposomes) in aqueous environments, which contribute to their potential utilization as gene delivery vectors. The first lipid-based gene delivery was demonstrated by Felgner et al. (1987), who utilized a liposome containing the cationic lipid DOTMA (1,2-di-*O*-octadecenyl-3-trimethylammonium propane) to deliver plasmid DNA into different eukaryotic cell lines. Cationic lipids are preferential to anionic or neutral lipids because they can adsorb easily and more efficiently onto cell membranes. In addition to liposomes, other lipid-based molecules such as lipidoids and gemini surfactants have sparked an interest as vehicles for gene delivery (Whitehead et al. 2014; Knapp et al. 2016; Akinc et al. 2008; Hait and Moulik 2015). Lipidoids are amino-alkyl-acrylate or acrylamide-based, cationic molecules with between one and seven side chains that have been shown to be more efficient at delivering siRNA than other lipids (Love et al. 2010). For example, Knapp et al. (2016) demonstrated how lipidoid-based NPs were able to successfully transfect mantle cell lymphoma cells with siRNAs to silence the anti-apoptotic protein Mcl-1 and induce cell death. Considering the difficulty in transfecting mantle cell lymphoma cells, this was a major accomplishment. Expanding upon their previous work, Knapp et al. (2018) demonstrated how lipidoid $306O_{13}$was able to deliver siRNA against three targets simultaneously with the aim of inducing cell death in mantle cell lymphoma cells, by downregulating the gene encoding cell cycle regulator Cyclin D1 and the genes encoding Bcl-2 and Mcl-1, which prevent apoptosis (cell death). This was the first study to attempt a multiplex gene silencing approach for mantle cell lymphoma, which was shown to be more effective than silencing just one or two of these genes.

5.5.5 Polymer-Based Nanomaterials

Polymer-based nanomaterials can be fabricated from both natural and synthetic polymers and display the largest variation in structure among the nanomaterials discussed so far, simply because of the many different polymers that can be used, the structural variation within them, and the opportunities that exist for additional chemical modification. For example, whether a polymer is linear or branched, the number of repeat units, the overall degree of polymerization (molecular weight), and the presence of anionic or cationic moieties on the polymer will influence its physicochemical properties.

Due to the negative charge of RNA and DNA at physiological pH, cationic polymers are able to readily interact with these nucleic acids, which makes them a natural choice for gene delivery. Some of the most commonly used synthetic polymers are polyethyleneglycol (PEG), poly-L-lysine (PLL), poly (dimethylaminoethyl methacrylate) (PDMAEMA) and PEI. PEGylation, surface modification with PEG, has been widely used as a process to increase the circulation time of various NPs by shielding them from protein absorption and opsonization (Section 5.4). While PEI has been shown to enhance endosomal escape of NPs (Varkouhi et al. 2011), unmodified PEI vectors are highly toxic to cells (Florea et al. 2002; Wang et al. 2011). To reduce the toxicity, PEI can be modified. For example, Ke et al. (2018) utilized carbamate-mannose-modified PEI (CMP) to deliver NF-κB shRNA into cancer stem cells to cause cell death. The CMP/shRNA nanocomplex exhibited a high transfection efficiency and low cytotoxicity in 4T1 cancer stem cells. In another modification, Amreddy et al. (2018) utilized polyamidoamine-dendrimer-PEI conjugated with folic acid to simultaneously deliver siRNA against mRNA encoding human antigen R (HuR, an RNA-binding protein necessary for the survival of cancer cells) and cis-diamine platinum (CDDP) to H1299 lung cancer cells. The group demonstrated that both siRNA and CDDP were released at the target cancer cells with minimal cellular toxicity, which can be attributed to the lower CDDP concentrations needed for the NP formulation.

In contrast to synthetic polymers, natural polymers have been investigated because of their biocompatibility, generally low immunogenicity, and renewability. The natural polymers that have sparked an interest include but are not limited to hyaluronic acid, chitosan (CS), cyclodextrins, PLL, alginate, and lignin. Cyclodextrins are polymers composed of alpha 1-4-D-glucose derived from starch. These natural polymers have a low immunogenicity, while their cationic nature facilitates the interaction with nucleic acids, which makes them excellent choices for gene delivery. Zhang et al. (2018) designed an adamantyl-terminated PEI (Ad_4-PEI) and polyaspartamide-modified cyclodextrin (Pasp-ss-CD) NP to target and deliver a plasmid encoding luciferase in HeLa and rat glioma cells (C6). The group demonstrated that this Pasp-SS-CD/Ad_4-PEI system had lower cytotoxicity and higher transfection efficiencies compared to the Ad_4-PEI system.

CS is another polysaccharide with potential for gene therapy. It is composed of randomly distributed *N*-acetyl D-glucosamine and D-glucosamine residues. CS is usually subjected to chemical modifications to reduce some of the inherent drawbacks, such as low solubility and swelling in aqueous environments. A successful combination of CS and protein for gene delivery was reported by

Karimi et al. (2013), who used a core-shell structure. The core was composed of albumin (Alb) and the shell was composed of CS, which can interact with DNA, creating novel Alb-CS-DNA NPs. These NPs were used to deliver a plasmid encoding shRNA against GL3 luciferase, into HeLa cells with 85% efficiency and minimal cytotoxicity. In a recent application of CS, Baghdan et al. (2018) designed lipochitoplexes (LCPs) (see Figure 5.4) composed of chitosan nanoparticles (CsNPs) encapsulated in liposomes consisting of dipalmitoylphosphatidylcholine and cholesterol to test the transfection efficiency of human embryonic kidney (HEK-293) cells. The LCP had a higher transfection efficiency and cell viability compared to non-encapsulated CsNPs and PEI, demonstrating how this chemical modification can enhance CS as a gene delivery system.

Lignin is an aromatic plant cell wall polymer formed from different hydroxycinnamyl alcohols and their related compounds, referred to as monolignols that provides hydrophobicity to water-conducting xylem vessels and structural support to fibers (Ralph et al. 2004). Variation in the properties of lignin among plants and tissues within plants is the result of variation in lignin subunit composition. Caicedo, Dempere, and Vermerris (2012) developed a protocol for the synthesis of lignin nanotubes using a sacrificial aluminum membrane template. Ten et al. (2014) demonstrated that these lignin nanotubes can deliver plasmid DNA into HeLa cells and that they were less cytotoxic than carbon nanotubes. The source of the lignin (plant species and extraction method) affected their elastic properties and interaction with the cells. While all formulations of lignin nanotubes were able to enter the cell, only some were able to penetrate the nucleus. A different use of lignin for delivery of plasmid DNA involves the formation of a co-polymer between residual lignin from commercial pulp and paper mills with the polymer PDMAEMA. Liu et al. (2015) grafted the hydrophilic PDMAEMA onto the lignin and demonstrated that in the presence of plasmid DNA, they formed NPs of 100–200 nm that could be used to deliver plasmid DNA into HeLa, Cos-7, and MDA-MB-231 cells. The properties of the lignin-PDMAEMA co-polymer depended on the chain length of the PDMAEMA, whereby co-polymers with shorter chains displayed a lower cytotoxicity and lower transfect efficiency than those with longer chains.

5.6 Conclusions and Future Perspectives

Nanotechnology has the potential to revolutionize every field in which it is applied. With our increasing understanding and ongoing research on how materials behave at the nanoscale, newly gained knowledge can be applied and tailored to novel applications in the field of electronics; environmental sciences; aerospace engineering; and as reviewed in this chapter, biology, medicine, and pharmacology.

Innovations in both nanotechnology and medicine enable rapid advances in methods for gene delivery. Although viral vectors are still the main gene delivery vector today, with the integration of nanotechnology, non-viral vector research has grown rapidly over the past few years. The key challenges for nanomaterials are to balance their environmental impact (e.g. biocompatibility, biodegradability) and efficiency of gene delivery, which is a function of size and target specificity. Two types of nanomaterial are of particular interest: Magnetic NPs and QDs, because of their roles in theranostics, and natural polymer-based nanomaterials, because of their biodegradability and low immunogenicity. At present, transfection with PEI is the reference against which newly developed transfection methods in cell culture systems are compared. Given PEI's toxicity to organisms, standardized assays will need to be designed that can be used to accurately and reliably compare different nanomaterial vectors in organisms. Additional research and testing will need to be conducted in different animal models to determine the potential interactions of new nanomaterial vectors with a human host.

The different types of nanomaterials available for use in gene delivery gives scientists and medical professionals a wide array of tools to use for different therapeutic options. As research in nanomaterials for diagnostic purposes and gene delivery advances, the medical sciences are expected to

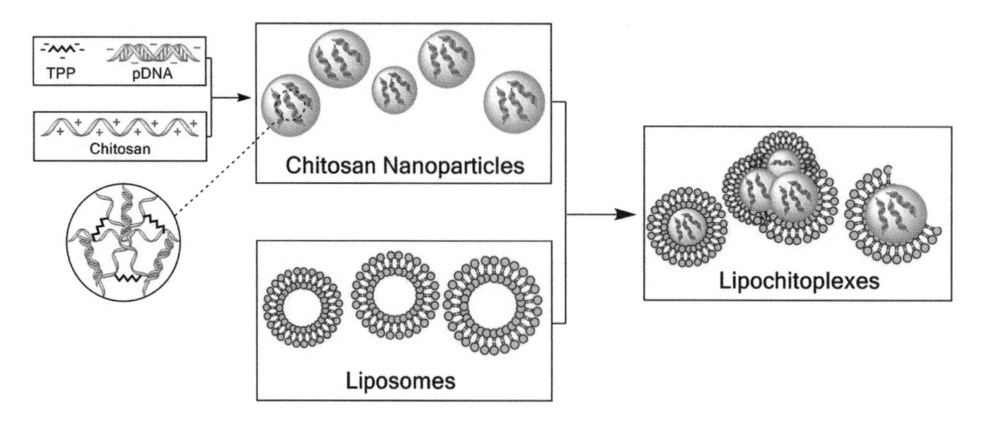

FIGURE 5.4 Scheme of CsNP and LCP formation. TPP = tripolyphosphate; pDNA = plasmid DNA. (Reproduced with permission from Baghdan et al. 2018.)

continue to improve as scientists are better able to diagnose, monitor, and treat different diseases. Because of the ability to tailor and synthesize nanomaterials for specific uses, nanomaterials are likely to be at the forefront of personalized medicine. We are, therefore, optimistic about the prospect of nanomaterial-based vectors for gene therapy in humans in the near future.

References

Akinc, A., A. Zumbuehl, M. Goldberg, et al. 2008. A combinatorial library of lipid-like materials for delivery of RNAi therapeutics. *Nature Biotechnology* 26 (5): 561–69. doi:10.1038/nbt1402.

Amreddy, N., A. Babu, J. Panneerselvam, et al. 2018. Chemo-biologic combinatorial drug delivery using folate receptor-targeted dendrimer nanoparticles for lung cancer treatment. *Nanomedicine: Nanotechnology, Biology, and Medicine* 14 (2): 373–84. doi:10.1016/j.nano.2017.11.010.

Baghdan, E., S. R. Pinnapireddy, B. Strehlow, K. H. Engelhardt, J. Schäfer, and U. Bakowsky. 2018. Lipid coated chitosan-DNA nanoparticles for enhanced gene delivery. *International Journal of Pharmaceutics* 535 (1–2): 473–79. doi:10.1016/j.ijpharm.2017.11.045.

Behnam, B., W. T. Shier, A. H. Nia, K. Abnous, and M. Ramezani. 2013. Non-covalent functionalization of single-walled carbon nanotubes with modified polyethyleneimines for efficient gene delivery. *International Journal of Pharmaceutics* 454 (1): 204–15. doi:10.1016/j.ijpharm.2013.06.057.

Brichkin, S. B., and V. F. Razumov. 2016. Colloidal quantum dots: Synthesis, properties and applications. *Russian Chemical Reviews* 85 (12): 1297–1312. doi:10.1070/RCR4656.

Bridge, A. J., S. Pebernard, A. Ducraux, A. L. Nicoulaz, and R. Iggo. 2003. Induction of an interferon response by RNAi vectors in mammalian cells. *Nature Genetics* 34 (3): 263–64. doi:10.1038/ng1173.

Caicedo, H. M., L. A. Dempere, and W. Vermerris. 2012. Template-mediated synthesis and bio-functionalization of flexible lignin-based nanotubes and nanowires. *Nanotechnology* 23: 105605. doi:10.1088/0957-4484/23/10/105605.

Cannon, P. M., and W. F. Anderson. 2003. Retroviral vectors for gene therapy. In: *Gene and Cell Therapy: Therapeutic Mechanisms and Strategies, Second Edition, Revised and Expanded*, Ed. N.S. Templeton. CRC Press, Boca Raton, FL, pp. 1–16.

Carrstensen, H., R. H. Müller, and B. W. Müller. 1992. Particle ize, surface hydrophobicity and interaction with serum of parenteral fat emulsions and model drug carriers as parameters related to RES uptake. *Clinical Nutrition* 11 (5): 289–97. doi:10.1016/0261-5614(92)90006-C.

Cartegni, L., and A. R. Krainer. 2002. Disruption of an SF2/ASF-dependent exonic splicing enhancer in SMN2 causes spinal muscular atrophy in the absence of SMN. *Nature Genetics* 30 (4): 377–84. doi:10.1038/ng854.

Cha, C., S. R. Shin, N. Annabi, M. R. Dokmeci, and A. Khademhosseini. 2013. Carbon-based nanomaterials: Multifunctional materials for biomedical Engineering. *ACS Nano* 7 (4): 2891–97. doi:10.1021/nn401196a.

Chen, F., E. B. Ehlerding, and W. Cai. 2014. Theranostic nanoparticles. *Journal of Nuclear Medicine* 55 (12): 1919–22. doi:10.2967/jnumed.114.146019.

Chiang, M.-Y., H. Chan, M. A. Zounes, S. M Freier, W. F. Lima, and C. F. Bennett. 1991. Antisense oligonucleotides inhibit intercellular adhesion molecule 1 expression by two distinct mechanisms. *The Journal of Biological Chemistry* 266 (27): 18162–71.

Choung, S., Y. J. Kim, S. Kim, H. O. Park, and Y. C. Choi. 2006. Chemical modification of siRNAs to improve serum stability without loss of efficacy. *Biochemical and Biophysical Research Communications* 342 (3): 919–27. doi:10.1016/j.bbrc.2006.02.049.

Dash, B. C., D. Thomas, M. Monaghan, et al. 2015. An injectable elastin-based gene delivery platform for dose-dependent modulation of angiogenesis and inflammation for critical limb ischemia. *Biomaterials* 65: 126–39. doi:10.1016/j.biomaterials.2015.06.037.

Deng, R., N. Shen, Y. Yang, et al. 2018. Targeting epigenetic pathway with gold nanoparticles for acute myeloid leukemia therapy." *Biomaterials* 167: 80–90. doi:10.1016/j.biomaterials.2018.03.013.

Eckstein, F. 2002. Developments in RNA chemistry, a personal view. *Biochimie* 84 (9): 841–48. doi:10.1016/S0300-9084(02)01459-1.

Elbashir, S. M., J. Harborth, W. Lendeckel, A. Yalcin, K. Weber, and T. Tuschl. 2001. Duplexes of 21 ± nucleotide RNAs mediate RNA interference in cultured mammalian cells." *Nature* 411 (6836): 494–98. doi:10.1038/35078107.

Fabbri, M., A. Paone, F. Calore, R. Galli, and C. M. Croce. 2013. A new role for microRNAs, as ligands of toll-like receptors. *RNA Biology* 10 (2): 169–74. doi:10.4161/rna.23144.

Felgner, P. L., T. R. Gadek, M. Holm, et al. 1987. Lipofection: A highly efficient, lipid-mediated DNA-transfection procedure. *Proceedings of the National Academy of Sciences USA* 84 (21): 7413–17. doi:10.1073/pnas.84.21.7413.

Fire, A., S. Xu, M. K. Montgomery, S. A. Kostas, S. E. Driver, and C. C. Mello. 1998. Potent and specific genetic interference by double-stranded RNA in *Caenorhabditis elegans*. *Nature* 391 (6669): 806–11. doi:10.1038/35888.

Florea, B. I., C. Meaney, H. E. Junginger, and G. Borchard. 2002. Transfection efficiency and toxicity of polyethylenimine in differentiated calu-3 and nondifferentiated COS-1 cell cultures. *AAPS PharmSci* 4 (3): 1–11. doi:10.1208/ps040312.

Fratoddi, I., I. Venditti, C. Cametti, and M. V. Russo. 2014. Gold nanoparticles and gold nanoparticle-conjugates for delivery of therapeutic molecules. Progress and challenges. *Journal of Materials Chemistry B* 2 (27): 4204–20. doi:10.1039/C4TB00383G.

Gaur, U., S. K. Sahoo, T. K. De, P. C. Ghosh, A. Maitra, and P. K. Ghosh. 2000. Biodistribution of fluoresceinated dextran using novel nanoparticles evading reticuloendothelial system. *International Journal of Pharmaceutics* 202 (1–2): 1–10. doi:10.1016/S0378-5173(99)00447-0.

Getz, T., J. Qin, I. L. Medintz, et al. 2016. Quantum dot-mediated delivery of siRNA to inhibit sphingomyelinase activities in brain-derived cells. *Journal of Neurochemistry* 139 (5): 872–85. doi:10.1111/jnc.13841.

Glover, D. J., H. J. Lipps, and D. A. Jans. 2005. Towards safe, non-viral therapeutic gene expression in humans. *Nature Reviews Genetics* 6 (4): 299–310. doi:10.1038/nrg1577.

Gordon, S. 2016. Phagocytosis: An immunobiologic process. *Immunity* 44 (3): 463–75. doi:10.1016/j.immuni.2016.02.026.

Graham, F. L., and L. Prevec. 1995. Methods for construction of adenovirus vectors. *Molecular Biotechnology* 3: 207–20. doi:10.1007/BF02789331.

Hait, S. K., and S. P. Moulik. 2015. Gemini surfactants : A distinct class of self-assembling molecules. Curr Sci Gemini Surfactants : 82: 1101–11.

Hamilton, A. J., and D. C. Baulcombe. 1999. A species of small antisense RNA in posttranscriptional gene silencing in plants. *Science* 286 (5441): 950–52. doi:10.1126/science.286.5441.950.

Haupenthal, J., C. Baehr, S. Kiermayer, S. Zeuzem, and A. Piiper. 2006. Inhibition of RNAse a family enzymes prevents degradation and loss of silencing activity of siRNAs in serum. *Biochemical Pharmacology* 71 (5): 702–10. doi:10.1016/j.bcp.2005.11.015.

He, L., and G. J. Hannon. 2004. MicroRNAs: Small RNAs with a big role in gene regulation. *Nature Reviews Genetics* 5 (7): 522–31. doi:10.1038/nrg1379.

Hola, K., Z. Markova, G. Zoppellaro, J. Tucek, and R. Zboril. 2015. Tailored functionalization of iron oxide nanoparticles for MRI, drug delivery, magnetic separation and immobilization of biosubstances. *Biotechnology Advances* 33 (6): 1162–76. doi:10.1016/j.biotechadv.2015.02.003.

Hwang, S., J. Nam, S. Jung, J. Song, H. Doh, and S. Kim. 2014. Gold nanoparticle-mediated photothermal therapy: Current status and future perspective. *Nanomedicine* 9 (13): 2003–22. doi:10.2217/nnm.14.147.

Iijima, S. 1991. Helical microtubules of graphitic carbon. *Nature* 354 (6348): 56–58. doi:10.1038/354056a0.

Iijima, S. 2002. Carbon nanotubes: Past, present, and future. *Physica B: Condensed Matter* 323 (1–4): 1–5. doi:10.1016/S0921-4526(02)00869-4.

Iijima, S. and T. Ichihashi. 1993. Single-shell carbon nanotubes of 1-Nm diameter. *Nature* 363: 603–5. doi:10.1038/363603a0.

Jahanshahi, M. and Z. Babaei. 2008. Protein nanoparticle: A unique system as drug delivery vehicles. *African Journal of Biotechnology* 7 (25): 4926–34. doi:10.4314/ajb.v7i25.59701.

Jeon, S., K. R. Hurley, J. C. Bischof, C. L. Haynes, and C. J. Hogan. 2016. Quantifying intra- and extracellular aggregation of iron oxide nanoparticles and its influence on specific absorption rate. *Nanoscale* 8 (35): 16053–64. doi:10.1039/C6NR04042J.

Jin, C., G. Fotaki, M. Ramachandran, B. Nilsson, M. Essand, and D. Yu. 2016. Safe engineering of CAR T cells for adoptive cell therapy of cancer using long-term episomal gene transfer. *EMBO Molecular Medicine* 8 (7): 702–11. doi:10.15252/emmm.201505869.

Jinek, M., K. Chylinski, I. Fonfara, M. Hauer, J. A. Doudna, and E. Charpentier. 2012. A programmable dual-RNA – guided. *Science* 337: 816–22. doi:10.1126/science.1225829.

Karimi, M., P. Avci, R. Mobasseri, M. R. Hamblin, and H. Naderi-Manesh. 2013. The novel albumin-chitosan core-shell nanoparticles for gene delivery: Preparation, optimization and cell uptake investigation. *Journal of Nanoparticle Research* 15: 1651. doi:10.1007/s11051-013-1651-0.

Ke, X., C. Yang, W. Cheng, and Y. Y. Yang. 2018. Delivery of NF-κB shRNA using carbamate-mannose modified PEI for eliminating cancer stem cells. *Nanomedicine: Nanotechnology, Biology, and Medicine* 14 (2): 405–14. doi:10.1016/j.nano.2017.11.015.

Knapp, C. M., J. He, J. Lister, and K. A. Whitehead. 2016. Lipidoid nanoparticle mediated silencing of Mcl-1 induces apoptosis in mantle cell lymphoma. *Experimental Biology and Medicine* 241: 1007–14. doi:10.1177/1535370216640944.

Knapp, C. M., J. He, J. Lister, and K. A. Whitehead. 2018. Lipid nanoparticle siRNA cocktails for the treatment of mantle cell lymphoma. *Bioengineering & Translational Medicine* 3: 138–47. doi:10.1002/btm2.10088.

Kumar, C. 2005. Gelatin nanoparticles and their biofunctionalization. *Biofunctionalization of Nanomaterials* 1 (M): 330–52. doi:10.1002/9783527610419.ntls0011.

Lam, J. K. W., M. Y. T. Chow, Y. Zhang, and S. W. S. Leung. 2015. siRNA versus miRNA as therapeutics for gene silencing. *Molecular Therapy - Nucleic Acids* 4 (9): 1–20. (Official Journal of the American Society of Gene & Cell Therapy). doi:10.1038/mtna.2015.23.

Lee, R. C., R. L. Feinbaum, and V. Ambros. 1993. The C. elegans heterochronic gene lin-4 encodes small RNAs with antisense complementarity to lin-14. *Cell* 75 (5): 843–54. doi:10.1016/0092-8674(93)90529-Y.

Liu, C. J., C. H. Wang, S. T. Chen, et al. 2010. Enhancement of cell radiation sensitivity by pegylated gold nanoparticles. *Physics in Medicine and Biology* 55 (4): 931–45. doi:10.1088/0031-9155/55/4/002.

Liu, X., H. Yin, Z. Zhang, B. Diao, and J. Li. 2015. Functionalization of lignin through ATRP grafting of poly(2-Dimethylaminoethyl Methacrylate) for gene delivery. *Colloids and Surfaces B: Biointerfaces* 125: 230–37. doi:10.1016/j.colsurfb.2014.11.018.

Lohcharoenkal, W., L. Wang, Y. C. Chen, and Y. Rojanasakul. 2014. Review article protein nanoparticles as drug delivery carriers for cancer therapy. *BioMed Research International* 2014: 4. doi:10.1155/2014/180549.

Love, K. T., K. P. Mahon, G. Christopher, K. A. Whitehead, W. Querbes, J. Robert, J. Qin, et al. 2010. Lipid-like materials for low-dose, in vivo gene silencing. *Proceedings of the National Academy of Sciences USA* 107 (21): 9915–9915. doi:10.1073/pnas.1005136107.

Lu, A. H., E. L. Salabas, and F. Schüth. 2007. Magnetic nanoparticles: Synthesis, protection, functionalization, and application. *Angewandte Chemie - International Edition* 46 (8): 1222–44. doi:10.1002/anie.200602866.

Mahajan, U. M., S. Teller, M. Sendler, et al. 2016. Tumour-specific delivery of siRNA-coupled superparamagnetic iron oxide nanoparticles, targeted against PLK1, stops progression of pancreatic cancer. *Gut* 65 (11): 1838–49. doi:10.1136/gutjnl-2016-311393.

Meng, Z., and M. Lu. 2017. RNA interference-induced innate immunity, off-target effect, or immune adjuvant?" *Frontiers in Immunology* 8: 1–7. doi:10.3389/fimmu.2017.00331.

Miller, D. G., M. A. Adam, and A. D. Miller. 1990. Gene transfer by retrovirus vectors occurs only in cells that are actively replicating at the time of infection. *Molecular and Cellular Biology* 10 (8): 4239–42. doi:10.1128/MCB.10.8.4239 (Updated).

Monfort, D. A., and P. Koria. 2017. Recombinant elastin-based nanoparticles for targeted gene therapy. *Gene Therapy* 24 (10): 610–20. doi:10.1038/gt.2017.54.

Morán, M. C., N. Rosell, G. Ruano, M. A. Busquets, and M. P. Vinardell. 2015. Gelatin-based nanoparticles as DNA delivery systems: Synthesis, physicochemical and biocompatible characterization. *Colloids and Surfaces B: Biointerfaces* 134: 156–68. doi:10.1016/j.colsurfb.2015.07.009.

Munk, M., R. De Souza Salomão Zanette, L. S. De Almeida Camargo, et al. 2017. Using carbon nanotubes to deliver genes to hard-to-transfect mammalian primary fibroblast cells. *Biomedical Physics and Engineering Express* 3 (4). doi:10.1088/2057-1976/aa7927.

Murray, C. B, and C. R Kagan. 2000. "Synthesis and characterization of monodisperse nanocrystals and close packed nanocrystal assemblies." *Annual Review of Materials Science* 30: 545–610.

Nakase, I., S. Kobayashi, and S. Futaki. 2010. Endosome-disruptive peptides for improving cytosolic delivery of bioactive macromolecules. *Biopolymers* 94 (6): 763–70. doi:10.1002/bip.21487.

Naldini, L., U. Blomer, F. H. Gage, D. Trono, and I. M. Verma. 1996. Efficient transfer, integration, and sustained long-term Expression of the transgene in adult rat brains injected with a lentiviral vector. *Proceedings of the National Academy of Sciences USA* 93 (21): 11382–88. doi:10.1073/pnas.93.21.11382.

Nie, S. 2010. Understanding and overcoming major barriers in cancer nanomedicine. *Nanomedicine* 5 (4): 523–28. doi:10.2217/nnm.10.23.

Oh, N., and J. H. Park. 2014. Endocytosis and exocytosis of nanoparticles in mammalian cells. *International Journal of Nanomedicine* 9: 51–63. doi:10.2147/IJN.S26592.

Paul, A., A. Hasan, H. A. Kindi, et al. 2014. Injectable graphene oxide/hydrogel-based angiogenic gene delivery system for vasculogenesis and cardiac repair. *ACS Nano* 8 (8): 8050–62. doi:10.1021/nn5020787.

Peng, L. H., Y. F. Huang, C. Z. Zhang, et al. 2016. Integration of antimicrobial peptides with gold nanoparticles as unique non-viral vectors for gene delivery to mesenchymal stem cells with antibacterial activity. *Biomaterials* 103: 137–49. doi:10.1016/j.biomaterials.2016.06.057.

Peracchia, M. T., S. Harnisch, H. Pinto-Alphandary, et al. 1999. Visualization of in vitro protein-rejecting properties of PEGylated stealth® polycyanoacrylate nanoparticles. *Biomaterials* 20 (14): 1269–75. doi:10.1016/S0142-9612(99)00021-6.

Petryayeva, E., and U. J. Krull. 2011. Localized surface plasmon resonance: nanostructures, bioassays and biosensing-a review. *Analytica Chimica Acta* 706 (1): 8–24. doi:10.1016/j.aca.2011.08.020.

Poiesz, B. J., F. W. Ruscetti, A. F. Gazdar, P. A. Bunn, J. D. Minna, and R. C. Gallo. 1980. Detection and isolation of type C retrovirus particles from fresh and cultured lymphocytes of a patient with cutaneous T-cell lymphoma. *Proceedings of the National Academy of Sciences USA* 77 (12): 7415–19. doi:10.1073/pnas.77.12.7415.

Qiu, Y., Y. Liu, L. Wang, et al. 2010. Surface chemistry and aspect ratio mediated cellular uptake of Au nanorods. *Biomaterials* 31 (30): 7606–19. doi:10.1016/j.biomaterials.2010.06.051.

Ralph, J., K. Lundquist, G. Brunow, et al. 2004. Lignins: Natural polymers from oxidative coupling of 4-hydroxyphenyl- propanoids. *Phytochemistry Reviews* 3 (1–2): 29–60. doi:10.1023/B:PHYT.0000047809.65444.a4.

Ran, F. A., P. D. Hsu, J. Wright, V. Agarwala, D. A. Scott, and F. Zhang. 2013. Genome engineering using the CRISPR-Cas9 system. *Nature Protocols* 8 (11): 2281–2308. doi:10.1038/nprot.2013.143.

Rao, D. D., J. S. Vorhies, N. Senzer, and J. Nemunaitis. 2009. siRNA vs. shRNA: Similarities and differences. *Advanced Drug Delivery Reviews* 61 (9): 746–59. doi:10.1016/j.addr.2009.04.004.

Re, F., M. Gregori, and M. Masserini. 2012. Nanotechnology for neurodegenerative disorders. *Maturitas* 73 (1): 45–51. doi:10.1016/j.maturitas.2011.12.015.

Robbins, P. D., and S. C. Ghivizzani. 1998. Viral vectors for gene therapy. *Pharmacology & Therapeutics* 80 (1): 35–47. doi:10.1016/S0163-7258(98)00020-5.

Rodriguez, A. 2004. Identification of mammalian microRNA host genes and transcription units *Genome Research* 14 (10a): 1902–10. doi:10.1101/gr.2722704.

Selby, L. I., C. M. Cortez-Jugo, G. K. Such, and A. P.R. Johnston. 2017. Nanoescapology: Progress toward understanding the endosomal escape of polymeric nanoparticles. *Wiley Interdisciplinary Reviews: Nanomedicine and Nanobiotechnology* 9: e1452. doi:10.1002/wnan.1452.

Silva, J. M., M. Z. Li, K. Chang, W. Ge, M. C. Golding, R. J. Rickles, D. Siolas, et al. 2005. Second-generation shRNA libraries covering the mouse and human genomes. *Nature Genetics* 37 (11): 1281–88. doi:10.1038/ng1650.

Siu, K. S., D. Chen, X. Zheng, et al. 2014. Non-covalently functionalized single-walled carbon nanotube for topical siRNA delivery into melanoma. *Biomaterials* 35 (10): 3435–42. doi:10.1016/j.biomaterials.2013.12.079.

Straarup, E. M., N. Fisker, M. Hedtjärn, et al. 2010. Short locked nucleic acid antisense oligonucleotides potently reduce apolipoprotein B mRNA and serum cholesterol in mice and non-human primates. *Nucleic Acids Research* 38 (20): 7100–11. doi:10.1093/nar/gkq457.

Sun, J., and Y. Jin. 2014. Fluorescent Au nanoclusters: Recent progress and sensing applications. *Journal of Materials Chemistry C* 2 (38): 8000–8011. doi:10.1039/C4TC01489H.

Ten, E., C. Ling, Y. Wang, A. Srivastava, L. A. Dempere, and W. Vermerris. 2014. Lignin nanotubes as vehicles for gene delivery into human cells. *Biomacromolecules* 15 (1): 327–38. doi:10.1021/bm401555p.

Thiel, K. W., and P. H. Giangrande. 2009. Therapeutic applications of DNA and RNA aptamers. *Oligonucleotides* 19 (3): 209–22. doi:10.1089/oli.2009.0199.

Tian, W., and Y. Ma. 2012. Insights into the endosomal escape mechanism via investigation of dendrimer–membrane interactions. *Soft Matter* 8 (23): 6378. doi:10.1039/c2sm25538c.

Urry, D. W. 1997. Physical chemistry of biological free energy transduction as demonstrated by elastic protein-based polymers. *The Journal of Physical Chemistry B* 101 (51): 11007–28. doi:10.1021/jp972167t.

Valizadeh, A., H. Mikaeili, M. Samiei, et al. 2012. Quantum dots: Synthesis, bioapplications, and toxicity. *Nanoscale Research Letters* 7: 480. doi:10.1186/1556-276X-7-480.

Varkouhi, A. K, M. Scholte, G. Storm, and H. J. Haisma. 2011. Endosomal escape pathways for delivery of biologicals. *Journal of Controlled Release* 151 (3): 220–28. doi:https://doi.org/10.1016/j.jconrel.2010.11.004.

Volpe, T. 2002. Regulation of heterochromatic silencing and histone H3 Lysine-9 by RNAi." *Science* 297 (5588): 1833–37. doi:10.1126/science.1074973.

Walters, R., R. P. Kraig, I. Medintz, et al. 2012. Nanoparticle targeting to neurons in a rat hippocampal slice culture model. *ASN Neuro* 4 (6): art:e00099. doi:10.1042/AN20120042.

Walters, R., I. L. Medintz, J. B. Delehanty, et al. 2015. The role of negative charge in the delivery of quantum dots to neurons. *ASN Neuro* 7 (4). doi:10.1177/1759091415592389.

Wang, L., J. Shi, H. Zhang, et al. 2013. Synergistic anticancer effect of RNAi and photothermal therapy mediated by functionalized single-walled carbon nanotubes." *Biomaterials* 34 (1): 262–74. doi:10.1016/j.biomaterials.2012.09.037.

Wang, P., L. Zhang, W. Zheng, et al. 2018. Thermo-triggered release of CRISPR-Cas9 system by lipid-encapsulated gold nanoparticles for tumor therapy. *Angewandte Chemie - International Edition* 57 (6): 1491–96. doi:10.1002/anie.201708689.

Wang, Y., M. Zheng, F. Meng, J. Zhang, R. Peng, and Z. Zhong. 2011. Branched polyethylenimine derivatives with reductively cleavable periphery for safe and efficient in vitro gene transfer. *Biomacromolecules* 12 (4): 1032–40. doi:10.1021/bm101364f.

Werengowska-Ciećwierz, K., M. Wisniewski, A. P. Terzyk, and S. Furmaniak. 2015. The chemistry of bioconjugation in nanoparticles-based drug delivery system. *Advances in Condensed Matter Physics* 2015. doi:10.1155/2015/198175.

Whitehead, K. A., J. R. Dorkin, A. J. Vegas, et al. 2014. Degradable lipid nanoparticles with predictable in vivo siRNA delivery activity. *Nature Communications* 5: 4277. doi:10.1038/ncomms5277.

Xu, C., D. Yang, L. Mei, et al. 2013. Encapsulating gold nanoparticles or nanorods in graphene oxide shells as a novel gene vector. *ACS Applied Materials and Interfaces* 5 (7): 2715–24. doi:10.1021/am400212j.

Ylä-Herttuala, S. 2012. Endgame: Glybera finally recommended for approval as the first gene therapy drug in the european union. *Molecular Therapy : The Journal of the American Society of Gene Therapy* 20 (10): 1831–32. doi:10.1038/mt.2012.194.

Zamore, P. D., T. Tuschl, P. A. Sharp, and D. P. Bartel. 2000. RNAi: Double-stranded RNA directs the ATP-dependent cleavage of mRNA at 21 to 23 nucleotide intervals. *Cell* 101 (1): 25–33. doi:10.1016/S0092-8674(00)80620-0.

Zhang, Y., Q. Jiang, B. Bi, L. Xu, J. Liu, R. Zhuo, and X. Jiang. 2018. A bioreducible supramolecular nanoparticle gene delivery system based on cyclodextrin-conjugated polyaspartamide and adamantyl-terminated polyethylenimine. *Journal of Materials Chemistry B* 6 (5): 797–808. doi:10.1039/C7TB02170D.

Zhang, Z., J. Wang, and C. Chen. 2013. Gold nanorods based platforms for light-mediated theranostics. *Theranostics* 3 (3): 223–38. doi:10.7150/thno.5409.

6

Bio-Inspired DNA Nanoswitches and Nanomachines: Applications in Biosensing and Drug Delivery

Arnaud Desrosiers and
Alexis Vallée-Bélisle
Université de Montréal

6.1 Introduction

Nature has evolved sophisticated biomolecular switches and nanomachines through billions of years of evolution that can change the shape or activity in response to various stimuli such as temperature, light, pH changes, small molecules, proteins and hormones. One example of such class of natural molecular switches is G protein-coupled receptors (GPCRs), a large protein family composed of more than 1,000 members [1,2]. These proteins can sense and respond to a plethora of extracellular stimuli to activate various biological outputs inside the cell (Figure 6.1). Such bio-nanosystems are of great inspiration to engineers interested in building nanosystems that produce useful output in response to specific chemical signal.

Nanoengineers have explored all types of chemistry in an attempt to recreate switches and nanomachines found in nature (see for example 2016 Nobel Prize [3–6]). While protein chemistry should, in principle, represent a material of choice to do so, it has revealed to be much more complicated than expected: using polypeptides to build useful switches from scratch remains extremely challenging [7–10]. Other types of chemistry have been also exploited with relatively good success (e.g. see 2016 chemistry Nobel Prize), but the switches and nanomachines engineered so far often remain simple proof of concept [3–6]. In recent years, DNA chemistry, with its high programmability and life-compatible features, has risen into a "happy medium" biopolymer that enables to rationally engineer a wide variety of useful nanomachines and molecular switches [11–18]. Our capacity to create specific DNA structures with defined thermodynamics has played a major role in propelling this research field into the spotlight [13,19,20]. Due to its simplicity, DNA chemistry has also enabled to recreate many highly complex signaling mechanism employed by natural switches and nanomachines (e.g. allosteric mechanism, population-shift regulation [21–26] (Figure 6.2)).

Artificial DNA switches are generally programmed and engineered to generate a measurable signal (e.g. fluorescent, electrochemical) or to trigger a precise activity such as drug release in response to a specific input stimulus. They typically alternate between two conformations often referred to "OFF" and "ON". These switches are now used as tools for biosensing (e.g. Kramer et al. 1996 molecular beacon for medical diagnostic [27]), fundamental research and drug delivery applications (see Ref. [12]) for a recent review on this field). In this chapter, we will briefly exemplify why DNA is an ideal polymer to build a wide range of useful nanotechnology. We will explain how we design and engineer DNA switches for useful biological applications such as biosensing and drug delivery. Finally, we will conclude with an outlook on the future of DNA switches and nanomachines in the field of medicine.

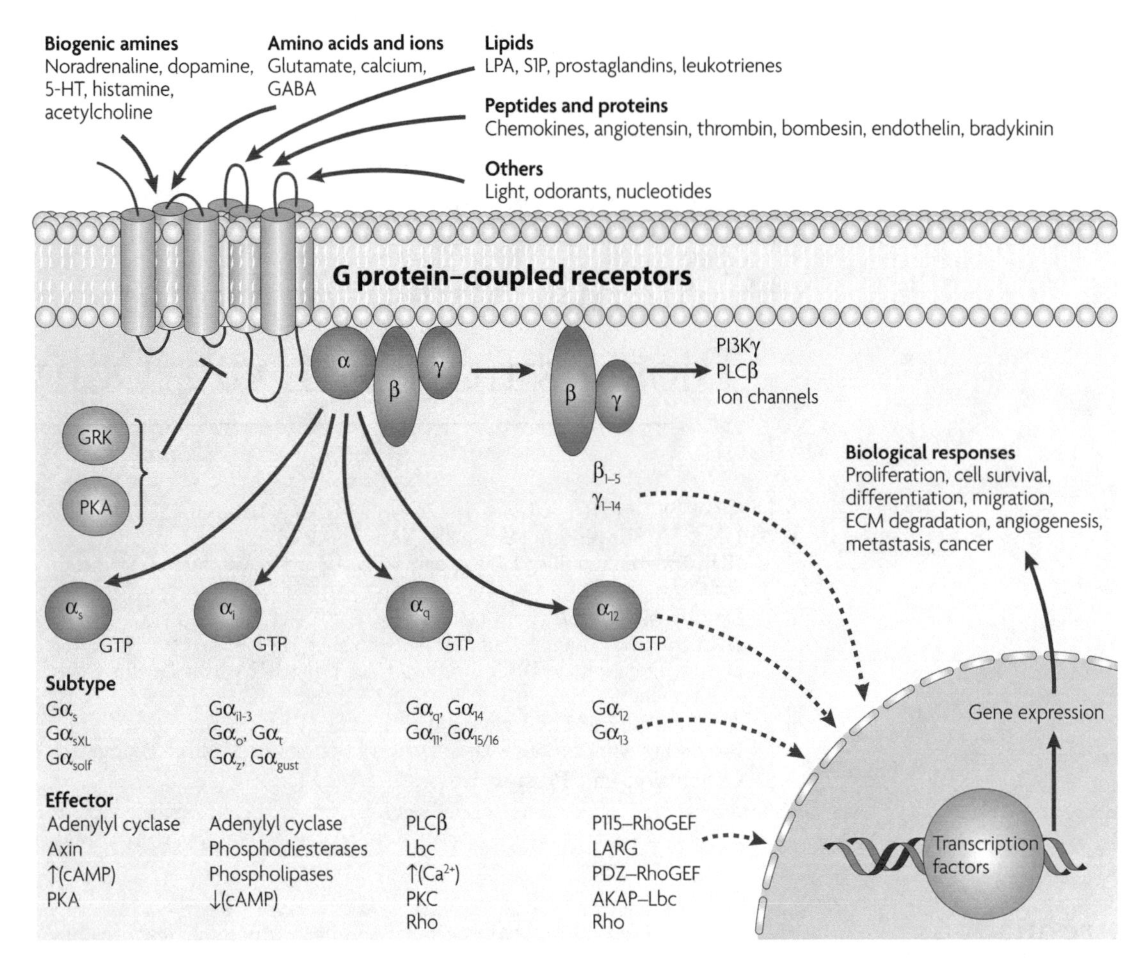

FIGURE 6.1 GPCRs are biomolecular switches that sense a wide variety of input stimuli and transduce this information into a wide variety of signal output. These represent a great inspiration for building nanoswitches or nanomachines for various applications. (Reprinted by permission from Springer customer service center gmbh: Springer Nature, Nature Reviews Cancer. Reference [2]. Copyright 2007.)

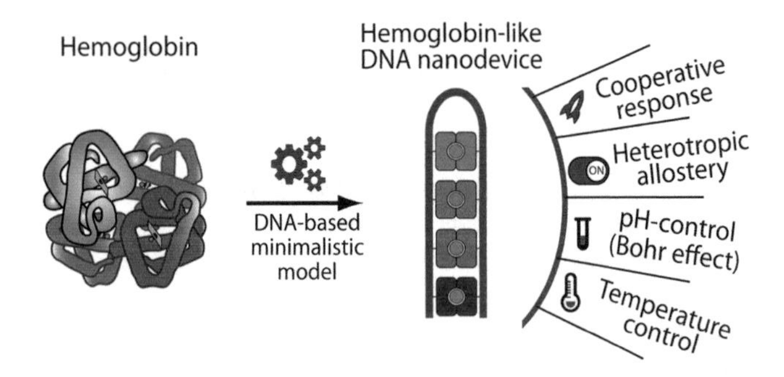

FIGURE 6.2 Bio-inspired DNA switch recreating hemoglobin's response behavior. (Adapted with permission from Ref. [26]. Copyright 2017 American Chemical Society.)

6.2 DNA: A Simple and Versatile Nanomaterial

Even if most natural switches are composed of proteins and to a lower extent of RNA, most recent advances in switch engineering in the field of nanotechnology have been carried out with DNA [13,19,20,28]. Although mainly studied for its genetic carrier role since its discovery by Watson and Crick, the immense potential of DNA programmability has been first applied in the 1980s and early 1990s to build specific 3D structures [29–34] with some displaying catalytic (e.g. DNAzyme) [35] or binding activities (e.g. DNA aptamers) [36,37]. In this section, we will go briefly over the three main reasons that make DNA an ideal programmable biopolymer for engineering nanosystems and will conclude with a brief overview on artificial automated DNA synthesis.

6.2.1 High Programmability

The first reason that makes DNA an ideal polymer is that its simple base pairing code, A-T and C-G, enables to predict its secondary structure and thermodynamics. This simple base pairing code, for example, was well exploited by software prediction tools such as *mfold* and NUPACK [38,39] to predict the folding energy of any DNA sequence into unimolecular or bimolecular structures in different conditions (e.g. salt concentration, temperature). This simple feature has fueled most of the advances in DNA nanotechnology [40–42] since the pioneering work of Seeman in the

1990s [43]. In addition to thermodynamic control, various strategies have been proposed to control DNA association kinetic to create more optimized DNA nanomachines [44,45]. Another feature highlighting the high programmability of DNA chemistry is that DNA sequences can be selected to bind almost any target of interest with high specificity and affinity (e.g. hydrogen ions, metal ions, small organics, proteins and even living cells, bacteria or virus) through natural evolution techniques like SELEX (see Section 6.3.1) [36,37].

6.2.2 Life-Compatible

Second, DNA switches are nanosized and already life-compatible, which allows the engineering of nanotechnological tools that can be directly applied to living systems. Most of the time, DNA requires very little modifications to resist living environment (e.g. mammalian cells or multicellular organisms) while still maintaining biocompatibility properties [46]. They can also be adapted easily to escape immunological response, which could lead to the development of therapeutic agents that could be used in medical settings [47].

6.2.3 Simple Automated Artificial Synthesis

The third and last reason that makes DNA an ideal polymer for nanotechnology is the simplicity of artificial DNA synthesis that produces DNA sequences in high yield and at a low cost (US$0.05–0.15 per nucleotide) [48]. While there are multiple different methods to synthesize oligonucleotides, the phosphoramidite approach [49] is the most often employed. Its adaptation on solid support allowed to build automated systems for faster and more efficient synthesis [50].

The automated method consists of four successive steps repeated for every nucleotide from the starting 3′ to 5′ end (i.e. in the other direction than DNA biosynthesis) (Figure 6.3): (i) detritylation, (ii) coupling, (iii) oxidation and (iv) capping. This synthesis method is usually performed on a column containing a controlled pore glass (CPG) solid support bearing a specific nucleotide load protected with a 5′-DMT ready for the first synthesis step. In the first step, the protective group 5′-DMT (4,4′-dimethoxytrityl) that protects the nascent DNA from unwanted coupling is removed leaving a 5′-OH on the solid support. Then, the next nucleotide phosphoramidite monomer from the DNA sequence is injected allowing its coupling to the DNA on solid support. The coupling step leaves a phosphite triester for the internucleotide linkage that is unstable and needs to be converted to the phosphate triester during the oxidation step. The last step, capping, enables the removal of 5′-OH unreacted DNA that is produced due to the impossibility of reaching a 100% coupling efficiency. If not removed, these sequences could keep reacting and introduce DNA sequences missing some

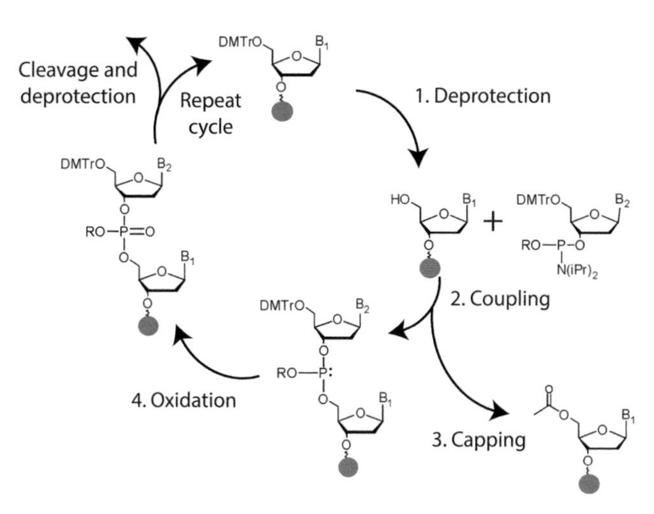

FIGURE 6.3 Automated organic synthesis of DNA oligonucleotide on solid support with the phosphoramidite method [50].

nucleotides. DNA oligonucleotides are then cleaved from the solid support and nucleobases are de-protected. Finally, DNA oligonucleotides are purified by various methods such as reverse-phase cartridge, high-performance liquid chromatography (HPLC) and denaturing electrophoresis.

Many research groups are ordering their specific oligonucleotides (modified or not) from companies (e.g. IDT Biosearch, IBA, Sigma Aldrich) by simply entering their sequence in the webserver. These are generally received within 2 days for unmodified oligonucleotides or in less than a month when special modifications are requested. Many academic laboratories also perform their own DNA synthesis using an automatic synthesizer and commercially available phosphoroamidite monomers. Commercially available phosphoroamidites monomers also include a wide diversity of modified nucleotide (aside normal A, T, C and G) with additional functional groups (e.g. amine, thiol, azide and alkyne), and many other more complex groups (e.g. fluorophore, quencher, cholesterol, tocopherol). It is also possible to work beyond commercially available phosphoroamidite monomers by simply synthesizing either the phosphoroamidites [51–53] or the CPG solid support [54–56] in house, which is routinely performed by various laboratories.

6.3 Engineering DNA Switches

DNA switches engineering is a key task in the field of DNA nanotechnology. Rational design of DNA switches can be summarized into four main steps: (i) a chemical input is first identified based on your research interest (e.g. what do you want to measure; what is the target stimulus you want your DNA switch to respond to); (ii) a DNA scaffold is selected and designed to provide a conformational change mechanism that is triggered by the selected input; (iii) a signal output is then mounted on the DNA switch to translate the input recognition and conformational change into a relevant output signal that will fit your specific application (e.g. fluorescence, current, drug release) and (iv) the DNA switch must be adapted to work in complex environment with specific chemical modifications (e.g. cellular uptake, nuclease resistance). These four steps will be discussed in detail in the following sections.

6.3.1 DNA Switches Input

DNA chemistry, like protein chemistry, can be adapted to respond specifically to a large variety of chemical inputs (Figure 6.4). Therefore, the first and most important step in DNA switch engineering is to identify the appropriate chemical input that suits the application of your choice. For example, if your application is biosensing, you should find an analyte of interest that can help you answer your scientific question (e.g. protons can be selected if pH measurement is key to your application). Another example, drug delivery DNA switches could involve a disease biomarker as the input signal to trigger the switch to release a specific drug. As such, we can divide the inputs used for DNA switch in two categories: chemical stimuli or physical phenomena. For example, one of the simplest physical change that can impact DNA folding is temperature. DNA unfolding due to temperature increase was one of the first DNA physical property studied extensively and is at the center of a myriad of scientific applications like polymerase chain reaction [57,58].

Melting temperature of DNA (Tm where 50% of DNA strands are in the single-stranded form) was also found to be dependent on the GC content of the DNA duplex since GC base pair is more stable than AT (that is because GC base pair forms three hydrogen bridges compared to two in AT base pair) or to salt concentration (an example of chemical stimuli) [59,60]. This important feature of DNA

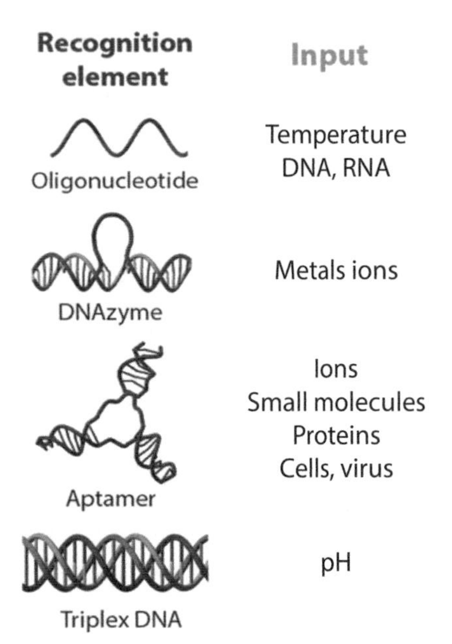

FIGURE 6.4 DNA structure can bind and/or respond to a plethora of input stimuli. (Adapted from Ref. [12] with permission from The Royal Society of Chemistry.)

allowed the engineering of various DNA switches like DNA nanothermometers [61–65].

Like natural GPCR, light can be used as physical stimuli to trigger DNA switches by the specific incorporation of functional groups bearing azo-benzene in the DNA strand that can switch between *cis* and *trans* conformation [66–69]. Another physical property of DNA, its natural conductance [70,71], can be selected as an input signal and successfully used to develop structure-switching mechanism by direct current flow through the DNA switch [72,73]. Electrochemical stimuli is another example of physical phenomena that can be employed to design a structure switching mechanism (e.g. Pb(II) reduction) [74].

On the other hand, chemical stimuli are based on molecules, ions or large biomolecules that bind various DNA structures, which can be used as inputs for DNA switches. The simplest chemical stimuli used to build DNA switch are hydrogen ions which allowed engineering of nanosized pH meter [75]. These DNA switches are based on specific DNA structures such as i-motif [76,77], poly dA helix [78] or parallel triplex [16], which need acidic pH to fold. Hence, a small variation in pH can result in the unfolding of the DNA switch, which allows the precise monitoring of pH variations at the nanoscale. These tools can be used to measure pH inside living cells [79,80] or to engineer drug delivery systems [81]. Similarly, other ions can be used for the folding of specific DNA structures like G-quadruplex that require potassium ion to form [82], which can be adapted into a simple potassium sensors [83] or into sophisticated DNA nanochannels [84]. Other G-quadruplex DNA structures are also of interest for their ability to bind specific metal ions such as Pb(II) [85], Tl(I) [86] and many other monovalent and divalent metal ions [87]. Small organic molecules, like members of the porphyrin family, can also bind various G-quadruplex structures [88,89]. DNA mismatch can also be used to detect metal ions like T:T mismatch for Hg(II) [90] or C:C mismatch for Ag(I) [91]. Nucleic acid sequences made of either DNA or RNA form perhaps one of the simplest class of molecular inputs that can be incorporated and easily programmed within the DNA switch (as it will be described in the next section). An example of DNA switch based on nucleic acid recognition is the molecular beacon, widely used for medical diagnosis since its conception by Tyagi and Kramer [27].

In the event that a specific input target of interest has no known DNA binding sequence, SELEX methods can be employed to identify nucleotide sequences that display high affinity for small organics [92], proteins [93,94] and even viruses [95] or whole bacteria [96] (mostly through surface proteins/sugars). Briefly, SELEX (short for systematic evolution of ligands by exponential enrichment) is a method developed in 1990 [37,97] in which specific nucleotide sequences (referred to as aptamer with typically 15–40 nucleotides) from random libraries are selected to bind specific targets with high affinity through several rounds of selection. This method was first developed for RNA selection but was then rapidly adapted to select DNA sequences [36]. In this method, one selection round typically consists in a simple binding experiment where DNA sequences bound to the target of interest are amplified and used for the next selection round.

6.3.2 Engineering Switching Mechanisms

Engineering switching mechanism represents, without a doubt, the most important step in the development of DNA switches. This mechanism involves a change in structure from conformation 1 to conformation 2 in the presence of the input selected (also often referred to as OFF and ON states, see output Section 6.3.3). This structural change from conformation 1 to conformation 2 will be adapted to produce a relevant signal output as it will be detailed in the next section. When engineering the switching mechanism, one should always keep in mind the desired signal output mechanism (e.g. fluorescence, electrochemistry, drug delivery). For example, signal output performance (signal-to-noise ratio) is often related to the magnitude of the conformational change that occurs in the presence of the selected input. Some DNA structures intrinsically go through a significant conformational change in the presence of the input. For instance, some G-quadruplex structures will go from a single-stranded DNA to a four-stranded quadruplex structure in the presence of the input (e.g. potassium ions) while other structures will only undergo mild-to-no conformational changes upon binding to their targets. To overcome this limitation and engineer a structure-switching mechanism in a specific DNA structure, one can take advantage of the population-shift model and stabilize an alternative non-binding structure within the original DNA sequence [42,98]. In this strategy, a slightly modified DNA sequence can be made to adopt a distinct conformation from the original one (i.e. a non-competent binding state) (Figure 6.5). This non-binding conformation will switch back and forth with the original conformation through chemical equilibrium in the absence of input molecule. Therefore, one can stabilize this non-binding conformation so that in the absence of input molecule, most of the DNA switches are in the non-binding conformation. Stabilizing the non-binding conformation also means there is now an energetic price required for the DNA switch to adopt its original conformation (to undergo a conformational switching). This energetic price is paid through binding of the input molecule, that form a more stable complex with the DNA switch than the non-binding conformation. This system is conveniently described using a three-state population-shift model: a binding-competent state and an input molecule bound state that both have the same conformation and a non-competent binding state that has a distinct conformation.

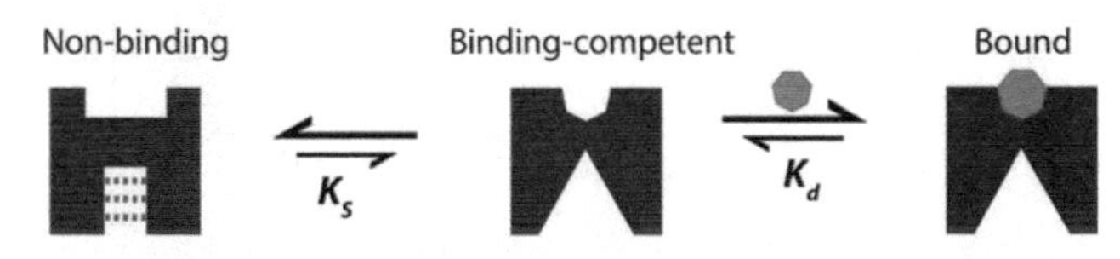

FIGURE 6.5 Population-shift mechanism: a non-binding DNA conformation shifts toward a binding-competent conformation upon binding to a specific chemical input. (Adapted from Ref. [12] with permission from The Royal Society of Chemistry.)

Many strategies can be used to engineer a non-binding state within the original DNA structure (Figure 6.6). A first approach consists in adding complementary nucleotides at both extremities of the original DNA sequence to force a stem-loop conformation (Figure 6.6a). This strategy was employed successfully by Kramer et al. in 1996 when they designed now famous molecular beacons that open upon binding to specific complementary nucleotides sequence [27]. Another strategy is to destabilize the original DNA structure by introducing mutations in the binding-competent conformation, which will favor an alternative non-binding conformation (Figure 6.6b) [99]. In the presence of input molecule, this equilibrium toward the non-binding conformation will shift back toward the binding-competent state to form a switch-input complex. It should be noted that mutation should be introduced with care in this strategy to avoid changing the intrinsic K_D and specificity of the DNA sequence with its input molecule. To avoid inserting mutations in the DNA sequence, a simpler strategy is to design a DNA sequence that is complementary to the switch sequence (Figure 6.6c). The addition of a complementary DNA will sequester the switch into a double-stranded DNA whose conformation will be different from the binding-competent

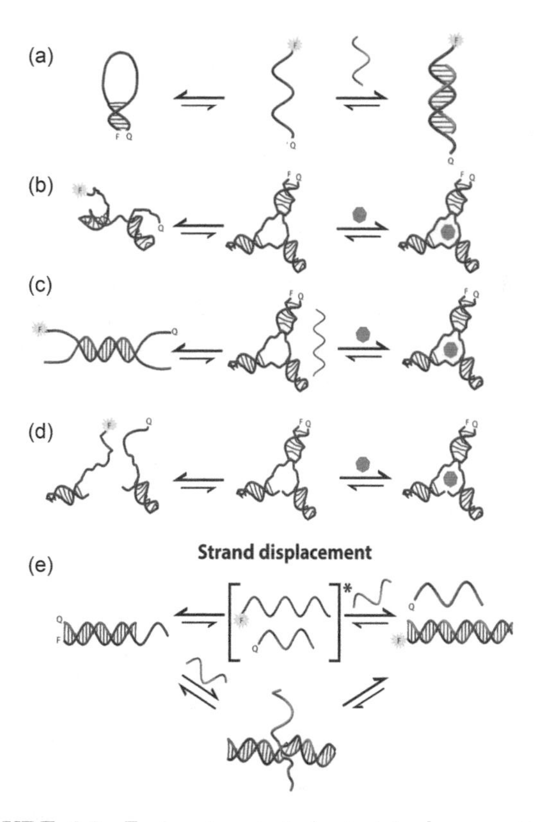

FIGURE 6.6 Engineering strategies to introduce a switching mechanism in DNA recognition elements. F: fluorophore; Q: quencher. (Adapted from Ref. [12] with permission from The Royal Society of Chemistry.)

conformation. Here, the chemical input will compete with the complementary DNA and the switch will undergo a structure switching mechanism upon input binding [100]. This approach is particularly useful when working with DNA aptamers that adopt a unique DNA conformation as it is easy to design a complementary DNA sequence [41,101]. Another way to avoid mutations is to divide the switch sequence in two parts that can dimerize to form the binding-competent state (Figure 6.6d) [98]. This strategy is also largely employed with DNA aptamer sequence whose structures can be roughly modeled using available computational tools like *mfold*. A last strategy, called strand displacement, can also be employed to build DNA switches (Figure 6.6e). In this strategy, the "switch" sequence is sequestered in a non-binding state with a complementary DNA strand like described previously except that there is a single-stranded terminal anchor (toehold domain) that is not bound. This toehold domain is free to hybridize to a part of an input DNA sequence (i.e. the input molecule in this case; see Ref. [100]). This process allows for strand exchange in the presence of the input molecule that results in a significant conformational change [45]. Here the input molecule is an invading DNA strand, but it is possible to adapt the strategy, so the invading input molecule can be replaced by proteins or small molecules. To achieve this, one can employ aptamer-bound toeholds [102,103], G-quadruplex structures [104] or even pH-sensitive DNA sequences [105].

The thermodynamics of the population-shift model controls the performance of the DNA switch. To understand this, let's consider the following two chemical equilibrium: K_S and K_D based on different conformations (C) of the switch:

$$C^{\text{non-binding}} \xrightleftharpoons{K_S} C^{\text{binding}}$$

$$C^{\text{input}} \xrightleftharpoons{K_D} C^{\text{binding}} + \text{input}$$

where K_D represents the affinity of the chemical input for the binding-competent conformation. However, the introduction of a non-binding conformation and, thus K_S, modifies the apparent binding affinity of the switch with the following relationship:

$$K_S = \frac{[\text{binding-competent state}]}{[\text{non-binding state}]}$$

$$K_D^{\text{apparent}} = K_D^{\text{instrinsic}} \cdot \left(\frac{1 + K_S}{K_S}\right)$$

This increase in apparent K_D represents the energetic price required to introduce the non-binding conformation (Figure 6.7a). As such, the K_S value should be chosen wisely since the lower its value, the higher the apparent K_D will be (Figure 6.7a). K_S may often be tuned to optimize the activity of the switch at the relevant concentration of the chemical input [21]. When the lowest detection limit of a switch is needed, K_S may be set between 0.1 and 1 as this maintains a low apparent K_D while between 90%–50% of the DNA switch goes from non-binding state to binding-competent and bound state in the presence of the input molecule, respectively (Figure 6.7b and c).

The high programmability nature of DNA is only fully exploited when we couple switching engineering strategies with DNA structure software prediction like freely available webserver *mfold* [38] and NUPACK [39]. These webservers are able to compute the folding and free energy of a DNA sequence at a specific temperature and salt concentration for unimolecular or bimolecular processes. Therefore, these tools are crucial to design non-binding state as they are able to predict the K_S with the Gibbs free energy:

$$\Delta G = -RT \cdot \text{In}(K_S)$$

For *mfold*, the unimolecular folding prediction is performed with the *mfold* application tool while the bimolecular folding is performed with DINAmelt application [106]. The K_S calculated by *mfold* can then be validated experimentally using urea [107] or temperature [108] denaturation curves.

6.3.3 Engineering Output Functions in DNA Switches

One of the last steps to engineer a functional DNA switch is to translate the conformational change induced by a chemical input into a relevant and measurable output signal.

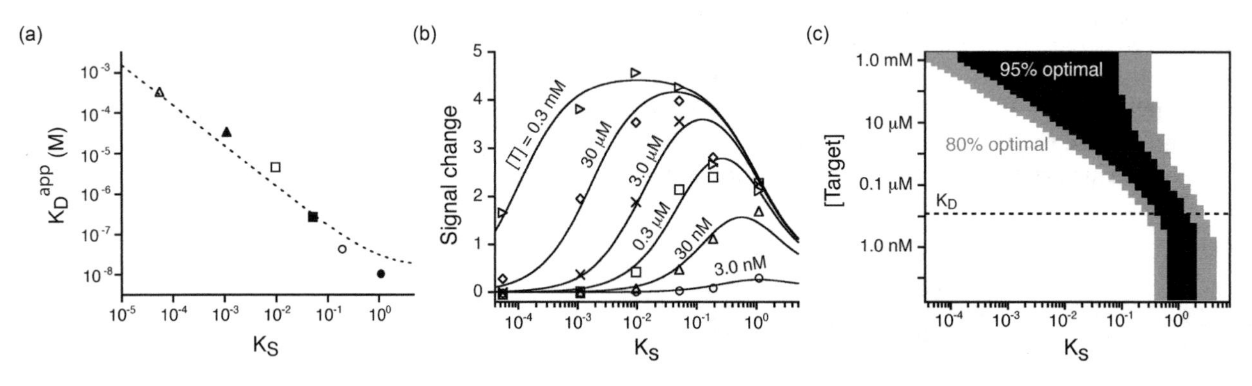

FIGURE 6.7 (a) Stabilizing a non-binding conformation to create a switching molecule with an equilibrium K_S decreases the apparent affinity K_D^{app} of the input molecule for the switch. (b) The amplitude of signal change induced by binding of the input molecule to the DNA recognition element depends on the K_S of the non-binding conformation. (c) A K_S between 0.1 and 1 is ideal to optimize the detection limit of the switch. (Reprinted with permission from Ref. [21]. Copyright 2009 National Academy of Sciences.)

There is a wide variety of output signaling mechanisms that can be discussed, but due to limited space, we will focus on the most commonly used strategies. The simplest and most widely used output signal implemented in DNA switch is fluorescence spectroscopy either by quenching or FRET (Förster resonance energy transfer) [109,110]. For instance, one of the first DNA switch developed, the molecular beacon by Kramer, employed fluorescence quenching as the output signal to monitor binding of a target DNA sequence (Figure 6.8a). To render the DNA switch fluorescent, a fluorophore (e.g. fluorescein, cyanine dyes and many others) is inserted at one of its extremities while a molecular quencher (i.e. a molecule that can absorb the light emitted by the fluorophore) or another fluorophore that can absorb light and re-emit at a higher wavelength is attached at the other extremity. A significant change in fluorescence signal can then be detected when the dyes are separated by a distance larger than their Förster radius (typically ~50Å) [111]. Since DNA switches alternate between two conformations, one can strategically place the fluorophore/quencher pair so that they are physically close (~20Å) in one conformation and far from each other (>50Å) in the other conformation. Binding of the input molecule will thus either bring fluorophore/quencher pair close or far from each other. Fluorescent switches can either be qualified as "signal OFF" or "signal ON" switches when the fluorescence decreases or increases upon input molecule addition. Fluorescence is often used for its high sensitivity and simplicity in biosensing applications although it cannot be used in medium that cannot transmit light (e.g. whole blood).

Electrochemistry is also often employed to monitor conformational changes within DNA switches (e.g. cyclic voltammetry (CV) or squarewave voltammetry (SWV)). In this approach, DNA is often covalently attached on an electrode surface via a sulfur linker, for example, if a gold surface is employed [115]. Other linkers may be used for other surface such as glass or glassy carbon [116]. Conformational changes are monitored by adding an electrochemical reporter (e.g. methylene blue or ferrocene) on the switch that undergoes redox reactions when located near the gold surface (Figure 6.8b). As described for fluorescence, this reporter needs to be strategically placed so that it is either brought closer (signal ON) or farther from the surface (signal OFF) in the presence of the input molecule. Classic electro-active molecules employed are methylene blue or ions like ferricyanide [117]). One great advantage of electrochemical output is their ability to signal efficiently directly in whole blood. Electrochemical DNA switches thus show much promises for point-of-care sensing applications in medical diagnostics (Figure 6.8c) [114,118–120].

Alternative signal output approaches like colorimetric changes employ nanoparticles to translate a DNA switch conformational change. In a colorimetric assay, for example, a visible color change can be detected upon nanoparticle aggregation in the presence of a specific input molecule [121,122]. To do so, single-stranded DNA switches are often used to prevent nanoparticles from aggregating. In the presence of their input molecule, these switches fold and decrease their affinity for the nanoparticles surface, which results in their aggregation. This approach can be adapted to a variety of input ranging from pH changes [123] to small organics and proteins [124]. Other approaches involve electrochemical plasmonic sensing systems [125] and surface-enhanced Raman spectroscopy [126,127].

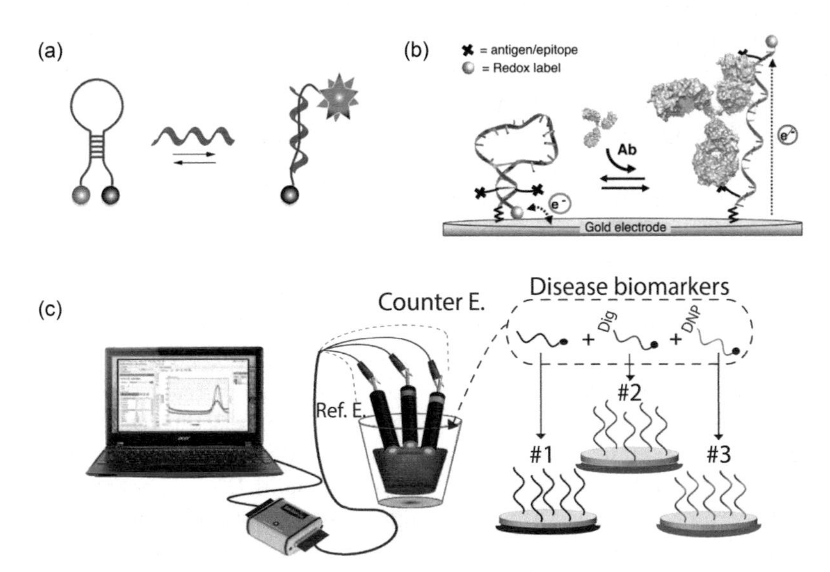

FIGURE 6.8 (a) Fluorescence output to monitor DNA switches requires a conformational change upon binding on the input molecule that brings a fluorophore/quencher pair far from each other. (b) Electrochemical output can be used to monitor DNA switches in which a redox element is brought far from an electrode surface upon binding of the input molecule. (c) Electrochemical output can be adapted to portable systems and multiplexed for point-of-care approaches. (Panel A is adapted with permission from Ref. [112]. Copyright 2010 American Chemical Society. Panel B is adapted with permission from Ref. [113]. Copyright 2012 American Chemical Society. Panel C is adapted with permission from Ref. [114]. Copyright 2015 American Chemical Society.)

6.3.4 Adapting DNA Switches for Complex Environment

An increasing proportion of DNA switches are used for applications related to living organism whether *in vitro* or *in vivo*. Such applications include biosensing inside living cells or drug delivery systems in living animals like mice. As a result, various strategies are required to adapt DNA switches for harsher environment (e.g. nuclease-rich environments) or simply to allow them to reach their final destination (e.g. cellular uptake) [47,128]. These modifications are not intended to impact the switching mechanism but rather to adapt DNA switches to the cellular environment. One of the biggest obstacle when employing DNA switches in living organism is the presence of nucleases that can easily degrade unmodified DNA (e.g. half-life of short unmodified DNA oligonucleotides is in the order of minutes in whole blood) [129]. To overcome this limitation, a myriad of chemical modifications can render DNA resistant to nuclease degradation with often little impact on the DNA thermodynamic and function (although it should be validated experimentally case by case). One common target for chemical modification is the DNA backbone structure in which internucleotide linkage can be modified using phosphorothioate [130], phosphonoacetate [131], morpholino oligomers [132] or methyl phosphonate [133] (Figure 6.9). Other chemical modification will target the nucleotide 2′ position of DNA to reduce nuclease degradation. Examples of 2′ chemical modifications include: 2′-fluoro [134], 2′O-methyl [135] and locked nucleic acid (LNA)-oligo [136] (Figure 6.9).

Some of these chemical modifications have already been approved by the FDA. For example, the phosphorothioate linkage is present in Nusinersen, an antisense oligonucleotides drug [137] used to treat spinal muscular atrophy [138]. Pegaptanib is an example of a 2′ modified oligonucleotide that is now used as a therapeutic agent to treat macular degeneration [139]. Another backbone modification, L-DNA (i.e. the mirror image of the natural form D-DNA) is also resistant to nuclease degradation and was successfully used to build DNA nanothermometers, which were used inside living cells [46].

In order to use DNA switches in living systems, it is also important to ensure that DNA switches are able to reach their final destination. Since cellular uptake of unmodified DNA is inefficient, this typically represents an obstacle for biosensing inside living cells. To overcome this limitation, various chemical modifications have been shown to greatly increase cellular uptake. For example, attaching a folic acid on a DNA switch enables it to target near cancer cells that overexpress folate receptors [140]. Hydrophobic chemical modifications such as cholesterol, uncharged DNA backbone or stearic acid are other examples of modifications that are used to increase cellular uptake of DNA oligonucleotides [141]. Targeting of oligonucleotide to specific cells can be achieved through specific cell surface receptors. One example of such chemical modification, N-acetyl galactosamine (GalNac), has helped to target oligonucleotides to a population of liver cells through binding specific cell surface receptors [141]. Attaching aptamers that target specific receptors (e.g. human transferrin receptor aptamer) is another simple strategy to transport DNA switches inside living cells [142]. Functionalization of DNA switches with protein like transferrin enables to target metabolic pathways (this was used to measure pH along endocytic transferrin pathway) [79]. Attaching specific signal peptide to DNA switches represents another strategy to target them to specific cell and organelle types like the nucleus (e.g. using the nuclear localization signal peptides) [143]. To this aim, researchers have recently developed an *in vivo* phage-display method in mice to select specific peptides that can localize to various organs (similar to homing peptide) [144,145]. Briefly, phage display is a selection method similar to SELEX in which phage bearing surface peptides are selected through binding interaction to a specific target to be then amplified which results in peptides with high affinity for the selected target [146]. Coupling of such method with DNA switches could allow to target them anywhere inside living mammals.

6.4 Applications of DNA Switches

In this section, we will describe two different downstream applications of DNA switches that highlight the high programmability and great potential of DNA nanotechnology. We will start by exploring with more detail the design and engineering of DNA switches as biosensing tools with applications in fundamental research and medical diagnosis (e.g. point-of-care approaches). We will then explore drug delivery systems in greater detail for applications as therapeutic agents in the medical field. For each application, we will integrate the four steps of DNA switches engineering in a single workflow toward downstream applications.

6.4.1 Biosensing Applications

Applications in Fundamental Research

DNA nanothermometers are DNA switches engineered to measure large or small temperature variations within nanosized systems [61]. Temperature variation is the obvious input stimuli selected for this application. A wide variety of DNA structures, however, can be selected since most of them will undergo temperature denaturation at a specific Tm. However, selecting DNA structures whose thermodynamic stability can be easily tuned will offer greater flexibility to build DNA nanothermometers tools that can be optimized for optimal temperature sensing. DNA stem-loop structure (e.g. molecular beacons) is an example of such programmable structure because it is possible to easily tune its folding energy by changing the GC/AT ratio of the stem structure [21]. One can rationally perform this by employing simulation software like *mfold* and change the stem sequence while keeping the loop sequence constant. For this application, a simple fluorescence output

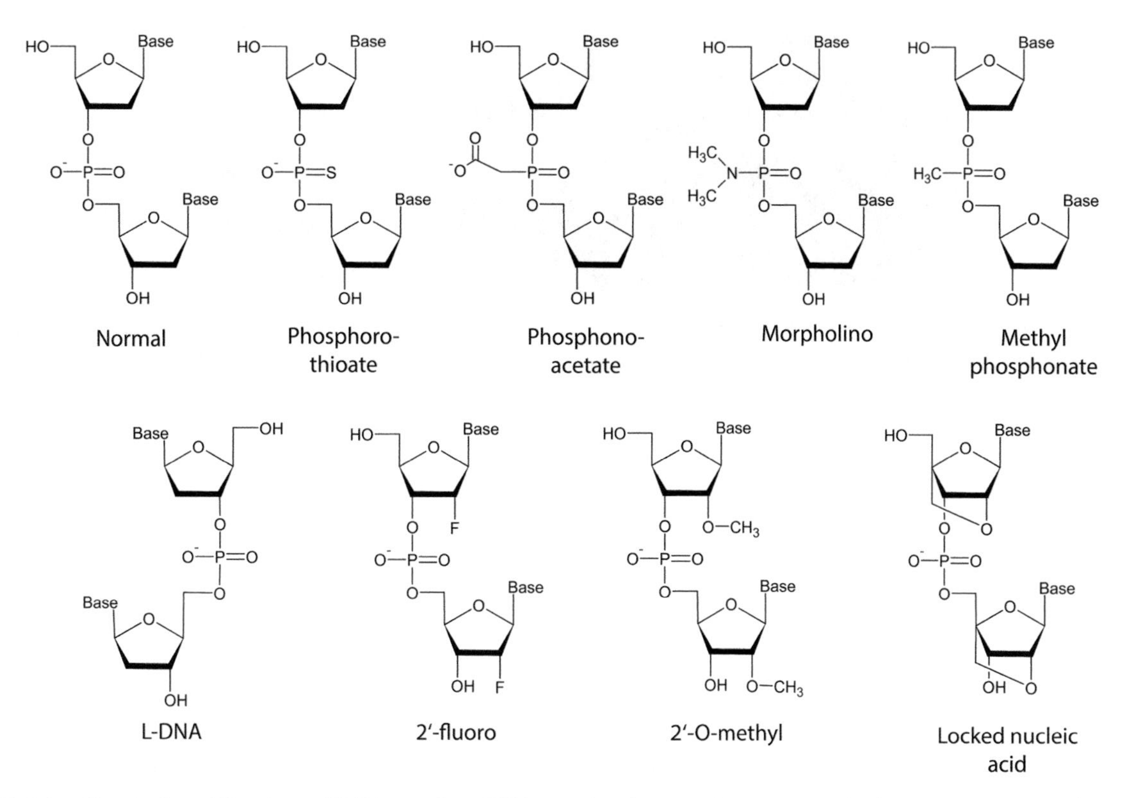

FIGURE 6.9 Chemical modifications of DNA to adapt DNA switches for complex environment (e.g. living organisms). All of these examples of modified nucleotides are commercially available from DNA synthesis companies such as Glen Research, Chemgenes, LINK and many others.

system (fluorophore/quencher pair) was implemented at the extremities of the stem structure whose opening due to increased temperature resulted in a fluorescence increase. To test this design, the GC content of the stem was varied from 0 to 5 GC base pair, which allowed to change the Tm from 40°C to 80°C (Figure 6.10a) [61]. Each nanothermometer has a linear range for a specific temperature interval (~15°C), but combining all nanothermometers together yielded a single linear range of ~50°C allowing extended temperature sensing (Figure 6.10b).

These DNA nanothermometers were employed in a real biological setting to monitor temperature inside living cells [46]. To do so, they were chemically modified to L-DNA which limited their binding to cellular proteins and conferred them with increased nuclease resistance. These nanothermometers were used to probe temperature increase in living cells after thermal therapy for cancer consisting in the irradiation of Pd nanosheets. Irradiation caused a temperature increase of up to 12°C that could be monitored by the DNA nanothermometers (Figure 6.11).

More cooperative DNA structures like DNA clamp switch (triple helix) were also adapted into nanothermometers, offering narrower temperature transition to monitor smaller temperature changes (~7°C) with enhanced precision

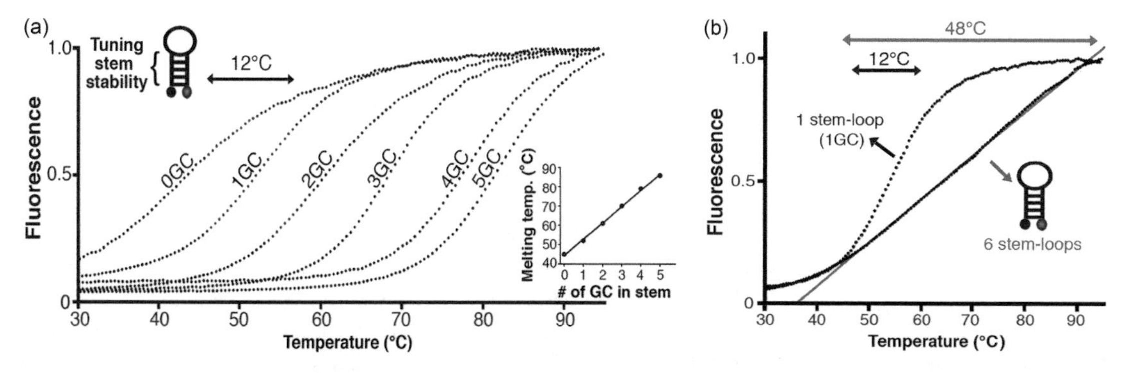

FIGURE 6.10 (a) Molecular beacons can be thermodynamically engineered into DNA nanothermometer tools with various temperature sensitivity. (b) Combining multiple DNA thermometers with different K_S results in an extended linear range to monitor extended temperature variations. (Adapted with permission from Ref. [61]. Copyright 2016 American Chemical Society.)

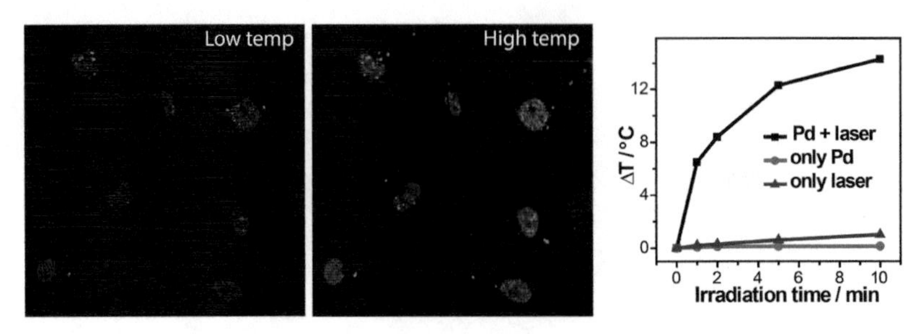

FIGURE 6.11 DNA nanothermometers were adapted through chemical modifications for temperature sensing inside living cells to monitor temperature increase caused by Pd nanosheets in real time. (Adapted with permission from Ref. [46]. Copyright 2012 American Chemical Society.)

(Figure 6.12). These ultrasensitive nanothermometers could be more effective to measure small temperature variations in living cells (e.g. to monitor metabolic activity).

DNA pH-sensitive switches are tools developed to monitor pH change at the nanoscale (i.e. nano pH meter), which resulted in various applications in biosensing and drug delivery. The selected input signal, pH variation, can be monitored, thanks to some DNA structures that are specifically stabilizes by hydrogen ions. As detailed in the section on DNA switch engineering, there is a wide variety of DNA structures whose folding is heavily dependent on specific protonation of nucleobase and hence on pH (e.g. i-motif [76,77], poly dA helix [78] or parallel triplex [16]). In particular, parallel triplex are composed of CGC or TAT triplets that display different pH sensitivities with TAT undergoing transition around pH 10 whereas CGC is around pH 7 (Figure 6.13a) [75]. A universal fluorescent DNA pH meter was therefore designed using a DNA triplex fold. For this application, Alexa Fluor 680 was employed as fluorophore due to their insensitivity to pH variation. This triplex thermometer can be programmed to respond to different pH variation by simply changing its TAT content versus CGC within the sequence (Figure 6.13b).

As described previously, these switches can be combined to extend the pH interval that can be monitored at the same time from 1.8 to 5.5 pH unit by combining three

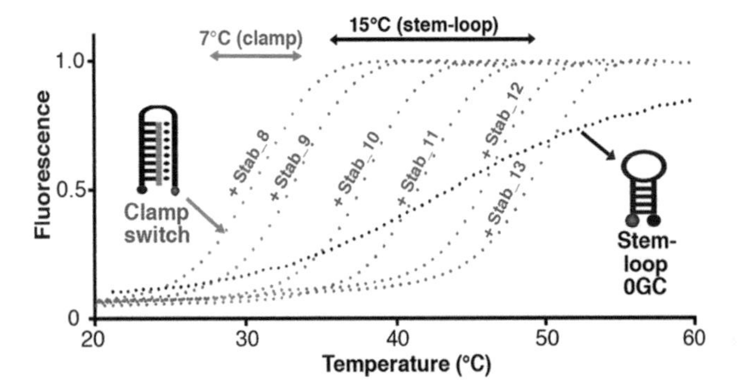

FIGURE 6.12 DNA clamp switch structures can be used to build nanothermometers with increased sensitivity to probe smaller temperature variations with high precision. (Adapted with permission from Ref. [61]. Copyright 2016 American Chemical Society.)

different DNA pH meters (Figure 6.14). Another important feature of these switches is their ability to be completely reversible [75].

DNA pH switches have also garnered interest for *in vitro* real-time pH sensing [79,80]. In this application, a pH-sensitive switch based on an i-motif DNA structure was covalently attached to transferrin, which enabled efficient switch internalization along the endocytic pathway of transferrin. Pulse chase of the DNA switch during time was performed to measure pH change through the endocytic pathway going, for example, from sorting endosome to recycling endosome with significant differences in pH distribution (Figure 6.15). This was also performed on the furin pathway using a specific DNA duplex domain. Interestingly, pH distribution determined by image processing of the DNA switches matched literature values [79]. This represents a unique method to probe pH value of cellular organelles in real time without having to lyse cells for analysis. Interestingly, these tools can be applied to study effects of toxin in real time as shown with Brefeldin A, an antibiotic known to cause extensive tubulation of the trans-Golgi network. These tools were also used on the more complex multicellular organism *Caenorhabditis elegans* to map spatiotemporal pH changes that was associated with endocytosis [147].

Biosensing for Medical Diagnostic Applications

Biosensing with DNA switches can also be used to build efficient diagnostic tools that can be employed in point-of-care setting. One of the earliest example of DNA switch application for diagnostic tools came from Kramer with fluorescent molecular beacons to detect specific DNA sequences (routinely employed in many medical facilities) [148–150]. However, fluorescence sensing is often not suitable for point-of-care approaches since it requires exhaustive sample handling. Another strategy based on electrochemical outputs directly in whole blood was then developed to overcome this limitation as described in the output section [115]. Recently, a DNA-based electrochemical biosensor exploited steric hindrance effect to detect biomarkers (Figure 6.16) [114,119,120]. In this strategy, a single-stranded capturing DNA is functionalized on a gold surface via Au-S chemistry and a complementary signaling DNA sequence bearing a redox element and a recognition element is added to the

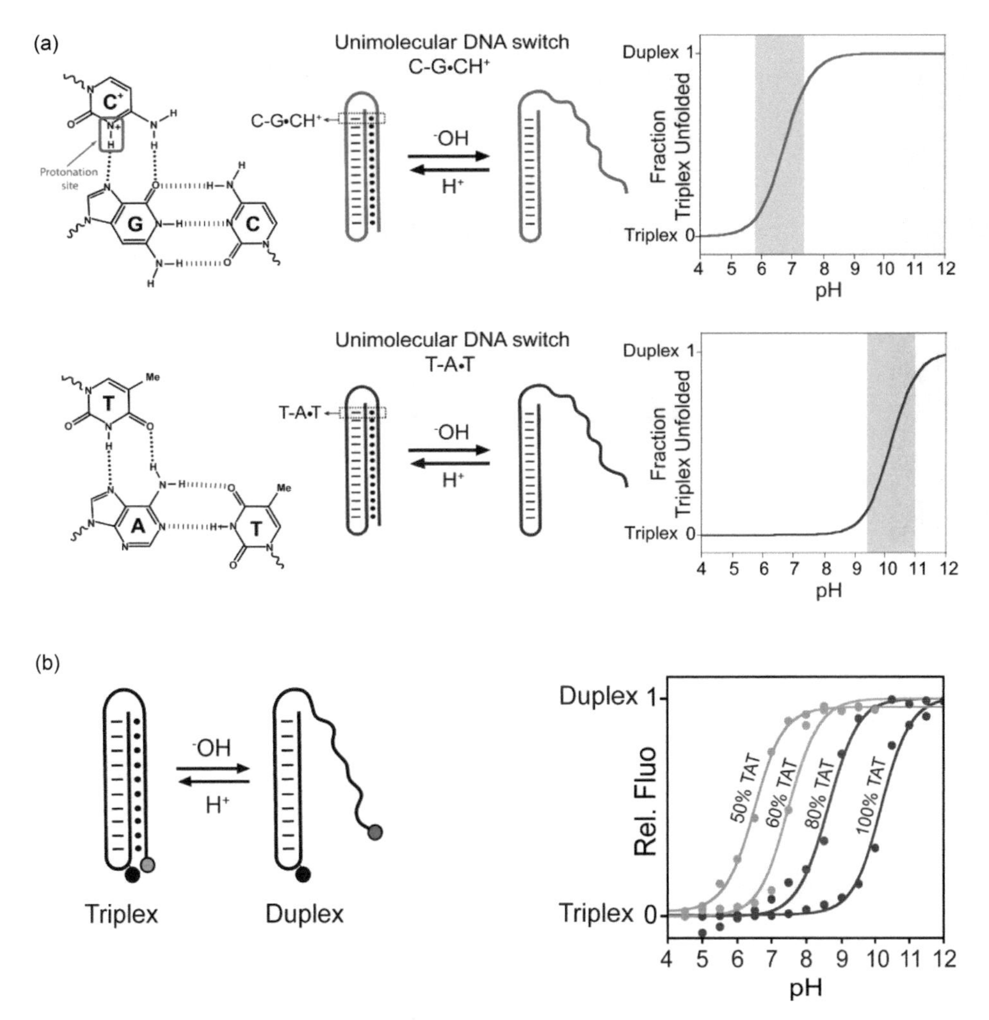

FIGURE 6.13 (a) Parallel triplex DNA structures were used to create pH-sensitive DNA switches to monitor pH variations (nano pH meter). (b) DNA pH meter has pH sensitivity that can be modulated by changing their DNA sequence with parallel triplex TAT or CGC. (Adapted with permission from Ref. [75]. Copyright 2014 American Chemical Society.)

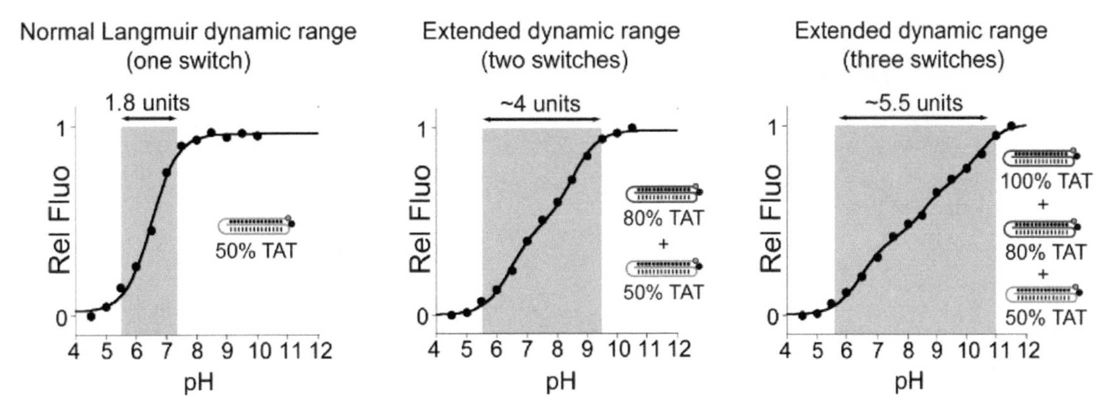

FIGURE 6.14 Combining DNA pH meter of different pH sensitivity can yield extended pH monitoring capacity. (Adapted with permission from Ref. [75]. Copyright 2014 American Chemical Society.)

blood sample. The specific biomarker (e.g. specific antibodies) can bind to the signaling DNA, which reduces the hybridization efficiency of the signaling DNA on the capturing DNA located on the surface of the electrode (>50%) due to steric hindrance. This results in a decrease of the measured current of the redox element (Figure 6.16).

This is an example of a signal-off DNA switch where the amplitude of the signal reduction is correlated with the molecular weight of the biomarker (i.e. steric hindrance effect) (Figure 6.17). This DNA switch can be easily adapted to bind other input molecules given that a small recognition element can be placed on the complementary DNA

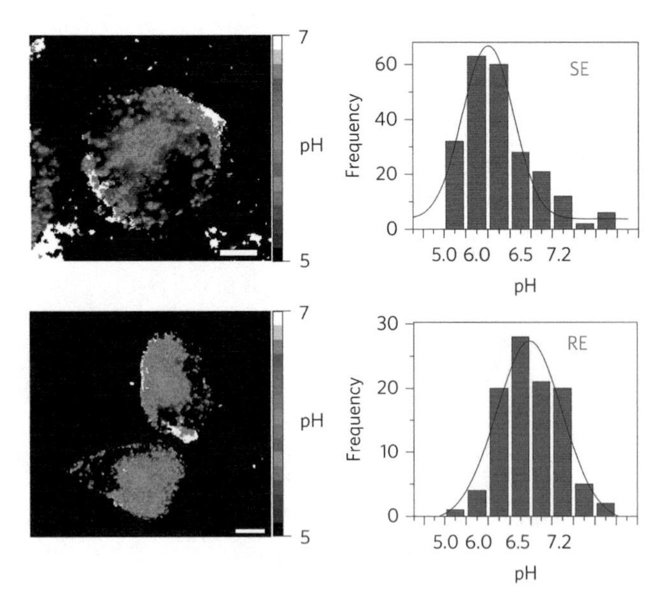

FIGURE 6.15 DNA pH meters were employed to map pH along endocytic pathways. (Adapted by permission from Springer customer service center gmbh: Springer Nature, Nature Nanotechnology. Reference [79]. Copyright 2013.)

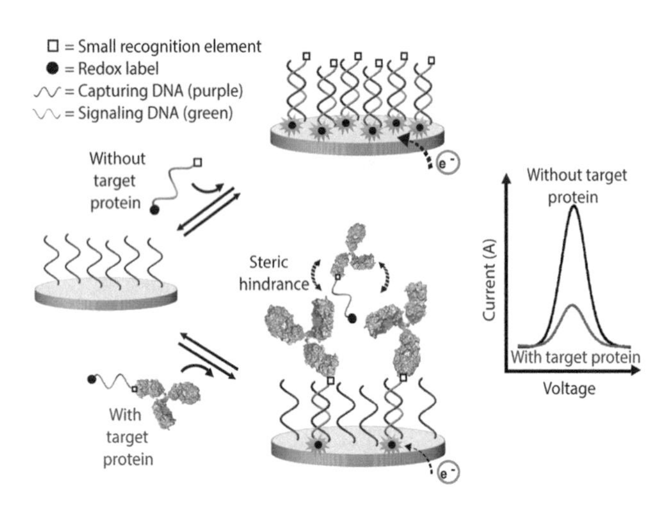

FIGURE 6.16 Steric hindrance was exploited to engineer a DNA switch for the detection of large biomarkers using gold electrodes. (Adapted with permission from Ref. [114]. Copyright 2015 American Chemical Society.)

(e.g. <30 kDa). This approach was tested directly in whole blood in a point-of-care format that employs an inexpensive potentiostat and provided result within 10 min [120].

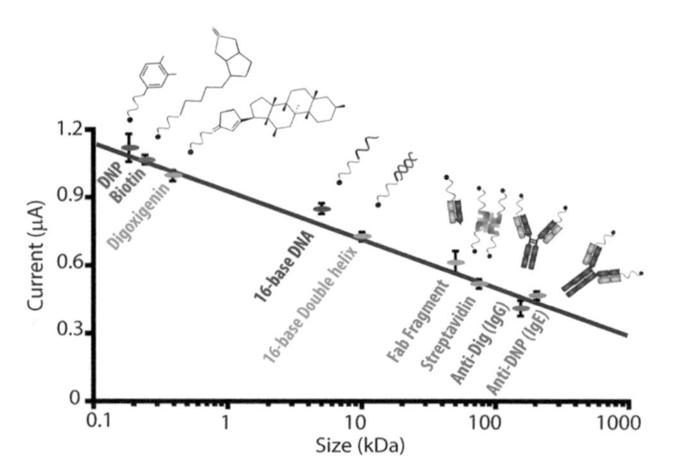

FIGURE 6.17 The steric hindrance effect on the DNA switch response was dependent on the molecular weight of the biomarker detected. (Adapted with permission from Ref. [114]. Copyright 2015 American Chemical Society.)

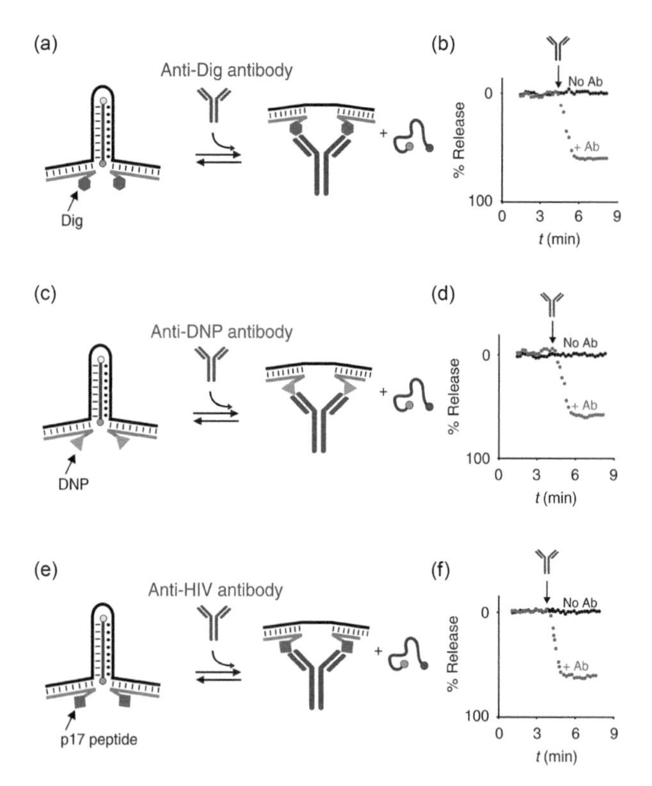

FIGURE 6.18 Antibody-induced drug-releasing nanomachines. DNA-based nanomachines programmed to release a nucleic acid-based cargo in response to: (a and b) anti-digoxigenin antibody; (c and d) anti-dinitrophenol antibody and (e and f) anti-HIV antibody. (Adapted with permission from Ref. [151]. Copyright 2017 American Chemical Society.)

6.4.2 Drug Delivery System for Medical Therapies

Drug delivery systems can take advantage of the high programmability of DNA. For example, DNA switches can be engineered to transport specific drug cargos that can be released in the presence of specific chemical input such as disease markers. For example, Ricci et al. 2017 have recently developed a DNA switch that can be programmed to release a nucleic acid drug cargo in the presence of specific antibodies (Figure 6.18) [151]. The switch–cargo complex consists of a DNA triplex clamp structure. This complex is thermodynamically engineered to be stable in the triplex form but unstable once the triplex unfolds into a DNA duplex, resulting in the release of the drug cargo [151].

The cargo release mechanism is based on the proximity of two recognition elements (i.e. hapten, epitopes) that specifically bind the input antibodies (e.g. digoxigenin binds anti-digoxigenin antibody). Upon binding the two antigens or epitopes on the DNA transporter, one single specific antibody provides the energy to destabilize the triplex structure by stretching the switch (Figure 6.18). This stretching mechanism has already been exploited to build DNA switches for sensing applications [113,152]. This DNA switch architecture is also highly programmable since the input biomarker can be changed to any other antibody by simply changing the hapten/epitope element on the switch scaffold (Figure 6.18c and e). The high programmability of this approach was also employed to build steric hindrance-based logic gates for drug delivery where drug cargo is released only in the presence of two different input biomarkers (i.e. two different antibodies) (Figure 6.19). These DNA switches were tested in blood serum and retained high specificity to its target antibody, which suggests that they could be useful to deliver nucleic acid-based drug in living organism.

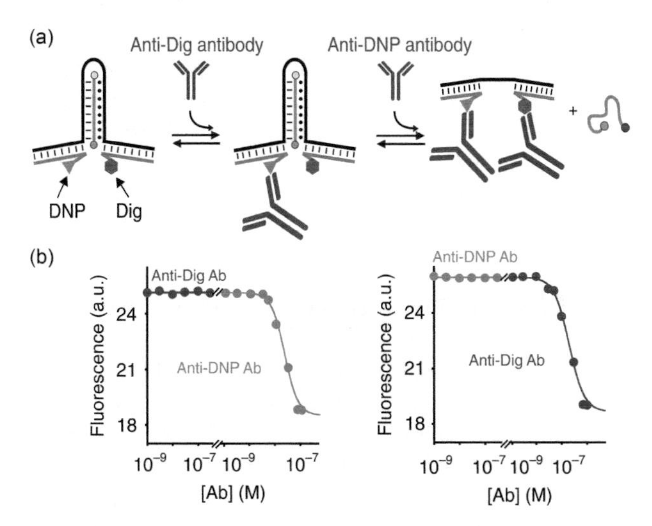

FIGURE 6.19 Antibody-induced drug-releasing nanomachines in response to the presence of two antibodies. The digoxigenin (Dig) and dinitrophenol (DNP) DNA switch were modified into a logic circuit requiring the presence of two different antibodies to induce drug cargo release. (Adapted with permission from Ref. [151]. Copyright 2017 American Chemical Society.)

There is a close proximity between DNA switches designed for biosensing and drug delivery systems since highly specific biosensing switches can be easily adapted to trigger the release of a drug cargo instead of simply providing a fluorescent output signal. Another example of such versatile switch is the adaptation of a DNA pH meter based on parallel triplex into a drug delivery switch triggered by acidic environments [81]. Acidic environments have been the input of a wide variety of drug delivery systems because tumor environment is usually more acidic due to specific tumor metabolism [153–155]. The switching mechanism employed is based on the addition of a pH-sensitive sequence at a distal position on a DNA-based recognition element in order to leave the high affinity and specificity of the DNA switch unchanged. The first design explored consisted of a simple stem-loop DNA that can transport a complementary nucleic acid drug hybridized to its loop (Figure 6.20a and b). When adding a parallel triplex-forming sequence to the DNA transporter (i.e. far from the loop), the triplex structure conformation is stabilized at lower pH (<6.0), therefore reducing the affinity of the transporter for its nucleic acid cargo (Figure 6.20c and d).

This design allowed to load single-stranded nucleic acid drugs on the loop structure of the DNA switch and trigger its release in acidic environment. This strategy was also adapted using more complex drug transporters such as the cocaine-binding aptamer where the same DNA sequence

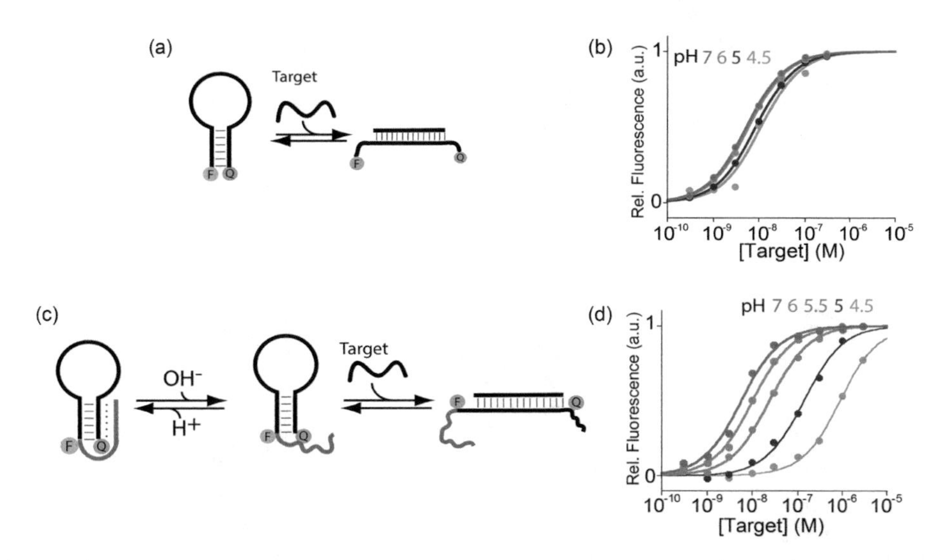

FIGURE 6.20 (a and b) The stem-loop DNA transporter is not sensitive to pH variation. (c and d) The addition of a triplex-forming, pH-sensitive DNA element renders the DNA transporter sensitive to pH variations. F: fluorophore; Q: quencher (Adapted with permission from Ref. [81]. Copyright 2015 American Chemical Society.)

sensitive to acidic environment was added at distal position from the recognition site of the aptamer (Figure 6.21a). The cocaine aptamer served as a mean to load the drug (i.e. cocaine) and the pH-sensitive sequence as the trigger to release cocaine. This cocaine transporter reacted as predicted by releasing cocaine in acidic environment (Figure 6.21b). This strategy is highly modular and could in principle be applied to a wide variety of DNA aptamers to transport and release a variety of therapeutic molecules.

Other smart drug transporters have been developed using DNA aptamers. For example, Krishnan et al. have recently employed a cyclic di-guanosine monophosphate (c-di-GMP)-binding aptamer to deliver fluorescent drug cargo in response to c-di-GMP, a second messenger in bacteria that regulates signal transduction and metabolism [156]. This DNA aptamer was engineered into a DNA switch that undergoes conformational change upon binding to c-di-GMP (Figure 6.22a) [157]. This DNA switch was then employed to build a small DNA cage that can transport fluorescent drugs. The DNA cage was designed to open upon the activation of the DNA switch with increasing concentration of c-di-GMP (Figure 6.22b). This approach is highly modular since it is possible to replace the DNA switch with any other DNA aptamers, and thus, these DNA cages could be in principle activated by any disease biomarker for which we possess a DNA aptamer. This approach also exploits novel properties of DNA cage like their intrinsic resistance to nucleases, which protect the drug cargo from being rapidly metabolized [158]. DNA cage has been successfully used in mammalian cells [158] and in the multicellular organism *C. elegans* [159] for bioimaging, suggesting that they could be eventually applied successfully for drug delivery applications in living organisms. This cage, sometimes referred to as DNA container, retains the programmability of DNA, and thus, it is possible to engineer them within the DNA switching mechanism to make a link between presence of the input and cage opening.

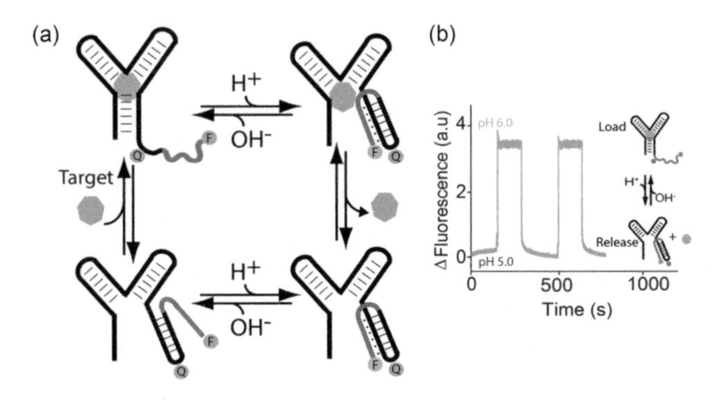

FIGURE 6.21 An aptamer-based pH-induced cocaine transporter that releases cocaine in response to low pH. F: fluorophore; Q: quencher (Adapted with permission from Ref. [81]. Copyright 2015 American Chemical Society.)

6.5 Conclusion and Perspective

DNA is a unique biopolymer that has enabled tremendous research applications in nanotechnology creating a field of its own: DNA nanotechnology. The high versatility of DNA arises from our capacity to predict its structure from the nucleotide sequence. Other biopolymers like proteins, for example, still lack this structure predictability, making it harder to develop them into new functional structures *de novo*. Another unique feature of DNA is its easy organic synthesis using automated synthesizer, which

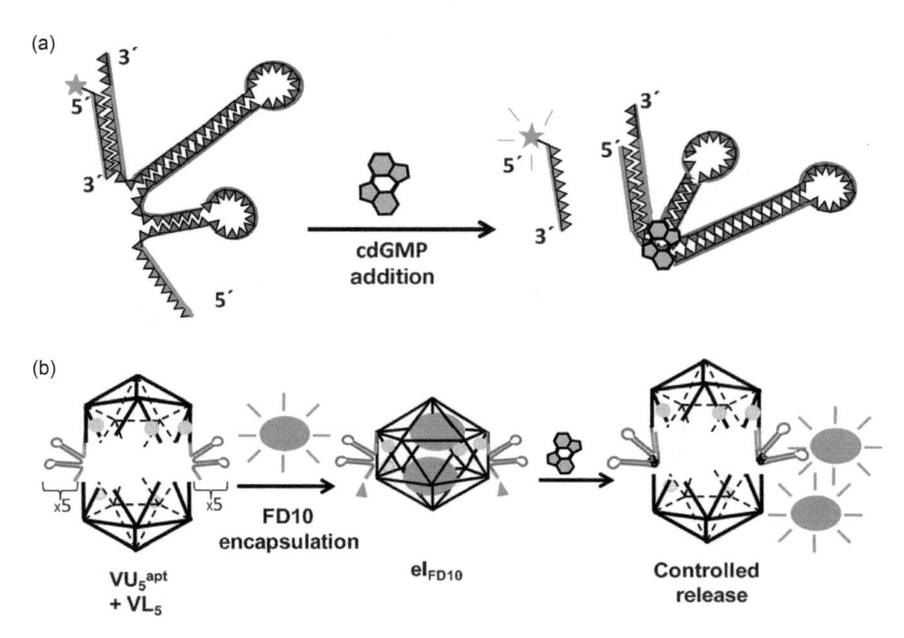

FIGURE 6.22 (a) A DNA aptamer was adapted into a DNA switch that undergoes conformational change upon binding to its input molecule c-di-GMP. (b) This DNA switch was inserted on a DNA cage whose opening, and drug cargo release, is triggered by activation of the DNA switch in the presence of c-di-GMP. (Reprinted with permission from Ref. [157]. Copyright 2013 John Wiley and Sons.)

drastically speeds up the progression in this field. These two features, along with the intrinsic biocompatibility of DNA, are at the center of the development of sophisticated nanotechnological tools that offer stimuli responsive capacity. One such bio-inspired example described in this review are DNA switches employed for biosensing and drug delivery applications. Throughout this chapter, we presented a summary of the step-by-step process to engineer DNA switches through multiple design and their adaptation into various applications. Selection of the input (e.g. disease marker) and the output signals (drug release, signaling mechanism) are the key elements to build DNA switches. The high programmability of DNA allows designing switches that recognize input ranging from light, to small ions, proteins or whole cells and that are able to translate this recognition event into a myriad of output signals from electrical current to fluorescence or drug release, for example. We also presented more than five different strategies to incorporate structure switching mechanisms within DNA switches. All of these strategies are aimed to engineer a non-binding conformation that corresponds to the switch inactive state. This non-binding conformation can switch to the bound-active conformation in the presence of the input through a population-shift model and generate the output signal implemented within the switch.

In addition to applications in biosensing and drug delivery, DNA switches have also found a lot of use in fundamental research and medicine. For example, DNA nanothermometer and nano pH meter have been developed to detect either temperature changes or pH variations at the nanoscale. These tools can be useful to monitor physical change within complex systems like living cells (e.g. mapping the pH of living cells in different conditions) or can also be adapted for drug delivery applications (e.g. pH-dependent drug delivery). Application of DNA switches in molecular diagnostics also represents a promising avenue in the medical field with the development of fast, inexpensive, point-of-care approaches that provide results within less than 10 min. These approaches often rely on DNA switches that provide electrochemical output in the presence of specific disease markers (e.g. HIV antibodies) and that can be employed directly in whole blood.

Although the future of DNA switches looks very promising, several challenges lie ahead before realizing this promise [160]. One example of such challenge is the need to create and develop DNA switches that can achieve sufficient specificity and selectivity. We believe that recent works seeking to expand the DNA code beyond the classic natural A, T, C and G base pairs with artificial nucleotides should greatly contribute to enhance the specificity of nucleic acid-based recognition elements [161,162]. Other challenges faced by DNA nanotechnology to move beyond laboratory-scale systems include improving accuracy (e.g. for DNA switch-based sensors), stability (to enable repeatability in DNA switches) and reproducibility (e.g. over industrial-scale production) [160]. To that end, an improvement and increased understanding of large-scale production of oligonucleotides to decrease cost and increase yield will be useful [163]. Finally, application of DNA switches in various medical fields for therapies will need a better understanding of intracellular uptake and trafficking mechanisms, DNA switches toxicology and their pharmacokinetic properties [164].

References

1. M. C. Lagerström, H. B. Schiöth, Structural diversity of G protein-coupled receptors and significance for drug discovery. *Nature Reviews Drug Discovery* **7**, 339 (2008).
2. R. T. Dorsam, J. S. Gutkind, G-protein-coupled receptors and cancer. *Nature Reviews Cancer* **7**, 79 (2007).
3. D. A. Leigh, Genesis of the nanomachines: The 2016 nobel prize in chemistry. *Angewandte Chemie International Edition* **55**, 14506–14508 (2016).
4. A. E. Friedman, J. C. Chambron, J. P. Sauvage, N. J. Turro, J. K. Barton, A molecular light switch for DNA: Ru (bpy) 2 (dppz) 2+. *Journal of the American Chemical Society* **112**, 4960–4962 (1990).
5. V. Balzani, A. Credi, F. M. Raymo, J. F. Stoddart, Artificial molecular machines. *Angewandte Chemie International Edition* **39**, 3348–3391 (2000).
6. N. Koumura, R. W. Zijlstra, R. A. van Delden, N. Harada, B. L. Feringa, Light-driven monodirectional molecular rotor. *Nature* **401**, 152 (1999).
7. D. Baker, A. Sali, Protein structure prediction and structural genomics. *Science* **294**, 93–96 (2001).
8. P. Bradley, K. M. Misura, D. Baker, Toward high-resolution de novo structure prediction for small proteins. *Science* **309**, 1868–1871 (2005).
9. L. Jiang *et al.*, De novo computational design of retro-aldol enzymes. *Science* **319**, 1387–1391 (2008).
10. B. Kuhlman *et al.*, Design of a novel globular protein fold with atomic-level accuracy. *Science* **302**, 1364–1368 (2003).
11. A. Desrosiers, A. Vallée-Bélisle. Nature-inspired DNA switches: applications in medicine. *Nanomedicine (Lond)* **12**(3), 175–179 (2016).
12. S. G. Harroun *et al.*, Programmable DNA switches and their applications. *Nanoscale* **10**, 4607–4641 (2018).
13. N. C. Seeman, DNA engineering and its application to nanotechnology. *Trends in Biotechnology* **17**, 437–443 (1999).
14. F. Wang, X. Liu, I. Willner, DNA switches: From principles to applications. *Angewandte Chemie International Edition* **54**, 1098–1129 (2015).
15. C.-H. Lu, B. Willner, I. Willner, DNA nanotechnology: From sensing and DNA machines to drug-delivery systems. *ACS Nano* **7**, 8320–8332 (2013).
16. Y. Hu, A. Cecconello, A. Idili, F. Ricci, I. Willner, Triplex DNA nanostructures: From basic properties

to applications. *Angewandte Chemie* **56**, 15210–15233 (2017).
17. L. Qian, E. Winfree, Scaling up digital circuit computation with DNA strand displacement cascades. *Science* **332**, 1196–1201 (2011).
18. C. G. Evans, E. Winfree, Physical principles for DNA tile self-assembly. *Chemical Society Reviews* **46**, 3808–3829 (2017).
19. N. C. Seeman, DNA nanotechnology: Novel DNA constructions. *Annual Review of Biophysics and Biomolecular Structure* **27**, 225–248 (1998).
20. A. V. Pinheiro, D. Han, W. M. Shih, H. Yan, Challenges and opportunities for structural DNA nanotechnology. *Nature Nanotechnology* **6**, 763 (2011).
21. A. Vallée-Bélisle, F. Ricci, K. W. Plaxco, Thermodynamic basis for the optimization of binding-induced biomolecular switches and structure-switching biosensors. *Proceedings of the National Academy of Sciences* **106**, 13802–13807 (2009).
22. F. Ricci, A. Vallée-Bélisle, A. J. Simon, A. Porchetta, K. W. Plaxco, Using nature's "tricks" to rationally tune the binding properties of biomolecular receptors. *Accounts of Chemical Research* **49**, 1884–1892 (2016).
23. F. Ricci, A. Vallée-Bélisle, K. W. Plaxco, High-precision, in vitro validation of the sequestration mechanism for generating ultrasensitive dose-response curves in regulatory networks. *PLoS Computational Biology* **7**, e1002171 (2011).
24. F. Ricci, A. Vallée-Bélisle, A. Porchetta, K. W. Plaxco, Rational design of allosteric inhibitors and activators using the population-shift model: In vitro validation and application to an artificial biosensor. *Journal of the American Chemical Society* **134**, 15177–15180 (2012).
25. A. J. Simon, A. Vallée-Bélisle, F. Ricci, H. M. Watkins, K. W. Plaxco, Using the population-shift mechanism to rationally introduce "Hill-type" cooperativity into a normally non-cooperative receptor. *Angewandte Chemie International Edition* **53**, 9471–9475 (2014).
26. D. Mariottini, A. Idili, A. Vallée-Bélisle, K. W. Plaxco, F. Ricci, A DNA nanodevice that loads and releases a cargo with hemoglobin-like allosteric control and cooperativity. *Nano Letters* **17**, 3225–3230 (2017).
27. S. Tyagi, F. R. Kramer, Molecular beacons: Probes that fluoresce upon hybridization. *Nature Biotechnology* **14**, 303 (1996).
28. N. C. Seeman, DNA in a material world. *Nature* **421**, 427 (2003).
29. P. W. Rothemund, Folding DNA to create nanoscale shapes and patterns. *Nature* **440**, 297 (2006).
30. J. Chen, N. C. Seeman, Synthesis from DNA of a molecule with the connectivity of a cube. *Nature* **350**, 631 (1991).
31. N. R. Kallenbach, R.-I. Ma, N. C. Seeman, An immobile nucleic acid junction constructed from oligonucleotides. *Nature* **305**, 829 (1983).
32. N. C. Seeman, Nucleic acid junctions and lattices. *Journal of Theoretical Biology* **99**, 237–247 (1982).
33. M. Churchill, T. D. Tullius, N. R. Kallenbach, N. C. Seeman, A Holliday recombination intermediate is twofold symmetric. *Proceedings of the National Academy of Sciences* **85**, 4653–4656 (1988).
34. R.-I. Ma, N. R. Kallenbach, R. D. Sheardy, M. L. Petrillo, N. C. Seeman, Three-arm nucleic acid junctions are flexible. *Nucleic Acids Research* **14**, 9745–9753 (1986).
35. J. Elbaz *et al.*, DNA computing circuits using libraries of DNAzyme subunits. *Nature Nanotechnology* **5**, 417 (2010).
36. A. D. Ellington, J. W. Szostak, Selection in vitro of single-stranded DNA molecules that fold into specific ligand-binding structures. *Nature* **355**, 850 (1992).
37. C. Tuerk, L. Gold, Systematic evolution of ligands by exponential enrichment: RNA ligands to bacteriophage T4 DNA polymerase. *science* **249**, 505–510 (1990).
38. M. Zuker, Mfold web server for nucleic acid folding and hybridization prediction. *Nucleic Acids Research* **31**, 3406–3415 (2003).
39. J. N. Zadeh *et al.*, NUPACK: Analysis and design of nucleic acid systems. *Journal of Computational Chemistry* **32**, 170–173 (2011).
40. R. Veneziano *et al.*, Designer nanoscale DNA assemblies programmed from the top down. *Science* **352**, 1534–1534 (2016).
41. R. Nutiu, Y. Li, Structure-switching signaling aptamers. *Journal of the American Chemical Society* **125**, 4771–4778 (2003).
42. A. Vallée-Bélisle, K. W. Plaxco, Structure-switching biosensors: Inspired by Nature. *Current Opinion in Structural Biology* **20**, 518–526 (2010).
43. M. R. Jones, N. C. Seeman, C. A. Mirkin, Programmable materials and the nature of the DNA bond. *Science* **347**, 1260901 (2015).
44. D. Y. Zhang, E. Winfree, Control of DNA strand displacement kinetics using toehold exchange. *Journal of the American Chemical Society* **131**, 17303–17314 (2009).
45. D. Y. Zhang, G. Seelig, Dynamic DNA nanotechnology using strand-displacement reactions. *Nature Chemistry* **3**, 103 (2011).
46. G. Ke *et al.*, L-DNA molecular beacon: A safe, stable, and accurate intracellular nano-thermometer for temperature sensing in living cells. *Journal of the American Chemical Society* **134**, 18908–18911 (2012).
47. S. Surana, A. R. Shenoy, Y. Krishnan, Designing DNA nanodevices for compatibility with the immune

system of higher organisms. *Nature Nanotechnology* **10**, 741 (2015).
48. S. Kosuri, G. M. Church, Large-scale de novo DNA synthesis: Technologies and applications. *Nature Methods* **11**, 499 (2014).
49. S. L. Beaucage, R. P. Iyer, Advances in the synthesis of oligonucleotides by the phosphoramidite approach. *Tetrahedron* **48**, 2223–2311 (1992).
50. D. A. Lashkari, S. P. Hunicke-Smith, R. M. Norgren, R. W. Davis, T. Brennan, An automated multiplex oligonucleotide synthesizer: Development of high-throughput, low-cost DNA synthesis. *Proceedings of the National Academy of Sciences* **92**, 7912–7915 (1995).
51. F. A. Aldaye, H. F. Sleiman, Guest-mediated access to a single DNA nanostructure from a library of multiple assemblies. *Journal of the American Chemical Society* **129**, 10070–10071 (2007).
52. Y.-H. M. Chan, B. van Lengerich, S. G. Boxer, Lipid-anchored DNA mediates vesicle fusion as observed by lipid and content mixing. *Biointerphases* **3**, FA17–FA21 (2008).
53. D. Musumeci, D. Montesarchio, Synthesis of a cholesteryl-HEG phosphoramidite derivative and its application to lipid-conjugates of the anti-HIV 5'TGGGAG3'Hotoda's sequence. *Molecules* **17**, 12378–12392 (2012).
54. M. J. Damha, P. A. Giannaris, S. V. Zabarylo, An improved procedure for derivatization of controlled-pore glass beads for solid-phase oligonucleotide synthesis. *Nucleic Acids Research* **18**, 3813–3821 (1990).
55. R. T. Pon, N. Usman, K. Ogilvie, Derivatization of controlled pore glass beads for solid phase oligonucleotide synthesis. *Biotechniques* **6**, 768–775 (1988).
56. S. Padmanabhan, J. E. Coughlin, R. P. Iyer, Microwave-assisted functionalization of solid supports: Application in the rapid loading of nucleosides on controlled-pore-glass (CPG). *Tetrahedron Letters* **46**, 343–347 (2005).
57. K. M. Ririe, R. P. Rasmussen, C. T. Wittwer, Product differentiation by analysis of DNA melting curves during the polymerase chain reaction. *Analytical Biochemistry* **245**, 154–160 (1997).
58. J. Lee, N. A. Kotov, Thermometer design at the nanoscale. *Nano Today* **2**, 48–51 (2007).
59. M. Frank-Kamenetskii, Simplification of the empirical relationship between melting temperature of DNA, its GC content and concentration of sodium ions in solution. *Biopolymers* **10**, 2623–2624 (1971).
60. C. Schildkraut, S. Lifson, Dependence of the melting temperature of DNA on salt concentration. *Biopolymers* **3**, 195–208 (1965).
61. D. Gareau, A. Desrosiers, A. Vallée-Bélisle, Programmable quantitative DNA nanothermometers. *Nano Letters* **16**, 3976–3981 (2016).
62. A. T. Jonstrup, J. FredsØe, A. H. Andersen, DNA hairpins as temperature switches, thermometers and ionic detectors. *Sensors* **13**, 5937–5944 (2013).
63. S. Ebrahimi, Y. Akhlaghi, M. Kompany-Zareh, Å. Rinnan, Nucleic acid based fluorescent nanothermometers. *ACS Nano* **8**, 10372–10382 (2014).
64. Y. Wu *et al.*, Novel ratiometric fluorescent nanothermometers based on fluorophores-labeled short single-stranded DNA. *ACS Applied Materials & Interfaces* **9**, 11073–11081 (2017).
65. R. Tashiro, H. Sugiyama, A nanothermometer based on the different π stackings of B-and Z-DNA. *Angewandte Chemie International Edition* **42**, 6018–6020 (2003).
66. H. Asanuma, T. Ito, T. Yoshida, X. Liang, M. Komiyama, Photoregulation of the formation and dissociation of a DNA duplex by using the cis–trans isomerization of azobenzene. *Angewandte Chemie International Edition* **38**, 2393–2395 (1999).
67. C. Dohno, K. Nakatani, Control of DNA hybridization by photoswitchable molecular glue. *Chemical Society Reviews* **40**, 5718–5729 (2011).
68. C. Dohno, S.-n. Uno, K. Nakatani, Photoswitchable molecular glue for DNA. *Journal of the American Chemical Society* **129**, 11898–11899 (2007).
69. H. Kang *et al.*, Single-DNA molecule nanomotor regulated by photons. *Nano Letters* **9**, 2690–2696 (2009).
70. J. Hihath, B. Xu, P. Zhang, N. Tao, Study of single-nucleotide polymorphisms by means of electrical conductance measurements. *Proceedings of the National Academy of Sciences of the United States of America* **102**, 16979–16983 (2005).
71. R. Korol, D. Segal, From exhaustive simulations to key principles in DNA nanoelectronics. *The Journal of Physical Chemistry C* **122**, 4206–4216 (2018).
72. L. Xiang *et al.*, Gate-controlled conductance switching in DNA. *Nature Communications* **8**, 14471 (2017).
73. K. Wang, DNA-based single-molecule electronics: From concept to function. *Journal of Functional Biomaterials* **9**, 8 (2018).
74. B. Y. Won, C. Jung, K. S. Park, H. G. Park, An electrochemically reversible DNA switch. *Electrochemistry Communications* **27**, 100–103 (2013).
75. A. Idili, A. Vallée-Bélisle, F. Ricci, Programmable pH-triggered DNA nanoswitches. *Journal of the American Chemical Society* **136**, 5836–5839 (2014).
76. H. A. Day, P. Pavlou, Z. A. Waller, i-Motif DNA: Structure, stability and targeting with ligands. *Bioorganic & Medicinal Chemistry* **22**, 4407–4418 (2014).
77. Y. Dong, Z. Yang, D. Liu, DNA nanotechnology based on i-motif structures. *Accounts of Chemical Research* **47**, 1853–1860 (2014).
78. S. Chakraborty, S. Sharma, P. K. Maiti, Y. Krishnan, The poly dA helix: A new structural motif for

high performance DNA-based molecular switches. *Nucleic Acids Research* **37**, 2810–2817 (2009).
79. S. Modi, C. Nizak, S. Surana, S. Halder, Y. Krishnan, Two DNA nanomachines map pH changes along intersecting endocytic pathways inside the same cell. *Nature Nanotechnology* **8**, 459 (2013).
80. S. Modi *et al.*, A DNA nanomachine that maps spatial and temporal pH changes inside living cells. *Nature Nanotechnology* **4**, 325 (2009).
81. A. Porchetta, A. Idili, A. Vallée-Bélisle, F. Ricci, General strategy to introduce pH-induced allostery in DNA-based receptors to achieve controlled release of ligands. *Nano Letters* **15**, 4467–4471 (2015).
82. A. Ambrus *et al.*, Human telomeric sequence forms a hybrid-type intramolecular G-quadruplex structure with mixed parallel/antiparallel strands in potassium solution. *Nucleic Acids Research* **34**, 2723–2735 (2006).
83. Z.-S. Wu, C.-R. Chen, G.-L. Shen, R.-Q. Yu, Reversible electronic nanoswitch based on DNA G-quadruplex conformation: A platform for single-step, reagentless potassium detection. *Biomaterials* **29**, 2689–2696 (2008).
84. X. Hou *et al.*, A biomimetic potassium responsive nanochannel: G-quadruplex DNA conformational switching in a synthetic nanopore. *Journal of the American Chemical Society* **131**, 7800–7805 (2009).
85. Z. Lin, Y. Chen, X. Li, W. Fang, Pb^{2+} induced DNA conformational switch from hairpin to G-quadruplex: Electrochemical detection of Pb^{2+}. *Analyst* **136**, 2367–2372 (2011).
86. M. Hoang, P.-J. J. Huang, J. Liu, G-quadruplex DNA for fluorescent and colorimetric detection of thallium (I). *ACS Sensors* **1**, 137–143 (2015).
87. D. Bhattacharyya, G. Mirihana Arachchilage, S. Basu, Metal cations in G-quadruplex folding and stability. *Frontiers in chemistry* **4**, 38 (2016).
88. D.-L. Ma *et al.*, Recent developments in G-quadruplex probes. *Chemistry & Biology* **22**, 812–828 (2015).
89. D. Zhao, Y. Fan, F. Gao, T.-M. Yang, "Turn-off-on" fluorescent sensor for (N-methyl-4-pyridyl) porphyrin-DNA and G-quadruplex interactions based on ZnCdSe quantum dots. *Analytica Chimica Acta* **888**, 131–137 (2015).
90. Y. Miyake *et al.*, $Mercury^{II}$-mediated formation of thymine–Hg^{II}–thymine base pairs in DNA duplexes. *Journal of the American Chemical Society* **128**, 2172–2173 (2006).
91. A. Ono *et al.*, Specific interactions between silver (I) ions and cytosine–cytosine pairs in DNA duplexes. *Chemical Communications* **39**, 4825–4827 (2008).
92. F. Pfeiffer, G. Mayer, Selection and biosensor application of aptamers for small molecules. *Frontiers in Chemistry* **4**, 25 (2016).
93. B. Deng *et al.*, Aptamer binding assays for proteins: The thrombin example—a review. *Analytica Chimica Acta* **837**, 1–15 (2014).
94. N. Hamaguchi, A. Ellington, M. Stanton, Aptamer beacons for the direct detection of proteins. *Analytical Biochemistry* **294**, 126–131 (2001).
95. P. K. Kumar, Monitoring intact viruses using aptamers. *Biosensors* **6**, 40 (2016).
96. A. Davydova *et al.*, Aptamers against pathogenic microorganisms. *Critical Reviews in Microbiology* **42**, 847–865 (2016).
97. A. D. Ellington, J. W. Szostak, In vitro selection of RNA molecules that bind specific ligands. *Nature* **346**, 818 (1990).
98. W. Zhou, P.-J. J. Huang, J. Ding, J. Liu, Aptamer-based biosensors for biomedical diagnostics. *Analyst* **139**, 2627–2640 (2014).
99. M. N. Stojanovic, P. De Prada, D. W. Landry, Aptamer-based folding fluorescent sensor for cocaine. *Journal of the American Chemical Society* **123**, 4928–4931 (2001).
100. Z.-X. Wang, An exact mathematical expression for describing competitive binding of two different ligands to a protein molecule. *FEBS Letters* **360**, 111–114 (1995).
101. Z. Tang *et al.*, Aptamer switch probe based on intramolecular displacement. *Journal of the American Chemical Society* **130**, 11268–11269 (2008).
102. Y. Xing, Z. Yang, D. Liu, A responsive hidden toehold to enable controllable DNA strand displacement reactions. *Angewandte Chemie International Edition* **50**, 11934–11936 (2011).
103. H. Zhang *et al.*, Assembling DNA through affinity binding to achieve ultrasensitive protein detection. *Angewandte Chemie International Edition* **52**, 10698–10705 (2013).
104. W. Tang *et al.*, DNA tetraplexes-based toehold activation for controllable DNA strand displacement reactions. *Journal of the American Chemical Society* **135**, 13628–13631 (2013).
105. A. Amodio *et al.*, Rational design of pH-controlled DNA strand displacement. *Journal of the American Chemical Society* **136**, 16469–16472 (2014).
106. N. R. Markham, M. Zuker, DINAMelt web server for nucleic acid melting prediction. *Nucleic Acids Research* **33**, W577–W581 (2005).
107. A. Idili, F. Ricci, A. Vallée-Bélisle, Determining the folding and binding free energy of DNA-based nanodevices and nanoswitches using urea titration curves. *Nucleic Acids Research* **45**, 7571–7580 (2017).
108. Y. You, A. V. Tataurov, R. Owczarzy, Measuring thermodynamic details of DNA hybridization using fluorescence. *Biopolymers* **95**, 472–486 (2011).
109. T. Schlichthaerle, M. T. Strauss, F. Schueder, J. B. Woehrstein, R. Jungmann, DNA nanotechnology

and fluorescence applications. *Current Opinion in Biotechnology* **39**, 41–47 (2016).
110. C. Feng, S. Dai, L. Wang, Optical aptasensors for quantitative detection of small biomolecules: A review. *Biosensors and Bioelectronics* **59**, 64–74 (2014).
111. R. Roy, S. Hohng, T. Ha, A practical guide to single-molecule FRET. *Nature Methods* **5**, 507 (2008).
112. F. Xia *et al.*, On the binding of cationic, water-soluble conjugated polymers to DNA: Electrostatic and hydrophobic interactions. *Journal of the American Chemical Society* **132**, 1252–1254 (2010).
113. A. Vallée-Bélisle, F. Ricci, T. Uzawa, F. Xia, K. W. Plaxco, Bioelectrochemical switches for the quantitative detection of antibodies directly in whole blood. *Journal of the American Chemical Society* **134**, 15197–15200 (2012).
114. S. S. Mahshid, S. B. Camiré, F. Ricci, A. Vallée-Bélisle, A highly selective electrochemical DNA-based sensor that employs steric hindrance effects to detect proteins directly in whole blood. *Journal of the American Chemical Society* **137**, 15596–15599 (2015).
115. C. Fan, K. W. Plaxco, A. J. Heeger, Electrochemical interrogation of conformational changes as a reagentless method for the sequence-specific detection of DNA. *Proceedings of the National Academy of Sciences* **100**, 9134–9137 (2003).
116. A. Kowalczyk *et al.*, Construction of DNA biosensor at glassy carbon surface modified with 4-aminoethylbenzenediazonium salt. *Biosensors and Bioelectronics* **26**, 2506–2512 (2011).
117. Y. Tang, B. Ge, D. Sen, H.-Z. Yu, Functional DNA switches: Rational design and electrochemical signaling. *Chemical Society Reviews* **43**, 518–529 (2014).
118. L. Yu, H. Lim, A. Maslova, I. Hsing, Rational design of electrochemical DNA biosensors for point-of-care applications. *ChemElectroChem* **4**, 795–805 (2017).
119. S. S. Mahshid, A. Vallée-Bélisle, S. O. Kelley, Biomolecular steric hindrance effects are enhanced on nanostructured microelectrodes. *Analytical Chemistry* **89**, 9751–9757 (2017).
120. S. S. Mahshid, F. Ricci, S. O. Kelley, A. Vallée-Bélisle, Electrochemical DNA-based immunoassay that employs steric hindrance to detect small molecules directly in whole blood. *ACS Sensors* **2**, 718–723 (2017).
121. C. A. Mirkin, R. L. Letsinger, R. C. Mucic, J. J. Storhoff, A DNA-based method for rationally assembling nanoparticles into macroscopic materials. *Nature* **382**, 607 (1996).
122. N. L. Rosi, C. A. Mirkin, Nanostructures in biodiagnostics. *Chemical Reviews* **105**, 1547–1562 (2005).
123. C. Chen, G. Song, J. Ren, X. Qu, A simple and sensitive colorimetric pH meter based on DNA conformational switch and gold nanoparticle aggregation. *Chemical Communications* **2008**, 6149–6151 (2008).
124. F. Xia *et al.*, Colorimetric detection of DNA, small molecules, proteins, and ions using unmodified gold nanoparticles and conjugated polyelectrolytes. *Proceedings of the National Academy of Sciences* **107**, 10837–10841 (2010).
125. A.-M. Dallaire, S. Patskovsky, A. Vallée-Bélisle, M. Meunier, Electrochemical plasmonic sensing system for highly selective multiplexed detection of biomolecules based on redox nanoswitches. *Biosensors and Bioelectronics* **71**, 75–81 (2015).
126. N. H. Kim, S. J. Lee, M. Moskovits, Aptamer-mediated surface-enhanced Raman spectroscopy intensity amplification. *Nano Letters* **10**, 4181–4185 (2010).
127. A. Barhoumi, N. J. Halas, Label-free detection of DNA hybridization using surface enhanced Raman spectroscopy. *Journal of the American Chemical Society* **132**, 12792–12793 (2010).
128. Y.-J. Chen, B. Groves, R. A. Muscat, G. Seelig, DNA nanotechnology from the test tube to the cell. *Nature Nanotechnology* **10**, 748 (2015).
129. T. L. Fisher, T. Terhorst, X. Cao, R. W. Wagner, Intracellular disposition and metabolism of fluorescently-labled unmodified and modified oligouncleotides microijjected into mammalian cells. *Nucleic Acids Research* **21**, 3857–3865 (1993).
130. S. D. Putney, S. J. Benkovic, P. R. Schimmel, A DNA fragment with an alpha-phosphorothioate nucleotide at one end is asymmetrically blocked from digestion by exonuclease III and can be replicated in vivo. *Proceedings of the National Academy of Sciences* **78**, 7350–7354 (1981).
131. D. Sheehan *et al.*, Biochemical properties of phosphonoacetate and thiophosphonoacetate oligodeoxyribonucleotides. *Nucleic Acids Research* **31**, 4109–4118 (2003).
132. R. M. Hudziak *et al.*, Resistance of morpholino phosphorodiamidate oligomers to enzymatic degradation. *Antisense and Nucleic Acid Drug Development* **6**, 267–272 (1996).
133. M. A. Reynolds *et al.*, Synthesis and thermodynamics of oligonucleotides containing chirally pure R P methylphosphonate linkages. *Nucleic Acids Research* **24**, 4584–4591 (1996).
134. T. Ono, M. Scalf, L. M. Smith, 2′-Fluoro modified nucleic acids: Polymerase-directed synthesis, properties and stability to analysis by matrix-assisted laser desorption/ionization mass spectrometry. *Nucleic Acids Research* **25**, 4581–4588 (1997).
135. J. K. Watts, G. F. Deleavey, M. J. Damha, Chemically modified siRNA: Tools and applications. *Drug Discovery Today* **13**, 842–855 (2008).
136. M. Frieden, H. F. Hansen, T. Koch, Nuclease stability of LNA oligonucleotides and LNA-DNA chimeras.

Nucleosides, Nucleotides and Nucleic Acids **22**, 1041–1043 (2003).

137. C. A. Stein, D. Castanotto, FDA-approved oligonucleotide therapies in 2017. *Molecular Therapy* **25**, 1069–1075 (2017).
138. E. W. Ottesen, ISS-N1 makes the first FDA-approved drug for spinal muscular atrophy. *Translational Neuroscience* **8**, 1–6 (2017).
139. E. S. Gragoudas, A. P. Adamis, E. T. Cunningham Jr, M. Feinsod, D. R. Guyer, Pegaptanib for neovascular age-related macular degeneration. *New England Journal of Medicine* **351**, 2805–2816 (2004).
140. S. Ko, H. Liu, Y. Chen, C. Mao, DNA nanotubes as combinatorial vehicles for cellular delivery. *Biomacromolecules* **9**, 3039–3043 (2008).
141. J. Winkler, Oligonucleotide conjugates for therapeutic applications. *Therapeutic Delivery* **4**, 791–809 (2013).
142. S. Saha, V. Prakash, S. Halder, K. Chakraborty, Y. Krishnan, A pH-independent DNA nanodevice for quantifying chloride transport in organelles of living cells. *Nature Nanotechnology* **10**, 645 (2015).
143. L. Liang *et al.*, Single-particle tracking and modulation of cell entry pathways of a tetrahedral DNA nanostructure in live cells. *Angewandte Chemie International Edition* **53**, 7745–7750 (2014).
144. M. G. Kolonin *et al.*, Synchronous selection of homing peptides for multiple tissues by in vivo phage display. *The FASEB Journal* **20**, 979–981 (2006).
145. R. Pasqualini, E. Ruoslahti, Organ targeting in vivo using phage display peptide libraries. *Nature* **380**, 364 (1996).
146. G. P. Smith, Filamentous fusion phage: Novel expression vectors that display cloned antigens on the virion surface. *Science* **228**, 1315–1317 (1985).
147. S. Surana, J. M. Bhat, S. P. Koushika, Y. Krishnan, An autonomous DNA nanomachine maps spatiotemporal pH changes in a multicellular living organism. *Nature Communications* **2**, 340 (2011).
148. A. S. Piatek *et al.*, Molecular beacon sequence analysis for detecting drug resistance in Mycobacterium tuberculosis. *Nature Biotechnology* **16**, 359 (1998).
149. S. Tyagi, D. P. Bratu, F. R. Kramer, Multicolor molecular beacons for allele discrimination. *Nature Biotechnology* **16**, 49 (1998).
150. S. A. Marras, F. R. Kramer, S. Tyagi, Multiplex detection of single-nucleotide variations using molecular beacons. *Genetic Analysis: Biomolecular Engineering* **14**, 151–156 (1999).
151. S. Ranallo, C. Prévost-Tremblay, A. Idili, A. Vallée-Bélisle, F. Ricci, Antibody-powered nucleic acid release using a DNA-based nanomachine. *Nature Communications* **8**, 15150 (2017).
152. S. Ranallo, M. Rossetti, K. W. Plaxco, A. Vallée-Bélisle, F. Ricci, A modular, DNA-based beacon for single-step fluorescence detection of antibodies and other proteins. *Angewandte Chemie* **127**, 13412–13416 (2015).
153. R. A. Gatenby, R. J. Gillies, Why do cancers have high aerobic glycolysis? *Nature Reviews Cancer* **4**, 891 (2004).
154. V. Estrella *et al.*, Acidity generated by the tumor microenvironment drives local invasion. *Cancer Research* **73**, 1524–1535 (2013).
155. J. Liu *et al.*, pH-sensitive nano-systems for drug delivery in cancer therapy. *Biotechnology Advances* **32**, 693–710 (2014).
156. R. Hengge, Principles of c-di-GMP signalling in bacteria. *Nature Reviews Microbiology* **7**, 263 (2009).
157. A. Banerjee *et al.*, Controlled release of encapsulated cargo from a DNA icosahedron using a chemical trigger. *Angewandte Chemie International Edition* **52**, 6854–6857 (2013).
158. A. S. Walsh, H. Yin, C. M. Erben, M. J. Wood, A. J. Turberfield, DNA cage delivery to mammalian cells. *ACS Nano* **5**, 5427–5432 (2011).
159. D. Bhatia, S. Surana, S. Chakraborty, S. P. Koushika, Y. Krishnan, A synthetic icosahedral DNA-based host–cargo complex for functional in vivo imaging. *Nature Communications* **2**, 339 (2011).
160. T. R. Fadel *et al.*, Toward the responsible development and commercialization of sensor nanotechnologies. *ACS Sensors* **1**, 207–216 (2016).
161. D. W. Drolet, L. S. Green, L. Gold, N. Janjic, Fit for the eye: Aptamers in ocular disorders. *Nucleic Acid Therapeutics* **26**, 127–146 (2016).
162. S. Gupta *et al.*, Chemically-modified DNA aptamers bind interleukin-6 with high affinity and inhibit signaling by blocking its interaction with interleukin-6 receptor. *Journal of Biological Chemistry* **289**, 8706–8719. doi:10.1074/jbc.M113.532580 (2014).
163. T. Tørring, K. V. Gothelf, DNA nanotechnology: A curiosity or a promising technology? *F1000prime Reports* **5**, 14 (2013).
164. S. Bamrungsap *et al.*, Nanotechnology in therapeutics: A focus on nanoparticles as a drug delivery system. *Nanomedicine* **7**, 1253–1271 (2012).

7

Therapeutic Benefits from Nanoparticles

Barbara Sanavio
Fondazione I.R.C.C.S. Istituto Neurologico Carlo Besta

Silke Krol
Fondazione I.R.C.C.S. Istituto Neurologico Carlo Besta
I.R.C.C.S. "S. De Bellis"—Ente Ospedaliero Specializzato in Gastroenterologia

7.1 The Vision

Nanomaterials in general have a great impact in all fields of research and daily life. The ability to manipulate matter at the nanoscale, that is, in the range between 1 and 100 nm,[1] impacted heavily first in physics and electronics, and it is now incorporated in our daily lives, from mobile cell phones, sunscreens, self-cleaning coatings of windows and surfaces by Lotus effect construction, sports equipment, transportation (nanocomposites), antibacterial fabrics, and Gecko tape. Nanomaterials exhibit size-dependent emergent properties compared to the parent bulk materials. On one hand, nanomaterials have high interfacial energy (because of their surface-to-volume ratio) that prompt higher reactivity (catalysis) and higher working surfaces. On the other hand, due to their size smaller than the wavelength of light (UV-visible: 400–700 nm), they can interact with light, and small differences can induce significant changes. It is the case for quantum dots or gold nanoparticles, for example, where solely tuning the size of the nanoparticle core (e.g. CdS) allows for tuning of the emission wavelength (color).

Due to their size which is usually 100–1,000 times smaller than a cell, nanotechnological applications in medicine have been growing steadily in the last 20 years and are set to provide unprecedented advantages in the clinical practice. In fact, nanomaterial often has a size comparable to most of the target biological entities and processes, which raised interest in biomedical applications of nanotechnology. Their impact was in two major areas of medicine: (i) in diagnosis either as nanomaterial for improved sensors (carbon nanotubes for glucose sensing) or as contrast agent for imaging (iron oxide (magnetic) nanoparticles for magnetic resonance imaging) and (ii) in therapy (as drug carrier or as drug itself (iron oxide nanoparticles are used for iron deficiency anemia as iron reservoir (Feraheme® (ferumoxytol)) or to magnetically induce hyperthermia in glioma treatment (MagForce®) (Moghimi et al. 2005, Arruebo et al. 2007, Etheridge et al. 2013, Pepic et al. 2014, Bobo et al. 2016). The highest expectations were on the dual use of nanomaterials as so-called theranostics (Lammers et al. 2010, Theek et al. 2014, Kunjachan et al. 2015).

The nanomedicine sector is largely dominated by research and application in drug delivery and *in vitro* diagnostics (Freitas 2005, Wagner et al. 2006, Sanhai et al. 2008, Kim et al. 2010, Ventola 2012a, 2012b, 2012c, Etheridge et al. 2013). Today, ~40–50 nanodrugs are already approved for use in humans (Moghimi et al. 2005, van der Meel et al. 2013, Pepic et al. 2014, Weissig et al. 2014, Weissig and Guzman-Villanueva 2015, Bobo et al. 2016, Ventola 2017). These can be simple systems such as proteins, antibodies, or hormones enwrapped by polyethylene glycol (PEG) to prolong their circulation time in blood and increase the probability to reach their target. Here one quite famous and controversial nanodrug is an antibody used for the treatment of macular degeneration (MD), Lucentis®.[2] But nanodrugs can

[1] Many definitions of nanomaterials are available in the literature (Balogh 2010a) and many includes restriction and derogation to accommodate materials that are naturally, unintentionally or intentionally produced with nanoscale properties for both classification and regulation purposes (we here refer to the EU definition).

[2] http://www.allaboutvision.com/conditions/lucentis-vs-avastin.htm

also be complex systems, which combine the programmed release in a specific environment with high drug loading and change of biodistribution to reduce toxicity to off-target organs. The most prominent example is Doxil®, a liposomal nanodrug approved in 1995 as an improved version for the cardiotoxic chemotherapeutic doxorubicin (Barenholz 2012).

The therapeutic benefits provided by nanomedicine encompass all the different stages of detection, prevention, treatment, and monitoring of a variety of medical conditions, with a considerable fraction of the applications devoted to cancer. Diagnostic applications (Hu et al. 2011) span from early detection of trace concentration of biomarkers in blood samples with unprecedented sensitivity, to miniaturization of analytical devices for point-of-care (PoC) patient-oriented, portable systems to facilitate therapeutic monitoring and patient follow-up (Sanavio and Krol 2015), contrast agents for diagnostic imaging (Cheng et al. 2006, 2012, Chen et al. 2016). Therapeutic approaches (Lammers et al. 2010, Kunjachan et al. 2015, Chen et al. 2016) range from formulation of poorly soluble drugs or delivery that crosses biological barriers (Krol 2012, Wong et al. 2012, Krol et al. 2013, Tosi et al. 2015), to tissue targeting, controlled/extended release (Bhaskar et al. 2010), or integrated sensors that monitor blood parameters, like in glycaemia, and release insulin accordingly without any need of intervention from the patient (Tüdős et al. 2001, Yager et al. 2008, Rusling et al. 2010).

7.1.1 Scope of This Chapter

Nanomaterials in medicine are used as smart driver and carriers to combine several or all of the following properties in one single unit: (i) targeting, (ii) controlled/extended drug release, (iii) solubility enhancement for hydrophobic drugs, (iv) high local drug concentrations (nanoparticles can work like a "sponge" binding high or even toxic concentrations[3] in a very small volume), (v) "stealth" properties (cannot be recognized by immune cells as foreign material), and (vi) visibility in clinically relevant imaging techniques. This multifunctionality makes them a promising tool in disease treatment or as smart component for novel biosensors.

In this chapter, we are going to explore the therapeutic benefits that are sought and achieved by these applications. But before we start, we need to establish some common working definition for nanoparticles and their properties that are useful in the following discussions. Next we will analyze the interaction of nanoparticles with biological processes or entities as this drives both its benefit but also its potential toxicity. Specific examples of commercially and clinically relevant nanoparticles are then presented (with a critical assessment of why so few others followed the same path).

[3]The toxic concentration for a body can be loaded easily on a particle as the total concentration is significantly lower and one particle should enter and kill or cure one cell.

7.2 The Science

Key concepts. Nanoparticle landscape classification: What is a nanoparticle, interfacial layers, scale-related properties, objects that are only statistically defined. Evolution of NPs in biological media. Tailor the nanoparticle to suit the need: therefore, each nanomaterial is made to fit a purpose and one size does not fit all.

Nanoparticles: Definition and characteristics. Nanomaterials are commonly classified depending on their material and dimensionality, and much effort is devoted to identify strategies for nanomaterial landscape classification (Xia et al. 2010, Arts et al. 2014, Tomalia and Khanna 2016, Castagnola et al. 2017), which also includes biological modification (Figure 7.1). In this chapter, we are going to focus on nanoparticles of different materials. Nanoparticles (NPs) are so-called 0D (0 dimensionality) objects, that is, objects with all three dimensions confined at the nanoscale,[4] in the range between 1 and 100 nanometers (nm) in size; they are usually represented by (but by no means limited to) a spherical shape. The main advantages of NPs include: size comparable to the target biological entity; size-dependent emergent properties compared to the parent bulk materials (below 100 nm); and high interfacial energy (surface-to-volume ratio). We already anticipated emergent properties like QD and plasmonic resonances of gold NPs that can be exploited for detection (Balogh 2010b).

One characteristic of nano-objects that is rarely underlined is that **a nanoproduct is an ensemble of objects whose composition in size and shape can only be specified statistically** (Sun et al. 2016). The final nanoparticle material is an ensemble of nanoparticles with certain size and shape distribution if it is prepared in a bottom-up technique.[5] It is possible to manipulate matter even at the subatomic scale with single-molecule techniques; however, this is not practical for most real-life applications outside more mature fields like nanoelectronics. Traditional drugs can be purified at a very high level, where all components (molecules) are chemically defined and identical at the atomic level. Nanomaterials, instead, cannot be purified in a traditional sense, as polydispersity is ultimately maintained at the atomic level. Even if it could be practically achieved, whether by synthetic approaches or by careful purification and separation techniques, it would be far too expensive to

[4]As compared to 1D object (at least two dimensions at the nanoscale, such as nanotubes and nanowires, 2D object (one dimensional on the nanoscale, such as thin films) and 3D objects, such as nanostructured composites.

[5]We will not discuss a top-down preparation for nanomaterials as this is typically not used for clinically relevant nanomaterials and is even farther from atomically identical NPs.

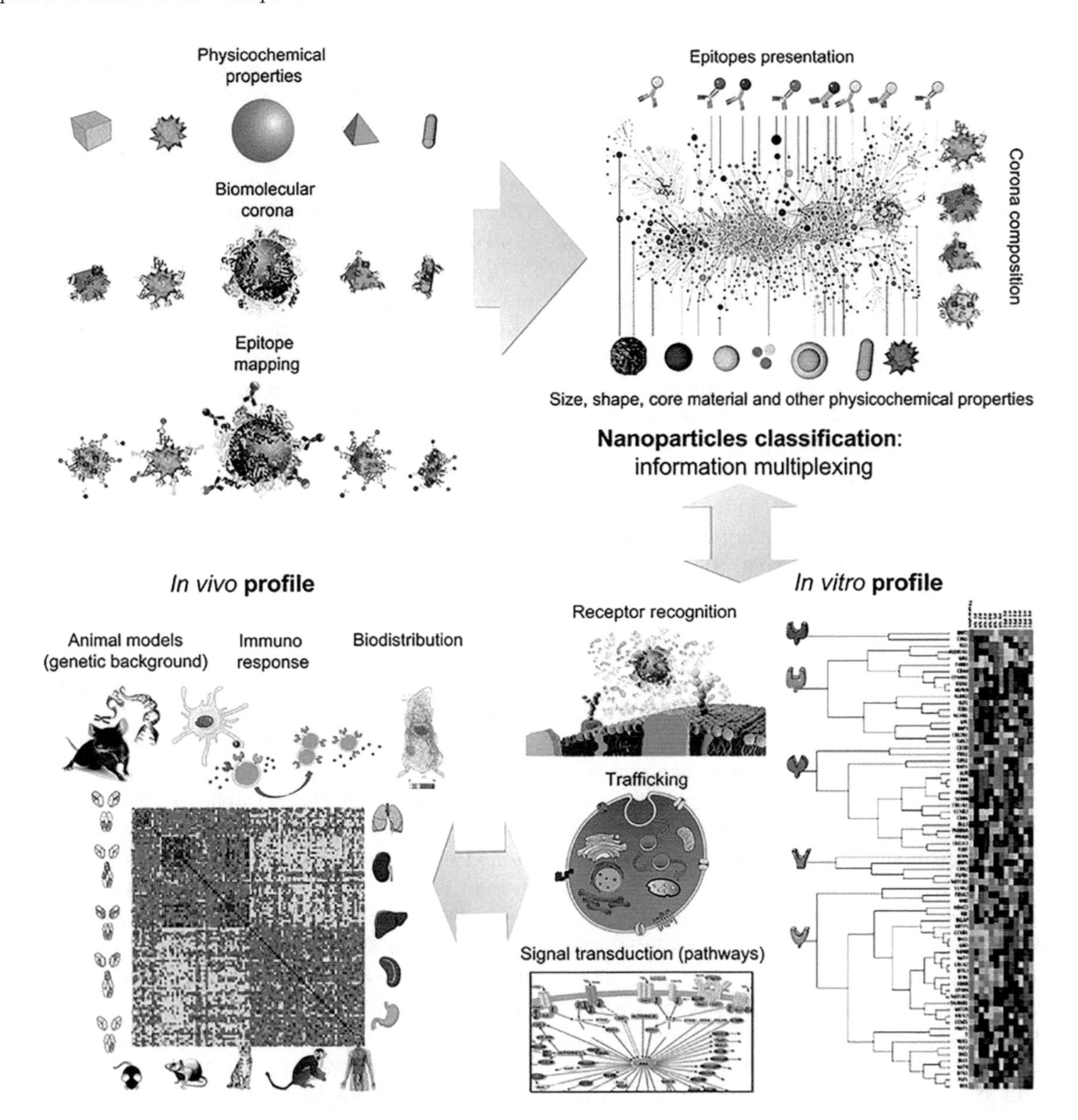

FIGURE 7.1 Schematic representation of the hypothesized workflow for nanomaterial bioclassification based on their intrinsic physicochemical properties (e.g. shape) and epitope presentation of biomolecular corona. Reprinted with permission from: (Castagnola et al. 2017), published by The Royal Society of Chemistry (Creative Commons Attribution-NonCommercial 3.0 Unported Licence).

ensure that all nanoparticles are equal at the atomic level, on commercial scales (Stone et al. 2010, Albanese et al. 2012, Ehmann et al. 2013, Pita et al. 2016).

Let's make an example: let's say we can synthesize a nanoparticle preparation where there is 1 impurity for every 1,000 NPs. So, 999 NPs are identical, and only 1 is different (the sample is 99.9% pure in a number distribution). If the 1 NP outside the distribution is 10 times larger in diameter by mass (proportional to volume), the sample is by weight 50% constituted by impurity. Nanomaterials evolve once they encounter biological media, and their evolution may vary according to their physicochemical properties, which ultimately depend on their size and shape. This example illustrates how much importance must be devoted to the proper characterization of the nanomaterial and the properties it exhibits. In real applications, NPs have a distribution of sizes and shapes that greatly exceeds two species. Does this difference in size affect or contribute to the biological activity of interest? Is it therefore really an impurity? Or shall we change paradigm and consider it as a feature of nanomaterials and analyze all results accordingly? *How can we take into account the fact that nanomaterials are intrinsically statistically defined? There is fortunately a growing branch of research that ensures that modern statistical methods (e.g. data science) are applied to the enormous set of data continuously produced in nanoscience* (Oomen et al. 2014, Sun et al. 2016). *This will help predicting structure–properties relationship while taking into account the "statistical" nature of the nanomaterial samples. If scientist were striving for unachievable level of purity and costly purification to provide experimental samples that matched the atomic level of precise selection of computational studies, in the future, Big Data science will help uncover structural relationship in "real life" nanoscience, with anticipated benefit for applications, like nanomedicines, where it is very difficult to achieve single-particle experimental precision.*

The **surface-to-volume ratio** of NPs increases with decreasing size. To make a practical example, let's consider 1 mm^3 of a cube of 1 mm side and 1 $mm^3$of cubes of many 1 nm side: the latter sample contains, at the same volume, 10^{18} object, with a total surface area that is 10^{12} times larger than the single 1 mm^3 object. Going small has obvious advantages in providing a larger surface for functionalization, drug loading, and receptor inclusion at the same volume (and mass) delivered. To be administered, an active molecule has to be formulated according to its properties and the route of administration. This means that the drug is not administered by itself but as a mixture with other compounds (excipients) that aims, among other things, at stabilizing, solubilizing, and prolonging the activity of the molecule. If the molecule is poorly soluble, there is a limit at the amount and/or the rate at which, once "encapsulated", it will be solubilized, for example, in the stomach, and it is possible that a higher dose has to be included in the formulation in order to achieve a therapeutic dose in the patient. If the drug is instead formulated through micronization, nanocrystals, or micellar/liposomial formulation, each formulation unit contains a small amount of drug that will be easier to solubilize, achieving therapeutic dosage faster and even reducing the total administered dose. Miniaturization of standard drug formulation was indeed implemented very early, long before modern nanotechnology started devising solutions for medicine.

As a consequence of the high exposed surface, the interface between the **surface** of these objects and the surrounding environment is extremely important in **defining the NP's characteristics and in affecting their properties**. The nanoparticle surface can consist of ions, inorganic, and organic molecules. The latter are commonly found in NPs intended for biomedical applications and may be named as surfactants, capping agents, surface ligands, or passivating agents. Their function is to allow for stabilization of the colloid and specific modifications for the target applications. This surface interacts with the surrounding (bulk) environment. The interface between the bulk properties of the surrounding environment and the proximal layer of the NP is usually called the ***interfacial layer****: the region of space comprising and adjoining the phase boundary within which the properties of matter are significantly different from the values in the adjoining bulk phases* (IUPAC 2014). It has specific properties and is closely associated with the respective nanomaterial. Indeed, the interfacial layer is considered as an integral part of the NPs. Careful tailoring and modification of this interfacial layers allows control over the stability, reactivity, interaction, transport, and fate of the nanoparticle throughout its life cycle (You et al. 2007, Alkilany and Murphy 2010, Park and Hamad-Schifferli 2010).

Details of the different techniques that allow proper characterization of NPs are outside the scope of this chapter. Many reviews and tutorials are available that describes how electron microscopy, light scattering, and other sophisticated techniques allow to assess (statistically) the properties of the NPs. It is good practice to use multiple techniques to describe a preparation.

Beside higher cargo delivering capability per volume of object, how can a nanoparticle about the size of protein or receptor of interest be useful in achieving therapeutic benefit? The major aims are: overcome biological barriers (e.g. gut to blood for oral delivery or blood–brain barrier (BBB) for brain diseases, (Wong et al. 2012, Barua and Mitragotri 2014, Blanco et al. 2015)), physicochemical issues (solubilizing hydrophobic drugs which otherwise cannot achieve concentrations high enough for a therapeutic affect (Riehemann et al. 2009)), reduce or avoid metabolic/distribution/toxicology drawbacks that traditional formulations cannot beat, and contemporarily providing therapy while diagnosing the disease (theranostics) (Theek et al. 2014, Kunjachan et al. 2015). To successfully achieve all of that, a deep understanding is needed of the molecular mechanisms at the basis of the disease and/or of the "obstacle" that nanotechnology may help overcoming: the BBB, the adsorption distribution and metabolism of the drug that need pharmacokinetic (PK) and pharmacodynamics (PD) improvements, the characteristics of the desired site of action, and the major side effects of the drug to be delivered (Garnett and Kallinteri 2006, Hagens et al. 2007, Fadeel and Garcia-Bennett 2010). In short, a profound knowledge of biology combined with a basic background in clinical (diagnostic tools) and medical requirements (potential treatment options) is required. Moreover, an understanding of how the proposed nanomaterials behave in biological settings (e.g. the human body) will help in predicting the desired properties or prevent undesired side effects and hence to tailor the design of the most suitable nanoparticle, meaning also physics and chemistry of nanomaterials are necessary. Nanomedicine, and especially translational nanomedicine, is indeed multidisciplinary and interdisciplinary and will be more successful the more cross-talk among fields is ensured.

Concluding from the literature, there is a very low conversion rate from the amount of "smart nanoparticles" at the research stage, to those which actually successfully made it to clinical trials or finally into clinical practice (Juliano 2013, Venditto and Szoka 2013). *Why is that?* With few exceptions, laboratory nanoparticle development was a lucky combination of different steps that fell in place together smoothly on the development pipeline (Etheridge et al. 2013). The development pipeline process starts with the design/identification of the most suitable nanotechnology platform for a specific drug/medical need, and progress through preclinical and clinical research, and eventually market. The process is highly regulated and may last up to 10 years, before the devised medicine makes it to the patient in the market. Many of the published research examples of smart nanoparticles are amazing advancements in the field but fail to incorporate in the design of the nanomaterial the "pipeline" for the clinical development that will follow. The rational formulation design of nanoparticles for therapy should include an assessment of the overall development process at the very early stage (Wacker 2013, Lee et al. 2015), so as to incorporate the appropriate design to fulfil the final goal, whether it has to do with enhanced solubilization, barrier trespassing, specific patient population, industrial scale up, and so on. That's what is called *smart by design*: you deal with the very issues that may affect the development and functioning of your therapeutic at an early developmental stage. Lately, there is also an increasing interest in including the issue of environmental impact right at the development stage: *where do the NPs go when they are excreted from the body? Are they biodegradable? How do they affect the environment? Will they accumulate in the food chain and at what level? Will the widespread use of that specific drug raise concern on any of these issues?*

How the physicochemical properties of nanomaterials can be exploited to achieve a therapeutic benefit?

Core material structure (such as crystallinity) **and composition**. According to the core materials, NPs can be broadly classified in organic NPs such as proteic, polymeric NPs, dendrimers (star-like polymers) and liposomes and carbon NPs like fullerenes, nanotubes, carbon dots, and graphene and inorganic NPs (metallic NPs such as gold, silver, iron oxide NPs, and QDs). All of them except the carbon NPs are in clinical use already or in clinical trials for approval as new nanodrugs. The core material may affect not only the behavior but also the toxicity of the NPs. The large surface area and surface atom of nanomaterials implicate that in certain cases it is highly accessible to oxygen or enzymes, and therefore, the surface is prone to corrosion (chemical changes or degradation), and therefore, leakage of the core material atoms or ions or entrapped drugs can occur. While this may be a devised property or an unwanted collateral effect, it must be taken into account and dealt with. QDs, for example, have excellent optical properties; however, leakage of heavy metal ions enhances toxicity. The same is true for silver nanoparticles, which have good antibiotic properties but also, here, a leakage of increased quantities of silver ions, from incomplete redox reaction, formerly incorporated in a metallic silver matrix, can cause toxicity (Christensen et al. 2010). The core has to be protected and sheltered to avoid this effect. The structure (crystallinity) of the core material can affect the properties of the nanoparticles, in terms of electronic and magnetic performances, for example. Or, NPs can be engineered to be hollow or porous to incorporate molecules in the core and release them slowly upon activation by changes in the environment (e.g. pH, ionic strength, temperature, light, magnetic/electromagnetic field).

Lipidic NPs and liposomes, polymeric NPs, proteic NPs, and gold and iron oxide NPs are the ones already available in the clinical practice or in clinical trials, either as therapeutics or as medical devices. Gold NPs are often chosen because of the inertial properties of gold in relation to biological system, although size-dependent accumulation in the body and a potential long-term adverse effect are still a matter of investigation. Liposomes were the first to enter the clinical practice because of the similarity to standard pharmaceutical production and regulation processes (see Doxil® story for further details).

Surface charge/chemical reactivity. The surface charge is fundamental to obtain colloidal stability (a balance between steric/repulsive forces) and influences interaction with the environment. There are accounts for enhanced toxicity of nanoparticles with a positive charge. However, reports where charge was changed keeping other variables fixed, it seems that toxicity of positively charged nanoparticles correlates with higher amounts of internalized NPs compared to negatively charged NPs. So, the higher toxicity could be a compound effect due to a higher dosage. Larger surface area offers larger possibilities for surface reactivity and functionalization, both engineered and passive (e.g. plasma protein binding) (Casals et al. 2010).

Shape and aspect ratio. The shape (spherical, cubic, cylindric,[6] stars, ellipsoid) of the nanoparticle affects its physical properties (e.g. optical and fluorescent properties). Shape can affect interaction with cells (like nanotubes that can be puncturing cell membranes). For synthetic and testing reasons, most nanoparticles are of spherical nature, although platelet-shaped nanoparticles have shown superior behavior when fluido-dynamics issues are taken into consideration (Ferrari 2008, Blanco et al. 2015).

Surface functionalization. Surface functionalization achieves both protection of the NPs, stealth properties (PEGylation to minimize protein adsorption and to prolong circulation of biological drug), and multimodality, which is the inclusion of different chemical entities to achieve different goals (targeting, therapy, imaging...) (Albanese et al. 2012, Pelaz et al. 2013).

7.2.1 Biology—Challenge or Trigger? Understanding Is the Key!

A day in the life of an NP. What happens to the NPs when they finally encounter the human body? The traditional ADMET (Administration, Distribution, Metabolism, Excretion, and Toxicity) approach can be used as a scheme for the study of potential NP therapeutics too. NPs, as any exogenous substance, evolve in biological (aqueous, protein-dense, and highly ionic) media. Insights on the fate of NPs can be obtained in *in vitro* and *in vivo* models of selected processes (Duncan and Gaspar 2011, Duncan and Richardson 2012).

[6]Aspect ratio measures the characteristic dimension of the material, for example, the ratio between length and diameter of a rod or tube.

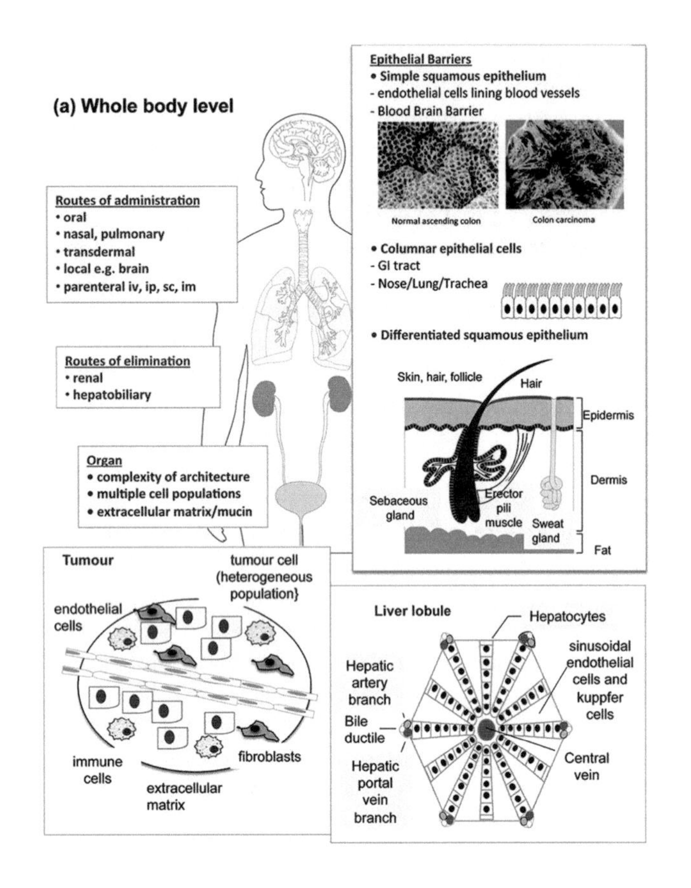

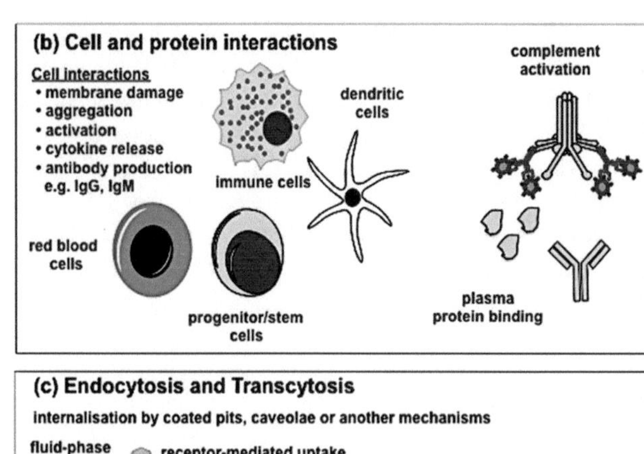

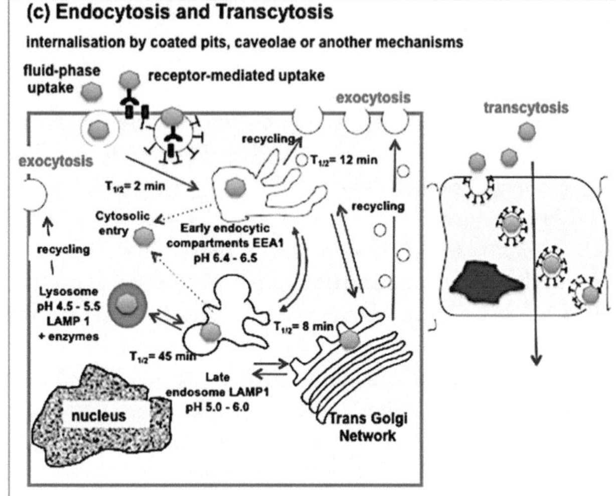

FIGURE 7.2 Illustration of pathophysiological complexity and biological barriers requiring consideration when designing nanomedicines for administration via different routes and with different pharmacological targets in mind. (Reprinted with permission from Duncan and Gaspar 2011. Copyright 2011 American Chemical Society.)

The study of the interactions of NPs and biological entities at different scales provides dual value information. On one side, we learn how the biological environment affects the NPs, and at the same time, we learn how the biological entity is affected by the NPs.

Let's proceed with a bird's-eye view of what the NPs journey may look like from the moment it enters the body to the moment it reaches its target (Duncan and Gaspar 2011, Duncan and Richardson 2012) (Figure 7.2). The fate of NPs will be influenced by the route of administration (oral, nasal, transdermal, local, parenteral, intravenous (iv), intraperitoneal (ip), intramuscular (im), subcutaneous (sc)[7]), the route of elimination (mainly renal (via kidneys and urine) or hepatobiliary (liver, bile, intestine, and feces), and the pathways of accumulation in the body. How adsorption is influenced by the barriers protecting our bodies? How the different epithelial barriers affect the NPs adsorption and distribution? The squamous epithelium will be the first to be investigated in case of transdermal administration, for example, but the role of endothelial cells in the passage of the therapeutic from the blood circulation to the tissues is of uttermost importance, especially if special endothelial barriers like the BBB have to be overcome to arrive at the target organ. In case of oral or nasal administration, the interaction with the columnar epithelial cells in the gastrointestinal tract, or the nose, lung, and tracheas, and how, for example, the mucous layer affects the nature of the NPs and the delivery need to be investigated.

Once the first barriers are crossed, the NPs have to travel through the organ-tissue architecture, through the extracellular matrix and through multiple cell populations. Here size matters! Both, for extravasation (release from blood vessels) and for mobility in the extracellular matrix between the single cells, the nanoparticle size is a limiting or restrictive factor. Even when they reach the target tissues, NPs have to discriminate somehow the healthy from the pathological state. Within the pathological tissue, some heterogeneity is expected. For example, tumor tissue presents a heterogeneous population of tumor cells, modified extracellular matrix, and other cells are recruited or affected by the tumor microenvironment (fibroblasts and immune cells, modification to the extracellular matrix, or the fenestration of the newly developed tumor vessel endothelium).

Intravenously injected NPs will travel through liver, lungs, and kidneys, whether or not they are primary target

[7]If you want to communicate with medical doctors, learn their lingo.

of therapy. The organ architecture will affect the amount of NPs that are retained in the organ, which remains in circulation and the fraction that is excreted, and the relative importance of different elimination and modification mechanisms is influenced by size, shape, and surface properties such as charge, stealth properties, antigenic features, and the proteins bound from the passing of barriers or in the blood (Schäffler et al. 2014, Bertoli et al. 2016).

At the cellular level, NP adsorption can influence and be influenced by membrane damage, aggregation/agglomeration state, activation of chemical reactivity, cytokine, and signaling molecule release (Duncan and Gaspar 2011, Canton and Battaglia 2012, Duncan and Richardson 2012). Interaction with cells from the immune system, such as dendritic cells and macrophages, may determine an inflammatory and/or an immune response, such as complement activation (CARPA,[8] (Szebeni 2001)) or elicit antibody production (Dobrovolskaia and McNeil 2007, Dobrovolskaia et al. 2009, Zolnik et al. 2010): approximately 30% of the population has antibodies against PEG, the major molecule for nanoparticle stealthing (Moghimi and Szebeni 2003). An example of inflammation is the reaction of the subcutaneous macrophages and dendritic cells to carbon microparticles contained in tattoo ink (Baranska et al. 2018): the tattoo spread over time as the immune cells feed on them, they die, and other cells recruited at the site perpetrate the same mechanism.

Endocytosis and transcytosis mechanisms, essential for epithelial and endothelial barriers trespassing, may be affected by the surface nature and size of the NPs (Alkilany and Murphy 2010, Canton and Battaglia 2012, Duncan and Richardson 2012). Internalization by coated pits, caveolae or other mechanism, and fluid phase/receptor-mediated uptake of NPs from cells may occur, and the pathways are influenced by size, surface charge (positively charged NPs are internalized at higher amount than negatively charged NPs, so the effective dose is higher and toxicity arise faster), and surface functionalization (ligands for specific receptors, protein corona, stealth polymers).

At the NP scale, all these steps are influenced by NP interactions with other molecules, like plasma proteins. As soon as the NP enters the bloodstream, for example, plasma proteins start adsorbing to the surface of the nanoparticle, in a process called *opsonization* (Owens and Peppas 2006). This can also partially happen in pegylated nanoparticles. Which proteins and in which relative amount they adsorb to the surface depends on the NP surface characteristics and on the relative abundance of the proteins in the blood. It is a phenomenon that was studied since the 1960s (*Vroman* effect, (Vroman 1962)) on surfaces of relevant biomedical devices and has been expanded and characterized on NPs recently (Casals et al. 2010, Monopoli et al. 2012, del Pino et al. 2014) and defined as protein corona. So, the "biological id card" of the nanomaterial, once it entered the bloodstream, is different from the initial "synthetic production id card" (Monopoli et al. 2012). It must be studied in detail, as it can affect how the original material behaves *in vivo*. As the protein corona changes the surface properties of the original NPs, the balance between steric and charge stabilization can be shifted. Will it still be sufficient to maintain functional NPs? Moreover, it can cover the originally bound targeting molecules and render them inefficient. It was recently found that the protein corona is only changed or released in the lysosomes of the cell (Bertoli et al. 2016). However, these questions can be addressed only *in vitro*. The binding of the protein corona can be influenced by size (Boselli et al. 2017) and surface properties (Bekdemir and Stellacci 2016), and its characteristics are differently depending on the nanomaterial: knowledge on the protein corona modifications can be exploited at your own advantage by passively targeting organs which contain the diseased cells (Schäffler et al. 2014).

How nanoparticle influences all these processes? Most of the answers come from toxicology studies on environmental exposure to nanoparticles and nanofibers (asbestosis and silicosis, indoor pollution by combustion, outdoor pollution by combustion, diesel exhaust particulate, volcanos, industries...). EMA[9] and FDA applications for nanomaterials testing follow regulation of drugs and/or devices (Hagens et al. 2007, Oomen et al. 2014), as the system is robust and provides a comprehensive regulatory framework to assess the safety of nanomaterials as therapeutics or medical devices. There is still a controversy if there is a difference of nanotoxicology to toxicology of drugs and if the testing provides a comprehensive understanding, as nanoparticles are neither bulk material nor single molecules and some specific toxicities, for example, small gold nanoparticles can induce genotoxicity (Xia et al. 2017) are not exactly covered. There is, however, in general heterogeneity in how toxicity studies are performed at the research and preclinical stage, as there is no consensus on specific requirements (see, for example, the variety of cells lines available for testing). In addition, the combinatorial variability of the end-point nanomaterials makes testing very complicated: one rule does not fit all. Moreover, the nanoparticles are also changing during the passage through the body and so, there is the question about the predictivity of *in vitro* tests to the *in vivo* situation. Other questions are: Are certain sets of tests suitable for nanoparticle of different sizes and shape? What about different core materials? Can the achieved results by different researchers be compared to each other? Can we extrapolate common trends that

[8]Cardiovascular distress syndrome due to complement bound to nanoparticles in the blood, also called infusion reaction. It's observed in 10% of patients treated with Doxil®.

[9]EMA-European Medicines agency: www.ema.europa.eu, FDA-Food and Drug Administration: http://www.fda.gov

would allow researchers to reach consensus on comprehensive guidelines? Initiatives like the material genome initiative, NIST, the EU NCL, are aiming to provide datasets and frameworks and technological access to researchers. Moreover, standardized guidelines are provided from the EU and US NCL (*Nanoparticle Characterization Laboratory, http://www.euncl.eu and https://ncl.cancer.gov/ respectively*), the latter dedicated to the characterization of nanomaterials since 2004.

Take home message: One rule does not fit all. The variety of nanomaterials, the continuous changes the material undergoes in different environments and time points (aging), and the variety of testing procedures and effects detected generated a large amount of data that makes it very difficult to extrapolate structure–activity relationship. The same definition of structure–activity relationship necessitates a new paradigm of analysis as the structure of a nanomaterial at the atomic level can be only statistically defined. Data science will be of utmost importance in the development of future nanotechnology, especially in defining structure/properties relationships (Sun et al. 2016).

7.2.2 Targeting

Since *Paul Ehrlich's* definition of "magic bullets", medicine has strived towards reaching and acting on the target pathological tissue, while sparing all other locations and therefore minimizing, if not avoiding at all, side effects. The possibility to engineer NPs with tailored properties opened avenues of attempts to targeted medicines, building also on the expertise of antibody therapies approved in the last decades. Through an antibody directed to specific features in the pathologic tissue, or through specific receptors, it is theoretically possible to select the target tissue.

Reviews of the recent literature (Duncan and Gaspar 2011, Ventola 2012b, van der Meel et al. 2013, Venditto and Szoka 2013, Bertrand et al. 2014, Bobo et al. 2016, Lammers et al. 2016, Torrice 2016) on targeted molecular medicines, however, revealed that – from the data available – targeting actually provides an **enhancement** of endocytosis/uptake at the cellular level at the targeted site but by no means provides a magic bullet that spares all other tissues. Since the late 1980s, large macromolecules (proteins, polymer-drug and polymer protein conjugates, liposomes) were known to accumulate in the tumor tissue via enhanced permeability and retention (EPR) effect (Matsumura and Maeda 1986, Maeda et al. 2000), a passive accumulation due to leaky tumor vessels (providing higher influx), and a reduced lymphatic clearance (Noguchi et al. 1998), depending on their molecular weight (which roughly correlate to size in macromolecules). Nanoformulated drugs show an improved accumulation by better exploiting the size selectivity of EPR. Compared to standard drug formulation, in addition, nanoparticles are bigger and highly loaded (Moghimi et al. 2005). In tumor tissue, the endothelium presents fenestrations that allow the passive permeation of <200 nm objects. By engineering cargo materials within this size, it is possible to achieve passive accumulation of nanoparticulate drugs at the site of interest, by exploiting a pathological modification to the endothelium. There is variability in the extent of fenestration within the tumor tissue and among different tumors: for example, nanoparticles designed for imaging could help in pre-screening, *via* diagnostic imaging, the tumor that will be more responsive to the nanomedicine treatment. However, the result is far beyond expectations. Recent estimates (Wilhelm et al. 2016) put the intratumor accumulation level as 0.7% of the iv administered (injected) dose of the nanotherapeutics. Although these values are astonishingly low, still the nanomedicine performs better than the free drug as the nanoformulation promotes the cellular uptake and therefore higher intracellular concentrations. In addition, nanotechnology can help deliver drugs to the patient that would be otherwise too toxic. The message is: although not yet ideal, the PK and PD of drugs as well as their organotoxicity now are more beneficial than what we could achieve without nanoparticles and offer a competitive advantage for the patients.

As mentioned before, it is possible to exploit size-dependent selection by the **EPR** effect at the target tissue. However, a major limitation for this approach is that tumors with a positive interstitial pressure (Heldin et al. 2004), in general therapy-resistant will also prevent the entry of nanoparticles. Here an additional treatment to reduce the interstitial pressure must be administered (such as collagenase), and again the nanoparticulate imaging is giving valuable information of the pressure before and after treatment with collagenase (Goodman et al. 2007, Hassid et al. 2008).

A novel approach that combines immune therapy with nanomaterials can be exemplified by trastuzumab. Trastuzumab (Herceptin®) binds to and inhibits human epidermal growth factor receptor (HER2, also known as ERBB2). Therefore, patients are screened for HER2-positive immunohistochemical staining in biopsies. From these selected patients, more than 50% respond well to trastuzumab treatment while only 10%–15% respond in the absence of molecular pre-selection of the patient subpopulation (Lammers et al. 2016). Trastuzumab bound to the drug-loaded nanoparticles should be able to deliver the drugs with higher selectivity in selected patients with HER2-positive tumors.

Another idea closer to the magic bullet was the use of magnetic focusing of iron oxide nanoparticles in specific tissue by placing an external magnet on the desired target area. However, also this approach showed limited success (Arruebo et al. 2007). The complexity of the different tissues, unspecific uptake in cells as well as physical parameters like flow rate and profile, and pore size in vessels and tissue requires a more elaborated and complex nanosystem, which usually is not suitable for industrial production.

Although it may seem a setback from the original idea of a magic bullet, it is an achievement that differential accumulation already allows to reduce side effects in off-target organs and it is a substantial improvement over the standard drug formulation. This is the example of Doxil® (Caelyx), the liposome formulation of doxorubicin.

In other instances, targeting is needed to overcome barriers: this topic merits a separate discussion and will be expanded in Section 7.2.5.

Most of the difficulties in targeting deal with the availability of a specific and selective target, which recognizes a feature unique to the tissue of interest, and that is extracellularly available. Or, the release of the drug may be engineered only upon interaction with specific intracellular targets, but NPs will have to first enter the cell and then act, therefore minimizing selective accumulation. Targeting may require "step-wise" closing-in onto the target. For example, a very selective neuronal feature is useless if the therapeutic agent cannot overcome the BBB, and multiple targeting agents may be installed on the same nanomaterial to first transit the BBB and reach the brain parenchyma and then preferentially reach the target cells. Therefore, the first characteristic for an effective targeting is: knowledge of the molecular features and mechanisms.

The design of nanomaterials allows to change the way we develop drug delivery: the nanopharmaceutical development process can be disease-driven, more than formulation-driven, in the sense that with a standard drug molecule, most of the action are limited by the solubility of the molecule itself. By choosing an appropriate nanomaterial, the same drug can be developed to devise the delivery system most appropriate for a specific pathology or subset of patient, in a quest of a more personalized medicine.

What has been achieved. Most of the knowledge of the effectiveness of targeting comes from targeted molecular medicine. A recent survey of the clinical trial data reported that far from providing a magic bullet, targeting does enhance uptake at the site of interest, therefore allowing for differential accumulation in the targeted tissue.

The first targeted nanomedicine to enter human clinical trials was BIND-014 (in 2014), an Accurin® family of proprietary polymeric nanoparticles, that aimed at delivering docetaxel as cytotoxic agent by targeting the prostate surface membrane antigen (PSMA). BIND-014 was generally well tolerated in Phase I trial (Von Hoff et al. 2016), with predictable and manageable toxicity and a unique PK profile different from conventional docetaxel and in multiple tumor types. It entered Phase 2, but results were less convincing than expected which put BIND pharmaceutical on the verge of bankruptcy and subsequent Pfizer acquisition.

In brief: *Nanomedicines: small size allows passive tumor targeting via EPR effect owing to the leaky endothelial structure in tumor vasculature; targeting allows ENHANCED endocytosis and accumulation in targeted tumor cells, by using antibody, substrate (like folate) for preferential targeting to the pathological cell population.*

7.2.3 Controlled Release

By acting on the drug formulation, it is possible to control the kinetic of drug release, so as to achieve predictable, prolonged level of the drug, both in the circulation or at the target site. Differently from traditional formulations, nanotechnology can help not only in improving and changing the physicochemical properties and therefore the PK and dynamic profile of the drug but can also make the formulation responsive to specific stimuli.

The controlled formulations available at the moment in the clinical practice are "nano-improvements" of a more traditional controlled formulation approach. For example, micellar formulation of hormone estradiol, Estrasorb® (Novavax), has been approved to achieve controlled delivery (controlled plasma concentration) in the cure of vasomotor symptoms in menopause (van der Meel et al. 2013). Alternatively, the use of polymeric formulations of active peptides can improve control over the clearance of the active compound and sustain prolonged activity, such as Copaxane® for multiple sclerosis (Bobo et al. 2016). DepoDur® is a liposomal morphine sulfate with extended release that has been approved for postoperative analgesia (Bobo et al. 2016). Similarly, nanocrystal formulations allow for faster solubilization and adsorption of pharmaceuticals: think at the 1 mm^3 cube example and how faster 10^{18} nm side cubes would solubilize (having much more surface exposed) than one single 1 mm side cube (Kim et al. 2010). Also, it is possible to exploit nanocrystal to improve cell recognition and adhesion, such as in the case of a variety of hydroxyapatite nanocrystals available as bone substitute that favors cell adhesion and bone regeneration (Bobo et al. 2016).

Nanotechnology offers the possibility to engineer nanoparticles to respond to stimuli of different nature: physical stimuli (UV light, laser radiation, temperature, ultrasounds, magnetic, electric and radiofrequency fields...), chemical stimuli (pI, ionic strength, redox conditions, specific gradient concentration), and biological stimuli (enzyme activity, bioactive molecule concentration changes...). Release of the drug or activity of the nanoparticle itself can be triggered by one of these factors. Visudyne®, for example, is a liposomal formulation of Verteporfin that allows for enhanced delivery and photosensitive release to diseased vessels in MD and ocular histoplasmosis (Morrow et al. 2007). Here, the specific site of action (the eye and specifically retina) is exploited to implement a controlled release mechanism (photosensitivity).

We can further extend the concept of controlled release to ***controlled activity***. Nanotherm® (MagForce) is an iron oxide-based treatment that is injected at the tumor site. By exposure to a oscillating magnetic field, the NPs absorb energy and release it as heat, generating a controlled hyperthermia treatment at the tumor site (Arruebo et al. 2007). Similarly, Auralase® achieve hyperthermia treatment by use of injected gold nanoparticle and photothermal (laser) therapy. Gold NPs absorb the laser excitation energy and release it as heat to initiate tumor necrosis by hyperthermia (Morrow et al. 2007).

Currently investigated at the research level are other kinds of stimuli-responsive nanoformulation/nanodevices. For example, changes in the glucose concentration are studied to allow for controlled expansion of a polymeric cage and subsequent drug release (insulin). Or, the acidic pH of the tumor microenvironment has been exploited to alter the surface charge or the surface of polymeric NPs in order to expose or activate secondary targeting agents to accumulate the nanoparticle or release the drug at the site of action (de las Heras Alarcón et al. 2005). Similarly, polymeric coatings like hyaluronic acid can be used as shielding agents that are then cleaved at the tissue sites by hyaluronidase enzymatic activity to expose the underlining active surface/drug (Zhou et al. 2018).

The examples at the research stage are numerous and have been reviewed elsewhere. Here we want to stress that fruitful multi-inter-disciplinary cross talks can foster enormous advances in these applications: as the interface between biological phenomenon and physicochemical properties of the nanomaterials is crucial for their efficacy, so it is the interface among the diverse knowledge areas that are needed to successfully develop such tailored nanomedicines.

7.2.4 Solubilizing Hydrophobic Drugs

As we mentioned, traditional approaches to drug formulation heavily rely on the physicochemical properties of the drug of interest. Differently, nanotechnology allows to devise new and innovative solutions to solve solubility issues.

Liposome, micellar and polymeric, or protein nanoparticle formulations have been implemented to improve drug solubility, protect unstable drug from early degradation or excessive plasma protein sequestration, or improve circulation time.

Protein, peptides, and aptamers, while not necessarily hydrophobic in nature, benefit from nanoformulation with pegylated polymer to gain improved stability and circulation and reduced immunogenicity. Other polymers like poly (D-L-lactide-coglycolide)) and poly(allylamine hydrochloride) helped in gaining augmented circulation time (respectively: Eligard®, for the controlled and prolonged delivery of leuprolide acetate for prostate cancer and Renagel® for chronic kidney disease) (Bobo et al. 2016).

Iron supplements in disease-related deficiency benefited from nanoparticle formulation to achieve higher IV administration doses and prolonged circulation through the use of iron oxide nanoparticles stabilized by polymeric coating such as pegylated polymers, silanes, polyglucose sorbitol carboxymethylether, and dextranes (Arruebo et al. 2007).

Good ol' pal: nanoformulations and liposomes. Pegylated liposomal formulation of doxorubicin was achieved by carefully exploiting the physicochemical properties of the drug (hydrophobicity, pH-dependent condensation in crystals) for optimal encapsulation into liposomes (Barenholz 2012). It is a textbook example of smart by design and cross-talk between disciplines. Both the size-dependent acquired tumoritropicity and the prolonged lifetimes helped improving the therapeutic index of doxorubicin, especially by mitigating the cardiac toxicity profile. This led to a subsequent extension of liposomal or polymeric/micellar formulation to other similarly toxic compound, such as daunorubicin, cytarabine, L-asparaginase, paclitaxel, irinotecan, vincristine, and mifamurtide (Morrow et al. 2007).

Ten years after the introduction of Doxil®, an albumin formulation of paclitaxel was approved (Abraxane®) (Morrow et al. 2007, Bobo et al. 2016). By exploiting the fact that a large portion of the low-solubility paclitaxel is bound to (human) serum albumin, this kind of formulation helps achieving improved solubility and circulation time in comparison with the standard paclitaxel formulation. Although protein NPs like Abraxane® are quite unstable in the circulation and do not provide a significant improvement in tumor-site accumulation. However, the therapeutic benefit derived from shorter infusion time and reduced need of corticosteroid administration to reduce immune response are a clear benefit for the patients.

Even if Doxil® and Abraxane® may be classified as incremental advances in formulation, more than a disruptive innovation in the implementation of nanotechnology in clinical practice, the therapeutic benefits for the patient in terms of reduced toxicity and improved quality of life during treatment (compared to the traditional formulations) are at the basis of their success.

7.2.5 Passing Barriers

As depicted in Section 7.2.1, many are the biological barriers that complicate the fate of nanoparticles and limit the number of nanomedicines that successfully translated into the clinics.

In a broader sense, biological barriers include all the biological phenomena that represent an obstacle to the effectiveness of the nanodrug. Therefore, opsonization, sequestration by the mononuclear phagocyte system, fluidodynamics, cellular uptake, and endosomal and lysosomal compartmentalization all contribute to the hurdle of NPs development (Blanco et al. 2015).

Ontak® was an FDA-approved diphtheria toxin-based recombinant fusion toxin for the treatment of human CD25+ cutaneous T cell lymphoma (CTCL), an early attempt to escape the lysosomal route of endocytosed drugs (van der Meel et al. 2013). It was withdrawn in 2014 due to boxed warning of visual acuity loss and non-sustainable purification issues in the recombinant bacterial production system. Lately, a new promising process based on a eukaryotic host was presented, reviving the interest in the compound.

Barriers in the body. Preserve, conquer, exploit. Barriers in the body usually have the function to preserve a tissue (organ or part thereof) from possible dangerous conditions and provide selective permeability (Barua and Mitragotri 2014, Blanco et al. 2015). The mucous and epithelial barriers are protecting the body from the external environment, while endothelial barriers such as the BBB, Sertoli-cell barrier in the testis, and placenta, are meant to provide protection to organs sensitive to homeostatic imbalances and exogenous chemicals by an extremely controlled transport into the tissue. Barriers evolve over time and disease, and therefore permeability can be impaired by medical conditions or exogenous factors. For example, ischemic events, epileptic seizures, and tumor can increase the BBB permeability (Krol 2012, Krol et al. 2013).

The BBB has attracted a lot of attention from the nanomedicine fields, as nanotechnology can provide new tools to selective trespassing of this barrier and allow an adequate quantity of mainly hydrophilic drugs to reach the brain parenchyma (Wong et al. 2012).

The BBB is an anatomo-functional barrier comprising different biological units (Figure 7.3). It is not the only barrier present in the brain, as a blood-cerebrospinal fluid (CSF) barrier is present in the choroid plexus, as well brain–CSF and CSF–brain barriers are described in the pia arachnoid and in the neuro ependyma, respectively (Abbott et al. 2006). The highly selective semipermeable BBB comprises different units, such as the endothelial cells, the adjacent pericyte, and the astrocyte processes contacting the vessels. However, the barrier function stems from a specific form of endothelial cells in the brain vasculature. Compared to the other capillaries throughout the body, the BBB endothelial cells do not present small fenestration. While other vessels are formed from several endothelial cells joined by gap junctions and tight junctions, the brain endothelium is formed by single cells connected to themselves by tight junction (Figure 7.3). In this way, the passage of microscopic object such as viruses or bacteria as well as large hydrophilic molecules is blocked, while passive entrance (diffusion regulated) of lipophilic and very small polar molecule (400 Da) through the endothelium is allowed. The endocytic mechanism is tightly regulated by receptors. Depending on the nature of the entering molecule, it can either by transported through the cells or excreted into to the vessel lumen by specific transporter on the membrane (e.g. P-glycoprotein).

Objects larger than 50 nm (Abbott et al. 2006, Banks 2016) do not pass an intact BBB, and nanoparticles much smaller than that have shown some success in traveling the barrier and reaching the parenchyma. However, the success rate is low, and specific strategy needs to be adopted.

Open sesame: the difficulties of selective opening. Traditional formulation for brain delivery resorted mainly in rendering the drug molecule of interest more lipophilic and therefore more prone to BBB passage. One example is to prepare a pro-drug, that is, modifying the drug with a specific hydrophobic chemical moiety. In particular, those which are known to have a specific shuttle in the BBB (transferrin or insulin) or traffic freely like caffeine. After entry and transport, the pro-drug is cleaved and activated. However, one of the limitations is that a more hydrophobic drug has lower solubility in blood.

Liposomal formulation present a more promising solution as they allow to entrap hydrophilic or hydrophobic

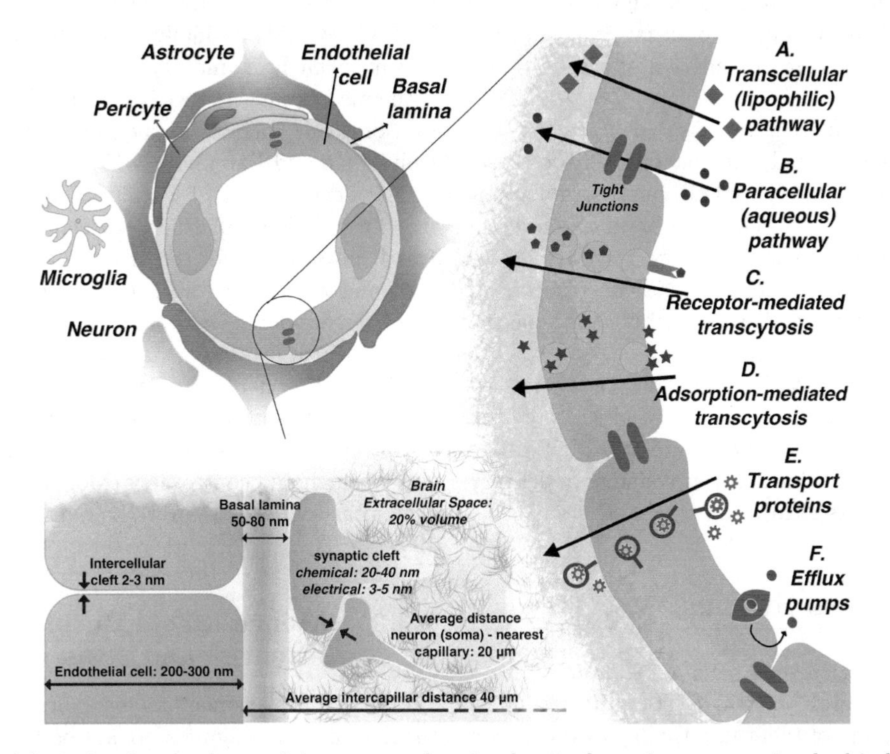

FIGURE 7.3 Blood–brain barrier. A scheme of the anatomo-functional unit, the major processes involved in barrier trespassing in a healthy BBB, and a highlight of the scales of the pathways traveled by NPs. Dimensions of anatomical and functional processes are repurposed from Krol, 2012 and reference therein.

molecules and hence deliver an increased amount of drug to the brain, compared to the standard drug (e.g. DaunoXome® (Lipoplatin), DepoCyt® (cytarabine), Ambisome®, Visudynev, and Doxil®/Caelyxv(doxorubicin). The use of specific receptors to trigger uptake at the BBB level has been promising. Transferrin-mediated uptake, as well as ApoE and albumin showed an increased rate of trespassing compared to non-targeted nanoparticles. On a molecular level, targets have been investigated to selectively exploit receptor-mediated endocytosis. OX26 is a rat monoclonal antibody specific for the transferrin receptor. By using OX-26 in conjugation to drug and/or nanocarriers, researchers have been able to enhance endothelial uptake and BBB passage of compounds (van der Meel et al. 2013). Similarly, ApoE (Apolipoprotein E) has been used to decorate liposomes and improve brain passage (Bhaskar et al. 2010). Albumin decoration is another protein functionalization of nanocarriers that showed increased transport to the brain (Krol et al. 2013). The greatest nanotechnology efforts are devoted to facilitate BBB passage in a non-invasive way. However, also the temporal BBB permeability changes can be exploited to increase drug accumulation in the brain. Different invasive techniques are available (Wong et al. 2012, Banks 2016) that alter the BBB permeability or bypass the BBB completely such as craniotomy-based delivery (e.g. during tumor resection), convection-enhanced delivery, and the use of microchip/wafer implant technology. Other invasive delivery strategies rely on temporal opening by focused ultrasound, where microbubbles as contrast agent can be used to force the endothelium open by a microexplosion, or the use of vasodilator like bradykinin that, by acting systemically, has an effect also on facilitating of the passage through the BBB (by modulation of caveolins and membrane channels).

Differently, nanoformulation for nasal delivery exploits the facilitated passage through the olfactory bulb of nasal formulation (Khan et al. 2017). Indeed, nasal formulations reach the circulation but also reach the brain parenchyma avoiding the BBB passage. Nasal delivery, however, has to overcome the mucous layer and the ciliated structure of the columnar epithelium and then deal with a very specific route of delivery that relies upon migration through the olfactory nerve, therefore limiting the regions accessible by this delivery route (Mitchell 2011). Nonetheless, there are promising NP-based applications that aim at exploiting the mucous layer for nanomedicine adsorption and then delivery (Wong et al. 2012, Banks 2016).

7.3 The Success Stories

Given the difficulties with several nanoformulations of drugs leading to their withdrawal, one will question the usefulness of nanodrugs and if any of them met the high expectation nanomedicine raised in the beginning. In the following chapter, we will briefly sketch two success stories of nanoformulated drugs and how they started new fields of research as in clinical use they showed unexpected and unforeseen side effects. The second story is to highlight the importance of cross-talk between science and industry.

7.3.1 Liposomes Doxil®

Doxil® was the first FDA-approved nanodrug, which hit the market in 1995. In Europe, it is sold under the name of Caelyx®. The nanoformulation of doxorubicin was necessary as the chemotherapeutic drug was highly efficient in treating tumor but could cause irreversible damage to the myocardium (heart muscle) if a cumulative concentration of doxorubicin exceeded 400 mg/m^2 (Rivankar 2014). This damage can, but must not, lead to a heart failure up to 10 years later and requires the annual checkup of patients who underwent doxorubicin treatment (Swain et al. 2003). Doxil®, for which doxorubicin was precipitated by an ammonium sulfate gradient inside a pegylated liposomal shell, should solve three of the major problems with doxorubicin: (i) decrease potentially lethal side effects by preventing the drug uptake by heart tissue; (ii) increase the drug concentration in the tumor site by EPR, and (iii) allow a controlled release. All three goals were achieved. Interestingly enough, the inventors, Prof. Barenholz and Prof. Gabizon, continued to study the potential mechanism of action of Doxil®. In 2012, Barenholz still was unsure about the mechanism by which the drug is released in the tumor environment or tumor cells nearly 20 years after approval and use in patients (Barenholz 2012). However, the mechanism for the increased entry of liposomal doxorubicin (EPR due to a size of 200 nm) and the reduced cardiac toxicity was clarified. Only in 2015, Barenholz finally was able to identify the mechanism, when he found that an increased concentration of ammonia in tumor tissues is responsible to trigger the solvation of doxorubicin from the liposome-enclosed precipitate (Silverman and Barenholz 2015).

The development of Doxil® opened the door for other nanoformulations of drugs especially with liposomes, but Doxil® also showed for the first time an unobserved toxicity, the infusion reaction, or, as it was later named the CARPA (complement-activation-related pseudoallergy), a hypersensitivity which could occur at the very first treatment with Doxil® or other nanodrugs or contrast agents (Szebeni 2001). The exact reason as well as the mechanism by which the symptoms develop spontaneously are still object to intensive research and controversy (Moghimi and Simberg 2017), even nearly 20 years after their discovery (Szebeni et al. 1999).

7.3.2 Science Meets Business! The Story of Lucentis® and Avastin®

One major achievement for nanoformulation of drugs was the binding of polymers to antibodies to stabilize them and hence increase their half-life time, increase their blood circulation time by avoiding immune recognition, and reduce adverse reaction to the antibody. One of the most successful examples is the antibody developed against wet MD, ranibizumab known under its trade name, Lucentis®. Wet MD is a form of age-related slow progressing blindness by vessels intruding the background of the eye. The development of anti-VEGF antibodies as inhibitor for neo-vascularization was a major breakthrough to delay the complete degeneration. Several anti-VEGF full-length, humanized monoclonal antibodies are known, one is the beforementioned ranibizumab and the other the anti-cancer-approved drug, bevacizumab (Avastin®). Recently a third one, aflibercept (brand name Eylea®) was been approved for treatment of wet MD. In all cases, the formulation in a matrix of polysorbate 20 is equal. As expected, also the efficacy (Mitchell 2011) and the drug-induced side effects are more or less the same for all drugs (Bevacizumab-Ranibizumab International Trials Group 2017) while studies in 2011 and 2012 indicated an elevated risk for bevacizumab (Mitchell 2011, Schmucker et al. 2012). So, what makes the difference? Lucentis® received approval in 2006 by the FDA as drug for wet MD while Avastin® received only approval for the treatment of cancer (Haddrill 2016) but never for MD. *So why were ophthalmologists pushing the off-label use of Avastin®?* Mainly because of the costs for the health system: while Avastin costs ca. 50$ per dose, Lucentis® costs ca. 2000$. The dimension of this collision of health care versus commercial interests is immense and led in 2017 to legal threat over *off-label* drug *use* by the interested companies selling ranibzumab in case patients are offered the choice for the off-label use of bevacizumab (Cohen 2017). This is an ongoing fight between pharma industry and clinicians, especially in Europe (industry complains to EU over French law pushing cheap eye drug 2015). The high costs for Lucentis® and Eylea® were dedicated to the process of approval by the FDA.

The Avastin®-Lucentis® story clearly shows the researchers developing nanomedicine have to consider not only the scientific aspects of their work but also the nanodrug if successful will have to fit into a clinical setting and must be commercially interesting. Both aspects need to be considered already in the planning of the nanodrug. If the drug is too complex, this presents an obstacle for scaling up and hence commercialization. If the nanodrug or theranostic tool is not fitting in existing imaging procedure in the hospital, it must exceed greatly the efficacy of the standard technique to justify the costs of a new imaging instrument. If these considerations are not included in the initial design phase of the nanosystems, there is a high risk that the developed system will be a great scientific achievement but one of the potential drug delivery systems or drugs which fail to make it into clinical trials or later in clinical practice.

7.4 Complexity and the Future

Nanomedicine, the development of nanomaterials for medical and/or clinical applications, involves a lengthy and expensive process, with an uncertain outcome. The nanodrugs aim to be as complex and flexible as the disease and the tissue they attempt to cure. A researcher interested in this field of study has to have a broad knowledge in various scientific fields such as biology, chemistry, physics as well as a basic knowledge of industrial production processes and clinical settings in which the drug or contrast agent or nanomaterial-based device will be used. While in the past it was required to become a specialist as a nanomedicine researcher, it is necessary to become a generalist once more.

However, the complexity of the diseases especially of tumors makes it necessary that also the drugs are complex. In the past, this was attempted by the use of a cocktail of different chemotherapeutic drugs. Nanodrug delivery systems with the possibility to combine multifunctionality and create a system as complex and as responsive to changes as the tumor itself promise to be the solution to a pressing problem to which by now we haven't found a convincing solution yet. What did we learn from the last 20 years of commercialized nanodrugs? That the process to fully understand the interaction of two complex systems (tumor and multi-component drug delivery system) is long, but nevertheless, nanodrugs can solve problems such as organotoxicity, by changing the PK and PD of the drug and improve the quality of life and expectation of cancer patients. We have to be careful to rise too high expectations as the development of a nanodrug has to take into consideration the interests of several stakeholders. However, the few successful examples showed that the efforts can result in an expected goal, to treat patients successfully and to improve their quality of life.

Abbreviations

BBB: Blood–Brain Barrier
NPs: Nanoparticles
PD: Pharmacodynamics
PoC: Point of Care
PK: Pharmacokinetics
UV: Ultraviolet

References

Abbott, N.J., Rönnbäck, L., and Hansson, E., 2006. Astrocyte–endothelial interactions at the blood–brain barrier. *Nature Reviews Neuroscience*, 7 (1), 41–53.

Albanese, A., Tang, P.S., and Chan, W.C.W., 2012. The effect of nanoparticle size, shape, and surface chemistry on biological systems. *Annual Review of Biomedical Engineering*, 14 (1), 1–16.

Alkilany, A.M. and Murphy, C.J., 2010. Toxicity and cellular uptake of gold nanoparticles: What we have learned so far? *Journal of Nanoparticle Research*, 12 (7), 2313–2333.

Arruebo, M., Fernández-Pacheco, R., Ibarra, M.R., and Santamaría, J., 2007. Magnetic nanoparticles for drug delivery. *Nano Today*, 2 (3), 22–32.

Arts, J.H.E., Hadi, M., Keene, A.M., Kreiling, R., Lyon, D., Maier, M., Michel, K., Petry, T., Sauer, U.G., Warheit, D., Wiench, K., and Landsiedel, R., 2014. A critical appraisal of existing concepts for the grouping of nanomaterials. *Regulatory Toxicology and Pharmacology*, 70 (2), 492–506.

Balogh, L.P., 2010a. Why do we have so many definitions for nanoscience and nanotechnology? *Nanomedicine: Nanotechnology, Biology and Medicine*, 6 (3), 397–398.

Balogh, L.P., 2010b. The nanoscopic range and the effect of architecture on nanoproperties. *Nanomedicine: Nanotechnology, Biology and Medicine*, 6 (4), 501–503.

Banks, W.A., 2016. From blood–brain barrier to blood–brain interface: New opportunities for CNS drug delivery. *Nature Reviews Drug Discovery*, 15 (4), 275–292.

Baranska, A., Shawket, A., Jouve, M., Baratin, M., Malosse, C., Voluzan, O., Manh, T.-P.V., Fiore, F., Bajénoff, M., Benaroch, P., Dalod, M., Malissen, M., Henri, S., and Malissen, B., 2018. Unveiling skin macrophage dynamics explains both tattoo persistence and strenuous removal. *Journal of Experimental Medicine*, 215, 1115–1133. doi:10.1084/jem.20171608.

Barenholz, Y. (Chezy), 2012. Doxil®—The first FDA-approved nano-drug: Lessons learned. *Journal of Controlled Release*, 160 (2), 117–134.

Barua, S. and Mitragotri, S., 2014. Challenges associated with penetration of nanoparticles across cell and tissue barriers: A review of current status and future prospects. *Nano Today*, 9 (2), 223–243.

Bekdemir, A. and Stellacci, F., 2016. A centrifugation-based physicochemical characterization method for the interaction between proteins and nanoparticles. *Nature Communications*, 7, 13121.

Bertoli, F., Garry, D., Monopoli, M.P., Salvati, A., and Dawson, K.A., 2016. The intracellular destiny of the protein corona: A study on its cellular internalization and evolution. *ACS Nano*, 10 (11), 10471–10479.

Bertrand, N., Wu, J., Xu, X., Kamaly, N., and Farokhzad, O.C., 2014. Cancer nanotechnology: The impact of passive and active targeting in the era of modern cancer biology. *Advanced Drug Delivery Reviews*, 66, 2–25.

Bevacizumab-Ranibizumab International Trials Group, 2017. Serious adverse events with Bevacizumab or Ranibizumab for age-related macular degeneration: Meta-analysis of individual patient data. *Ophthalmology Retina*, 1 (5), 375–381.

Bhaskar, S., Tian, F., Stoeger, T., Kreyling, W., de la Fuente, J.M., Grazú, V., Borm, P., Estrada, G., Ntziachristos, V., and Razansky, D., 2010. Multifunctional Nanocarriers for diagnostics, drug delivery and targeted treatment across blood-brain barrier: Perspectives on tracking and neuroimaging. *Particle and Fibre Toxicology*, 7 (1), 3.

Blanco, E., Shen, H., and Ferrari, M., 2015. Principles of nanoparticle design for overcoming biological barriers to drug delivery. *Nature Biotechnology*, 33 (9), 941–951.

Bobo, D., Robinson, K.J., Islam, J., Thurecht, K.J., and Corrie, S.R., 2016. Nanoparticle-based medicines: A review of FDA-approved materials and clinical trials to date. *Pharmaceutical Research*, 33 (10), 2373–2387.

Boselli, L., Polo, E., Castagnola, V., and Dawson, K.A., 2017. Regimes of biomolecular ultrasmall nanoparticle interactions. *Angewandte Chemie International Edition*, 56 (15), 4215–4218.

Canton, I. and Battaglia, G., 2012. Endocytosis at the nanoscale. *Chemical Society Reviews*, 41 (7), 2718.

Casals, E., Pfaller, T., Duschl, A., Oostingh, G.J., and Puntes, V., 2010. Time evolution of the nanoparticle protein corona. *ACS Nano*, 4 (7), 3623–3632.

Castagnola, V., Cookman, J., de Araújo, J.M., Polo, E., Cai, Q., Silveira, C.P., Krpetić, Ž., Yan, Y., Boselli, L., and Dawson, K.A., 2017. Towards a classification strategy for complex nanostructures. *Nanoscale Horizons*, 2 (4), 187–198.

Chen, G., Roy, I., Yang, C., and Prasad, P.N., 2016. Nanochemistry and nanomedicine for nanoparticle-based diagnostics and therapy. *Chemical Review*, 116, 2826–2885.

Cheng, M., Cuda, G., Bunimovich, Y., Gaspari, M., Heath, J., Hill, H., Mirkin, C., Nijdam, A., Terracciano, R., and Thundat, T., 2006. Nanotechnologies for biomolecular detection and medical diagnostics. *Current Opinion in Chemical Biology*, 10 (1), 11–19.

Cheng, Z., Zaki, A.A., Hui, J.Z., Muzykantov, V.R., and Tsourkas, A., 2012. Multifunctional Nanoparticles: Cost Versus Benefit of Adding Targeting and Imaging Capabilities. *Science*, 338, 903–910.

Christensen, F.M., Johnston, H.J., Stone, V., Aitken, R.J., Hankin, S., Peters, S., and Aschberger, K., 2010. Nano-silver - feasibility and challenges for human health risk assessment based on open literature. *Nanotoxicology*, 4 (3), 284–295.

Cohen, D., 2017. CCGs face legal threat for offering off-label drug for wet AMD. *BMJ*, 359, doi:10.1136/bmj.j5021.

de las Heras Alarcón, C., Pennadam, S., and Alexander, C., 2005. Stimuli responsive polymers for biomedical applications. *Chemical Society Reviews*, 34 (3), 276–285.

del Pino, P., Pelaz, B., Zhang, Q., Maffre, P., Nienhaus, G.U., and Parak, W.J., 2014. Protein corona formation around nanoparticles: From the past to the future. *Materials Horizons*, 1 (3), 301–313.

Dobrovolskaia, M.A., Germolec, D.R., and Weaver, J.L., 2009. Evaluation of nanoparticle immunotoxicity. *Nature Nanotechnology*, 4 (7), 411–414.

Dobrovolskaia, M.A. and McNeil, S.E., 2007. Immunological properties of engineered nanomaterials. *Nature Nanotechnology*, 2 (8), 469–478.

Duncan, R. and Gaspar, R., 2011. Nanomedicine(s) under the Microscope. *Molecular Pharmaceutics*, 8 (6), 2101–2141.

Duncan, R. and Richardson, S.C.W., 2012. Endocytosis and intracellular trafficking as gateways for nanomedicine delivery: Opportunities and challenges. *Molecular Pharmaceutics*, 9 (9), 2380–2402.

Ehmann, F., Sakai-Kato, K., Duncan, R., Pérez de la Ossa, D.H., Pita, R., Vidal, J.-M., Kohli, A., Tothfalusi, L., Sanh, A., Tinton, S., Robert, J.-L., Silva Lima, B., and Amati, M.P., 2013. Next-generation nanomedicines and nanosimilars: EU regulators' initiatives relating to the development and evaluation of nanomedicines. *Nanomedicine*, 8 (5), 849–856.

Etheridge, M.L., Campbell, S.A., Erdman, A.G., Haynes, C.L., Wolf, S.M., and McCullough, J., 2013. The big picture on nanomedicine: The state of investigational and approved nanomedicine products. *Nanomedicine: Nanotechnology, Biology and Medicine*, 9 (1), 1–14.

Fadeel, B. and Garcia-Bennett, A.E., 2010. Better safe than sorry: Understanding the toxicological properties of inorganic nanoparticles manufactured for biomedical applications. *Advanced Drug Delivery Reviews*, 62 (3), 362–374.

Ferrari, M., 2008. Nanogeometry: Beyond drug delivery. *Nature Nanotechnology*, 3 (3), 131–132.

Freitas, R.A., 2005. What is nanomedicine? *Nanomedicine: Nanotechnology, Biology and Medicine*, 1 (1), 2–9.

Garnett, M.C. and Kallinteri, P., 2006. Nanomedicines and nanotoxicology: Some physiological principles. *Occupational Medicine*, 56 (5), 307–311.

Goodman, T.T., Olive, P.L., and Pun, S.H., 2007. Increased nanoparticle penetration in collagenase-treated multicellular spheroids. *International Journal of Nanomedicine*, 2 (2), 265–274.

Haddrill, M., 2016. Lucentis vs. Avastin: A macular degeneration treatment controversy.

Hagens, W.I., Oomen, A.G., de Jong, W.H., Cassee, F.R., and Sips, A.J.A.M., 2007. What do we (need to) know about the kinetic properties of nanoparticles in the body? *Regulatory Toxicology and Pharmacology*, 49 (3), 217–229.

Hassid, Y., Eyal, E., Margalit, R., Furman-Haran, E., and Degani, H., 2008. Non-invasive imaging of barriers to drug delivery in tumors. *Microvascular Research*, 76 (2), 94–103.

Heldin, C.-H., Rubin, K., Pietras, K., and Ostman, A., 2004. High interstitial fluid pressure - an obstacle in cancer therapy. *Nature Reviews. Cancer*, 4 (10), 806–813.

Hu, Y., Fine, D.H., Tasciotti, E., Bouamrani, A., and Ferrari, M., 2011. Nanodevices in diagnostics: Nanodevices in diagnostics. *Wiley Interdisciplinary Reviews: Nanomedicine and Nanobiotechnology*, 3 (1), 11–32.

Industry complains to EU over French law pushing cheap eye drug, 2015. *Reuters*, 1 Sep.

IUPAC (International Union of Pure and Applied Chemistry), 2014. Surface layer (or interfacial layer). *In: IUPAC. Compendium of Chemical Terminology, 2nd ed. (the 'Gold Book')*. Compiled by A. D. McNaught and A. Wilkinson. Blackwell Scientific Publications, Oxford (1997). XML on-line corrected version: http://goldbook.iupac.org (2006-) created by M. Nic, J. Jirat, B. Kosata; updates compiled by A. Jenkins. ISBN 0-9678550-9-8. https://doi.org/10.1351/goldbook.

Juliano, R., 2013. Nanomedicine: is the wave cresting? *Nature Reviews Drug Discovery*, 12 (3), 171–172.

Khan, A.R., Liu, M., Khan, M.W., and Zhai, G., 2017. Progress in brain targeting drug delivery system by nasal route. *Journal of Controlled Release*, 268, 364–389.

Kim, B.Y.S., Rutka, J.T., and Chan, W.C.W., 2010. Nanomedicine. *New England Journal of Medicine*, 363 (25), 2434–2443.

Krol, S., 2012. Challenges in drug delivery to the brain: Nature is against us. *Journal of Controlled Release*, 164 (2), 145–155.

Krol, S., Macrez, R., Docagne, F., Defer, G., Laurent, S., Rahman, M., Hajipour, M.J., Kehoe, P.G., and Mahmoudi, M., 2013. Therapeutic benefits from nanoparticles: The potential significance of nanoscience in diseases with compromise to the blood brain barrier. *Chemical Reviews*, 113 (3), 1877–1903.

Kunjachan, S., Ehling, J., Storm, G., Kiessling, F., and Lammers, T., 2015. Noninvasive imaging of nanomedicines and nanotheranostics: Principles, progress, and prospects. *Chemical Reviews*, 115 (19), 10907–10937.

Lammers, T., Kiessling, F., Ashford, M., Hennink, W., Crommelin, D., and Storm, G., 2016. Cancer nanomedicine: Is targeting our target? *Nature Reviews. Materials*, 1 (9), 16069.

Lammers, T., Kiessling, F., Hennink, W.E., and Storm, G., 2010. Nanotheranostics and image-guided drug delivery: Current concepts and future directions. *Molecular Pharmaceutics*, 7 (6), 1899–1912.

Lee, B.K., Yun, Y.H., and Park, K., 2015. Smart nanoparticles for drug delivery: Boundaries and opportunities. *Chemical Engineering Science*, 125, 158–164.

Maeda, H., Wu, J., Sawa, T., Matsumura, Y., and Hori, K., 2000. Tumor vascular permeability and the EPR effect in macromolecular therapeutics: a review. *Journal of Controlled Release*, 65(1–2), 271–284.

Matsumura, Y. and Maeda, H. 1986. A new concept for macromolecular therapeutics in cancer chemotherapy: mechanism of tumoritropic accumulation of proteins and the antitumor agent smancs. *Cancer Research*, 46(12 Pt 1), 6387–6392.

Mitchell, P., 2011. A systematic review of the efficacy and safety outcomes of anti-VEGF agents used for treating neovascular age-related macular degeneration: Comparison of ranibizumab and bevacizumab. *Current Medical Research and Opinion*, 27 (7), 1465–1475.

Moghimi, S.M., Hunter, A.C., and Murray, J.C., 2005. Nanomedicine: Current status and future prospects. *The FASEB Journal*, 19 (3), 311–330.

Moghimi, S.M. and Simberg, D., 2017. Complement activation turnover on surfaces of nanoparticles. *Nano Today*, 15, 8–10.

Moghimi, S.M. and Szebeni, J., 2003. Stealth liposomes and long circulating nanoparticles: Critical issues in pharmacokinetics, opsonization and protein-binding properties. *Progress in Lipid Research*, 42 (6), 463–478.

Monopoli, M.P., Åberg, C., Salvati, A., and Dawson, K.A., 2012. Biomolecular coronas provide the biological identity of nanosized materials. *Nature Nanotechnology*, 7 (12), 779–786.

Morrow, K.J., Bawa, R., and Wei, C., 2007. Recent advances in basic and clinical nanomedicine. *Medical Clinics of North America*, 91 (5), 805–843.

Noguchi, Y., Wu, J., Duncan, R., Strohalm, J., Ulbrich, K., Akaike, T., and Maeda, H., 1998. Early phase tumor accumulation of macromolecules: a great difference in clearance rate between tumor and normal tissues. *Japanese Journal of Cancer Research: Gann*, 89 (3), 307–314.

Oomen, A.G., Bos, P.M.J., Fernandes, T.F., Hund-Rinke, K., Boraschi, D., Byrne, H.J., Aschberger, K., Gottardo, S., von der Kammer, F., Khnel, D., Hristozov, D., Marcomini, A., Migliore, L., Scott-Fordsmand, J., Wick, P., and Landsiedel, R., 2014. Concern-driven integrated approaches to nanomaterial testing and assessment – report of the NanoSafety cluster working group 10. *Nanotoxicology*, 8 (3), 334–348.

Owens, D.E. and Peppas, N.A., 2006. Opsonization, biodistribution, and pharmacokinetics of polymeric nanoparticles. *International Journal of Pharmaceutics*, 307 (1), 93–102.

Park, S. and Hamad-Schifferli, K., 2010. Nanoscale interfaces to biology. *Current Opinion in Chemical Biology*, 14 (5), 616–622.

Pelaz, B., Charron, G., Pfeiffer, C., Zhao, Y., de la Fuente, J.M., Liang, X.-J., Parak, W.J., and del Pino, P., 2013. Interfacing engineered nanoparticles with biological systems: Anticipating adverse nano-bio interactions. *Small*, 9 (9–10), 1573–1584.

Pepic, I., Hafner, A., Lovric, J., and Perina Lakos, G., 2014. Nanotherapeutics in the EU: An overview on current state and future directions. *International Journal of Nanomedicine*, 9, 1005–10023.

Pita, R., Ehmann, F., and Papaluca, M., 2016. Nanomedicines in the EU—regulatory overview. *The AAPS Journal*, 18 (6), 1576–1582.

Riehemann, K., Schneider, S.W., Luger, T.A., Godin, B., Ferrari, M., and Fuchs, H., 2009. Nanomedicine-challenge and perspectives. *Angewandte Chemie International Edition*, 48 (5), 872–897.

Rivankar, S., 2014. An overview of doxorubicin formulations in cancer therapy. *Journal of Cancer Research and Therapeutics*, 10 (4), 853–858.

Rusling, J.F., Kumar, C.V., Gutkind, J.S., and Patel, V., 2010. Measurement of biomarker proteins for point-of-care early detection and monitoring of cancer. *The Analyst*, 135 (10), 2496–2511.

Sanavio, B. and Krol, S., 2015. On the slow diffusion of point-of-care systems in therapeutic drug monitoring. *Frontiers in Bioengineering and Biotechnology*, 3, 20.

Sanhai, W.R., Sakamoto, J.H., Canady, R., and Ferrari, M., 2008. Seven challenges for nanomedicine. *Nature Nanotechnology*, 3 (5), 242–244.

Schäffler, M., Sousa, F., Wenk, A., Sitia, L., Hirn, S., Schleh, C., Haberl, N., Violatto, M., Canovi, M., Andreozzi, P., Salmona, M., Bigini, P., Kreyling, W.G., and Krol, S., 2014. Blood protein coating of gold nanoparticles as potential tool for organ targeting. *Biomaterials*, 35 (10), 3455–3466.

Schmucker, C., Ehlken, C., Agostini, H.T., Antes, G., Ruecker, G., Lelgemann, M., and Loke, Y.K., 2012. A safety review and meta-analyses of bevacizumab and ranibizumab: Off-label versus goldstandard. *PLoS One*, 7 (8), e42701.

Silverman, L. and Barenholz, Y., 2015. In vitro experiments showing enhanced release of doxorubicin from Doxil® in the presence of ammonia may explain drug release at tumor site. *Nanomedicine: Nanotechnology, Biology, and Medicine*, 11 (7), 1841–1850.

Stone, V., Nowack, B., Baun, A., van den Brink, N., von der Kammer, F., Dusinska, M., Handy, R., Hankin, S., Hassellöv, M., Joner, E., and Fernandes, T.F., 2010. Nanomaterials for environmental studies: Classification, reference material issues, and strategies for physico-chemical characterisation. *Science of the Total Environment*, 408 (7), 1745–1754.

Sun, B., Fernandez, M., and S. Barnard, A., 2016. Statistics, damned statistics and nanoscience–using data science to meet the challenge of nanomaterial complexity. *Nanoscale Horizons*, 1 (2), 89–95.

Swain, S.M., Whaley, F.S., and Ewer, M.S., 2003. Congestive heart failure in patients treated with doxorubicin: A retrospective analysis of three trials. *Cancer*, 97 (11), 2869–2879.

Szebeni, J., 2001. Complement activation-related pseudoallergy caused by liposomes, micellar carriers of intravenous drugs, and radiocontrast agents. *Critical Reviews in Therapeutic Drug Carrier Systems*, 18 (6), 567–606.

Szebeni, J., Fontana, J.L., Wassef, N.M., Mongan, P.D., Morse, D.S., Dobbins, D.E., Stahl, G.L., Bünger, R., and Alving, C.R., 1999. Hemodynamic changes induced by liposomes and liposome-encapsulated hemoglobin in pigs: A model for pseudoallergic cardiopulmonary reactions to liposomes. Role of complement and inhibition by soluble CR1 and anti-C5a antibody. *Circulation*, 99 (17), 2302–2309.

Theek, B., Rizzo, L.Y., Ehling, J., Kiessling, F., and Lammers, T., 2014. The theranostic path to personalized nanomedicine. *Clinical and Translational Imaging*, 2 (1), 67–76.

Tomalia, D.A. and Khanna, S.N., 2016. A systematic framework and nanoperiodic concept for unifying nanoscience: Hard/soft nanoelements, superatoms, meta-atoms, new emerging properties, periodic property patterns, and predictive mendeleev-like nanoperiodic tables. *Chemical Reviews*, 116 (4), 2705–2774.

Torrice, M., 2016. Does nanomedicine have a delivery problem? *ACS Central Science*, 2 (7), 434–437.

Tosi, G., Vandelli, M.A., Forni, F., and Ruozi, B., 2015. Nanomedicine and neurodegenerative disorders: So close yet so far. *Expert Opinion on Drug Delivery*, 12 (7), 1041–1044.

Tüdős, A.J., Besselink, G.A.J., and Schasfoort, R.B.M., 2001. Trends in miniaturized total analysis systems for point-of-care testing in clinical chemistry. *Lab on a Chip*, 1 (2), 83–95.

Venditto, V.J. and Szoka, F.C., 2013. Cancer nanomedicines: So many papers and so few drugs! *Advanced Drug Delivery Reviews*, 65 (1), 80–88.

Ventola, C.L., 2012a. The nanomedicine revolution. *Pharmacy and Therapeutics*, 37 (9), 512–525.

Ventola, C.L., 2012b. The nanomedicine revolution: Part 2: Current and future clinical applications. *P & T: A Peer-Reviewed Journal for Formulary Management*, 37 (10), 582–591.

Ventola, C.L., 2012c. The nanomedicine revolution: Part 3: Regulatory and safety challenges. *P & T: A Peer-Reviewed Journal for Formulary Management*, 37 (11), 631–639.

Ventola, C.L., 2017. Progress in nanomedicine: Approved and investigational nanodrugs. *Pharmacy and Therapeutics*, 42 (12), 742–755.

Von Hoff, D.D., Mita, M.M., Ramanathan, R.K., Weiss, G.J., Mita, A.C., LoRusso, P.M., Burris, H.A., Hart, L.L., Low, S.C., Parsons, D.M., Zale, S.E., Summa, J.M., Youssoufian, H., and Sachdev, J.C., 2016. Phase I study of PSMA-targeted docetaxel-containing nanoparticle BIND-014 in patients with advanced solid tumors. *Clinical Cancer Research: An Official Journal of the American Association for Cancer Research*, 22 (13), 3157–3163.

van der Meel, R., Vehmeijer, L.J.C., Kok, R.J., Storm, G., and van Gaal, E.V.B., 2013. Ligand-targeted particulate nanomedicines undergoing clinical evaluation: Current status. *Advanced Drug Delivery Reviews*, 65 (10), 1284–1298.

Vroman, L., 1962. Effect of adsorbed proteins on the wettability of hydrophilic and hydrophobic solids. *Nature*, 196 (4853), 476–477.

Wacker, M., 2013. Nanocarriers for intravenous injection–the long hard road to the market. *International Journal of Pharmaceutics*, 457 (1), 50–62.

Wagner, V., Dullaart, A., Bock, A.-K., and Zweck, A., 2006. The emerging nanomedicine landscape. *Nature Biotechnology*, 24 (10), 1211–1217.

Weissig, V. and Guzman-Villanueva, D., 2015. Nanopharmaceuticals (part 2): Products in the pipeline. *International Journal of Nanomedicine*, 10, 1245–1257.

Weissig, V., Pettinger, T.K., and Murdock, N., 2014. Nanopharmaceuticals (part 1): Products on the market. *International Journal of Nanomedicine*, 9, 4357–4373.

Wilhelm, S., Tavares, A.J., Dai, Q., Ohta, S., Audet, J., Dvorak, H.F., and Chan, W.C.W., 2016. Analysis of nanoparticle delivery to tumours. *Nature Reviews Materials*, 1 (5), 16014.

Wong, H.L., Wu, X.Y., and Bendayan, R., 2012. Nanotechnological advances for the delivery of CNS therapeutics. *Advanced Drug Delivery Reviews*, 64 (7), 686–700.

Xia, Q., Li, H., Liu, Y., Zhang, S., Feng, Q., and Xiao, K., 2017. The effect of particle size on the genotoxicity of gold nanoparticles. *Journal of Biomedical Materials Research Part A*, 105 (3), 710–719.

Xia, X.-R., Monteiro-Riviere, N.A., and Riviere, J.E., 2010. An index for characterization of nanomaterials in biological systems. *Nature Nanotechnology*, 5 (9), 671–675.

Yager, P., Domingo, G.J., and Gerdes, J., 2008. Point-of-care diagnostics for global health. *Annual Review of Biomedical Engineering*, 10 (1), 107–144.

You, C.-C., Chompoosor, A., and Rotello, V.M., 2007. The biomacromolecule-nanoparticle interface. *Nano Today*, 2 (3), 34–43.

Zhou, L., Wang, H., and Li, Y., 2018. Stimuli-responsive nanomedicines for overcoming cancer multidrug resistance. *Theranostics*, 8 (4), 1059–1074.

Zolnik, B.S., González-Fernández, Á., Sadrieh, N., and Dobrovolskaia, M.A., 2010. Minireview: Nanoparticles and the immune system. *Endocrinology*, 151 (2), 458–465.

8

Nanoprobes for Early Diagnosis of Cancer

Pengfei Xu
Jining Medical University

Zhen Cheng
Stanford University

8.1 Introduction

8.1.1 Cancer and Early Detection

Cancer is a major public health problem worldwide. In 2012, over 14 million cancer diagnoses were reported, and about 8.2 million people are dying due to this disease worldwide. It is estimated that the cancer death may increase up to 11.0 million in 2030.[1] According to WHO estimates, cancer now causes more deaths than all coronary heart disease or stroke. The continuing global demographic and epidemiologic transitions signal an ever-increasing cancer burden over the next decades, particularly in low- and middle-income countries. Cancer continues to present a major, yet unmet challenge to global healthcare.

Therefore, many researches focus on how to reduce the mortality rate of cancer. In fact, survival of a cancer patient depends heavily on early detection. Early diagnosis enables timely treatment, significantly improves patient outcomes, and is essential for successful therapy.[2–4] For example, non–small-cell lung patients who start therapy in the early stages of the disease (stage I) have an 80% overall 5-year survival rate. However, patients in the advanced stages (III and IV) of the disease have only 5%–15% and <2% 5-year survival rates, respectively. Consequently, early diagnosis is crucial for improving cancer patient prognosis and reducing the mortality rate.

Unfortunately, most cancers at early stage are difficult to be detected due to the lack of adequate diagnostic techniques.[5–7] At present, clinical detection of cancer primarily relies on imaging techniques or the morphological analysis of tissues (histopathology) or cells that are suspected to be diseased (cytology). Traditional imaging techniques for cancer detection, such as ultrasound (US), computed tomography (CT) and magnetic resonance imaging (MRI), have low sensitivity and also are limited in their ability to differentiate between benign and malignant lesions.[8,9] Histopathology, generally based on taking a biopsy of a suspected tumor, is typically used to probe the malignancy of tissues that are identified through alternative imaging techniques, such as CT or MRI, and it may not be used alone for early diagnosis of cancer. Due to the lack of resolution to detect nascent sub-mm-sized tumors, these traditional diagnostic methods are not very powerful methods when it comes to cancer detection at early stage. Therefore, the development of technology that is specific and reliable for detecting cancers at early stage is of utmost importance.

8.2 Imageable Nanoprobes for Early Cancer Detection

Nanomaterials and nanotechnology have facilitated major advancements in the development of powerful diagnostic techniques. Nanomaterials exhibit unique physical and chemical properties that make them excellent scaffolds for the fabrication of novel imaging nanoprobes. As a result of the properties afforded by the highly tunable size and shape-dependent features of nanomaterials, their large surface-to-volume ratios, and their conjugations with biomolecules with desired specificity or assemblies with desired signal transduction mechanisms, nanomaterial-based probes can markedly improve the imaging sensitivity and specificity, which is important for the early diagnosis of cancers.[7,10] Numerous nanoprobes possessing unique properties have been developed and extensively studied.

8.2.1 Key Nanoprobe Properties

Passive Targeting of Nanoprobes

The physicochemical properties of nanoprobes dictate not only their signaling characteristics but also their transport and biodistribution properties *in vivo*. Termed the enhanced permeation and retention (EPR) effect, this process represents a premier mechanism by which nanoprobes passively, yet preferentially, accumulate at a target cancer site. This permeation is due to rapidly, dysfunctionally grown blood vessels characterized by aberrant tortuosity and abnormalities in the basement membrane and the lack of pericytes lining endothelial cells. The incomplete tumor vasculature results in leaky vessels with gap sizes of 100 nm–2 μm depending upon the tumor type.

If nanoprobes can evade immune surveillance and circulate for a long period, the EPR effect will be applicable and high local signals of nanoprobes can be achieved at the tumor site. To this end, at least three properties of nanoprobes are particularly important. (i) The ideal nanoprobe size should be somewhere between 10 and 100 nm. Indeed, for efficient extravasation from the fenestrations in leaky vasculature, nanocarriers should be much less than 400 nm. On the other hand, to avoid the filtration by the kidneys, nanoprobes need to be larger than 10 nm; and to avoid the specific capture by the liver, nanoprobes need to be smaller than 100 nm. (ii) The charge of the nanoprobes should be neutral or anionic for efficient evasion of the renal elimination. (iii) The nanoprobes must be hidden from the reticuloendothelial system (RES), which destroys any foreign material through opsonization followed by phagocytosis.[11]

Active Targeting of Nanoprobes

Active targeting uses peptides and other biomolecules as targeting agents to selectively and successfully transport nanoprobes to cancerous tissue. Generally, passive targeting facilitates the efficient localization of nanoprobes in the tumor interstitium but cannot further promote their uptake by cancer cells. However, in terms of active targeting, specific interactions between the targeting agents on the surface of nanoprobes and receptors expressed on the tumor cells can facilitate nanoprobe's internalization by triggering receptor-mediated endocytosis. For instance, folate targeting is a classic example of active targeting, as it has been extensively tried and tested over the past years. Folate receptor is overexpressed in a variety of cancer types such as ovarian carcinomas, osteosarcomas, and non-Hodgkin's lymphomas.[12] Nanoprobes conjugated to folate have greater chances of being internalized to a substantial extent, wherein the folate receptors are highly overexpressed.

By taking advantages of passive and active targeting, a variety of nanoprobes have been designed and studied for tumor imaging using different modalities as elaborated below.

8.2.2 Magnetic Nanoprobes

MRI is one of the most widely used and powerful imaging modalities for cancer detection. In *in vivo* MRI, the signal intensity is inherently related to the tissue characteristics, such as the proton density and relaxation time (T_1, spin–lattice relaxation and T_2, spin–spin relaxation), and it does not always generate sufficient contrast in a clinical setting. To increase contrast, various contrast agents (UCAs), which can vary the relaxation time and produce hyperintense or hypointense signals in shorter times, are administered prior to the scanning.

There are many strategies to develop contrast agents, but the use of nanoparticles as MRI probes has several advantages over conventional imaging agents. First, the loadability is one of the advantages where the concentration of the imaging agent can be controlled in each nanoparticle during the development process. Second, the tunability of the surface of the nanoparticles is another advantage which can potentially extend the circulation time of the magnetic nanoprobes in the blood or actively target tumor within the body.

T_1 MRI Nanoprobes

In a clinical situation, T_1 contrast agents are usually preferred for better clarity because the dark signal produced by T_2 contrast agents is sometimes confused with some endogenous conditions, such as calcification, air, hemorrhage, and blood clots.[13] Gd(III) is the most clinically used metal ion in paramagnetic T_1 contrast agents. It possesses an electron spin of 7/2 and, hence, seven unpaired electrons promoting spin relaxation due to flipping spins and rotational motion.[14] The Gd(III)-based nanoprobes, including Gd(III) containing inorganic crystalline nanoparticles and Gd(III) chelate-grafted nanoparticles, have been extensively studied in preclinical and clinical applications.

Nanomaterials gadolinium oxide (Gd_2O_3), gadolinium fluoride (GdF_3), and inorganic fluoride nanoparticles $KGdF_4$ and $NaGdF_4$ have been explored as promising T_1 contrast agents for detection of cancer.[15–18] For example, in a study performed by Hou and co-workers,[19] they constructed differently sized $NaGdF_4$-PEG-mAb nanoprobes with excellent binding specificity to epidermal growth factor receptor (EGFR). EGF and the aberrant activation of EGFR, such as overexpression of EGFR in tumor cells, play key roles in cell proliferation, cell motility, cell adhesion, invasion, cell survival, and angiogenesis, which result in the development and progression of various tumors. Thus, EGFR can work as a biomarker for early detection of tumor.[20] $NaGdF_4$ nanocrystals conjugated to anti-EGFR monoclonal antibody (mAb) via "click" reaction both on the maleimide residue on nanoparticle surface and on thiol group from the partly reduced anti-EGFR mAb. As shown in Figure 8.1, this probe exhibits excellent biocompatibility and presents superior contrast agent properties than Gd-DTPA (Magnevist). Moreover, $NaGdF_4$-PEG-mAb probe displays promising tumor-specific targeting ability and enhanced MRI contrast effects.

Gd(III) chelates coated or immobilized on nanoparticles is another way to develop T_1 MRI nanoprobes for cancer imaging. Kataoka et al. constructed a new T_1 MRI nanoprobe by incorporating Gd-DTPA into

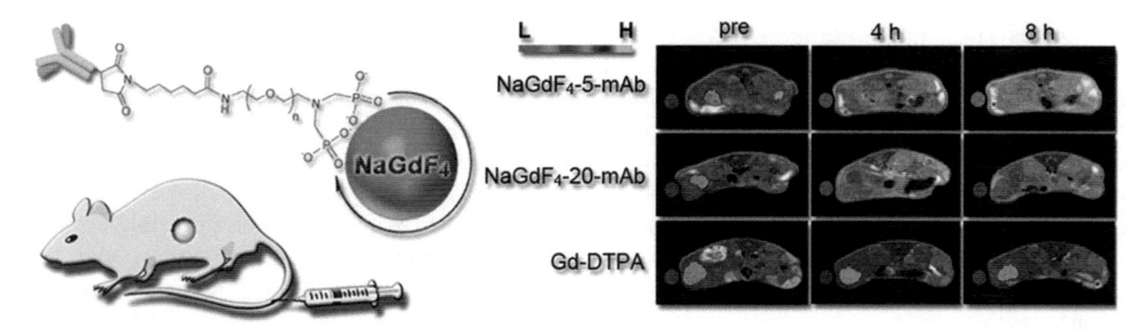

FIGURE 8.1 T_1-weighted MR images of tumor-bearing mice acquired before and at different time points after the intravenous injections of Gd-DTPA or NaGdF4–PEG-mAb probes formed by using $NaGdF_4$-5 and $NaGdF_4$-20, respectively. (Reprinted with permission from Ref. [19]. Copyright 2013 American Chemical Society.)

DACHPt-loaded micelles utilizing the reversible complex formation between DACHPt and Gd-DTPA. This T_1 MRI nanoprobe has many advantages for MRI, such as slow molecular reorientation, decreased mobility, and increased relaxivities. Animal experiments revealed that Gd-DTPA/DACHPt-loaded micelles accumulated effectively in subcutaneous murine colon carcinoma and orthotopic human pancreatic adenocarcinoma models and specifically enhanced the signal at the tumor site for prolonged time.[21]

In a recent work, Hyeon et al. developed a pH-responsive "smart" magnetic nanoprobes (PMNs) for early-stage diagnosis of tumors (Figure 8.2). Compared with the physiological tissue (pH ~7.4), one commonality among tumors is acidity; the microenvironment usually has a pH of ~6.8 and endo/lysosomes experience even lower pH values of 5.0–5.5.[22–24] In neutral conditions, PMNs display high r_2 relaxivity because of the clustering of nanoparticles, preventing an effective T_1 contrast effect. However, in the tumor conditions, pH-sensitive ligands become protonated and hydrophilic, resulting in disassembly of extremely small iron oxide nanoparticles (ESIONs). The positive T_1 MR contrast effect, which is quenched at pH 7.4, is recovered at pH 5.5. When injected intravenously, PMNs enable sensitive diagnosis of small tumor (as small as 3 mm in diameter) with a positive contrast, demonstrating that PMNs can be used for early-stage diagnosis of tumors.[25]

T_2 MRI Nanoprobes

T_2 contrast agents (or negative contrast agents) decrease the MR signal intensity of the regions they are delivered to. Iron oxide nanoparticles (IONPs) have been used as T_2 contrast agents for more than 25 years. These IONPs can be ferromagnetic or superparamagnetic, depending on the size of the core of the nanoparticle. Numerous T_2 MRI nanoprobes have been reported for early cancer detection.[14] For instance, ultra-small superparamagnetic iron-oxide nanoparticles (USPIONs) of <10 nm in hydrodynamic diameter were tested for tumor-specific MRI targeting. In this study, the USPIONs were stabilized by 4-methylcatechol and were coupled with a cyclic arginine-glycine-aspartic acid (cRGD) peptide. The

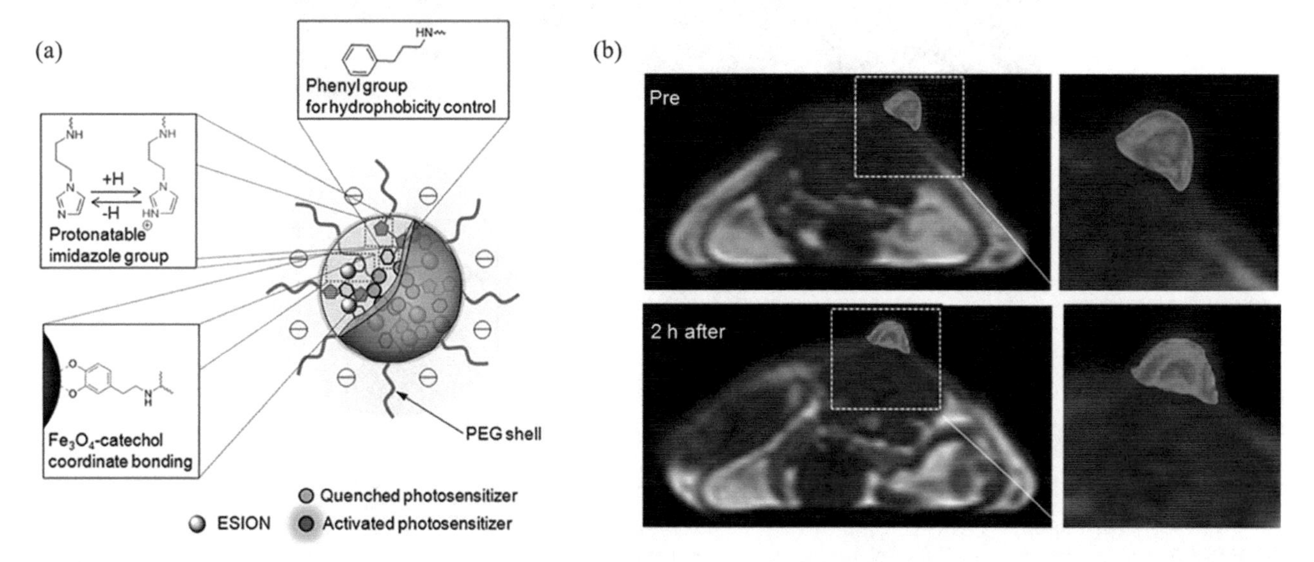

FIGURE 8.2 (a) Schematic representation of tumor pH-recognizable strategy using PMNs; (b) *in vivo* tumor imaging using PMNs. (Reprinted with permission from Ref. [25]. Copyright 2014 American Chemical Society.)

peptide RGD binds the $\alpha_V\beta_3$-integrin, a cell adhesion molecule that is overexpressed in tumor vasculature and invasive tumor cells.[26] After the administration of the RGD nanoparticles, the tumor MRI signal intensity decreased by 40%.[27] Other T_2 MRI nanoprobes with potential applications for early diagnosis of tumors include gold-iron oxide (Au-Fe_3O_4) nanoparticles, metallic ion nanoparticles, porous hollow Fe_3O_4 nanoparticles, and Fe-based alloy nanoparticles, such as iron-cobalt (FeCo) and iron-platinum (FePt) nanoparticles.[14]

T_1-T_2 MR Nanoprobes

Conventional MRI contrast agents are mostly effective only in a single imaging mode: either T_1 or T_2. They frequently result in ambiguities in early diagnosis of tumors, because tumors are small at early stage. The combination of simultaneously strong T_1 and T_2 contrast effects in a single contrast agent could be a new breakthrough, since it can potentially provide more accurate MRI *via* self-confirmation with better differentiation of normal and tumor regions. In a work of Yang et al., they developed a Gd-labeled superparamagnetic Fe_3O_4 NPs conjugated with RGD peptides (Fe_3O_4@SiO_2(Gd-DTPA)-RGD NPs) for MRI of cancer (Figure 8.3).[28] This nanoprobe efficiently reduces both T_1 and T_2 relaxation times. MRI data clearly indicates that the dual-mode Fe_3O_4@SiO_2(Gd-DTPA)-RGD NPs have high binding specificity to overexpress $\alpha_v\beta_3$-integrin cancer cells through receptor-mediated delivery pathway. The T_1-weight positive and T_2-weighted negative enhancement in the dual-mode MRI could significantly improve the diagnosis accuracy for early detection of cancer.

8.2.3 Radioactive Nanoprobes

Single-photon emission computed tomography (SPECT) and positron emission tomography (PET) are the two major radionuclide imaging modalities in the field of nuclear medicine. Unlike MRI, which mainly provides detailed anatomical images, PET and SPECT can measure chemical changes that occur before macroscopic anatomical signs of a tumor that are observed, making them well suited for early cancer detection. The basic principle behind PET involves

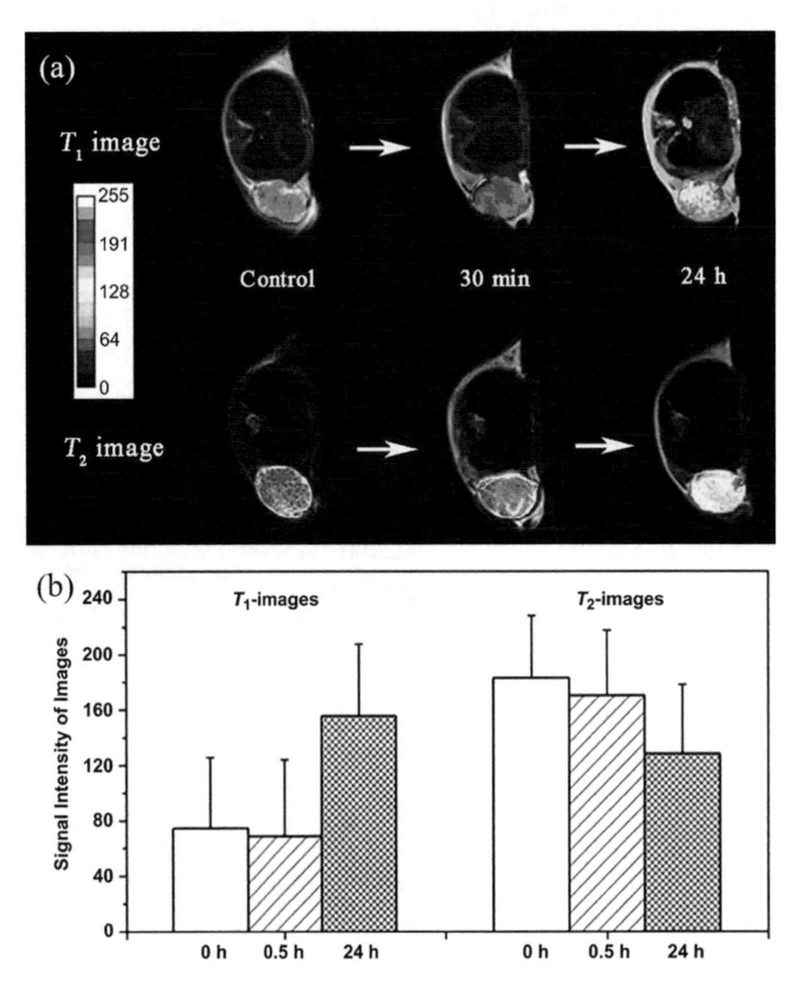

FIGURE 8.3 (a) T_1-weighted and T_2-weighted magnetic resonance images of tumor injected with Fe_3O_4@SiO_2(Gd-DTPA)-RGD NPs; (b) signal intensity analysis for T_1- and T_2-weighted MR images in tumor. (Reprinted with permission from Ref. [28]. Copyright 2011 Elsevier.)

the use of PET radionuclides, such as ^{11}C ($t_{1/2} = 20.3$ min), ^{18}F ($t_{1/2} = 110$ min), ^{64}Cu ($t_{1/2} = 12.7$ h), ^{68}Ga ($t_{1/2} = 67.7$ min), ^{86}Y ($t_{1/2} = 14.7$ h), and ^{89}Zr ($t_{1/2} = 78.4$ h), which emit a positron (β^+) when the radionuclides decays. This positron undergoes annihilation with an electron releasing two 511-keV γ-rays 180° apart. A dedicated camera with multiple detectors is used for collecting these γ photons; and the detection signal is then processed electronically to generate an image showing the localization of the radiotracer in the body.[29] Unlike PET radionuclides, the radionuclides used in SPECT can directly release γ photons upon decay, which include ^{99m}Tc ($t_{1/2} = 6.01$ h), ^{111}In ($t_{1/2} = 2.80$ days), ^{123}I ($t_{1/2} = 13.2$ h), ^{67}Ga ($t_{1/2} = 3.26$ days), ^{131}I ($t_{1/2} = 8.02$ days), and ^{125}I ($t_{1/2} = 59.4$ days). Compared to PET, SPECT generally has lower sensitivity because its imaging relies on physical collimators that absorb photons that are not within a small angular range.

Nanoparticle-based platforms offer new strategies to design specific radioactive probes to realize early detection of cancers. Various nanoparticles offering a unique size and physiochemical properties can be used in designing of radioactive nanoprobes. After the choice of the nanomaterial, the two key design considerations are: (i) which radiolabel to use and (ii) how to incorporate the radiolabel into the nanoparticle. The choice of radiolabel is determined by the physical characteristics of the radioisotope such as the positron energy (which dictates its travel distance in tissue and thus the spatial resolution of the resultant image), the decay half-life, and the efficacy of the radiolabeling approach.[30] Nanomaterial radiolabeling strategies must optimally be robust, safe, rapid, and efficient, which can be classified into two main categories. The first type involves incorporation of a radioactive element into a nanosized cluster. The second type involves attaching a radioactive element to a nanoparticle. This method is versatile and can incorporate various radioelements of choice into a ligand on the NP surface.

^{18}F is the most widely used PET radioisotope for cancer imaging because of its nearly ideal nuclear properties, including high-positron emission abundance (β^+, 97%), low positron energy (0.635 MeV) resulting in limited positron migration (<2 mm), and suitable physical half-life (109.7 min).[31,32] In a recent work, Schirrmacher and coauthors used the silicon-fluoride-acceptor (SiFA) ^{18}F-labeling strategy to label ^{18}F to the four novel polymeric core-shell nanoparticles NP1-NP4. Four nanoparticles in the sub-100 nm range with distinct hydrodynamic diameters of 20 nm (NP1), 33 nm (NP2), 45 nm (NP3), and 72 nm (NP4), respectively, were synthesized under size-controlled conditions. The SiFA-labeling building block acted as an initiator for the polymerization of polymer P1. The nanoparticles were radiolabeled with ^{18}F through simple isotopic exchange (IE) and analyzed *in vivo* in a murine mammary tumor model (EMT6). The nanoparticle size of 33 nm showed the highest tumor accumulation of SUV_{mean}0.97 (=4.4% ID/g) after 4 h p.i. (Figure 8.4) through passive diffusion based on the EPR effect.[33]

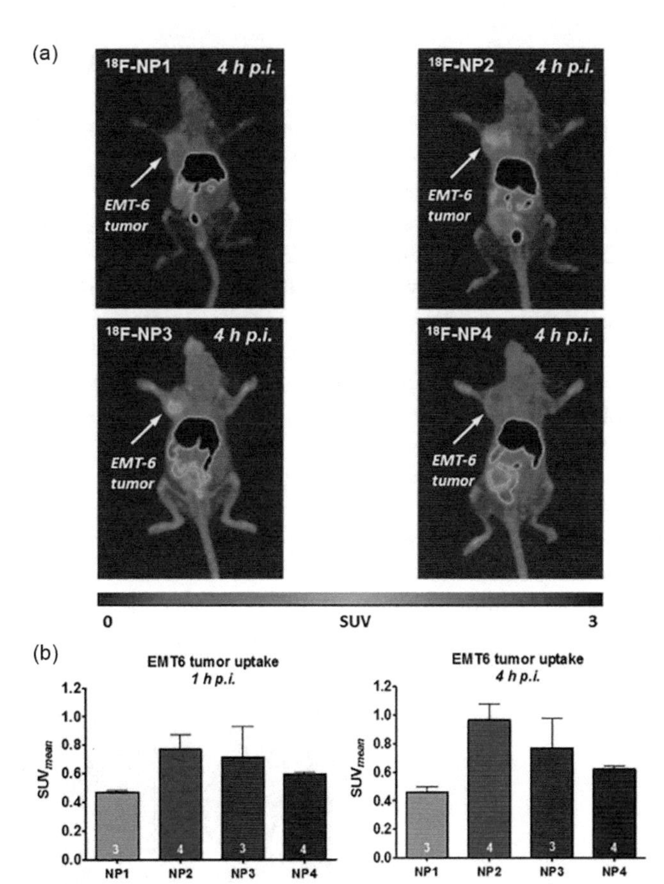

FIGURE 8.4 (a) Representative static PET images (MIP, maximum intensity projection) of ^{18}F-labeled NP1, NP2, NP3, and NP4 in EMT6 tumor-bearing mice at 4 h p.i. (summed 20-min scan). (b) SUV_{mean} values for tumor uptake after 1 h (left) and 4 h p.i. (right). Data are shown as mean ± SEM from n-experiments (# in bar). (Reprinted with permission from Ref. [33]. Copyright 2017 American Chemical Society.)

Among various radioisotopes studied, ^{64}Cu ($t_{1/2} = 12.7$ h, EC 45%, β^- 37.1%, β^+ 17.9%) is unique as it decays by three different routes, namely, electron capture (EC) and β^- and β^+ decays. Due to simultaneous emission of both β^+ and β^- particles, this radioisotope holds promise toward development of PET imaging probes for noninvasive visualization of diseases and can also be used in targeted radiotherapy.[34] Anderson et al. demonstrated the development of ^{64}Cu-labeled somatostatin-targeted liposomes for the detection of neuroendocrine tumors (NET). The PEGylate ^{64}Cu liposomes with the targeting moiety showed higher tumor uptake in NET xenograft model (NCI-H727) (high tumor-to-muscle ratio) compared to untargeted liposomes. The targeted liposomes showed faster accumulation times relative to the untargeted liposomes in NE tumors allowing for earlier detection.

Chemokine receptors (CCRs) and their ligands regulate cell motility, invasiveness, and survival. In some tumors, the expression of CCRs (e.g., CCR5) is upregulated and is associated with promoted cancer progression. Therefore, CCR5 can be a useful biomarker for the early detection of cancer.

To image CCR5, Xia and colleagues developed DAPTA (aSTTTNYT)-modified PdCu@Au core–shell tripods.[35] The CCR5 expression in 4T1 tumors can be visualized using a radiolabeled version of the nanoprobe, DAPTA-$Pd^{64}Cu$@Au, with higher uptake and tumor-to-muscle accumulation ratios compared to non-targeted nanoprobes and free DAPTA-blocked nanoprobes (Figure 8.5).

Li and colleagues developed chelator-free ^{64}CuS NPs by adding a trace amount of $^{64}CuCl_2$ into an aqueous solution of $^{nat}CuCl_2$ and Na_2S in the presence of PEG.[36] The specificity of the resulting ^{64}CuS NPs can be controlled by adjusting the ratio of $^{64}Cu/^{nat}Cu$ in the reaction mixture. PEGylated ^{64}CuS NPs exhibited significant uptake and retention in U87 human glioblastoma xenografts in mice after intravenous (i.v.) injection as a result of their small size (~11 nm in average diameter) and their prolonged blood circulation time.

^{99m}Tc is the most widely used SPECT radionuclide. Radiolabeling with ^{99m}Tc has usually been accomplished using two different methods: direct labeling and labeling using a bifunctional chelator. In the case of IONPs, direct labeling with ^{99m}Tc was performed by Madru et al.[37] However, other metallic or polymeric NPs have been modified by both the direct labeling approach and ligand systems for labeling with ^{99m}Tc. Torchilin and colleagues have utilized DTPA-polylysyl-N-glutaryl-phosphatidylethanol-amine for stable loading of ^{99m}Tc. The liposomes were designed to be tumor specific by conjugation of mAb 2C5 to the liposomes. These tumor-specific liposomes were found to be effective imaging agents displaying almost six times more uptake in tumor cells compared to nontargeted liposomes.[38]

8.2.4 Optical Nanoprobes

With the improvements in current optical imaging instrumentations and the advances in design and development of imaging probes, optical imaging has been one of the fastest growing fields for early detection of cancers.

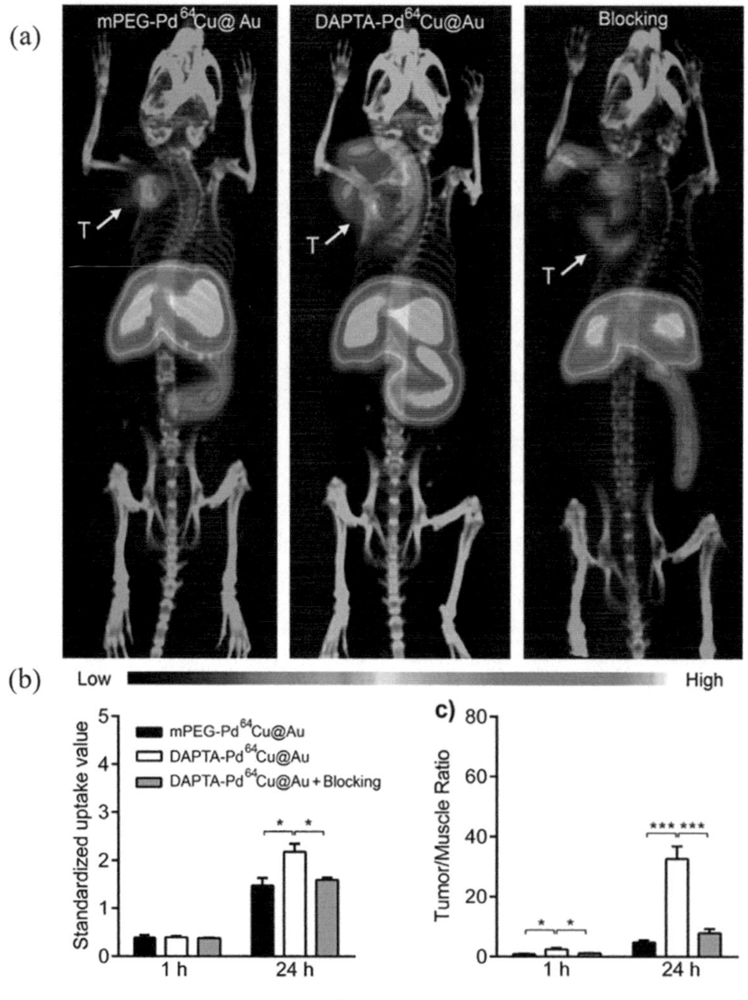

FIGURE 8.5 (a) PET/CT images of mice injected with the ^{64}Cu-doped core–shell tripods: (left) nontargeted tripods whose surface were covered by PEG, (middle) CCR5-targeted tripods, and (right) CCR5-targeted tripods with blocking by the nonradiative counterpart. The images were recorded at 24 h post injection (3.7 MBq per mouse). T, tumor; L, liver. (b) The SUV in tumor region obtained from the PET/CT images taken at different points post injection. (c) Tumor-to-muscle SUV ratios obtained from the PET/CT images taken at different time points. $^{*}p < 0.05$; $^{***}p < 0.001$. (Reprinted with permission from Ref. [35]. Copyright 2016 American Chemical Society.)

Luminescence Imaging Nanoprobes

Luminescence is an umbrella term for entities that emit light upon energetic excitation without heating, including photoluminescence, bioluminescence, chemiluminescence, and phosphorescence, mechanoluminescence. Luminescent materials are one of the most promising optical materials proposed for cancer diagnostic implementation. For example, highly fluorescent semiconductor nanocrystals, quantum dots (QDs), with their unique size-dependent physical and chemical properties have been suggested as ideal candidates for the early detection of cancers.[39–41] QDs hold many superior properties for this purpose than organic fluorophores, for example, strong resistance to photobleaching, continuous absorption spectra covering UV to near-infrared (NIR) region, high quantum efficiency, narrow emission spectra, large effective Stokes shifts, long fluorescence lifetime, and excellent multiphoton emission.[42] For cancer detection, QDs should be surface functionalized with specific recognition agents for target detection, such as primary or secondary antibody (Ab), streptavidin, peptides, proteins, and oligonucleotides.[43–48] For example, highly fluorescent CdTeS-alloyed QDs functionalization with folate were synthesized. The QDs were first synthesized by a hydrothermal method and coated with N-acetyl-L-cysteine (NAC) as both bioactive ligand and sulfur source for biocompatibility and biological stability. After folate-PEG decoration of the CdTeS-alloyed QDs, the conjugated NIR QDs displayed good biocompatibility, excellent sensitivity, and specificity for optical imaging of tumors overexpressing the folate receptor.[49] Another striking example is the biocompatible peptide-coated (RGD) Ag_2S Qdots, displaying an outstanding 9:1 tumor-to-muscle ratio,[50] which is higher than 5.5:1 ratio for the radiolabeled peptide alone.[51] Intriguingly, Ag_2S Qdots are one of the few nanomaterials that have been imaged in the second NIR window (~1,000–1,700 nm) *in vivo*.[52] Ag_2S Qdots have many advantages suited for early-stage tumor diagnosis including deep tissue penetration, high sensitivity, and elevated spatial and temporal resolution owing to their high emission efficiency in the unique NIR-II imaging window.[53]

Similar to Qdots, upconversion nanoparticles (UCNPs) are extremely bright and photostable, and they are another type of promising candidates for optical imaging of cancer. Compared with traditional down-conversion fluorescence imaging, the NIR light-excited upconversion luminescence (UCL) imaging relying on UCNPs exhibits improved tissue penetration depth, higher photochemical stability, and is free of auto-fluorescence background. Therefore, more and more attention has been paid to investigate cancer imaging using UCNPs.[54–56]

Resonance Energy Transfer Nanoprobes

The most well-known and widely used form of optical RET (nonradiative transfer of virtual photons from donor to nearby acceptor) in imaging is fluorescence resonance energy transfer (FRET). FRET probes are often known as “smart” or “activatable” probes because they do not simply travel through the body always “on”-rather, they can be activated to produce signal (“on” state) when and where one wants or be quenched (“off” state) otherwise; this substantially decreases nonspecific background signal.[57] FRET occurs when the distance between donor and acceptor molecules is close. Nanomaterials are generally excellent tools for spatially constraining molecules that must, in order to be effective, cooperate by localizing within a given distance of one another. Thus, many FRET-nanoprobes have been constructed and applied for *in vivo* tumor imaging. As shown in Figure 8.6, Lee and colleagues prepared a self-assembled activatable polymeric nanomaterial consisting of a dye (Cy5.5) linked to a quencher (BHQ-3) *via* an enzyme-cleavable peptide with selectivity toward several matrix metalloproteases (MMPs, several of which tend to be overexpressed in cancer and other diseases).[57] The nanomaterial precursors were synthesized by generating polymeric amphiphiles *via* hydrophobic modification (5β-cholanic acid) of glycol chitosan. By sonicating the solution with the activatable peptide, a self-assembled nanoprobe was formed. Intratumoral injection of these MMP-nanoprobes provides a distinct fluorescent readout of the MMP activity in the imaged region (tumor site).

Raman Imaging Nanoprobes

Raman imaging is a viable alternative to fluorescent imaging of cancers.[58] Raman imaging presents several advantages for the cancer diagnostics community. In particular, Raman

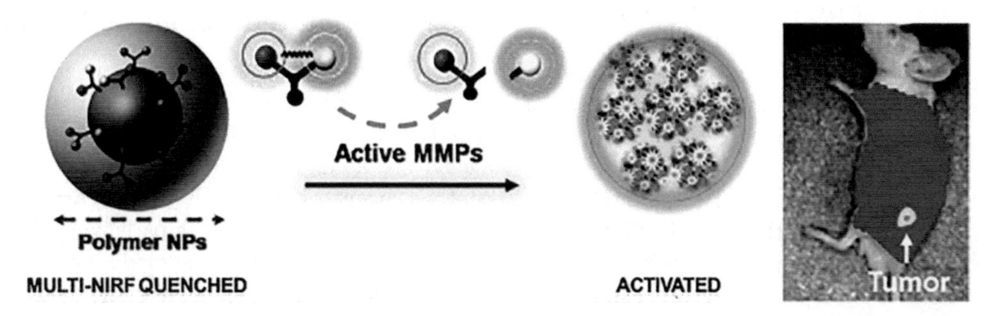

FIGURE 8.6 Schematic depicts a new protease activatable strategy based on a polymer nanoparticle platform. (Reprinted with permission from Ref. [57]. Copyright 2009 American Chemical Society.)

tools have the ability to provide sensitive, quantitative, and chemically specific information about important biological components in cellular and tissue milieu.[58]

However, due to the low signal efficiency, *in vivo* Raman imaging in a reasonable time frame is challenging (but possible for small regions),[59] unless nanomaterials are used for signal amplification in deep tissues. By applying nanomaterial-enabled surfaced-enhanced Raman scattering (SERS) imaging, researchers have detected nanoprobes up to 1–5 cm deep in tissues at concentrations as low as 1.5 fM. In this work, SERS-enhanced gold nanostars have been developed. The SERS-nanostars feature a star-shaped gold core, a Raman reporter resonant in the NIR spectrum, and a primer-free silication method. In mouse models of pancreatic cancer, breast cancer, prostate cancer, and sarcoma, SERRS-nanostars enabled accurate detection of macroscopic malignant lesions as well as microscopic disease. High sensitivity and broad applicability, in conjunction with their inert gold-silica composition, render SERRS-nanostars, a promising imaging agent for more precise cancer imaging.[60]

Photoacoustic Imaging Nanoprobes

Owing to their increased tissue penetration depth and fairly high spatial resolution, photoacoustic imaging (PAI) has attracted lots of research attention for cancer detection. PAI harvests the advantages of both optical and acoustic imaging by directing (laser) light pulses into a sample and receiving acoustic information in the form of US to create images. Nonetheless, the photoacoustic signal intensity mainly depends on the optical absorption intensity and the photothermal conversion efficiency of contrast agents. Thus, nanomaterials, which can be highly engineered for optimal absorption in the visible or NIR with very high cross sections, are favored in PAI applications. Recent development on nanochemistry has led to nanoprobes with 100-fold or more improved sensitivities so that PAI can now detect nanomaterial concentrations at picomolar levels.[61]

Due to their strong and tunable optical absorption that results from the surface plasmon resonance (SPR) effect, AuNPs have been widely used as PAI contrast agents. In addition, gold forms strong gold-thiolate bonds that enable covalent surface modifications for optimizing biocompatibility (e.g. polyethylene glycol (PEG) functionalization), stability (e.g., silica encapsulation), and active targeting. It was reported that AuNPs could be utilized to detect human breast cancer tumor xenografts at a wavelength of 532 nm, B16 melanomas at 778 nm,[62,63] and U87 brain tumors.[64]

Recently, some of the developments in PAI probes involve smart or activatable nanomaterials that, like the FRET activatable probes, are off until they are turned on by molecules of interest. These molecules are proteins or other molecules that signal tumoral state. In a recent work by Pu,[65] they have developed an activatable PA nanoprobe (SON_{50}) based on semiconducting oligomer with amplified signals for *in vivo* imaging of pH. The probe design takes advantage of nanodoping to simultaneously achieve signal amplification and pH sensing in a single particle unity wherein a bifunctional dye (pH-BDP) effectively creates both intraparticle PET and intramolecular protonation. Upon variation of pH from 7.4 to 5.5, the PA brightness of the nanoprobe (SON_{50}) substantially amplified by ∼3.1-fold at 680 nm, while its ratiometric PA signal (PA680/PA750) is able to increase by ∼3.1-fold. Due to the high PA brightness and efficient accumulation in tumors, SON_{50} has the potential to detect tumors in early stage.

8.2.5 Acoustic Nanoprobes

US imaging has been extensively used in clinical diagnostic applications based on its noninvasive, cost-effective, and portable features.[66] Typical UCAs include microbubbles (MBs) that are composed of an inner gaseous core with a thin shell coating. However, MBs are too big for imaging of tumor cell surface biomarker, because the leaky vasculature of solid tumors allows only nanosized or smaller objects to pass through. An alternative strategy is to deliver fluorocarbons as smaller nanodroplets to tumor sites and then use US to induce phase transition into nanobubbles (NBs), which will coalesce transiently into larger MBs. However, the random transient nature of NB generation and coalescence makes the use of nanodroplets as UCAs for tumor imaging unreliable.

Recently, several nanoscale echo-contrast agents which can improve the performance of US-based diagnostic imaging, such as perfluorocarbon (PFC)-filled nanoemulsion or nanodroplets with sizes ranging between 250 and 500 nm, have been reported.[67–69] For example, in a work performed by Zhang et al., they have developed a type of pH-sensitive, polymersome-based, PFC-encapsulated ultrasonographic nanoprobe, capable of maintaining at 178 nm during circulation and increasing to 437 nm at the acidic tumor microenvironment. Its small size allowed efficient tumor uptake. At the tumor site, the nanoparticle swells, resulting in lowering of the vaporization threshold for the PFC, efficient conversion of nanoprobes to echogenic nano/MBs for ultrasonic imaging.[70]

8.2.6 Computed Tomography Nanoprobes

X-ray CT has long been a common technique used in clinical imaging. However, its soft-tissue contrast is quite poor; therefore, to improve soft-tissue contrast, extrinsic contrast agents are used. Nanomaterials can make compelling CT contrast agents because they can carry a high-contrast payload (necessary for CT compared with the more sensitive PET/SPECT nuclear imaging modalities which require only trace amounts of probes), circulate for long periods of time, and can easily be fabricated with the requisite electron-dense contrast materials.

Iodine-Based Nanoprobes

With the ability of inducing stronger X-ray attenuation, iodine has historically been the atom of choice for CT. Recently, tremendous nanoparticulate CAs based on iodine for CT have been explored, including nanosuspensions, nanoemulsions, microspheres, liposomes, micelles, polymeric particles, nanospheres, and nanocapsules.[71] For example, Vandamme et al. have formulated PEGylated nanoemulsions with two iodinated oils (i.e., iodinated monoglyceride and iodinated castor oil), which were endowed with very high iodine concentration. Owing to stronger X-ray attenuation properties, this iodine-based nanoprobe provides strong X-ray contrast enhancement.[72]

Lanthanide-Based Nanoprobes

Given their high atomic numbers, lanthanide-based contrast agents as CT probes have also been explored. Free lanthanide ions are extremely toxic. So most lanthanide-based contrast agents are used in the form of chelate complexes. Many nanoparticle lanthanide CT agents are designed as multimodal imaging media. As an example, nanoprobes containing a rare-earth core (consisting of a mixture of lanthanides such as Gd, Yb, Er, Tm, and Yttrium) and conjugated with additional X-ray attenuating lanthanide materials were explored as multimodal (CT/MRI/upconversion fluorescence) nanoparticulate contrast media.[73]

Gold Nanoprobes

The atomic number of gold is much higher than that of the currently used CT contrast material iodine, and therefore, gold can induce stronger X-ray attenuation.[74] Gold provides about 2.7 times greater contrast per unit weight than iodine.[75] In addition, the small size of iodine molecules allows only very short imaging times owing to rapid clearance by the kidneys. In contrast, gold nanoparticles can be designed so as to overcome biological barriers and remain confined to the intravascular space for prolonged times. What's more, specific targeting could be achieved through the conjugation of nanoparticles to a variety of ligands, including antibodies, peptides, aptamers, or small molecules that possess high affinity toward unique molecular signatures, such as cancer cells.

Hainfeld et al. demonstrated molecular imaging of cancer with actively targeted CT contrast agents.[76] They showed that gold nanoparticles can enhance the visibility of millimeter-sized human breast tumors in mice, and that active tumor targeting (with anti-HER2 antibodies) is 1.6-fold more efficient than passive targeting. They also showed that the specific uptake of the targeted gold nanoparticles in the tumor's periphery was 22-fold higher than in surrounding muscle tissue. In another study, Chanda et al. reported bombesin functionalized gold nanoparticles for selective detection of prostate, breast, and small-cell lung carcinoma.[77]

8.2.7 Multimodal Nanoprobes

Each imaging modality has its own unique advantages along with intrinsic limitations, such as insufficient sensitivity or spatial resolution, which make it difficult to obtain accurate and reliable information for tumor detection.[78] To overcome the individual limitations of these modalities, taking advantage of their individual strengths, they can be combined in so-called multimodal imaging approaches.[79] Multimodal imaging is a powerful method that may provide more reliable and accurate information for early cancer detection.

Nanoprobes for Radionuclide-MR Imaging

MRI is among the best noninvasive imaging techniques used in clinic with advantages of excellent temporal and spatial resolution and long effective imaging window. However, MRI is less sensitive than radionuclide imaging when used to monitor small tissue lesions, molecular activity, and cellular activities. The radionuclide imaging techniques have a very high sensitivity down to the pico-molar level, while its spatial resolution is not high. Therefore, dual-modal imaging combining PET (or SPECT) and MRI provides extremely sensitive and high-resolution images. A very good example is to use IO-based nanoprobe for simultaneous dual PET and MRI of prostate cancer.[80] In this study, prostate-specific membrane antigen (PSMA) peptides were conjugated on the surface of IONPs where the 1,4,7,10-tetraazacyclododecane-1,4,7,10-tetraacetic acid (DOTA) chelators were also incorporated for ^{68}Ga labeling. PET/MR dual images were successfully obtained by using an adequate amount of radioactivity and cold NPs to adjust the sensitivity of each modality (Figure 8.7). MR images showed a high uptake of the nanoprobe by the PSMA-positive tumor with high resolution but were limited in providing quantitative information. PET images also showed specific uptake by the PSMA-positive tumor and furthermore could provide quantitative information. The success of this PET/MR imaging approach may allow for earlier tumor detection with a high degree of accuracy and provide insight into the molecular mechanisms of cancer with comprehensive information. Except for using ^{68}Ga, ^{18}F and ^{64}Cu and ^{124}I can also be labeled on the surface of IONPs.[81,82]

Nanoprobes for CT-MR Imaging

CT and MRI have been regarded as two most widely used noninvasive diagnosis techniques. In particular, CT renders high spatial and density resolution imaging of hard tissues, while MR offers superior vivid tomographic information of soft tissues. Therefore, the combination of CT and MR imaging would provide more comprehensive and accurate diagnostic information by imaging different tissues. Heterostructured nanocrystals composed of radiopaque elements and magnetic nanoparticles can be an effective approach for bimodal CT-MRI probes.[83] One of the advantages of heterostructured nanoparticles is that the contrast effect for each modality can be controlled by

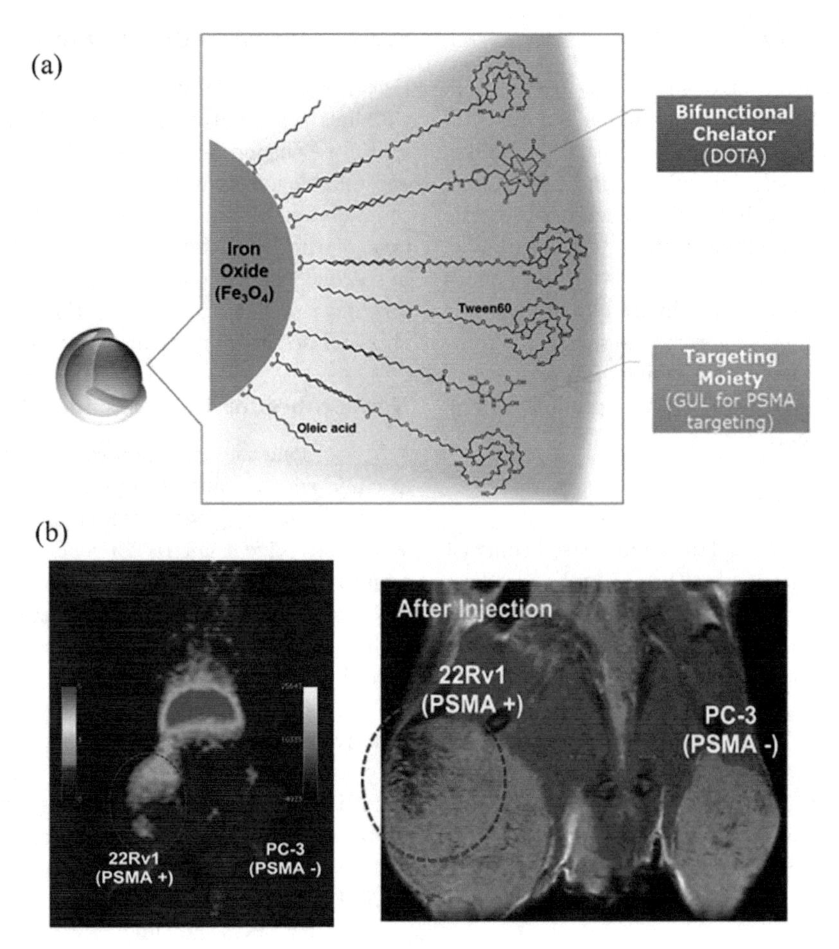

FIGURE 8.7 (a) Diagram of encapsulated nanoprobe. (b) *In vivo* micro PET and MRI image results. (Reprinted with permission from Ref. [80]. Copyright 2015 Elsevier.)

changing the sizes of the corresponding nanoparticles. For example, Jon et al. constructed a novel hybrid nanoprobe by thermal decomposition of Fe-oleate and Au-oleylamine complexes.[84] Furthermore, using a nano-emulsion method, the nanoparticles are coated with amphiphilic poly(DMA-r-mPEGMA-r-MA) to impart water-dispersity and anti-biofouling properties. An *in vitro* phantom study shows that the hybrid nanoprobe has high CT attenuation and a good MR signal. Intravenous injection of the hybrid nanoprobe into hepatoma-bearing mice results in high contrast between the hepatoma and normal hepatic parenchyma in both CT and MRI modal. These results suggest that the hybrid nanoparticles may be useful as CT/MRI dual contrast agents for *in vivo* detection of hepatoma.

In another work performed by Shi et al., they have successfully synthesized lactobionic acid (LA)-modified PEI-entrapped AuNPs (Gd-Au PENPs-LA probe) for targeted CT/MR dual modal imaging.[85] This nanoprobe displays good X-ray attenuation properties, good r_1 relaxivity, and hepatocellular carcinoma targeting specificity. In *in vivo* study, Gd-Au PENPs-LA probe could accumulate in the tumor site and be cleared out of the body at 96 h post injection. With the versatile advantages, Gd-Au PENPs-LA probe has the potential for accurate diagnosis of hepatocellular carcinoma. With the advances of nanotechnology, more nanoprobes have been developed for dual-mode CT/MRI applications, including nanoparticles (NPs), Gd-based G_8 dendrimers, FePt Gd-loaded Au NPs, Fe_3O_4@TaOx core@shell NPs, and lanthanide fluoride-based upconversion NPs.

Nanoprobes for Optical-MR Imaging

Another approach to bimodal imaging is to combine MR and optical imaging capabilities. In this combination, MR offers detailed anatomic imaging, while the optical component offers the option of real-time optical imaging. Three primary approaches are currently used to develop these probes: (i) biocompatible molecules that are inherently paramagnetic yet also have optical properties, (ii) IONPs functionalized with fluorescent molecules, and (iii) nanomaterials specifically designed for both MR and optical imaging. Extensive dual-function optical-MR probes have been developed for cancer detection.[86] In one paradigm, dendrimers with multiple Gd-DTPA chelates have been proposed as viable MR agents. Substituting one of the amine bound Gd-chelates on the dendrimer for a fluorophore creates a dual MR/optical imaging agent, with the virtue that the sensitivity differences between MR and optical are compensated by the relative number of attached molecules (Gd-chelates > optical fluorophores by >100:1).

Nanoprobes for Optical-Radionuclide Imaging

As an excellent complement to nuclear imaging techniques, near infrared fluorescent (NIRF) imaging displays properties of high spatial resolution, high sensitivity, and low risk to the living subjects by using nonionizing radiation. PET/SPECT can be combined with NIRF imaging, allowing increased depth penetration and provide complimentary information. Cai et al. constructed a novel QD whose surface was successfully conjugated with RGD peptides and DOTA chelators through amide coupling.[87] The resulting QDs (DOTA-QD-RGD) were radiolabeled with ^{64}Cu for PET/NIRF imaging. The tumor-targeting efficacy of the dual-functional QD was quantitatively evaluated in integrin $\alpha V\beta_3$ expressing U87MG tumor xenografts. Compared to nontargeted ^{64}Cu-DOTA-QD,^{64}Cu-DOTA-QD-RGD showed better tumor uptake at various imaging time points.

Another study was performed by Liang et al. to use streptavidin nanoparticle-based complexes as SPECT/NIRF tumor imaging probes.[88] The NPs consisted of three biotinylated components, including a Cy5.5 fluorophore for NIRF imaging, DOTA chelator for ^{111}In radiolabeling, and anti-HER2 Herceptin for tumor targeting. These components were assembled via streptavidin. The imaging results showed that the resulting ^{111}In-DOTA/Cy5.5/Herceptin NPs possess favorable *in vivo* biodistribution, high tumor accumulation, and strong tumor-to-normal tissue contrast. The authors summarize that this nanoprobe may have potential for effective detection of cancer in early stage.

Nanoprobes for Ultrasound-MR Imaging

Bimodal US-MRI agent can be prepared *via* the incorporation of IONPs within UCA.[89] For example, nanoparticle-embedded MBs are prepared *via* one-pot emulsion polymerization of poly(butyl cyanoacrylate) along with the oil-in-water encapsulation of nanoparticles.[90] The hybrid imaging agents show strong contrast in US and MRI. Furthermore, the MR contrast effect is enhanced after US-induced bubble destruction, demonstrating a triggerable MRI capability.

8.3 Conclusion and Future Prospects

Over the past decade, development of nanostructure-based probes has made remarkable progress. With sufficient specificity and sensitivity, many types of nanoprobes for different imaging modalities have been designed and applied for early detection and theranostic of tumors. Despite recent progress and promising results, many researches remain in the pre-clinical stage, and there are still huge challenges in translation of these researches from pre-clinical studies to clinical applications. Therefore, during the development of nanoprobes for cancer detection, several points should be carefully considered. First, biocompatibility should be kept in the forefront to choose the components for NP fabrication. From the materials point of view, the safety issues of nanoprobes depending on physicochemical characteristics (e.g., surface charges, shape, size, functional group, and coating materials) need to be thoroughly investigated. Second, target specificity of nanoprobes is very important. To obtain successful results in imaging, sufficient amounts of imaging agents should be specifically localized in the target site. Third, in order to design of multimodal nanoprobes, the doses of each imaging agent should be carefully considered to match the sensitivity of different modalities.

Nanoprobes are changing the face of what is possible in imaging by fostering superior sensitivity, depths, spatial resolutions, and contrast. Merging imaging techniques and nanotechnology appears to be the promising way to design "smart", targeted nanoprobes for accurate detection of cancers in early stage. Although much future work remains to be done, the prospects are exciting, and the success may be just around the corner.

References

1. Ferlay, J.; Soerjomataram, I.; Dikshit, R.; Eser, S.; Mathers, C.; Rebelo, M.; Parkin, D. M.; Forman, D.; Bray, F., Cancer incidence and mortality worldwide: sources, methods and major patterns in GLOBOCAN 2012. *International Journal of Cancer* **2015,** *136* (5), E359–86.
2. Laget, S.; Broncy, L.; Hormigos, K.; Dhingra, D. M.; BenMohamed, F.; Capiod, T.; Osteras, M.; Farinelli, L.; Jackson, S.; Paterlini-Brechot, P., Technical insights into highly sensitive isolation and molecular characterization of fixed and live circulating tumor cells for early detection of tumor invasion. *PLoS One* **2017,** *12* (1), e0169427.
3. Souza, E. S. V.; Chinen, L. T.; Abdallah, E. A.; Damascena, A.; Paludo, J.; Chojniak, R.; Dettino, A. L.; de Mello, C. A.; Alves, V. S.; Fanelli, M. F., Early detection of poor outcome in patients with metastatic colorectal cancer: Tumor kinetics evaluated by circulating tumor cells. *OncoTargets and Therapy* **2016,** *9*, 7503–13.
4. Wang, H.; Stoecklein, N. H.; Lin, P. P.; Gires, O., Circulating and disseminated tumor cells: Diagnostic tools and therapeutic targets in motion. *Oncotarget* **2017,** *8* (1), 1884–1912.
5. Rudin, M.; Weissleder, R., Molecular imaging in drug discovery and development. *Nature reviews. Drug discovery* **2003,** *2* (2), 123–31.
6. Ferrari, M., Cancer nanotechnology: Opportunities and challenges. *Nature Reviews Cancer* **2005,** *5* (3), 161–71.
7. Weissleder, R., Molecular imaging in cancer. *Science* **2006,** *312* (5777), 1168–71.

8. Xing, L.; Todd, N. W.; Yu, L.; Fang, H.; Jiang, F., Early detection of squamous cell lung cancer in sputum by a panel of microRNA markers. *Modern Pathology: An Official Journal of the United States and Canadian Academy of Pathology, Inc* **2010,** *23* (8), 1157–64.
9. Choi, Y. E.; Kwak, J. W.; Park, J. W., Nanotechnology for early cancer detection. *Sensors* **2010,** *10* (1), 428–55.
10. Fu, X.; Chen, L.; Choo, J., Optical nanoprobes for ultrasensitive immunoassay. *Analytical Chemistry* **2017,** *89* (1), 124–137.
11. Smith, B. R.; Gambhir, S. S., Nanomaterials for in vivo imaging. *Chemical Reviews* **2017,** *117* (3), 901–86.
12. Qiu, L.; Dong, C.; Kan, X., Lymphoma-targeted treatment using a folic acid-decorated vincristine-loaded drug delivery system. *Drug Design, Development and Therapy* **2018,** *12*, 863–72.
13. Ahmed, H. U.; Kirkham, A.; Arya, M.; Illing, R.; Freeman, A.; Allen, C.; Emberton, M., Is it time to consider a role for MRI before prostate biopsy? *Nature Reviews Clinical Oncology* **2009,** *6* (4), 197–206.
14. Estelrich, J.; Sanchez-Martin, M. J.; Busquets, M. A., Nanoparticles in magnetic resonance imaging: From simple to dual contrast agents. *International Journal of Nanomedicine* **2015,** *10*, 1727–41.
15. Taylor, A.; Wilson, K. M.; Murray, P.; Fernig, D. G.; Levy, R., Long-term tracking of cells using inorganic nanoparticles as contrast agents: Are we there yet? *Chemical Society Reviews* **2012,** *41* (7), 2707–17.
16. Terreno, E.; Castelli, D. D.; Viale, A.; Aime, S., Challenges for molecular magnetic resonance imaging. *Chemical Reviews* **2010,** *110* (5), 3019–42.
17. Park, J. Y.; Baek, M. J.; Choi, E. S.; Woo, S.; Kim, J. H.; Kim, T. J.; Jung, J. C.; Chae, K. S.; Chang, Y.; Lee, G. H., Paramagnetic ultrasmall gadolinium oxide nanoparticles as advanced T1 MRI contrast agent: Account for large longitudinal relaxivity, optimal particle diameter, and in vivo T1 MR images. *ACS Nano* **2009,** *3* (11), 3663–9.
18. Anishur Rahman, A. T.; Majewski, P.; Vasilev, K., Gd_2O_3 nanoparticles: Size-dependent nuclear magnetic resonance. *Contrast Media & Molecular Imaging* **2013,** *8* (1), 92–5.
19. Hou, Y.; Qiao, R.; Fang, F.; Wang, X.; Dong, C.; Liu, K.; Liu, C.; Liu, Z.; Lei, H.; Wang, F.; Gao, M., $NaGdF_4$ nanoparticle-based molecular probes for magnetic resonance imaging of intraperitoneal tumor xenografts in vivo. *ACS Nano* **2013,** *7* (1), 330–8.
20. Taniguchi, K.; Uchida, J.; Nishino, K.; Kumagai, T.; Okuyama, T.; Okami, J.; Higashiyama, M.; Kodama, K.; Imamura, F.; Kato, K., Quantitative detection of EGFR mutations in circulating tumor DNA derived from lung adenocarcinomas. *Clinical Cancer Research: An Official Journal of the American Association for Cancer Research* **2011,** *17* (24), 7808–15.
21. Cabral, H.; Nishiyama, N.; Kataoka, K., Supramolecular nanodevices: From design validation to theranostic nanomedicine. *Accounts of Chemical Research* **2011,** *44* (10), 999–1008.
22. Gallagher, F. A.; Kettunen, M. I.; Day, S. E.; Hu, D. E.; Ardenkjaer-Larsen, J. H.; Zandt, R.; Jensen, P. R.; Karlsson, M.; Golman, K.; Lerche, M. H.; Brindle, K. M., Magnetic resonance imaging of pH in vivo using hyperpolarized 13C-labelled bicarbonate. *Nature* **2008,** *453* (7197), 940–3.
23. Lee, E. S.; Kim, D.; Youn, Y. S.; Oh, K. T.; Bae, Y. H., A virus-mimetic nanogel vehicle. *Angewandte Chemie* **2008,** *47* (13), 2418–21.
24. Kim, B. H.; Lee, N.; Kim, H.; An, K.; Park, Y. I.; Choi, Y.; Shin, K.; Lee, Y.; Kwon, S. G.; Na, H. B.; Park, J. G.; Ahn, T. Y.; Kim, Y. W.; Moon, W. K.; Choi, S. H.; Hyeon, T., Large-scale synthesis of uniform and extremely small-sized iron oxide nanoparticles for high-resolution T1 magnetic resonance imaging contrast agents. *Journal of the American Chemical Society* **2011,** *133* (32), 12624–31.
25. Ling, D.; Park, W.; Park, S. J.; Lu, Y.; Kim, K. S.; Hackett, M. J.; Kim, B. H.; Yim, H.; Jeon, Y. S.; Na, K.; Hyeon, T., Multifunctional tumor pH-sensitive self-assembled nanoparticles for bimodal imaging and treatment of resistant heterogeneous tumors. *Journal of the American Chemical Society* **2014,** *136* (15), 5647–55.
26. Desgrosellier, J. S.; Cheresh, D. A., Integrins in cancer: Biological implications and therapeutic opportunities. *Nature Reviews Cancer* **2010,** *10* (1), 9–22.
27. Xie, J.; Chen, K.; Lee, H. Y.; Xu, C.; Hsu, A. R.; Peng, S.; Chen, X.; Sun, S., Ultrasmall c(RGDyK)-coated Fe_3O_4 nanoparticles and their specific targeting to integrin alpha(v)beta$_3$-rich tumor cells. *Journal of the American Chemical Society* **2008,** *130* (24), 7542-3.
28. Yang, H.; Zhuang, Y.; Sun, Y.; Dai, A.; Shi, X.; Wu, D.; Li, F.; Hu, H.; Yang, S., Targeted dual-contrast T1- and T2-weighted magnetic resonance imaging of tumors using multifunctional gadolinium-labeled superparamagnetic iron oxide nanoparticles. *Biomaterials* **2011,** *32* (20), 4584–93.
29. Rahmim, A.; Zaidi, H., PET versus SPECT: Strengths, limitations and challenges. *Nuclear Medicine Communications* **2008,** *29* (3), 193–207.
30. Sun, X.; Cai, W.; Chen, X., Positron emission tomography imaging using radiolabeled inorganic nanomaterials. *Accounts of Chemical Research* **2015,** *48* (2), 286–94.
31. Pimlott, S. L.; Sutherland, A., Molecular tracers for the PET and SPECT imaging of disease. *Chemical Society Reviews* **2011,** *40* (1), 149–62.

32. Smith, G. E.; Sladen, H. L.; Biagini, S. C.; Blower, P. J., Inorganic approaches for radiolabelling biomolecules with fluorine-18 for imaging with positron emission tomography. *Dalton ransactions* **2011,** *40* (23), 6196–205.
33. Berke, S.; Kampmann, A. L.; Wuest, M.; Bailey, J. J.; Glowacki, B.; Wuest, F.; Jurkschat, K.; Weberskirch, R.; Schirrmacher, R., (18)F-Radiolabeling and in vivo analysis of SiFA-derivatized polymeric core-shell nanoparticles. *Bioconjugate Chemistry* **2018,** *29* (1), 89–95.
34. Niccoli Asabella, A.; Cascini, G. L.; Altini, C.; Paparella, D.; Notaristefano, A.; Rubini, G., The copper radioisotopes: A systematic review with special interest to ^{64}Cu. *BioMed Research International* **2014,** *2014*, 786463.
35. Pang, B.; Zhao, Y.; Luehmann, H.; Yang, X.; Detering, L.; You, M.; Zhang, C.; Zhang, L.; Li, Z. Y.; Ren, Q.; Liu, Y.; Xia, Y., (6)(4)Cu-Doped PdCu@Au tripods: A multifunctional nanomaterial for positron eEmission tomography and image-guided photothermal cancer treatment. *ACS Nano* **2016,** *10* (3), 3121–31.
36. Zhou, M.; Zhang, R.; Huang, M.; Lu, W.; Song, S.; Melancon, M. P.; Tian, M.; Liang, D.; Li, C., A chelator-free multifunctional [^{64}Cu]CuS nanoparticle platform for simultaneous micro-PET/CT imaging and photothermal ablation therapy. *Journal of the American Chemical Society* **2010,** *132* (43), 15351–8.
37. Madru, R.; Kjellman, P.; Olsson, F.; Wingardh, K.; Ingvar, C.; Stahlberg, F.; Olsrud, J.; Latt, J.; Fredriksson, S.; Knutsson, L.; Strand, S. E., 99mTc-labeled superparamagnetic iron oxide nanoparticles for multimodality SPECT/MRI of sentinel lymph nodes. *Journal of Nuclear Medicine: Official Publication, Society of Nuclear Medicine* **2012,** *53* (3), 459–63.
38. Silindir, M.; Erdogan, S.; Ozer, A. Y.; Dogan, A. L.; Tuncel, M.; Ugur, O.; Torchilin, V. P., Nanosized multifunctional liposomes for tumor diagnosis and molecular imaging by SPECT/CT. *Journal of Liposome Research* **2013,** *23* (1), 20–7.
39. Rakovich, A.; Rakovich, T.; Kelly, V.; Lesnyak, V.; Eychmuller, A.; Rakovich, Y. P.; Donegan, J. F., Photosensitizer methylene blue-semiconductor nanocrystals hybrid system for photodynamic therapy. *Journal of Nanoscience and Nanotechnology* **2010,** *10* (4), 2656–62.
40. Rakovich, A.; Savateeva, D.; Rakovich, T.; Donegan, J. F.; Rakovich, Y. P.; Kelly, V.; Lesnyak, V.; Eychmuller, A., CdTe quantum dot/dye hybrid system as photosensitizer for photodynamic therapy. *Nanoscale Research Letters* **2010,** *5* (4), 753–60.
41. Rousserie, G.; Sukhanova, A.; Even-Desrumeaux, K.; Fleury, F.; Chames, P.; Baty, D.; Oleinikov, V.; Pluot, M.; Cohen, J. H.; Nabiev, I., Semiconductor quantum dots for multiplexed bio-detection on solid-state microarrays. *Critical Reviews in Oncology/Hematology* **2010,** *74* (1), 1–15.
42. Bruchez Jr., M.; Moronne, M.; Gin, P.; Weiss, S.; Alivisatos, A. P., Semiconductor nanocrystals as fluorescent biological labels. *Science* **1998,** *281* (5385), 2013–6.
43. Michalet, X.; Pinaud, F. F.; Bentolila, L. A.; Tsay, J. M.; Doose, S.; Li, J. J.; Sundaresan, G.; Wu, A. M.; Gambhir, S. S.; Weiss, S., Quantum dots for live cells, in vivo imaging, and diagnostics. *Science* **2005,** *307* (5709), 538–44.
44. Pinaud, F.; King, D.; Moore, H. P.; Weiss, S., Bioactivation and cell targeting of semiconductor CdSe/ZnS nanocrystals with phytochelatin-related peptides. *Journal of the American Chemical Society* **2004,** *126* (19), 6115–23.
45. Wu, X.; Liu, H.; Liu, J.; Haley, K. N.; Treadway, J. A.; Larson, J. P.; Ge, N.; Peale, F.; Bruchez, M. P., Immunofluorescent labeling of cancer marker Her2 and other cellular targets with semiconductor quantum dots. *Nature Biotechnology* **2003,** *21* (1), 41–6.
46. Goldman, E. R.; Balighian, E. D.; Mattoussi, H.; Kuno, M. K.; Mauro, J. M.; Tran, P. T.; Anderson, G. P., Avidin: A natural bridge for quantum dot-antibody conjugates. *Journal of the American Chemical Society* **2002,** *124* (22), 6378–82.
47. Shemetov, A. A.; Nabiev, I.; Sukhanova, A., Molecular interaction of proteins and peptides with nanoparticles. *ACS Nano* **2012,** *6* (6), 4585–602.
48. Sukhanova, A.; Venteo, L.; Devy, J.; Artemyev, M.; Oleinikov, V.; Pluot, M.; Nabiev, I., Highly stable fluorescent nanocrystals as a novel class of labels for immunohistochemical analysis of paraffin-embedded tissue sections. *Laboratory Investigation; A Journal of Technical Methods and Pathology* **2002,** *82* (9), 1259–61.
49. Xue, B.; Deng, D. W.; Cao, J.; Liu, F.; Li, X.; Akers, W.; Achilefu, S.; Gu, Y. Q., Synthesis of NAC capped near infrared-emitting CdTeS alloyed quantum dots and application for in vivo early tumor imaging. *Dalton Transactions* **2012,** *41* (16), 4935–47.
50. Tang, R.; Xue, J.; Xu, B.; Shen, D.; Sudlow, G. P.; Achilefu, S., Tunable ultrasmall visible-to-extended near-infrared emitting silver sulfide quantum dots for integrin-targeted cancer imaging. *ACS Nano* **2015,** *9* (1), 220–30.
51. Yapp, D. T.; Ferreira, C. L.; Gill, R. K.; Boros, E.; Wong, M. Q.; Mandel, D.; Jurek, P.; Kiefer, G. E., Imaging tumor vasculature noninvasively with positron emission tomography and RGD peptides labeled with copper 64 using the bifunctonal chelates DOTA, oxo-DO3a. and PCTA. *Molecular Imaging* **2013,** *12* (4), 263–72.

52. Hong, G.; Robinson, J. T.; Zhang, Y.; Diao, S.; Antaris, A. L.; Wang, Q.; Dai, H., In vivo fluorescence imaging with Ag_2S quantum dots in the second near-infrared region. *Angewandte Chemie* **2012,** *51* (39), 9818–21.
53. Zhang, Y.; Hong, G.; Zhang, Y.; Chen, G.; Li, F.; Dai, H.; Wang, Q., Ag_2S quantum dot: a bright and biocompatible fluorescent nanoprobe in the second near-infrared window. *ACS Nano* **2012,** *6* (5), 3695–702.
54. Xiong, L.; Chen, Z.; Tian, Q.; Cao, T.; Xu, C.; Li, F., High contrast upconversion luminescence targeted imaging in vivo using peptide-labeled nanophosphors. *Analytical Chemistry* **2009,** *81* (21), 8687–94.
55. Wang, M.; Mi, C. C.; Wang, W. X.; Liu, C. H.; Wu, Y. F.; Xu, Z. R.; Mao, C. B.; Xu, S. K., Immunolabeling and NIR-excited fluorescent imaging of HeLa cells by using NaYF(4):Yb,Er upconversion nanoparticles. *ACS Nano* **2009,** *3* (6), 1580–6.
56. Yu, X. F.; Sun, Z.; Li, M.; Xiang, Y.; Wang, Q. Q.; Tang, F.; Wu, Y.; Cao, Z.; Li, W., Neurotoxin-conjugated upconversion nanoprobes for direct visualization of tumors under near-infrared irradiation. *Biomaterials* **2010,** *31* (33), 8724–31.
57. Lee, S.; Ryu, J. H.; Park, K.; Lee, A.; Lee, S. Y.; Youn, I. C.; Ahn, C. H.; Yoon, S. M.; Myung, S. J.; Moon, D. H.; Chen, X.; Choi, K.; Kwon, I. C.; Kim, K., Polymeric nanoparticle-based activatable near-infrared nanosensor for protease determination in vivo. *Nano Letters* **2009,** *9* (12), 4412–6.
58. Kong, K.; Kendall, C.; Stone, N.; Notingher, I., Raman spectroscopy for medical diagnostics: From in-vitro biofluid assays to in-vivo cancer detection. *Advanced Drug Delivery Reviews* **2015,** *89*, 121–34.
59. Jermyn, M.; Mok, K.; Mercier, J.; Desroches, J.; Pichette, J.; Saint-Arnaud, K.; Bernstein, L.; Guiot, M. C.; Petrecca, K.; Leblond, F., Intraoperative brain cancer detection with Raman spectroscopy in humans. *Science Translational Medicine* **2015,** *7* (274), 274ra19.
60. Harmsen, S.; Huang, R.; Wall, M. A.; Karabeber, H.; Samii, J. M.; Spaliviero, M.; White, J. R.; Monette, S.; O'Connor, R.; Pitter, K. L.; Sastra, S. A.; Saborowski, M.; Holland, E. C.; Singer, S.; Olive, K. P.; Lowe, S. W.; Blasberg, R. G.; Kircher, M. F., Surface-enhanced resonance Raman scattering nanostars for high-precision cancer imaging. *Science Translational Medicine* **2015,** *7* (271), 271ra7.
61. de la Zerda, A.; Bodapati, S.; Teed, R.; May, S. Y.; Tabakman, S. M.; Liu, Z.; Khuri-Yakub, B. T.; Chen, X.; Dai, H.; Gambhir, S. S., Family of enhanced photoacoustic imaging agents for high-sensitivity and multiplexing studies in living mice. *ACS Nano* **2012,** *6* (6), 4694–701.
62. Zhang, Q.; Iwakuma, N.; Sharma, P.; Moudgil, B. M.; Wu, C.; McNeill, J.; Jiang, H.; Grobmyer, S. R., Gold nanoparticles as a contrast agent for in vivo tumor imaging with photoacoustic tomography. *Nanotechnology* **2009,** *20* (39), 395102.
63. Kim, C.; Cho, E. C.; Chen, J.; Song, K. H.; Au, L.; Favazza, C.; Zhang, Q.; Cobley, C. M.; Gao, F.; Xia, Y.; Wang, L. V., In vivo molecular photoacoustic tomography of melanomas targeted by bioconjugated gold nanocages. *ACS Nano* **2010,** *4* (8), 4559–64.
64. Zhang, Y. S.; Wang, Y.; Wang, L.; Wang, Y.; Cai, X.; Zhang, C.; Wang, L. V.; Xia, Y., Labeling human mesenchymal stem cells with gold nanocages for in vitro and in vivo tracking by two-photon microscopy and photoacoustic microscopy. *Theranostics* **2013,** *3* (8), 532–43.
65. Miao, Q.; Lyu, Y.; Ding, D.; Pu, K., Photoacoustic imaging: Semiconducting oligomer nanoparticles as an activatable photoacoustic probe with amplified brightness for in vivo imaging of pH (Adv. Mater. 19/2016). *Advanced Materials* **2016,** *28* (19), 3606.
66. Zheng, Y.; Wang, L.; Krupka, T. M.; Wang, Z.; Lu, G.; Zhang, P.; Zuo, G.; Li, P.; Ran, H.; Jian, H., The feasibility of using high frequency ultrasound to assess nerve ending neuropathy in patients with diabetic foot. *European Journal of Radiology* **2013,** *82* (3), 512–7.
67. Matsunaga, T. O.; Sheeran, P. S.; Luois, S.; Streeter, J. E.; Mullin, L. B.; Banerjee, B.; Dayton, P. A., Phase-change nanoparticles using highly volatile perfluorocarbons: Toward a platform for extravascular ultrasound imaging. *Theranostics* **2012,** *2* (12), 1185–98.
68. Teng, Z.; Wang, R.; Zhou, Y.; Kolios, M.; Wang, Y.; Zhang, N.; Wang, Z.; Zheng, Y.; Lu, G., A magnetic droplet vaporization approach using perfluorohexane-encapsulated magnetic mesoporous particles for ultrasound imaging and tumor ablation. *Biomaterials* **2017,** *134*, 43–50.
69. Rapoport, N.; Nam, K. H.; Gupta, R.; Gao, Z.; Mohan, P.; Payne, A.; Todd, N.; Liu, X.; Kim, T.; Shea, J.; Scaife, C.; Parker, D. L.; Jeong, E. K.; Kennedy, A. M., Ultrasound-mediated tumor imaging and nanotherapy using drug loaded, block copolymer stabilized perfluorocarbon nanoemulsions. *Journal of Controlled Release: Official Journal of the Controlled Release Society* **2011,** *153* (1), 4–15.
70. Zhang, L.; Yin, T.; Li, B.; Zheng, R.; Qiu, C.; Lam, K. S.; Zhang, Q.; Shuai, X., Size-modulable nanoprobe for high-performance ultrasound imaging and drug delivery against cancer. *ACS Nano* **2018,** *12* (4), 3449–60.
71. Lusic, H.; Grinstaff, M. W., X-ray-computed tomography contrast agents. *Chemical Reviews* **2013,** *113* (3), 1641–66.

72. Attia, M. F.; Anton, N.; Chiper, M.; Akasov, R.; Anton, H.; Messaddeq, N.; Fournel, S.; Klymchenko, A. S.; Mely, Y.; Vandamme, T. F., Biodistribution of X-ray iodinated contrast agent in nano-emulsions is controlled by the chemical nature of the oily core. *ACS Nano* **2014,** *8* (10), 10537–50.
73. Zhang, G.; Liu, Y.; Yuan, Q.; Zong, C.; Liu, J.; Lu, L., Dual modal in vivo imaging using upconversion luminescence and enhanced computed tomography properties. *Nanoscale* **2011,** *3* (10), 4365–71.
74. Xu, C.; Tung, G. A.; Sun, S., Size and concentration effect of gold nanoparticles on X-ray attenuation as measured on computed tomography. *Chemistry of Materials: A Publication of the American Chemical Society* **2008,** *20* (13), 4167–9.
75. Tkachenko, A. G.; Xie, H.; Liu, Y.; Coleman, D.; Ryan, J.; Glomm, W. R.; Shipton, M. K.; Franzen, S.; Feldheim, D. L., Cellular trajectories of peptide-modified gold particle complexes: Comparison of nuclear localization signals and peptide transduction domains. *Bioconjugate Chemistry* **2004,** *15* (3), 482–90.
76. Hainfeld, J. F.; O'Connor, M. J.; Dilmanian, F. A.; Slatkin, D. N.; Adams, D. J.; Smilowitz, H. M., Micro-CT enables microlocalisation and quantification of Her2-targeted gold nanoparticles within tumour regions. *The British Journal of Radiology* **2011,** *84* (1002), 526–33.
77. Chanda, N.; Kattumuri, V.; Shukla, R.; Zambre, A.; Katti, K.; Upendran, A.; Kulkarni, R. R.; Kan, P.; Fent, G. M.; Casteel, S. W.; Smith, C. J.; Boote, E.; Robertson, J. D.; Cutler, C.; Lever, J. R.; Katti, K. V.; Kannan, R., Bombesin functionalized gold nanoparticles show in vitro and in vivo cancer receptor specificity. *Proceedings of the National Academy of Sciences of the United States of America* **2010,** *107* (19), 8760–5.
78. Willmann, J. K.; van Bruggen, N.; Dinkelborg, L. M.; Gambhir, S. S., Molecular imaging in drug development. *Nature Reviews Drug Discovery* **2008,** *7* (7), 591–607.
79. Lee, D. E.; Koo, H.; Sun, I. C.; Ryu, J. H.; Kim, K.; Kwon, I. C., Multifunctional nanoparticles for multimodal imaging and theragnosis. *Chemical Society Reviews* **2012,** *41* (7), 2656–72.
80. Moon, S. H.; Yang, B. Y.; Kim, Y. J.; Hong, M. K.; Lee, Y. S.; Lee, D. S.; Chung, J. K.; Jeong, J. M., Development of a complementary PET/MR dual-modal imaging probe for targeting prostate-specific membrane antigen (PSMA). *Nanomedicine* **2016,** *12* (4), 871–9.
81. Glaus, C.; Rossin, R.; Welch, M. J.; Bao, G., In vivo evaluation of ^{64}Cu-labeled magnetic nanoparticles as a dual-modality PET/MR imaging agent. *Bioconjugate Chemistry* **2010,** *21* (4), 715–22.
82. Cui, X.; Belo, S.; Kruger, D.; Yan, Y.; de Rosales, R. T.; Jauregui-Osoro, M.; Ye, H.; Su, S.; Mathe, D.; Kovacs, N.; Horvath, I.; Semjeni, M.; Sunassee, K.; Szigeti, K.; Green, M. A.; Blower, P. J., Aluminium hydroxide stabilised $MnFe_2O_4$ and Fe_3O_4 nanoparticles as dual-modality contrasts agent for MRI and PET imaging. *Biomaterials* **2014,** *35* (22), 5840–6.
83. Lee, N.; Yoo, D.; Ling, D.; Cho, M. H.; Hyeon, T.; Cheon, J., Iron oxide based nanoparticles for multimodal imaging and magnetoresponsive therapy. *Chemical Reviews* **2015,** *115* (19), 10637–89.
84. Kim, D.; Yu, M. K.; Lee, T. S.; Park, J. J.; Jeong, Y. Y.; Jon, S., Amphiphilic polymer-coated hybrid nanoparticles as CT/MRI dual contrast agents. *Nanotechnology* **2011,** *22* (15), 155101.
85. Li, D.; Yang, J.; Wen, S.; Shen, M.; Zheng, L.; Zhang, G.; Shi, X., Targeted CT/MR dual mode imaging of human hepatocellular carcinoma using lactobionic acid-modified polyethyleneimine-entrapped gold nanoparticles. *Journal of Materials Chemistry B* **2017,** *5* (13), 2395–401.
86. Kobayashi, H.; Longmire, M. R.; Ogawa, M.; Choyke, P. L., Rational chemical design of the next generation of molecular imaging probes based on physics and biology: Mixing modalities, colors and signals. *Chemical Society Reviews* **2011,** *40* (9), 4626–48.
87. Cai, W.; Chen, K.; Li, Z. B.; Gambhir, S. S.; Chen, X., Dual-function probe for PET and near-infrared fluorescence imaging of tumor vasculature. *Journal of Nuclear Medicine: Official Publication, Society of Nuclear Medicine* **2007,** *48* (11), 1862–70.
88. Liang M.; Liu X.; Cheng D.; Liu G.; Dou S.; Wang Y.; Rusckowski M.; Hnatowich D. J., Multimodality nuclear and fluorescence tumor imaging in mice using a streptavidin nanoparticle. *Bioconjugate Chemeistry* **2010,** *1* (7): 1385–8.
89. Andrasi, M.; Gaspar, A.; Kovacs, O.; Baranyai, Z.; Klekner, A.; Brucher, E., Determination of gadolinium-based magnetic resonance imaging contrast agents by micellar electrokinetic capillary chromatography. *Electrophoresis* **2011,** *32* (16), 2223–8.
90. Liu, Z.; Lammers, T.; Ehling, J.; Fokong, S.; Bornemann, J.; Kiessling, F.; Gatjens, J., Iron oxide nanoparticle-containing microbubble composites as contrast agents for MR and ultrasound dual-modality imaging. *Biomaterials* **2011,** *32* (26), 6155–63.

9

Anti-Arthritic Potential of Gold Nanoparticle

Jayeeta Sengupta, Sourav Ghosh, and Antony Gomes
University of Calcutta

9.1 Introduction

Arthritis is a global problem which is present in almost all countries in the world, affecting people from all age groups, particularly the elderly population. It's one of the oldest known diseases which has been recorded since the birth of human civilization and still remains a major problem to date. Arthritis affects not only people's day-to-day to life, working conditions, and lifestyle but also mental health. Though arthritis is a very common disease, affecting people irrespective of their age, gender, race, or geographical location, quite a lot of the disease progression and treatment strategy are yet to be understood. As many as 100 types of different arthritis have been recorded to date, which imposes the further challenge for its treatment and diagnosis. Its wider occurrence, severity, challenges in treatment and number of people affected every day make arthritis one of the most important areas of research. Therefore, search for alternative therapy and technology to deal with arthritis is always an area of vital importance for society and scientific research.

The introduction of nanotechnology in biomedical science is a phenomenal event of the present decade since it has proven to be a potential tool for diagnosis, imaging, drug delivery, and therapeutics. The possibility to manipulate and tailor the physical, chemical, and biological properties of matter at nanometer scale and molecular level has open doors for researchers to use nanoparticles for a wide array of applications in health and diseases. The unique properties exhibited by nanoparticles owing to the size, shape, surface charge, aggregation and agglomeration profile, and conjugation possibility make these dwarf particles some excellent candidates for application in biomedical sciences. The ease with which they can interact at the molecular level with cells and biomolecules gives them a great possibility to be used for therapeutic purpose. Therefore, using nanotechnology to find a solution for arthritis has been the latest trend in arthritis research.

Among a variety of nanoparticles explored till date, gold nanoparticles, in particular, are being explored for a vast range of applications in the field of medical science owing to their unique physical and chemical properties at colloidal state. The surface plasmon resonance phenomenon associated with gold nanoparticle enables them to generate strong electromagnetic fields on its surface, thereby enhancing their absorbing and scattering capacities. This makes them as potential tools to be applied for targeted delivery, drug delivery, imaging and diagnosis, theranostics, and molecular therapies. The other attractive features for counting them among potential candidates are their excellent surface modulation and functionalization capacity, easy synthesis, and ability to absorb light at near-infrared region making them useful in diagnosis and therapy. They are like those "magic wands" which has brought forward a plethora of possibilities in diagnosing and treating diseases. Apart from showing promising results in cancer therapy, diagnosis, photodynamic therapy, as contrast agents, and as biosensors, they have been also evolved out to be equally potent enough to be used as the anti-arthritic agent.

The use of gold nanoparticle as the anti-arthritic agent is not new; it dates back to thousands of years and has been well documented in folk medicine and ancient medicinal books. Recent advances in nanoscience and arthritis research have yielded scientific validations of gold nanoparticle possessing anti-arthritic potential, thereby confirming the validity of the ancient use of gold nanoparticle to treat arthritis. In this book chapter, we will discuss in detail the disease arthritis, disease progression and the related problems, the role of nanotechnology in biomedical science, the use of nano-based materials in different areas of medical science, gold nanoparticles, and its use in medical science and anti-arthritic potential of the gold nanoparticle. This would let us have a better insight into the understanding of the technology and its miracle power to change the scenario of medical science, with special reference to arthritis.

9.2 Arthritis: Arthritis Pathophysiology and Challenges in Its Treatment

Arthritis is one of the oldest diseases of the civilization, with first-known fossil dated back around 4500 B.C. in the Native Americans of Tennessee, USA. It has been mentioned in the ancient Greek and Hindu mythology, and the first written reference on arthritis was found in the *Charaka Samhita* (an Indian holistic book). There are three main categories of arthritic conditions: rheumatoid arthritis, osteoarthritis, and gouty arthritis, which can further be classified into around 100 types of arthritic conditions by the modern medicinal system. Rheumatoid arthritis mainly affects the limb joints, whereas osteoarthritis generally affects the hips, fingers, and knees. Pain is the main symptom of almost every type of arthritis, and the other symptoms include the morning stiffness, difficulty in moving joints, muscle weakness, weight loss, and fatigue.

9.2.1 Pathophysiology of Rheumatoid Arthritis

The risk factors leading to rheumatoid arthritis include genetic as well as non-genetic factors. Women have a higher risk of the disease, probably due to hormonal changes during pregnancy, lactating period, and use of hormonal contraceptives. The family history of rheumatoid arthritis, smoking (active as well as passive), occupational hazard (such as exposure to silica), the diet with high caffeine, infection (such as Epstein–Barr virus), and mucosal inflammation remain the major risk factors of rheumatoid arthritis. Periodontal diseases leading to the inflammation of gums, bone destruction, and collagen matrix destruction are associated with the pathogenesis of rheumatoid arthritis. *Porphyromonasgingivalis*, a bacteria causing periodontal disease, has an enzyme that can cause citrullination of proteins, leading to the pathogenesis of rheumatoid arthritis.

It is thought that the main risk factor for rheumatoid arthritis is the genetic factors. Antigen-presenting cells of the immune system possess major histocompatibility (MHC) antigens (class II) at the cell surface, encoded by a gene HLA-DR4. It is estimated that a conserved sequence in HLA-DR4 gene contributes around 30% of total genetic risks for rheumatoid arthritis. This conserved and specific sequence in HLA-DR4 gene causes a slight change in class II MHC antigens, leading to the activation of T lymphocytes. The other genetic risk factors contributing to the initiation of the pathogenesis of rheumatoid arthritis include peptidyl arginine deiminase-4, STAT4, PTNP22, and CTLA4. Peptidyl arginine deiminase-4 causes increased citrullination of peptides, which unfolds the peptide (change in 3D structure) due to the loss of positively charged arginine residues. It makes the self-protein an antigen, producing specific anti-citrullinated peptide antibodies (rheumatoid arthritis-specific autoantibody). A polymorphism in peptidyl arginine deiminase-4 gene causes increased citrullination of peptides. Anti-citrullinated peptide antibodies can be identified up to 12–15 years before the commencement of clinical symptoms of rheumatoid arthritis, indicating a preclinical period of the pathogenesis of the disease. STAT4, PTNP22, and CTLA4 are involved in the activation of tumor necrosis factor receptor and T-cell activation, leading to the initiation of rheumatoid arthritis pathogenesis.

When T lymphocyte interacts with MHC on antigen-presenting cells, it may be activated, or show tolerance to the antigen, or undergo programmed cell death, depending on a second signal through appointing additional cellular receptors. The CD24 molecule on the T-cell surface may act as this second signal of costimulation. Activated T cells proliferate and secrete additional cytokines such as interleukin-2, interleukin-4, tumor necrosis factor, and interferon-γ. Interleukin-2 amplifies the proliferation of T cells. Very early inflamed synovium has been found to have an unexpected T helper cell profile, along with increased expression of interleukin-4, interleukin-5, and interleukin-13. When the diseased is established, synovial T cells produce low amounts of interferon-γ, interleukin-10, and tumor necrosis factor-α, while interleukin-2 and interleukin-4 become virtually absent. Interleukin-17 is produced spontaneously in synovial cells of rheumatoid arthritis patients. Interleukin-23 promotes the survival and proliferation of T helper cells. The receptor for interleukin-17 is expressed all over the synovium and exerts pleiotropic effects. Thus, T-cell-derived interleukin-17 (subtypes 17A and 17F) promotes monocyte-dependent interleukin-1 and tumor necrosis factor-α production and induces osteoclast differentiating factor RANKL (receptor activator of nuclear factor kappa-B ligand), stimulating synovial fibroblasts to express interleukin-6, interleukin-8, granulocyte colony-stimulating factor, prostaglandin E_2, and matrix metalloproteinases.

B lymphocytes are the source of the anti-citrullinated peptide antibodies and rheumatoid factors, contributing to the pathogenesis of rheumatoid arthritis through complement activation and immune complex formation in the synovium. They act as antigen-presenting cells and expresses costimulatory molecules that can activate T cell. They also produce and respond to cytokines and chemokines promoting leukocyte infiltration into the synovium, synovial hyperplasia, formation of ectopic lymphoid structures, and angiogenesis. Interaction with T cells and some soluble cytokines (which increase the differentiation and proliferation of B cells) activate B cells. Activated B cells and plasma cells (fully differentiated B cells) are found in the synovium of rheumatoid arthritis patients. B cells are involved in the production of cytokines and cellular interactions, and act as antigen-presenting cells to T cells. Being an autoimmune disease, rheumatoid arthritis is characterized by the presence of autoantibodies (including anti-citrullinated peptide antibodies and rheumatoid factors). Rheumatoid factors are autoantibodies directed against native antibodies (mostly immunoglobulin M, sometimes IgG or IgA) recognizing the Fc portion of the immunoglobulins.

Effector cells mediate maximum damage in rheumatoid arthritis, through the production of cytokines and other mediators. Macrophages act as the mastermind in the pathogenesis of the disease, producing pro-inflammatory cytokines including interleukin-1, interleukin-6, interleukin-8, tumor necrosis factor, and granulocyte-macrophage colony-stimulating factor, which further stimulates the macrophages and amplifies local and systemic inflammation, leading to synovial matrix destruction. These cytokines also activate other cells in the synovium (including osteoclast and fibroblasts) and cells in the liver (for production of C-reactive protein), contributing to the pathogenesis of rheumatoid arthritis. Activated macrophages also produce leukotrienes, prostaglandins, NO, and many other pro-inflammatory substances causing systemic and local inflammation. Local actions of macrophages include recruitment and potentiation of inflammatory cells, production of matrix-degrading proteolytic enzymes, and cytokine-induced differentiation of synovial cells and angiogenesis, whereas macrophage systemically intensifies rheumatoid arthritis, produces tumor necrosis factor, and activates circulating monocytes. Fibroblasts in the synovium secrete interleukin-6, interleukin-8, granulocyte-macrophage colony-stimulating factor, proteases, and collagenase, contributing to the pathogenesis of the disease. Neutrophils are the most abundant cell types found in the synovium of rheumatoid arthritis. Interleukin-8, leukotriene, and activation of complement pathway recruit neutrophils in the synovium, causing the generation of reactive oxygen species that depolymerize hyaluronic acid. Reactive oxygen species also inactivates protease inhibitors, causing protein denaturation at the synovium. Interleukin-1 and tumor necrosis factor activate chondrocytes to produce a significant amount of proteases causing the destruction of cartilage matrix.

An imbalance between pro-inflammatory cytokines and anti-inflammatory cytokines exerts one of the most profound roles in the pathogenesis of rheumatoid arthritis. Interleukin-1, interleukin-6, and tumor necrosis factor have major contributions in initiation and progression of the disease. These cytokines act by stimulating the same cells (autocrine function), stimulating neighboring cells (paracrine function), or stimulating cells at distant sites (endocrine function). Cytokines activate B cells, promote the synthesis and secretion of other pro-inflammatory cytokines, increase the synthesis of adhesion molecules, stimulate the osteoclasts, and upregulate the matrix metalloproteinases, prostaglandins, leukotrienes, and C-reactive proteins. Interleukin-8, interleukin-15, interleukin-17, and interleukin-23 are involved in cellular recruitment, T-cell proliferation, osteoclast activation, and helper T cell differentiation.

9.2.2 Pathophysiology of Osteoarthritis

Like rheumatoid arthritis, osteoarthritis also has genetic and non-genetic risk factors. Family history epidemiological studies have shown that genetics have around 50% or more impact on the pathogenesis of osteoarthritis. The human chromosomes 2q, 9q, 11q, and 16p, as well as other genes including AGC1, CRTM, CRTL, ER alpha, IGF-1, TGF beta, VDR, and genes for collagens II, IX, and XI, have proven to have profound effects on osteoarthritis. Age, gender (13% females and 10% male of age 60 years or above), weight, obesity, poor muscle strength, lifestyle, body posture (squatting and kneeling), knee injury, extensive use of joints, etc. are the major risk factors of the disease.

The summative effects of risk factors (intrinsic and extrinsic) initiate the commonest chronic joint disorder, osteoarthritis. It occurs when chondrocytes fail to maintain the homeostasis between production and degradation of extracellular matrix components on articular cartilage. Chondrocytes are influenced by growth factors, cytokines, and physical stimulations. The pro-inflammatory cytokines (interleukin-1β, interleukin-6, interleukin-15, interleukin-17, interleukin-18, tumor necrosis factor α) and anti-inflammatory cytokines (interleukin-4, interleukin-10, interleukin-13) are the most contributing pathophysiologic factors in osteoarthritis. Pro-inflammatory cytokines destroy the joint cartilages, slower the rate of formation of extracellular matrix key components (including collagen fiber, proteoglycan, aggrecan), and release proteolytic enzymes that decompose joint cartilages. They also help in recruiting immune cells at the articular cartilage, producing inflammatory prostaglandins, cyclooxygenases, phospholipases, reactive oxygen species, and NO. Anti-inflammatory cytokines inhibit the synthesis and actions of pro-inflammatory cytokines, inhibit apoptotic cell death of chondrocytes, reduce the synthesis of proteases, promote proteoglycan synthesis, and stimulate the production of growth factors.

Obesity leads to increased adipokine and leptin expression in the human body, promoting proteases, pro-inflammatory cytokines, reactive oxygen species, and NO. Adipokine and leptin receptors are found in the cell membrane of chondrocytes, subchondral osteoblasts, and synoviocytes, activation of which leads to the secretion of proteases (especially matrix metalloproteinases), exerting the destruction of joint tissue. Increased body mass index is directly correlated with the pathogenesis of osteoarthritis.

9.2.3 Challenges in the Treatment of Arthritis

Treatments of rheumatoid arthritis and osteoarthritis have many limitations and side effects, including frequent administration, high doses, drug resistance, high cost, bacterial infections, and systemic side effects. Nearly 30% of patients show low responsiveness to the available therapeutics against rheumatoid arthritis and osteoarthritis. The current drugs rarely lead to full recovery of the patients suffering from rheumatoid arthritis. Absence of tissue specificity of the drugs causes serious adverse systemic side effects. The drugs can be administered by intra-articular injection to increase the accumulation of drugs at the inflamed joints and to decrease its systemic side effects, but it often leads to increase risk of infection and poor compliance. The local administration of drugs in the poly-arthritic patients is not a suitable choice. The control of joint inflammation and bone damage is necessary in the management of rheumatoid arthritis and osteoarthritis. The limitations and side effects of the present therapeutics lead to search for alternative therapies against arthritis with increased efficacy and decreased limitations.

9.3 Nanotechnology and Nanomedicine: An Alternative Therapeutic Option

Nanotechnology is the next revolution in the history of scientific innovations and advancements since it is an interdisciplinary field which bridges various disciplines, from physics to chemistry to biology, from engineering to medicine, thereby establishing new trends of technological era. It is of no surprise that nanotechnology is considered a key to the 21st century, as it holds excellent promises in sectors of biotechnology and medicine. The explosive growth of nanoscale technology and its applications in a broader spectrum brings forward a new paradigm of development for the next-generation products. This upcoming technology is still a virgin area to be explored by the researchers, but it has been already proved that it has great potential for revolutionizing the ways in which materials and products are created. Moreover, the range and nature of accessible functionalities where it can be used are extremely promising. Due to its multidisciplinary overview, the opportunities for nanotechnology development are vast and represent enormous prospect to design and innovate smarter devises, more precise solutions to meet a wide arena of human needs. Nanoparticles are known to exist in diverse shapes such as cubical, triangular, spherical, pentagonal, shells, rod-shaped, ellipsoidal, and so forth. Nanoparticles by themselves and when used as building blocks to construct complex nanostructures such as nano-chains, nano-wires, nano-fibers, nano-clusters, and nano-aggregates find use in a wide variety of applications in the fields of electronics, chemistry, biotechnology, and medicine. From its wider application perspective and interdisciplinary content, it becomes an interesting area of research, attracting researchers from various disciplines. A simple way to define is "Nanotechnology" is manipulation of matter at nanoscale level, i.e., one billionth of a meter. In fact, to be precise, we can consider it as the scale of technology and not type of technology.

Major innovations in nanotechnology have brought forward possibilities of application in the field of biomedical science. Use of nanomaterials at this scale-level provides a high range of flexibility in the modifications as per desired requirements to use it in the biomedical field. Nanotechnology and its advancement have to lead to the production of a number of commercially based nano-products which are used for therapeutics. In fact, the present popularity and development in the field of nanomedicine have attracted researchers to investigate on several nanoparticle-based products which can be employed in therapeutic application. The field of nanotechnology has been attractive and promising which can be easily understood by the fact that a very good number of nanoparticle-based products are under clinical trial. Thus, there has been a marked increase in the research and development of nanoscale-based products with distinctive physical and chemical features to be employed as potential drug clues in the field of oncology, cardiology, immunology, neurology, endocrinology, ophthalmology, pulmonary, orthopedics, and dentistry. The major areas of biomedical application are as follows.

9.3.1 Imaging and Diagnosis

One of the important aspects of medical diagnosis is molecular imaging which helps in early diagnosis of a disease, disease stage and progression, and therapeutic monitoring which mainly relies on probes specifically designed to detect the biological and cellular phenomenon at the molecular level. An important application of nanotechnology is in the field of imaging. Inherited properties such as superior photostability, narrow range of emission, broad excitation wavelength, multiple possibilities of modification, luminescence, and surface plasmon resonance of nanomaterials make these dwarf particles as potential candidates to be employed as probes for imaging purpose. Nano-based probes have shown an edge over existing single molecule-based contrast agents by having advantages such as better contrast, longer circulating time, and higher number of particulate assembly for better detection. These properties give the opportunity for imaging to be done in multiple ways and techniques and hence be efficient for molecular-level diagnosis and drug delivery. Nanomaterials such as quantum dots which are highly light absorbing, luminescent semiconductor nanocrystals have been extensively used in the field of optical imaging. Fluorescence resonance energy transfer analysis (FRET), gene technology, fluorescent labeling of cellular proteins, cell tracking, pathogen and toxin detection, and in vivo animal imaging have been greatly influenced by nanotechnology. Further, in future, potential imaging contrast agents with nanoparticles can be created over the time which will be multifunctional, more efficient, biocompatible, of better efficacy, specificity, and multiple targets oriented. Since nanotechnology is an emerging science and every day more and more new particles are synthesized with different properties, it is for sure medical imaging with nanotechnology will continue to develop.

9.3.2 Drug Delivery Systems

The employment of nanotechnology in the field of drug delivery has changed the whole scenario and perspective of pharmaceutics and therapeutics. Miniature drug delivery systems are formed by using engineered nanoparticles which can precisely target specific locations to deliver drug and hence gives hope in the field of target-oriented therapy. Using nanoparticles as drug delivery agents has some added advantages which have been scientifically documented. Nanoparticles as drug delivery agents are not only effective for targeted drug release but also monitored release along with two or more drugs to give combined effects, lesser side effects, prolonged half-life of the drug in the systemic solution, and better solubility. The potential nano-candidates used as drug delivery systems are dendrimers, fullerenes and fullerites, gold nanoshells, ceramic, polymeric, metallic, liposomes, and magnetic particle systems. The basic principle underlying the design of these drug delivery systems is entrapment or encapsulation of the drug within the nanostructure which are not rejected by the immune system from which drug can be released uniformly over an extended period of time. The carried therapeutic agent can be DNA, plasmid DNA, proteins, peptides, low-molecular-weight compounds, contrast dyes, anticancer, antiviral, and anti-bacterial agents, and has better efficacy, stability, and cost-effectiveness. The research outcomes till date give us hope that drug delivery using nanotechnology gathers more momentum and opens new doors for the treatment of various diseases.

9.3.3 Bio-nanoconjugation

Another arena in which nanotechnology has a great influence in the field of biomedical science is bio-nanoconjugation. Bio-nanoconjugation is tailored manipulative systems operating at the molecular level involving the combination of functionalities of biomolecules and non-biologically derived molecular species for specialized use such as biomarkers, biosensors, bio-imaging agents, and masking of immunogenic moieties and targeted drug delivery systems used in disease diagnosis and treatment. To mention a few bio-nano conjugates are quantum dots as fluorescent biological labels, gold nanoparticle bio conjugate-based colorimetric assay, gold nanoshell polymer composite photothermally triggered drug delivery system, core/shell fluorescent magnetic silica-coated composite nanoparticles, bio-conjugated silica and colloidal gold nanoparticles, and many more. These can be proven to be miniature magic tools for a wide array of applications.

9.3.4 Biosensors

Nanomaterials exhibit unique physical and chemical properties which make them promising for designing efficient sensing devices, especially biosensors. The nano-based biosensors can come up with added advantages of better sensitivity, intensified performance, and having very low detection values. Nanoparticles can be used in a variety of bioanalytical formats in the form of quantification tags, such as the optical detection of quantum dots and the electrochemical detection of metallic nanoparticles, encoded nanoparticles as substrates for multiplexed bioassays, such as striped metallic nanoparticles, nanoparticles that leverage signal transduction, for example, in colloidal gold-based aggregation assays, and functional nanoparticles that exploit specific physical or chemical properties of nanoparticles to carry out novel functions, such as the catalysis of a biological reaction. Metal, oxide, and semiconductor nanoparticles are employed in the construction of biosensors. The edge of using nanoparticles in designing biosensors includes the immobilization of biomolecules, the catalysis of electrochemical reactions, the enhancement of electron transfer between electrode surfaces and proteins, the labeling of biomolecules, and even acting as the reactant. With advancement in biosensors, nanomaterials have become an important component in making bioanalytical devices. Since a lot of benefits are provided by the employment of these particles in making biosensors, they can be considered as an alternative option for bioanalysis. With the availability of long line-up of these engineered particles with distinct properties, there is a huge possibility of multiple bio-analytic devices to be constructed in coming years which can form a new era of modified biosensors.

9.3.5 Therapeutic Use

The biggest revolution brought by nanotechnology is in the area of therapeutics. With the introduction of nanomaterials, the medical science has seen the impossible to be possible. Nanoscience and nanotechnology comes with the perquisites of better drug efficacy, targeted delivery, sustainable and slow drug release, biocompatibility, minimal toxicity, interaction at molecular level, and better cell-to-cell interaction. The technology has showed unique problem-solving capabilities in treating various disease conditions by its unique nanolevel physicochemical properties.

Different important metals, such as iron, iron oxides, copper, cobalt, zinc, gold, and silver, have been potentially shown to find its application in biomedical science. Zinc and zinc oxide nanoparticles are extensively in cancer therapy. Cobalt nanoparticles have been also used as drug delivery agents and in cancer therapy. For therapeutic approaches, another big milestone in the field of nanoscience is "magnetic nanoparticles" which can be functionalized accordingly and therefore of prime focus for researchers today. The versatility of magnetic nanoparticles makes these unique particles as extremely potential candidates for theranostic such as finding application for targeted drug delivery, magnetic resonance imaging (MRI) contrast agent, biosensors, tissue engineering, and magnetic transfections. Magnetic NPs such as Fe, Co, and Pt have been used as for MRI and MPI applications. In fact, the unique capacity of magnetic nanoparticles to bind to drugs, proteins, enzymes, antibodies, and nucleotides makes them more efficient as drug-carrying vehicles and further adding to their targeted therapy being directed by the magnetic field. Magnetic nanoparticles at this nanoscale level exhibit the property of super-paramagnetism, i.e., not retaining the magnetism effect after the magnetic field. The most promising candidate under this category is iron oxide nanoparticles which have attracted a great deal of interest in biomedical research and clinical applications, termed as SPIONS (superparamagnetic iron oxide nanoparticles) owing to their biocompatibility, proper surface architecture, and conjugated targeting ligands/proteins. SPIONs have been successfully employed in both in vitro magnetic separation and in vivo applications such as MRI, gene delivery (GD), chelation therapy, and nanomedicine. Nanoparticles in stem cell therapy involve the use of magnetic nanoparticles and quantum dot-based applications which can track and guide transplanted stem cells and control functions of cellular signals, thereby finding its application in the fields of atherosclerosis, thrombosis, and vascular biology. The technology has also found its application in the field of tissue engineering, and synthesis of new nanobiomaterials can be useful in tissue regeneration and tissue replacement. A lot of experimental evidences have been reported on self-healing properties of nano-based materials along with their use in tissue regeneration and wound healing. With the generation of new nanomaterials every day with tailored functionalities, there is a possibility to use these materials as prosthetics, implants, or artificial organs. Researchers have already got encouraging results in using nanomaterials as implants, and it will be soon that these nano-based materials will be part of our healthcare system. A number of research studies are being conducted throughout the world to use nanomaterials in artificial organs, which will be a great leap for medical science.

Cancer nanotechnology holds promise in the field of cancer diagnosis and therapy since these particles can be modified using nanotechnology to engineered vehicles having unique therapeutic properties which can penetrate tumors deeply with a high-level specificity and target-oriented therapy. The key factors which contribute to the anticancer attribute of the nanoparticles are its size range, surface charge, surface functionalization, loading capacity, targeting ligands, enhanced permeability, circulation time, and retention. One of the biggest achievements is being able to load nanoparticles with anticancer drugs such as doxorubicin, paclitaxel, and 5-fluorouracil to target tumor cells. Presently, it has been also possible to use loading of both imaging and therapeutic agent for theranostic applications. Using nanoparticles for photodynamic therapy is another emerging platform in cancer treatment. The National Cancer Institute has identified the effectiveness of nanotechnology research in the field of cancer providing an extraordinary, paradigm-changing opportunity to make significant breakthroughs in cancer diagnosis and treatment. Nanotechnology provides the interface between living cells with engineered particulate matter which holds promises from tissue engineering to sensors implanted within human tissues to obtain real-time information on biological processes and functions. Nanoparticles have been also used to evaluate microbial drug resistance and determining the minimum inhibitory concentration of an antibiotic in less amount of time. Therefore, it is being evident that in recent year, nanoparticles are the most abundantly used resource in biomedical field.

9.4 Nanogold: History, Synthesis, and Applications in Biomedical Science

Though nanotechnology is a very recent technology, but the concept of it dates back to thousands of years before. The concept of reduction in particle size has been documented in "Charaka Samhita" a Sanskrit script on Ayurveda in Indian subcontinent. In Ayurveda, there is description on the preparation of "Bhasmas" which are based on the concept of nanoparticle preparation and have been known to be used in treating various ailments. For gold nanoparticle, the history of use not only limits to Indian subcontinent but has been documental evidence of Chinese and Arabic using colloidal gold since 2500 B.C. In Indian Ayurvedic medicine, colloidal gold known as "*Swarna Bhasma*" (gold ash) has been used during the Vedic period for rejuvenation and treating different ailments such as bronchial asthma, rheumatoid arthritis, diabetes mellitus, and nervous system diseases.

In present times, gold nanoparticle is being synthesized by the following methods: chemical method, Turkevich method, Brust–Schiffrin method, electrochemical method, seeding growth method, and biological method. Among these, the Turkevich method is one of the oldest and well-known methods designed by Turkevich in 1951, which is based on the reduction of gold salt ($HAuCl_4$) by citrate in water. Chemical reduction method involves two phases: first reduction by chemical compounds and then followed by stabilization phase. Brust–Schiffrin method could make thermally and air-stable gold nanoparticle particles which are of controlled size and low dispersity. Seed growth technique is mainly for obtaining gold nanoparticles of other shapes rather than only spherical shapes; in this technique, gold salt is reduced by strong reducing agent to produce seed particles followed by addition to a metal salt in presence of weak reducing agent and a structure directing agent to prevent further nucleation and accelerating anisotropic growth of gold nanoparticle particles. With all the above methods mentioned in detail, the biggest drawbacks remain to be the production of toxic by-products, and the gold nanoparticles produced by these techniques may not be biocompatible for biological applications. Therefore, approaches to produce nontoxic and biocompatible gold nanoparticle particles came into the picture. Particles were therefore produced using biological compounds, phytochemicals, and microorganisms. The particles were mostly prepared in concept with "green chemistry". The other methods such as digestive ripening, ultrasonic waves, microwaves laser ablation, solvothermal method, and electrochemical and photochemical reduction were also explored to make gold nanoparticle. In recent years with advancement in nanotechnology research, there has been increased interest on drug discovery strategies using gold nanoparticle. Gold nanoparticles are of interest due to its distinct properties exhibited at colloidal state which includes easy synthesis, synthesis of different shapes and sizes, and unique optical and photothermal properties. The unique property of surface plasmon resonance gives them the unique capacity of binding with peptides and protein which demonstrates that they can act as probes to study cancer cells since they selectively accumulate in tumor cells, exhibiting bright scattering. The SPR effect provides these particles with enhanced radiating properties which makes them useful in areas of cancer therapy, biomedical imaging and diagnosis, controlled drug delivery, and gene therapy. It has been proved gold nanoparticle can be used as better X-ray contrast agents than other iodine-based agents. It has been documented that gold nanoparticle can have potential therapeutic applications in the treatment of B-chronic lymphocytic leukemia. Gold nanoparticle has anti-angiogenic properties and thereby can inhibit the functions of angiogenic factors. Evidences of gold nanospheres, nanorods, nanoshells, and nanocages, capable of killing bacteria and cancer cells when irradiated with focused laser pulses of suitable wavelength, have been documented. Oligonucleotide-capped gold nanoparticles have been reported for polynucleotide or protein (such as p53, a tumor suppressor gene) detection. Gold nanoparticles have also been used in other assays such as immunoassay, protein assay, and detection of cancer cells. There are many studies documented in the literature about gold particles being used as drug delivery vehicles. PEG-coated gold nanoparticle loaded with TNF-α was constructed to maximize the tumor damage and minimize the systemic toxicity of TNF-α. It has been witnessed that intracellular uptake of gold nanoparticles of various shapes and sizes depends on the physical dimensions of the particles; therefore, particles for successful drug delivery should be chosen keeping in mind these factors. The effective role of the gold nanoparticle as an anti-oxidative agent has also been established, by inhibiting the formation of ROS and scavenging free radicals, thus increasing the antioxidant defense enzymes and creating a sustained control over hyperglycemic conditions. Therefore, it has been evident that gold nanoparticles have extensive and diverse application in the field of medical science. The diversity of areas in which gold nanoparticle use has been till date makes it one of the most lucrative study areas of today which can provide answers to many unsolved medical challenges.

9.5 Anti-arthritic Potential of Gold Nanoparticle

The advantages of using gold nanoparticle in arthritis include the following: (i) it controls the release of anti-arthritic drugs at a dose- and time-dependent manner, which requires less frequent administration of the drug; (ii) drugs that are insoluble in aqueous medium are made more bioavailable, improving the delivery at the target site and reducing the systemic side effects; (iii) it makes the peptide drugs to be available for more time in the bloodstream (increases the half-life); (iv) it co-delivers the targeting agent and drugs; and (v) it may combine the anti-arthritic drug and nano-sized diagnostic tool to increase the efficacy and decrease the side effect of the drug.

9.5.1 Mechanism of Action of Gold Nanoparticle against Arthritis

There are two targeting strategies (passive and active) for the treatment of rheumatoid arthritis and osteoarthritis using nanotechnology. Passive targeting strategies allow the drug (encapsulated in nanocarrier) to be accumulated locally at the inflammatory target site through systemic administration, relieving adverse reactions in the other parts of the body. Rheumatoid arthritis results in inflammatory infiltration and abnormal vessels (increased endothelial gaps) at the affected sites, causing increased permeability (wide gap of up to 700 nm at the inter-endothelial junctions) and retention. It causes the nanoparticles of a specific size to reach the site of inflammation through the leakage in the endothelial vasculature. Passive targeting in the treatment of rheumatoid arthritis using nanoparticles includes the loading of drugs (such as disease modifying anti-rheumatic drugs, glucocorticoids, and non-steroidal anti-inflammatory drugs) in nanoparticles. Active targeting strategies of rheumatoid arthritis are more specific, which bind to the receptors and mediate their effects. The drug (and nanoparticle)–receptor interaction causes the remediation of rheumatoid arthritis actively. Use of nanoparticle causes increased therapeutic efficacy of the drug through the high-affinity binding ability to the receptors of cells (including T cells, macrophages, and vascular endothelial cells) involved in the pathogenesis of rheumatoid arthritis. It causes the increase in bioaccumulation of the drug at the target site (i.e., at the inflammatory joints) and reduces the drug retention at the normal tissue level. Therefore, use of nanotechnology may decrease the dose of drugs and side effects, and increase the efficacy.

Being an autoimmune disease, rheumatoid arthritis causes an inflammatory environment in the body by activating T cells and B cells, leading to the activation of antigen-presenting cells that recognize auto-antigens. Nanoparticles generally target the antigen-presenting cells or the lymphocytes, and block the autoimmunity by the destruction of regulatory T cells and B cells. The therapeutically important inorganic nanoparticles (including gold nanoparticle) have a basic structure: a central metallic nanoparticle core and an outer protective coating of drugs/biologically active components. The central metallic nanoparticle core is mostly responsible for the physical properties of the drug–nanoparticle conjugate, whereas the outer coating increases the nanoparticle solubility, increases the half-life of the nanoparticle, decreases the clearance, protects the core from degradation, and exerts their therapeutic actions in rheumatoid arthritis.

9.5.2 Gold Nanoparticle in Arthritis

Gold nanoparticles are easy to synthesize in bulk, increase the activity of the drugs, and show potent applications in drug delivery and imaging vehicle in rheumatoid arthritis. They show a strong affinity towards amine and thiol groups, allowing them to attach with anti-arthritic agents having those active groups. Till date, some drugs (such as methotrexate, prednisolone, methylprednisolone hemisuccinate, betamethasone hemisuccinate, diclofenac sodium, ethyl cellulose, and tocilizumab) have been conjugated with the gold nanoparticle to increase the efficacy of the drugs and to decrease their side effects. Lower doses of these drugs showed superior therapeutic index compared to conventional treatments. Some drugs that contain gold (not as the nanoparticle but in cationic form) are in the market (such as sodium aurothiomalate and monomeric neutral auranofin) to treat rheumatoid arthritis, although there are many side effects of using those in long run. Cationic properties of gold in these drugs cause bone marrow damage, dermatitis, nausea, and other side effects, which colloidal gold nanoparticle (with neutral charge) does not cause.

As discussed before, rheumatoid arthritis is associated with angiogenesis at the inflamed joints. Therapeutically, gold nanoparticles can bind to vascular endothelial growth factor, mediate phosphorylation of its receptor, and increase intracellular calcium release, exerting their anti-angiogenic effects. It helps in reducing inflammation and protects the arthritic joints from immune cell recruitment. Intra-articular administration of colloidal gold nanoparticles can reduce pristine-, collagen- and mycobacterium-collagen-induced arthritis in the experimental animal model.

Gold nanoparticles have been used as contrast agents for imaging of joints and bones in rheumatoid arthritis and osteoarthritis. Photoacoustic tomography (using gold nanoparticle and other contrasting agents, such as etanercept) is useful in monitoring anti-rheumatic drug delivery in inflamed joints. It is a non-ionizing, non-invasive, highly sensitive imaging of drug delivery in good resolution (both temporal and spatial) at the joints.

Many studies have shown the association of oxidative free radical generation in the pathogenesis of rheumatoid arthritis and osteoarthritis, although its specific role in the progression of arthritis is still not clear. Gold nanoparticles can act as antioxidant, as well as the prooxidant, depending on the size and capping molecule of the nanoparticles. Gold nanoparticles synthesized by herbs and herbal extracts have been shown to have anti-oxidative properties, which may reduce the progression of rheumatoid arthritis and osteoarthritis. They can reduce oxidative free radicals by catalyzing the cleavage of hydrogen peroxide to water and oxygen, and by converting superoxide anion to nontoxic by-products.

Significant researches have been done to establish the anti-arthritic property of gold nanoparticle (with or without capping agent) in animal models, although the models may not reproduce the actual features of human arthritic conditions. These can help to study the normal immune responses and inflammatory sequences during the pathogenesis of rheumatoid arthritis and osteoarthritis or may serve as a vehicle to test anti-arthritic drugs (known or unknown) in experimental animal models.

9.5.3 Bio-Distribution of Gold Nanoparticle in the Treatment of Arthritis

Bio-distribution of gold nanoparticle in experimental animals is mediated by mononuclear phagocyte system or reticuloendothelial system. It is dependent on the size and capping agent of nanoparticles. Gold nanoparticles have a high tendency to accumulate in ht cells of the innate immune system, making them advantageous to be used in the treatment of rheumatoid arthritis and osteoarthritis. They can target macrophages, neutrophils, and dendritic cells of the immune system, altering the immune response and ameliorating the inflammatory progression.

9.5.4 Retention and Clearance of Gold Nanoparticle After Treatment of Arthritis

The retention of gold nanoparticle in tissue level has been a great concern to the therapeutics, as they show the size-dependent clearance from the tissue level and from the body.

It has been found that surface modulation of gold nanoparticle with certain drugs can increase the retention time in the body, making the drug available for long time. Some drugs can increase the retention of nanoparticle in the targeted organ and decrease their accumulation in normal tissue, making them effective for targeted drug delivery and less harmful for normal organs. It has been observed that the clearance of gold nanoparticle after administration is governed by immune cells of macrophage family, as well as spleen. Macrophages can engulf larger nanoparticles more effectively than smaller ones. The spleen can eliminate gold nanoparticles of size over 200 nm effectively. The nanoparticles below the size of 10–15 nm can be cleared by the kidneys' normal filtration process.

9.6 Challenges to Take the Research Studies from Bench to Clinics

The promises and the hopes generated by nanoscience are really overwhelming. It shows a new generation of material and technology which can be applied to almost all domains of medical science. The biggest challenge still remains taking these research products from bench to clinic and to make them a part of our day-to-day life. Most of these anti-arthritic gold nanoparticle materials and strategies have to go miles before being implemented in practical dimensions. A major concern also lies in the fact about the safety and acceptability of them in our body. Issues such as toxicity, immune reactions, biocompatibility, bio-clearance, and teratogenic effects have to be kept in mind while thinking of using these products for therapeutic purpose. With so many types of nanomaterials existing, with varied physicochemical properties, it becomes difficult in screening the nanoparticles. The challenge becomes more because every day a new particle is discovered with a set of new properties. And the biggest problem is the interaction of nanomaterial with the living system which is very less deciphered and understood. Therefore, a complete understanding and thorough investigation of the nanoparticle of interest is highly recommended. From development of the drug to its clinical trials, each and every step is a crucial one to determine the drug's fate.

9.7 Conclusion

Every coin has two sides, and for a new technology to be impregnated in biomedical science, it takes a lot of cross-evaluations and cross-validations. This is also pertinent for the use of nanoparticles and nano-based approaches for treating arthritis. With so many promising drug clues, positive scientific outcomes, and researches on using nanoparticle-based arthritis management, it is sure nanoparticles are new hopes for the treatment option of this disease. Among the nano-based therapeutic options explored till date, it is clear that gold nanoparticle is one of the most prospective aspirants in this field. The gold nanoparticle-based anti-arthritic solutions can make the future scenario of arthritic research beam with new possibilities. If the factors such as toxicology, biocompatibility, adverse side effects, half-life period, release profile, and efficacy of the gold nanoparticles can be taken care of, then the day is not far when gold nanoparticles will be in mainstream medicine for treating arthritis. Considerations can be also given to the drug options which combines gold nanoparticles with other bioactive compounds which can act as multi action drugs for treating arthritis and management of its symptoms. This book chapter gives a very brief overview on arthritis and gold nanoparticles as anti-arthritic agents; however, there is room for other nanoparticles which can be of equal or better therapeutic potential. We need to explore more about them as well. At the same time, initiatives should also be taken to decipher more about these "magic wands" gold nanoparticles. We can definitely expect to have gold nanoparticles to be used as new-generation medicine targeting patient-specific requirements and conditions.

Further Reading

Boisselier, E., & Astruc, D. (2009). Gold nanoparticles in nanomedicine: Preparations, imaging, diagnostics, therapies and toxicity. *Chemical Society Reviews*, *38*(6), 1759–1782.

Brown, C. L., Bushell, G., Whitehouse, M. W., Agrawal, D. S., Tupe, S. G., Paknikar, K. M., & Tiekink, E. R. (2007). Nanogold-pharmaceutics (i) The use of colloidal gold to treat experimentally-induced arthritis in rat models; (ii) Characterization of the gold in Swarnabhasma, a microparticulate used in traditional Indian medicine. *Gold Bulletin*, *40*(3), 245–250.

Davis, M. E., & Shin, D. M. (2008). Nanoparticle therapeutics: An emerging treatment modality for cancer. *Nature Reviews Drug Discovery*, *7*(9), 771.

Dwivedi, P., Nayak, V., & Kowshik, M. (2015). Role of gold nanoparticles as drug delivery vehicles for chondroitin sulfate in the treatment of osteoarthritis. *Biotechnology Progress*, *31*(5), 1416–1422.

Herizchi, R., Abbasi, E., Milani, M., & Akbarzadeh, A. (2016). Current methods for synthesis of gold nanoparticles. *Artificial Cells, Nanomedicine, and Biotechnology*, *44*(2), 596–602.

Holzinger, M., Le Goff, A., & Cosnier, S. (2014). Nanomaterials for biosensing applications: A review. *Frontiers in Chemistry*, *2*, 63.

Leonavičienė, L., Kirdaitė, G., Bradūnaitė, R., Vaitkienė, D., Vasiliauskas, A., Zabulytė, D., ... & Mackiewicz, Z. (2012). Effect of gold nanoparticles in the treatment of established collagen arthritis in rats. *Medicina (Kaunas)*, *48*(2), 91–101.

Luo, X., Morrin, A., Killard, A. J., & Smyth, M. R. (2006). Application of nanoparticles in electrochemical sensors and biosensors. *Electroanalysis*, *18*(4), 319–326.

Mahmoudi, M., Sant, S., Wang, B., Laurent, S., & Sen, T. (2011). Superparamagnetic iron oxide nanoparticles (SPIONs): Development, surface modification and applications in chemotherapy. *Advanced Drug Delivery Reviews, 63*(1–2), 24–46.

Pascarelli, N. A., Moretti, E., Terzuoli, G., Lamboglia, A., Renieri, T., Fioravanti, A., & Collodel, G. (2013). Effects of gold and silver nanoparticles in cultured human osteoarthritic chondrocytes. *Journal of Applied Toxicology, 33*(12), 1506–1513.

Prosperi, D., Colombo, M., Zanoni, I., & Granucci, F. (2017). Drug nanocarriers to treat autoimmunity and chronic inflammatory diseases. *Seminars in Immunology*, 34, 61–67.

Sahoo, S. K., Misra, R., & Parveen, S. (2017). Nanoparticles: A boon to drug delivery, therapeutics, diagnostics and imaging. In L. P. Balogh (ed.) *Nanomedicine in Cancer* (pp. 73–124). Pan Stanford, Singapore.

Sarkar, P. K., & Chaudhary, A. K. (2010). Ayurvedic Bhasma: The most ancient application of nanomedicine. *Journal of Scientific and Industrial Research, 69*, 901–905.

Shubayev, V. I., Pisanic II, T. R., & Jin, S. (2009). Magnetic nanoparticles for theragnostics. *Advanced Drug Delivery Reviews, 61*(6), 467–477.

Thanh, N. T., & Green, L. A. (2010). Functionalisation of nanoparticles for biomedical applications. *Nano Today, 5*(3), 213–230.

Wang, Q., & Sun, X. (2017). Recent advances in nanomedicines for the treatment of rheumatoid arthritis. *Biomaterials Science, 5*, 1407–1420.

Yang, M., Feng, X., Ding, J., Chang, F., & Chen, X. (2017). Nanotherapeutics relieve rheumatoid arthritis. *Journal of Controlled Release, 252*, 108–124.

Yih, T. C., & Al-Fandi, M. (2006). Engineered nanoparticles as precise drug delivery systems. *Journal of Cellular Biochemistry, 97*(6), 1184–1190.

10

Nanovations in Neuromedicine for Shaping a Better Future

Jyotirekha Das and
G. K. Rajanikant
National Institute of Technology Calicut

10.1 Introduction

Neurological disorders are an important cause of disability and death worldwide, and their burden has increased substantially over the past 25 years (GBD 2015). However, the complexity of the brain and associated disorders, as well as restricted access of biologics to the central nervous system (CNS) owing to the presence of blood–brain barrier (BBB) and the blood–cerebrospinal fluid barrier (BCSFB), has limited successful development of safe and effective diagnostics and therapeutics for the identification and treatment of various neurological diseases. The quest for exploring the neurological disease mechanism and fostering theranostics equipped for effective diagnosis and treatment has prompted the use of nanotechnology in neuroscience and has paved the way for a new interdisciplinary field, nanoneuromedicine that deals with the application of nanotechnology in neuroscience. This chapter is an attempt to recognize the progress made in understanding the hurdle posed by the complex physiology of the nervous system and novelties in the design of advanced nanomaterials to overcome the difficulties.

10.2 Answers to the Challenges

10.2.1 Blood–Brain Barrier (BBB)

The BBB is the largest interface between blood and CNS and is responsible for maintaining the brain homeostasis. It is a highly selective physical barrier that controls the molecular traffic through the brain and the systemic endothelium circulation. Bio-inspired surface modification to regulate the specific composition of the outer corona of nanocarriers allows them to permeate through the BBB (mainly by endocytosis mediated by the receptor) and deliver their payload into the brain tissue (Figure 10.1). The potentials of nanoparticle to penetrate the BBB have been utilized for developing targeted therapies against a range of neuropathological conditions starting from brain tumors (its delineation and enhanced radio-sensitization) to neurodegenerative diseases such as Parkinson's disease (PD), Huntington's disease, Alzheimer's disease (AD), and cerebral palsy. Despite the development of nano-based platforms for CNS, therapeutics is a recent phenomenon; nano-sized viral pathogens have successfully crossed the BBB and

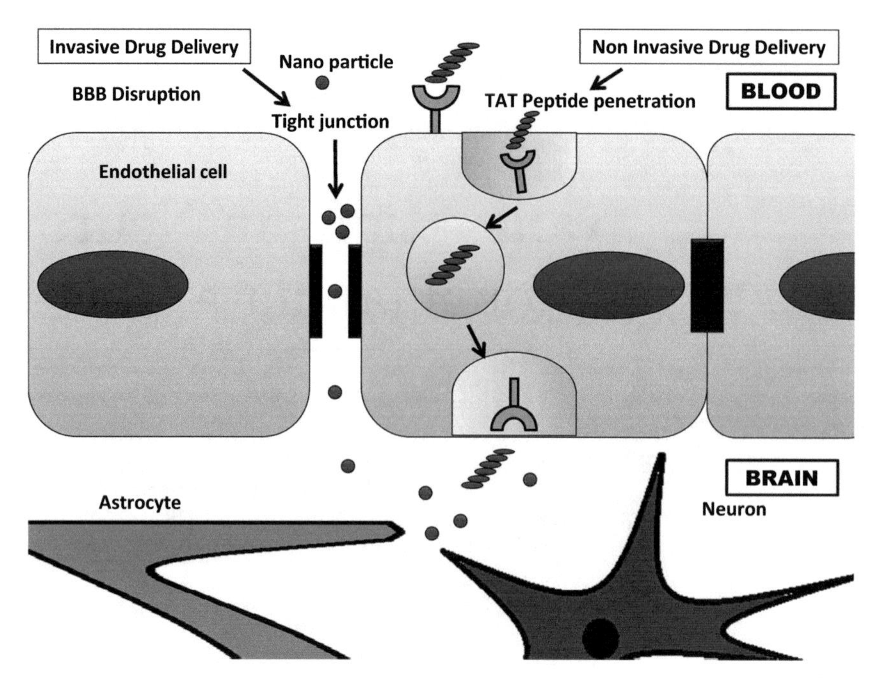

FIGURE 10.1 Schematic diagram illustrating the modes of transportation of nanoparticles across the BBB.

have been located in the brain tissue for ages. For example, particles of human immunodeficiency virus (HIV) with an average size of 120 nm are internalized across the BBB by adsorptive transcytosis (Banks et al. 2001, Mishra et al. 2011). The stealth mechanism applied by viruses such as HIV to cross the BBB has inspired the scientists for developing functionalized nano-vehicles with cell-penetrating transactivator of transcription (TAT) peptides. These have been successfully used to deliver ritonavir drug across the BBB in brain parenchyma (Rao et al. 2008). The trans-endothelial leukocytes migration caused by increased permeability and brain endothelial hypertrophy owing to cytokines secretions such as interleukins, TNF-α (tumor necrosis factor), and interferon-γ is another biological phenomenon extensively exploited in various nanoparticles-based delivery strategies (Persidsky et al. 1997, Williams et al. 2012).

10.2.2 Blood–Cerebrospinal Fluid Barrier (BCSFB)

The other barrier system within the CNS is the BCSFB formed by the choroid plexus (CP). CP forms the principal interface between the cerebrospinal fluid (CSF) and the systemic circulation. It consists of fenestrated capillaries linked to the adjacent CP cells by tight junctions further limiting the paracellular mode of hydrophilic substance diffusion (Engelhardt and Sorokin 2009). However, the trans-endothelial electrical resistance (TEER) value of CBSF is significantly lower than BBB suggesting a higher degree of permeability as compared to BBB. Large substances like peptides have higher efficiency in crossing the BCSF as compared to BBB by paracellular diffusion owing to its limited tight junctions and reduced exocytosis (Johanson et al. 2011). Additionally, CP also produces CSF secreted into the lateral, third, and fourth ventricles playing a major role in maintaining the CSF homeostasis (Abbott et al. 2006). The expression of a variety of transporters aids the CP barrier and secretory functions to allow detailed ion regulation and nutrient content of CSF, as well as waste products removal, and limit the entry of potential neurotoxic compounds (Engelhardt and Sorokin 2009). Several carriers were identified by using quantitative gene analysis *in vivo* biotinylation, Western blot analysis, and immunohistochemistry in the CP. These transporters include peptide transporters (PEPT2) (Smith et al. 2004), organic anion transporters (Oat 3, Oat 2) (Tachikawa et al. 2012, Choudhuri et al. 2003), organic anion polypeptide transporters (Oatp 1a1, Oatp 1c4, Oatp1a6, Oatp2a1, Oatp4a1) (Johanson et al. 2011), organic cation transporters (Oct 3), monocarboxylate transporters (MCT 3) (Philp et al. 2001), multidrug resistance proteins (Mrp 1, Mrp4) (Roberts et al. 2008), and amino acid transporters (Lat1) (Roberts et al. 2008). Additionally, BCRP (Halwachs et al. 2011), P-gp (Rao et al. 1999), and nucleoside transporters (Xia and Wang 2007) are expressed at the BCSF barrier

10.3 Medical Application of Nanotech in CNS

10.3.1 Nanotech-Based Approach for the Diagnosis of CNS

Diagnosis and imaging techniques have also become more precise and steadfast with the application of nanoparticles (NPs). NPs have helped in the advancement of new diagnostic and imaging techniques for brain function evaluation and CNS disease analysis. Brain imaging nanoparticles can be classified into two major divisions:

nanosized imaging agents for, e.g., superparamagnetic iron oxide (SPIO) (Kremer et al. 2007, Manninger et al. 2005, Neuwelt et al. 2007, Bourrinet et al. 2006) and trimetallic nitride endohedral metallofullerene nanoparticles (Fatouros et al. 2006) and carriers functionalized with gadolinium as the imaging agent. There is growing attention regarding the multiple functioning of NPs for therapy, imaging, and diagnosis (Figure 10.2). However, dendrimers and quantum dots (QDs) have been the most investigated NPs for brain imaging and diagnosis with their elevated structural adaptability and great carrying capacity (Tables 10.1 and 10.2).

Quantum Dots

QDs have emerged as a revolutionary imaging technology with high brightness, long-term photostability, and size-tunable narrow emission spectra. It is composed of a metalloid crystalline core and a shell that shields the core. Monoclonal antibodies (MAbs) (mouse MAb 528, H-11, and H199.12) coupled with QDs have been used to map human glioblastoma multiforme with fluorescence microscopy (Arndt-Jovin et al. 2009). In this study, different glioblastoma cell lines were tested with and without QD-conjugated EGF. They have found that QD-conjugated EGF was easily taken up by the cells in contrast to EGF alone. However, the level of their uptake varied consistently with the wild-type and mutant EGFR expression in the cell lines. The orthotopic glioblastomas demonstrated in this study could also reveal uptake QD-EFG after transplantation into mouse brains (Arndt-Jovin et al. 2009). The feasibility of anti-EGFR conjugated QDs in intraoperative diagnosis is confirmed by another study (Wang et al. 2007).

Dendritic Nanoparticles

Dendrimers have been extensively studied for drug delivery since its formulation. Currently, dendrimer-based brain imaging and diagnosis are gaining considerable attention. In this regard, hydrophilic dendritic manganese has been formulated from diethylenetriamine penta-acetic acid (Bertin et al. 2009). This complex increases hydrophilicity and subsequently helps build the adequacy of the contrast which is dependent on the paramagnetic species concentration. A contrast in the images following the intraperitoneal injection (i.p.), in both the longitudinal and transverse relaxation time group, was observed (Bertin et al. 2009). This complex can be used to probe neurological disorders as the uptake of Mn(II) by a neuron reveals its physiological activity and function. Despite the relative instability in the manganese complex *in vivo*, this can be used for the diagnosis of neurological disorders such as AD and PD.

Iron Oxide Nanoparticles

Magnetic iron oxide nanoparticles have been widely explored for different biomedical applications including drug targeting and imaging tools. A multimodal

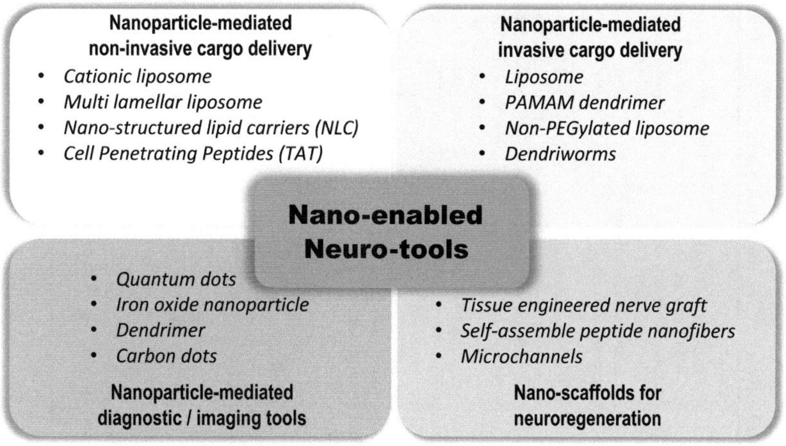

FIGURE 10.2 Diagnostic and therapeutic applications of nano-enabled neuro-tools.

TABLE 10.1 Nanoparticles Used for the Diagnosis of Various Neurological Disorders

Type of Nanoparticles/Nanomaterials	Type of Study	Use	Model	Reference
MnO nanoparticles	Preclinical	Imaging of brain gliomas	Glioma mice model	Sintov et al. (2016)
Iron oxide nanoparticles	Preclinical	MRI contrast agents to brain tumor and antiretroviral drug delivery	Rat tumor model	
Ultrasmall superparamagnetic iron oxide nanoparticles (ferumoxytol and ferumoxtran-10)	Clinical	Brain tumor imaging	Rat tumor model	
Gold nanoparticles	Preclinical	Prion disease therapy and brain tumor image	Mice model	
Gadolinium metallofullerene nanoparticle	Preclinical	Brain tumor drug delivery and longitudinal imaging	Mice glioma model	
Ultrasmall gadolinium oxide nanoparticles	Preclinical	Image brain cancer cells *in vivo* with MRI	Glioma cell line GL-261	

TABLE 10.2 Nanomaterials Used for the Delivery of Therapeutic Cargo to the Diseased Brain

Nanomaterials	Description	Advantage	Application	Inference	References
Nanocapsules and nanospheres	Nanocapsules represent a nanoparticulate system wherein an oil-filled cavity is enclosed by a thin polymeric envelope, whereas nanospheres are fabricated by microemulsion polymerization technique constituting a solid core with dense polymeric matrix	They offer higher drug stability and protection against systemic degradation, and also allow surface modification, evading reticuloendothelial system (RES) recognition.	Delivery of indomethacin Thioflavin-T Clioquinol d-Penicillamine	Neuroprotective Diagnostic Inhibition of Aβ aggregation Dissolution of Aβ	Kanwar et al. (2012)
Nanogels	These are prepared by the emulsification solvent evaporation technique, wherein the ionic and nonionic polymers combine to form a network of cross-linked polymers.	They swell in the presence of water and are often used to deliver oligonucleotides, small interfering RNA (siRNA), and DNA with an encapsulation efficiency of 40%–60%	Delivery of Oligonucleotides Cholesterol-bearing Pullulan	Promising for oligonucleotide delivery Inhibition of Aβ aggregation	Kanwar et al. (2012)
Carbon nanotubes and nanofibers	Carbon nanotubes, an attractive mode of delivery with unique structural and electrical properties, are classified as single-walled carbon nanotubes and multiwalled carbon nanotubes, with the diameters ranging from 1 nm and 10–100 nm, respectively	The application of carbon nanotubes for stimulation of the CNS has gained momentum in *in vitro* neuronal circuit models, widening the scope and understanding of CNS electrophysiology nanofibers are safe in terms of preparation, entail less risk of pollution, and are unique for designing neural prosthetics	Deliver of Nerve growth factor Biosensor Acetylcholine	Neuronal outgrowth Therapeutic Therapeutic	
Nano-micelles	They represent a core–shell structural design with the core being hydrophobic and the shell represented by a hydrophilic polymer block	The core encapsulates 20%–30% of hydrophobic drugs and prevents unwanted drug release and loss, whereas the shell offers stability and non-selective interactions	Delivery of phospholipids	Neuroprotective with Aβ aggregation	
Nano-liposomes	These are the vesicular structures consisting of aqueous compartments internally surrounded by either uni- or multilamellar lipid bilayers, thus facilitating the encapsulation of increased drug content	They also mask the drugs from RES recognition and subsequent removal	Delivery of curcumin	Diagnostic and therapeutic for AD	

fluorescent-magnetic iron oxide nanoparticle tagged with rhodamine or congo red was able to detect Aβ40 fibrils, the main constituent of amyloid plaques in the AD (Skaat and Margel 2009). This approach promises an early diagnosis of the AD. During multiple sclerosis (MS), macrophage invasion occurs at an early stage. To study the pathophysiology of MS tracking, this macrophage invasion is essential. A group of workers has developed very small superparamagnetic iron oxide particles (VSOP) to detect the infiltration. These particles were able to detect the subtle infiltration of macrophages in murine experimental autoimmune encephalomyelitis (EAE), an animal model of MS (Tysiak et al. 2009).

Carbon Dots

The use of carbon dots (CDs) for imaging is advantageous over other nanoparticles due to its unique fluorescent properties with good biocompatibility and absence of optical blinking and photobleaching. CDs with glycine as a precursor were prepared with a size of 3–4 nm which can have the strongest emission of 600 nm, making it suitable for *in vivo* tumor imaging (Ruan et al. 2014a). Owing to the enhanced permeability and retention effect, these CDs could accumulate in the brain tumor site. To enhance the imaging effect, CDs were conjugated with angiopep-2 which showed higher uptake by C6 glioma cells when compared with unconjugated CDs (Ruan et al. 2014b). Additionally, it improved brain tumor targeting effect and exhibited a twofold higher intensity in brain tumor than the normal CDs. The glioma-to-normal brain ratio of the angiopep-2-modified CDs displayed considerably higher value when compared with unmodified CDs. However, additional modification with a ligand might enhance the brain tumor targeting efficiency. In this context, CDs prepared using glucose and 1-aspartic acid (CD-Asp) as precursors showed considerable C6 glioma cell selectivity when compared to CDs prepared with other precursors such as glucose alone (CD-G), glucose and glutamic acid (CD-Glu), and 1-aspartic acid (CD-A) (Zheng et al. 2015). The cellular uptake and G/N ratio of CD-Asp are much higher than any other CDs (1.42, 0.91, 0.86, and 0.88 for CD-Asp, CD-G, CD-A, and CD-Glu, respectively). The presence of glucose and aspartic acid on the surface of the CDs may contribute to this high uptake level as both glucose transporter and ASCT2 are highly expressed on the BBB and/or brain tumor cells (Luciani et al. 2004, McAllister et al. 2001).

10.3.2 Nanocarriers and CNS-Targeted Delivery

Invasive Methods for CNS Delivery

Disruption of the BBB

Ample amount of research has been made since the past years using several modifications to open up the BBB momentarily. Hyperosmotic solutions like mannitol or urea were used to momentarily open up the tight junctions enabling the drugs to cross the BBB. Additionally, biochemical disruption of the BBB was also studied using several vasoactive agents such as histamine and bradykinin to increase the permeability (Tajes et al. 2014). Despite the success of these strategies in animal models, it also holds certain drawbacks. The impact of the drawbacks of neuronal damage, microembolism, and alteration in glucose uptake is of greater magnitude. Moreover, this approach directly impedes the defense mechanism of the brain allowing the entry of the chemicals and toxins (Garbayo et al. 2014, Tajes et al. 2014).

Direct Implantation to the Brain

To avoid the entry of toxins across the BBB, the simple method for the delivery of drug is local implantation to the brain. However, the drug diffusion efficiency depends on several other factors such as the location of the administration, polarity, tissue affinity, and molecular mass.

Stereotaxy can be a way of drug delivery in specific brain areas without damaging the surrounding, but the dosage regulation after implantation is a major drawback (Garbayo et al. 2014, Riley and Lozano 2007). However, intrathecal or intraventricular delivery could be promising theoretically, in which case the drug is delivered to the CSF bypassing the BBB. However, this administration hasn't gained successful experimental results, and it is also accompanied with several risks and need of infusion due to CSF replacement (Blasberg et al. 1975, Patel et al. 2009).

Non-invasive Methods for CNS Delivery

Nasal Delivery

Nasal delivery is one of the most primitive non-invasive ways of drug delivery. It transports the drug directly to the brain via the olfactory and trigeminal nerve pathway enabling the drug to bypass the BBB (Djupesland et al. 2014). The nanoparticles follow the transcellular pathway for nasal drug delivery (Ambikanandan Misra et al. 2012). This drug transportation could be enhanced with the addition of a mucoadhesive polymer like chitosan into the nasal formulation which increases the mucosal contact time (Wen 2011). With this reference, Nano-structured lipid carriers (NLC) coated with chitosan were developed for intranasal administration. This novel approach showed safe and effective drug delivery to the brain while maintaining the dosage concentration (Gartziandia et al. 2015). Following this, many other mucoadhesive polymers were studied, and amidst others, lectins were found to be promising bioadhesive delivery system to improve drug absorption on nasal mucosa (Clark et al. 2000).

Cell-Penetrating Peptides (CPPs)

Cell-penetrating peptides (CPPs) have been explored widely and gained a lot of attention in recent years. These are short amphipathic and cationic that rapidly internalizes across the cell membrane unlike most other peptides (Zou et al. 2013). The first CPP was derived from the human immunodeficiency virus (HIV-1), described as the TAT commonly used to improve the drug access to the brain (Kanazawa et al. 2013). Another study has made an attempt to modify this TAT with suitable NLC to increase the rate of transport to the brain (Gartziandia et al. 2016). Further, studies have confirmed maximum therapeutic effect with minimal systemic side effects on the modification of CPPs with nanomaterials (Kanazawa et al. 2013, Qin et al. 2011).

Drug Delivery Systems (DDSs)

As mentioned above, drug delivery systems (DDSs) were raised to reach the brain as a solution. They have some advantages in overcoming certain drug limitations such as lack of selectivity, poor solubility, and multidrug resistance development (Garbayo et al. 2014). There are currently several types of DDS available for biomedical applications with appropriate physicochemical characteristics to cross the BBB or to improve brain access. They have positive characteristics such as high chemical and biological stability, and the ability to incorporate both hydrophilic and hydrophobic molecules, and can also be administered through a variety of routes (Re et al. 2012).

10.4 Neuroregeneration

Neuroregeneration was first revealed by Dr. Robert Sperry while performing an experiment on a frog's split-brain research. This experiment proved the neuronal regeneration and rewiring neural circuits which bestowed Dr. Sperry with a Nobel Prize in 1981. Unlike humans, many other animals such as goldfish, lampreys, and salamanders possess this regeneration property. However, in humans, the neural stem cells obtained from specific brain areas (dentate gyrus, olfactory bulb, and cerebellum) and the spinal have shown regeneration property *in vitro*. The neurons actively generated from the sub-ventricular region migrated to the olfactory bulb to form interneurons. Conversely, those generated in the dentate gyrus add to memory formation in the hippocampus (Ming and Song 2005). Axon regeneration is another phenomenon which occurs upon axonal damage. It reconnects the neurons to restore functional loss following the damage; however, it is limited to native CNS's inhibitory signals (Sandvig et al. 2004). These inhibitory signals are mainly generated by the glial cells (mostly astrocytes) of the CNS. The astrocytes inhibit neuronal regeneration by forming an impenetrable barrier accompanied with the cellular debris and myelin. Conversely, the Schwann cells play a major role in nerve regeneration by phagocytic infiltration and clearing the debris created by their CNS counterparts. This is aided by the physiological access to peripheral nerves involved the axons sprouting from a nearby node of Ranvier. These axons sprout extends to form new axons under the guidance of neurotrophins secreted by the Schwann cells and the extracellular matrix surrounding the axons prior to damage. This regeneration occurs at the rate of 2–5 mm/day, thereby restricting the natural healing if the gap exceeds 6 mm which will require surgical interventions (Schmidt and Leach 2003).

10.4.1 Nanostructures Facilitating Neuroregeneration

The field of regenerative neuroscience has been revolutionized upon the intervention of nanotechnology. The functional recovery after axonal regrowth is poor particularly in case of large spaces in between the nerve stumps. In these cases, the ends are linked by the autologous nerve grafts. However, the major disadvantage of these grafts is the requisite surgical intervention on the second nerve to harvest the necessary tissue. The development of tissue-engineered nerve grafts (TENGs) holds the solution to this. TENGs, for example, NeuraGenTM (Integra Life-Sciences) and NeuroMendTM (Stryker Orthopaedics) are hollow tubes that can connect two nerve stumps and currently available for clinical use. These collagen-based nerve conduits restrict foreign cell types to access the nerve while guiding the regenerating axons and the Schwann cells across the lesion. To actively facilitate the axonal regeneration, TENGs are modified with the combination of physical and biochemical cues frequently represented as nanofibers. The advantages of nanofibers for neural regeneration are as follows: tailoring them to meet specific requirements of a particular lesion type, fabricating with a variety of biocompatible and degradable substances, and also their large surface area boosts the biochemical cues presentation.

Electrospinning is a common method for the fabrication of nanofibers. Hence, the fibers formed have a wide range of dimensions that can be regulated easily through polymer solution variations, strength of the electric field, and the use of field pattern in fabrication process. Most commonly, synthetic materials are used for the fabrication of electrospun nanofibers for nerve regeneration. However, the nanofiber scale possesses immense importance as a regenerative scaffold. A natural material like chitosans was also used for the fabrication of nanofiber scaffolds; however, an exhibition of better biocompatibility limits their use as nerve grafts (Wang et al. 2008).

Self-assembly is the major alternative of electrospinning which comprises the design of mainly peptide molecules, assembling into non-covalent bonding and polymeric structures directly in their target areas (Zhang 2003). Self-assembled peptide nanofibers (SAPNs) have a close association with the natural environment of the extracellular matrix as it is characterized by the small diameter of 5–10 nm. In addition to that SAPNs are non-toxic as it is made out of self-assembling peptides (sapeptides). These sapeptides are also modified to support adhesion and facilitate neurite outgrowth; furthermore, the functionalization of motifs also facilitates neural stem cell proliferation to a greater degree (Gelain et al. 2006).

Nanostructures exploited in neuronal regeneration area are not restricted to nanofibers alone. Microchannels with a diameter similar to that of individual axons are produced by membrane buckling on to nano-thin silicon membranes (Cavallo et al. 2014). These channels can guide neurite outgrowth and also can be engineered to facilitate network formation. One of these approaches is the combination of nanofiber scaffolds with the stem cell therapy. Neural stem cells (NSCs) in combination with scaffolds led to superior axonal regeneration and a functional recovery as compared to NSC and scaffold alone (Teng et al. 2002). Certain nanomaterials are well suited for the combination of neural scaffolding with electrical stimulation. Electrical stimulation has revealed not only the extension of neurites from the cells but also increase in neurite length and number.

10.5 Neural Stimulation

Neural stimulation is a fundamental technique used to restore neural functions and disordered neural circuits in neurological disorders. It has multiple applications such as restoring auditory, bladder, visual, and limb functions and, moreover, treatment of PD, dystonia, tremor, epilepsy, obsessive-compulsive disorder, and depression (Cogan 2008). The conventional electrical stimulation has several limitations as the biocompatibility of the implanted electrodes as well as the surgery-induced trauma. In contrast, the non-invasive electrical stimulation undergoes even poorer resolution with the high power which can cause complications to the intermediate tissues. These limitations are addressed by using light, magnetic fields, or ultrasound to directly stimulate neurons in the contactless way (Wells et al. 2005). However, these techniques are constrained by poor spatial resolution which is highly depended on the ultrasound frequency. Conversely, for the stimulation of deep neural tissue like in the brain requires low ultrasound frequency for deep tissue penetration which again leads to low spatial resolution (Menz et al. 2013). High spatial resolution neural stimulation is required for the clinical diagnosis and treatment of neurological diseases and neuroscience research (Menz et al. 2013). Neural stimulation occurs through two mechanisms proposed: the thermal effect on the cell membrane changes the membrane capacity and/or activates temperature-gated ion channels of the transient receptor potential vanilloid (TRPV) channels (Paviolo et al. 2014). To achieve this, the unique properties of the nanomaterials are to be explored, and surface modification can be done.

10.6 Distribution of Nanomaterials in the Brain

The knowledge of the biological fate of the nanoparticles inside the brain is vital to maximize their effect on therapy and minimize their adverse effects. The pharmacokinetic and toxicity profiles of the nanomaterials are entirely dependent on their biological distribution. If the nanoparticle is trapped within the carriers for the sufficiently long time it can alter their toxicity profile. There are different factors affecting the distribution of the nanoparticles in the brain including pharmaceutical factors such as particle size, shape, surface charge, administration routes and targeting ligands, and biological factors such as chronobiology and disease conditions.

10.6.1 Size

The particle size is a key parameter that decides the energy and impact of nanomaterials in the living being and furthermore their entrance of the cerebrum tissue. Most commonly, bigger particles over 2 μm are caught by pulmonary capillary vessels. Smaller nanoparticles with a hydrodynamic distance across under 5.5 nm are quickly discharged by the kidney. In the advancement of therapeutics against CNS diseases, 20 nm, sufficiently expansive to escape renal glomerular filtration, is believed to be the adequately small size (Jo et al. 2015). In general, nanoparticles of smaller size can easily enter the brain compared to the bigger ones. In a study, it has been demonstrated that the NPs at 70nm delivered more drugs into the brain as compared to the NPs of 170, 220, and 345 nm (Gao and Jiang 2006). Likewise, in another study, NPs of similar size (85 nm) and surface properties showed the similar result in drug delivery (Nance et al. 2012).

10.6.2 Surface Charge

The fate of the nanoparticles is also depended on the surface charge for its administration and distribution into the cerebrum. The positively charged NPs can get easily internalized as compared to the neutral and negatively charged particles. For example, cationic bovine serum egg whites have been altered on the surface of nanoparticles and polymersome to upgrade their brain delivery. However, the non-specific interaction of the cationic surface-charged nanoparticles leads to its accumulation within the biological membranes causing unwanted side effects, thereby limiting their applications.

10.6.3 Shape

The shape of the particle plays an important role as a carrier of therapeutic agents and affects their fate not only on the circulation but also in the cell. The studies revealed the advantages of the rod-shaped nanoparticles to exhibit higher specific and lower non-specific accumulation underflow at the target brain endothelium when compared with its spherical counterparts (Kolhar et al. 2013). In another study, it has been found that the biconcave geometry of iron oxide (Fe_3O_4) showed higher penetration and localized at the nuclei; however, the nanotube geometry was restricted mostly in the cytoplasm (Chaturbedy et al. 2015).

10.6.4 Targeting Ligands

Surface modification of the NPs with specific targeting molecules (e.g., ligand or antibody) is equally essential in addition to the particle size, surface, and shape. These particles can easily recognize their targets located in the brain capillary endothelium or inside the brain parenchyma which thereby plays an important role in modulating the brain distribution of NPs. This is one of the promising strategies to overcome the BBB and deliver the therapeutics to the target. For example, apolipoproteins E and B can specifically bind to LRP-1 that expressed on the BBB.

10.6.5 Administration Routes

The distribution of nanomaterials is definitely affected by its route of administration. This is the major challenge overcame with the advent of nanotechnology. Invasive approaches face major hurdles in the form of poor diffusion rate. The diffusion rate is exponentially dependent on the distance from the site of administration. Thereby, in the area with close proximity to the injection site, drug accumulation leads to neurotoxicity. The drugs crossing the BBB are easily accessible by the brain cells because of brain's high vascular density. Thereby, the intravenous administration of drugs generally gets widely distributed inside the brain.

10.6.6 Chronobiology

In addition to the other factors, chronobiology is another important factor influencing the uptake of the nanoparticles into the brain. For example, a circadian phase-dependent pain reaction was exhibited by the mice following the intravenous injection of dalargin-loaded polysorbate 80-coated poly(butylcyanoacrylate) (PBCA) nanoparticles denoting the transport activity of endo- and excitotoxic nanoparticle got enhanced during the rest time or suppressed during the active phase (Kreuter 2015).

10.6.7 Disease Conditions

Disease condition is the major factor in determining the distribution of NPs inside the brain. Various pathological diseases compromise the rigidity of the BBB leading to the enhancement of BBB and increase permeability for NPs. In case of a brain tumor, the BBB remains intact at the early stage, but in the later stage when the tumor size increases, it forms a blood–brain tumor barrier (BBTB). With further increase in the tumor size, the BBB and BBTB get impaired and endothelial gaps are formed in the microvessels of the tumors leading to an enhanced permeability of the nanoparticles accumulation inside the tumor tissue (Zhan and Lu 2012). In case of different neurological diseases such as AD, PD, stroke, ischemia, and brain tumors, the integrity of the BBB gets compromised, thus giving an advantage for drug delivery.

10.7 Elimination of Nanomaterials in Brain

Enormous researches have been made in the area of targeted drug delivery and increased the half-life of a drug. However, there is a dearth of research in the area of elimination of these nanoparticles from the body which is very crucial to study the efficacy and safety of the drug. There are proposed four routes through which these nanoparticles get eliminated from the brain. These routes are as follows: (i) metabolizing enzymes mediated extracellular degradation, (ii) neurons or glial cells mediated intracellular degradation, (iii) convective bulk flow mediated CSF circulation entry and then eliminated to the bloodstream or cervical lymphatics, and (iv) finally efflux from brain to blood. Nanoparticles like PEGylated hyaluronic acid nanocarrier undergo extracellular degradation by the metabolic enzymes like hyaluronidases and aminopeptidase present in the brain parenchyma (Choi et al. 2011). In addition, microglia-mediated intracellular degradation in the lysosome, glymphatic pathway, brain-to-blood efflux, and lymphatic system-mediated clearance of nanomaterials are also hypothesized to be involved in eliminating nanomaterials form the brain. Pharmacological factors such as matrix biodegradability and deformability, ligand targeting, surface charge, route of administration, and size along with biological factors like conscious state and disease condition affect the elimination of nanomaterials.

10.7.1 Deformability and Biodegradability of the Matrix

The particle stability in the biological system is clearly determined by its deformability and biodegradable property. This in turn also affects the elimination of the NPs in the brain. Biodegradable liposome, reconstituted lipoproteins, deformable micelle (Sonali et al. 2015), PBCA or poly(isohexyl cyanoacrylate) and PLA or its copolymer (lactide-*co*-glycolide) (PLGA) and HSA are the some of the nanoparticle candidates which can be easily eliminated from the brain by the process of deformation or degradation (Wohlfart et al. 2012). On the other hand, in the non-biodegradable particles such as silver NPs, gold NPs, and cerium oxide NPs, the elimination occurs via the CSF flow, or brain-to-blood efflux could be the major mechanism for elimination (Prades et al. 2012).

10.7.2 Size and Surface Charge

The size and surface charge are the key factors for the entry of a particle inside the brain. It is crucial to direct the particle distribution in the brain and also affect the elimination of the nanoparticles from the brain. In a previous research work, it has been validated that lower molecular weighted dextran could easily pass the CSF flow when compared with the high molecular weight (Iliff et al. 2012). Therefore, from this, it could be considered that CSF-mediated delivery is also size dependent.

10.7.3 Targeting Ligands

The ligands mediating brain delivery of nanomaterials can be focused likewise to influence their elimination from the CNS. Biomimetic nanocarrier made up of 1, 2-dimyristoyl-sn-glycero-3-phosphocholine and ApoA-I that targets the scavenger receptor class B type I receptor prompts lysosome escape and direct cytosolic medicate delivery (Zhang et al. 2009). In contrast, the clearance of ApoE-functionalized liposome targeting the low-density lipoprotein receptor was probably by intracellular lysosomal degradation (Bana et al. 2014).

10.7.4 Conscious State

Metabolic homeostasis is maintained and ensured by the critical function of the sleep; 60% of the increase in the interstitial space has been demonstrated to increase by natural sleep or anesthesia. This further resulted in the striking increase in the exchange of CSF and ISF while increasing the rate of Aβ clearance during the sleep (Xie et al. 2013). Thus, sleep and anesthesia must also have a major role in the clearance of nanomaterials from the brain.

10.7.5 Disease Conditions

Brain tumors and various other malignant tumors are known to possess the increased amount of different metabolic enzymes like hyaluronidases. Thereby, the PEG-conjugated hyaluronic acid nanoparticle degradation in the site of the tumor and its unique hyaluronic acid degradation pathway allows the use as a carrier for tumor targeting drug delivery.

10.7.6 Brian Regional Distribution

Membrane-bound pyroglutamate aminopeptidase distribution in the brain is not found to be homogenous in nature. There is a difference of tenfold between the two regions of the brain. The activity of this enzyme was observed the highest at the olfactory bulb, whereas in the cervical part of the spinal cord had the lowest effect. The differential distribution of this compound is steady with the hypothesis that it is in charge of extracellular degradation of neuroactive peptides (Vargas et al. 1987). Such site-specific circulation profile of metabolic enzyme in the brain can likewise initiate site-specific degradation of certain nanomaterials.

10.8 Personalized Medicine

Numerous research interventions have been made on neurological disorders; however, the underlying mechanisms and treatments are yet to be delineated. One of the reasons is that tissue samples from patient's brain are often difficult to obtain. In addition to that, there are many hurdles while establishing an animal model for neurological diseases including incapability of summarizing the pathophysiological characteristics of a specific disorder. The introduction of induced pluripotent stem cells (iPSCs) has emerged as a landmark in recent neurological research and drug development (Takahashi and Yamanaka 2006). The advantage of iPSCs is the ability to reserve epigenetic features of the individual patients (Kim et al. 2010). Since iPSCs are obtained directly from the patient, the treatment and diagnosis provide specific information on the pattern of mutations, duration, and severity of the disease of the patient. Thereby, this is a more reliable platform to study the disease progression and reaction to new drug treatments on a patient when compared with modeled disease phenotype (Mattis and Svendsen 2011).

Recent advances made in the iPSCs have successfully generated human iPSCs in somatic cells by forced expression of transcription factors provides new opportunities for regenerative medicines and *in vitro* disease modeling (Saha and Jaenisch 2009). iPSCs generated from diseased patient have been demonstrated to the model-specific pathogenesis of genetically inherited diseases, to facilitate cell replacement therapy, and also to screen candidate drugs. The application of nanoparticles in patient-specific iPSCs is yet to be studied. However, NPs can be considered as a strong weapon in reprogramming the iPSCs based on their nanoscaled structure and tremendously large surface area. A study has been carried out to show the efficiency of mesoporous silica nanoparticles (MSNs) to be efficiently internalized by iPSCs without causing cytotoxicity (Chen et al. 2013). Therefore, NPs can be used as an ideal vector for stem cell labeling gene delivery and potential drug carrier for inducing patient-specific differentiation and subsequent personalized techniques.

10.9 Nanotoxicity

Nanoparticles are being used extensively and in an enormous amount since the day of its invention. It is found in all forms integrated into clothing, plastic wares, food products, and electrical appliances (Becheri et al. 2008, Serpone et al. 2007). Their applications have touched all the fields of science starting from its use in medical imaging and diagnosis, pharmaceuticals, clinical therapy, and drug delivery to biomedical and healthcare fields (Bakry et al. 2007, Bianco et al. 2005, Gooding 2005, Katz and Willner 2004, Liu et al. 2009). The increased production of synthetic NPs has led to the increase in the level of exposure to the public and the environment which is expected to increase at a higher rate in the coming years (Table 10.3). To study the effect of these NPs on human health and environment, many types of research have been made since the past decades under a new field of science termed as nanotoxicology. This field focuses on the understanding of the properties held by these NPs and to modify or formulate these properties towards reducing the toxicity level of NPs.

The nervous system is a complex structure with unique functional characteristics, including different cell types with numerous functions. Neurotoxicology is an important part to study in neurosciences. Neurotoxicology is a field of science arose from the integration of pharmacology, experimental psychology, toxicology, psychopharmacology,

TABLE 10.3 Cellular Mechanisms of Neurotoxicity Induced by Various Nanoparticles

Nanoparticles	Description	Toxicity	References
Titanium oxide	Titanium oxide (TiO_2) is available in the form of nanocrystals or nanodots exhibiting magnetic properties	Oxidative stress Apoptosis and autophagy Immune mechanism Activated signaling pathways	Jiang and Gao (2017)
Iron oxide	Iron oxide nanoparticles (IONs) are of particular interest because of their unique intrinsic magnetic properties, so-called superparamagnetism. This feature, together with their high colloidal stability, makes them very attractive for a broad range of uses	Reactive oxygen species (ROS) production iron dysregulation is involved in the pathogenesis of several neurodegenerative diseases including PD and AD. Toxic to neuronal-type cells, this gives rise to concerns regarding the potential neurotoxic effects of exposure to IONs, due to their increasing usage in the treatment of CNS malignancies	
Silver NP	Silver nanoparticles (AgNPs) are increasingly used in various fields, including medical, food, healthcare, consumer, and industrial purposes, due to their unique physical and chemical properties. These include optical, electrical, thermal, and high electrical conductivity, and biological properties	Mitochondria, oxidative stress, inflammation, and cell death Interactions with cellular calcium and N-methyl-D-aspartate (NMDA) glutamate receptors AgNP-induced neurodegeneration	
Gold NP	Rapid emergence of gold nanoparticle (AuNP) technology, which holds great promise for future applications because of their large volume-specific surface areas and high diverse surface activities compared with bulk gold	Alterations in microglial activation. AuNP crossed BBB and accumulated in the neural tissues AuNP damaged cognition, size-dependent, and entered hippocampus Impaired brain GPx, generated 8-OHdG, caspase-3, and Hsp70, IFN-γ, DNA damage DNA damage in the cerebral cortex. Chronic exposure caused more damage Proinflammatory response influenced the integrity of the BBB	Jiang and Gao (2017)
Manganese NP	Manganese oxide nanoparticles (MnO_2 NPs) can be utilized for advanced materials in batteries, as well as other applications, such as water treatment and imaging contrast agents	Accumulation of manganese in the olfactory bulb, CNS, and other peripheral tissues, especially lung, induce a dose-dependent decrease in spatial learning, memory ability, and spontaneous activity; overt morphological changes in hippocampi; cellular degeneration and death; and a significant increase in Hsp70 mRNA levels of hippocampal cells in rats	Jiang and Gao (2017)
Silica NP	Silicon dioxide nanoparticles, also known as silica nanoparticles, are promising for biological applications owing to their excellent biocompatibility, low toxicity, thermal stability, facile synthetic route, and large-scale synthetic availability	Elevation of intracellular ROS and thiobarbituric acid reactive species (TBARS), as well as depletion of Glutathione (GSH), was seen in PC12 cells exposed to SiO2 NPs. Nano-SiO_2 exposure diminished the ability of neurite extension in response to nerve growth factor (NGF) in treated PC12 cells	Jiang and Gao (2017)
Carbon nanotube (CNT)	A CNT is a tube-shaped material, made of carbon, having a diameter measuring on the nanometer scale. A nanometer is one-billionth of a meter, or about 10,000 times smaller than a human hair. CNT is unique because the bonding between the atoms is very strong and the tubes can have extreme aspect ratios	Passing the olfactory neurons into the olfactory bulb will then induce an inflammatory response by activating microglial cells Alternation of hippocampal CA1 neuron excitability and consequently induced abnormal neuronal action potential with high firing frequency and a spiked half-width, which is considered as an early diagnostic phenomenon of neuronal dysfunctional diseases	

and the area of studying the structural and functional changes in the nervous system while undergoing chemical or other environmental influences (Barbosa et al. 2015). An expansive number of compounds have indicated ability in causing neurotoxicity, including industrial chemicals, metals, regular toxins, solvents, pesticides, and pharmaceutical medications. The nervous system is especially sensitive to toxic substances due to certain natural qualities, for example, dependence on aerobic metabolism for a steady oxygen supply, in the occurrence of axonal transport, or the events of neurotransmission. A neurotoxic impact can be the immediate change of the neuronal structure, or it can be the aftereffect of a course of impacts because of glial activation and glial-neuronal interactions. Additionally, a neurotoxic impact can be expressed promptly or years after the affront. Neurotoxicity can be permanent or reversible, and it can influence the entire nervous system or just parts of it. In a general point of view, neurotoxicity can be seen by neurotoxicants in pathways including causing neuropathy, hence harming the entire neuron, focusing on the axon, causing axonopathy, activating myelopathy, and influencing neurotransmission (Costa et al. 2008). Various chemicals may cause toxicity that leads to the loss of neurons (neuronopathy), either by necrosis or by apoptosis. Such neuronal loss is irreversible and may bring about a global encephalopathy; when just subpopulations of neurons are influenced, it brings about the loss of specific functions.

10.9.1 Mechanisms of Nanotoxicity

Nanotoxicity has been an integral possession of nanoparticles since the introduction of synthetic nanoparticles and the predominant mechanism underlying nanotoxicity is ROS. A small amount of oxygen (O_2) escapes during the process of ATP synthesis without reducing the coupled proton and electron transfer reaction completely which leads to the production of superoxide anion (O_2^-) radicals or other oxygen-containing radicals. Subsequently, ROS including superoxide anion (O_2^-), hydroxyl radicals ($^{\cdot}OH$), singlet oxygen ($^1O_2$), and hydrogen peroxide (H_2O_2) are produced as by-products of intracellular oxidative metabolism, which usually occurs in the mitochondria (Halliwell 1991, Turrens 2003). Overproduction of ROS induces oxygen species (OS) causes cellular dysfunction by the generation of protein radicals, DNA-strand breaks, lipid peroxidation, and also a modification to nucleic acids (Bagchi et al. 1995, Cabiscol et al. 2010, Griveau et al. 1995). The level of ROS generated depends on the physicochemical properties of the NPs including the size, surface area, and reactivity.

ROS production is seen related to many neurodegenerative diseases such as Huntington's disease, AD, and PD. As compared to other organs, the high O_2 consumption rate, relative paucity of antioxidant enzymes, and high content of easily peroxidizable unsaturated fatty acids make the brain to be highly vulnerable to OS damage. When the brain gets continuous exposure to the NPs, simultaneously it also gets exposed to the ROS production. Generally, microglia mediates the production of OS, which gets activated under the external stimuli (e.g., chemicals, xenobiotics). The neurotoxicity of NPs usually starts via ROS production following the release of OS, thus upregulating the mitogen-activated cytokines protein (MAP) kinases to activate MAP signaling pathway. This is followed by the high secretion of inflammation-related cytokines inside the brain leading to brain injury.

10.9.2 Reducing Exposure and Neurotoxicity

The study of nanoparticle related toxicity is still at its initial stage; thereby, the assays to determine their effects also remain at the developmental level. The advancement of Nanotechnological modification has led the entry of NPs through a number of routes, thus crossing the BBB. However, the interaction of NPs with the biological system especially at the nanophysical–biology interface remains a mystery. Therefore, the adoption of the clause precautionary principles has become the necessity under the Control of Substances Hazardous to Health Regulations (COSHH) when the toxicity level is unknown (Calliess and Stockhaus 2012). Additionally, it is essential to carry out an assessment of the work carried out with nanomaterials with risk management procedures applied (Health and Safety Executive 2005). The emerging use of NPs with evidence of the level of toxicity, there is an urgent need in defining the hazard profile and risk management strategies to incorporate a fine balance between over-regulation and maintaining a cutting edge (Doak et al. 2012).

There are no criteria to develop health surveillance strategies for the use of NPs as their remains debate to create appropriate dose matrices. On the basis of laboratory evidence, there is an increasing concern regarding the possible hazards posed due to the exposure of NPs. Therefore, there is an urgent need to develop a novel approach for early detection of NP-induced toxicity. To start with further research is necessary to understand the NP-mediated mechanisms of action to develop and validate early biomarkers for NP-induced effects.

10.10 Nanotech: Brain and Future Ethical Considerations

The social and ethical implication of nanotechnology is applied in the variety of areas including economics, health and environmental, technological and educational, moral, and philosophical pointed out by M. Roco (2003). These ethical issues are again subdivided into different areas based on the types of research. In some studies, it has been considered in the areas of equity, privacy, environment, security, and metaphysical questions concerning human–machine interaction (Mnyusiwalla et al. 2003). However, others consider it to be primarily privacy and control,

TABLE 10.4 Ethical Issues on the Use of Nanoparticles

Key Ethical Issues	General Description	Applications to the Brain	References
Health risk	Possible toxicity risk of nanoparticle exposure to human health	The ability of the nanoparticles to enter the BBB may potentially put human health at risk	Valerye (2013)
Environmental risk	Possible toxicity risk to the environmental destruction	Human elimination of the nanoparticles into the environment puts a potential risk to the environment	
Privacy risk	Privacy of an individual may be compromised upon implementation of nanotechnology	The use of neuroimaging and nano-enabled implants may lead the access to personal information	
Security risk	Implementation of nanotechnology in a security system may also lead to terrorism or warfare	It may improve biochemical terrorism or also the creation of super soldiers may lead to better warfare	
Catastrophe risk	Development of nanotechnology may also lead to catastrophic consequences	It may lead to widespread and irreversible damage to the human brain	
Identity	The individual identity might be compromised with the implementation of nanotechnology	The brain plays a major role in an individual's identity. Implementation of nanotechnology can potentially alter that identity	Valerye (2013)
Human enhancement	Nanotechnology can be used to enhance the human functions both physically and cognitively	The use of neuronal implantations and psychopharmaceuticals may alter or enhance brain functioning for enhancement of human beings	
Human–machine interfacing	Nanotechnology may integrate machine components in humans to the point of making trans-humans	It may improve the brain biotic/abiotic hybridization and integration	

longevity, and runaway nanobots (Moor and Weckert 2004); or issues of safety and environment, nanomedicine, individual privacy, and justice (Litton 2007); or even a much wider range of ethical issues, including environmental issues, workforce issues, privacy issues, national and international political issues, intellectual property issues, and human enhancement (Lewenstein 2005). These ethical issues are categorized into three main groups: social justice, safety, and transformation of humanity (Keiper 2007) Table 10.4. In contrary, a few other researchers divide them into four main groups: questions of human and personal identity, risk assessment, the justice of the distribution of risk and benefits, and the possibility of human and personal enhancement (Lenk and Biller-Adorno 2007).

10.11 Conclusion

In this chapter, we have made an attempt to study the uses of nanotechnology in clinical neurology and in basic neuroscience with a special attention on the effect of nanomaterials and their viability in bypassing the neurophysiological barriers. The advent of nanotechnology and its applications to address the neurodegenerative diseases seems to change the future of neuroscience. It is quite evident in this chapter that the nanotechnology-based solutions for therapy, diagnosis, and imaging have their firm roots in the natural phenomenon of transportation across the BBB. The nano-formulated particles such as HIV-TAT peptide, receptor-based translocation via a GLUT-1 receptor, activated leukocytes have been effective in the treatment of neurological disorders by offering brain target delivery and a sustained release profile. Additionally, it has also exhibited more specificity and sensibility than the current diagnostic criteria for an accurate and early diagnosis of the neurological disorders such as AD and PD. The fate of the nanoparticles inside the brain is yet to be explored completely. We have made an attempt to address this area and highlighted the importance of different pharmacological factors and biological factors determining the distribution and elimination of the nanoparticles in the brain. We have emphasized on the fact that for safe and efficient applications of nanotechnology in neurological disorders, more investigations on the fate of nanomaterial should be conducted.

However, the toxicological and ethical concerns towards the application of nanomaterials cannot be ignored. Extensive research interventions have been made in the area of neurotoxicity of nanomaterials. The corona of the nanoparticle and its biomolecular composition determines the toxicity of the nanoparticle *in vivo*. The neurotoxicity is also attributed to the size and the ROS produced by the nanoparticles. Therefore, extensive study is necessary on the toxicology and biocompatibility of the nanoparticles both *in vitro* and *in vivo* before its clinical use. Moreover, nanomaterials used in brain implants raise ethical concerns on capacity of the patient experiencing brain disease to provide an informed consent during the clinical trials. However, such limitations ought not to thwart innovations and development of nano-neurotechnology. Considering the safety and ethical issues, nanotechnology has a potential to help elucidate the intricate mechanisms involved in neuro-pathophysiology and can address the challenge of the human nervous system by developing highly efficient theranostic solutions.

References

Abbott, N.J., Ronnback, L., Hansson, E., 2006. Astrocyte-endothelial interactions at the blood-brain barrier. *Nat. Rev. Neurosci.* 7(1), 41–53.

Ambikanandan Misra, G.K., 2012. Drug delivery systems from nose to brain. *Curr. Pharm. Biotechnol.* 13(12), 2355–2379.

Arndt-Jovin, D.J., Kantelhardt, S.R., Caarls, W., de Vries, A.H., Giese, A., Jovin Ast, T.M., 2009. Tumor-targeted quantum dots can help surgeons find tumor boundaries. *IEEE Trans. Nanobiosci.* 8, 65–71.

Bagchi, D., Bagchi, M., Hassoun, E.A., Stohs, S.J., 1995. In vitro and in vivo generation of reactive oxygen species, DNA damage and lactate dehydrogenase leakage by selected pesticides. *Toxicology* 104, 129–140.

Bakry, R., Vallant, R.M., Najam-ul-Haq, M., Rainer, M., Szabo, Z., Huck, C.W., Bonn, G.K., 2007. Medicinal applications of fullerenes. *Int. J. Nanomed.* 2, 639.

Bana, L., Minniti, S., Salvati, E., Sesana, S., Zambelli, V., Cagnotto, A. et al., 2014. Liposomes bi-functionalized with phosphatidic acid and an ApoE-derived peptide affect Aβ aggregation features and cross the bloodbrain-barrier: Implications for therapy of Alzheimer disease. *Nanomedicine* 10, 1583–1590.

Banks, W.A., Freed, E.O., Wolf, K.M., Robinson, S.M., Franko, M., Kumar, V.B., 2001. Transport of human immunodeficiency virus type 1 pseudoviruses across the blood-brain barrier: Role of envelope proteins and adsorptive endocytosis. *J. Virol.* 75, 4681–4691.

Barbosa, D.J., Capela, J.P., de Lourdes Bastos, M., Carvalho, F., 2015. In vitro models for neurotoxicology research. *Toxicol. Res.* 4, 801–842.

Becheri, A., Durr, M., Nostro, P.L., Baglioni, P., 2008. Synthesis and characterization of zinc oxide nanoparticles: Application to textiles as UV-absorbers. *J. Nanopart. Res.* 10, 679–689.

Bertin, A., Steibel, J., Michou-Gallani, A.I., Gallani, J.L., Felder-Flesch, D., 2009. Development of a dendritic manganese-enhanced magnetic resonance imaging (MEMRI) contrast agent: Synthesis, toxicity (in vitro) and relaxivity (in vitro, in vivo) studies. *Bioconjug. Chem.* 20, 760–767.

Bianco, A., Kostarelos, K., Prato, M., 2005. Applications of carbon nanotubes in drug delivery. *Curr. Opin. Chem. Biol.* 9, 674–679.

Blasberg, R.G., Patlak, C., Fenstermacher, J.D., 1975. Intrathecal chemotherapy: Brain tissue profiles after ventriculocisternal perfusion. *J. Pharmacol. Exp. Ther.* 195(1), 73–83.

Bourrinet, P., Bengele, H.H., Bonnemain, B., Dencausse, A., Idee, J.M., Jacobs, P.M., Lewis, J.M., 2006. Preclinical safety and pharmacokinetic profile of ferumoxtran-10, an ultrasmall superparamagnetic iron oxide magnetic resonance contrast agent. *Invest. Radiol.* 41, 313–324.

Cabiscol, E., Tamarit, J., Ros, J., 2010. Oxidative stress in bacteria and protein damage by reactive oxygen species. *Int. Microbiol.* 3, 3–8.

Calliess, C., Stockhaus, H., 2012. Precautionary principle and nanomaterials: REACH revisited. *J. Eur. Environ. Plan. Law* 9, 113–135.

Cavallo, F., Huang, Y., Dent, E.W., Williams, J.C., Lagally, M.G., 2014. Neurite guidance and threedimensional confinement via compliant semiconductor scaffolds. *ACS Nano* 8, 12219–12227.

Chaturbedy, P., Kumar, M., Salikolimi, K., Das, S., Sinha, S.H., Chatterjee, S. et al., 2015. Shape-directed compartmentalized delivery of a nanoparticleconjugated small-molecule activator of an epigenetic enzyme in the brain. *J. Control. Release* 217, 151–159.

Chen, W., Tsai, P.H., Hung, Y., Chiou, S.H., Mou, C.Y., 2013. Nonviral cell labeling and differentiation agent for induced pluripotent stem cells based on mesoporous silica nanoparticles. *ACS Nano* 7, 8423–8440.

Choi, K.Y., Yoon, H.Y., Kim, J.H., Bae, S.M., Park, R.W., Kang, Y.M. et al., 2011. Smart nanocarrier based on PEGylated hyaluronic acid for cancer therapy. *ACS Nano* 5, 8591–8599.

Choudhuri, S., Cherrington, N.J., Li, N., Klaassen, C.D., 2003. Constitutive expression of various xenobiotic and endobiotic transporter mRNAs in the choroid plexus of rats. *Drug Metab. Dispos.* 31(11), 1337–1345.

Clark, M.A., Hirst, B.H., Jepson, M.A., 2000. Lectin-mediated mucosal delivery of drugs and microparticles. *Adv. Drug Deliv. Rev.* 43(2–3), 207–223.

Cogan, S.F., 2008. Neural stimulation and recording electrodes. *Annu. Rev. Biomed. Eng.* 10, 275–309.

Costa, L.G., Giordano, G., Guizzetti, M., Vitalone, A., 2008. Neurotoxicity of pesticides: A brief review. *Front. Biosci.* 13, 1240–1249.

Djupesland, P.G., Messina, J.C., Mahmoud, R.A., 2014. The nasal approach to delivering treatment for brain diseases: An anatomic, physiologic, and delivery technology overview. *Ther. Deliv.* 5(6), 709–733.

Doak, S., Manshian, B., Jenkins, G., Singh, N., 2012. In vitro genotoxicity testing strategy for nanomaterials and the adaptation of current OECD guidelines. *Mutat. Res.* 745, 104–111.

Engelhardt, B., Sorokin, L., 2009. The blood-brain and the blood-cerebrospinal fluid barriers: Function and dysfunction. *Semin. Immunopathol.* 31(4), 497–511.

Fatouros, P.P., Corwin, F.D., Chen, Z.J., Broaddus, W.C., Tatum, J.L., Kettenmann, B. et al., 2006. In vitro and in vivo imaging studies of a new endohedral metallofullerene nanoparticle. *Radiology* 240, 756–764.

GBD, 2015 Neurological disorders collaborator group. Global, regional, and national burden of neurological disorders during 1990-2015: A systematic analysis for the Global Burden of Disease Study 2015. *Lancet Neurol.* 16, 877–897.

Jo, D.H., Kim, J.H., Lee, T.G., Kim, J.H., 2015. Size, surface charge, and shape determine therapeutic effects of nanoparticles on brain and retinal diseases. *Nanomedicine* 11(7), 1603–1611.

Gao, K., Jiang, X., 2006. Influence of particle size on transport of methotrexate across blood–brain barrier by polysorbate 80-coated polybutylcyanoacrylate nanoparticles. *Int. J. Pharm.* 310, 213–219.

Garbayo, E., Estella-Hermoso de Mendoza, A., Blanco-Prieto, M.J., 2014. Diagnostic and therapeutic uses of nanomaterials in the brain. *Curr. Med. Chem.* 21(36), 4100–4131.

Gartziandia, O., Egusquiaguirre, S.P., Bianco, J., Pedraz, J.L., Igartua, M., Hernandez, R.M. et al., 2016.

Nanoparticle transport across in vitro olfactory cell monolayers. *Int. J. Pharm.* 499(1–2), 81–89.

Gartziandia, O., Herran, E., Pedraz, J.L., Carro, E., Igartua, M., Hernandez, R.M., 2015. Chitosan coated nanostructured lipid carriers for brain delivery of proteins by intranasal administration. *Colloids Surf. B* 134, 304–313.

Gelain, F., Bottai, D., Vescovi, A., Zhang, S., 2006. Designer self-assembling peptide nanofiber scaffolds for adult mouse neural stem cell 3-dimensional cultures. *PLoS One* 1, e119.

Gooding, J.J., 2005. Nanostructuring electrodes with carbon nanotubes: A review on electrochemistry and applications for sensing. *Electrochim. Acta* 50, 3049–3060.

Griveau, J.F., Dumont, E., Renard, P., Callegari, J.P., Le Lannou, D., 1995. Reactive oxygen species, lipid peroxidation and enzymatic defence systems in human spermatozoa. *J. Reprod. Fertil.* 103, 17–26.

Halliwell, B., 1991. Reactive oxygen species in living systems: Source, biochemistry, and role in human disease. *Am. J. Med.* 91, S14–S22.

Halwachs, S., Lakoma, C., Schafer, I., Seibel, P., Honscha, W., 2011. The antiepileptic drugs phenobarbital and carbamazepine reduce transport of methotrexate in rat choroid plexus by down-regulation of the reduced folate carrier. *Mol. Pharmacol.* 80(4), 621–629.

Health and Safety Executive, 2005. Control of substances hazardous to health: The control of substances hazardous to health regulations 2002 (as amended): Approved code of oractice and guidance. *HSE Books*, Sudbury.

Iliff, J.J., Wang, M., Liao, Y., Plogg, B.A., Peng, W., Gundersen, G.A. et al., 2012. A paravascular pathway facilitates CSF flow through the brain parenchyma and the clearance of interstitial solutes, including amyloid beta. *Sci. Transl. Med.* 4, 111r–147r.

Jiang, X., Gao, H., 2017. *Neurotoxicity of Nanomaterials and Nanomedicine.* Elsevier.

Johanson, C.E., Stopa, E.G., McMillan, P.N., 2011. The blood-cerebrospinal fluid barrier: Structure and functional significance. *Methods Mol. Biol.* 686, 101–131.

Kanazawa, T., Akiyama, F., Kakizaki, S., Takashima, Y., Seta, Y., 2013. Delivery of siRNA to the brain using a combination of nose-to-brain delivery and cell-penetrating peptide-modified nano-micelles. *Biomaterials* 34(36), 9220–9226.

Kanwar, J.R., Sun, X., Punj, V., Sriramoju, B., Mohan, R.R., Zhou, S.F. et al., 2012. Nanoparticles in the treatment and diagnosis of neurological disorders: Untamed dragon with fire power to heal. *Nanomedicine* 8, 399–414.

Katz, E., Willner, I., 2004. Biomolecule-functionalized carbon nanotubes: Applications in nanobioelectronics. *ChemPhysChem* 5, 1084–1104.

Keiper, A., 2007. Nanoethics as a discipline? *New Atlantis* 16, 55–67.

Kim, D.S., Lee, J.S., Leem, J.W., Huh, Y.J., Kim, J.Y., Kim, H.S. et al., 2010. Robust enhancement of neural differentiation from human ES and iPS cells regardless of their innate difference in differentiation propensity. *Stem Cell Rev. Rep.* 6, 270–281.

Kolhar, P., Anselmo, A.C., Gupta, V., Pant, K., Prabhakarpandian, B., Ruoslahti, E., Mitragotri, S., 2013. Using shape effects to target antibody-coated nanoparticles to lung and brain endothelium. *Proc. Natl. Acad. Sci. USA* 110, 10753–10758.

Kremer, S., Pinel, S., Vedrine, P.O., Bressenot, A., Robert, P., Bracard, S., Plenat, F., 2007. Ferumoxtran-10 enhancement in orthotopic xenograft models of human brain tumors: An indirect marker of tumor proliferation? *J. Neurooncol.* 83, 111–119.

Kreuter, J., 2015. Influence of chronobiology on the nanoparticle-mediated drug uptake into the brain. *Pharmaceutics* 7, 3–9.

Lenk, C., Biller-Adorno, N., 2007. Nanomedicine: Emerging or re-emerging ethical issues? A discussion of four ethical themes. *Med. Health Care Philos.* 10, 173–184.

Lewenstein, B.V., 2005. What counts as a 'social and ethical issue' in nanotechnology? *HYLE* 11, 5–18.

Litton, P., 2007. "Nanoethics"? What's new? *Hastings Center Rep.* 37, 22–25.

Liu, Z., Tabakman, S., Welsher, K., Dai, H., 2009. Carbon nanotubes in biology and medicine: In vitro and in vivo detection, imaging and drug delivery. *Nano Res.* 2, 85–120.

Luciani, A., Olivier, J.C., Clement, O., Siauve, N., Brillet, P.Y., Bessoud, B. et al., 2004. Glucose-receptor MR imaging of tumors: Study in mice with PEGylated paramagnetic niosomes. *Radiology* 231, 135–142.

Manninger, S.P., Muldoon, L.L., Nesbit, G., Murillo, T., Jacobs, P.M., Neuwelt, E.A., 2005. An exploratory study of ferumoxtran-10 nanoparticles as a blood-brain barrier imaging agent targeting phagocytic cells in CNS inflammatory lesions. *Am. J. Neuroradiol.* 26, 2290–2300.

Mattis, V.B., Svendsen, C.N., 2011. Induced pluripotent stem cells: A new revolution for clinical neurology? *Lancet Neurol.* 10, 383–394.

McAllister, M.S., Krizanac-Bengez, L., Macchia, F., Naftalin, R.J., Pedley, K.C., Mayberg, M.R. et al., 2001. Mechanisms of glucose transport at the blood–brain barrier: An in vitro study. *Brain Res.* 904, 20–30.

Menz, M.D., Oralkan, O., Khuri-Yakub, P.T., Baccus, S.A., 2013. Precise neural stimulation in the retina using focused ultrasound. *J. Neurosci.* 33, 4550–4560.

Ming, G., Song, H., 2005. Adult neurogenesis in the mammalian central nervous system. *Annu. Rev. Neurosci.* 28, 223–250.

Mishra, A., Lai, G.H., Schmidt, N.W., Sun, V.Z., Rodriguez, A.R., Tong, R. et al., 2011. Translocation of HIV TAT peptide and analogues induced by multiplexed membrane and cytoskeletal interactions. *Proc. Natl. Acad. Sci. USA* 108, 16883–16888.

Mnyusiwalla, A., Daar, A.S., Singer, P.A., 2003. Mind the gap: Science and ethics in nanotechnology. *Nanotechnology* 14: R9–R13.

Moor, J., Weckert, J., 2004. Nanoethics: Assessing the nanoscale from an ethical point of view. In: *Discovering the Nanoscale*, eds. D. Baird, A. Nordmann, and J. Schummer. Ios Press, Amsterdam, pp. 301–310.

Nance, E.A., Woodworth, G.F., Sailor, K.A., Shih, T.Y., Xu, Q., Swaminathan, G. et al., 2012. A dense poly(ethylene glycol) coating improves penetration of large polymeric nanoparticles within brain tissue. *Sci. Transl. Med.* 4, 119r–149r.

Neuwelt, E.A., Varallyay, C.G., Manninger, S., Solymosi, D., Haluska, M., Hunt, M.A. et al., 2007. The potential of ferumoxytol nanoparticle magnetic resonance imaging, perfusion, and angiography in central nervous system malignancy: A pilot study. *Neurosurgery* 60, 601–611.

Patel, M.M., Goyal, B.R., Bhadada, S.V., Bhatt, J.S., Amin, A.F., 2009. Getting into the brain: Approaches to enhance brain drug delivery. *CNS Drugs* 23(1), 35–58.

Paviolo, C., Thompson, A.C., Yong, J., Brown, W.G., Stoddart, P.R., 2014. Nanoparticle-enhanced infrared neural stimulation. *J. Neural Eng.* 11, 065002.

Persidsky, Y., Stins, M., Way, D., Witte, M.H., Weinand, M., Kim, K.S. et al., 1997. A model for monocyte migration through the blood-brain barrier during HIV-1 encephalitis. *J. Immunol.* 158, 3499–3510.

Philp, N.J., Yoon, H., Lombardi, L., 2001. Mouse MCT3 gene is expressed preferentially in retinal pigment and choroid plexus epithelia. *Am. J. Physiol. Cell Physiol.* 280(5), C1319–C1326.

Prades, R., Guerrero, S., Araya, E., Molina, C., Salas, E., Zurita, E. et al., 2012. Delivery of gold nanoparticles to the brain by conjugation with a peptide that recognizes the transferrin receptor. *Biomaterials* 33, 7194–7205.

Qin, Y., Chen, H., Zhang, Q., Wang, X., Yuan, W., Kuai, R. et al., 2011. Liposome formulated with TAT-modified cholesterol for improving brain delivery and therapeutic efficacy on brain glioma in animals. *Int. J. Pharm.* 420(2), 304–312.

Rao, K.S., Reddy, M.K., Horning, J.L., Labhasetwar, V., 2008. TAT-conjugated nanoparticles for the CNS delivery of anti-HIV drugs. *Biomaterials* 29, 4429–4438.

Rao, V.V., Dahlheimer, J.L., Bardgett, M.E., Snyder, A.Z., Finch, R.A., Sartorelli, A.C. et al., 1999 Choroid plexus epithelial expression of MDR1 P glycoprotein and multidrug resistance-associated protein contribute to the blood-cerebrospinal-fluid drugpermeability barrier. *Proc. Natl. Acad. Sci. USA* 96(7), 3900–3905.

Re, F., Gregori, M., Masserini, M., 2012. Nanotechnology for neurodegenerative disorders. *Maturitas* 73(1), 45–51.

Riley, D., Lozano, A., 2007. The fourth dimension of stereotaxis: Timing of neurosurgery for Parkinson disease. *Neurology* 68(4), 252–253.

Roberts, L.M., Black, D.S., Raman, C., Woodford, K., Zhou, M., Haggerty, J.E. et al., 2008. Subcellular localization of transporters along the rat blood-brain barrier and bloodcerebral-spinal fluid barrier by in vivo biotinylation. *Neuroscience* 155(2), 423–438.

Roco, M.C., 2003. Broader societal issues of nanotechnology. *J. Nanopart. Res.* 5, 181–189.

Ruan, S., Qian, J., Shen, S., Chen, J., Zhu, J., Jiang, X. et al., 2014a. Fluorescent carbonaceous nanodots for noninvasive glioma imaging after angiopep-2 decoration. *Bioconjug. Chem.* 25(12), 2252–2259.

Ruan, S., Qian, J., Shen, S., Zhu, J., Jiang, X., He, Q., Gao, H., 2014b. A simple one-step method to prepare fluorescent carbon dots and their potential application in non-invasive glioma imaging. *Nanoscale* 6(17), 10040–10047.

Saha, K., Jaenisch, R., 2009. Technical challenges in using human induced pluripotent stem cells to model disease. *Cell Stem Cell* 5, 584–595.

Sandvig, A., Berry, M., Barrett, L.B., Butt, A., Logan, A., 2004. Myelin-, reactive glia-, and scarderived CNS axon growth inhibitors: Expression, receptor signaling, and correlation with axon regeneration. *Glia* 46, 225–251.

Schmidt, C.E., Leach, J.B., 2003. Neural tissue engineering: strategies for repair and regeneration. *Annu. Rev. Biomed. Eng.* 5, 293–347.

Serpone, N., Dondi, D., Albini, A., 2007. Inorganic and organic UV filters: Their role and efficacy in sunscreens and suncare products. *Inorg. Chim. Acta* 360, 794–802.

Sintov, A.C., Velasco-Aguirre, C., Gallardo-Toledo, E., Araya. E., Kogan. M.J., 2016. Metal nanoparticles as targeted carriers circumventing the blood–brain barrier. *Int. Rev. Neurobiol.* 130, 199–227.

Skaat, H., Margel, S., 2009. Synthesis of fluorescent-maghemite nanoparticles as multimodal imaging agents for amyloid-beta fibrils detection and removal by a magnetic field. *Biochem. Biophys. Res. Commun.* 386, 645–649.

Smith, D.E., Johanson, C.E., Keep, R.F., 2004. Peptide and peptide analog transport systems at the blood-CSF barrier. *Adv. Drug Deliv. Rev.* 56(12), 1765–1791.

Tachikawa, M., Ozeki, G., Higuchi, T., Akanuma, S.I., Tsuji, K., Hosoya, K.I., 2012. Role of the bloodcerebrospinal fluid barrier transporter as a cerebral clearance system for prostaglandin E(2) produced in the brain. *J. Neurochem.* 343(3), 608–616.

Tajes, M., Ramos-Fernandez, E., Weng-Jiang, X., Bosch-Morató, M., Guivernau, B., Eraso-Pichot, A. et al., 2014. The blood-brain barrier: Structure, function and therapeutic approaches to cross it. *Mol. Membr. Biol.* 31(5), 152–167.

Takahashi, K., Yamanaka, S., 2006. Induction of pluripotent stem cells from mouse embryonic and adult fibroblast cultures by defined factors. *Cell* 126, 663–676.

Teng, Y.D., Lavik, E.B., Qu, X., Park, K.I., Ourednik, J., Zurakowski, D., Langer, R., Snyder, E.Y., 2002. Functional recovery following traumatic spinal cord injury mediated by a unique polymer scaffold seeded with neural stem cells. *Proc. Natl. Acad. Sci. USA* 99, 3024–3029.

Turrens, J.F., 2003. Mitochondrial formation of reactive oxygen species. *J. Physiol.* 552, 335–344.

Tysiak, E., Asbach, P., Aktas, O., Waiczies, H., Smyth, M., Schnorr, J., Taupitz, M., Wuerfel, J., 2009. Beyond blood brain barrier breakdown: In vivo detection of occult neuroinflammatory foci by magnetic nanoparticles in high field MRI. *J. Neuroinflammation.* 6, 20.

Valerye, M., 2013. *Nanotechnology, the Brain, and the Future: Ethical Considerations Nanotechnology*, vol 3. Springer, New York.

Vargas, M., Mendez, M., Cisneros, M., Joseph-Bravo, P., Charli, J.L., 1987. Regional distribution of the membrane-bound pyroglutamate amino peptidase-degrading thyrotropin-releasing hormone in rat brain. *Neurosci. Lett.* 79, 311–314.

Wang, J., Yong, W.H., Sun, Y., Vernier, P.T., Koeffler, H.P., Gundersen, M.A., Marcu, L., 2007. Receptor-targeted quantum dots: Fluorescent probes for brain tumor diagnosis. *J. Biomed. Opt.* 12, 044021.

Wang, W., Itoh, S., Matsuda, A., Aizawa, T., Demura, M., Ichinose, S., Shinomiya, K., Tanaka, J., 2008. Enhanced nerve regeneration through a bilayered chitosan tube: The effect of introduction of glycine spacer into the CYIGSR sequence. *J. Biomed. Mater. Res. A* 85, 919–928.

Wells, J., Kao, C., Mariappan, K., Albea, J., Jansen, E.D., Konrad, P. et al., 2005. Optical stimulation of neural tissue in-vivo. *Opt. Lett.* 30, 504–506.

Wen, M.M., 2011. Olfactory targeting through intranasal delivery of biopharmaceutical drugs to the brain: Current development. *Discov. Med.* 11(61), 497–503.

Williams, D.W., Eugenin, E.A., Calderon, T.M., Berman, J.W., 2012. Monocyte maturation, HIV susceptibility, and transmigration across the blood brain barrier are critical in HIV neuropathogenesis. *J. Leukoc. Biol.* 91, 401–415.

Sonali, Agrawal, P., Singh, R.P., Rajesh, C.V., Singh, S., Vijayakumar, M.R., Pandey, B.L., Muthu, M.S., 2016. Transferrin receptor-targeted vitamin E TPGS micelles for brain cancer therapy: preparation, characterization and brain distribution in rats. *Drug Deliv.* 23(5), 1788–1798.

Wohlfart, S., Gelperina, S., Kreuter, J., 2012. Transport of drugs across the blood–brain barrier by nanoparticles. *J. Control. Release* 161, 264–273.

Xia, L.Z.M., Wang, J., 2007. Nucleoside transporters: CNTs and ENTs. In: *Drug Transportes: Molecular Characterization and Role in Drug Disposition*, ed. GMME You. John Wiley & Sons, Inc, Hoboken, NJ, pp. 171–200.

Xie, L., Kang, H., Xu, Q., Chen, M.J., Liao, Y., Thiyagarajan, M., O'Donnell, J., Christensen, D.J., Nicholson, C., Iliff, J.J., Takano, T., Deane, R., Nedergaard, M., 2013. Sleep drives metabolite clearance from the adult brain. *Science* 342, 373–377.

Zhan, C., Lu, W., 2012. The blood–brain/tumor barriers: Challenges and chances for malignant gliomas targeted drug delivery. *Curr. Pharm. Biotechnol.* 13, 2380–2387.

Zhang, S., 2003. Fabrication of novel biomaterials through molecular self-assembly. *Nat. Biotechnol.* 21, 1171–1178.

Zhang, Z., Cao, W., Jin, H., Lovell, J.F., Yang, M., Ding, L. et al., 2009. Biomimetic nanocarrier for direct cytosolic drug delivery. *Angew. Chem.* 48, 9171–9175.

Zheng, M., Ruan, S., Liu, S., Sun, T., Qu, D., Zhao, H. et al., 2015. Self-targeting fluorescent carbon dots for diagnosis of brain cancer cells. *ACS Nano* 9(11), 11455–11461.

Zou, L.L., Ma, J.L., Wang, T., Yang, T.B., Liu, C.B., 2013. Cell-penetrating peptide-mediated therapeutic molecule delivery into the central nervous system. *Curr. Neuropharmacol.* 11(2), 197–208.

11

Magnetic Particle Hyperthermia

M. Angelakeris
Aristotle University

11.1 Introduction

11.1.1 History and Background

Nowadays cancer has become a major public health problem and one of the leading causes of morbidity and mortality in the world. Apart from the fact that the disease has many etiologies and pathological manifestations, the lack of understanding the mechanisms underlying tumor development limits the development of effective new therapies. Although chemotherapy, radiation therapy, and surgery are still the principle therapies for treating cancer up to date, alternative milder approaches such as therapeutic hyperthermia are currently explored either as synergetic modules or, more recently, as stand-alone therapy schemes with quite impressing results.

The question of how to cure a disease goes back in 500 B.C., when Hippocrates, the father of medicine stated, "What medicine cannot cure, iron cures; what iron cannot cure, fire cures; what fire does not cure, is to be considered incurable". When medicine fails in cure, iron comes on the scene. Iron, practically refers to a surgery where the malignant region is removed. Eventually, fire i.e. a heat protocol is the ultimate attempt to burn a so-far incurable region.

Generally, hyperthermia is elevated body temperature due to failed thermoregulation that occurs when a body produces or absorbs more heat than it dissipates. The therapeutic role of hyperthermia is often misinterpreted, since hyperthermia is often met either as a relaxing event in hot spring baths or as a non-medical treatment with doubtful beneficial results in pain and discomfort situations due to increased blood circulation.

A systematic attempt to control body temperate against diseases appeared in the late 19th century, when Dr. William Coley, an innovative New York oncologist surgeon, induced a fever in the body of a cancer patient to stimulate the immune response as a treatment for cancer. Named after him, Coley's toxins were a mixture consisting of killed bacteria of species *Streptococcus pyogenes* and *Serratia marcescens*.[1] Coley's toxins acted as infection simulators, since cancer is not recognized as enemy but as normal body tissue. The role of Coley's toxins was to induce a real fever due to the "simulated" infection. Real fever not only means heating of the body but also higher activity of the immune system. More macrophages are driven to "defense mode" instead of "repair mode" i.e. become more active and lead cancer cells to destruction. Thus, fever is a prerequisite for the success of a Coley's toxin therapy. Until the early 1950s, more than 1,000 patients were treated over 40 years. Unfortunately, Coley was forced to stop his work, interrupting this field for about half a century, since it was not highly accepted. For example, Cancer Research UK[2] stated that people with cancer who take Coley's toxins alongside with conventional cancer treatments, or who use it as a substitute for those treatments, risk seriously harming their health since "... available scientific evidence does not currently support claims that Coley's toxins can treat or prevent cancer...". American Cancer Society[3] reported on Coley's toxins that more research would be needed to determine what benefit, if any, this therapy might have for people with cancer. The history of cancer immunotherapy consisted of extremely great expectations after peculiar case reports of enormous success, followed by decreasing levels of enthusiasm as the results of controlled clinical trials were available. At present, again a high level of anticipation for the use of immune stimulation seems to be more promising than the incremental improvements using previous immunotherapies.[4]

Hyperthermia differs from fever in the fact the body's temperature set point remains unchanged while the heat generation is driven by an external stimulus such as microwave radiation, capacitive or inductive coupling of radio frequency (RF) fields, electrodes, ultrasound, lasers. There are two distinct application schemes of hyperthermia with respect to its localization, with the first one to be whole-body hyperthermia aiming to increase the systemic temperature up to ~41.8°C by a heat bath. On the other hand, regional hyperthermia is the specific local temperature increase in the region, 41°C–45°C for several minutes, with respect to standard temperature of the human body and is therapeutically useful over a relatively broad temperature range where different (mainly apoptotic) mechanisms of cell damaging initiate with increasing temperature. An "extreme" hyperthermia scenario is thermoablation, where temperatures between 46°C and 56°C are applied, aiming to kill cancer cells by direct cell necrosis, coagulation or carbonization.[5]

Necrosis is an "uncontrollable" form of cell injury that results in the premature death of cells following an unregulated digestion of cell components concluding to autolysis. In other words, cellular death due to necrosis does not follow the apoptotic signal transduction pathway, but rather various receptors are activated and result in the loss of cell membrane integrity and an unmanageable release of products of cell death into the extracellular space. This initiates in the surrounding tissue an inflammatory response, which attracts leukocytes and nearby phagocytes which attempt to eliminate the dead cells by phagocytosis. Probably, excessive collateral damage to surrounding tissues, inhibiting the healing process, will occur, due to the release of microbial damaging substances by the leukocytes. Thus, untreated necrosis results in a build-up of decomposing dead tissue and cell debris at or near the site of the cell death. It is often necessary to remove necrotic tissue surgically, a procedure known as debridement. Eventually, necrosis is almost always detrimental and can be fatal while apoptosis, a naturally occurring programmed and targeted process of cellular death, usually achieved in hyperthermia protocols often provides beneficial effects to the organism. The challenge of this cancer therapy lies in controlling the heating effect specifically to the local tumor site so as to not harm the nearby healthy cells. To this end, magnetic hyperthermia has emerged as one of the most promising approaches for heat localization.

Magnetic nanoparticles (MNPs) dispersed in several media are currently a subject of basic and applied research due to their unusual and eventually exploitable important properties that lead to wide variety of applications in technological[6–8] and biomedical fields[9,10] such as biological object manipulation,[11] genetic engineering,[12] bind biomolecules,[13] cell separation,[14] cell signaling,[15] gene delivery and gene therapy,[16] contrast enhancement in magnetic resonance imaging (MRI),[17,18] site-specific drug delivery[19,20] and magnetic particle hyperthermia (MPH).[21–23]

11.1.2 Evolution and Applicability

MPH (alternatively named as magnetic fluid hyperthermia: MFH) is the regional, magnetically originated, heat release due to the presence of MNPs in the malignant region. First attempts, to incorporate macroscopic magnetic implants as heat generators, goes back in the 1950s[24] while gradually, nanoscale entities such as MNPs are up to date routinely used to deliver heat cargo at specific sites.[25–27] MNPs generate heat in an alternating magnetic field as a result of hysteresis and/or relaxational loss, resulting in heating of the tissue in which MNPs are accumulated.[28,29] With the development of precise methods for synthesizing functionalized MNPs,[30,31] MNPs with "adequate" surfaces, which have high tumor specificity, have been developed as heating elements for magnetic hyperthermia.[32] Furthermore, there is renewed interest in magnetic hyperthermia as a treatment modality for cancer, especially when it is combined with other more traditional therapeutic approaches such as the co-delivery of anticancer drugs or photodynamic therapy.[33] The current development of magnetic hyperthermia is heavily focused on two aspects, namely the composition of the nanoparticles (where reproducibility and scalability are consistently found to be hard to achieve), and the instrumentation needed for applying external fields to generate the magnetic hyperthermia, together with measuring the resultant heat deposition in tissues. Regarding the instrumentation, the clear majority of the devices are purpose-built designs intended for *in vitro* testing or at most pre-clinical *in vivo* testing. Clinical scale appliances are very much the exception to date and are being developed by companies such as Magforce GmbH in Berlin[34,35] and Resonant Circuits Ltd in London.[36]

The incorporation of MNPs in hyperthermia protocols has many advantages. First, it is a mild, least-invasive scheme since the frequencies and intensities of magnetic fields generally utilized pass harmlessly through the body. Additionally, its performance is effectively and externally guided, while its effect may be delivered down to the cellular level. Second, MNPs can be targeted through cancer-specific binding agents making the treatment much more selective and effective. Since MNPs are only a few tens of nanometer in size, they may find easy passage into several tumors whose pore sizes are in 380–780 nm range. Moreover, specific MNPs can also effectively cross blood–brain barrier (BBB) and, hence, can be used for treating brain tumors. Third, with the possibility to obtain stable colloids using MNPs, they can be administered through several drug delivery routes. Compared to macroscopic implants, MNPs provide much more efficient and homogeneous treatment. Fourth, MNP-based hyperthermia treatment may induce antitumoral immunity while it may appear as a branch of multi-therapeutic approaches for treating many diseases. For example, a multifunctional combination of MPH with other assisting toxic agents (mostly irradiation or chemotherapy drugs) often succeeds in reliable damage of tumor cells where isolated treatments are proved inefficient.[37,38]

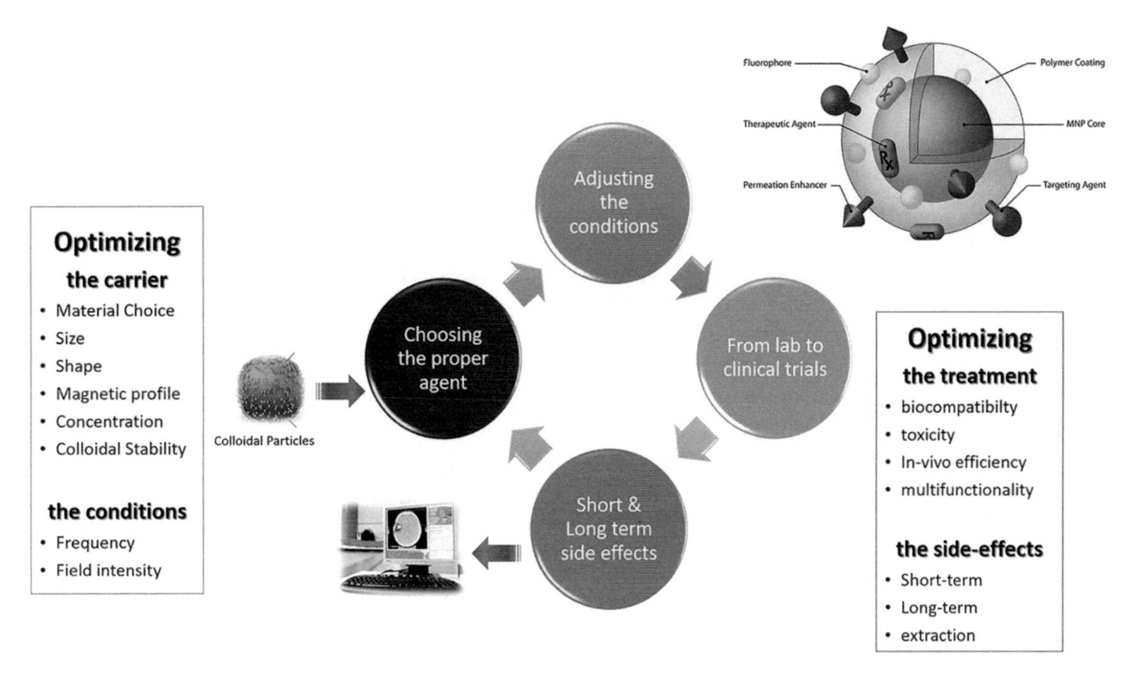

FIGURE 11.1 An application scheme with magnetic nanoparticles (MNPs) comprises of four stages: Stage 1 refers to the optimum properties of the agent with respect to material selection, dimensions. Stage 2 refers to optimum field conditions to be applied to control agent's performance. Stages 3 and 4 correspond to biomedical applicability steps necessary to check agent's performance *in vitro*, *in vivo* and eventually lead to clinical trials. Top-right image[40] is a schematic representation of a multifunctional MNP comprised of a magnetic core (MNP core) where diverse probes with desirable functions and properties are attached.

MPH directly addresses the issues raised by the modern medicine trends such as theranostics, a term coming from the prefix of "Therapeutics" and the suffix of "Diagnostics", used to describe a simultaneously applied diagnostic therapy scheme. Its application refers to individual patients – to test them for possible reactions when taking a new medication and to tailor a treatment for them based on personalized test results. A major challenge in theranostics for the 21st century is to detect disease biomarkers non-invasively at early stages of disease progression while effectively tune personalized medical treatment considering specific (patient) genetic and (disease) phenotypic characteristics. MNPs acting as vehicles may carry different cargos/probes[39] and deliver them at the cellular level provided specific functionalization steps are undertaken as shown in Figure 11.1.

11.2 Background

11.2.1 Underlying Physics

In MPH, as more generally in magnetically driven hyperthermia, a magnetic substance is subjected to an alternating magnetic field within the bottom window (kilohertz) of radio-wave frequencies. The influence of the magnetic field on any magnetic material is straightforward, the magnetic atoms will attempt to follow the external field variations provided the external magnetic field is strong enough to penetrate the material and manipulate the magnetic moments. Namely, you need a strong enough system of magnetic atoms and an even stronger magnetic field to control them. A key issue in all the magnetically driven therapies is to tune the interaction of an external magnetic field drive with the internalized magnetic entities to get the maximum efficiency with the minimum dosage levels. In other words, the magnetic nanomaterials should possess as high as possible magnetization and size to be effectively treated by as small as possible in magnitude magnetic fields. It is unambiguous that the ferromagnetic (FM) materials are the strongest materials with respect to magnetic features; provided that the external magnetic field surpasses their coercivity, it can easily manipulate them.

11.2.2 Nanomagnetism

Specifically, in MPH, the magnetic substance is a colloidal solution of MNPs. The hysteresis loop of magnetic materials is characterized mainly by typical material-dependent parameters: Saturation magnetization M_S is the maximum magnetization that may be achieved by increasing the external magnetic field. The field value required to achieve this is termed saturation field H_s. After removing the external field, a remanent magnetization M_R is, in general, left over. In order to remove that remanence, a reversed external field must be applied, which is called coercivity H_C. Since nanoparticles possess diameters in the nanometer range, their collective magnetic features such as saturation magnetization, coercivity and magnetic anisotropy may be distinctly different from their bulk counterparts. All these parameters are important for the heat output of nanoparticles and may vary considerably for different particle types.

One of the most controversial issues in MNPs is the observed variation of the saturation magnetization with particle size. A general rule is that the saturation magnetization decreases with the decrease of particle size. To explain this reduction, there are several arguments in favor of either surface origin (random canting of the surface spins) or finite size effects. Surface effects dominate the coercivity values of the smallest particles at low temperatures. Thus, the anisotropy increases as the volume is reduced due to the contribution of surface anisotropy. As particle size goes smaller, fewer magnetic atoms compose the MNP, and there are critical sizes where MNPs originating from a multidomain (MD) magnetic structure turn to a single-domain (SD) one and eventually to a superparamagnetic structure (SPM) as shown in Figure 11.2, with dramatically different magnetic features.

Superparamagnetism is a term that corresponds to certain conditions of size and temperature where the magnetic atoms within a nanoparticle no longer can interact with each other, resulting in isolated magnetic moments, as in the case of typical paramagnetic materials. Therefore, coercivity attenuates, while when size of nanoparticles gradually begins to increase, they again start to interact and return to SD particles with increasing coercivity up to a diameter, where their breakdown to multiple magnetic domains is energetically favored and their coercivity begins to decrease, having a direct impact on the shape of the hysteresis loop (Figure 11.2, top graphs). These magnetic transitions are dependent on the magnetic hardness of the ingredient material as depicted in Figure 11.2 where the effective anisotropy and room temperature saturation magnetization values appear next to corresponding bar charts as collected from Angelakeris[41] and references therein. For MPH, it is a prerequisite to know the magnetic status of the magnetic MNPs since the heating mechanism differs for particles possessing or not hysteresis.

In every MNP system, a critical temperature exists, named blocking temperature (T_B), where the SPM $\Leftrightarrow$ SD transition occurs. For the temperature region below the blocking temperature, $T < T_B$, as the temperature

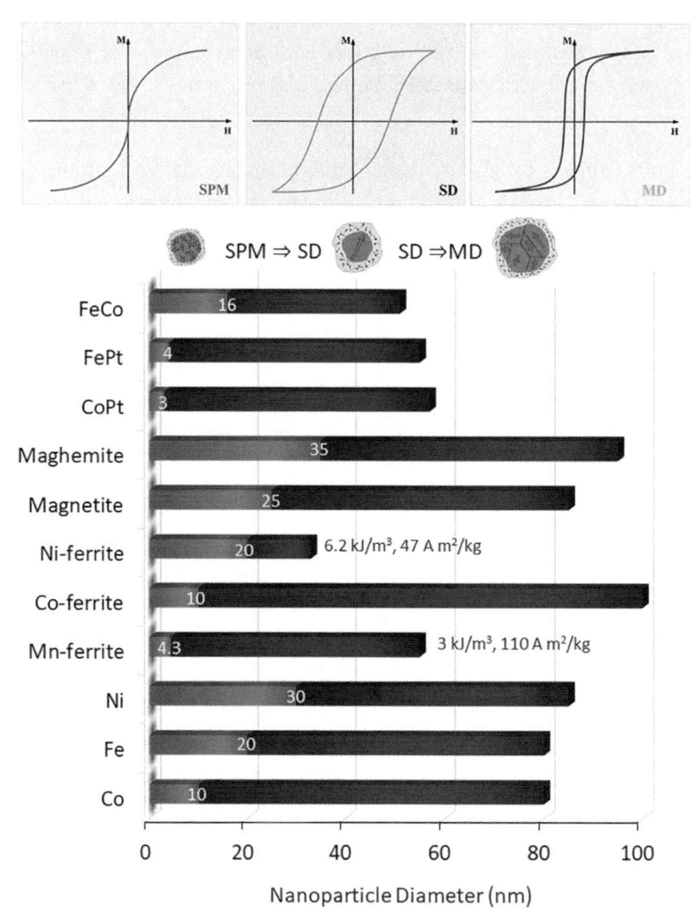

FIGURE 11.2 Nanomagnetic size effects: Main graph shows the magnetic transitions occurring as MNPs grow in diameter. D_1 (shown in the graph at transition area between leftmost bars) corresponds to transition from (SPM) superparamagnetism to (SD) ferromagnetism-SD particles while D_2 (rightmost edge of bars) corresponds to the critical diameter above which the formation of MDs is energetically favored. Top graphs outline the collective magnetic features exhibited by the hysteresis loops in each one of the three regions. Next to the bar charts, effective anisotropy and room temperature saturation magnetization values are given as collected from Angelakeris.[41] and references therein.

gets smaller, thermal energy decreases, the nanoparticles become unpinned and align with the applied field increasing the sample's net magnetization showing typical hysteretic features. At T_B, superparamagnetic particles become thermally unstable and the magnetization decreases for $T > T_B$ leading eventually to almost immeasurable coercivity and remanence.

11.2.3 Heating Mechanisms under an External Magnetic Field

Heating of magnetic substances under an external alternating magnetic field may be related to several physical mechanisms either of magnetic or non-magnetic origin. In order to compare different setups (due to different particles and/or conditions), the amount of heat (W) produced during one cycle of the alternating field is related to the mass of particles, measured in Joules per gram of magnetic material. The corresponding specific loss power P is obtained by multiplying W with the frequency f taking into account that the losses also depend on the frequency and amplitude (H) of the alternating field: $P(f, H) = Wf$. Accordingly, these measurements must be cautiously selected to optimize heating efficiency of the characterized material, as expressed by the specific loss power index (SLP) in *Watts per gram*, which refers to the amount of energy converted into heat (W) per time (Δt) and mass (m_{magn}) of the magnetic material[42]:

$$\text{SLP} = \frac{P}{m_{\text{magn}}} = \frac{W}{\Delta t \cdot m_{\text{magn}}} = C\frac{m_{\text{sol}}}{m_{\text{magn}}}\frac{\Delta T}{\Delta t}$$

where C is the specific heat of the solution, m_{sol} is the solution mass and $\Delta T/\Delta t$ the slope of the heating curve under magnetic field extracted from experimental data.[43] Alternatively, in many reports, SLP is mentioned as SAR arising from specific absorption ratio, in analogy to the measure of rate at which energy is absorbed by the human body when exposed to an RF electromagnetic field. Since SAR practically refers to the amount of heat that is eventually absorbed and may be correlated with the physiochemical features of the surrounding tissue, the use of the term SLP is more reasonable as it is strictly referring to the energy losses from the MNPs towards their environment.

To optimize heating efficiency, SLP has to be maximized, by solving the hyperthermia trilemma. As shown in Figure 11.3: Left panel, three issues must be addressed successfully: *Materials*: dimensions and magnetic profile, *Conditions*: field amplitude and frequency and *Dosage* corresponding to which concentration is optimum with respect to colloidal stability and minimum toxicity. The theoretical predictions for MNPs suitable for magnetic hyperthermia[44] are given in Figure 11.3: Right panel, together with, the schematic representations of the SLP dependencies on MNPs diameter (D), saturation magnetization (M_s) and effective anisotropy (K_{eff}).[45]

Let's discuss briefly the major mechanisms responsible for energy loss (electromagnetic energy to heat conversion) when a substance enters an alternating magnetic field, shown schematically in Figure 11.4. We may start from the non-magnetic origin mechanisms i.e. Eddy current loss and viscous loss apparent to all particle systems. We will conclude with magnetic relaxation and hysteresis loss dominating in SPM and FM nanoparticles, respectively.

Eddy Current Loss

Eddy current generation is a consequence of the law of induction, providing an essential heating effect in electrically conducting materials. The high electrical conductivity of metals yields considerable currents in a closed circuit

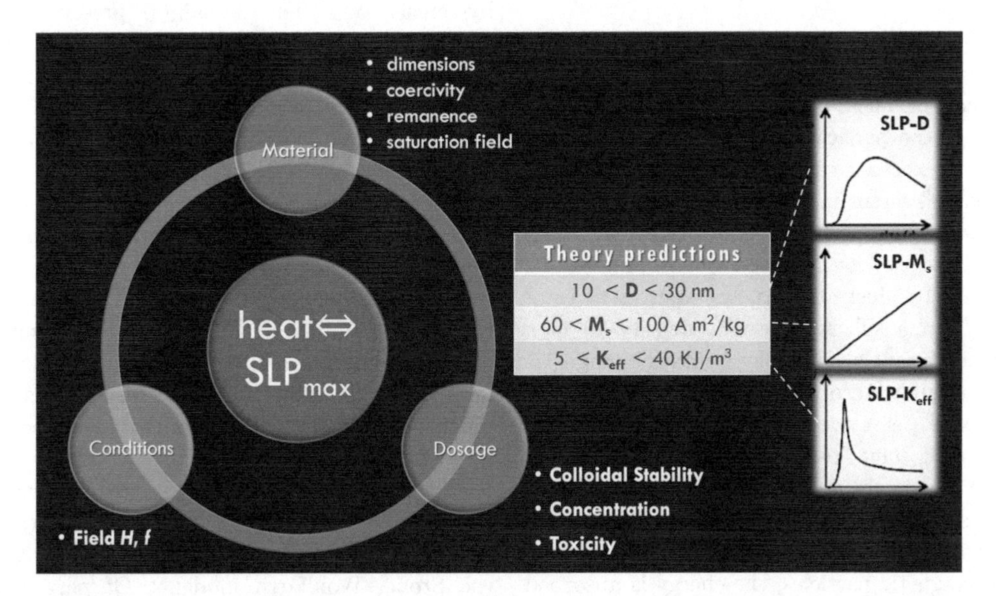

FIGURE 11.3 Left panel: The hyperthermia trilemma, showing which parameters must be fine-tuned to maximize energy conversion. Right panel: Theory predictions for the optimum material parameters for MPH together with the corresponding dependencies of SLP on size (D), magnetization (M_s) and anisotropy (K_{eff}).

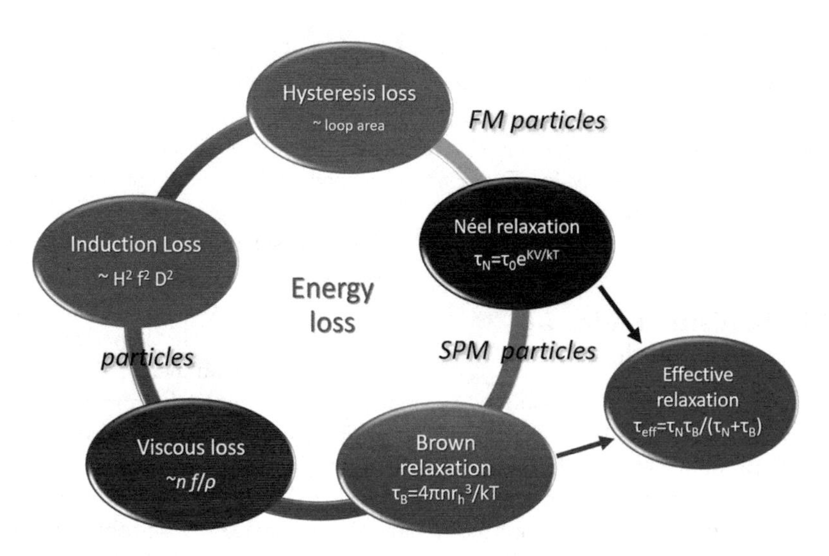

FIGURE 11.4 Energy loss mechanisms appearing when a particle is subjected to an AC magnetic field. Left side corresponds to non-magnetic origin loss such as eddy current loss (induction loss) and viscous loss while right side to magnetic hysteresis and magnetic relaxation, dominating in MNP systems.

surrounding an alternating magnetic flux. An alternating magnetic field induces eddy currents in the tissues which are undesirable in MPH as they occur non-selectively both in tumor tissues as well as in the surrounding healthy tissues. Eddy current-induced heating of a material of diameter D under an AC field (H, f) is proportional to $H^2f^2D^2$. Since this type of heating relies on the electrical conductivity of materials, specifically for magnetic oxide nanoparticles, widely used, is rather low. Secondly, the induced electrical voltage is rather weak due to the extremely small dimensions of possible eddy current loops within a nanoparticle. So, a practical induction heating effect does not occur, not even in good electrically conducting metal nanoparticles.

Viscous Loss

If the dispersion medium of the nanoparticles is a liquid, the alternating magnetic field may also cause oscillating or rotating motions of the particles. In cases where the field amplitude does not exceed a certain critical value, magnetization of the particles remains essentially unaffected and the particle reacts as permanent magnet with a certain mass of inertia giving rise to losses caused by friction in the surrounding liquid. To illustrate this, the stationary rotation (frequency f) of a sphere (mass m, density ρ) in a liquid (viscosity coefficient η) forced by a rotating magnetic field may be investigated. According to the theory of viscous friction,[46] the losses per cycle are proportional to $\eta f/\rho$. Interestingly, these specific losses depend neither on the size of the sphere nor on the field amplitude, provided that the amplitude is sufficiently strong to overcome the torque exerted by the viscous friction.

In systems of magnetic particles, the energy is absorbed from an alternating magnetic field and is transformed into heat via at least one of the following magnetic origin mechanisms:

- Reversal of magnetization inside a magnetic particle.
- Rotation of the magnetic particle in a fluid suspension relative to the surroundings.

With respect to nanoparticle's magnetic profile, we encounter a dominant mechanism e.g. magnetic hysteresis loss for FM nanoparticles and magnetic relaxation loss for SPM nanoparticles, respectively.

Magnetic Hysteresis Loss

As shown in Figure 11.2, in the case of ferromagnetic or ferriMNPs where hysteresis is apparent, during the alternating field application, MNPs outline their hysteresis loop. The domain(s) present in the material will change their orientation after every half cycle. The power consumed by the magnetic domains for changing the orientation after every cycle is called hysteresis loss: $W_{\text{hyst}} = \frac{\mu_o}{\rho} \oint M_H(H)\,dH$. This energy is delivered at the surroundings of the MNPs and practically renders them as hyperthermia agents.

Hysteresis losses may be determined in a well-known manner by integrating the area of hysteresis loop, a measure of energy dissipated per cycle of magnetization reversal depending strongly on the field amplitude as well as the magnetic prehistory. Above a critical particle size domain, walls exist and hysteresis losses of those MD particles depend in a complicated manner on the type and configuration of wall pinning centers given by the particle structure. With decreasing particle size, a transition to SD particles occurs, leading to the simplest process of coherent magnetization reversal as described by the Stoner–Wohlfarth model.[47] Of particular interest is the coercivity since below that field strength, no reversal of the magnetization and accordingly no hysteresis losses may occur. Additionally, magnetic particles below about a certain diameter (D_1 in Figure 11.2) become superparamagnetic, the maximum of hysteresis losses may be expected for SD MNPs, possessing the maximum coercivity.[41]

Magnetic Relaxation Loss

Since magnetic hysteresis is, in principle, a non-equilibrium process, after shutting off the external field, the macroscopic magnetization will tend to vanish within a typical relaxation time τ. Although such a relaxation time expands up to millions of years in magnetic minerals, in the case of MNPs, it may vary down to very low values down to nanoseconds. The reason for this reversal of the magnetic moments in a magnetic material after vanishing of the aligning external magnetic field is the thermal agitation kT. For magnetic SD nanoparticles, the energy barrier against reversal of the magnetization is given by KV where K is the magnetic anisotropy and V is the particle volume.

In such a case, heating is achieved via the reversal of the magnetic moments due to the overcome or an energy barrier proportional to the anisotropy energy KV (V particle volume) separating two stable magnetic configurations of opposite magnetization.[48] Then, the so-called Néel relaxation time of the system is determined by the ratio of anisotropy energy KV to thermal energy kT (Néel, 1949) 48: $\tau_N = \tau_0 \cdot e^{\frac{KV}{kT}}$ with $\tau_0 \sim 10^{-9}$.

As nanoparticle size decreases, these barriers decrease and the probability of jumps of the spontaneous magnetization due to thermal activation processes increases. For example, when MNPs enter the SPM region in Figure 11.2, they experience vanishing of remanent magnetization, coercivity as well as hysteresis losses. The loss power density for a magnetic material in a small amplitude (H) alternating magnetic field is given in linear approximation by (Landau and Lifschitz, 1960)[49]: $P(f, H) = \mu_0 \pi \chi''(f) H^2 f$ where $\chi''(f)$ is the imaginary part of magnetic susceptibility: $\chi''(f) = \chi_0 \varphi(1 + \varphi^2)$, $\varphi = f\tau_N$, $\chi_0 = \mu_0 M_s^2 V/kT$.

Except the Néel losses, caused by the reversal of magnetic moments inside the particles, a further loss type may arise due to particle rotation in case of liquid suspensions of magnetic particles. This is called Brownian relaxation, and its characteristic time (τ_B) is given by (Debye, 1929) 48: $\tau_B = 4\pi\eta r_h^3/kT$, where r_h refers to the hydrodynamic radius which may dramatically be different than the magnetic radius. The dependence of loss power density is different for Néel and Brownian relaxations, with the effective relaxation time in co-existence, given by: $\tau_{\text{eff}} = \tau_N\, \tau_B/(\tau_N + \tau_N)$. For smaller diameters, Brownian relaxation is ruling while there is a critical size, specific for each material, above which Néel relaxation prevails. Moreover, the effect of Brownian relaxation dramatically suppresses when MNPs are transferred from a liquid suspension to a more dense matrix such as tissue environment.

Accordingly, relaxation effects may be observed if the measurement frequency is smaller than the characteristic relaxation frequency of the particle system. Comparing different types of magnetic particles (SPM, SD, MD), with respect to their specific loss power, differences by orders of magnitude may arise, due to differences of particle size, shape and microstructure.[49]

11.2.4 Magnetic Particle Hyperthermia Experiment and Evaluation

A typical magnetic hyperthermia setup is shown in Figure 11.5. Its main components are: an AC magnetic field generator with the corresponding coil where the sample

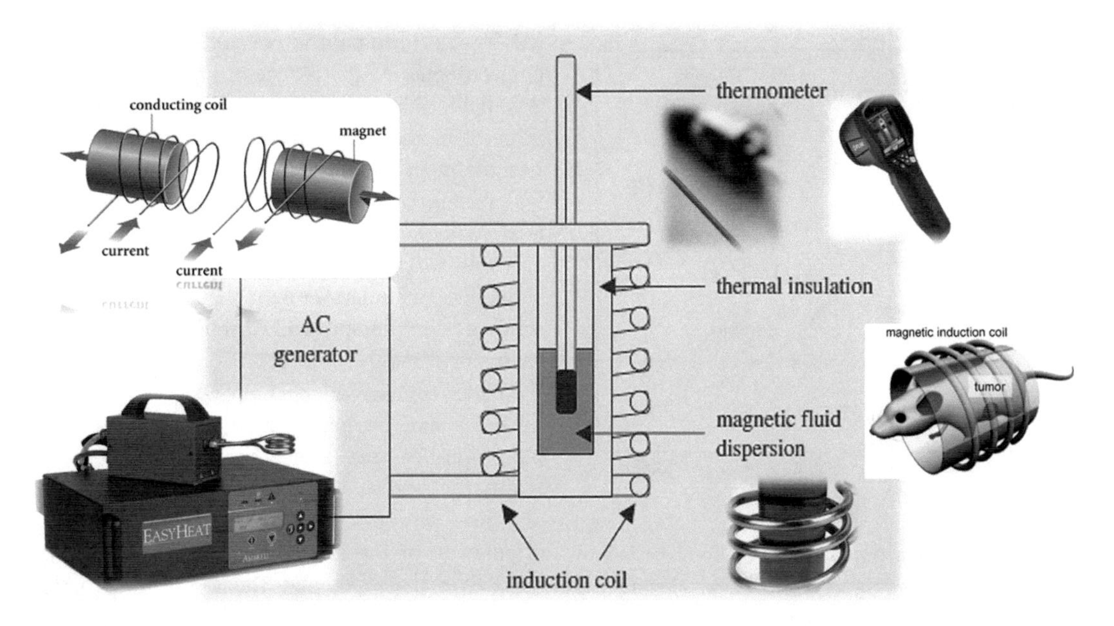

FIGURE 11.5 Schematics of a typical magnetic hyperthermia setup. Components: AC magnetic field generator, induction coil of variable diameter to support the "sample" either liquid suspension, culture dish or malignant region and thermometer to accurately measure point or regional temperature variations.

under study is put in and a precise temperature-measuring system based on an optic fiber for point measurements and/or a thermal camera useful for more macroscopic regional temperature recording as in the case for *in vitro* and *in vivo* studies. A rapidly reversing (kilohertz to megahertz) magnetic field is easily generated by changing the polarity of a powerful alternating current.

A typical hyperthermia sequence is shown in Figure 11.6 (bottom curve) recorded under non-adiabatic conditions, meaning that, the sample under study, despite the surrounding thermal insulation, shown in Figure 11.5, it experiences the thermal variations due to the lower temperature air ambient environment and due to the higher temperature induction coil (as shown by the thermal camera inset photos). The experimental curves under non-adiabatic conditions, where thermal losses occur naturally due to the lower surrounding temperature, are substantially different from the ideal ones corresponding to zero thermal losses.[50,51] Thus, together with the non-magnetic origin mechanisms, the need to accurately estimate the temperature variation solely due to the application of AC magnetic field on the MNPs arises.

In an effort to perform an error-free SLP evaluation to mimimize overestimation or underestimations due to concurrent non-magnetic heat exchanges, certain steps should be considered: (i) Coil heating effect: Heating of the coil surface becomes pronounced and may lead to misinterpretation (despite cooling water circulating through the copper coil). A gradual increase in the temperature of the coil surface may influence the measurement considerably, despite use of an internal insulating spiral coil and Teflon sample tube. For this reason, a background signal, measured for pure solvent under similar conditions, should be subtracted to eliminate any contribution to sample temperature from coil heating. This signal may not vary much with field, but its contribution to the overall signal differs substantially based on the overall temperature rise. Therefore, this procedure may lead to considerable changes of the overall signal depending on its relative magnitude. This correction is very important for measurements with low temperature rises, i.e. small sample volumes and/or low solution concentrations, where a small nanoparticle effect may be overestimated because of external heating and misinterpretation of the parameters of MPH.[52,53] (ii) Environmental losses: Although the use of thermal insulation around the sample tube reduces heat losses, the heat transfer remains significant as the "plateau" occurrence in the heating curve suggests. The modified law of cooling, $\frac{dQ}{dt} = mC\frac{dT}{dt} + UA(T - T_e)$ where Q is the thermal energy, m the solution mass, U the heat transfer coefficient, A the effective surface area, T the measured temperature in the solution and T_e the environment temperature, provides a "corrected" estimate of the energy rate including both the heating contribution due to MNPs and the cooling contribution due to losses to environment.[50,51] If we divide by mC, we find $\frac{dT'}{dt} = \frac{dT}{dt} + k(T - T_e)$. The term dT/dt is the magnetic origin temperature rate, dT'/dt is the ultimate temperature rate and $k = UA/mC$ is the cooling constant, estimated by exponentially fitting the cooling stage of the experimental curve.

Alternatively, SLP index may be derived from other data manipulation methods, such as the initial slope and Box-Lucas,[54] or non-calorimetric measurements following magnetometry experimental and theoretical approaches.[55–57]

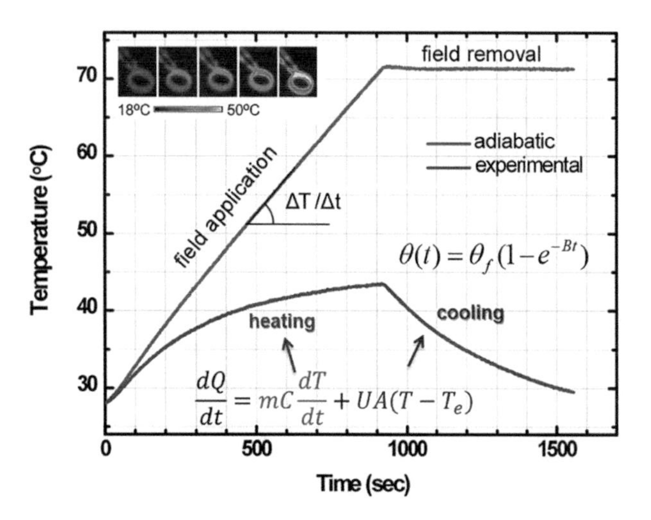

FIGURE 11.6 Typical hyperthermia sequence (bottom curve) composed of a temporal recording of temperature during AC magnetic field application (heating) and after AC field shut down (cooling). Top curve corresponds to ideal adiabatic conditions where temperature increases linearly during AC field application and remains stable after its removal.

11.2.5 Biomedical Applicability Constraints

MNPs should remain intact and not degrade within cellular environments to deliver successfully their heat cargo under AC field. The toxicity of MNPs is part of the general issue of biomedical applicability of nanomaterials; a multi-parameter problem comprising of materials and morphological parameters such as composition, degradation, oxidation, size, shape, surface area and structure.[58] Nanoparticles due to their multivalency and multifunctionality pose challenge for understanding their pharmacokinetics because different components will have different features that affect their performance, toxicity, distribution, clearance.[59]

When compared to micron-sized particles, nano-sized particles can be generally more toxic because they have larger surface area (hence, more reactive), for a given mass, to interact with cell membranes and deliver toxicity. They are also retained for longer periods in the body (more circulation or larger clearance time) and, in principle, can be delivered deeper into the tissue due to their size. The 3Ds: Dose, Dimensions and Durability provide the set of parameters to be fine-tuned in order to have optimum performance together with minimum side effects within a biological environment.[41]

A major ingredient is a biocompatible surface that will promote a long blood half-life and minimum toxicity together with colloidal stability over a wide pH range. Depending on the treatment, particles should evade or allow uptake by reticuloendothelial system (RES) and/or participate on specific biomolecule interactions. Thus, a typical first step is to coat the MNP's outer surface with amphiphilic or anti-biofouling polymers to enhance biocompatibility of the nanoparticles. Rapid clearance of nanoparticles from circulation can substantially reduce their biomedical functionality. Active clearance of nanoparticles is mainly due to their recognition by macrophages of the RES. Nanoparticles have a large surface-to-volume ratio and tend to adsorb plasma proteins (opsonization), which are easily recognized by macrophages making them vulnerable to rapid clearance before reaching their target.

Despite, Fe ions' unambiguous domination in biocompatibility against other metal ions, additional shells (such as citric acid or dextran) should be included to guarantee that MNPs will continue their undivided and unobstructed performance in time. Generally, MNPs, seen as foreign materials by the cells, generate a quite standard reactive oxygen species (ROS) reaction reaching maximal levels after 24 h and decreasing to near control levels over a period of about 72 h.[60] The level of induced ROS depends not only on the total amount of MNPs internalized but on the stability of the coating against intracellular degradation.

Although it is generally stated that the typical range of magnetic hyperthermia fields (10 kHz to 1 GHz, 1–100 mT) are safely tolerated by living tissues, a maximum field-frequency product for patient discomfort, $Hf \leq 4.85 \times 10^8$ Am^{-1}s^{-1}, called the Atkinson–Brozovich limit, initially proposed in 1984, based on micron-sized magnetic implants is still under discussion.[61] Ever since, MNPs are routinely incorporated in MPH and up-to-date practice has shown that such a limitation is rather stringent and should be reconsidered by accounting both the nanoscale character of the particles and the treatment itself.[62,63] Therefore, a less rigid criterion, one order of magnitude greater ($Hf \leq 5 \times 10^9$ Am^{-1} s^{-1}), was more recently proposed[64] while relevant case studies of different materials[65] and numerical calculation of an optimization criterion[66] are on the same trend. The results of the abovementioned exhaustive experiments and analysis performed have shown that, in all of the considered cases, the allowable value of *Hf*, is much larger (~two orders of magnitude) than Atkinson–Brozovich criterion allowing us to use more effective manipulation of the nanocarriers and thus decrease their dosage scheme.

11.3 Up to Date

11.3.1 Systems under Study

MNPs provide a unique versatile platform for modern technologies since they can be remotely and non-invasively employed not only as heating agents but also as imaging probes, carrier vectors and smart actuators. To start with, we need MNPs with several well-defined and reproducible structural, physical and chemical features, while specific-application nanoparticle design imposes several additional constraints. Since an external magnetic field is the usual drive of magnetism-based application schemes, we should have to maximize field effect by tuning particle collective magnetic features.

MNPs of different phases, sizes and shapes are currently studied as enhanced performance hyperthermia agents due to their magnetic features, affected by individual particles' intrinsic features and collective phenomena where particles as parts of an ensemble, within a colloidal suspension, interact with each other. Different formations of particles include single-phase particles such as Fe_3O_4[49] or other ferrite types such as $CoFe_2O_4$, $MnFe_2O_4$ or $NiFe_2O_4$.[67–69] An additional tool to manipulate the effective magnetic anisotropy of a system is to introduce shape anisotropy by changing the shape of the MNPs[70] by routine synthetic controls. Exchange anisotropy introduction may arise in core-shell structures comprised of a soft magnetic material as a core and a harder magnetic material as a shell or vice versa with such systems appearing with much enhanced heating efficiency due to anisotropy tuning.[71,72]

Nanoparticles will eventually appear with suppressed magnetic features, depending on surrounding conditions/features/interactions, thus their configuration in 2D or even 3D-oriented assemblies may provide a powerful pathway to control their macroscopic magnetic response on demand. Such assemblies have a direct impact on effective anisotropy, susceptibility and hysteresis losses. Mobility of the MNPs within the colloidal dispersion is a crucial factor that affects the assembly formation success rate, since diverse effects, such as magnetic, electrostatic, viscous, gravitational and molecular interactions are involved.[73] Eventually, MNPs may form via tunable aggregation effects either multi-core particle clusters[74] or oriented arrays.[75,76]

The choice of the MNPs to be employed not only as hyperthermia agents but more generally as biomedical probes has to do with four aspects: (i) material, (ii) size, (iii) shape and (iv) formation (i.e. core-shell, multi-core, chains). Since an external magnetic field is the drive of MNPs, we should have to maximize field effect by tuning MNP's collective magnetic features. Magnetism both at the nanoscale level (single particles) and also macroscopic collective magnetic features (ensembles) will be dramatically different for the different formations while their collective magnetic features are visualized in magnetic hysteresis loops.

11.3.2 Self-Regulated Hyperthermia

MNPs possessing low Curie temperatures (T_C) offer the possibility for self-regulated heating of cancer cells, where the T_C acts as an upper limit to heating to prevent damage to neighboring healthy tissues. Once they reach the Curie temperature, M_s drops to zero and heating stops. Thus, if the Curie temperature is fixed by judicious selection of particle composition and size, hyperthermia response is eventually fixed below a certain temperature level preventing excessive temperatures.[77]

11.3.3 Focused Hyperthermia

A typical problem of the application of hyperthermia is the difficulty to locate the heat without damaging potentially healthy surrounding tissues. The control of temperature rise in MPH may be achieved by the incorporation of an external DC magnetic field concurrently with the AC field. Thus, a different regional field map appear and the energy dissipation of MNPs tuned by the co-existence of both fields, providing typical SLP values where DC field is not present and suppressed SLP values where DC and AC fields are in opposite directions. The combination of AC and DC fields may be a feasible method not only to optimize and regionally focus MPH effect to specific malignant sites but also to combine two magnetic particle modalities such as MPH and magnetic particle imaging.[78,79]

11.3.4 Multiple Pulses Hyperthermia

Versatile hyperthermia protocols may be attempted to further improve MNP's *in vitro* performance. Manganese and cobalt ferrites combined with magnetite and covered by citric acid, synthesized by a high-yield "green" approach, possess remarkable structural and magnetic properties and display a great potential as MPH agents.[80] Their low inherent toxicity, ensured by magnetite and citric acid, allows them to be internalized and delivers their cargo (heat) inside target cells. In this study, such MNPs were used in combination with a versatile *in vitro* hyperthermia protocol consisting of two hyperthermia runs separated by a 48-h incubation stage with the latter run resulting in a maximized cytotoxicity effect on SaOS-2 cells. Additional variation of *in vitro* hyperthermia protocol by altering AC single pulses (continuous hyperthermia) with multiple pulses (intermittent MNPs exposure to magnetic field) led to further enhancement of the cytotoxic effect specifically on primary human osteogenic sarcoma cells. On the contrary, two reference healthy cell lines (C2–C12 and 3T3-L1 fibroblastlike preadipocytes) managed to survive the multiple-pulse sequences fairly well. Consequently, a pulsed magnetic field combined with the enhanced heating efficiency of binary MNPs can lead to a promising tumor-selective MPH protocol.[81]

11.3.5 Multifunctionality Combinatory Schemes

The case of Fe/MgO is a novel proposal since nearly spherical metallic iron nanoparticles with a primary diameter of about 75 nm (±16%), comprising of a biocompatible MgO layer and bcc Fe core may be exploited as contrast agents in MRI[82] and as potential hyperthermia agents in tumor therapy.[50] Their concentration-dependent cytotoxity profile outlines that, MNPs are adequately internalized by the different cell lines (3T3 normal cell lines, cancer breast lines: MDA-MB-231, SkBr3 and MCF7) and concurrently manage to deliver heat at malignant sites and serve as contrast agents.

The simultaneous drug release during MPH is an issue under study. Attachment of drug via heat-breaking bonds with MNPs or encapsulation of drugs within temperature-sensitive structures renders MPH not only as heating modality for cancer treatment but also as trigger to ignite drug delivery upon request.[83] For example, iron oxide nanoparticles have been embedded in different biocompatible and biodegradable polymeric matrices in order to study their structural, magnetic features and AC magnetic heating efficiency. The anticancer Taxol drug was also encapsulated in the same polymeric matrices. During MPH application, the polymer matrix, reaching adequate temperatures (close to its decomposing point: up to 42°C), started to degrade, allowing the release of anticancer drug at specific sites, providing a twofold therapy scheme together with the application of MPH performed by the iron oxide nanoparticles.[84]

11.3.6 Synergy with Other Modalities

MPH, as the hyperthermia in general, is a type of medical modality for cancer treatment using the biological effect of artificially induced heat. MPH may play an additional intriguing role in amplifying immune response in the body against cancer while decreasing the immune suppression and immune escape of cancer (Figure 11.7).[85]

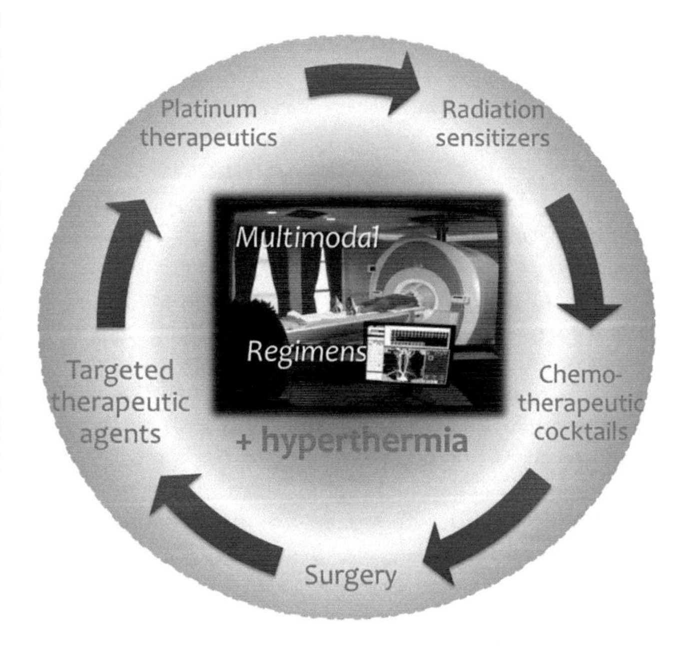

FIGURE 11.7 Synergy of hyperthermia with cancer treatment modalities is an extra "hand" in modern cancer therapy. Since hyperthermia is the least-invasive modality with respect to both materials and conditions, it may repeatedly be applied at any stage in-between combinatory treatment schemes to induce at least "fatigue" at cancer cells and not only facilitate the effects of other modalities but decrease their dosage levels as well.

11.3.7 Current Trends and Challenges

So far, the anticancer effect of hyperthermia alone has not yet been adequately exploited because deep heating techniques and devices to augment heat effects selectively in cancer tissues are difficult in practical terms. Yet, if jointly applied, hyperthermia inhibits the repair of damaged cancer cells together with chemotherapy or radiotherapy.[86] Radiation and chemotherapy induce DNA damage via double strand breaks and/or oxygen radicals. Hyperthermia increases oxygenation levels in cancer cells and, thus, hinders the ability of cancer cells to sense DNA destruction, impedes DNA repair mediators. These perceptions indicate that hyperthermia has potential for cancer therapy in conjunction with immunotherapy, chemotherapy, radiotherapy and surgery.

However, there are major issues to be tackled prior to its general clinical approval or even its reintroduction as a stand-alone effective cancer treatment. The hydrodynamic diameter of the MNPs with a biological system may be dramatically different than the magnetically "active" diameter. Accurate measurement of something as fundamental as the hydrodynamic diameter becomes a tricky operation because of effects of protein corona formation, adding layers of complexity. Hydrodynamic diameter plays an important role in true effectiveness and potential side effects of a particular non-medical system based on particle uptake by cells and clearance from the biological system which will govern its approval and adoption in the clinic. Specifically, the hydrodynamic diameter will dictate the blood flow rate, the infusion route and the circulation time while the ferrofluid concentration may require adjustments since the magnetic ingredient may become a small part of a much bigger entity.

Tumor heterogeneity refers to sub-populations of cells, with distinct genotypes and phenotypes, divergent biological behaviors, within a primary tumor and its metastases, or between tumors of the same histopathological subtype: intra- and inter-tumor, respectively. Thus, the physiological parameters of malignant cells having to do with tissue depth and perfusion, tumor volume and finally with the strength carrier/binding in varying environment are also a key factor in the effectiveness of a treatment.

It is important that nanoparticle formulations can overcome the main biological barriers that prevent them from reaching their targets. However, intravenous injection of nanomaterials introduces new concerns such as dosage, distribution and circulation times making their use and development similar to pharmaceuticals. Possible changes in behavior upon injection and interactions with cells such as specific binding and endocytosis may have dramatic impacts. These interactions can also result in nanoparticle agglomeration or regions of high concentration with inter-particle interactions leading to altered properties. The overall size of the nanoparticles (hydrodynamic size), surface charge and functionalization play a large role in their distribution and circulation time; however, these parameters may change upon interaction with blood constituents. So, special design architecture of MNPs-based biomedical probes is required to avoid rapid opsonization and subsequent macrophage phagocytosis and to manipulate the slow accumulation through the leaky vasculature in a variety of lesions.

Among the side effects is the metabolism of the metallic iron metabolism (a major component in various MPH agents, not to mention the more toxic Co and Ni ions), which may induce oxidative stress through harmful ROS and the use of high-frequency magnetic fields which may result in stimulation of peripheral or skeletal muscles, possible cardiac stimulation and arrhythmia and potentially to non-specific inductive heating of tissues.

MPH is a scheme against cancer, exploiting the fact that many tumors thrive in a hypoxic environment in which the oxygenation of the tumor is much lower than in normal tissue. Thus, tumors cannot dissipate heat as quickly as healthy tissue, they can get hotter than that tissue if enough heat is applied. Up to date, hyperthermia at relatively low levels – as in the early clinical use of thermal medicine – ends up increasing the amount of blood flow and oxygenation of the tumor, making it more vulnerable to radiation and chemotherapy. On the other hand, if adequately functionalized, MPH may play an additional role as an independent, powerful, least-invasive technique that directly and efficiently treats cancer.

References

1. Kramer M. G., Masner M., Ferreira F. A., Hoffman R. M. 2018. Bacterial therapy of cancer: Promises, limitations, and insights for future directions. *Frontiers in Microbiology* 9:16.
2. www.cancerresearchuk.org.
3. www.cancer.org.
4. Sell S. 2017. Cancer immunotherapy: Breakthrough or "deja vu, all over again"? *Tumor Biology* 39(6):1010428317707764.
5. Cortie M. B., Cortie D. L., Timchenko V. 2018. Heat transfer from nanoparticles for targeted destruction of infectious organisms. *International Journal of Hyperthermia* 34(2):157–1676.
6. Majetich S. A., Jin Y. 1999. Magnetization directions of individual nanoparticles. *Science* 284(5413): 470–473.
7. Reiss G., Hütten A. 2005. Magnetic nanoparticles: Applications beyond data storage. *Nature Materials* 4(10):725–726.
8. Zhang Y. X., Xu S., Luo Y., Pan S., Ding H., Li G. 2011. Synthesis of mesoporous carbon capsules encapsulated with magnetite nanoparticles and their application in waste water treatment. *Journal of Materials Chemistry* 21:3664–3671.
9. Wu W., Wu Z., Yu T., Jiang C., Kim W. S. 2015. Recent progress on magnetic iron oxide nanoparticles: Synthesis, surface functional strategies and

biomedical applications. *Science and Technology of Advanced Materials* 16: 023501.
10. Pankhurst Q., Thanh N., Jones S., Dobson J. 2009. Progress in applications of magnetic nanoparticles in biomedicine. *Journal of Physics D* 42:224001.
11. Liu Y., Gao Y., Xu C. 2013. Using magnetic nanoparticles to manipulate biological objects. *Chinese Physics B* 22:097503.
12. Sakamoto J. H., van de Ven A. L., Godin B., Blanco E. et al. 2010. Enabling individualized therapy through nanotechnology. *Pharmacological Research* 62:57–89.
13. Tseng P., Judy J. W., Di Carlo D. 2012. Magnetic nanoparticle-mediated massively parallel mechanical modulation of single-cell behavior. *Nature Methods* 9:1113–1119.
14. Plouffe B. D., Murthy S. K., Lewis L. H. 2015. Fundamentals and application of magnetic particles in cell isolation and enrichment: A review. *Reports on Progress in Physics* 78:016601.
15. Bonnemay L., Hoffmann C., Gueroui Z. 2014. Remote control of signaling pathways using magnetic nanoparticles. *Wiley Interdisciplinary Reviews: Nanomedicine and Nanobiotechnology* 7(3):342–354. doi: 10.1002/wnan.1313.
16. Dobson J. 2006. Gene therapy progress and prospects: Magnetic nanoparticle-based gene delivery. *Gene Therapy* 13:283–287.
17. Corot C., Robert P., Idée J. M., Port M. 2006. Recent advances in iron oxide nanocrystal technology for medical imaging. *Advanced Drug Delivery Reviews* 58:1471–1504.
18. Yang F., Li Y., Chen Z., Zhang Y., Wu J., Gu N. 2009. Superparamagnetic iron oxide nanoparticle-embedded encapsulated microbubbles as dual contrast agents of magnetic resonance and ultrasound imaging. *Biomaterials* 30:3882–3890.
19. Yang L., Cao Z., Sajja H. K., Mao H. et al. 2008. Development of receptor targeted magnetic iron oxide nanoparticles for efficient drug delivery and tumor imaging. *Journal of Biomedical Nanotechnology* 4:439–449.
20. Estelrich J., Escribano E., Queralt J., Busquets M. 2015. Iron oxide nanoparticles for magnetically-guided and magnetically-responsive drug delivery. *International Journal of Molecular Sciences* 16:8070–8101.
21. Ling-Yun Z., Jia-Yi L., Wei-Wei O., Dan-Ye L. et al. 2013. Magnetic mediated hyperthermia for cancer treatment: Research progress and clinical trials. *Chinese Physics B* 22:108104.
22. Dutz S., Hergt R. 2014. Magnetic particle hyperthermia–a promising tumour therapy? Nanotechnology 25:452001.
23. Périgo E. A. Hemery G., Sandre O. Ortega D. et al. 2015. Fundamentals and advances in magnetic hyperthermia. *Applied Physics Reviews* 2:041302.
24. Gilchrist R. K., Medal R., Shorey W. D., Hanselman R. C. et al. 1957. Heating of lymph nodes. *Annals of Surgery* 146:596–606.
25. Jordan A., Scholz R., Maier-Hauff K., Johannsen M. et al. 2001. Presentation of a new magnetic field therapy system for the treatment of human solid tumors with magnetic fluid hyperthermia. *Journal of Magnetism and Magnetic Materials* 225:118–126.
26. Kozissnik B., Bohorquez A. C., Dobson J., Rinaldi C. 2013. Magnetic fluid hyperthermia: Advances, challenges, and opportunity. *International Journal of Hyperthermia* 29:706–714.
27. Hilger I. 2013. In vivo applications of magnetic nanoparticle hyperthermia. *International Journal of Hyperthermia* 29:828–834.
28. Rosensweig R. E. 2002. Heating magnetic fluid with alternating magnetic field. *Journal of Magnetism and Magnetic Materials* 252:370–374.
29. Neuberger T., Schopf B., Hofmann H., Hofmann M., von Rechenberg B. 2005. Superparamagnetic nanoparticles for biomedical applications: Possibilities and limitations of a new drug delivery system. *Journal of Magnetism and Magnetic Materials* 293:483–496.
30. Grüttner C., Müller K., Teller J., Westphal F. 2013. Synthesis and functionalization of magnetic nanoparticles for hyperthermia applications. *International Journal of Hyperthermia* 29:777–789.
31. Ito A., Shinkai M., Honda H., Kobayashi T. 2005. Medical applications of functionalized magnetic nanoparticles. *Journal of Bioscience and Bioengineering* 100:1–11.
32. Balivada S., Rachakatla R. S., Wang H., Samarakoon T. N. et al. 2010. A/C magnetic hyperthermia of melanoma mediated by iron(0)/iron oxide core/shell magnetic nanoparticles: A mouse study. *BMC Cancer*, 10:119–127.
33. Johannsen M., Gneueckow U., Thiesen B., Taymoorian K. et al. 2007. Thermotherapy of prostate cancer using magnetic nanoparticles: Feasibility, imaging and three-dimensional temperature distribution. *Urology* 52:1653–1662.
34. Gneveckow U., Jordan A., Scholz R., Bruss V. et al. 2004. Description and characterization of the novel hyperthermia- and thermoablation-system MFH 300F for clinical magnetic fluid hyperthermia. *Medical Physics* 31:1444–1451.
35. Johannsen M., Thiesen B., Wust P., Jordan A. 2010. Magnetic nanoparticle hyperthermia for prostate cancer. *International Journal of Hyperthermia* 26:790–795.
36. Resonant Circuits Ltd., London, www.resonantcircuits.com.
37. Thorat N. D., Bohara R. A., Noor M. R., Dhamecha D. et al. 2017. Effective cancer theranostics with polymer encapsulated superparamagnetic nanoparticles: Combined effects of magnetic

hyperthermia and controlled drug release. *ACS Biomaterials Science and Engineering* 3(7):1332–1340.
38. Liang X., Fang. L., Li X., Zhang X., Wang F. 2017. Activatable near infrared dye conjugated hyaluronic acid based nanoparticles as a targeted theranostic agent for enhanced fluorescence/CT/photoacoustic imaging guided photothermal therapy. *Biomaterials* 132:72–84.
39. Heinz H., Pramanik C., Heinz O, Ding Y. 2017. Nanoparticle decoration with surfactants: Molecular interactions, assembly, and applications. *Surface Science Reports* 72:1–58.
40. www.cd-bioparticles.com/t/Properties-and-Applications-of-Magnetic-Nanoparticles_55.html.
41. Angelakeris M. 2017. Magnetic nanoparticles: A multifunctional vehicle for modern theranostics. *Biochimica et Biophysica Acta* 1861:1642–1651.
42. Zhang L.-Y., Gu H. C., Wang X.-M. 2007. Magnetite ferrofluid with high specific absorption rate for application in hyperthermia. *Journal of Magnetism and Magnetic Materials* 311:228–233.
43. Samaras T., Regli P., Kuster N. 2000. Electromagnetic and heat transfer computations for non-ionizing radiation dosimetry. *Physics in Medicine and Biology*, 45:2233.
44. Habib H., Ondeck C. L., Chaudhary P., Bockstaller M. R., McHenry M. E. 2008. Theory of magnetic fluid heating with an alternating magnetic field with temperature dependent materials properties for self-regulated heating. *Journal of Applied Physics* 103:07A307.
45. Yoo D., Lee J.-H., Shin T.-H., Cheon J. 2011. Theranostic magnetic nanoparticles. *Accounts of Chemical Research* 44:863.
46. Landau L. D., Lifshitz E. M. 1959 (reprinted 1975). *Fluid Mechanics* (Translated from Russian by J. B. Sykes and W. H. Reid). Pergamon Press (Oxford, New York, Toronto, Sydney, Paris, Braunschweig), pp. xii, 536.
47. Cullity B. D., Graham, C. D. 2008. *Introduction to Magnetic Materials*. John Wiley & Sons, Inc.: Hoboken, NJ.
48. Hergt R., Andra W. Magnetic hyperthermia and thermoablation. In: *Magnetism in Medicine: A Handbook*, Second Edition, pp. 550–570. doi: 10.1002/9783527610174.ch4f.
49. Bakoglidis K., Simeonidis K., Sakellari D., Stefanou G., Angelakeris M. 2011. Size-dependent mechanisms in AC magnetic hyperthermia response of iron-oxide nanoparticles. *IEEE Transactions on Magnetism* 48:1320.
50. Chalkidou K., Simeonidis M., Angelakeris T., Samaras C. et al. 2011. *In vitro* application of Fe/MgO nanoparticles as magnetically mediated hyperthermia agents for cancer treatment. *Journal of Magnetism and Magnetic Materials* 323:775.
51. Simeonidis K., Martinez-Boubeta C., Balcells Ll., Monty C. et al. 2013. Fe-based nanoparticles as tunable magnetic particle hyperthermia agents. *Journal of Applied Physics* 114(10):103904.
52. Huang S., Wang S.-Y., Gupta A., Borca-Tasciuc D.-A., Salon S. J. 2012. On the measurement technique for specific absorption rate of nanoparticles in an alternating electromagnetic field. *Measurement Science and Technology* 23:035701.
53. Makridis A., Curto S., van Rhoon G. C., Samaras T., Angelakeris M. 2019. A standardisation protocol for accurate evaluation of specific loss power in magnetic hyperthermia. *Journal of Physics D: Applied Physics* 52(25). doi:10.1088/1361-6463/ab140c.
54. Wildeboer R. R., Southern P., Pankhurst Q. A. 2014. On the reliable measurement of specific absorption rates and intrinsic loss parameters in magnetic hyperthermia materials. *Journal of Physics D: Applied Physics* 47(49): 495003.
55. Rosensweig R. E. 2002. Heating magnetic fluid with alternating magnetic field. *Journal of Magnetism and Magnetic Materials* 252:370–374.
56. Hergt R., Dutz S., Müller R., Zeisberger M. 2006. Magnetic particle hyperthermia: Nanoparticle magnetism and materials development for cancer therapy. *Journal of Physics: Condensed Matter* 18:S2919.
57. Carrey J., Mehdaoui B., Respaud M. 2011. Simple models of dynamic hystereis loop calculations of magnetic single-domain nanoparticles: Application to magnetic hyperthermia optimization. *Journal of Applied Physics* 109:083921.
58. Lei L., Ling-Ling J., Yun Z., Gang L. 2013. Toxicity of superparamagnetic iron oxide nanoparticles: Research strategies and implications for nanomedicine. *Chinese Physics B* 22(12):1–10.
59. Yildirimer L., Thanh N. T. K., Loizidoua M., Seifalian A. M. 2011. Toxicological considerations of clinically applicable nanoparticles. *Nano Today* 6: 585–607.
60. Soenen S. J., Rivera-Gil P., Montenegro J.-M., Parak W. J., De Smedt S. C., Braeckmans K. 2011. Cellular toxicity of inorganic nanoparticles: Common aspects and guidelines for improved nanotoxicity evaluation. *Nano Today* 6:446–465.
61. Atkinson W. J., Brezovich I. A., Chakraborty D. P. 1984. Usable frequencies in hyperthermia with thermal seeds. *IEEE Transactions on Biomedical Engineering* 31(1):70–75.
62. Hergt R., Dutz S. 2007. Magnetic particle hyperthermia-biophysical limitations of a visionary tumour therapy. *Journal of Magnetism and Magnetic Materials* 311(1 SPEC. ISS):187–192.
63. Ortega Ponce D., Pankhurst Q. 2012. Magnetic hyperthermia. In: P. O'Brien (Ed.) *Nanoscience*. Royal Society of Chemistry: Cambridge, pp. 60–88.

64. Hergt R., Dutz S., Müller R., Zeisberger M. 2006. Magnetic particle hyperthermia: Nanoparticle magnetism and materials development for cancer therapy. *Journal of Physics: Condensed Matter* 18(38):S2919–S2934.
65. Mamiya H. 2013. Recent advances in understanding magnetic nanoparticles in AC magnetic fields and optimal design for targeted hyperthermia. *Journal of Nanomaterials* 2013:752973.
66. Bellizzi G., Bucci O. M., Chirico G. 2016. Numerical assessment of a criterion for the optimal choice of the operative conditions in magnetic nanoparticle hyperthermia on a realistic model of the human head. *International Journal of Hyperthermia* 32(6):688–703.
67. Makridis A., Topouridou K., Tziomaki M., Sakellari D. et al. 2014. In vitro application of Mn-ferrite nanoparticles as novel magnetic hyperthermia agents. *Journal of Materials Chemistry B* 2:8390.
68. Stefanou G., Sakellari D., Simeonidis K., Kalabaliki T. et al. 2014. Tunable AC magnetic hyperthermia efficiency of Ni ferrite nanoparticles. *IEEE Transactions on Magnetics* 50(12):6872577.
69. Simeonidis K., Liebana Vinas S., Wiedwald U., Ma Z. et al. 2016. A versatile large-scale and green process for synthesizing magnetic nanoparticles with tunable magnetic hyperthermia features. *RSC Advances* 6:53107.
70. Martinez-Boubeta C., Simeonidis K., Makridis A., Angelakeris M. et al. 2013. Learning from nature to improve the heat generation of iron-oxide nanoparticles for magnetic hyperthermia applications. *Scientific Reports* 3:1652.
71. Angelakeris M., Li Z.-A., Hilgendorff M., Simeonidis K. et al. 2015. Enhanced biomedical heat-triggered carriers via nanomagnetism tuning in ferrite-based nanoparticles. *Journal of Magnetism and Magnetic Materials* 381:179–187.
72. Liebana Vinas S., Simeonidis K., Li Z.-A., Ma Z., Myrovali E. et al. 2016. Tuning the magnetism of ferrite nanoparticles. *Journal of Magnetism and Magnetic Materials* 415:20–23.
73. Myrovali E., Maniotis N., Makridis A., Terzopoulou A. et al. 2016. Arrangement at the nanoscale: Effect on magnetic particle hyperthermia. *Scientific Reports* 6, 37934.
74. Sakellari D., Brintakis K., Kostopoulou A., Myrovali E. et al. 2016. Ferrimagnetic nanocrystal assemblies as versatile magnetic particle hyperthermia mediators. *Materials Science and Engineering C* 58:187–193.
75. Serantes D., Simeonidis K., Angelakeris M., Chubykalo-Fesenko O. et al. 2014. Multiplying magnetic hyperthermia response by nanoparticle assembling. *The Journal of Physical Chemistry C* 118:5927–5934.
76. Simeonidis K., Puerto Morales M., Marciello M., Angelakeris M. et al. 2016. In-situ particles reorientation during magnetic hyperthermia application: Shape matters twice. *Scientific Reports* 6:38382.
77. McNerny K. L., Hudgins D. M., Brown III L. W., Yoon S. D. et al. 2012. Annealing of amorphous $Fe_xCo_{100-x}Fe_xCo_{100-x}$ nanoparticles synthesized by a modified aqueous reduction using $NaBH_4$. *Journal of Applied Physics* 107:09A312.
78. Murase K., Takata H., Takeuchi Y, Saito S. 2013. Control of the temperature rise in magnetic hyperthermia with use of an external static magnetic field. *Physical Medicine* 29(6):624–630.
79. Bauer L. M., Situ S. F., Griswood M. A., Samia A. C. 2016. High-performance iron oxide nanoparticles for magnetic particle imaging – guided hyperthermia (hMPI). *Nanoscale* 8:12162–12169.
80. Makridis A., Topouridou K., Tziomaki M., Sakellari D. et al. 2014. In vitro application of Mn-ferrite nanoparticles as novel magnetic hyperthermia agents. *Journal of Materials Chemistry B* 2:8390.
81. Makridis A., Tziomaki M., Topouridou K., Yavropoulou M. P. et al. 2016. A novel strategy combining magnetic particle hyperthermia pulses with enhanced performance binary ferrite carriers for effective in vitro manipulation of primary human osteogenic sarcoma cells. *International Journal of Hypethermia* 32(7):778–785.
82. Martinez-Boubeta C., Balcells Ll., Cristòfol R., Sanfeliu C. et al. 2010. Self-assembled multifunctional Fe/MgO nanospheres for magnetic resonance imaging and hyperthermia. *Nanomedicine: Nanotechnology, Biology, and Medicine* 6(2):362–370.
83. Kumar S. S. R, Mohammad F. 2011. Magnetic nanomaterials for hyperthermiabased therapy and controlled drug delivery. *Advanced Drug Delivery Reviews* 63:789–808.
84. Filippousi M., Altantzis T., Stefanou G., Betsiou M. et al. 2013. Polyhedral iron oxide core-shell nanoparticles in a biodegradable polymeric matrix: Preparation, characterization and application in magnetic particle hyperthermia and drug delivery. *RSC Advances* 3:24367.
85. Ito A., Honda H, Kobayashi T. 2006. Cancer immunotherapy based on intracellular hyperthermia using magnetite nanoparticles: A novel concept of "heat-controlled necrosis" with heat shock protein expression. *Cancer Immunology, Immunotherapy* 55:320–328.
86. Yagawa Y., Tanigawa K., Kobayashi Y., Yamamoto M. 2017. Cancer immunity and therapy using hyperthermia with immunotherapy, radiotherapy, chemotherapy, and surgery. *Journal of Cancer Metastasis and Treatment* 3:218–230.

12

Graphene Applications in Biology and Medicine

Stefano Bellucci
INFN Laboratori Nazionali di Frascati

12.1 Introduction

In agreement with International Union of Pure and Applied Chemistry (IUPAC), the recommended definition for graphene is the following: a single layer of carbon atoms of the graphitic structure, whose conformation can be described in analogy to a polycyclic hydrocarbon aromatic measuring almost infinite. It is further specified by the IUPAC itself that previous definitions such as graphitic layer, carbonaceous layer, or carbonaceous sheet have also been used, and sometimes still are, corresponding to that of graphene. However, since the term 'graphite' refers to the organization of the carbon chemical element in planar leaflet structures, within which each atom binds to three others in a hexagonal cell morphology, in turn organized into one regular three-dimensional structure, it is not correct to use for a single layer terms referring to the graphite, which instead implies a three-dimensional regular structure. The term 'graphene' should only be used when reactions, structural relationships, or other properties of a single layer are discussed [1].

The exact history of graphene and how it appeared in the scientific landscape is quite fascinating. Theoretically, as an integral part of various three-dimensional materials, graphene was studied since 1940. In 1947, Philip Wallace [2] wrote a pioneering treatise on the electronic characteristics of graphite that aroused great interest in the exploration of the material, but only with the work Novoselov et al. [3,4] and Zhang et al. [5] interest in graphene grew thanks to its unusual properties. In 2004, Novoselov et al. reported the techniques of development and processing of a sample, in which microscopic monolayers of graphene crystalline on silicon support (silicon dioxide and silicon dioxide) could be observed. Subsequently, this technique was globally adopted as a protocol for the production of samples of a single layer of graphene, allowing the development of subsequent studies on conductivity and chemical–physical characteristics of the material. As a consequence, in 2010 the Nobel Prize in physics was awarded jointly to Andre Geim and Konstantin Novoselov for the innovative experiments concerning the two-dimensional graphene material. However, as pointed out by de Heer (as reported in [6] by Eugenie Samuel Reich), there is a common misconception regarding the 2004 article by Geim and Novoselov: in his letter, addressed to the Nobel Prize committee, the latter highlights that most of the scientific publications incorrectly mention the 2004 article as the one presenting the 'scotch tape method' to the world and the unique electronic characteristics of graphene. In fact, de Heer points out that these discoveries were not reported in reference to the single graphene layer of 2004, but only in an article of the following year always by the same authors. It should also be noted that in reality, graphene was identified and characterized in numerous articles prior to that of 2004, in which reference was made to thin graphitic films or even occasionally to monolayers.

Dreyer et al. [7] have elegantly presented, in the following years, a detailed account of the synthesis and various characterizations of graphene; the following page shows schematically a representation of the history of the preparations, purifications, and characterizations of the material: we note how the term 'graphene' was proposed in 1962 by H. P. Boehm et al. [8,9] who later reported his observations on the allotrope in 1986, demonstrating beyond doubt the existence of the latter. It is also noted that, again in an article of 1962, Boehm et al. isolated reduced graphene oxide (GO) with heteroatomic contaminations and from this, they observed a significant reduction of electrical conduction compared to the pristine graphene produced by means of the 'scotch tape method'. The reason why we do not assign particular historical importance to these discoveries is that despite having the work of Boehm et al. reported the existence of the graphene before 2004, these failed to describe the peculiar properties of the material and limited themselves to a blind observation of the experimental data. Therefore, the 2005 article by Novoselov and Geim can be considered the first concerning the isolation of pristine graphene (i.e., a single layer of graphene not showing heteroatomic contaminations) that presented its unique characteristics in the world, triggering the period of gold of the material in the universe of chemistry and physics.

Since the pioneering articles of 2004/2005 have been highlighted, unique characteristics of this material and a significant number of other methodologies have been reported about its manufacture. Graphene has therefore sincerely captured the imagination of scientists in all corners of the world and is today an extensive and vibrant area of research: its use has led to an improvement in the performance of devices in a wide range of technological fields and a better understanding of basic physico-chemical processes [10].

12.2 Biological Applications of Graphene and Related Materials

Graphene is considered the best and most enduring monolayer capable of free existence. Its two-dimensional structure and the presence of delocalized π electrons can be exploited for the loading of drugs for hydrophobic interactions and π–π stacking. Furthermore, the availability of a large surface area (2,600 m^2/g) allows for a high density of bio functionalizations through covalent and non-covalent modifications. Several in vivo studies on the specificity of graphene have confirmed its potential for the replacement and implementation of materials currently used for bio-sensors and drug delivery [11].

12.2.1 Drug Delivery

Since its discovery graphene has shown excellent potential as a transport molecule (*carrier*) in *drug delivery* research. The high and defined surface area increases the opportunities for a targeted transfer from the administration site to the target site: polymer modifications and conjugation techniques lead, moreover, to an increase in biocompatibility. Many studies have been conducted, in recent years, on the transport of anticancer drugs, genes, and peptides through graphene and related materials: the simple physisorption, for π–π, interactions, can be used to load several hydrophobic drugs that, through the following functionalization with antibodies, can lead to the selective destruction of cancer cells. Thanks to its small size, intrinsic optical properties, large surface area, low cost, and non-covalent functional interaction with aromatic compounds, graphene has encouraging features for the nanocarrier approach. The extended molecular surface and interactions π–π or hydrophobic in particular, as can be seen in the references to the studies reported on the following page, contribute to the possibility of a high degree of loading of poorly soluble molecules, without compromising their potentiality or therapeutic efficiency. We also see how the use of graphene is extended to completely different fields, with extremely promising results in the biomedical field, with possible and future therapeutic application.

Liu et al. developed one of the first works in this field by synthesizing GO functionalized with polyethylene glycol (PEG) loaded with a camptothecin analogue (CPT), SN38. The NGO-PEG-SN38 complex exhibited good water solubility while maintaining the potentiality and efficiency of the loading.

The complex also showed high cytotoxicity in HCT-116 cells, about a thousand times higher than the free drug: camptothecin is a cytotoxic quinolinic alkaloid that has the ability to inhibit the activity of the enzyme DNA-topoisomerase I. The CPT binds itself to the covalent I-DNA mouse complex with the formation of a highly stabilized ternary structure: this assembly leads to the non-rewinding of the DNA with consequent cellular apoptosis. The CPT, in particular, binds the enzyme and the DNA through the hydrogen bond: the most important part in the structure is the E-ring which interacts with three different H-bridges with the enzyme itself (Figure 12.1). The hydroxyl group at position 20 forms a hydrogen bond with the side chain of the enzyme at an aspartic acid residue (Asp533); the lactone is bound by two H-bridges to the amine group of Arg364. Camptothecin, in particular, is selectively cytotoxic for the cell in the S phase of DNA replication and its property is, in the first place, the result of the conversion of a single-stranded fragment into a double-stranded fragment when the replication fork coincides with the breaking complex formed by DNA and CPT. In another study, the same group investigated the selective transport of rituxan (a specific monoclonal antibody to the CD20 protein, found primarily on the surface of B cells of the immune system) conjugated with PEG-NGO. In both cases, non-covalent interactions π–π are exploited for drug loading on the surface of the PEG-NGO complex and for pH-dependent release of the same [12].

Joo et al. reported studies of GO, loaded with doxorubicin (DOX) again via interactions π–π, and how this shows

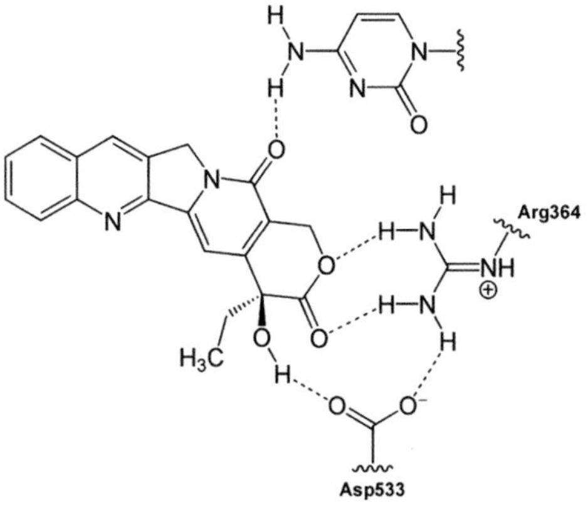

FIGURE 12.1 Interaction of CPT and the covalent complex topo I-DNA.

a drug release in specific cell sites as a result of glutathione (GSH) triggering. Another research group reported as GO loaded with DOX, exhibiting a greater ability to release to an acidic pH (=5.3) due to the reduction of interactions between the drug and the carrier: it is in fact known that the pH of the cellular tumor environment is more acidic than healthy one, and this evidence has been exploited to obtain a targeted drug release at the target cell. The GO-DOX complex showed increased cell toxicity and promising tumor inhibition with a mortality range of 66%–91%. Other chemotherapeutic drugs, such as Paclitaxel and Methotrexate, loaded on GO for π–π stacking and amide bonds, have shown surprising effects in the treatment of lung cancer and breast cancer, which resulted in an inhibition of tumor growth between 66% and 90% [13].

GO, loaded with a second generation of photosensitizers, chlorine e6 (Ce6), has led to greater accumulation in tumor cells compared to previous treatments, allowing greater effectiveness in photodynamic therapy (PDT).

Graphene-based materials (GFNs) have been conjugated with a series of biopolymers such as gelatin and chitosan, acting as functionalizing agents for subsequent pharmacological application. Natural biopolymers are biocompatible and biodegradable, and have low immunogenicity that can greatly reduce the toxic effect of graphene. Gelatin has been successfully used as a reducing and functionalizing agent for loading DOX onto graphene nanosheets (GS): the gelatin-GS complex showed a greater loading capacity compared to the usual carriers due to the large surface area and the high interaction π. The tinnitus gelatin-GS-DOX complex also exhibited high toxicity to MCF-7 cells for endocytosis. Chitosan, a linear cationic polysaccharide, obtained by alkaline deacetylation of chitin and composed of D-glucosamine and *N*-acetyl-D-glucosamine bound by bonds β (1–4), was used, in combination with graphene, for the loading of various compounds including ibuprofen, camptothecin, and 5-fluoroacyl. Frog et al. used GO functionalized with chitosan to transport ibuprofen (IBU), 5-fluoroacyl (5-FU), and CPT. The 5-FU showed a lower loading capacity due to the relatively hydrophilic character of the compound, to less interaction π–π and in the presence of di-amide groups. In a subsequent study, Bao et al. synthesized a chitosan-GO-CPT complex that showed characteristics of higher toxicity, compared to pure CPT, for HepG2 and HeLa cell lines.

The conjugation of iron oxide nanoparticles with GFNs makes the latter superparamagnetic and can be useful in transport applications. Yang et al. have prepared a hybrid and superparamagnetic GO by the addition of iron oxide nanoparticles (Fe_3O_4) for precipitation methods followed by the loading of DOX. The magnetic hybrid showed a good aqueous dispersion before and after the loading with DOX with the formation of agglomerates in acid solution and subsequent redispersion in basic solution. This pH-dependent release of GO- Fe_3O_4 nanoparticles can be explored and optimized for the development of controllable release systems. In Table 12.1, the applications of graphene-based systems in the field of the drug delivery mentioned above are summarized and the descriptions of the results obtained [12].

Drug Delivery: Release Controlled by Endogenous Stimuli

The release of a molecule in an area of interest plays an important role in the field of drug delivery. Recently, graphene-based drug delivery systems (*DDSs*), responding to various endogenous stimuli such as pH, redox potential, and specific biomolecules, have been widely used to increase therapeutic efficacy and reduce unwanted effects of the drug used (Table 12.2).

Release Mediated by pH Variation

DDSs sensitive to extreme pH variations, such as those occurring in diseases such as ischemia, infections, inflammation, and cancer, have been extensively studied in order

TABLE 12.1 Summary and Applications of Drug Delivery Systems Based on Graphene

Complex Examined	Drug	In Evidence
NGO-PEG-RB-DOX	DOX	Selective and pH-dependent release obtained with rituxan (CD20 + antibody) as a target agent
PEG-BPEI-rGO-DOX	DOX	Endosomal intracellular transport of photothermally induced DOX
PEG-GO-DOX	DOX	In vivo and in vitro release of DOX, dependent on the concentration of glutathione in the absence of fluorescent molecules
DOX-GO-CHI-FA	DOX	Accelerated pH-dependent release in acidic conditions
NGO-SS-PEG-DOX	DOX	Rapid intracellular DOX release from GO composites in the presence of significant GSH concentrations in a tumor environment (HeLa cells)
GO-FA-βCD-DOX	DOX	In vivo study of DOX loading on GO-folic acid complexes β cyclodextrin. Good in vitro cytocompatibility and inhibition of tumor growth
NGO-PEG-SN38	SN38	Transport of anticancer molecule insoluble in water, with greater efficiency compared to CPT

TABLE 12.2 Summary of the Modalities of Release in Different Models of Drug Delivery

Complex Examined	Drug	Stimulus Utilized
GO-PEG	DOX, Ce6	Light/pH
GO-Pluronic F127	DOX	pH
GO-IONP	DOX	Magnetic field/pH
GO-Au	DOX	Light/pH/redox
GO-Ag	DOX	Light
GO-CuS	DOX	Light/pH
GO-acido ialuronico (HA)	DOX, Ce6, ICG	Light/pH
GO-lipid	DOX	pH
GO-DNA	DOX	ATP
GO-silica	DOX	Light/pH
GO-PDEA	CPT	pH
GO-peptide-FA	CPT	pH
GO-transferrina	Dihydroartemisinin	pH
GO-PEG	DOX	pH
GO-acido sulfonico	DOX e CPT	pH
GO-gelatin	DOX, MTX	pH
GO-chitosan	DOX, CPT	pH
GO-clorotossina	DPX	pH
GO-PEG-BPEI	DOX	Light
GO-SS-mPEG	DOX	Redox

to implement easily controllable systems. Since the tumor microenvironment is more acid when compared to healthy tissue, the search for pH-dependent systems has been explored for effective use in cancer therapy. In acidic conditions, hydrophobic loads like doxorubicin can be protonated, which reduces the amount of interactions π–π and of the hydrophobic ones between the molecule under examination and the surface of the graphene, realizing a pH-dependent system. In one of the first works in this sense, reported by Dai et al., the GO was functionalized with polyethylene glycol (*PEG*) and studied as a two-dimensional nano-carrier for loading various substances. In this work, an antibody (anti-CD20, rituxan) was conjugated with the PEG-GO system for a targeted and specific transport dependent on pH variation: starting from this study, various surface loading used for the realization of a release model depends on the hydrogen ion concentration. For example, Pluronic F127 was used to make PF127-GO nanocomposites that exhibited a high loading capacity (289% w/w) and pH-controlled release; similar characteristics have also been observed for lipid functionalizing lipid with DOX.

In order to increase the therapeutic efficacy and to reduce the side effects related to the administration of the drug, various systems based on graphene have been used: graphene sheets conjugated with a peptide (chlorotoxin) (CTX-GO) have been prepared and used for the transport of DOX for non-covalent CTX-GO-DOX interactions. Chlorotoxin or CTX is a peptide of 36 amino acids that is found, together with other neurotoxins, in the venom of the yellow scorpion (Leiurus quinquestriatus), a scorpion of the Buthidae family. This toxin blocks the chlorine-dependent ion channels, acting as a neurotoxin: this fact, together with the fact that chlorotoxin exceeds the blood–brain barrier (BEE) and binds to the tumor cells of the gliomas, has suggested that the same can be usefully used in the treatment of the same tumor forms. The release of DOX proved to be pH dependent and showed good diffusion properties. In a subsequent study, Depan et al. used folic acid conjugated with chitosan to modify nano-GO later used to transport DOX; in a recent work, nano-GO functionalized with dihydroartemisinin (DHA) and transferrin was used in the development of a controlled-release chemotherapeutic drug: in this case, a significant increase in tumor specificity was observed. In addition, hyaluronic acid (HA) was used for the modification of nano-graphene, aimed at the transport of an anti-tumor drug by means of endocytosis-mediated HA receptors.

Lastly, in the last few years, non-neutral nano-carriers, in which the surface charge can be modified from negative to positive by pH lowering inducing the loading or release of a drug, have received great interest in the field of DDSs. In a recent work by Yang, Feng, and Liu, variable-load GO was developed: 2,3-dimethylmaleic (DA) and poly-allylamine (PAH) were used together to combine this reversible change to combine PEG-GO obtaining a nano-compound GO-PEG-DA. It has been studied how this ternary compound exhibits strongly stable negative charges under a physiological pH (approximately 7.0), but these fillers are rapidly converted into positive under weakly acidic conditions (pH 6.8), at which the process of loading DOX onto GO-PEG-DA has been significantly increased. As a result, the GO-PEG-DA/DOX complex within the tumor microenvironment (pH $>$ 6.8) showed greater efficacy in the destruction of drug-resistant MCF-7/ADR cells, which are unlikely to be attacked in the presence of free DOX under the same pH conditions.

In summary, nano-graphene-based DDSs sensitive to pH changes were extremely promising for increasing the effectiveness of the usual cancer treatment drugs.

Redox Stimulus-Mediated Release

It is well known that the cellular redox environment is strictly controlled by the level of glutathione (GSH): GSH is a tripeptide with antioxidant properties, consisting of cysteine and glycine, bound by a normal peptide bond, and glutamate, which is instead linked to cysteine with an atypical peptide bond between the carboxylic group of the glutamate side chain and the cysteine amino group (Figure 12.2); glutathione is a strong antioxidant, certainly one of the most important among those that the body is able to produce. Relevant is its action against both free radicals and molecules such as hydrogen peroxide, nitrites, nitrates, benzoates, and others. The essential element for its correct functioning is the NADPH. This molecule is a derivative of vitamin PP (nicotinic acid) with the function of oxidative–reductive cofactor of the enzyme glutathione reductase (or GSR). This enzyme regenerates reduced glutathione from the oxidized molecule (or GSSG) through the electrons transferred from NADPH to GSSG. A decrease in GSH levels always leads to a consequent increase in the possibility of oxidative stress, while an excess of GSH in the cytoplasm increases the antioxidant capacity: the presence of glutathione could be exploited as a stimulus for the release of substances from DDSs.

In a paper by Shi et al., a coating of PEG was used for the modification of nano-GO (NGO) by formation of disulfide bridges, leading to the formation of an NGO-SS-mPEG complex. This innovative system has been used for the transport of DOX by interaction π-π and showed the ability to be introduced into the cellular environment by endocytosis: in the presence of the cytoplasmic GSH concentration, the disulfide bridge of the NGO-SS-mPEG complex is rapidly reduced leading to the release of the loaded drug. In another work, NGO-Ag nanocomposites were prepared for intracellular drug delivery monitored by Raman scattering (SERS) and fluorescence spectroscopy. Doxorubicin is directly bound to the NGO-Ag nanocomposite for formation

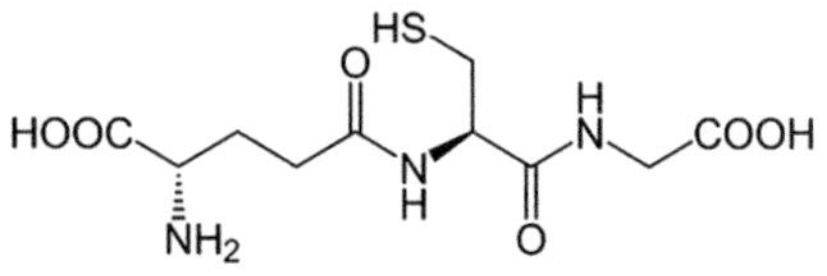

HS, HOOC, NH_2, O, N, H, COOH

FIGURE 12.2 Structure of the tripeptide glutathione.

of disulfide bridges, which can then be broken down by intracellular GSH leading to diffusion of the loading. In addition to the possibility of redox-mediated release from molecules following superficial changes, in a subsequent work it was established that the degradability characteristics of the GO can be regulated by the redox sensitivity of the superficial coating: it has been discovered that GO without any surface coating, although proving to be toxic for macrophage activity, can be gradually degraded through oxidative inducing enzymes such as horseradish peroxidase (HRP); at the same time, GO coated with biocompatible macromolecules, such as PEG or bovine serum albumin (BSA), does not show evident cellular toxicity but is degraded with difficulty in the organism. Therefore, to obtain functionalized and biocompatible GO, which can undergo enzymatic degradation, the latter has been conjugated with PEG by reversible disulfide bridges, thus obtaining GO-SS-PEG with negligible toxicity and considerable degradability. It is thus seen that a surface coating responsive to redox reactions can not only be used for the synthesis of intelligent DDSs, but also to mark and influence the biodegradability characteristics of the graphene itself.

Release Mediated by Biomolecules

In addition to the release from pH-dependent DDSs and redox balances, transport systems have been studied and developed in which the release mechanism is linked to the specific presence of a specific biological molecule. In a recent work by Gu et al., adenosine-5′- triphosphate (ATP), the main energetic molecule of cellular metabolism, has been chosen as a target for the control of the release capacity by nano-carrier of GO. In this work, a hybrid nano-aggregate GO-DNA was prepared containing a single strand of DNA1, DNA2, the aptamer of ATP (the aptamers are nucleic acids having the property of binding to a molecule or a protein) and GO, and the latter used as a nano-platform for loading the drug. It has been seen that the individual strands of DNA1 and DNA2 together with the aptamer of the ATP can cross-link with each other on the surface of the GO, effectively inhibiting the release of DOX from the nanosheets. In the presence of ATP, however, the interaction between the latter and the aptamer can induce the dissociation of the GO-DNA aggregate, promoting the release of DOX from the nanosheets.

Drug Delivery: Release Controlled by Exogenous Stimuli

In addition to endogenous stimuli, there are a number of external physical impulses potentially useful for controlling DDSs such as light, magnetic fields, and temperature. Differently from what was discussed for endogenous stimuli (which were present within the same cellular environment), DDSs that respond to this type of stress can show or exercise amplified therapeutic functions only under specific signals applied to the cellular environment from outside (Table 12.2).

Release Mediated by Electromagnetic Radiation

By photothermal therapy (PTT), we mean the heating, in which heat generated by appropriate nanoparticles, following irradiation by near-infrared radiation (NIR). To date, a wide variety of organic and inorganic compounds, including nano-graphene, have been investigated as effective photothermal agents for direct tumor cell ablation; on the other hand, unlike high temperature heating (e.g., >50°C), a mild warming, which elevates the temperature of the tumor to 43°C–45°C and does not induce certain cell death, it has been discovered to be useful to increase the loading capacity of drugs (absorbers in NIR) and their subsequent release, for a more effective cancer therapy. In a series of works by different authors, nano-graphene and its derivatives have been reported as effective nano-carriers for the transport of a number of aromatic molecules. In a 2011 work by Yang, Feng, and Liu, these show how a photosensitizer, chlorine 6 (Ce6), can be effectively loaded on the surface of nGO-PEG for interactions π–π and hydrophobic interactions. These have also noted how a mild photothermal heating induced by a laser radiation of 808 nm can greatly increase the loading of Ce6 by nGO-PEG, without, interalia, inducing evident cytotoxicity at the cellular level and also increasing the efficacy of photothermal therapy against the tumor itself. In a subsequent work, reduced nano-graphene functionalized with PEG was used for the transport of resveratrol (RV), forming NrGO-PEG/RV: under NIR irradiation for a limited period of time, the RV released by the complex grew significantly, contributing, consequently, to an increased apoptosis. Therefore, as nano-carriers with strong NIR absorption, the graphene and its derivatives have proved to be promising DDSs mediated by electromagnetic radiation: in particular, a mild heat generated by photothermal effect can lead to a significant increase in the control of the concentration of absorbed molecules and subsequently released, thus leading to the reduction of side effects currently present in healthy tissues.

Release Mediated by Magnetic Fields

In the past few years, various nanocomposites based on graphene with peculiar magnetic properties have been used for the realization of controlled delivery drug delivery. Iron oxide nanoparticles (IONPs) decorated with GO (GO-IONP) were first used by Yang et al. as nano-carriers for the release of DOX mediated by pH variations: it was then discovered that cancer cells, incubated with GO-IONP-PEG-DOX under a magnetic field, showed a high loading of DOX, while a small absorption had been highlighted for the same cell culture in the absence of the applied field, thus demonstrating the effectiveness of the field in the elimination of cells following induced absorption.

Release Mediated by Temperature Variation

In addition to responses due to light and magnetic field, temperature variations have shown to be useful for the controlled release of molecules of biological interest. Therapy refers to the use of heat as a therapeutic tool for the treatment of diseases, such as tumors. Generally, in cancer therapy, heat is applied with the aim of increasing the temperature of the tissue by only a few degrees, in order to exploit the increased sensitivity of tumors to ionizing radiation and some drugs. Hyperthermia is a treatment performed under the temperature range roughly between 41°C and 47°C. At these temperatures, greater sensitivity to heat of tumors was observed experimentally compared to healthy tissues: when higher temperatures are applied, higher than about 50°C, the treatment is called thermotherapy; this catalyzes the rapid destruction of the fabric. However, at these temperatures, there is no difference in the sensitivity to heat between healthy tissue and neoplastic tissue; for this reason, thermotherapy must be applied accurately and in the right position because when the tissue is heated, it necrotizes. The poly (*N*-isopropylacrylamide) (PNIPAM), one of the most known thermosensitive polymers with a Lower Critical Solution Temperature (LCST) easily modifiable in water, has been thoroughly used as a material responding to variations of temperature. We recall that the LCST can be defined as the critical temperature below which the components of a mixture become fully soluble in all compositions. It is generally pressure-dependent, increasing directly proportionally to the pressure itself. In the case of polymeric solutions, the LCST depends on the degree of polymerization, on the size, and on the composition and architecture of the polymer. PNIPAM can also be used to functionalize GO through click-chemistry, obtaining GO-PNIPAM nanocomposites, subsequently loaded with IBU or CPT, which show dependent temperature release profiles [14].

12.3 Toxicity of Graphene and Related Materials

As already seen, the GFNs range in shape, size, surface area, number of layers, side dimensions, chemical surface, hardness, density of defects, and purity; all these properties significantly influence the interactions of GFNs with biological systems. Generally, GFNs with limited dimensions, sharp edges, and rough surfaces are introduced into cells more easily when compared with larger and more regular members. Within this family, the monolayer graphene has the maximum surface area allowed as each atom lies on a plane, providing an extremely high loading and functionalization capacity. For biological molecules, the members of the more stratified GFNs result in a lower adsorption capacity: the lateral dimensions, which range in a range between 10 nm and 100 μm, influence cellular uptake modalities, renal disposal, and other biological interactions. Finally, since graphene is possible for different synthesis modes, for example, mechanical exfoliation or processing of graphite intercalation compounds, it is inevitable that GFNs contain impurities, such as chemical additives or interlayer residues, which may include nitrates, sulfates, and peroxides.

12.3.1 Toxicity *In Vitro* on Mammal's Cells

An initial screening of new in vitro toxicity materials generally uses several cell lines. Literature data suggest that exposure to GFNs may result in cytotoxicity and/or genotoxicity in mammalian cells.

Graphene

A comparative study measuring mitochondrial toxicity and cell membrane integrity in neuronal cells has suggested that the biological activity of graphene and single-walled carbon nanotubes (SWCNTs) strongly depends on their shape. Following a 24 h exposure, the metabolic activity of PC12 cells decreases in a variable manner: graphene leads to high toxicity at low concentrations and low toxicity at high concentrations, even more than compared to SWCNTs. The highest concentration of graphene used in these studies (100 μg/mL) significantly increases the release of LDH (a total LDH level higher than normal is found in diseases such as myocardial infarction, pulmonary infarction, acute viral hepatitis, toxic hepatitis, shock condition, severe anemia, muscular dystrophy, diabetes, renal failure, cirrhosis hepatic, leukemia, and neoplasms; decreased values are found in subjects exposed to ionizing radiation) and the generation of reactive oxygen species (ROS). In addition, caspase-3 activation (there are two types of caspases: initiator caspases (caspase-2, caspase-8, caspase-9, caspase-10) that cut off inactive forms of other caspases called effector (caspase-3), caspase-6, caspase-7) activating them, the effector caspases in turn will cut precise protein substrates, giving rise to the apoptotic process) suggests a time-dependent increase in the apoptotic process at a concentration equal to or greater than 10 μg/mL. Yuan et al. have compared the potential cytotoxicity of graphene and SWCNTs on the HepG2 cell line: overall, a concentration of 1 μg/mL of both nanomaterials led to the different expression of 37 proteins involved in cell metabolism, redox regulation, cytoskeletal formation, and cell growth. An interesting discovery has been that graphene and SWCNTs produce different pathways of expression of calcium-binding proteins, thus indicating a different mode of action. Finally, pristine graphene has been identified as responsible for increased ROS concentration and apoptotic processes of macrophages of RAW 264.7 cell line, important for the innate immunity system.

Graphene Oxide (GO)

The GO is the member of the graphene family whose toxicity has been most investigated. Although the first toxicity studies did not show cell loading or effects on the morphology, viability and integrity of the membrane in cells affected by adenocarcinoma are influenced by exposure to GO. This is in fact able to induce oxidative stress at a concentration equal to or greater than 10 μg/mL. In a subsequent study, Hu et al. using the same cell line, they reported a cytotoxicity directly proportional to the concentration of the product, which can be strongly reduced by incubation with 10% of fetal bovine serum, due to the great capacity of protein absorption by the GO. Subsequently, the toxicity, genotoxicity, and mechanism of action of the GO were studied in a variety of animal and plant cell lines, including normal and immortalized cells, immune cells, stem cells, and blood flow components. In studies including immortalized cells, the toxicity of GO was studied with the HepG2 line: in this case, a decrease in fluorescence intensity was observed starting from the concentration of 4 μg/mL, which indicates possible damage to the plasma membrane; the loss of structural integrity of the plasma membrane is associated with a strong interaction of the GO with the double phospholipid layer. The use of TEM and SEM (electronic scanning microscope) has shown that the GO has the ability to penetrate through the membrane, leading to an alteration of the cell morphology and an increase in the number of cells subject to apoptosis. Concerning the mechanism of interaction, the authors concluded that damage to the plasma membrane and oxidative stress play a crucial role in the cytotoxicity of the component. Yuan et al. subsequently evaluated the toxicity of GO and oxidized SWCNTs in HepG2 cells: similar to their previous study, a concentration of 1 μg/mL oxidized GO and SWCNTs lead to an alteration of the expression of proteins involved in metabolic pathways, cytoskeletal formation, and cell proliferation, with a much less pronounced action of the GO compared to that of SWCNTs. Furthermore, a lower reduction in proliferation rate, a slightly modified cell cycle, and a high concentration of intracellular ROS were observed in cells treated with GO, suggesting that GO has lower toxicity in HepG2 cells. The induction of cytotoxicity, genotoxicity, and oxidative stress was also studied in pulmonary fibroblasts: the MMT assay indicated a significant decrease in cell viability and an increase in toxicity following prolonged treatment, as well as the possibility of apoptosis at concentrations of 100 μg/mL; DNA damage has been identified for all tested concentrations including that of 1 μg/mL. The MTT assay, where the acronym indicates the 3-(4,5-dimetiltiazol-2-yl)-2,5-diphenyltetrazolium bromide compound, is a standard colorimetric assay for the measurement of the activity of enzymes that reduce the MTT at formazan (Figure 12.3) giving the substance a blue/violet color in the experiment; this occurs predominantly in the mitochondria, and the assay can be used to determine the cytotoxicity of drugs or other types of chemically active and potentially toxic substances. In fact, the mitochondrial enzyme succinate dehydrogenase is active only in living cells, and its function consists in cutting the tetrazolium ring of MTT (which appears yellow) with the formation, consequently, of formazan (which is a blue salt).

Reduced Graphene Oxide (rGO)

In the first studies of reduced GO toxicity on three different cell lines, it has been reported that the latter has less accentuated toxicity and therefore greater biocompatibility when compared with SWCNTs. The diacetate fluorescein test showed significant cytotoxicity effects for rGOs with an average lateral size of 11 nm, even at the lowest concentration of 1 μg/mL and following an hour of exposure; rGOs with an average lateral size of 3.8 μm on the other hand showed lower cytotoxicity compared to systems with dimensions of 91 and 418 nm. Assays for the estimation of RNA flow from the cellular environment, indirect indicators of membrane damage, have confirmed a response strongly dependent on the size and shape of the reduced GO (rGO) with hMSCs. The rGO of smaller size showed an outflow of RNA higher than that of a larger size; moreover, the rGO showed ROS levels 13–26 times higher than the control sample, thus suggesting the involvement of oxidative stress in the cytotoxic mechanism. In genotoxic studies, after an hour exposure of rGO having an average lateral size from 11 to 91 nm, increases in the frequency of DNA damage and chromosomal aberrations at concentrations of 0.1 and 1.0 μg/mL have been observed. Using the MTT test, Hu et al. have found that nanosheets of rGO with an average thickness of 4.6 μg reduce cell viability from 47% to 15% at concentrations, respectively, of 20 and 85 μg/mL.

Functionalized Graphene Nanomaterials

Many of the GFNs tend to aggregate into physiological solution due to electrostatic interactions and nonspecific binding with proteins. Thus, the development of functionalized GFNs led to increased solubility and biocompatibility, and consequently reduced cytotoxicity and

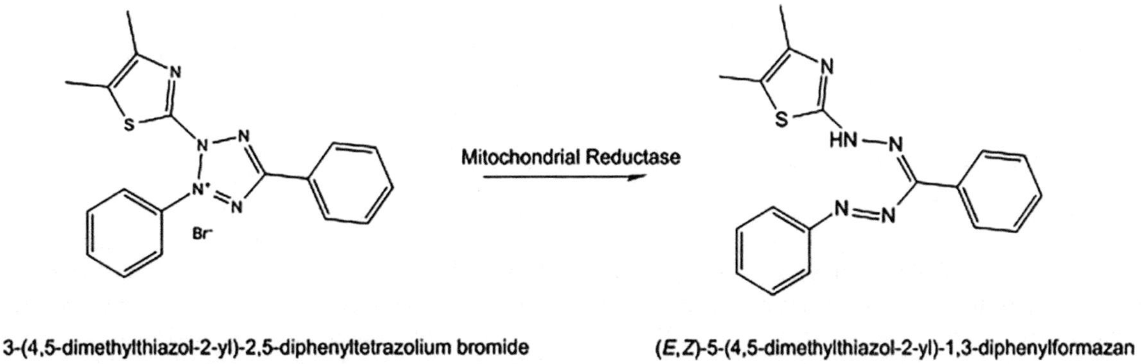

Mitochondrial Reductase

3-(4,5-dimethylthiazol-2-yl)-2,5-diphenyltetrazolium bromide

(*E,Z*)-5-(4,5-dimethylthiazol-2-yl)-1,3-diphenylformazan

FIGURE 12.3 Schematics of the reduction of the MTT to formazan.

genotoxicity. As said, two main methods are used for the synthesis of functionalized compounds: covalent interactions and non-covalent physisorption. The molecules and polymers used for the modification of GFN include different types of aliphatic and aromatic amines, amino acids, silanes, or enzymes. Non-covalent functionalization methods use hydrophobic interactions, π–π interactions, van der Waals forces, and electrostatic bonds, and these seem to be more versatile than covalent methods. Studies on covalent and non-covalent functionalization have shown a different decrease in toxicity and intensity of side effects in the members of GFNs.

In a study by Sasidharan et al., the prinine graphene toxicity was compared and functionalized in monkey renal epithelial cells, RAW 264.7 rat macrophages, and primary components of the human blood stream. In monkey cells, the internalization of functionalized graphene within cells has not shown any short-term toxicity, while the accumulation of pristine graphene on the cell membrane leads to ROS-mediated apoptosis. Similarly, pristine graphene is mainly thought to be on the surface of RAW 264.7 macrophages, leading to an impressive reduction in cell viability and a consistent increase in intracellular ROS in 24% of cells at a concentration of 75 μg/mL. For comparison, only 4% of cells treated with functionalized graphene showed ROS generation, with non-toxic effects at concentrations greater than 75 μg/mL and also as a result of high cellular loading. Finally, the treatment of mononuclear cells from peripheral blood with pristine graphene produced a high expression of IL-8 and IL-6 (thanks to the secretion of interleukins, the cells of the immune system can regulate the activity of other cells, triggering one of the most important mechanisms of cellular communication at the level of the immune system, their action can be autocrine, paracrine, and, in rare cases, endocrine) compared to treatment with functionalized graphene, indicating a greater inflammatory capacity. These results suggest that the superficial functionalization of pristine graphene can prevent most of its toxicity. Unlike GO and rGO, which cause a strong aggregation response in the platelets, the amino-functionalized graphene has no stimulating effects on human platelets; the intravenous administration of functionalized graphene does not lead to an increased lysis of erythrocytes or other diseases in rats. These results indicate how appropriately functionalized graphene can be potentially safe for in vivo biomedical applications. Functionalization, however, does not always lead to complete elimination of GFNs toxicity.

12.3.2 Toxicity *In Vivo* on Mammal's Cells

Information on in vivo toxicity is critical in case the GFNs are to be used for a system of *drug delivery*. The toxicity of GO was studied by administration in guinea pigs: no problems were found in rats, exposed intravenously, at low GO concentrations (0.1 mg) and medium ones (0.25 mg), whereas a high dose (0.4 mg) leads to a chronic toxicity. A substantial proportion of subjects died from suffocation within 1–7 days of administration due to blockage of the respiratory tract for the formation of agglomerates of GO. The maximum accumulation of GO occurs mainly in the lungs, followed by the liver and kidneys; the hepatic gathering indicates that the GO is basically eliminated by excretion into the bile, as only a small amount of material has concentrated in the kidneys. A similar study has also shown that GO is rapidly subtracted from the bloodstream and then accumulated in the liver and lungs, with the larger oxide (1–5 μm) concentrated in the airways and the thinner one (110–500 nm) retained in the liver. Also in this case, superficial changes significantly modulate the toxicity of graphene in vivo: a series of toxicological tests performed using different routes of administration (intravenous, oral, and intraperitoneal) for graphene and graphene functionalized with PEG were conducted on BALB/c rats. One hour after the administration of 20 mg/kg, nanosheets of PEG-graphene are distributed in a series of different organs; 3 days later, PEG-graphene is fundamentally concentrated in the reticuloendothelial system, including liver and kidney. Toxicological studies on nanosheets of PEG-graphene have not reported cases of deaths or significant weight loss, over a period of 90 days after treatment. The biochemistry of the bloodstream and hematological analyzes have not identified any changes in the sensitive markers of liver and kidney including alanine aminotransferase, aspartate aminotransferase, and alkaline phosphatase. In addition, no obvious systemic damage was found, except for discoloration in the liver and kidney, due to the accumulation of PEG-graphene in the first 20 days of treatment.

Recently, Yang et al. investigated the biodistribution and potential toxicity of GO and a series of PEG-based derivatives with different sizes and surface coatings, following oral and intraperitoneal administration in BALB/c rats of a dose of 4 mg/kg. No marked loading at the tissue level was observed following oral administration, indicating a limited intestinal absorption of these nanomaterials; on the contrary, as a result of intraperitoneal treatment, the researchers observed a greater accumulation of PEG-GO derivatives, but not GO, in the reticuloendothelial system, including liver and kidney. Similar to other studies, histological examinations of dissected organs and hematological analyzes have revealed negligible changes in animals, although the nanomaterial persists within the organism for over 3 months. These results therefore suggest that the characteristics of in vivo toxicity depend to a considerable extent on the methods of administration.

A subsequent study investigated problems related to the inhalation of four carbon-based nanomaterials, i.e. MWCNTs, graphene, graphene nano platelets (GNP) and carbon-black nanoparticulate matter, in adult Wistar rats. The rats were exposed to atmospheres containing 0.1, 0.5, or 2.5 mg/m^3 of MWCNT or 0.5, 2.5, or 10 mg/m^3 of graphene, GNP, and carbon-black for 6 h/day for five consecutive days. No undesirable effects were observed following exposure of GNP or carbon-black; on the contrary, subjects exposed to a concentration of 2.5 mg/m^3 of MWCNTs and graphene had a higher number than the norm of lymphocytes, cytokines, and an increased number activity of Y-glutamyl-transpeptidase, LDH, and alkaline phosphatase. Microgranulomas were also observed at the pulmonary level, with a more intense response provided by MWCNTs [15].

12.4 Conclusions

Graphene is a two-dimensional sheet consisting of a single layer of sp_2hybridized carbon atoms with exceptional electrical, mechanical, and thermal properties. From the day of its discovery, research on this material has grown exponentially, exploring the properties and the different fields of use: these range from electronics, to the development of optical systems, to photoconductive materials for solar cells or possible applications in the pharmaceutical and biomedical field. The focus of the present paper has been the exploration of the graphene potential for the loading of drugs by hydrophobic interactions and π–π stacking, as well as bio functionalizations through covalent and non-covalent modifications. A for a high density of the latter is allowed for, thanks to the availability of a large surface area (200 m^2/g) in graphene. We have reviewed *in vivo* studies on the specificity of graphene, strongly suggesting the possibility of implementing graphene-based solutions as a replacement for materials currently used for bio-sensors and drug delivery applications.

Acknowledgment

The participation of Alessio Di Tinno to the early stages of this work is gratefully acknowledged.

References

1. E. Fitzer, K.-H. Kochling, H.-P. Boehm, and H. Marsh, *Pure Appl. Chem.*, vol. 67, pp. 473–506, 1995.
2. P. R. Wallace, *Phys. Rev.*, vol. 71, pp. 622–634, 1947.
3. K. S. Novoselov, A. K. Geim, S. Morozov, Y. Zhang, and D. Jiang, *Science*, vol. 306, pp. 666–669, 2004.
4. K. S. Novoselov, D. Jiang, F. Schedin, S. Morozov, A. K. Geim, and T. J. Booth, *Proc. Natl. Acad. Sci. U.S.A.*, vol. 102, pp. 10451–10453, 2005.
5. Y. Zhang, Y.-W. Tan, and P. K. Stormer, *Nature*, vol. 438, pp. 201–204, 2005.
6. E. S. Reich, *Nature*, vol. 468, p. 486, 2010.
7. D. R. Dreyer, R. S. Ruoff, and C. W. Bielawski, *Angew. Chem. Int. Ed.*, vol. 49, pp. 9336–9344, 2010.
8. H.-P. Boehm, A. Clauss, G. O. Fischer, Z. Hofmann, and Z. Naturforsch, *Anorg. Chem. Org. Chem. Biochem. Biophys. Biol.*, vol. 17, pp. 150–153, 1962.
9. H.-P. Boehm, R. Setton, and E. Stumpp, *Carbon*, vol. 24, pp. 241–245, 1986.
10. A. Dale, C. Brownson, and C. E. B. Banks, *The Handbook of Graphene Electrochemistry*. Springer, London.
11. S. Goenka, V. Sant, and S. Sant, *J. Controlled Release*, vol. 173, pp. 75–88, 2014.
12. D.-J. Lim, M. Sim, L. Oh, K. Lim, and H. Park, *Arch. Pharmacal Res.*, vol. 37, no. 1, pp. 43–52, 2014.
13. L. M. Viculis, J. J. Mack, and R. B. Kaner, *Science*, vol. 299, no. 5611, pp. 1361–1361, 2003.
14. K. Yang, L. Feng, and Z. Liu, *Adv. Drug Delivery Rev.*, vol. 105, pp. 228–241, 2016.
15. X. Guo and N. Mei, *J. Food Drug Anal.*, vol. 22, no. 1, pp. 105–115, 2014.

13

Biofunctional Three-Dimensional Nanofibrous Surface for Tissue Engineering and Apoptotic Carcinogenic Approach

Lucas B. Naves
CAPES Foundation, Ministry of Education of Brazil
University of Minho
National University of Singapore

Luis Almeida
University of Minho

Seeram Ramakrishna
National University of Singapore
Jinan University

13.1 Introduction

One of the most aggressive types of skin cancer is melanoma. The melanoma cancer is originated from the malignant transformation of melanocytes. It has a low survival rate, is easy to relapse, and has a notorious high multidrug resistance (MDR). In 2014, only in the United States, it has been reported that nearly 76,100 people were diagnosed with new cases of melanoma, and an estimation of 9,710 was expected to die. The trend worldwide is to use more and more drugs and nanotechnologies, for biomedical applications [1].

Nanotechnology is the field of applied science focused on the design, development, fabrication, characterization, and application of devices and materials in nanoscale. It has been shown a potentially significant impact of nanotechnology on healthcare by delivering changes in drug delivery, disease monitoring, and diagnosis, regenerative medicine, implants, as well as research tools for biomedical science and drug delivery. The use of nanotechnology in the healthcare environment can create new alternative treatments, which are more efficient due to the target compounds minimizing the side effects and, consequently, minimizing the drug dosage and the treatment cost either to the patient or to the government. In addition to that, nanotechnology has been applied to most conventional melanoma therapies.

In the last few years, the researchers have been focused their research on discovering new alternative approaches to treating cancer using nanotechnology. This approach has demonstrated that nanodelivery of drugs for chemotherapy, immunotherapy, target therapy, and photodynamic therapy has increased the treatment efficacy. Nanofibers are now often used for the biomedical and sustained drug delivery systems (DDS). Nanofibers are ideal for this approach due to their characteristics, and their dimensions are similar to the components of the native extracellular matrix (ECM), possibility to mimic the fibrillar structure, which can provide essential cues for cellular survival and organization functions.

This chapter mainly demonstrates the melanoma skin cancer treatment and tissue engineering using different techniques, namely, electrospinning and 3D printing, both techniques for three-dimensional macro/nanostructure processing for biocompatible medical applications.

13.2 Skin

Skin is the largest organ in the human body: in adults, it represents 8% of their total body mass and has the surface area of about 1.8 m^2. The skin is mainly composed of three different layers, namely, the epidermis, dermis, and hypodermis, as shown in Figure 13.1. These layers are mainly composed of nerves and blood vessels. All these layers are responsible for protection from the risk that might be posed by thermoregulation and surrounding environment [2]. The main two layers are the epidermis and dermis. Epidermis is the outer layer, surrounded by extracellular lipid matrix, corneocytes and keratinocytes (which produce keratin), both packed into the extracellular lipid matrix (arranged in bilayers): some studies reported this arrangement as "brick and mortar", separated from the dermis by basement membrane and melanocytes (which provide pigmentation for the skin). The epidermis is responsible for promoting a protective barrier against bacteria or even harmful toxins. The dermis is located just under the epidermis, formed by a variety of connective tissues, such as nerves, lymphatic system, and many types of cells and blood vessels [3]. The dermis is formed by ECM, formed by glycosaminoglycans (GAGs), collagen, and elastin. This functions foremost as a barrier, as a regulator for water retention and heat loss, and as a sensory organ, preventing pathogens from entering the body. The primary type of cell found in the dermal layer is the fibroblast; these cells are responsible for synthesizing ECM proteins and enzymes and

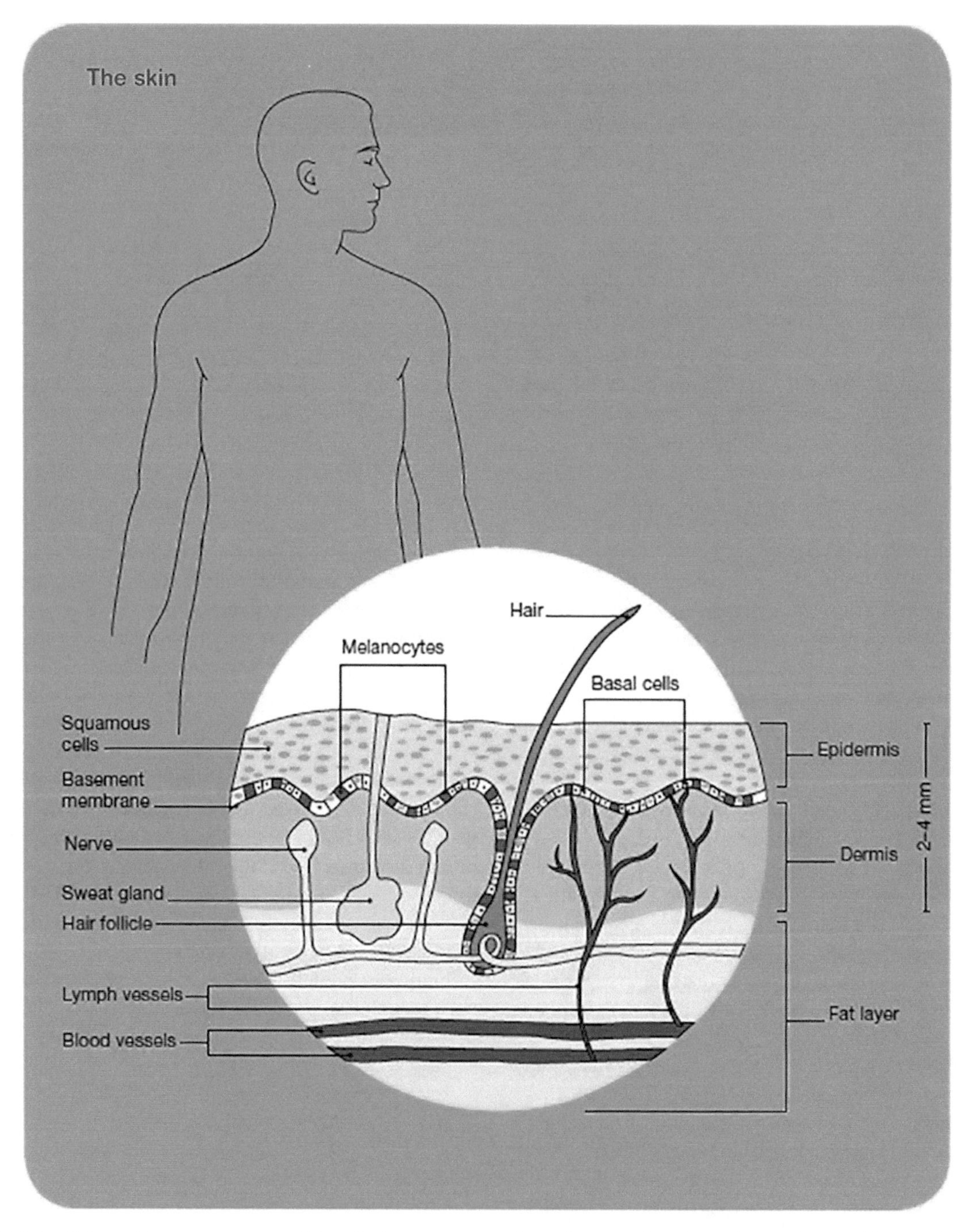

FIGURE 13.1 Schematic representation of the skin layers, namely, the epidermis, dermis, and hypodermis. (Source: Ref. [5], p. 7.)

activate the response to wound-healing process by protease and collagenases. The innermost layer is called hypodermis, which is responsible for thermoregulation and mechanical properties of the skin [4]. In addition to that, skin is an attractive model to test novel drugs and regenerative medicine, with emphasis on skin tissue regeneration for chronic or acute wounds. The skin surface is a potential route for local and systemic drugs delivery, using either nanoparticles or electrospun mats. The delivery of the drugs can also be reached into opened follicles of the hair. One of the primary and most important functions of skin in mammalian is to provide a protective barrier against UV radiation, fungi, bacteria, and any nanoparticle that might come from the external and natural environment [1].

Topical DDSs have been investigated and practiced since the end of the 1970s, by using either nanoparticles or transdermal patches to be administrated as readily available skin localized therapy [6].

The pH of the skin surface is directly influenced by several factors such as anatomical site, sweat, gender, and hydration [7]. Usually, the natural skin pH is around 4.2–5.6 [8]. The pH of the skin surface is very important for the degradation of drugs when approaching it through the transdermal pathway, and it also supports nanoparticles penetration. Electrostatic force can be decreased by a solution with lower pH [9].

A natural process that occurs in the human skin is the desquamation, which happens in the squamous cell (SC) layers. The whole process of renewing the skin surface is completed in approximately 14 days. This process is very important for our protection, once that the corneocytes can provide the elimination of certain matters such as pathogens, solid particles, or even cancer cell. It is important to keep in mind that the desquamation process may vary from one person to another depending on anatomy, malaise, and age [10,11].

13.3 Skin Cancer

Our body is constantly making new cells, aiming growth and proliferation, replacing dead cells, healing injuries, and replacing worn-out tissue. The cells normally multiply and die in an orderly way. The cancer is formed by cell disease, when cells divide and proliferate in an abnormal way, which may lead to lymph or blood fluid in the body, thus forming a lump called tumor. Briefly, the types of skin cancer referred to as nonmelanocytic skin cancer (NMSC) are basal cell carcinoma (BCC) and squamous cell carcinoma (SCC). Cutaneous malignant melanomas (CMs) appear in the literature also reported as melanoma or malignant melanoma [12]. Skin cancer like many other types of cancers has increased the incidence significantly with age and environmental etiologies (UV exposure), reflecting long latency between cancer establishment and carcinogen exposure. Many cases of skin cancer remain not reported, and the incidence of these types of cancer continues increasing dramatically. Table 13.1 shows a substantial underestimation of the epidemiological data.

13.4 Melanoma

Melanoma skin cancer is one of the most common types of cancer. It is known that in women, it is the seventh leading cancer and, in men, the fifth in the United States. Annually, 800,000 people are expected to develop melanoma skin cancer. Over the last few years, the incidence of melanoma skin cancer has increased substantially, especially in women, although it is most common among Caucasians with green or blue eyes and blondish [22]. Approximately one in 100 newborns will at some stage of their lives develop malignancy in their lifetime [23].

This type of cancer is the most aggressive skin cancer and represents only 3% of all diagnosed skin cancer in the United States. The high rate of patient's death is related to advanced melanoma metastasis, which usually occurs several months up to years after the primary melanoma diagnosis. At an early stage, the melanoma tumor can be removed, therefore promoting a survival rate up to 99%. When the patients are diagnosed with the late stage of melanoma, cancer may have a metastatic effect, thus spreading the malignancy cells to the other parts of the body.

When the rate of melanoma skin cancer is compared to different races, the rate is distinguished. It is ten times

TABLE 13.1 Overview of Epidemiological Data of BCC, SCC, and CM

Highlights of Epidermal Data		
BCCs	SCCs	CMs
Two to three million cases of NMSCs occur worldwide each year [13] Only in the United States, 1.3 million cases each year [14] Rates are expected to double in the next 30 years [14] NMSC mortality rate is lower in Nordic countries and higher in women and men in southern European countries such as Italy, Portugal, Greece, and Spain [15]		Sixty-five thousand people die per year worldwide [13]
Highest rates in elderly men and increasing incidence in young women [16]	Increasing rates, although it varies depending on the country [17]	Incidence rates are at least 16 times greater in Caucasians than African-Americans and 10 times greater than Hispanics [18]
In the United States, 30% of all new cancers diagnosed are BCC [19]	SCC has higher possibility to invade other tissue and cause more death than BCC [20]	132 new cases worldwide will occur each year [12]

Source: Adapted from Ref. [21].

greater in Caucasians than in Hispanics and 16 times greater than in African-American. It has been reported that people with darker skin have higher mobility and fatality, when compared to other skin tons, once they go undiagnosed for a while. In African-American, the incidence rate is lower, because this group has increased epidermal melanin, resulting in a sun photoprotection factor (SPF) of up to 13.4 in African-American skin. Another study suggested that the epidermal melanin can filter twice as much as UVB radiation in African-Americans as does in Caucasians.

In the literature is reported the incidence and correlation of UV radiation and melanoma rates. The UV radiation of sun exposure may cause irreversible damage; both UVA and UVB can damage directly and indirectly the DNA. Once these mutations occur in the DNA, it leads to the modification of the DNA, which will further develop skin cancer [24]. There are five important factors that might affect the incidence of UVR levels reaching the earth's surface such as

1. Clouds and others: at higher cloud-cover densities, there are lower UV levels. On the one hand, pollutants, haze, and fog can decrease UV incidence by 10%–90%, and on the other hand, sand, metals, and snow can reflect it up to 90% [25];
2. Shades: the shadows may protect the UVR exposure for up to 90% according to different types of shade; for example, dense foliage had the highest protection, while an umbrella at the beach shows low levels of solar radiation protection;
3. Seawater: it can reflect UV rays up to 15%;
4. Latitude: living closer to equator zones increases the incidence of solar radiation, thus maximizing the probability of skin cancer incidence. The lower the latitude, the higher the UVR incidence; on the other words, the UVR must travel a shorter distance through ozone-rich portions of the atmosphere, and in turn, more UVR is emitted [26];
5. Altitude: every 1,000 m increasing in elevation, the UVR intensity increases by 10%–12% [25].

In addition to the external process that may lead to skin cancer, the use of artificial UV tanning can also be linked to skin cancer development. Only in the United States, approximately 28 million Americans are reported to use the artificial UV tanning. The exposure to the tanning bed and sunlamps has been warned to be carcinogenic by the National Institute of Environmental Health Science (NIEHS). The effects of UV exposure either natural or artificial may take 20 years to result in skin cancer [27].

Fortunately, this paradigm has changed due to the advances done in new pathways and targeting in DDSs, which may play a major impact on the development of immunotherapy and target therapy for melanoma cancer. Most common type of drugs used for the treatment of melanoma metastatic effect is ipilimumab and vemurafenib, both approved by the U.S. Food and Drug Administration (FDA), although both therapies still present their limitations. These drugs play a different role targeting in melanoma cells. Ipilimumab can achieve durable benefits to the target cells by blocking the immune suppression of T cells which is induced by cytotoxic T lymphocyte antigen 4 (CTLA-4); however, around 80%–85% of the patients do not respond to this therapy. Vemurafenib can achieve rapid tumor regression, targeting melanoma harboring BRAFV600E mutations. The negative side of this drug is that patients might have drug resistance after 6 months of treatment. In 2015, Wang and Yun presented the melanoma network environment affecting the development of melanoma focusing the tissue hypoxia, macrophages, and stromal fibroblasts [28]. It is only possible to develop new drugs and strategies for melanoma therapy, prognosis, and diagnosis through a better understanding of how tumor microenvironment can directly affect the melanoma cancer progression.

13.5 Tissue Engineering

It is known that tissue engineering (TE) is a multidisciplinary field that integrates and encompasses engineering, biology, and chemistry. The main goal for TE approach is to create and mimic functional substitutes for damaged tissue. In the last few years, the need for regenerative medicine has increased, due to the increasing demand for surgeries, resulted from aging, disease, and athletic activities [29]. TE has emerged as an alternative to conventional approaches to repairing and restoring tissue functions, such as allograft, autograft, and xenograft. Macromolecules, either natural or synthetic based, are of great interest in TE field, once they possess a wide range of properties and tunability to match the desired functionality of specific target [30].

To allow the biological function, mass transport (permeability and diffusion), and mechanical integrity, it is crucial to control some important characteristics such as architecture and porosity of engineering scaffolds [31]. Some methods have been employed in order to develop constructive TE scaffolds, including electrospinning, gas foaming, molding [32], and particulate leaching [33]. Each of these approaches has some limitations when trying to mimic the biological function of natural tissue, because it is difficult to finely control the scaffold porosity, dimensions, and architecture [30].

13.6 Nanofibers Introduction

Nanofibers are widely used in various medical and biomedical applications such as TE, cell therapy, regenerative medicine, drug delivery, and cancer therapy which is the main goal of this doctoral thesis. These fibers have a diameter range of 1–100 nm; when the diameter is higher than 100 nm, they can be called either electrospun mats or scaffolds developed by electrospinning technique. Thus, the scaffold properties have been proven to be much more efficient than any other system for molecular and cellular applications, as compared to their micro–macro scale. We can distinguish some functional properties, such as high aspect ratio, quantum confinement effects, large surface area, and fast-absorbing biomolecules which provide abundant binding and adhesion to cell receptors. The strong cell–matrix interaction allows developing cell engineering, organs, and tissues [34]. Electrospun scaffolds possess among all their characteristics surface morphology, mechanical strength, structural integrity, chemical functionalities, and porosity; these characteristics can be tailored, according to their employed application by modulating the fiber orientation, material composition, dimension, and alignment. Monophasic electrospun mats have numerous advantages; however, some biomedical applications require composite scaffolds owing to superior functional and structural properties [35].

Several studies have been carried out on nanofibers composites, demonstrating that for biomedical and biological application, nanofibers composites are a better option than their monophasic nanofibers. When comparing chitosan nanofibers to blended polycaprolactone (PCL)/chitosan, the cell proliferation increased (>50%) and obtained better mechanical properties to the composite nanofibers polycaprolactone (PCL)/chitosan than the monophasic chitosan nanofibers system [36]. Nanofiber composites are engineered materials composed of two or more distinct phases combined to impart new desirable, chemical, physical, or even biological properties, processing bulk of different forms than those of any of the constituent phases.

Nanofibers composites can be fabricated using various techniques, such as self- ensemble, phase separation, template synthesis, and electrospinning. Among all these techniques, electrospinning appears to be the most widely used method, due to its ease of fabrication, tunable structure, and the possibility of large-scale fabrication. One of the most important characteristics is that the resultant fiber is well suitable for a various biomedical applications. Figure 13.2 shows a schematic representation of electrospinning apparatus.

The fabrication of nanofibers to develop scaffolds by electrospinning method is made by using an electrostatically driven jet of charged polymer melt or polymer solution [37]. Briefly, by applying an optimized electrical potential to the spinneret, the droplet located at the edge of the spinneret tip, polymer solution, gets electrified. There is an accumulation of charge on the surface of the droplet, which will be subsequently deformed into the Taylor cone. The droplet deformation is caused by

1. Coulombic force, which is exerted by the external electrical field applied;
2. Electrostatic repulsion within the charge of the surface droplet and the Coulombic force.

The threshold occurs, when the electric field applied surpasses the critical value, the polymer electrostatic force at this stage tends to exceed the viscoelastic force and surface tension, consequently, is formed a very fine jet of charged polymer, which will be ejected from the tip of the Taylor cone.

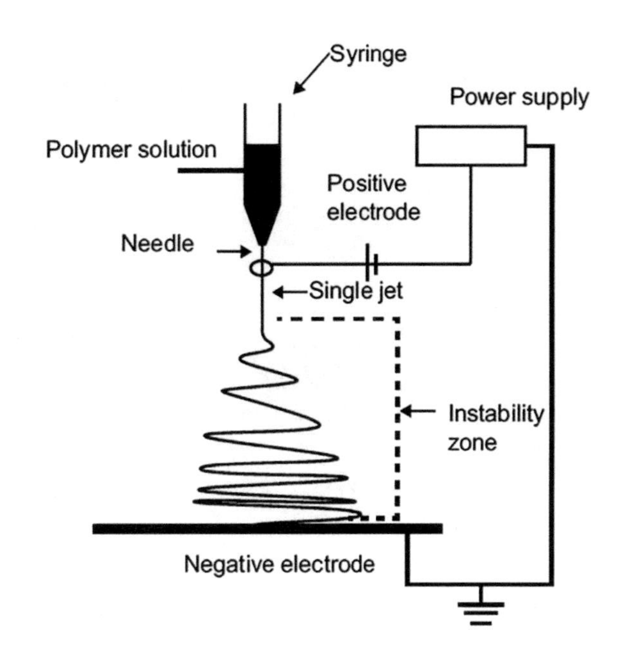

FIGURE 13.2 Schematic of electrospinning apparatus. (Source: adapted from Ref. [38], p. 119.)

As shown in Figure 13.2, the negative electrode, also called fiber collector, is the direction which the polymer jet will move towards. The instability zone is the distance from the Taylor cone, and the collector of the fibers is where will happen a rapid evaporation of the solvents molecules. Due to the mutual repulsion, while in transit, the different polymer strands in the jet get separated, this phenomenon is known as splaying, rising to a series of ultra-fine dry fibers. The fibers are collected at the fiber collector (grounded metallic target), and the range size may vary from nanometer to micrometers.

13.7 Nanofibers Composites Properties

Without compromising their volume fraction, nanofiber composites, when compared to conventional composites, have demonstrated to have a significantly large surface area. This large surface area can be useful to compensate the imperfection bonding between the fiber matrix interphases; therefore, even when prepared with the same volume fraction, nanofiber composites offer greater strength than conventional composites. These composite structures could be subject to surface either modification or treatment to impart or enhance new functional properties within them. Figure 13.3 shows some of the key characteristics of the nanofibers composites, which are very tunable, and it will basically depend upon the application and the specific need of the nanofibers composites. As an example, we can mention the poly(lactic-*co*-glycolic acid) (PLGA) nanofiber biodegradability which can be easily fine-tuned by just manipulating the concentrations of lactic or glycolic.

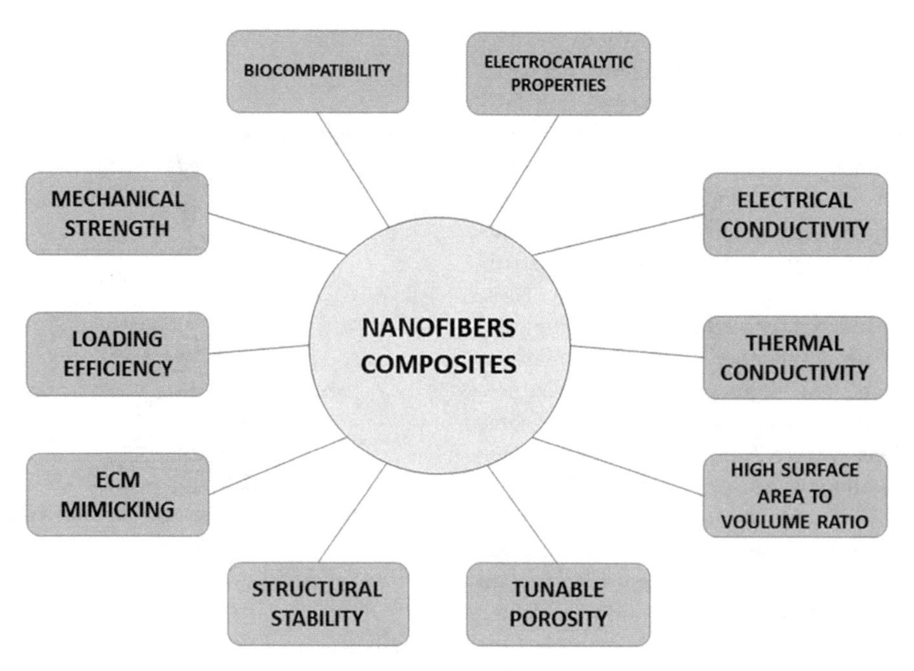

FIGURE 13.3 Schematic of the key nanofibers composites. (Source: adapted from Ref. [35], p. 6.)

The interaction between the surrounding matrix and the nanofibers is one of the important factors that determine the composite properties interactions. It has been demonstrated that these interface properties have the capability to control broadly properties, such as bonding strengths, bonding types, and dislocation densities. The surface structure of each nanofiber will determine each nanofiber composites interface; therefore, every single electrospun mats surface will have a unique property.

13.8 Surface Modifications

The biofunctionalization of fibrous scaffolds is essential, the surface of electrospun fibers can be incorporated with either bioactive molecules or bioactive agents or even coated for use in advanced biological application with therapeutic properties. Surface modification of the electrospun fibers is very important for their use as antimicrobial or bioactive scaffold material.

There are many natural polymers that can be employed, having unique biological functions improving the characteristics during the fabrication of electrospun scaffolds. In some cases, the release of the drug of hydrophobic therapeutic drugs can be directly blended into the electrospinning polymer solution, to fabricate the fibrous scaffolds. When using synthetic biodegradable polymers, they all are hydrophobic in nature, so, when developing electrospun scaffolds aiming cell adhesion to the surface of the scaffold, the surface modification is imperative, changing the surface property from hydrophobic to hydrophilic, thus enabling the cell adhesion, proliferation, and growth on the surface of the mats.

In a study of surface modification reported by Yao and colleagues [39], they used poly(4-vinyl-*N*-hexyl pyridinium bromide) to treat the polyurethane fibers. The polyurethane fibers were treated with the plasma treatment to produce peroxide and oxide groups to the mats surface, then the fibers were immersed in a solution of 4-vinyl pyridine monomer. Polyurethane fibers were then exposed to UV radiation, leading to the production of poly(4-vinyl pyridine), which were grafted into the polyurethane fibers. Following these steps, the fibers were immersed into heptane solution containing hexyl bromide, thus giving to the mats antimicrobial properties. In this study, the authors reported that after surface modification, the fibers had excellent antimicrobial activity against Gram-negative *Escherichia coli* as well as Gram-positive *Staphylococcus aureus*. Therefore, these fibers might have a potential application on protective textiles, filters, and biomedical devices.

In a similar study, Ma and co-workers [40] used poly(methacrylic acid) as graft polymerization, to modify the surface of the polyethylene terephthalate (PET). First, they treated the PET fibers with formaldehyde to introduce the hydroxyl functional group. The oxidation of hydroxyl groups using Ce (IV) formed free radicals. The following step was the grafting of polymerization of methacrylic acid. Finally, by using a coupling agent, poly(methacrylic acid)-grafted PET is grafted with gelatin. They could show that gelatin is covalently attached to poly(methacrylic acid) grafted on PET fibers. In Figure 13.4, it is possible to observe the better endothelial cells (EC) growing on tissue culture plate (a), and better cell adhesion and proliferation in (c) than (b), due to the surface modification of the electrospun mats by grafting gelatin, and (d), the AFM – Atomic Force Microscopy images of the EC growing on the surface of the surface-medicated mats.

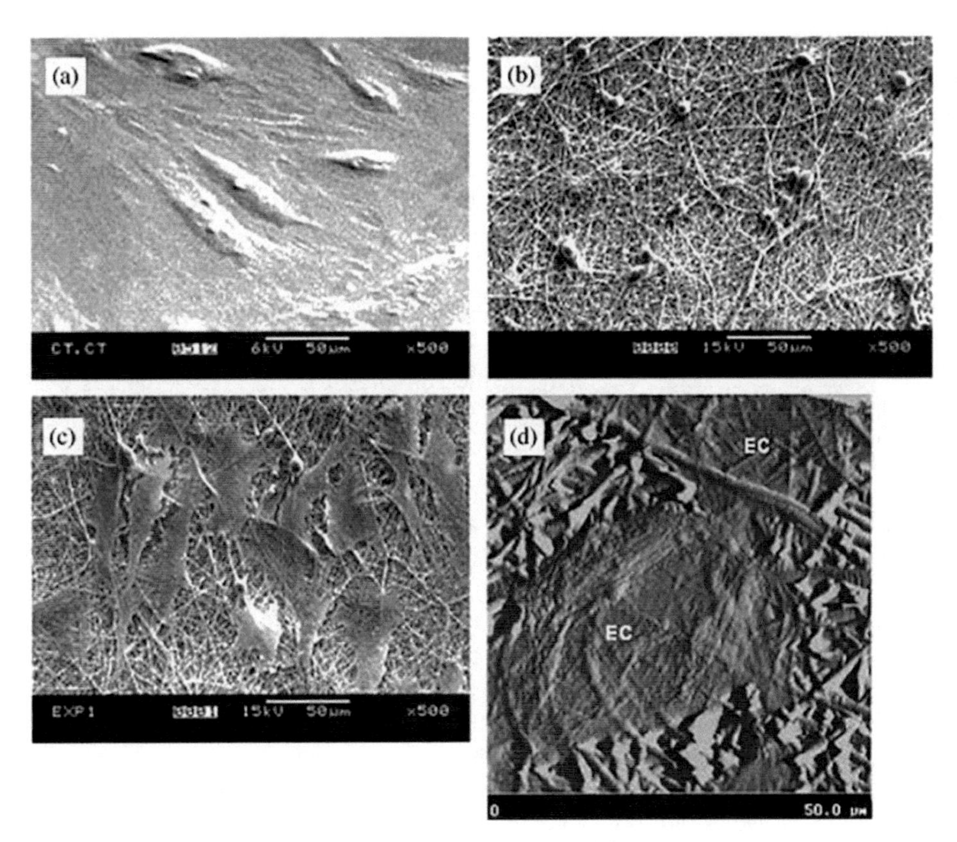

FIGURE 13.4 (a) ECs cultured in TCP, scanning electron microscope (SEM) image, (b) pure PET NFM, (c) gelatin-grafted PET NFM, and (d) ECs on the gelatin-modified PET NFM-AFM image. (Source: Ref. [40], p. 2533.)

13.9 Physicochemical Characterization at Materials Level

One of the major features for the characterization of electrospun mats is the fiber morphology, including diameter, alignment, shape (curved or straight; rounded or flat) and uniformity (uniform fibers of beaded). The morphology of prepared electrospun nanofibers can be fast screened by optical or light microscopy, to observe a nanofiber composite or a thin layer of nanofibers. Even though light microscopy is more convenient, it cannot achieve the high level of resolution and magnification as an electron microscope can. The typical resolution of the light microscope is about 200 nm. In addition to that, by using the light microscope it is not possible to visualize the three-dimensional shape of nanofibers. Therefore, most of the studies are made using SEM for morphological characterization of electrospun composites [41].

The SEM technique is described in the literature as a technique that uses electrons, instead of visible light to produce images of the samples. SEM has a wavelength much smaller than visible light; considering the electrons, for this reason, this technique enables high-quality images with resolution less than 1 nm. In addition to that, the SEM technique has a good depth of field thanks to a small aperture and large work distance. The sample is scanned by a focused beam of electrons; the various signals at the sample surface, which are produced by the interaction of the sample and the bean, can further be collected, revealing some important sample information such as surface topography, morphology, and chemical composition. The secondary electrons (SE), X-rays characteristics, and backscattered electrons (BSE) are the most useful signals detected by SEM. The images of SEM can have the contrast property useful for the qualitative analysis of the sample, once the contrast is dominated by the so-called edge effect: the secondary electron at the edge leads to increased brightness at the location which is being beamed. To reveal the chemical composition of nanofibers composites, BSE mode is the more used tool. The atoms with greater atomic numbers (larger atoms), due to their greater cross-sectional area, have a higher probability of producing an elastic collision, resulting in more BSE production [41]. So, when analyzing electrospun composite fibers, the polymeric fiber matrix appears less bright than metal particles or doped ceramic. The SE images, generally, have higher resolution than BSE images, as BSE is emitted from the depth of the sample.

Using SEM micrographs image analysis software, alignment and fiber diameter can be measured from microscopy images. The measurement of fiber diameters is made measuring multiple fibers of the sample, and the report result is described with $\pm$, meaning the standard deviation, depicted as histograms of fiber diameter distribution [42].

13.10 3D Printing

Bioprinting is an emerging field that is having a revolutionary impact on medical technology. It offers great precision for the spatial placement of cells, drugs, proteins, genes, and biocompatibility to molecules and active particles to better guide tissue regeneration and formation. This emerging technology appears to be promising for advancing tissue or organs fabrication, aiming cancer or disease modeling, research investigation, drug testing, and transplantation [43].

The term "bioprinting" can be defined as the spatial patterning of living cells and other biologics by assembling and stacking them using computer-aided technology in which the approach of layer-by-layer deposition is used to develop living tissue and organs analogs for regenerative medicine, biological studies, TE, and pharmacokinetics [44]. With the development of bioprinting, four different approaches can be employed, such as laser, extrusion, acoustic, and inkjet based [45]. Current technology enables the development of tissue constructs or partial organs that do not require substantial vascularization, as well as mini-tissue models mimicking the biological of natural counterparts for cancer studies and pharmaceutical testing [45].

Three-dimensional printing technique for the development of biocompatible scaffolds has fielded the interest of many scientists worldwide, specializing in TE. The key role of 3D printing approach for TE application is to provide and mimic the microenvironment which intricate properties compared to the native ECM; by doing so, it is possible to potentiate an improved mode of tissue engineering, thus seeded or infiltrating stem cells dedicated to regenerate a specific type of tissue. In order to achieve optimal tissue regeneration, it is important that scaffolds mimic the ECM as closely as possible [46]. The ECM ineffective tissue formation is crucial, once that it is responsible for basic cellular functions such as proliferation, migration, and differentiation.

In order to avoid the limitation of conventional techniques for scaffolds development, 3D printing provides a more versatile and controlled technology with capability needed to create biomimetic scaffolds to promote the formation of functional tissue. The scaffolds can be produced in nanoscale precision with reproducibly fabricated, depending upon the 3D printing technology, which is assisted by computer-aided designs (CADs). All scaffolds developed by 3D printing technique work on the basis of assistive manufacturing where the designed structure is completed in a layer-by-layer process through the means of thermal fixation, light-assisted polymerization, or chemical binders. Aiming the creation of scaffolds with similar properties to those of ECM, the 3D printing technology must be able to yield quality scaffolds with the ability to support cellular growth, biodegradability, high mechanical strength, interconnectivity, and controlled pore size [47]. Three-dimensional printing technologies have continually made advances in order to improve scaffolds for TE purpose and have demonstrated promising results in many preclinical studies. In Table 13.2, we present a comparative list of 3D printing technologies and their preclinical progress.

13.11 Biomedical Application of Three-Dimensional Scaffolds

Cell-laden scaffolds in vitro or bioprinting living tissue have been studied thoroughly. It has been reported that tissues that do not need vascularization, including cartilage, skin, and blood vessel, have been grown, as well as thin tissue [81]. However, in situ, bioprinting is promising due to the capacity of the porous tissue analogs to be able to engraft with endogenous tissue and later regenerate new tissue. The idea of in situ laser bioprinting has been performed to test on mouse model the feasibility of printing nanohydroxyapatite particles [81], and inkjet-based bioprinting of skin cells has been tested for burn wounds. With the hope of repairing major wound defects, the idea of bioprinting skin cells (keratinocytes and fibroblasts) has been considered feasible for transitioning technology to clinical setting [43].

To take the in situ bioprinting technology into a robust state, further systematic research is required. There are major limitations associated with its engineering, biomaterial, and biological aspects, such as requirement for highly effective extrudable bionic enabling the rapid solidification in living body, without the need of solidifier as chemical crosslink or ultraviolet light, there is also the need of more advanced robotic bioprinters, which can be scanning the defect body site, while promoting interactive uses interface for surgeons. It is important to mention that sometimes, in situ bioprinting can increase the surgery cost and duration.

When tissue constructs can be engineered and built inside explants, it can be considered ex vivo, the bioprinting transitional stage, see Figure 13.5a [82]. When developing tissue analogs directly of the defected model (Figure 13.5b), in situ bioprinting is a promising approach, which may mimic the injured body site as a new way to develop technology for human body use in the near future. The use of in situ bioprinting can avoid limited cells activity in vitro and prevent risk associated with contamination. In addition to that, in situ bioprinting can eliminate the need for progenitor cells or differentiation of stem in vitro, and this last step can be time-consuming and not cost-effective. The alternative to using bioprinting progenitor or stem in situ can enable the precise deposition of cytokines, cells, or genes inside the defect, unlike manual interventions as traditional scaffolds which can suffer some considerable changes regarding deformation, contraction, or swelling. In the in situ bioprinting, the cells are exposed to the natural environment with growth factor which is a key factor for differentiation into the desired lineage, so it can be applied to the various body sites, such as extremity injuries, craniofacial defects, and deep dermal injuries. In Figure 13.5c, we can see an example that might be useful for humans application of in

TABLE 13.2 3D Printing Techniques Used to Print Scaffolds for TE Application

Printing Method	Advantages	Disadvantages	Preclinical Process
Electrospinning	Cell printing [48] Speed of fabrication Low shear stress (bioelectrospraying) [49] Soft TE [50]	Non-uniform pore sizes [51] Random orientation of fibers [52] High voltage requirement (1–30 kV) [53]	Rabbit/vascular tissue [54] Mouse/biocompatibility [55] Rat/bone [56]
Direct 3D printing/inkjet	It is not necessary to support for overhang or complex structures Versatile in terms of usable materials	Low mechanical strength prints compared to laser sintering Potential toxicity due to incompletely removed binders Time-consuming (postprocessing)	Mouse/bone [57] Rabbit/bone [58] Rat/bone [59]
With electrospinning bioplotting	Soft tissue application [60] Prints viable cells	It requires support structures for printing complex shapes Limitation on nozzle size [61]	Mouse/skin Mouse/tooth regeneration [62] Mouse/cartilage [63] Rabbit/trachea [64] Rabbit/cartilage [65] Rat/cartilage [62]
Stereolithography	Smooth surface finish Speed of fabrication [66] Very high resolution [67]	Support system is necessary for overhang and intricate objects Materials must be photopolymers [68] Expensive [67]	Pig/tendon [69] Rabbit/trachea [70] Rat/bone [71]
Selective laser sintering	Fine resolution Powder bed provides support for complex structure Provides scaffolds with high mechanical strength	Expensive and time-consuming High temperature required, up to 140°C [72] Limitation of the materials, once it is needed to be heat and shrinkage resistant	Mouse/skin [73] Mouse/bone [74] Mouse/heart [73] Rat/bone Rat/heart
Indirect 3D printing	Material versatility casting once mold is obtained Good for preproduction/prototyping	Long production times Mold required for casting Low accuracy Required proprietary waxes for biocompatibility [75]	Mouse/tooth regeneration [76] Rat/bone [77]
Fused deposition modeling	Relatively inexpensive Low cytotoxicity versus direct 3D printing	Low resolution [67] Required support structure for overhangs and complex shapes Postprocess may be necessary Non-biodegradable materials are used Limitation on materials, often is used thermoplastic materials [78]	Rat/bone [79] Swine/bone [80]

situ bioprinting, which Temple and colleagues demonstrate the application of bioprinting for large calvarial defects that need a great microvascularization along with structural support, envisioning stem cells (loaded in blends of hydrogel, such as fibrin and collagen) and bioprinting a frame using osteoconductive polymers, such as blend of hydroxyapatite, copolymers, and polycaprolactone; these combinations together can drive multiple cells differentiation, including endothelial and osteoblast cells for angiogenesis and bone regeneration, respectively [82].

Researchers are focusing their studies in the fabrication of scale-up tissue with a high volume of oxygen consumption rates such as liver, pancreas, or cardiac tissue; however, it is still remaining a challenge. One of the major steps to be overcome is associated with the integration of the network of vascular hierarchical veins and arteries down to capillaries. In order to bioprint thick organs, future protocols should be developed. Once it is difficult to print capillaries at submicron scale with the technology available, as an alternative it can be developed by printing microvasculature and then, the body interaction could create the capillaries. Aiming this approach, it is possible to think two different alternatives that are shown in Figure 13.6: (i) bioprint directly the vasculature network in a tubular shape, and (ii) by using indirect bioprinting, it is possible to utilize fugitive ink which is removed by de-crosslinking induced thermally, by doing so the vascular network is left behind [83].

In Figure 13.6a,b, Lee and co-workers demonstrate that endothelial spouting generated a considerable increase in the permeability of soft tissue construct [84]. Advanced vasculogenesis and angiogenesis have been developed in microfluid devices, as result it was possible to use several cells in cancer metastasis studies [85]. Figure 13.6c–e demonstrates the hybrid formation of vasculature in tandem with tissue strands, where fibroblasts cells fuse quickly to each other, forming the tissue around the vasculature [86]. Not only vascularization should be taken into consideration when bioprinting large-scale tissue constructs for transplantation, but we should also take into consideration the functionality post-transplantation and the anastomosis to the circulatory system. The vascular network should be bioprinted and designed in a manner that it can possess some important properties such as high patency rate to support occlusion-free circulation, sufficient intactness of endothelium to prevent thrombosis, and good mechanical properties to satisfy burst pressure and suture retention [87].

13.12 In Vitro Biocompatibility with Three-Dimensional Scaffolds

The success of three-dimensional scaffolds employed for biomedical application is principally due to its functional

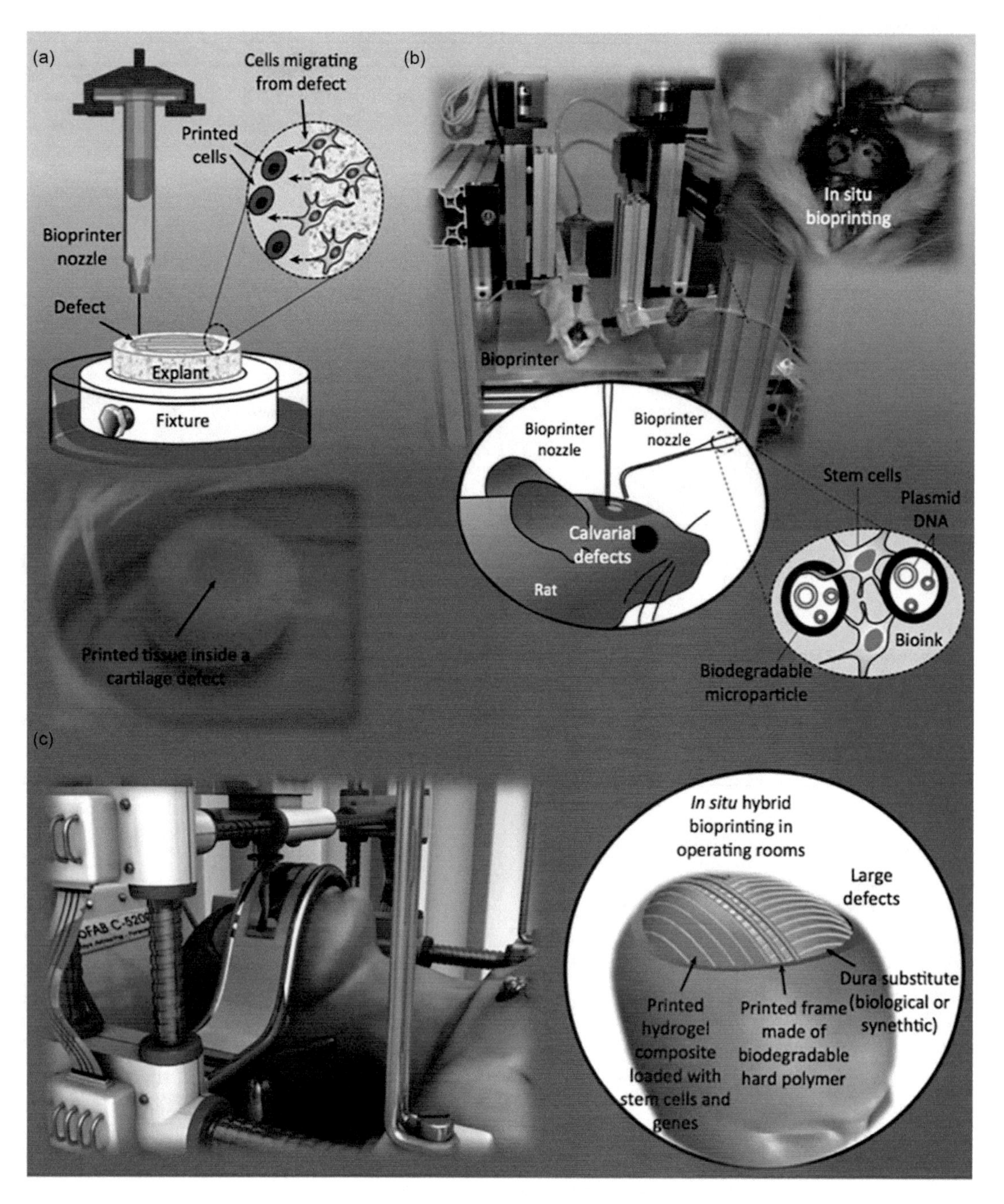

FIGURE 13.5 In situ bioprinting. (a) Bioprinting cells into explants in situ can be considered an intermediate step towards, (b) bioprinting tissue constructs directly into animal models, where DNA can also be printed for bioprinting mediated gene therapy to transfect and differentiate printed stem cells into multiple lineages. This will have a great impact on (c) transitioning in situ bioprinting technology into operating rooms for humans in the near future; for example, large, deep calvarial defects can be regenerated on a synthetic or biological Dura substitute, where a hard polymeric biodegradable frame can be printed along with stem cells and genes loaded in hydrogels to generate the vascularized bone tissues. (Image in (c) reproduced courtesy of Christopher Barnatt, published in [43], with permission from Elsevier.)

properties, e.g., the design of the device and the characteristics of the materials, as well the biological material interaction with the device. The biomaterials can be surface functionalized with biomolecules, such as peptides, polysaccharides, and protein. The biological properties of the scaffolds are directly related to the density and distribution of the ligands, and biomaterials surface. See Table 13.3.

The cell/surface interaction is influenced by microscale surface features; some new technological advances enable the physical surface modification, which is directly related to the cell behavior and protein absorption. The protein adsorbed onto biomaterials surface constitutes important anchor sites for cell attachment. The study of cell attachment to the scaffolds surface is covered by the complementary ligands

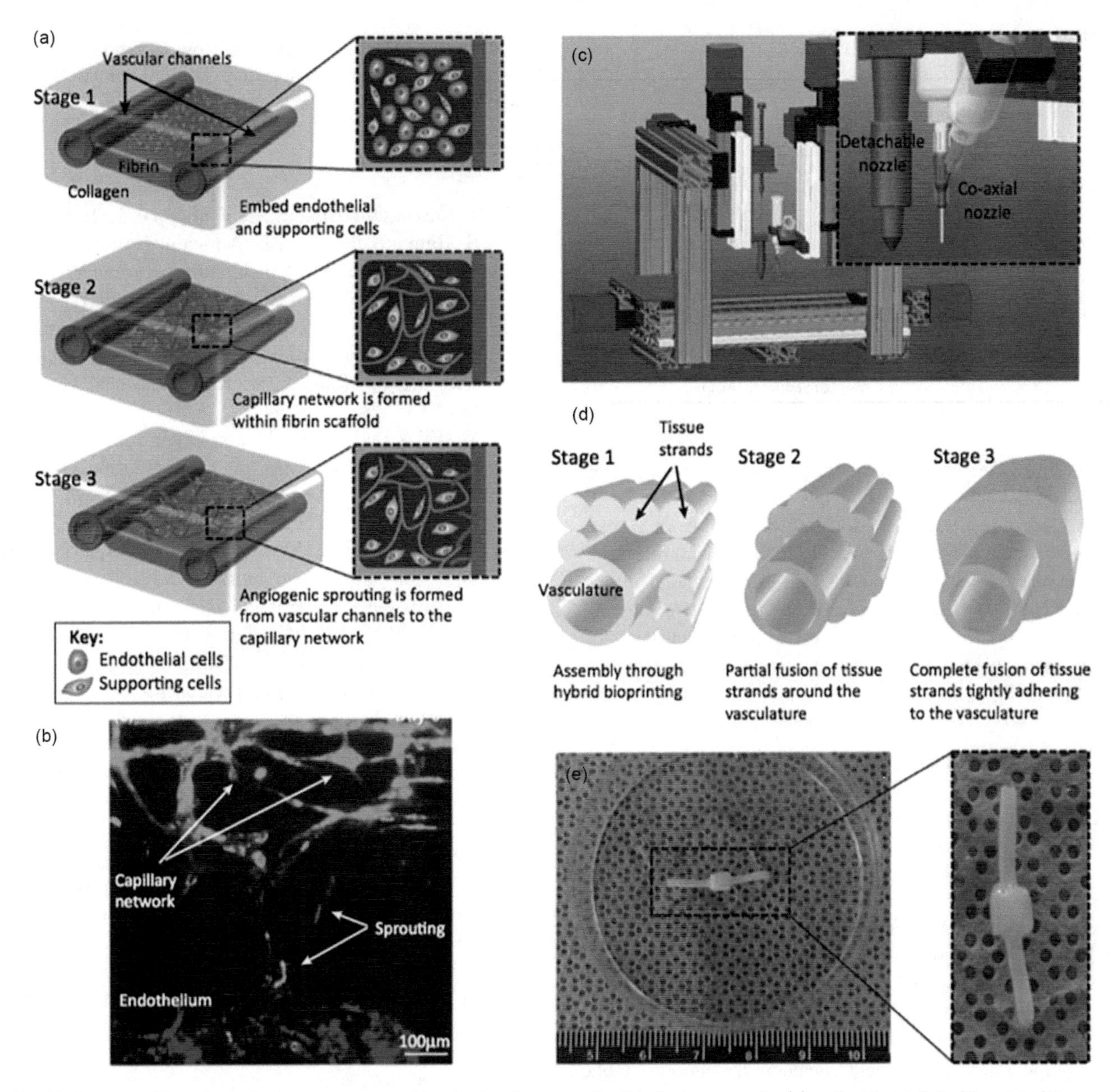

FIGURE 13.6 Vascularized tissue construct bioprinting in vitro. In the first approach, (a) a fugitive ink is bioprinted to create vascular channels inside a collagen support structure along with deposition of endothelial and fibroblast cells in a fibrin scaffold, resulting in capillary network formation in fibrin scaffold followed by endothelial sprouting from vascular channel to capillary network in 9 days, where (b) a immunohistochemistry image shows integration of large sprouts from the parental vascular channel (dark gray) to the capillary network. In the second approach, (c) a hybrid bioprinting approach can be applied using a Multi-Arm BioPrinter, where (d) the tissue assembly is created by placing the vasculature in the middle, resulting in fusion and maturation of scaffold-free tissue strands around the vasculature, further resulting in (e) the tight fusion of the fibroblast tissue to the vasculature in 1 week [3]. The co-axial nozzle used in that study can enable continuous printing of a single lumen vasculature throughout the fabrication of larger-scale hybrid tissue constructs. (Source Ref. [43], with permission from Elsevier.)

on the biomaterial surface and the analysis of the formation of interaction between the cell surface receptors, as well as the analysis of cell response in terms of changes of cell morphology and cell behavior during the attachment event on the surface of the biomaterial scaffold [89].

An important strategy for TE is to take together the bifunctional, chemical, and topographic surface modification. The biocompatibility is reported as the interaction of tissue or living system with a finished component material or medical device. Traditional cell culture system may be of value for in vitro testing for biocompatibility prior to the preclinical experiments [90]. Taking into consideration the biocompatibility, it is not only the absence of a toxic effect but also a positive influence in terms of biofunctionality, as in case of wound-healing scaffolds. The cell culture approach in two dimensions has been routinely utilized worldwide,

TABLE 13.3 Optimal Peptide Surface Functionalization by Covalent Grafting [88]

Peptide Characteristic	Functionalization Criteria
Preserved biorecognition	To prevent activity in vivo, peptide biofouling by protein absorption should be avoided by the combined use of antifouling molecules
Bioactivity	To recreate natural peptide conformation, it can be employed flanking amino acids
Selective	Control conformation of peptides by neighboring residues
Surface distribution	Peptide efficacy can be increased by surface clustered distribution
Surface density	Cell migration shows a bell-shaped peptides concentration response, while cell attachment shows a sigmoidal increase as function of peptide concentration

so now the researchers are focusing their efforts to test the biocompatibility in vitro using three-dimensional cell culture environments, and the insights obtained from 3D cell culture are more accurate for testing biomaterials, although its development and performance still require a multidisciplinary expertise. The relevance of in vitro testing models is increased by the approach foresees the preparation of 3D tissue-engineered models of human tissue for in vitro biocompatibility screening.

13.13 Conclusion

This chapter has approached the melanoma skin cancer, which has been increased the rate of mortality annually. It is crucial for the scientist to develop a new approach to drug delivery as target cancer therapy. We presented an overview regarding the biomedical development of scaffolds which has been explored over the past few years such as electrospinning technique and 3D bioprinting. With the help of emerging new technologies, we can expect in the near future to be able to perform 3D in situ scaffolds which mimic the carcinogenic microenvironment, achieving a more reliable output from the study, while avoiding animals suffering in preclinical trials. A great bio mimic for TE would be in situ bioprinting, with microvascularization along with structural support, envisioning stem cells in a combination that can drive multiple cells differentiation, possessing vasculogenesis and angiogenesis in microfluid devices, resulting in the possibility of using in several cells and cancer metastasis studies. The three-dimensional biomimetic scaffolds described in this chapter are expected to play a key role in TE, with multiple applications such as engineering new scaffolds, drug delivery profile, and tissue regeneration.

Acknowledgment

Authors wish to acknowledge financial support from CAPES Foundation for the Ph.D. Grant with Process Number 13543/13-0, Brazilian Ministry of Education, Brazil. Authors also like to thank FEDER funds through the Competitivity Factors Operational Programme—COMPETE and national funds through FCT—Foundation for Science and Technology (POCI-01-0145-FEDER-007136).

References

1. L. B. Naves, C. Dhand, J. R. Venugopal, L. Rajamani, S. Ramakrishna, and L. Almeida, Nanotechnology for the treatment of melanoma skin cancer, *Prog. Biomater.*, vol. 6, no. 1–2, pp. 13–26, 2017.
2. L. Naves, The contribution of fashion design to the development of alternative medical clothing, University of Beira Interior, MSc Thesis, Portugal, 2013.
3. P. M. Elias and G. K. Menon, Structural and lipid biochemical correlates of the epidermal permeability barrier, *Adv. Lipid Res.*, vol. 24, pp. 1–26, 1991.
4. E. Boughton and S. V. Mclennan, Biomimetic scaffolds for skin tissue and wound repair. In: A. J. Ruys (ed.) *Biomimetic Biomaterials: Structure and Applications*. Woodhead Publishing Limited, Cambridge, pp. 153–180, 2013.
5. C. Grove, *Understanding Skin Cancer*. Cancer Council Australia, Sydney, pp. 1–44, 2016.
6. S. D. Roy, M. Gutierrez, G. L. Flynn, and G. W. Cleary, Controlled transdermal delivery of fentanyl: Characterizations of pressure-sensitive adhesives for matrix patch design, *J. Pharm. Sci.*, vol. 85, no. 5, pp. 491–495, 1996.
7. S. S. Tinkle et al., Skin as a route of exposure and sensitization in chronic beryllium disease, *Environ. Health Perspect.*, vol. 111, no. 9. pp. 1202–1208, 2003.
8. M. H. Schmid-Wendtner and H. C. Korting, The pH of the skin surface and its impact on the barrier function, *Skin Pharmacol. Physiol.*, vol. 19, no. 6. pp. 296–302, 2006.
9. R. J. Murphy, D. Pristinski, K. Migler, J. F. Douglas, and V. M. Prabhu, Dynamic light scattering investigations of nanoparticle aggregation following a light-induced pH jump, *J. Chem. Phys.*, vol. 132, no. 19, p. 194903, 2010.
10. M. B. Reddy, R. H. Guy, and A. L. Bunge, Does epidermal turnover reduce percutaneous penetration? *Pharm. Res.*, vol. 17, no. 11, pp. 1414–1419, 2000.
11. L. B. Naves, C. Dhand, J. R. Venugopal, L. Rajamani, S. Ramakrishna, and L. Almeida, Nanotechnology for the treatment of melanoma skin cancer, *Prog. Biomater.*, vol. 6, pp. 1–14, 2017.
12. D. L. Narayanan, R. N. Saladi, and J. L. Fox, Ultraviolet radiation and skin cancer, *Int. J. Dermatol.*, vol. 49, no. 9, pp. 978–986, 2010.
13. J. L. Narayanan, D. L. Saladi, and R. N. Fox, Ultraviolet radiation and skin cancer, *Int. J. Dermatol.*, vol. 49, pp. 978–986, 2010.

14. J. S. Rhee, B. A. Matthews, M. Neuburg, B. R. Logan, M. Burzynski, and A. B. Nattinger, The skin cancer index: Clinical responsiveness and predictors of quality of life, *Laryngoscope*, vol. 117, pp. 399–405, 2007.
15. P. Boyle, J. F. Dor, P. Autier, and U. Ringborg, Cancer of the skin: A forgotten problem in Eufrope, *Ann. Oncol.*, vol. 15, no. 1, pp. 5–6, 2004.
16. L. O'Driscoll et al., Investigation of the molecular profile of basal cell carcinoma using whole genome microarrays, *Mol. Cancer*, vol. 5, p. 74, 2006.
17. A. Lomas, J. Leonardi-Bee, and F. Bath-Hextall, A systematic review of worldwide incidence of nonmelanoma skin cancer, *Br. J. Dermatol.*, vol. 166, no. 5, pp. 1069–1080, 2012.
18. C. Battie, M. Gohara, M. Vershoore, and W. Roberts, Skin cancer in skin of color, *J. Drugs Dermatol.*, vol. 12, no. 2, pp. 194–198, 2013.
19. L. Rittié et al., Differential ErbB1 signaling in squamous cell versus basal cell carcinoma of the skin, *Am. J. Pathol.*, vol. 170, no. 6, pp. 2089–2099, 2007.
20. B. Suárez et al., Occupation and skin cancer: The results of the HELIOS-I multicenter case-control study, *BMC Public Health*, vol. 7, no. 1, p. 180, 2007.
21. M. C. F. Simões, J. J. S. Sousa, and A. A. C. C. Pais, Skin cancer and new treatment perspectives: A review, *Cancer Lett.*, vol. 357, no. 1, pp. 8–42, 2015.
22. J. Scotto, T. R. Fears, and J. F. Fraumeni, Incidence of non-melanoma skin cancer in the United States, 1982.
23. R. S. Kopf, A. W. Salopek, T. G. Slade, J. Marghoob, and A. A. Bart, Techniques of cutaneous examination for the detection of skin cancer, *Cancer Suppl.*, vol. 72, no. 2, pp. 684–690, 1995.
24. O. N. Soehnge, H. Ouhtit, and A. Ananthaswany, Mechanisms of induction of skin cancer by UV radiation, *Front. Biosci.*, vol. 2, pp. 538–551, 1997.
25. W. H. Organization, Ultraviolet radiation and health. [Online]. Available: www.who.int/uv/uv_and_health/en/index.html. [Accessed: 15-Jul-2016].
26. M. R. Lautenschlager, S. Wulf, and H. C. Pittelkow, *Photoprotection.* Lancet, no. 370 (9586) pp. 528–537, 2007.
27. M. Chen, Y. T. Dubrow, R. Zheng, T. Barnhill, R. L. Fine, and J. Berwick, Sunlamp use and the risk of cutaneous malignant melanoma: A population based case-control study in Connecticut, USA, *Int. J. Epidemiol*, vol. 27, pp. 758–765, 1998.
28. C. Wang et al., Enhanced bioavailability and anticancer effect of curcumin-loaded electrospun nanofiber: In vitro and in vivo study, *Nanoscale Res. Lett.*, vol. 10, no. 1, p. 439, 2015.
29. J. L. Nichol, N. L. Morozowich, and H. R. Allcock, Biodegradable alanine and phenylalanine alkyl ester polyphosphazenes as potential ligament and tendon tissue scaffolds, *Polym. Chem.*, vol. 4, no. 3, pp. 600–606, 2013.
30. R. J. Mondschein, A. Kanitkar, C. B. Williams, S. S. Verbridge, and T. E. Long, Polymer structure-property requirements for stereolithographic 3D printing of soft tissue engineering scaffolds, *Biomaterials*, vol. 140, pp. 170–188, 2017.
31. N. Annabi et al., Controlling the porosity and microarchitecture of hydrogels for tissue engineering, *Tissue Eng. Part B Rev.*, vol. 16, no. 4, pp. 371–383, 2010.
32. J. M. Kemppainen and S. J. Hollister, Tailoring the mechanical properties of 3D-designed poly(glycerol sebacate) scaffolds for cartilage applications, *J. Biomed. Mater. Res. Part A*, vol. 94, no. 1. pp. 9–18, 2010.
33. P. Zhang, Z. Hong, T. Yu, X. Chen, and X. Jing, In vivo mineralization and osteogenesis of nanocomposite scaffold of poly(lactide-co-glycolide) and hydroxyapatite surface-grafted with poly(l-lactide), *Biomaterials*, vol. 30, no. 1, pp. 58–70, 2009.
34. S. Yang, F. Murugan, R. Wang, and S. Ramakrishna, Electrospinning of nano/micro scale poly(l-lactic acid) aligned fibers and their potential in neural tissue engineering, *Biomaterials*, vol. 26, no. 12, pp. 2603–2610, 2005.
35. S. Ramalingam and M. Ramakrishna (eds.) Introduction to nanofiber composites. In: *Nanofiber Composites for Biomedical Applications*, 1st edn. Elsevier Woodhead Publishing, Chennai, India, pp. 3–29, 2017.
36. S. Prabhakaran, M. P. Venugopal, J. R. Chyan, T. T. Hai, L. B. Chan, C. K. Lim, and A. Y. Ramakrishna, Electrospun biocomposite nanofibrous scaffolds for neural tissue engineering, *Tissue Eng. Part A*, vol. 14, pp. 1787–1797, 2008.
37. R. Murugan and S. Ramakrishna, Design strategies of tissue engineering scaffolds with controlled fiber orientation, *Tissue Eng.*, vol. 13, no. 8, pp. 1845–1866, 2007.
38. K. Wei, H. R. Kim, B. S. Kim, and I. S. Kim, Electrospun metallic nanofibers fabricated by electrospinning and metallization. In: *Nanofibers: Production, Properties and Functional Applications.* InTech, London, pp. 117–134, 2011.
39. C. Yao, X. Li, K. G. Neoh, Z. Shi, and E. T. Kang, Surface modification and antibacterial activity of electrospun polyurethane fibrous membranes with quaternary ammonium moieties, *J. Memb. Sci.*, vol. 320, pp. 259–267, 2008.
40. Z. Ma, M. Kotaki, T. Yong, W. He, and S. Ramakrishna, Surface engineering of electrospun polyethylene terephthalate (PET) nanofibers towards development of a new material for blood vessel engineering, *Biomaterials*, vol. 26, pp. 2527–2536, 2005.

41. A. Polini and F. Yang, Physicochemical characterization of nanofiber composites. In: S. Ramalingam, M. Ramakrishna (eds) *Nanofiber Composites for Biomedical Applications*, 1st edn. Woodhead Publishing Series in Biomaterials, Chennai, India, pp. 97–116, 2017.
42. W. Ji et al., Fibrous scaffolds loaded with protein prepared by blend or coaxial electrospinning, *Acta Biomater.*, vol. 6, no. 11, pp. 4199–4207, 2010.
43. I. T. Ozbolat, Bioprinting scale-up tissue and organ constructs for transplantation, *Trends Biotechnol.*, vol. 33, no. 7, pp. 395–400, 2015.
44. F. Guillemot, V. Mironov, and M. Nakamura, Bioprinting is coming of age: Report from the International Conference on bioprinting and biofabrication in Bordeaux (3B'09), *Biofabrication*, vol. 2, no. 1, pp. 1–7, 2010.
45. F. Guillemot et al., Laser-assisted bioprinting to deal with tissue complexity in regenerative medicine, *MRS Bull.*, vol. 36, no. 12, pp. 1015–1019, 2011.
46. Y. Tan et al., 3D printing facilitated scaffold-free tissue unit fabrication, *Biofabrication*, vol. 6, no. 2, pp. 1–9, 2014.
47. A.-V. Do, B. Khorsand, S. M. Geary, and A. K. Salem, 3D printing of scaffolds for tissue regeneration applications, *Adv. Healthc. Mater.*, vol. 4, no. 12, pp. 1742–1762, 2015.
48. S. N. Jayasinghe, Cell electrospinning: A novel tool for functionalising fibres, scaffolds and membranes with living cells and other advanced materials for regenerative biology and medicine, *Analyst*, vol. 138, no. 8, pp. 2215–2223, 2013.
49. S. N. Jayasinghe, Bio-electrosprays: From bio-analytics to a generic tool for the health sciences, *Analyst*, vol. 136, no. 5, pp. 878–890, 2011.
50. K. E. Ng et al., Bio-electrospraying primary cardiac cells: In vitro tissue creation and functional study, *Biotechnol. J.*, vol. 6, no. 1, pp. 86–95, 2011.
51. C. L. Casper, J. S. Stephens, N. G. Tassi, D. B. Chase, and J. F. Rabolt, Controlling surface morphology of electrospun polystyrene fibers: Effect of humidity and molecular weight in the electrospinning process, *Macromolecules*, vol. 37, no. 2, pp. 573–578, 2004.
52. W. J. Li, C. T. Laurencin, E. J. Caterson, R. S. Tuan, and F. K. Ko, Electrospun nanofibrous structure: A novel scaffold for tissue engineering, *J. Bio. Mater. Res.*, vol. 60, no. 4, pp. 613–621, 2002.
53. H. Niu and T. Lin, Fiber generators in needleless electrospinning, *J. Nanomater.*, vol. 2012, pp. 1–13, 2012.
54. B. W. Tillman, S. K. Yazdani, S. J. Lee, R. L. Geary, A. Atala, and J. J. Yoo, The in vivo stability of electrospun polycaprolactone-collagen scaffolds in vascular reconstruction, *Biomaterials*, vol. 30, no. 4, pp. 583–588, 2009.
55. S. L. Sampson, L. Saraiva, K. Gustafsson, S. N. Jayasinghe, and B. D. Robertson, Cell electrospinning: An in vitro and in vivo study, *Small*, vol. 10, no. 1. pp. 78–82, 2014.
56. S. Srouji, D. Ben-David, R. Lotan, E. Livne, R. Avrahami, and E. Zussman, Slow-release human recombinant bone morphogenetic protein-2 embedded within electrospun scaffolds for regeneration of bone defect: In vitro and in vivo evaluation, *Tissue Eng. Part A*, vol. 17, no. 3–4, pp. 269–277, 2011.
57. M. T. Poldervaart et al., Prolonged presence of VEGF promotes vascularization in 3D bioprinted scaffolds with defined architecture, *J. Control. Release*, vol. 184, no. 1, pp. 58–66, 2014.
58. J.-H. Shim et al., Three-dimensional printing of rhBMP-2-loaded scaffolds with long-term delivery for enhanced bone regeneration in a rabbit diaphyseal defect, *Tissue Eng. Part A*, vol. 20, no. 13–14, pp. 1980–1992, 2014.
59. F. Tamimi et al., Osseointegration of dental implants in 3D-printed synthetic onlay grafts customized according to bone metabolic activity in recipient site, *Biomaterials*, vol. 35, no. 21, pp. 5436–5445, 2014.
60. F. Pati et al., Printing three-dimensional tissue analogues with decellularized extracellular matrix bioink, *Nat. Commun.*, vol. 5, pp. 2014–2016, 2014.
61. I. T. Ozbolat, H. Chen, and Y. Yu, Development of 'multi-arm bioprinter' for hybrid biofabrication of tissue engineering constructs, *Robot. Comput. Integr. Manuf.*, vol. 30, no. 3, pp. 295–304, 2014.
62. N. E. Fedorovich et al., Biofabrication of osteochondral tissue equivalents by printing topologically defined, cell-laden hydrogel scaffolds, *Tissue Eng. Part C Methods*, vol. 18, no. 1, pp. 33–44, 2012.
63. T. Xu et al., Hybrid printing of mechanically and biologically improved constructs for cartilage tissue engineering applications, *Biofabrication*, vol. 5, no. 1, pp. 1–10, 2013.
64. J. W. Chang et al., Tissue-engineered tracheal reconstruction using three-dimensionally printed artificial tracheal graft: Preliminary report, *Artif. Organs*, vol. 38, no. 6, pp. E95–E105, 2014.
65. C. H. Lee, J. L. Cook, A. Mendelson, E. K. Moioli, H. Yao, and J. J. Mao, Regeneration of the articular surface of the rabbit synovial joint by cell homing: A proof of concept study, *Lancet*, vol. 376, no. 9739, pp. 440–448, 2010.
66. J. R. Tumbleston et al., Continuous liquid interface of 3D objects, *Science*, vol. 347, no. 6228, pp. 1349–1352, 2015.
67. B. C. Gross, J. L. Erkal, S. Y. Lockwood, C. Chen, and D. M. Spence, Evaluation of 3D printing and its potential impact on biotechnology and the chemical sciences, *Anal. Chem.*, vol. 86, no. 7, pp. 3240–3253, 2014.

68. J. Stampfl et al., Photopolymers with tunable mechanical properties processed by laser-based high-resolution stereolithography, *J. Micromech. Microeng.*, vol. 18, no. 12, pp. 1–9, 2008.
69. X. Li, J. He, W. Bian, Z. Li, D. Li, and J. G. Snedeker, A novel silk-TCP-PEEK construct for anterior cruciate ligament reconstruction: An off-the shelf alternative to a bone-tendon-bone autograft, *Biofabrication*, vol. 6, no. 1, pp. 1–11, 2014.
70. J. H. Park, J. W. Jung, H. W. Kang, Y. H. Joo, J. S. Lee, and D. W. Cho, Development of a 3D bellows tracheal graft: Mechanical behavior analysis, fabrication and an in vivo feasibility study, *Biofabrication*, vol. 4, no. 3, pp. 1–10, 2012.
71. J. W. Lee, K. S. Kang, S. H. Lee, J. Y. Kim, B. K. Lee, and D. W. Cho, Bone regeneration using a microstereolithography-produced customized poly(propylene fumarate)/diethyl fumarate photopolymer 3D scaffold incorporating BMP-2 loaded PLGA microspheres, *Biomaterials*, vol. 32, no. 3, pp. 744–752, 2011.
72. F. H. Liu, Synthesis of biomedical composite scaffolds by laser sintering: Mechanical properties and in vitro bioactivity evaluation, *Appl. Surf. Sci.*, vol. 297, pp. 1–8, 2014.
73. R. Gaebel et al., Patterning human stem cells and endothelial cells with laser printing for cardiac regeneration, *Biomaterials*, vol. 32, no. 35, pp. 9218–9230, 2011.
74. V. Keriquel et al., In vivo bioprinting for computer- and robotic-assisted medical intervention: Preliminary study in mice, *Biofabrication*, vol. 2, no. 1, pp. 1–8, 2010.
75. E. Sachlos, N. Reis, C. Ainsley, B. Derby, and J. T. Czernuszka, Novel collagen scaffolds with predefined internal morphology made by solid freeform fabrication, *Biomaterials*, vol. 24, no. 8, pp. 1487–1497, 2003.
76. C. H. Park et al., Biomimetic hybrid scaffolds for engineering human tooth-ligament interfaces, *Biomaterials*, vol. 31, no. 23, pp. 5945–5952, 2010.
77. J. P. Temple et al., Engineering anatomically shaped vascularized bone grafts with hASCs and 3D-printed PCL scaffolds, *J. Biomed. Mater. Res. Part A*, vol. 102, no. 12, pp. 4317–4325, 2014.
78. B. Bhushan and M. Caspers, An overview of additive manufacturing (3D printing) for microfabrication, *Microsyst. Technol.*, vol. 23, no. 4, pp. 1117–1124, 2017.
79. P. F. Costa, C. Vaquette, Q. Zhang, R. L. Reis, S. Ivanovski, and D. W. Hutmacher, Advanced tissue engineering scaffold design for regeneration of the complex hierarchical periodontal structure, *J. Clin. Periodontol.*, vol. 41, no. 3, pp. 283–294, 2014.
80. J. Jensen et al., Surface-modified functionalized polycaprolactone scaffolds for bone repair: In vitro and in vivo experiments, *J. Biomed. Mater. Res. Part A*, vol. 102, no. 9, pp. 2993–3003, 2014.
81. F. P. W. Melchels, M. A. N. Domingos, T. J. Klein, J. Malda, P. J. Bartolo, and D. W. Hutmacher, Additive manufacturing of tissues and organs, Prog. Polym. Sci., vol. 37, no. 8, pp. 1079–1104, 2012.
82. X. Cui, K. Breitenkamp, M. G. Finn, M. Lotz, and D. D. D'Lima, Direct human cartilage repair using three-dimensional bioprinting technology, *Tissue Eng. Part A*, vol. 18, no. 11–12, pp. 1304–1312, 2012.
83. D. B. Kolesky, R. L. Truby, A. S. Gladman, T. A. Busbee, K. A. Homan, and J. A. Lewis, 3D bioprinting of vascularized, heterogeneous cell-laden tissue constructs, *Adv. Mater.*, vol. 26, no. 19. pp. 3124–3130, 2014.
84. V. K. Lee, A. M. Lanzi, H. Ngo, S. S. Yoo, P. A. Vincent, and G. Dai, Generation of multi-scale vascular network system within 3D hydrogel using 3D bio-printing technology, *Cell. Mol. Bioeng.*, vol. 7, no. 3, pp. 460–472, 2014.
85. I. K. Zervantonakis, S. K. Hughes-Alford, J. L. Charest, J. S. Condeelis, F. B. Gertler, and R. D. Kamm, Three-dimensional microfluidic model for tumor cell intravasation and endothelial barrier function, *Proc. Natl. Acad. Sci. U. S. A.*, vol. 109, no. 34, pp. 13515–13520, 2012.
86. Y. Yu, Y. Zhang, and I. T. Ozbolat, A hybrid bioprinting approach for scale-up tissue fabrication, *J. Manuf. Sci. Eng.*, vol. 136, no. 6, p. 061013, 2014.
87. C. Quint, Y. Kondo, R. J. Manson, J. H. Lawson, A. Dardik, and L. E. Niklason, Decellularized tissue-engineered blood vessel as an arterial conduit, *Proc. Natl. Acad. Sci. U.S. A.*, vol. 108, no. 22, pp. 9214–9219, 2011.
88. U. Hersel, C. Dahmen, and H. Kessler, RGD modified polymers: Biomaterials for stimulated cell adhesion and beyond, *Biomaterials*, vol. 24, no. 24, pp. 4385–4415, 2003.
89. A. A. Khalili and M. R. Ahmad, A review of cell adhesion studies for biomedical and biological applications, *Int. J. Mol. Sci.*, vol. 16, no. 8, pp. 18149–18184, 2015.
90. A. Pizzoferrato et al., Cell culture methods for testing biocompatibility, *Clin. Mater.*, vol. 15, no. 3, pp. 173–190, 1994.

14

Environmental Nanoresearch Centers

Tonya R. Pruitt, Matthew Y. Chan, and Aaron J. Prussin II
Virginia Tech

Jeffrey M. Farner
McGill University

Arielle C. Mensch
Pacific Northwest National Laboratory

Michael F. Hochella, Jr.
Virginia Tech
Pacific Northwest National Laboratory

14.1 Introduction and Basic Concepts

Human life well-being is profoundly tied to the quality of the environment. The Indus Valley Civilization, in the northwestern region of South Asia, had some understanding of this some four thousand years ago, as there is clear archeological evidence of sewage systems and other relevant environmental technologies in this remarkable society (Schladweiler Jon, 2002). In the United States, the federal government formally recognized the importance of environmental quality by establishing a cabinet-level independent agency, the US Environmental Protection Agency, as well as passing laws that were designed to control and regulate practices that lead to pollution, such as the Clean Water Act and the Clean Air Act (Library of Congress, Environmental Policy Division, 1973; United States House Committee on Interstate and Foreign Commerce, 1963). Technology meta-narratives for the past few decades describe a convergence in which knowledge formerly self-contained in disciplinary silos spills over and blends with other distinct disciplines (Jeong et al., 2015). The fields of environmental sciences and engineering are not exempt from this trend, especially at the quantum and molecular scales (Bottero and Wiesner, 2007). Nanoscience and nanotechnology are the discovery and application of phenomenon and objects precisely at this scale, at the 1–100 nm range (Roco et al., 2000). It is well known that materials at this size range exhibit novel properties and exhibit behavior not observed at the bulk scale. These novel properties and the ongoing convergence of nanoscience and nanotechnology to environmental science and engineering have unlocked new possibilities in related research.

The most obvious intersection between nanotechnology and environmental engineering is the application of the former to latter. Environmental remediation using nanotechnology is a broad example resulting from this convergence. Conventional remediation methods generally rely on materials and processes that can adsorb/absorb and remove contaminants, chemically (or even more ideally: catalytically) degrade contaminants, or detect trace but severely harmful contaminants in complex system for further treatment (Khin et al., 2012). Nanomaterials can be exceptional adsorbents because they have very high surface area-to-volume ratios. They can also exhibit higher than typical catalytic activities because of their high curvature and potentially high density of localized surface-reactive chemical species. Both of these properties, as well as the possibility of nanoparticles surface modifications, enable the design of effective sensors for trace contaminants – including biological contaminants such as pathogens – in complex environments (Wujcik et al., 2014). The reduced size of nanomaterials also enables effective deployment of nano-enabled remediation technologies due to their relatively high mobility.

Another related application that has seen a large amount of ongoing effort is in drinking water treatment. The United Nations ranked access to clean water and sanitation as one of their Sustainable Development Goals, linking clean and safe water to the prosperity of human society as well as protection of the planet (United Nations General Assembly, 2015). Because the goal of drinking water treatment is to improve the quality of a source water to acceptable standards for drinking, typically water treatment involves multiple unit operations. These include but are not limited to: coagulation and flocculation (the formation of larger and denser "flocs" of particulate matter); sedimentation to remove "flocs" via gravity; and filtration and disinfection to kill/remove potential pathogens. There are opportunities in each step to supplement or even supersede existing unit operation technologies with nanotechnology. Issues that can be addressed by nanotechnology include filtration membranes that can foul due to failure of upstream unit operations, large quantities of particulates in source water, and costly maintenance of systems. As a result, there is now a great deal of effort to develop "self-cleaning" reverse-osmosis filtration membranes, for example, by embedding titanium oxide nanoparticles into the membranes such that the nano-enabled membranes exhibit antifouling properties from the resulting photocatalytic and superhydrophobic properties of these nanomaterials (Pendergast and Hoek, 2011; Madaeni and Ghaemi, 2007). Upstream of filtration, chemicals and other additives coagulate larger particulate matter for gravitational removal. This process is not perfect as certain contaminants, such as polymerized metal ionic hydroxyl complexes, do not form large-enough flocs to be removed, leading to the aforementioned filtration failure downstream (Ambashta and Sillanpää, 2010). Magnetic nanoparticles, such as nano-magnetite, can be used as recoverable adsorbents for removal of these metal ions before they form complexes, after which the nanoparticles can be easily removed via a relatively weak magnetic field and recovered (Liu et al., 2008). Beyond municipal water treatment applications, there are vast opportunities for applying nanotechnology to point-of-use water treatment methods, especially in developing countries. Conventional disinfection methods can be inaccessible and/or cost-prohibitive for developing communities, who may rely on solar disinfection that can be insufficient to render water for safe consumption (Acra et al., 1980; Clasen, 2009). Nanotechnology can help overcome this shortcoming by enhancing the disinfection process via titanium oxide nanoparticles or functionalized fullerenes, both of which exhibit photocatalytic reactivity (Brame et al., 2011; Lee et al., 2009; Dunlop et al., 2002).

The novel properties of nanomaterials are essential to advancing environmental applications; however, it is a double-edged sword. The implication of nanotechnology and its associated nanomaterials can have unpredictable impacts on the environment (Hannah and Thompson, 2008). In the aforementioned examples of applying nanotechnology to solve environmental challenges, the applications themselves often involve exposing or releasing nanomaterials to humans and the environment. Additionally, nanotechnology is becoming ubiquitous in consumer products, food products, textile and building materials, and other such products with limited risk management (Vance et al., 2015; Bouwmeester et al., 2009; Patra and Gouda, 2013; Pacheco-Torgal and Jalali, 2011; Shatkin et al., 2017). While nanomaterials associated with related technologies are often grouped into a single category, their chemical and physical properties, as well as how they would behave in the environment, are vastly diverse. To date, there is limited standardization on the type of measurement assays needed for the necessary risk assessment of the various nanomaterials utilized (Shatkin and Ong, 2016).

Due to these reasons, there is a need to pursue work in determining the diverse environmental implications of different nanomaterials and related applications. Composition (e.g. carbon, metals, and oxides), exposure route (e.g. surface water, air, soil, and plant intake), and preparation (e.g. nanoparticle surface medications), among other factors that can affect the environmental behavior and fate of nanomaterials. Because of the aforementioned high surface area-to-volume ratio, changes to nanoparticle surface properties can have a high impact on their environmental behavior not observed in bulk materials. For example, gold nanoparticles exhibit very different transport behavior in groundwater depending on their surface coatings (Chan and Vikesland, 2014). In the air, the high surface reactivity of fullerenes has significant implications to their behavior and chemical fate in the ozone layer, and this is in addition to conventional concerns for other particulate or aerosolized matter in the atmosphere (Tiwari et al., 2014). In the event of nanomaterials accumulating in soil, there are ongoing studies of the nanomaterials trophic transfer between animals, such as monitoring gold nanoparticle transformation as they moved up the food chain of first being consumed by earthworms, which in turn are consumed by bullfrogs. There is also research work on the plant uptake of silver nanoparticles from the soil and the location of their bioaccumulation (Unrine et al., 2012; Stegemeier et al., 2017b). Beyond exposure and fate studies (as well as toxicology and ecotoxicology studies not mentioned here), there is a need of understanding the cradle-to-grave lifecycle of nanomaterials with simulated environment such as mesocosms and constructed wetlands (Ferry et al., 2009; Lowry et al., 2012).

Nanotechnology's environmental applications and implications must be informed by the advancement of fundamental science at the nanoscale. For example, there needs to be an understanding of the fundamental crystallization and degradation processes of nanominerals in nature in order to elucidate the transformation of engineered nanomaterials in the environment. For example, the investigation into microbiologically assisted formation of sphalerite/wurtzite nanominerals provides insight into the origin and fate of metal sulfide nanoparticles in the environment (Xu et al., 2016). To produce engineered nanomaterials with the desired functionality, scientists must utilize specific chemical mechanisms or other preparation tools such as protein templates. It is necessary to have a full understanding of the nanoscale structure and properties of these protein templates in order to prepare, for example, gold nanoparticles with tunable morphologies (Roth et al., 2016). For biological exposure, scientists are gaining a better understanding of nanoparticle-eukaryotic cell interactions by evaluating that interface with synthetic supported lipid bilayers (Melby et al., 2016).

The robust advancement of nanoscience and its associated underlying knowledge framework is necessary for the continued development of new environmental nanotechnology applications as well as further elucidating implications of nanotechnology to the environment. Despite the innate interdisciplinary nature of nanoscience, disciplinary silos remain. The examples of research mentioned above are the tip of the iceberg as it is simply impossible to list all facets of environmental nanoresearch. An environmental nanoresearch center must be able to converge expertise, make available necessary facilities and instruments, and provide a culture and climate that is conducive to environmental nanoscience discovery and environmental nanotechnology research. Herein we describe the mechanics of an environmental nanotechnology research center, as adding an environmental component makes them different from traditional nanotechnology/nanoscience centers.

14.2 Environmental Nanoresearch Centers

As evident from the research described in the previous section, environmental nanoresearch is often quite different from other types of nanoscience and engineering described in this handbook. This results in substantial differences in centers that promote and facilitate this environmental nanoscience and technology. From sample acquisition to storage and analysis, the unique needs of an environmental nanoresearcher shape the environment, analytical tools, and culture of an environmental nanoresearch center.

14.2.1 How Environmental Nanoscience Laboratories and Centers Work

Samples for environmental nanoresearch are acquired in one of two ways: (i) directly from the natural environment or (ii) developed in the laboratory to simulate the natural environment. Consequently, the environmental nanoresearch laboratory tends to be structured differently from other nanoresearch laboratories, which frequently focus on design and fabrication. Traditional clean rooms have little use as environmental samples are rarely "clean" and frequently are contaminated with natural materials.

For samples collected directly from the environment, environmental nanoresearch centers require specialized collection, storage, and processing facilities and tools. Environmental samples can be collected in glass or plastic bottles and vials in a variety of shapes and sizes. These sterile containers protect samples during transport from contamination and frequently from UV light. Air samples are frequently collected through filtration onto filter membranes or impingement into liquid (Lodge Jr., 1988). Once in the lab, samples can be stored in refrigerators, freezers, or other environmental controlled chambers (e.g. temperature, relative humidity, light). For long-term storage, environmental samples are frequently kept in $-80°C$ freezers to halt any chemical or biochemical changes from occurring after collection. Desiccators, frequently used for semiconductor components in traditional nanoresearch facilities, can also be used for environmental nanoresearch in which moisture should be removed from sample storage. Other samples may even be used in mesocosms, microcosms, or experimental chambers, discussed below.

To simulate the natural environment in a laboratory setting, unique mesocosms, microcosms, and experimental chambers are needed. A mesocosm, literally "medium world", is an outdoor experimental setup that allows researchers to study the natural environment while controlling certain conditions. Samples derived from mesocosms serve as a middle ground between samples acquired in the field and those developed in a full laboratory setting. A basic mesocosm could be produced by developing a greenhouse-enclosed ecosystem for research. A good example of an environmental nanoresearch mesocosm is the Center for the Environmental Implications of NanoTechnology (CEINT) Mesocosm Facility at Duke University in Durham, NC. This facility was designed to simulate a wetland ecosystem, and it features boxes containing soil, water, aquatic and terrestrial plants, and aquatic life (Center for the Environmental Implications of NanoTechnology, 2018). The exact contents of a box can be modified to simulate different wetland environments, and by using multiple identical boxes, researchers quickly have a means of comparison when introducing nanomaterials into the environment. A few research questions addressed by this mesocosm include "How does the nutrient level in the environment influence nanomaterial movement through the environment and biological or ecological effects?", "How are these effects influenced by nanoparticle size?", and "Can we design functional assays for laboratory studies that are capable of predicting these realistic environmental responses?" (Center for the Environmental Implications of NanoTechnology, 2018) and examples of research completed by CEINT are discussed later in this chapter.

Similar to mesocosms, laboratory experimental chambers and microcosms ("miniature world") seek to simulate natural conditions while holding certain factors under control. While the chambers can vary wildly depending on the natural environment or conditions they are designed to simulate, they can generally be divided into aqueous chambers, aerosol chambers, terrestrial chambers, or a combination thereof (Tiwari et al., 2014). Even within these categories, the chambers can vary drastically depending on the study parameters. For example, an aerosol chamber may be a closed system with a particular gaseous makeup where nano-aerosols can be introduced and monitored over time. Another simulated aerosol system could be designed to simulate incidental nanoparticles created while cooking. Additionally, rotating drums are frequently used to study the effect that different environmental conditions (e.g. relative humidity and temperature) have on the viability of biological nanoparticles (e.g. viruses). An aqueous experimental chamber could be a fish tank containing continuously circulating water and clams to simulate a river. A hot water heater connected to a complex series of pipes can be an aqueous system constructed to simulate the water system in a house and designed to study if nanomaterials are transformed in such an environment. A terrestrial chamber could be as simple as a tray with soil and plants of interest. A combination chamber could simulate soil runoff into a stream or could be a complex combination of soil, water, and air designed to mimic a particular natural environment. Introducing a relevant nanoparticle to any of these example systems could lead to studies on the fate of nanomaterials in the environment such as how the nanomaterials move through the environment, how they interact with the environment, if they transformed during this interaction, if bio-uptake occurs, and, if so, how the nanomaterials interact with plants or other organisms.

Environmental nanoresearch laboratories have other facility needs. First, the laboratory needs facilities to synthesize relevant nanomaterials to introduce into the mesocosms and experimental chambers described above. These synthesis facilities could be used to produce manufactured nanomaterials that can be inadvertently released in the environment, those that may be intentionally released as an active agent in environmental remediation, ones that simulate naturally occurring nanomaterials, or nanomaterials to be used in nano-enabled detection devices. In all of these cases, laboratory production in carefully controlled conditions results in nanomaterials that are much more customized to a particular research need (in terms of precise compositions and more uniform sizes and shapes) than equivalent commercially available nanomaterials.

Environmental samples are often heterogeneous mixtures that contain many different components. Frequently, the nanoparticle of interest is a tiny fraction of the collected sample. Therefore, to increase the chances of finding the particle of interest, the first step of analysis is often mechanical or chemical separation. The separation method is dependent on the makeup of the sample and the feature of interest. Filtration, sedimentation, and centrifugation are common mechanical separation techniques, whereas chromatography, distillation, and extraction are common forms of chemical separation. Though each of these techniques requires common laboratory equipment, their frequent use in environmental nanoresearch laboratories often results in designated bench space contrary to traditional nanoresearch centers. Additionally, high-velocity ultracentrifuges prove to be especially useful in separating out nanoparticles from environmental samples and are becoming essential components of environmental nanoresearch centers. Further, ultracentrifuges are critical for purifying biological nanoparticles, such as viruses (Hermens et al., 1999; Reimer et al., 1967).

Depending on the research focus of the laboratory, other highly specific research tools may be necessary. For example, a lab that is studying sewage treatment may need a bioreactor or bio-culturing facilities for the production of bacteria, viruses, or pathogens commonly found in human waste. Researchers developing nanotechnology-enabled sensors will need to produce the environmental target of the sensor from heavy metals or other environmental contaminants to DNA, proteins, or infectious organisms. Laboratory animal cages and associated supplies could be needed for researchers interested in health implications of engineered or incidental nanoparticles released into the environment. The use of computational tools has rapidly taken off in environmental nanotechnology research centers, especially when studying biological agents. Metagenomics is a bioinformatics tool that can be used to study the microbiome or collection of all viruses, bacteria, fungi, etc. in an environmental sample (Handelsman, 2004). Understanding the microbiome and what organisms we are exposed to has become of great interest to researchers recently, due to the impact on human health (Gomez-Alvarez et al., 2012; National Academies of Sciences Engineering and Medicine et al., 2017; Vikesland and Raskin, 2016). Computational tools can also be used for modeling the transport and fate of nanomaterials in the atmosphere (Kumar et al., 2011).

14.2.2 Sample Collection and Analytical Tools

Typical nanoscience and engineering research centers often include clean rooms, fabrication, patterning, and lithography tools. Due to the focus on analysis of environmental samples instead of fabrication, environmental nanoresearch centers typically feature a wide variety of characterization tools. These tools range from small portable devices designed to gather or analyze samples at the environmental source all the way to complex, environmentally tuned electron microscopes.

Field deployable equipment is critical for environmental nanoresearch centers as they allow samples to be analyzed at the study site. These tools are specifically designed to be portable, sturdy, and to work in harsh environmental conditions. These devices generally are not intended to measure nanoparticles but are able to gather useful information on particles in the micron size range. For example, aerosol particle concentrations can be obtained in the field easily using equipment such as an AeroTrak or DustTrak. Researchers are also able to measure PM1, PM2.5, and PM10 aerosol size fractions in the field using relatively inexpensive optical counters. For aqueous samples, portable spectrophotometers may be used to determine if a particular particle is present and at what concentration. Portable X-ray fluorescence (XRF) tools provide preliminary elemental and chemical analysis information for soil samples.

When collecting samples in the environment, it is important to collect other relevant data about the environment that the samples are being drawn from. For aerosol samples collected outdoors, meteorological data is recorded (i.e. temperature, relative humidity, wind speed, wind direction, solar radiation, etc.). Meteorological information can be obtained using portable weather stations. When aerosol samples are collected indoors, researchers typically measure temperature, relative humidity, and the air exchange rate. The air exchange rate can be measured using a tracer gas or examining the decay in CO_2 concentration in a building. For aqueous samples, recording the temperature, flow rate, reduction potential, dissolved oxygen, conductivity, and turbidity are common measurements. Taking these environmental measurements at the source can be just as essential as measuring the concentration of the particle of interest as interactions with these environmental factors may affect features of the particle.

Environmental nanoresearch centers tend to have a range of tools to gather a wide variety of data. Environmental samples often contain much more than the nanoparticle of interest. Though researchers tend to have an idea of what might be in a sample, the first step is often to determine the composition of a sample, then separate out particles of interest, and finally, fully characterize those particles. A Scanning Mobility Particle Sizer (SMPS) and Aerodynamic Particle Sizer (APS) are routinely used to measure the size distribution of aerosolized nanoparticles (Wang and Flagan, 1990). In addition to providing an aerosol particle concentration, SMPS and APS are also able to provide a mass and volume concentration. For liquid samples or those that can be suspended in solution, high-performance liquid chromatography (HPLC) can be used to separate, identify, and quantify sample components. Ultraviolet-visible (UV-Vis) spectrophotometers, common in both teaching and research laboratories, are frequently used in environmental nanoresearch to determine composition and concentration typically of liquid samples. Dynamic light scattering (DLS) provides complementary information on the size distribution and diameter of particles in solution. For trace element analysis of soil samples, specialized forms of spectrometry including secondary ion mass spectrometry (SIMS), inductively coupled plasma mass spectrometry (ICP-MS), and inductively coupled plasma atomic emission spectrometry (ICP-AES) can detect elements ranging from parts per billion to parts per quadrillion. Instruments using Brunauer–Emmett–Teller (BET) theory are useful for characterizing the surface area of solid materials.

Raman spectroscopy, which measures laser scattering (Rayleigh and Raman scattering) to measure vibrational excitation, provides structural information for molecule identification. While traditional Raman spectroscopy works best with pure compounds, there are multiple advanced Raman spectroscopy techniques that expand its capabilities to mixtures and impure samples. Discussing the numerous advanced Raman spectroscopy techniques is beyond the scope of this chapter; however, it would be remiss to not mention the notable advances of surface-enhanced Raman spectroscopy (SERS) for analysis of nanoparticles in aqueous environmental samples. SERS is an invaluable tool for environmental nanoresearch for several reasons: (i) the surface enhancement in SERS can generally be provided by various nanomaterials such as gold and silver nanoparticles (Fleischmann et al., 1974). This allows for the design of nanotechology-based environmental sensors using SERS (Wei et al., 2015); (ii) Raman spectroscopy often suffers from low signals, but surface enhancements in SERS have potential to boost signal by 10^{10} times (Le Ru et al., 2007). For the aforementioned example on environmental nanoresearch for nanotechnology-based sensors, this high enhancement is imperative to track trace environmental pollutants. Finally, both conventional Raman spectroscopy as well as SERS has the advantage of requiring very little sample preparation, which preserves analytes of interest as they would appear in the environment.

X-ray diffraction (XRD) is a technique frequently used to determine the atomic and molecular structure of a crystal. Many materials of interest for environmental nanoresearchers form crystals including minerals, metals, and even proteins and other biological molecules. Though originally designed for identification of crystalline structures in powders, new X-ray diffractometers are capable of measuring atomic structural data on amorphous materials and those in liquid suspensions. Additionally, materials can be studied at environmental conditions with temperature and pressure control. Use of XRD is frequently a preliminary step to determine sample makeup before imaging via electron microscopy.

Scanning electron microscopy (SEM) is a common analytical and characterization method used in nanoresearch laboratories. SEM requires high vacuum and non-conducting specimens to be coated with a thin layer of a conductive material such as gold or carbon in order to acquire information on the sample's surface topography and composition. Aptly named, an environmental scanning electron microscope (E-SEM) is specially designed to handle the non-homogenous and often delicate samples obtained in environmental research by operating with low vacuum in the specimen chamber and removing the conductivity requirement. These adaptations allow researchers to observe specimens in a more natural state, work with gas and liquid samples or those containing multiple phases, and even perform *in situ* studies involving hydrating, dehydrating, or heating/cooling samples. Similar to E-SEM, environmental transmission electron microscopes (E-TEMs) allow researchers to capture high resolution images of samples under environmentally relevant pressure and perform *in situ* studies. Specialized sample holders can be used in traditional TEMs to allow for similar applications as an E-TEM.

As with the specially needed laboratory facilities, specialized analytical tools may be necessary for specific research areas. This will depend on the individual needs of a laboratory and the research questions that are being asked. For example, if an environmental nanoresearch team is focused on viruses in the environment, then a molecular biology setup will be required. Each environmental nanoresearch center should tailor their infrastructure to their specific needs.

14.2.3 The Scientific and Engineering Culture

Interdisciplinary research and collaboration are hallmarks to environmental nanoresearch teams. The natural environment is inherently complex, and, accordingly, research studying how natural, incidental, and engineered nanoparticles interact with and within the environment is complicated. There is a great diversity of environmental nanoresearch questions, and a wide variety of expertise is needed to address them. Therefore, environmental nanoresearch, including far reaching implications, frequently occurs at the intersection of disciplines as diverse as physics and a variety of social sciences.

These factors result in a diverse group of researchers, focusing on a broad array of topics within the overarching theme of environmental nanoresearch. For example, at Virginia Tech, the environmental nanoresearch group has faculty and students from a wide variety of departments in three different colleges, one school, and a research institute – the College of Engineering (civil and environmental engineering, materials science engineering, computer science), the College of Science (chemistry, geosciences), the College of Natural Resources and the Environment (sustainable biomaterials), the School of Biomedical Engineering and Sciences, and the Institute for Critical Technology and Applied Science (ICTAS).

This interdisciplinary and collaborative culture provides a strong foundation for environmental nanoresearch centers. With the breadth of disciplines under the environmental nanoresearch umbrella, it is evident that no researcher could be an expert in all areas. Bringing together researchers with varied backgrounds brings different perspectives together to solve unique problems. Even as researchers may be working on very different topics, they are able to share techniques and best practices. For example, even though working with environmental soil samples and those from aqueous environments are very different, there are similarities involved in both collecting the samples and how they can be analyzed based on what information the researcher is interested in.

14.2.4 Summary

When compared to traditional nanoresearch centers, those focusing on environmental nanoresearch tend to feature distinctive laboratory facilities, specialized analytical instrumentation, and a robust interdisciplinary and collaborative culture. With samples collected from the natural environment or developed in mesocosms, microcosms, and environmental chambers, the facilities necessary for environmental nanoresearch are often quite different from the clean rooms of nanofabrication research centers. From handheld portable environmental sensors to specially designed E-SEMs and E-TEMs, environmental nanoresearch centers use a variety of analytical tools to collect a wide range of information in environmentally relevant ways. Sharing laboratory space, instrumentation, techniques, and best practices is a way of life for environmental nanoresearchers. Interdisciplinary research teams and a culture of collaboration are commonplace within environmental nanoresearch centers as they frequently work on a breadth of research topics that span multiple disciplines.

In the following section, three different environmental nanoresearch centers will be described by those who have been a part of them not only to demonstrate the variety between centers, but also the similarities in environment, instrumentation, and culture previously described.

14.3 Examples of Environmental Nanoresearch Centers

14.3.1 Center for the Environmental Implications of NanoTechnology (CEINT)

CEINT, based at Duke University, is a National Science Foundation (NSF) and US Environmental Protection Agency funded entity whose research is primarily focused on the implications of nanoparticles entering and interacting with the environment (Wiesner et al., 2009, 2011). Initially funded in 2008 with a $15 million, 5-year grant, CEINT is comprised of research groups from seven primary institutions (Duke University, Carnegie Mellon University, Howard University, Virginia Tech, Stanford University, the University of Kentucky, and Baylor University) and includes key collaborations with researchers in the United Kingdom (ENPRA) and France (iCEINT). The research aims of the center are divided into three themes:

1. Transport and transformation
2. Cellular and organismal responses
3. Ecosystem responses

These themes are investigated in terms of manufactured nanomaterials (e.g. silver, TiO_2) (Gorka et al., 2015; Jassby et al., 2012), naturally occurring nanomaterials, and incidental nanomaterials (e.g. byproducts of engine combustion and automotive traffic) (Tiwari et al., 2016; Yang et al., 2016). Modeling and risk forecasting efforts are applied to each of the themes and nanomaterials (Therezien et al., 2014; Hendren et al., 2015a; Money et al., 2012; Dale et al., 2015a,b). Given that the focus of CEINT centers on environmental interactions, significant efforts have been made in identifying nanoparticles in environmental or complex matrices (Kim et al., 2010; Schierz et al., 2012; Badireddy et al., 2012) and investigating their interactions with plants and other organisms (Judy et al., 2011; Starnes et al., 2015; Pitt et al., 2018; Geitner et al., 2017; Stegemeier et al., 2017a). Recently, the center has pushed to develop an understanding of results across various experimental conditions by utilizing the power of big data through the CEINT NanoInformatics Knowledge Commons (NIKC) (Hendren et al., 2015b; Karcher et al., 2018; Powers et al., 2015; Robinson et al., 2016).

For any center spanning such large physical distances, meaningful collaborations on research can be difficult to carry out. CEINT attempts to combat this using a number of techniques: by holding annual meetings, live streaming research seminars and symposiums, an online presence that includes a list of instrumentation available at each institution and jobs postings, and by supporting graduate student travel between sites.

The CEINT Mesocosm Facility also provides a unique platform for inter-institutional collaboration. Thirty mesocosms (1 m × 4 m, and ~1 m deep) enable CEINT to study the impact of nanomaterials at the ecosystem level while fully exposed to natural environmental conditions. Although the mesocosm facility is covered and heated in winter to prevent freezing, at all other points, it remains exposed to the North Carolina weather. While day-to-day maintenance of the mesocosms is performed by researchers at Duke, experiments are planned, begun, and concluded by members from across CEINT institutions. These long-term studies bring together researchers from varied disciplines (e.g. ecologists, biologists, environmental engineers, geoscientists) and have investigated aggregation, aging, plant uptake, and organismal impact of nanoparticles (Stegemeier et al., 2017a; Espinasse et al., 2018; Lowry et al., 2012). Samples are shared and shipped within the center to research labs best equipped for their analysis. In addition to the large-scale mesocosm experiments, complimentary studies have been performed within CEINT employing terrestrial only microcosms and aquatic only mesocosms (Colman et al., 2013; Auffan et al., 2014; Wright et al., 2018; Bone et al., 2014). Ultimately, these semi-controlled environmental studies help to explain how understanding NP behavior in lab-scale experiments might be applied to realistic environmental exposures.

14.3.2 Center for Sustainable Nanotechnology (CSN)

The Center for Sustainable Nanotechnology (CSN) is a National Science Foundation Division of Chemistry funded by Centers for Chemical Innovation. The funding for the CSN is administered through the University of Wisconsin–Madison, where its director is based. Unlike other centers, the CSN does not operate out of one physical building. Instead, the CSN is a geographically diffuse team of multidisciplinary scientists distributed across 14 different institutions, including Augsburg University, Boston University, the Connecticut Agriculture Experiment Station, Johns Hopkins University, Northwestern University, Pacific Northwest National Laboratory, Tuskegee University, University of Illinois Urbana-Champaign, University of Iowa, University of Maryland – Baltimore County, University of Minnesota, University of Wisconsin – Madison, University of Wisconsin – Milwaukee, and an independent evaluator at Georgia Institute of Technology. The scientists share the goal of using fundamental chemistry to enable the development of nanotechnology in a sustainable manner, for societal benefit.

The CSN takes a chemistry-focused approach to study the nano-bio interface, in which equal emphasis is placed on understanding both the surface chemistry of different engineered nanomaterials and the surface chemistry of different organisms that nanomaterials may encounter in the environment. The team utilizes both experimental and computational approaches to develop a molecular-level understanding of how engineered nanomaterials interact with different biological systems. They accomplish this by varying the complexity of both the nanoparticles and model organisms that they use in a controlled manner. Nanoparticles with controllable surface chemistries, such as diamond (Robinson et al., 2018; Zhang et al., 2017) and gold (Liang et al., 2018; Lohse et al., 2013; Torelli et al., 2015; Zheng et al., 2018), as well as, more technologically relevant nanomaterials, such as complex metal oxides (Doğangün et al., 2017; Gunsolus et al., 2017; Hang et al., 2016; Huang et al., 2017) and quantum dots, are used to understand their impact on biological systems. These biological systems range in complexity from model cell membranes (Melby et al., 2016; Mensch et al., 2017; Troiano et al., 2015) to single-celled bacteria (Buchman et al., 2018; Feng et al., 2015; Jacobson et al., 2015) and multicellular organisms (Bozich et al., 2014; Dominguez et al., 2015; Qiu et al., 2015). The mechanistic insights gained from these studies allow the CSN to predict the impact of different nanomaterials based on intrinsic chemical and physical properties. Having this insight will ultimately allow for the design of more sustainable nanoparticles with retained functionality but reduced negative environmental consequences.

The CSN shares funding, research interests, and instrumentation, but the majority of interaction occurs virtually via e-mail, online meetings, and phone calls due to the distributed nature of the collaboration. To manage a large, distributed center, the CSN follows a shared governance structure where leadership roles are spread throughout the center; however, the center director is ultimately responsible for the CSN's research and broader impacts. An associate director assists the director, while the CSN's executive committee, or management team, advises the center director on all aspects of management. The executive committee of the CSN is comprised of the center director, the associate director, a managing director, a director of education and outreach, a student/postdoc board liaison, and a theme leader (faculty member) from each of the six scientific research themes of the CSN (Nanoparticles, Transformations and Impact on Biological Systems, Analytical/Computational Tools and Methodologies, Green Chemistry, Model Cell Surfaces, and Organisms: Diversity of Molecular Interactions). In addition to its science themes, the CSN also has a strong emphasis on its integrative activities, which include professional development, informal science communication, broadening participation, and innovation. The centerpiece of the CSN's informal science communication effort is Sustainable Nano (sustainable-nano.com), a blog aimed at communicating science related to sustainability and nanotechnology to the general public written by the students, postdocs, faculty, and staff associated with the CSN (Bishop et al., 2014).

14.3.3 Virginia Tech National Center for Earth and Environmental Nanotechnology Infrastructure (NanoEarth)

The Virginia Tech National Center for Earth and Environmental Nanotechnology Infrastructure (NanoEarth) is a node of the National Nanotechnology Coordinated Infrastructure (NNCI), an NSF-funded network of 16 centers spread through the United States serving as user facilities for cutting edge nanotechnology research. While many NNCI nodes have a materials science and engineering focus, NanoEarth is specifically designed to support researchers who work with nanoscience – and nanotechnology – related aspects of the Earth and environmental sciences/engineering at local, regional, and global scales, including the land, atmospheric, water, and biological components of these fields. NanoEarth supports researchers across academia, government, and industry by providing advanced tools and expertise to guide nanotechnology research and propel environmental solutions. In addition to facilities at Virginia Tech, the capabilities of NanoEarth are significantly enhanced by a partnership with the Environmental Molecular Sciences Laboratory (EMSL) at Pacific Northwest National Laboratory (PNNL). PNNL, operated by Battelle, is a United States Department of Energy (DOE) national laboratory, managed by DOE's Office of Science. EMSL within PNNL is a national scientific user facility funded and managed by DOE's Office of Biological and Environmental Research.

Users from across the country, and even internationally, submit access requests, which provide a brief project description, explanation of the research question, overview of the methods, and a description of what NanoEarth instrumentation and staff assistance are needed. After application review, a plan is developed and users either come to NanoEarth facilities to work with staff or mail samples to be analyzed remotely.

NanoEarth represents a strong push from the NSF to support Earth and environmental nanoresearch by ensuring that infrastructure and expertise are available. All projects supported through NanoEarth staff are exempted from technical assistance fees and users only pay for instrument usage time. Additionally, NanoEarth is able to provide need-based seed funding to support travel to the facilities and instrument usage time for innovative Earth and environmental nanoresearch.

Complimenting NanoEarth's expert environmental nanoresearch support is a robust education and outreach program (Figure 14.1) featuring three signature initiatives: (i) MUNI (Multicultural and Underrepresented Nanoscience Initiative), (ii) *Pulse of the Planet* radio programs and podcasts, and (iii) a unique focus on innovation and entrepreneurship. NanoEarth's MUNI program seeks to support underserved populations by providing access to NanoEarth facilities and expertise free of charge to researchers and also providing workshops and other educational activities for students of all academic levels. To increase public awareness of impactful geoscience and environmental nanoresearch, NanoEarth sponsors *Pulse of the Planet* radio programs, which reaches over 200,000 listeners per broadcast. Widely known and multiple award-winning radio producer, Jim Metzner, produces ten NanoEarth sponsored shows per year featuring NanoEarth members, users, and other well-known environmental nanoscientists explaining environmental nanoscience and engineering topics of general interest to this very broad listening audience. NanoEarth supports innovation and entrepreneurship through an industry speaker series, lectures intended to broaden the horizons of students, and the NanoTechnology Entrepreneurship Challenge (NTEC). Students participating in NTEC vie for funding, mentorship, and business support services by pitching nanotechnology-based business concepts focused on global sustainability challenges.

(a)

(b)

FIGURE 14.1 Education and outreach are frequently key components of environmental nanoresearch centers. Pictured: Community college students from the City University of New York (a) synthesize different sizes of gold nanoparticles and (b) learn about using the SIMS for trace element analysis of environmental samples while visiting NanoEarth facilities.

14.4 Going Forward

CEINT, CSN, and NanoEarth have shown that Earth- and environment-based nanoscience and technology centers can exist independently and flourish. Just as importantly, we know now that these Earth- and environment-based centers are respected contributors among the massive and still rapidly growing mainstream nanoscience and technology communities, among them, the great chemistry-, physics-, materials science-, and electrical engineering-based nano-centers around the world. All of these centers are now enjoying and in fact reveling in a revolution similar to the molecular biology and the information technology revolutions in achieving their promises to improve society. Nanoscience and nano-engineering are, overall, intellectually rich, technology driven, and continually growing. One only needs to look at the fact that participants in the nanoscience and technology revolution are currently producing about 150,000 scholarly publications per year, this number having increased by over two orders of magnitude over the last three decades, with funding in the many billions of US dollars per year, accompanied by sales of nano-enabled products in the trillions of US dollars per year (sources: Scopus, Elsevier's citation database, and a number of academic news articles).

Certainly, given this enormous backdrop, the number of nano-related papers in Earth and environmental sciences is still considered somewhat small as measured by the percentage of nano-related papers in each general scientific field normalized by the total papers in that same field in 2016, as follows: roughly 1% in the geosciences, 4% in the biological sciences, 5% in the environmental sciences, vs. what is considered a "healthy" 10%–20% in the chemical-, materials-, and physics-based sciences (sources: Scopus and ProQuest citation databases). This indicates that Earth and environmental nano-based sciences still have some distance to go to "catch up" with more mainstream nanoscience-impacted fields of science and technology.

Like the other great revolutions in science and engineering, the nano-revolution is expected to continue onward for literally centuries, always being fundamentally relevant to the way we live, the products we buy, our understanding of this planet, and new technological evolutions that will depend on it in the future. The grand challenges that the environmental sciences face in our rapidly changing world, which can be decisively advanced by applying modern nanoscience and technology, include a better understanding of the geosphere and biosphere that have evolved in the presence of naturally occurring nanosized materials, but now with exposure levels and types of nanomaterials (natural, incidental, and engineered) increasing radically over the last two centuries, with many consequences currently seen and unseen. Examples include metal oxide nanoparticles that have significant utility in a variety of industries, including pharmaceutical applications. It is clear that these therapeutic approaches are effective, but we need to understand the implications and consequences of the persistence of these nanomaterials in the environment after they have served their intended function. Nanoplastics as an unexpected environmental contaminant is another great example of the need of Earth and environmental nanoscience. And there still remain emerging, newly discovered, and presumably as yet undiscovered environmental nano-contaminants. For example, widespread titanium suboxide incidental nanoparticles produced during industrial coal burning were recently discovered (Yang et al., 2017). These particles are now widespread in the Earth system, and they likely have environmental contaminant consequences. Ironically, although hard to conceive of at this time, geo-engineering practices to mitigate global warming in the future will definitely involve nanoscience and technology. Nano-geo solutions have already been implemented to restrict the movement of environmental contaminants (including nuclear wastes) in vadose zones, aquifers, soils, and surface waters.

References

Acra, A., Karahagopian, Y., Raffoul, Z. & Dajani, R. 1980. Disinfection of oral rehydration solutions by sunlight. *The Lancet*, 316, 1257–1258.

Ambashta, R. D. & Sillanpää, M. 2010. Water purification using magnetic assistance: A review. *Journal of Hazardous Materials*, 180, 38–49.

Auffan, M., Tella, M., Santaella, C., Brousset, L., Paillès, C., Barakat, M., Espinasse, B., Artells, E., Issartel, J., Masion, A., Rose, J., Wiesner, M. R., Achouak, W., Thiéry, A. & Bottero, J.-Y. 2014. An adaptable mesocosm platform for performing integrated assessments of nanomaterial risk in complex environmental systems. *Scientific Reports*, 4, 5608.

Badireddy, A. R., Wiesner, M. R. & Liu, J. 2012. Detection, characterization, and abundance of engineered nanoparticles in complex waters by hyperspectral imagery with enhanced darkfield microscopy. *Environmental Science and Technology*, 46, 10081–10088.

Bishop, L. M., Tillman, A. S., Geiger, F. M., Haynes, C. L., Klaper, R. D., Murphy, C. J., Orr, G., Pedersen, J. A., Destefano, L. & Hamers, R. J. 2014. Enhancing graduate student communication to general audiences through blogging about nanotechnology and sustainability. *Journal of Chemical Education*, 91, 1600–1605.

Bone, A. J., Matson, C. W., Colman, B. P., Yang, X., Meyer, J. N. & Di Giulio, R. T. 2014. Silver nanoparticle toxicity to Atlantic killifish (Fundulus heteroclitus) and Caenorhabditis elegans: A comparison of mesocosm, microcosm, and conventional laboratory studies. *Environmental Toxicology and Chemistry*, 34, 275–282.

Bottero, J.-Y. & Wiesner, M. R. 2007. *Environmental Nanotechnology: Applications and Impacts of Nanomaterials*. McGraw-Hill Professional, New York.

Bouwmeester, H., Dekkers, S., Noordam, M. Y., Hagens, W. I., Bulder, A. S., De Heer, C., Ten Voorde, S. E. C. G., Wijnhoven, S. W. P., Marvin, H. J. P. & Sips, A. J. A. M. 2009. Review of health safety aspects of nanotechnologies in food production. *Regulatory Toxicology and Pharmacology*, 53, 52–62.

Bozich, J. S., Lohse, S. E., Torelli, M. D., Murphy, C. J., Hamers, R. J. & Klaper, R. D. 2014. Surface chemistry, charge and ligand type impact the toxicity of gold nanoparticles to Daphnia magna. *Environmental Science: Nano*, 1, 260–270.

Brame, J., Li, Q. & Alvarez, P. J. J. 2011. Nanotechnology-enabled water treatment and reuse: Emerging opportunities and challenges for developing countries. *Trends in Food Science and Technology*, 22, 618–624.

Buchman, J. T., Rahnamoun, A., Landy, K. M., Zhang, X., Vartanian, A. M., Jacob, L. M., Murphy, C. J., Hernandez, R. & Haynes, C. L. 2018. Using an environmentally-relevant panel of Gram-negative bacteria to assess the toxicity of polyallylamine hydrochloride-wrapped gold nanoparticles. *Environmental Science: Nano*, 5, 279–288.

Center for the Environmental Implications of Nanotechnology. 2018. CEINT mesocosm history Center for the Environmental Implications of NanoTechnology [Online]. Available: https://ceint.duke.edu/research/mesocosm/history [Accessed].

Chan, M. Y. & Vikesland, P. J. 2014. Porous media-induced aggregation of protein-stabilized gold nanoparticles. *Environmental Science and Technology*, 48, 1532–1540.

Clasen, T. F. 2009. *Scaling Up Household Water Treatment among Low-Income Populations*. World Health Organization, Geneva.

Colman, B. P., Arnaout, C. L., Anciaux, S., Gunsch, C. K., Hochella, M. F., Jr., Kim, B., Lowry, G. V., Mcgill, B. M., Reinsch, B. C., Richardson, C. J., Unrine, J. M., Wright, J. P., Yin, L. & Bernhardt, E. S. 2013. Low concentrations of silver nanoparticles in biosolids cause adverse ecosystem responses under realistic field scenario. *Plos One*, 8, e57189.

Dale, A. L., Casman, E. A., Lowry, G. V., Lead, J. R., Viparelli, E. & Baalousha, M. 2015a. Modeling nanomaterial environmental fate in aquatic systems. *Environmental Science and Technology*, 49, 2587–2593.

Dale, A. L., Lowry, G. V. & Casman, E. A. 2015b. Stream dynamics and chemical transformations control the environmental fate of silver and zinc oxide nanoparticles in a watershed-scale model. *Environmental Science and Technology*, 49, 7285–7293.

Doğangün, M., Hang, M. N., Machesky, J., Mcgeachy, A. C., Dalchand, N., Hamers, R. J. & Geiger, F. M. 2017. Evidence for considerable metal cation concentrations from lithium intercalation compounds in the nano–bio interface gap. *The Journal of Physical Chemistry C*, 121, 27473–27482.

Dominguez, G. A., Lohse, S. E., Torelli, M. D., Murphy, C. J., Hamers, R. J., Orr, G. & Klaper, R. D. 2015. Effects of charge and surface ligand properties of nanoparticles on oxidative stress and gene expression within the gut of Daphnia magna. *Aquatic Toxicology*, 162, 1–9.

Dunlop, P. S. M., Byrne, J. A., Manga, N. & Eggins, B. R. 2002. The photocatalytic removal of bacterial pollutants from drinking water. *Journal of Photochemistry and Photobiology A: Chemistry*, 148, 355–363.

Espinasse, B. P., Geitner, N. K., Schierz, A., Therezien, M., Richardson, C. J., Lowry, G. V., Ferguson, L. & Wiesner, M. R. 2018. Comparative persistence of engineered nanoparticles in a complex aquatic ecosystem. *Environmental Science and Technology*, 52, 4072–4078.

Feng, Z. V., Gunsolus, I. L., Qiu, T. A., Hurley, K. R., Nyberg, L. H., Frew, H., Johnson, K. P., Vartanian, A. M., Jacob, L. M., Lohse, S. E., Torelli, M. D., Hamers, R. J., Murphy, C. J. & Haynes, C. L. 2015. Impacts of gold nanoparticle charge and ligand type on surface binding and toxicity to Gram-negative and Gram-positive bacteria. *Chemical Science*, 6, 5186–5196.

Ferry, J. L., Craig, P., Hexel, C., Sisco, P., Frey, R., Pennington, P. L., Fulton, M. H., Scott, I. G., Decho, A. W., Kashiwada, S., Murphy, C. J. & Shaw, T. J. 2009. Transfer of gold nanoparticles from the water column to the estuarine food web. *Nature Nanotechnology*, 4, 441–444.

Fleischmann, M., Hendra, P. J. & Mcquillan, A. J. 1974. Raman spectra of pyridine adsorbed at a silver electrode. *Chemical Physics Letters*, 26, 163–166.

Geitner, N. K., O'brien, N. J., Turner, A. A., Cummins, E. J. & Wiesner, M. R. 2017. Measuring nanoparticle attachment efficiency in complex systems. *Environmental Science and Technology*, 51, 13288–13294.

Gomez-Alvarez, V., Revetta, R. P. & Domingo, J. W. S. 2012. Metagenomic analyses of drinking water receiving different disinfection treatments. *Applied and Environmental Microbiology*, 78, 6095–6102.

Gorka, D. E., Osterberg, J. S., Gwin, C. A., Colman, B. P., Meyer, J. N., Bernhardt, E. S., Gunsch, C. K., Digulio, R. T. & Liu, J. 2015. Reducing environmental toxicity of silver nanoparticles through shape control. *Environmental Science and Technology*, 49, 10093–10098.

Gunsolus, I. L., Hang, M. N., Hudson-Smith, N. V., Buchman, J. T., Bennett, J. W., Conroy, D., Mason, S. E., Hamers, R. J. & Haynes, C. L. 2017. Influence of nickel manganese cobalt oxide nanoparticle composition on toxicity toward Shewanella oneidensis MR-1: Redesigning for reduced biological impact. *Environmental Science: Nano*, 4, 636–646.

Handelsman, J. 2004. Metagenomics: Application of genomics to uncultured microorganisms. *Microbiology and Molecular Biology Reviews*, 68, 669–685.

Hang, M. N., Gunsolus, I. L., Wayland, H., Melby, E. S., Mensch, A. C., Hurley, K. R., Pedersen, J. A., Haynes, C. L. & Hamers, R. J. 2016. Impact of nanoscale lithium Nickel Manganese Cobalt Oxide (NMC) on the bacterium shewanella oneidensis MR-1. *Chemistry of Materials*, 28, 1092–1100.

Hannah, W. & Thompson, P. B. 2008. Nanotechnology, risk and the environment: A review. *Journal of Environmental Monitoring*, 10, 291–300.

Hendren, C. O., Lowry, G. V., Unrine, J. M. & Wiesner, M. R. 2015a. A functional assay-based strategy for nanomaterial risk forecasting. *Science of the Total Environment*, 536, 1029–1037.

Hendren, C. O., Powers, C. M., Hoover, M. D. & Harper, S. L. 2015b. The nanomaterial data curation initiative: A collaborative approach to assessing, evaluating, and advancing the state of the field. *Beilstein Journal of Nanotechnology*, 6, 1752–1762.

Hermens, W. T. J. M. C., Brake, O. T., Dijkhuizen, P. A., Sonnemans, M. A. F., Grimm, D., Kleinschmidt, J. A. & Verhaagen, J. 1999. Purification of recombinant adeno-associated virus by iodixanol gradient ultracentrifugation allows rapid and reproducible preparation of vector stocks for gene transfer in the nervous system. *Human Gene Therapy*, 10, 1885–1891.

Huang, X., Bennett, J. W., Hang, M. N., Laudadio, E. D., Hamers, R. J. & Mason, S. E. 2017. Ab initio atomistic thermodynamics study of the (001) surface of $LiCoO_2$ in a water environment and implications for reactivity under ambient conditions. *The Journal of Physical Chemistry C*, 121, 5069–5080.

Jacobson, K. H., Gunsolus, I. L., Kuech, T. R., Troiano, J. M., Melby, E. S., Lohse, S. E., Hu, D., Chrisler, W. B., Murphy, C. J., Orr, G., Geiger, F. M., Haynes, C. L. & Pedersen, J. A. 2015. Lipopolysaccharide density and structure govern the extent and distance of nanoparticle interaction with actual and model bacterial outer membranes. *Environmental Science and Technology*, 49, 10642–10650.

Jassby, D., Farner Budarz, J. & Wiesner, M. 2012. Impact of aggregate size and structure on the photocatalytic properties of TiO_2 and ZnO nanoparticles. *Environmental Science and Technology*, 46(24), 13270–13277.

Jeong, S., Kim, J.-C. & Choi, J. Y. 2015. Technology convergence: What developmental stage are we in? *Scientometrics*, 104, 841–871.

Judy, J. D., Unrine, J. M. & Bertsch, P. M. 2011. Evidence for biomagnification of gold nanoparticles within a terrestrial food chain. *Environmental Science and Technology*, 45, 776–781.

Karcher, S., Willighagen, E. L., Rumble, J., Ehrhart, F., Evelo, C. T., Fritts, M., Gaheen, S., Harper, S. L., Hoover, M. D., Jeliazkova, N., Lewinski, N., Marchese Robinson, R. L., Mills, K. C., Mustad, A. P., Thomas, D. G., Tsiliki, G. & Hendren, C. O. 2018. Integration among databases and data sets to support productive nanotechnology: Challenges and recommendations. *NanoImpact*, 9, 85–101.

Khin, M. M., Nair, A. S., Babu, V. J., Murugan, R. & Ramakrishna, S. 2012. A review on nanomaterials for environmental remediation. *Energy and Environmental Science*, 5, 8075–8109.

Kim, B., Park, C.-S., Murayama, M. & Hochella, M. F., Jr. 2010. Discovery and characterization of silver sulfide nanoparticles in final sewage sludge products. *Environmental Science and Technology*, 44, 7509–7514.

Kumar, P., Ketzel, M., Vardoulakis, S., Pirjola, L. & Britter, R. 2011. Dynamics and dispersion modelling of nanoparticles from road traffic in the urban atmospheric environment: A review. *Journal of Aerosol Science*, 42, 580–603.

Lee, J., Mackeyev, Y., Cho, M., Li, D., Kim, J.-H., Wilson, L. J. & Alvarez, P. J. J. 2009. Photochemical and antimicrobial properties of novel C60 derivatives in aqueous systems. *Environmental Science and Technology*, 43, 6604–6610.

Le Ru, E. C., Blackie, E., Meyer, M. & Etchegoin, P. G. 2007. Surface enhanced Raman scattering enhancement factors: A comprehensive study. *The Journal of Physical Chemistry C*, 111, 13794–13803.

Liang, D., Hong, J., Fang, D., Bennett, J. W., Mason, S. E., Hamers, R. J. & Cui, Q. 2018. Analysis of the conformational properties of amine ligands at the gold/water interface with QM, MM and QM/MM simulations. *Physical Chemistry Chemical Physics*, 20, 3349–3362.

Library of Congress, Environmental Policy Division. 1973. *A Legislative History of the Water Pollution Control Act Amendments of 1972, together with a Section-by-Section Index*. United States Government Publishing Office, Washington, DC.

Liu, J.-F., Zhao, Z.-S. & Jiang, G.-B. 2008. Coating Fe_3O_4 magnetic nanoparticles with humic acid for high efficient removal of heavy metals in water. *Environmental Science and Technology*, 42, 6949–6954.

Lodge Jr., J. P. 1988. *Methods of Air Sampling and Analysis*, CRC Press, Bosa Roca, FL.

Lohse, S. E., Eller, J. R., Sivapalan, S. T., Plews, M. R. & Murphy, C. J. 2013. A simple millifluidic benchtop reactor system for the high-throughput synthesis and functionalization of gold nanoparticles with different sizes and shapes. *ACS Nano*, 7, 4135–4150.

Lowry, G. V., Espinasse, B. P., Badireddy, A. R., Richardson, C. J., Reinsch, B. C., Bryant, L. D., Bone, A. J., Deonarine, A., Chae, S., Therezien, M., Colman, B. P., Hsu-Kim, H., Bernhardt, E. S., Matson, C. W. & Wiesner, M. R. 2012. Long-term transformation and fate of manufactured Ag nanoparticles in a simulated large scale freshwater emergent wetland. *Environmental Science and Technology*, 46, 7027–7036.

Madaeni, S. S. & Ghaemi, N. 2007. Characterization of self-cleaning RO membranes coated with TiO_2 particles under UV irradiation. *Journal of Membrane Science*, 303, 221–233.

Melby, E. S., Mensch, A. C., Lohse, S. E., Hu, D., Orr, G., Murphy, C. J., Hamers, R. J. & Pedersen, J. A. 2016. Formation of supported lipid bilayers containing phase-segregated domains and their interaction with gold nanoparticles. *Environmental Science: Nano*, 3, 45–55.

Mensch, A. C., Hernandez, R. T., Kuether, J. E., Torelli, M. D., Feng, Z. V., Hamers, R. J. & Pedersen, J. A. 2017. Natural organic matter concentration impacts the interaction of functionalized diamond nanoparticles with model and actual bacterial membranes. *Environmental Science and Technology*, 51, 11075–11084.

Money, E. S., Reckhow, K. H. & Wiesner, M. R. 2012. The use of Bayesian networks for nanoparticle risk forecasting: Model formulation and baseline evaluation. *Science of the Total Environment*, 426, 436–445.

National Academies of Sciences Engineering and Medicine. 2017. *Microbiomes of the Built Environment: A Research Agenda for Indoor Microbiology, Human Health, and Buildings*. National Academies Press, Washington, DC.

Pacheco-Torgal, F. & Jalali, S. 2011. Nanotechnology: Advantages and drawbacks in the field of construction and building materials. *Construction and Building Materials*, 25, 582–590.

Patra, J. K. & Gouda, S. 2013. Application of nanotechnology in textile engineering: An overview. *Journal of Engineering and Technology Research*, 5, 104–111.

Pendergast, M. M. & Hoek, E. M. V. 2011. A review of water treatment membrane nanotechnologies. *Energy and Environmental Science*, 4, 1946–1971.

Pitt, J. A., Kozal, J. S., Jayasundara, N., Massarsky, A., Trevisan, R., Geitner, N., Wiesner, M., Levin, E. D. & Di Giulio, R. T. 2018. Uptake, tissue distribution, and toxicity of polystyrene nanoparticles in developing zebrafish (Danio rerio). *Aquatic Toxicology*, 194, 185–194.

Powers, C. M., Mills, K. A., Morris, S. A., Klaessig, F., Gaheen, S., Lewinski, N. & Hendren, C. O. 2015. Nanocuration workflows: Establishing best practices for identifying, inputting, and sharing data to inform decisions on nanomaterials. *Beilstein Journal of Nanotechnology*, 6, 1860–1871.

Qiu, T. A., Bozich, J. S., Lohse, S. E., Vartanian, A. M., Jacob, L. M., Meyer, B. M., Gunsolus, I. L., Niemuth, N. J., Murphy, C. J., Haynes, C. L. & Klaper, R. D. 2015. Gene expression as an indicator of the molecular response and toxicity in the bacterium Shewanella oneidensis and the water flea Daphnia magna exposed to functionalized gold nanoparticles. *Environmental Science: Nano*, 2, 615–629.

Reimer, C. B., Baker, R. S., Vanfrank, R. M., Newlin, T. E., Cline, G. B. & Anderson, N. G. 1967. Purification of large quantities of influenza virus by density gradient centrifugation. *Journal of Virology*, 1, 1207–1216.

Robinson, M. E., Ng, J. D., Zhang, H., Buchman, J. T., Shenderova, O. A., Haynes, C. L., Ma, Z., Goldsmith, R. H. & Hamers, R. J. 2018. Optically detected magnetic resonance for selective imaging of diamond nanoparticles. *Analytical Chemistry*, 90, 769–776.

Robinson, R. L. M., Lynch, I., Peijnenburg, W., Rumble, J., Klaessig, F., Marquardt, C., Rauscher, H., Puzyn, T., Purian, R., Åberg, C., Karcher, S., Vriens, H., Hoet, P., Hoover, M. D., Hendren, C. O. & Harper, S. L. 2016. How should the completeness and quality of curated nanomaterial data be evaluated? *Nanoscale*, 8, 9919–9943.

Roco, M. C., Weber, T. A., Henhart, M. P., Kalil, T. A., Trew, R., Murday, J. S., Genther Yoshida, P., Casassa, M. P., Shull, R. D., Thomas, I. L., Price, R., Valentine, B. G., John, R. R., Lacombe, A., Murphy, E., Kirkpatrick, K. S., Daum, M. M., Matsumura, M., Porter, J., Radzanowski, D., Hirschbein, M., Krabach, T., Mucklow, G. H., Meyyappan, M., Schloss, J. & Kousvelari, E. 2000. National nanotechnology initiative: The initiative and its implementation plan, NSTC/NSET report. National Science and Technology Council, Committee on Technology, Subcommittee on Nanoscale Science, Engineering and Technology, Washington, DC.

Roth, K. L., Geng, X. & Grove, T. Z. 2016. Bioinorganic interface: Mechanistic studies of protein-directed nanomaterial synthesis. *The Journal of Physical Chemistry C*, 120, 10951–10960.

Schierz, A., Parks, A. N., Washburn, K. M., Chandler, G. T. & Ferguson, P. L. 2012. Characterization and quantitative analysis of single-walled carbon nanotubes in the aquatic environment using near-infrared fluorescence spectroscopy. *Environmental Science and Technology*, 46, 12262–12271.

Schladweiler Jon, C. Tracking down the roots of our sanitary sewers. *Pipeline Division Specialty Conference 2002, 2002/03/31/2002*, Cleveland, Ohio, United States: American Society of Civil Engineers.

Shatkin, J. A. & Ong, K. J. 2016. Alternative testing strategies for nanomaterials: State of the science and considerations for risk analysis. *Risk Analysis*, 36, 1564–1580.

Shatkin, J. A., Ong, K. & Ede, J. 2017. Minimizing risk: An overview of risk assessment and risk management of nanomaterials. In: *Metrology and Standardization of Nanotechnology*. Wiley-Blackwell. doi:10.1002/9783527800308.ch24.

Starnes, D. L., Unrine, J. M., Starnes, C. P., Collin, B. E., Oostveen, E. K., Ma, R., Lowry, G. V., Bertsch, P. M. & Tsyusko, O. V. 2015. Impact of sulfidation on the bioavailability and toxicity of silver nanoparticles to Caenorhabditis elegans. *Environmental Pollution*, 196, 239–246.

Stegemeier, J. P., Avellan, A. & Lowry, G. V. 2017a. Effect of initial speciation of copper- and silver-based nanoparticles on their long-term fate and phytoavailability in freshwater wetland mesocosms. *Environmental Science and Technology*, 51, 12114–12122.

Stegemeier, J. P., Colman, B. P., Schwab, F., Wiesner, M. R. & Lowry, G. V. 2017b. Uptake and distribution of silver in the aquatic plant Landoltia punctata (Duckweed) exposed to silver and silver sulfide nanoparticles. *Environmental Science and Technology*, 51, 4936–4943.

Therezien, M., Thill, A. & Wiesner, M. R. 2014. Importance of heterogeneous aggregation for Np fate in natural and engineered systems. *Science of the Total Environment*, 485–486, 309–318.

Tiwari, A. J., Ashraf-Khorassani, M. & Marr, L. C. 2016. C60 fullerenes from combustion of common fuels. *Science of the Total Environment*, 547, 254–260.

Tiwari, A. J., Morris, J. R., Vejerano, E. P., Hochella, M. F., Jr. & Marr, L. C. 2014. Oxidation of C_{60} aerosols by atmospherically relevant levels of O_3. *Environmental Science and Technology*, 48, 2706–2714.

Torelli, M. D., Putans, R. A., Tan, Y., Lohse, S. E., Murphy, C. J. & Hamers, R. J. 2015. Quantitative determination of ligand densities on nanomaterials by X-ray photoelectron spectroscopy. *ACS Applied Materials and Interfaces*, 7, 1720–1725.

Troiano, J. M., Olenick, L. L., Kuech, T. R., Melby, E. S., Hu, D., Lohse, S. E., Mensch, A. C., Dogangun, M., Vartanian, A. M., Torelli, M. D., Ehimiaghe, E., Walter, S. R., Fu, L., Anderton, C. R., Zhu, Z., Wang, H., Orr, G., Murphy, C. J., Hamers, R. J., Pedersen, J. A. & Geiger, F. M. 2015. Direct probes of 4 nm diameter gold nanoparticles interacting with supported lipid bilayers. *The Journal of Physical Chemistry C*, 119, 534–546.

United Nations General Assembly 2015. Transforming our world: The 2030 Agenda for sustainable Development; Resolution Adopted by the General Assembly on 25 September 2015. A/Res/70/1.

United States House Committee on Interstate and Foreign Commerce. 1963. *Clean Air Act*. United States Government Publishing Office, Washington, DC.

Unrine, J. M., Shoults-Wilson, W. A., Zhurbich, O., Bertsch, P. M. & Tsyusko, O. V. 2012. Trophic transfer of Au nanoparticles from soil along a simulated terrestrial food chain. *Environmental Science and Technology*, 46, 9753–9760.

Vance, M. E., Kuiken, T., Vejerano, E. P., Mcginnis, S. P., Hochella, M. F., Jr., Rejeski, D. & Hull, M. S. 2015. Nanotechnology in the real world: Redeveloping the nanomaterial consumer products inventory. *Beilstein Journal of Nanotechnology*, 6, 1769–1780.

Vikesland, P. J. & Raskin, L. 2016. The drinking water exposome. *Environmental Science: Water Research and Technology*, 2, 561–564.

Wang, S. C. & Flagan, R. C. 1990. Scanning electrical mobility spectrometer. *Aerosol Science and Technology*, 13, 230–240.

Wei, H., Abtahi, S. M. H. & Vikesland, P. J. 2015. Plasmonic colorimetric and Sers sensors for environmental analysis. *Environmental Science: Nano*, 2, 120–135.

Wiesner, M. R., Lowry, G. V., Casman, E., Bertsch, P. M., Matson, C. W., Di Giulio, R. T., Liu, J. & Hochella, M. F., Jr. 2011. Meditations on the ubiquity and mutability of nano-sized materials in the environment. *ACS Nano*, 5, 8466–8470.

Wiesner, M. R., Lowry, G. V., Jones, K. L., Hochella, M. F., Jr., Di Giulio, R. T., Casman, E. & Bernhardt, E. S. 2009. Decreasing uncertainties in assessing environmental exposure, risk, and ecological implications of nanomaterials. *Environmental Science and Technology*, 43, 6458–6462.

Wright, M. V., Matson, C. W., Baker, L. F., Castellon, B. T., Watkins, P. S. & King, R. S. 2018. Titanium dioxide nanoparticle exposure reduces algal biomass and alters algal assemblage composition in wastewater effluent-dominated stream mesocosms. *Science of the Total Environment*, 626, 357–365.

Wujcik, E. K., Wei, H., Zhang, X., Guo, J., Yan, X., Sutrave, N., Wei, S. & Guo, Z. 2014. Antibody nanosensors: A detailed review. *RSC Advances*, 4, 43725–43745.

Xu, J., Murayama, M., Roco, C. M., Veeramani, H., Michel, F. M., Rimstidt, J. D., Winkler, C. & Hochella, M. F., Jr. 2016. Highly-defective nanocrystals of ZnS formed via dissimilatory bacterial sulfate reduction: A comparative study with their abiogenic analogues. *Geochimica et Cosmochimica Acta*, 180, 1–14.

Yang, Y., Chen, B., Hower, J., Schindler, M., Winkler, C., Brandt, J., Giulio, R., Ge, J., Liu, M., Fu, Y., Zhang, L., Chen, Y., Priya, S. & Hochella, M. F., Jr. 2017. Discovery and ramifications of incidental Magnéli phase generation and release from industrial coal-burning. *Nature Communications*, 8, 194.

Yang, Y., Vance, M., Tou, F., Tiwari, A., Liu, M. & Hochella, M. F., Jr. 2016. Nanoparticles in road dust from impervious urban surfaces: Distribution, identification, and environmental implications. *Environmental Science: Nano*, 3, 534–544.

Zhang, Y., Fry, C. G., Pedersen, J. A. & Hamers, R. J. 2017. Dynamics and morphology of nanoparticle-linked polymers elucidated by nuclear magnetic resonance. *Analytical Chemistry*, 89, 12399–12407.

Zheng, Z., Saar, J., Zhi, B., Qiu, T. A., Gallagher, M. J., Fairbrother, D. H., Haynes, C. L., Lienkamp, K. & Rosenzweig, Z. 2018. Structure–property relationships of amine-rich and membrane-disruptive poly(oxonorbornene)-coated gold nanoparticles. *Langmuir*, 34, 4614–4625.

15

Iron Nanoparticles in Environmental Technology

Lenka McGachy, Radek Škarohlíd, and Marek Martinec
University of Chemistry and Technology Prague

15.1 Introduction

In recent history, nanomaterials have attracted attention due to their excellent properties in many fields, including environmental applications. Principally, the key features required for the use of any nanomaterials in environmental applications, especially groundwater remediation, are (i) high removal efficiency for contaminants, (ii) sufficient mobility in the subsurface, (iii) adequate reactive lifetime, and (iv) low toxicity. Due to environmental compatibility and high reactivity, the nanoscale zero valent iron (nZVI) particles are the most widely used nanomaterials in environmental applications. Research has shown that nZVI may be efficient in removal of many reducible environmental pollutants such as chlorinated organic contaminants and heavy metals (Masciangioli and Zhang, 2003; Sun et al., 2006; Wang and Zhang, 1997; Liu and Lowry, 2006; Liu et al., 2005a; Wu et al., 2005; Xu et al., 2005). The progress in nZVI technology can be traced back to the studies by Gillham (1994), who discovered that bulk ZVI can effectively reduce halogenated organic molecules in groundwater and launched the expansion of iron permeable reactive barriers (PRBs – vertical trenches filled with granular ZVI and build in the flow of the subsurface contaminant plume). The idea of using smaller particles of ZVI that may be effective in treating contaminants more rapidly in situ came simultaneously with identifying the technical difficulties of PRB (Borda et al., 2009). The nZVI technology was first tested for groundwater remediation by Wang and Zhang at Lehigh University, USA (Wang and Zhang, 1997). Therefore, iron nanoparticles entered the sector of contaminated site remediation along with the commercialization of nanomaterials over two decades ago. Since that time they have been documented in over more than 50 pilot-scale or full-scale nZVI applications in the subsurface remediation worldwide. The benefit of using nZVI consists especially in its flexibility to in situ applications. Moreover, the proportion of atoms located at the surface increases with decreasing particle size, and that raises the tendency to adsorb, interact, and react with other atoms, molecules, and complexes (Crane and Scott, 2012). Therefore, in comparison with submicron iron particles (used in PRB), the nZVI particles have larger surface areas relative to their volumes and, thus, higher surface reactivity (Zhang, 2003; Li and Zhang, 2006). The high surface reactivity and the potential to degrade the contaminants directly in the subsurface suggest that nZVI can accelerate contaminant removal and reduce the time and cost of remediation. However, there are critical aspects such as nZVI aggregation and fate in the subsurface that have not been fully addressed. Here, focus is placed on recent advances in the use of nZVI technology for the subsurface remediation.

15.2 Synthesis of nZVI

In general, as with other nanomaterials, nZVI can be formed by top-down and bottom-up strategies based on the prevailing mechanisms either physical or chemical in laboratory or commercial scale (Cao and Wang, 2011; Li et al., 2006a). Principally, the top-down approaches make use of bulk material transformation that provides nanoscale particles. In turn, the bottom-up approaches are based on piecing together individual molecules, which lead to the formation of nanoscale clusters. Concise description of the most utilized synthesis methods is provided in subsequent chapters and clearly summarized in Table 15.1.

15.2.1 Liquid-Phase Chemical Reduction: Borohydride Reduction

The underlying principle of liquid-phase reduction is a redox reaction between a strong reductant, commonly sodium borohydride ($NaBH_4$), and metallic ions in liquid solution of $FeCl_3 \cdot 6H_2O$ and $FeSO_4 \cdot 7H_2O$. Because of the simplicity and adjustable nature of the method allowing researchers to modify diverse properties, such as surface stabilization, nZVI emulsification, control of size, size distribution, and core-shell morphology (Carpenter et al., 2003; Li et al., 2003; Ponder et al., 2000, 2001), borohydride reduction has been the most widely studied and applied method within academia (Glavee et al., 1995; Choe et al., 2000; Elliott and Zhang, 2001; Kanel et al., 2005; Liu et al., 2005a; Zhang, 2003; Li et al., 2006a; Philippini et al., 2006; Wang and Zhang, 1997). However, unmodified nZVI particles with the size of tens to hundreds of nanometers produced by this method significantly polydisperse and have high tendency to agglomerate (Nurmi et al., 2005; Sun et al., 2006).

The noticeable fact is that borohydride reduction is relatively expensive mainly due to the high cost of reagents. Mentioned altogether with the production of large volumes of hydrogen gas and wastewater and also several separation steps excludes this method from industrial applications (Li et al., 2009; Yan et al., 2013).

15.2.2 Gas-Phase Reduction

In 2006, Toda Kogyo Corp. developed commercial product RNIP (**R**eactive **N**anoscale **I**ron **P**articles) that is manufactured by gas-phase reduction. RNIP is a crystalline type of nZVI produced by reduction of goethite ($\alpha - FeOOH$) or hematite ($\alpha - Fe_2O_3$) particles with H_2 at elevated temperature (350°C–600°C). RNIP consists of $Fe_3O_4Fe_3O_4$ and $\alpha - Fe^0$ with the weight content of Fe no less than 65% and average particle size of 50–300 nm and specific surface area (SSA) of 7–55 m^2/g. Particles are coated by polyacrylic or polymaleic acid (Li et al., 2006a; Nurmi et al., 2005).

NANOFER products manufactured by NANOIRON Ltd. are made by a gas-phase reduction of nanosized ferrihydrite at 600°C (Filip et al., 2007). NANOFER products exhibit an average size of 50 nm, an average surface area of 20–25 m^2/g, and a high content of iron and are offered on the market as powder or aqueous dispersion (NANOIRON Ltd., 2018). The suspensions are considerably reactive with a significant tendency to agglomerate (Honetschlägerová et al., 2015).

15.2.3 High-Energy Ball Milling

The high-energy ball milling method uses steel balls as the mechanical grinding media to disintegrate iron grains into nanoparticles. In 1998, both Del Bianco et al. (1998) and Malow et al. (1998) published studies where they successfully incorporated this technique. Li et al. (2009) produced nZVI particles of size 20 nm and surface area of 39 m^2/g by their technique modification. More recently, as the market leader of nZVI implementation for environmental applications, Golder Associates Inc. manufacture considerable quantities of nZVI particles by planetary ball milling. Despite the fact that the technique is straightforward and uncomplicated in its nature, it is energy consuming (Crane and Scott, 2012) and the nanoparticles produced are of an irregular shape and have a high energy surface and, thus, display a significant tendency to agglomerate (Stefaniuk et al., 2016). Carbothermal reduction described by Hoch et al. (2008) and Bystrzejewski (2011) lies in the reduction of iron oxide nanoparticles or iron(II/III) salts by gaseous reducing agents, such as H_2, CO, CO_2, resulting from carbon-based materials' decomposition under elevated temperatures (>600°C). Endothermic reaction with only gaseous by-products and the inexpensive carbon-based input materials (carbon black, biochar, etc.) make this method promising for future commercial production.

Ultrasound is the method utilizing the application of ultrasound combined with chemical reduction (Jamei et al., 2013, 2014) or physical methods (Tao et al., 1999). Jamei et al. (2014) showed possibility to obtain different morphologies based on ultrasound frequency and also achieved smaller nanoparticle size and greater SSA without application of ultrasounds.

15.2.4 Innovative Methods

Electrolysis was also investigated as a method for nZVI synthesis. This method is considered by many authors as extremely simple, rapid, and cheap in comparison with chemical reduction (Crane and Scott, 2012; Stefaniuk et al., 2016; Yoo et al., 2007). The main constraining factor is the application of an effective method to disperse nanoparticles generated at the cathode.

VeruTek and the US EPA developed the so-called nZVI green synthesis based on reducing Fe^{2+} by polyphenolic plant extracts, which are produced by heating plant extracts, such as coffee, green tea, and sorghum bran (Crane and Scott, 2012; Hoag et al., 2009). Showing the possibility to use the method under ambient temperature, the crucial advantage is the potential in situ application eliminating offsite manufacture associated also with storage, transport, and other peripheral remote manufacturing costs.

TABLE 15.1 Summary of Reported Methods for the Synthesis of nZVI

Method	Short Description	Pros	Cons	Diameter (nm)	Surface Area (m^2/g)	References
Lithography grinding	Break down bulk iron materials	Inexpensive method	Limited control over particle size distribution and morphology	N.A.	N.A.	Shan et al. (2009)
Precision milling method	The rotary chamber with steel beads	Elimination of toxic reagents, short processing time, low energy consumption		10–50	39.0	Li et al. (2009)
Borohydride reduction	Reduction of the iron salts using reducing agent	Simple and easy to use in any laboratory	The use of toxic reducing agent	1–100	33.5	Wang and Zhang (1997)
Carbothermal reduction	Fe^{2+} are reduced to nZVI at elevated temperatures with the use of thermal energy in the presence of gaseous reducing agents	Spherical iron particles, cheap reducing agent $-H_2$, CO_2, CO	Not well known	20–150	130 (C was used as a matrix and created Fe^0/C had better properties)	Hoch et al. (2008)
Ultrasound method	Application of ultrasound waves and reducing agent	The creation of small nanoparticles	The use of toxic reducing agent	10	34.0–42.0	Jamei et al. (2013)
Electrochemical method	Reduction of the iron salt in the presence of the electrodes and electrical	Inexpensive method	Tendency to form nZVI clusters	1–20	25.4	Chen et al. (2004)
Green synthesis	Biosynthesis of nanoparticles using plant extracts	Replacing toxic reducing agent	Irregular shape	20–120	5.8	Kuang et al. (2013)

Source: Stefaniuk et al. (2016)

15.3 Colloidal Stability

One of the greatest challenges of nZVI technology is the successful transport of iron nanoparticles to the impacted zone in the subsurface. nZVI mobility is affected by various processes that occur immediately after its production and continue during the transport and storage. The primary size of nZVI particles ranges between 50 and 100 nm (U. S. Navy, 2007; Borda et al., 2009). Iron nanoparticles then due to high surface energy and strong magnetic interactions start to aggregate and form secondary particles with sizes approaching the micron scale (Saleh et al., 2007; Phenrat et al., 2007) (see Figure 15.1). Moreover, the presence of ions naturally occurring in groundwater, particularly bivalent cations such as Mg^{2+} an Ca^{2+}, further affects the stability of iron nanoparticles and increases their rates of aggregation (Wiesner and Bottero, 2007). It is believed that these secondary particles do not possess the properties that are expected from "true" nanosized particles.

The rapid aggregation of nZVI in an aqueous environment results from the magnetic attractive forces between particles due to their intrinsic magnetic moments. Moreover, nanoparticle's aggregation is affected by their surface charge, which depends on mineral phases present on the particle surface (Reinsch et al., 2010). Nonmagnetic nanoparticles, such as hematite, aggregate less than magnetite and much less than Fe^0 (Phenrat et al., 2008). Therefore, the Fe^0 content and the magnetic potential of oxide presented on the particle surface significantly affect the stability of nZVI against aggregation.

To improve nZVI dispersion and prevent aggregation, different strategies have been investigated including coating the nanoparticles with a stabilizer (e.g. soluble polymers), conjugating with a solid support, and dispersing the particles in oil–water emulsions (Phenrat et al., 2008; Kanel and Choi, 2007; Kanel et al., 2008).

Surface coating improving colloidal stability of nZVI may be achieved by two mechanisms: electrostatic and steric stabilization. In this regard, several stabilizers including butyl methacrylate (Sirk et al., 2009), carboxymethyl cellulose (He et al., 2010; Phenrat et al., 2008), guar gum (Tiraferri et al., 2008), poly(styrene sulfonate) (Hydutsky et al., 2007; Phenrat et al., 2008), polyacrylic acid (Schrick et al., 2004; Kanel and Choi, 2007; Yang et al., 2007), polyaspartate (Phenrat et al., 2008), polyoxyethylene sorbitan monolaurate (Kanel et al., 2007), polymethylmethacrylate (Sirk et al., 2009), polymethacrylic acid (Sirk et al., 2009), polyvinyl alcohol-co-vinyl acetate-co-itaconic acid (Sun et al., 2007), triblock copolymers (Saleh et al., 2005), xanthan gum (Comba and Sethi, 2009), starch (Yang et al., 2007), and silica (Honetschlägerová et al., 2015) have been used. The coating can be achieved by adding a stabilizer into an iron precursor solution prior to the particle synthesis (in situ, pre-synthesis method) or by mixing or vibrating (sonication) (Sakulchaicharoen et al., 2010; O'Carroll et al., 2013) the synthesized nZVI with stabilizer solutions (post-synthesis method). The stabilized particles exhibit mobility up to several meters, thereby influencing the radius of nZVI and thus the efficiency of nZVI technology (Kocur et al., 2013).

The dispersion stability of nZVI can be achieved also by immobilization of nZVI on a solid support. nZVI has been successfully affixed to a variety of materials including polyvinylidene fluoride membranes, resins, polystyrene resins, silica, carbon, carbon nanotubes, activated carbon, and clay (Xiao et al., 2015; Sheng et al., 2016; Kim et al., 2013). Anchoring of nZVI to a solid support is usually achieved by using carboxylic, hydroxylic, or amine groups as chelating sites for the nanoparticles (or their precursor ions). The mechanism of interaction between contaminants and nZVI-based materials is still not fully clear, but it is expected that the main mechanisms are adsorption and reduction. The nZVI-based material's removal capacity depends on physicochemical properties of the solid support and conditions such as pH, temperature, ionic strength, and concentration of contaminants (Bhowmick et al., 2014; Uzum et al., 2009; Wang et al., 2014; Zou et al., 2016). The adsorptive capacities of nZVI-based materials can be further increased by the formation of iron oxides and hydroxides on their surfaces (Li et al., 2015b, 2016) (Table 15.2).

The colloidal stability of nZVI can be further attained by encapsulation of nZVI inside oil vesicles (Li et al., 2003; Quinn et al., 2005) that improve the particle delivery to the source of the contamination. So far, emulsification has

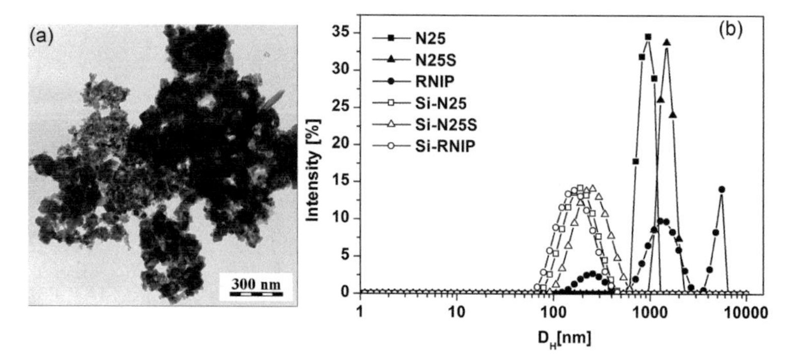

FIGURE 15.1 (a) Image from transmission electron microscope (TEM) – agglomerated commercial product NANOFER 25. (b) Particle size distribution of original and silica coated nZVI, as determined by dynamic light scattering (DLS) - (N25 - NANOFER 25, N25S - NANOFER 25S) (Honetschlägerová et al., 2015).

TABLE 15.2 Summary of Studies on Surface Modication of nZVI

Stabilizing Agent	Method of Stabilization	References
I. Surface-Coated nZVI		
Polyacrylic acid (PAA)	Coating before or after synthesis	Wei et al. (2010), Schrick et al. (2004), and Jiemvarangkul et al. (2011)
Carboxymethyl cellulose (CMC)	Coating before synthesis	He et al. (2007, 2010), Bennett et al. (2010), and Kanel et al. (2008)
Polyvinyl alcohol-co-vinyl acetate-co-itaconic acid (PV3A)	Coating before or after synthesis	Sun et al. (2007)
poly(methacrylic acid)-block-poly(methyl methacrylate)-block-poly(styrenesulfonate) (PMAA–PMMA–PSS) triblock copolymers	Coating after synthesis	Saleh et al. (2007)
Polystyrene sulfonate (PSS)	Coating after synthesis	Phenrat et al. (2008, 2009a)
Polyaspartate	Commercial nZVI	Phenrat et al. (2008)
Silica-water-glass	Commercial nZVI	Honetschlägerová et al. (2018a)
Maleic acid-based polymer	Commercial nZVI	Phenrat et al. (2011)
Natural biopolymers: Guar gum Xanthan gum Calcium alginate Starch	Coating after synthesis	Tiraferri et al. (2008), He and Zhao (2005), Bezbaruah et al. (2009), Comba and Sethi (2009), and Dalla Vecchia et al. (2009)
II. Supported nZVI		
Functionalized membranes	Membranes are soaked in iron salt solutions and followed by reduction with borohydride	Meyer et al. (2004) and Xu et al. (2005)
Carbon supports	Granular activated carbon or porous carbon supports. Ions of iron are anchored before the reduction	Choi et al. (2008), Schrick et al. (2004), and Zhan et al. (2011)
Silica supports	Porous or bulk silica fabricated from aerosol-based processes or sol–gel.	Zhan et al. (2008) and Zhu et al. (2006)
Clay supports	Zeolites and pillared interlayered clays (PILCs) (e.g. bentonite).	Wang et al. (2010b), Chen et al. (2011), and Shi et al. (2011)
III. Emulsifed nZVI		
Water-in-oil-in-water (W/O/W) emulsion	Emulsification after synthesis. nZVI is dispersed in aqueous droplets of diameter 1–20 mm surrounded by an oil membrane	Quinn et al. (2005)
Oil-in-water (O/W) emulsion	Emulsification after synthesis. nZVI is dispersed in O/W droplets of diameter 1–2 μm	Berge and Ramsburg (2009)

Source: Adopted from Yan et al. (2013).

been performed by using commercially available food-grade vegetable oil and surfactant (e.g. SPAN™, TWEEN™, oleic acid).

The enhanced stability and mobility of nZVI is usually at the expense of reduced reactivity. The surface coatings, supported forms or emulsification, create a barrier of physically or chemically bonded macromolecules on the particle surface (Phenrat et al., 2009b). The decrease in reactivity is usually observed with commercial products, mainly RNIP (Chatterjee et al., 2010; Phenrat et al., 2009b; Saleh et al., 2007; Tratnyek et al., 2011). However, post-synthetized silica-coated NANOFER 25 and RNIP prepared by Honetschlägerová et al. exhibited higher removal efficiency of chlorinated ethenes compared to their commercial precursors (Honetschlägerová et al., 2018a).

15.4 Structure and Reactivity

15.4.1 Core-Shell Structure

The chemistry of nZVI including its corrosion, stabilization, contaminant adsorption, and redox transformation is controlled by the structure of nZVI especially by the composition of its surface. Iron nanoparticles have typical core-shell structure. Depending on the way of synthesis, the core is created by amorphous or crystalline $\alpha - Fe^0$ and iron oxide/hydroxide shell. The shell is a thin layer that forms spontaneously during synthesis and undergoes continuous evolution during the whole lifetime of nZVI particles. The minerals commonly presented in the shell are magnetite (Fe_3O_4), maghemite (Fe_2O_3, $\gamma - Fe_2O_3$), and goethite (FeO (OH)) (Reinsch et al., 2010).

Nurmi et al. (2005) investigated the structure of two types of nZVI particles. X-ray diffraction (XRD) of commercial particles prepared by gas-phase reduction showed $\alpha - Fe^0$ and Fe_3O_4, with the metal to oxide proportion of 70%–30%. Iron nanoparticles prepared by borohydride reduction consisted of a nearly single-crystal Fe^0 core with a polycrystalline oxide shell. Honetschlägerová et al. (2015) investigated composition of three commercial nZVI products: RNIP-10APS (Toda Kogyo corp., Japan) and NANOFER 25 (N25) and NANOFER 25S (N25S) (NANO-IRON Ltd., Czech Republic). The XRD examination of N25 and N25S indicated two phases, $\alpha-Fe$ and $Fe_3O_4/\gamma-Fe_2O_3$. N25 contained 98 ± 1.5 wt% of α-Fe and 2 ± 1.5 wt% of $Fe_3O_4/\gamma - Fe_2O_3$ (mean ± SD, $n = 3$). N25S contained 92 ± 1.5 wt% of α-Fe and 8 ± 1.5 wt% of $Fe_3O_4/\gamma - Fe_2O_3$ (mean ± SD, $n = 3$). The same two phases were indicated also for RNIP, which showed α-Fe content of 28 ± 1.5 wt% and $Fe_3O_4/\gamma-Fe_2O_3$ content of 78 ± 1.5 wt% (mean ± SD, $n = 3$).

15.4.2 Why Is nZVI Highly Reactive?

nZVI exhibits qualitatively different reactivity than larger ZVI particles due to quantum and surface size effects (Roduner, 2006). The quantum effects change energy and electronic structure of ≈10 nm metal particles or 10–150 nm oxides particles (Simonet and Valcárcel, 2008). The surface effects include large specific surface-to-volume ratio, which enhance mass transfer between the bulk environment and the nanoparticle surface and improve the adsorption capacity and reactivity (Wang et al., 2009, 2010a). The size of the SSA as well as the number of surface atoms is inversely proportional to the size of nZVI particles (Tratnyek and Johnson, 2006). Because the surface atoms accumulate high binding energy, nanoparticles tend to strongly interact with the environment (Li et al., 2006b). nZVI provides a greater number of particles per unit of mass and a larger surface area resulting in greater density of intrinsic reactive sites on a smaller scale. The literature suggests that the SSA of nZVI, larger by up to three orders of magnitude than the SSA of microscale ZVI, is the main reason of its enhanced reactivity (Nurmi et al., 2005; Wang and Zhang, 1997).

Noubactep et al. (2012) assumed that the reactivity of nZVI also comes from the ability of nZVI particles to release a huge amount of electrons in a very short time due to the large number of particles and the low number of layers in a nanoparticle (Noubactep et al., 2012). As the particle size decreases, the number of the layers also decrease, which are given by a dimension of Fe lattice.

As a result of its high reactivity, the nZVI exhibits three phenomena (Tratnyek and Johnson, 2006): (i) the faster degradation rate of contaminant microscale ZVI, (ii) the proved degradation of contaminants that do not react with the microscale ZVI, and (iii) a shift in distribution of breakdown products to less toxic by-products and more harmless final products compared to microscale ZVI.

Aside from the effect on size of SSA, the reactivity of nZVI and, therefore, its degradation rate and degradation efficiency are strongly dependent on the core-shell structure (Macé et al., 2006; Nurmi et al., 2005), which is influenced by preparation, particle structure, and composition including agglomeration and surface coating (O'Carroll et al., 2013; Phenrat et al., 2009b; Nurmi et al., 2005). The core-shell structure of nZVI significantly changes due to particle aging, groundwater chemistry, and environmental conditions (Liu et al., 2005b; O'Carroll et al., 2013). The reactivity of nZVI can be enhanced by the application of a noble metal to integrate its catalytic effect (Lien and Zhang, 2001) or by appropriate surface-coating process (Sakulchaicharoen et al., 2010).

15.4.3 nZVI Aqueous Chemistry

In aqueous environment, nZVI primarily reacts with dissolved oxygen (DO) and water resulting in electrochemical/corrosion reaction that oxidizes the iron (Eqs. 15.1 and 15.2) (Sun et al., 2016; Lefevre et al., 2016). Primary product of Fe^0 corrosion is ferrous iron (Fe^{2+}) that can be further oxidized (Eqs. 15.3 and 15.4) to ferric iron (Fe^{3+}).

$$\mathbf{Fe^0 + 2H_2O \rightarrow Fe^{2+} + H_2 + 2OH^-} \quad (15.1)$$

$$\mathbf{2Fe^0 + 2H_2O + O_2 \rightarrow 2Fe^{2+} + 4OH^-} \quad (15.2)$$

$$\mathbf{4Fe^{2+} + 4H^+ + O_2 \rightarrow 4Fe^{3+} + 4H_2O} \quad (15.3)$$

$$\mathbf{4Fe^{2+} + 4H^+ + O_2 \rightarrow 4Fe^{3+} + 4H_2O} \quad (15.4)$$

Overall, the ratio of DO, H^+, and H_2O to metallic iron surface determines the rate of Fe^0 oxidation. Moreover, the oxidation of Fe^{2+} to Fe^{3+} will increase the degree of Fe^0 oxidation (Le Chatelier's principle) (Crane and Scott, 2012; Li et al., 2006b; Noubactep et al., 2012). During the oxidation of Fe^0, H^+ is consumed and hydroxide anions are released; therefore, an increase of pH by two to three units was recorded in laboratory studies as a sign of nZVI reactivity (Liu and Lowry, 2006; Matheson and Tratnyek, 1994; Zhang, 2003). Significant hydrogen production and consumption of DO can rapidly induce strong reduction conditions reaching the redox potential between -500 and -900 mV, which is favorable for reduction of contaminants (Crane and Noubactep, 2012).

15.4.4 nZVI Aging

The reactive nature of nZVI is significantly affected by structure that changes during nZVI lifetime due to the oxidation or aging process. Since the shell is initially porous, contaminant transformation and/or removal can occur either at the Fe^0 core or within the thin surface oxide shell. However, contaminant transport across the shell is limited by diffusion. Increasing quantities of corrosion products result in lowering the porosity of the shell, increase crystallinity, and changes into less conductive iron oxides/hydroxides, which limits Fe^0 – contaminant interactions (Noubactep, 2008). The nZVI aging depends greatly on environmental conditions. In water and especially in the presence of DO, nZVI is rapidly oxidized which leads to the development of multilayered oxide shell and a rapid loss in the reactivity (Crane and Scott, 2012). Surface oxidation of nZVI particles always occurs even under highly anerobic conditions. After the injection of nZVI into the groundwater, Fe^{2+} and Fe^{3+} are temporarily present and then quickly oxidize to insoluble iron species (Lefevre et al., 2016). Under oxic conditions, high oxidation rates cause formation of lepidocrocite (FeOOH) or goethite (αFeOOH) (Kumar et al., 2014a). However, under anoxic conditions, lower oxidation rates lead to the formation of maghemite (γFe_2O_3) or magnetite (Fe_3O_4), as the end product of nZVI oxidation due to the soluble iron dehydroxylation (Stumm and Morgan, 2013; Lefevre et al., 2016). These minerals form many structures around the nZVI core including aggregates of spheres, needles, or board shapes of magnetite, lepidocrocite, and goethite, with individual sizes between 20 nm and 1 μm (Stumm and Morgan, 2013; Greenlee et al., 2012; Kumar et al., 2014b; Lefevre et al., 2016) (Figure 15.2). Among other common groundwater constituents, nitrites (Sohn et al., 2006) or carbonates (Hua et al., 2018) strongly affect nZVI aging.

Factors Affecting nZVI Reactivity

Solution chemistry including pH, DO, and water constituents is one of the most important factors that affect the nZVI reactivity under natural conditions. Furthermore, nZVI/contaminant concentration, temperature, and

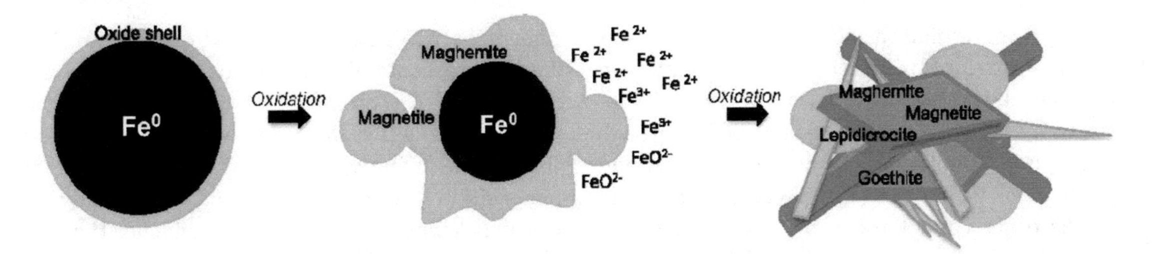

FIGURE 15.2 nZVI oxidation process in the environment (Lefevre et al., 2016).

presence of indigenous microbes can significantly affect the reactivity of nZVI.

pH strongly influences the rate of nZVI corrosion (Bae and Hanna, 2015; O'Carroll et al., 2013). Many researchers have revealed that with decreasing pH, the kinetic constants for reductive removal of organic contaminants (Song and Carraway, 2005; Matheson and Tratnyek, 1994) and inorganic contaminants (Jiang et al., 2015; Guan et al., 2015; Alowitz and Scherer, 2002) increase. In summary, the high removal efficiency of nZVI at low pH could be caused by the acceleration of iron corrosion, while at high pH values ($pH > 8$), the removal efficiency of nZVI is lower because of mineral precipitation at the Fe^0 core surface, which inhibits the mass transfer (Dong et al., 2010; Bae and Lee, 2014). However, low pH could also reduce the nZVI performance due to a fast dissolution of nZVI particles.

Another parameter that can affect nZVI performance is DO. So far, there is no general agreement about the effect of DO on the nZVI performance (Huang and Zhang, 2005). Some researchers found that the presence of DO could lower the removal efficiency of nZVI due to the passivation of nZVI and/or its ability to compete for electrons. Others stated that DO may increase nZVI performance due to iron corrosion (Sun et al., 2016).

Furthermore, co-solutes present in water can affect nZVI performance. Many studies had showed that typical anions such as Cl^-, SO_4^{2-}, NO_3^-, CO_3^{2-}, HCO_3^-, SiO_3^{2-}, ClO_4^-; cations such as Fe^{2+}, Co^{2+}, Ni^{2+}, Ca^{2+}, Mg^{2+}; and natural organic matter (NOM) would significantly influence the performance of nZVI. The effect of co-existing solutes on the nZVI performance is shown in Table 15.3.

The effect of temperature on heterogeneous reaction of nZVI can determine reaction kinetics. As reported in Lien and Zhang (2007), the reaction rate of perchloroethene (PCE) increases by factor of almost two when temperature growth from 15°C to 25°C (Lien and Zhang, 2007).

The reactivity of nZVI can be affected also by indigenous microorganisms in the subsurface. Conflicting conclusions

TABLE 15.3 Effect of Co-existing Solutes on the nZVI Performance

Co-existing Solutes	Effect	Mechanism	References
Cl^-	Enhance the performance of nZVI	Breakdown the protective oxide film; causing corrosion	Devlin and Allin (2005) and Hernandez et al. (2004)
	Deteriorate the performance of nZVI	Competitive sorption with respect to perchlorate or nitrate reduction	Hwang et al. (2015) and Moore et al. (2003)
SO_4^{2-}	Enhance the performance of nZVI	Increase the surface reactivity or sorption capacity; breakdown the protective oxide film	Hwang et al. (2015) and Moore et al. (2003)
	Deteriorate the performance of nZVI	Iron consumption. Alternation of surface properties by Fe anion complexes	Liu et al. (2007) and Yu et al. (2013)
NO_3^-	Enhance the performance of nZVI	Not clearly described	Liu et al. (2007) and Su et al. (2014)
	Deteriorate the performance of nZVI	Competes for reactive sites with target contaminants. Generates passive oxide film	Klausen et al. (2001) and Xu et al. (2012)
HCO_3^-	Enhance the performance of nZVI	Not clearly described but depends on concentration and exposure time	Bi et al. (2009) and Agrawal et al. (2002)
	Deteriorate the performance of nZVI	Passivation caused by carbonate-bearing minerals. Passivation of reactive sites by forming Fe anion	Klausen et al. (2001) and Kober et al. (2002)
SiO_3^{2-}	Deteriorate the performance of nZVI	Block reactive sites by forming a protective layer. Change the speciation of the surface	Su and Puls (2001) and Klausen et al. (2001)
ClO_4^-	Potential deteriorating or enhancing effects	Reduction by nZVI in long term or at elevated temperatures	Lim and Zhu (2008) and Im et al. (2011)
Fe^{2+},Co^{2+},Ni^{2+},Mg^{2+}, Mn^{2+},Zn^{2+},Cu^{2+},Pb^{2+}	Enhance the performance of nZVI	Enhance electron transfer. Form bimetallic systems. Depassivate Fe^0	Bae and Hanna (2015) and Tang et al. (2014)
Ca^{2+}, Mg^{2+}	Deteriorate the performance of nZVI	Masking the reactive sites on the Fe^0 surface by precipitate accumulation	Dong et al. (2013) and Mak et al. (2009)
NOM	Enhance the performance of nZVI	Enhance adsorption	Tratnyek et al. (2001) and Xie and Shang (2005)
	Deteriorate the performance of nZVI	Block reactive sites; decrease mass transfer to/from the reactive surfaces; change the surface electrostatic and reductive potential	Johnson et al. (2009) and Tsang et al. (2009)

Source: Sun et al. (2016)

have been reported concerning the effect of microorganisms on nZVI. Recent study by Honetschlägerová et al. revealed that the presence of iron reducing bacteria (IRB) inhibited nZVI reactivity towards trichloroethene (TCE) through iron corrosion caused by microbial respiration on iron and to cell sorption to nZVI surface (Honetschlägerová et al., 2018b). One the other hand Park et al. (2010) examined IRB metabolism in microcosms containing nZVI. This work suggested that nZVI did not exert serious impact on the microbial population. However, this research did not take into account the presence of a contaminant.

15.4.5 Nature of nZVI Reactivity with Aqueous Contaminants

The core–shell nature of nZVI serves many functional properties. The nZVI Fe^0 core acts as an electron donor and reductant and the oxide shell facilitates sorption and surface complexation while allowing electrons to pass into the metal core (Figure 15.3). The reactivity of core-shell structured nZVI, thus reaction kinetics, is controlled by the kinetics of its core oxidation and the transport of electrons through the oxide shell (Nurmi et al., 2005).

Fe^0 is a moderately strong oxidant (Stumm and Morgan, 1996) ($E^0 = 0.44V$) that can reduce chlorinated aliphatic hydrocarbons (CAHs), metals cations, and other materials such as azo dyes (Fan et al., 2009; Lin et al., 2008; Shirin and Balakrishnan, 2011; Luo et al., 2013), organophosphates (Ambashta et al., 2011), nitroamines (Naja et al., 2008), or nitroaromatics (Zhang et al., 2009, 2010; Yin et al., 2012).

Reductive Dehalogenation

Being the most commonly detected pollutants in the subsurface, the chemical transformation of CAHs by nZVI has been well investigated. The mechanism of chlorinated ethanes and

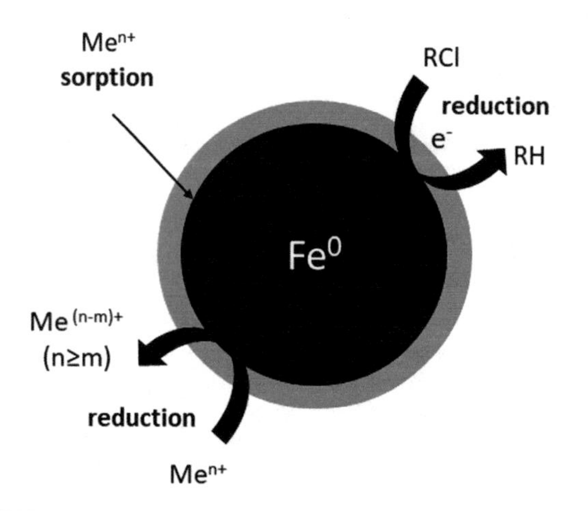

FIGURE 15.3 A core-shell structure for iron nanoparticles in aqueous solution. The core is made of metallic iron while the shell consists mostly of iron oxides and hydroxides. Thus iron nanoparticles exhibit characteristics of both iron oxides (e.g. as a sorbent) and metallic iron (e.g. as a reductant). (Adopted from Sun et al., 2006.)

ethenes by nZVI in aqueous environment involves adsorption of the contaminants at the iron surface followed by breaking of carbon-halogen bonds. Because the adsorption and reductive dehalogenation is a surface-mediated process, the reactivity of nZVI is determined by surface composition (Wang et al., 2010a). Especially impurities, alloying elements, adsorbates can influence the reactive sites (Johnson et al., 1996).

When Fe^0 is oxidized (Eq. 15.2), released electrons cleave chlorides (Cl^-) from halogenated organic molecules (RCl) to produce dehalogenated hydrocarbons (RH) and chlorides according to:

$$RCl + Fe^0 + H^+ \rightarrow RH + Fe^{2+} + Cl^- \quad (15.5)$$

$$C_2Cl_4 + 5Fe^0 + 6H^+ \rightarrow C_2CH_6 + 5Fe^{2+} + 4Cl^- \quad (15.6)$$

The degree of dehalogenation depends on available electrons. Equation (15.6) describes example of the reductive dehalogenation of PCE by nZVI to final ethane (Li et al., 2006b). Chlorinated ethenes such as PCE are probably the most targeted contaminants in nZVI technology. Reductive dehalogenation of chlorinated ethenes by nZVI is expected to proceed via the reaction pathways including reductive α- and β-dihalo/elimination, hydrogenolysis, and hydrogenation which proceed simultaneously (Arnold and Roberts, 2000; Liu et al., 2005a). The reaction products formed during the reductive dehalogenation by nZVI must be considered. Some reaction products such as vinyl chloride (VC) are more toxic and more mobile than the parent compounds. The formation of degradation products is highly affected by the structure of particles and the synthesis method. RNIP degraded chlorinated ethenes by the governing pathway of β-elimination while Fe^{BH} is assumed to mainly proceed via hydrogenolysis (Liu et al., 2005a,b). Doping a small amount of a noble metal (e.g. Pd, Pt, Ag) on nZVI surface can significantly improve the reactivity. Trace amount of Pd on nZVI surface (0.05–0.1 wt%) acts as a catalyst and greatly increases the reaction rate (k_{sa}, L m^2/d) of nZVI ($\approx$ 1.2–3.0 $\times$ 10^{-2}) and leads to little or no accumulation of chlorinated intermediates (Lien and Zhang, 2007; Sakulchaicharoen et al., 2010; Lien and Zhang, 2001). Unfortunately, the reactivity of bimetallic nZVI is significantly reduced under the natural conditions, the accelerated iron oxidation, and entrapment of active noble metal sites underneath an iron oxide shell (Yan et al., 2010).

Metal Cations

nZVI has demonstrated effective reduction or removal ability for various types of metallic ions such as As^{3+},As^{5+},Cr^{6+},Pb^{2+},Fe^{3+},Cu^{2+},Cd^{2+},Co^{2+},Zn^{2+},Hg^{2+}, and Ni^{2+} (Kanel et al., 2005, 2006; Li and Zhang, 2006, 2007; Ponder et al., 2000; Zou et al., 2016; Ulucan-Altuntas et al., 2017; Rathor et al., 2017; Li et al., 2017). Two possible mechanisms have been documented: reduction of the contaminants to a more inoffensive form and surface sorption. Fe^0 acts as an electron donor for many metal ions (Zhang et al., 2011; Song and Carraway, 2005) and influences their valence and physicochemical properties (Li et al., 2015a; Song and Carraway, 2005; Zou et al., 2016; Liu et al., 2015). Metal ions such as Cd^{2+} and Zn^{2+} with E^0 close to, or more negative than, $E^0(Fe^0)$ are removed predominantly via sorption/surface complex formation. In comparison, the removal of metal ions, such as Cu^{2+} and Hg^{2+}, with E^0 much more positive than $E^0(Fe^0)$, which occurs mainly through reduction (Li et al., 2015b, 2016).

Phosphates and Nitrites

Recent studies show that also phosphates can be easily removed from water using nZVI. Generally, processes of precipitation and adsorption contribute to the removal of phosphates. At low pH, the nZVI is easily dissolved in the aqueous solution. In presence of DO, oxidation of Fe^{2+} results in the formation of Fe^{3+}, which can bind with phosphates. In addition, the redox reaction between nZVI particles and water increases the pH and promotes the formation of $Fe(OH)_2$ and $Fe(OH)_3$, which can adsorb phosphate by incorporation (Wen et al., 2014; Zhang et al., 2017). nZVI seems to be also promising for removal of nitrates from the solution. Recent studies show that nitrates can be successfully removed from wastewaters using nZVI especially at low pH (Siciliano, 2015; Babaei et al., 2015; Moradi et al., 2017; Wang et al., 2016). In the first step, nitrates are reduced by nZVI to nitrites. In the next step, nitrites are transformed to ammonium.

15.5 Transport in Porous Media

The most critical factor of nZVI application in the subsurface remediation is the ability to access a contaminant in the subsurface. It has been shown that nZVI does not transport easily in saturated porous media and is not delivered for more than few centimeters from an injection well (Saleh et al., 2007; Schrick et al., 2004; Honetschlägerová et al., 2016). The delivery of nZVI in the subsurface is analogous to filtration in porous media. nZVI injected into the subsurface can colloid or attach to soil grain. The iron oxide shell possesses a high affinity for adsorption onto surfaces of soil grains (Iler, 1959). Natural geochemical conditions (pH and ionic strength) can destabilize nanoparticles and promote aggregation. Formation of aggregates allows plugging of the porous media, thus affecting the mobility of nZVI due to the straining and gravitational sedimentation. Transport of nZVI can be facilitated by reducing their affinity to the surfaces of the soil grains and by increasing their colloidal stability. nZVI stabilized by negatively charged carriers are repelled more by the negatively charged soil grains and quickly migrate through saturated porous media. Furthermore, nZVI can colloid with pure nonaqueous phase liquids (NAPLs). While nZVI filtration by straining and attachment to soil grain is undesirable due to the affecting nZVI mobility, interactions with entrapped NAPLs are desirable. The mobility of nZVI will also depend on hydrodynamic conditions of the media (pore size, porosity, flow velocity, and degree of mixing or turbulence). At a high pore-water velocity, the residence time of the nanoparticles at the collector surface may be too short to allow deposition to occur. Low deposition efficiency will result in longer transport distance. In general, the hydrogeochemistry of the system, along with the properties of nZVI, will determine the transportability of nZVI at a specific site. Therefore, adequate site characterization is essential to determine whether the particles can infiltrate the remediation source zone. However, it is difficult to predict how the various interactions between nZVI and collector will affect their transport.

15.6 Field Implementation of nZVI

The attractiveness of nZVI is its potential to be used in situ. nZVI can be injected directly into contaminated subsurface using a well (Figure 15.4a) and degrade the contaminants directly in the subsurface. The injection of nZVI is conceptually possible at almost any location and depth of the terrestrial subsurface. Principally, there are two approaches of in situ nZVI remediation. First approach represents remediation of contaminant plume by injecting low mobile nZVI forming "reactive treatment zone" (Figure 15.4c) and second one treats static contaminant plume (NAPLs) and involves injecting of mobile nZVI (Figure 15.4b), which is thereafter transported down gradient to the wider area (Crane and Scott, 2012; Tratnyek and Johnson, 2006) of the subsurface.

Since the first field nZVI application by Elliott and Zhang (2001), other 50+ full-scale or pilot field applications have been conducted worldwide with a variety of success, with most of them in the United States and the remainder in Europe (Czech Republic and Germany). In general, the most frequent case is remediation of groundwater contaminated by chlorinated aliphatic hydrocarbons using nZVI injection by gravity feed or low pressure (Karn et al., 2009; Mueller et al., 2012). While nZVI applied in Europe in all cases was standard surface-modified, in United States, it was only 60%, and 40% were bimetallic nZVI and emulsified nZVI (Mueller et al., 2012). Based on their summary of pilot-scale and field-scale applications of nZVI, Stefaniuk et al. (2016) show that effectiveness of pollutant degradation varied from 40% to 100%. The pilot test in Tuhnice in the Czech Republic led to the creation of strongly reducing conditions and a sharp decrease in contaminant concentrations. However, the concentration of the main contaminant dichlorethene (DCE) remained low for 1 month and thereafter increased to original concentration (Honetschlägerová et al., 2010). At the Kuřivody site (Czech Republic), the application of nZVI led to permanent reduction of the original total contaminants' concentration by about 75%–95% (Mueller et al., 2012).

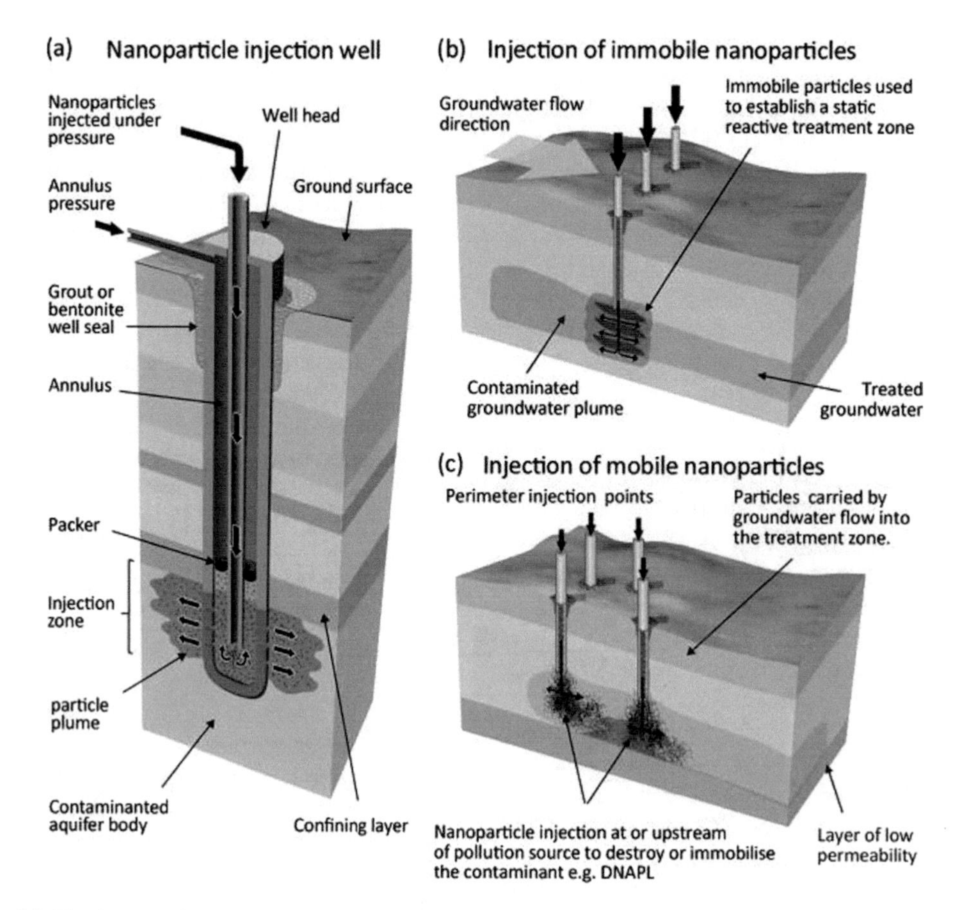

FIGURE 15.4 (a) The layout of a typical nanoparticle injection well. The technology is very similar to that used for injection of CO_2 into subterranean storage reservoirs. Right: the use of nanoparticle injection technology for treating contaminant bodies that are (b) mobile and (c) immobile. The location and number of particle injection wells are site specific and tailored to achieve the best possible remediation (Crane and Scott, 2012).

15.7 Laboratory Tests

nZVI environmental stability and mobility and contamination issues are of significant importance for nZVI field application. Conceptually, the passivation layer reacts and evolves with time. In addition, nZVI also reacts with surrounding environment with water constituents and contaminants. The distinction of these interactions can be useful in predicting the fate of nZVI in the subsurface environment. nZVI characterization may help in indicating these interactions. Moreover, preliminary lab assessments including batch and column experiments using groundwater and soil from a contaminated site can help in estimating the nZVI injection dose, the number of injection boreholes, and the total amount of nZVI to be applied.

15.7.1 Characterization

When characterizing nZVI information about particle size, size distribution, and surface area; surface and bulk composition and chemical state and colloidal stability have to be gathered (Nurmi et al., 2005; Honetschlägerová et al., 2015). Direct TEM or scanning electron microscopy (SEM) usually performs coarse particle analysis, while particle size distribution is obtained with dynamic light scattering (DLS). However, DLS results may be altered by the presence of aggregates. Furthermore, sedimentation profiles obtained by monitoring the optical absorbance of suspensions are also a part of particle size analysis. Shell composition is analyzed using X-ray photoelectron spectroscopy (XPS). X-ray powder diffraction (XRD) provides the crystallinity of nanoparticles and X-ray fluorescence spectrometry (WD-XRF) examines elemental composition and the presence of impurities. To evaluated colloidal stability of nZVI suspension, ζ-potential is measured.

15.7.2 Batch Experiments

The reactivity of nZVI with target contaminant is a critical issue of nZVI application. nZVI is not only a selective agent, it also reacts with other groundwater compartments. To stimulate the reactivity of nZVI under the natural conditions, batch experiments are usually performed. Batch experiments are conducted by adding a certain amount of solid into solution containing specific concentrations of contaminant and nZVI with a specific solid and liquid ratio. The tested batch microcosms are conducted in glass vials (Qian et al., 2019; Dong et al., 2018;

Honetschlägerová et al., 2018a,b) or a conical flask (Eljamal et al., 2018). To set up conditions similar to those of natural groundwater, the batch microcosms can be incubated on a shaker, in darkness, at temperatures similar to those in the subsurface (8°C–15°C) (Honetschlägerová et al., 2018b).

15.7.3 Column Experiments

To predict the behavior of nZVI in the subsurface including mobility, interactions with soil, and reactivity under the flow conditions, column experiments are usually used. The basic apparatus is a vertical cylindrical column, in which the bottom is provided with a suitable filtration membrane. The material of a column using for the column experiments with nZVI is mainly glass (Schrick et al., 2004; Koch et al., 2005; Tiraferri and Sethi, 2009), plexiglas (Vecchia et al., 2009), or stainless steel (Phenrat et al., 2009a; Saleh et al., 2008). Columns are packed wet usually with spherical silica sand (d_{50} = 300–690 nm) as model porous media (Vecchia et al., 2009; Phenrat et al., 2009a; Saleh et al., 2008). Ionic strength and ionic composition have a significant impact on the subsurface behavior of nZVI (Schrick et al., 2004; Saleh et al., 2007, 2008). Typical concentration of monovalent cations (e.g. Na^+, K^+) in groundwater is 1–10 mM and divalent cations (e.g. Ca^{2+}, Mg^{2+}) is 0.1–2 mM (Busenberg et al., 2000). Therefore, columns are, prior the injection of nZVI, usually flushed with background electrolyte solution (10–12 pore volumes) to remove background turbidity and to provide a uniform collector surface charge (Saleh et al., 2008). A peristaltic pump is commonly used to feed both background electrolyte solution and nZVI into a column. Figure 15.5 shows setup of lab-scale experiments and pilot test.

15.8 Interactions of nZVI with the Living Organisms

The activity of bacteria in aquifer sediment can be enhanced or inhibited depending on the dose of nZVI. Thanks to high reactivity of nZVI, these nanoparticles could be toxic for living organisms. Recent studies revealed that nZVI particles can be adsorbed on bacterial cell membranes and penetrate them (Honetschlägerová et al., 2018b). These nanoparticles can cause changes in cell membrane structure and inhibit nutrient intake and mobility, which could cause death of the microorganism (Xiu et al., 2010; Xie et al., 2017). Also, formation of reactive oxygen species may cause damage to DNA, peroxidation of lipids (Stefaniuk et al., 2016; Stohs and Bagchi, 1995), and cell death. In contrast, few studies indicate that nZVI could have a positive effect on the microorganisms. For example, it was observed that the presence of nZVI increased the reduction of nitrates by microbes (Shin and Cha, 2008; Kotchaplai et al., 2017). Other authors observed stimulation of the Gram-positive bacteria growth in the soil (Nemecek et al., 2014). Anoxic corrosion of nZVI produces hydrogen that may biostimulate bacterial growth (Liu and Lowry, 2006).

Not only microorganisms but also plants can be negatively affected by nZVI mainly due to blocking of nutrient and water uptake by roots, thanks to deposition of nZVI particles on its surface (El-Temsah and Joner, 2012; Ma et al., 2013). It has been reported that nZVI inhibits growth of young leaves and promotes the decay of older leaves (Ma et al., 2013; Ghosh et al., 2017). In contrast, other studies show that low concentrations of nZVI can stimulate growth of some plants e.g. Typha latifolia and Oryza sativa (Guha et al., 2018).

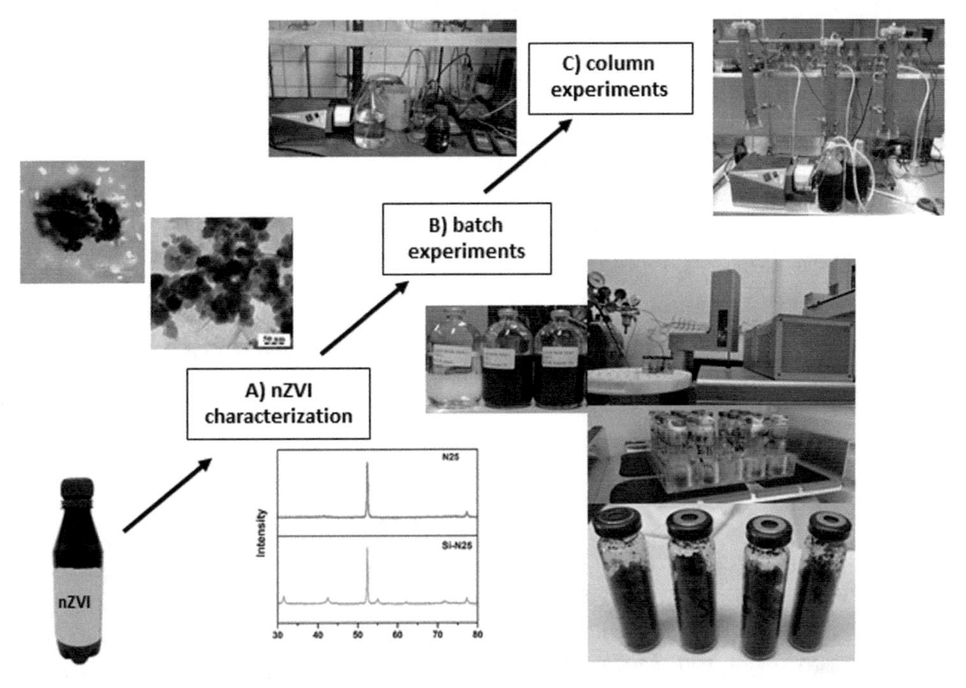

FIGURE 15.5 Setup of lab-scale tests.

A little is known about nZVI toxicity on mammalian cells, but few studies show that nZVI was lethal for bronchial epithelium human cells and that nZVI can be neurotoxic for neuron microglia in rodent (Sevcu et al., 2012; Phenrat et al., 2009c; Stefaniuk et al., 2016).

15.9 nZVI in Non-subsurface Remediation Applications

nZVI can be used not only for a remediation of contaminated groundwater, but many studies have shown that nZVI is effective also in removing viruses and bacteria (Shi et al., 2012; You et al., 2005), disinfection by-products (DBPs), and DBP precursors (NOM) (Kristiana et al., 2011) or phenols (Shimizu et al., 2012; Segura et al., 2012). Other studies suggested special use of nZVI nanoparticles for H_2S removal from gas (Su et al., 2018), enhancement of germination and growth in aromatic rice cultivar (Guha et al., 2018), enhancement of biodiesel production from glucose-fed activated sludge microbial cultures by addition of nZVI and $FeCl_3$ (Huang et al., 2016), removal of medicament residues (Wu et al., 2018; Song et al., 2017), and many more.

15.10 Future Challenges

nZVI as a subject matter has been well documented. There are number of laboratory studies concerning nZVI structure, stability, mobility in porous media, and reactivity towards different groups of contaminants together with documented pilots or full-scale nZVI applications. However, this outstanding status of nZVI does not decrease the need to address various issues such as connecting the laboratory results to field implementation, developing models to predict the mobility of nZVI in the natural environment, and tackling hydro-chemical and biological interactions of nZVI in the subsurface. The complex nature of subsurface environment and distribution of contaminants are difficult to reproduce in laboratory settings. Missing knowledge of specific interactions between nZVI and the contaminated subsurface ecosystems is the basic knowledge gap, which limits results of every remediation. From an engineering point of view, there is a need to develop decision-making tools and techniques for optimal application of nZVI into a contaminated subsurface. To address these challenges, the close collaboration between research community and the remediation industry is highly desired. Moreover, nZVI reactivity offers more applications out of the remediation sector.

References

Agrawal, A., Ferguson, W. J., Gardner, B. O., Christ, J. A., Bandstra, J. Z. & Tratnyek, P. G. 2002. Effects of carbonate species on the kinetics of dechlorination of 1,1,1-trichloroethane by zero-valent iron. *Environmental Science and Technology*, 36, 4326–4333.

Alowitz, M. J. & Scherer, M. M. 2002. Kinetics of nitrate, nitrite, and Cr(VI) reduction by iron metal. *Environmental Science and Technology*, 36, 299–306.

Ambashta, R. D., Repo, E. & Sillanpaa, M. 2011. Degradation of tributyl phosphate using nanopowders of iron and iron-nickel under the influence of a static magnetic field. *Industrial and Engineering Chemistry Research*, 50, 11771–11777.

Arnold, W. A. & Roberts, A. L. 2000. Pathways and kinetics of chlorinated ethylene and chlorinated acetylene reaction with Fe(O) particles. *Environmental Science and Technology*, 34, 1794–1805.

Babaei, A. A., Azari, A., Kalantary, R. R. & Kakavandi, B. 2015. Enhanced removal of nitrate from water using nZVI@MWCNTs composite: Synthesis, kinetics and mechanism of reduction. *Water Science and Technology*, 72, 1988–1999.

Bae, S. & Hanna, K. 2015. Reactivity of nanoscale zero-valent iron in unbuffered systems: Effect of pH and Fe(II) dissolution. *Environmental Science and Technology*, 49, 10536–10543.

Bae, S. & Lee, W. 2014. Influence of riboflavin on nanoscale zero-valent iron reactivity during the degradation of carbon tetrachloride. *Environmental Science and Technology*, 48, 2368–2376.

Bennett, P., He, F., Zhao, D. Y., Aiken, B. & Feldman, L. 2010. In situ testing of metallic iron nanoparticle mobility and reactivity in a shallow granular aquifer. *Journal of Contaminant Hydrology*, 116, 35–46.

Berge, N. D. & Ramsburg, C. A. 2009. Oil-in-water emulsions for encapsulated delivery of reactive iron particles. *Environmental Science and Technology*, 43, 5060–5066.

Bezbaruah, A. N., Krajangpan, S., Chisholm, B. J., Khan, E. & Bermudez, J. J. E. 2009. Entrapment of iron nanoparticles in calcium alginate beads for groundwater remediation applications. *Journal of Hazardous Materials*, 166, 1339–1343.

Bhowmick, S., Chakraborty, S., Mondal, P., Van Renterghem, W., Van Den Berghe, S., Roman-Ross, G., Chatterjee, D. & Iglesias, M. 2014. Montmorillonite-supported nanoscale zero-valent iron for removal of arsenic from aqueous solution: Kinetics and mechanism. *Chemical Engineering Journal*, 243, 14–23.

Bi, E. P., Bowen, I. & Devlin, J. F. 2009. Effect of mixed anions (HCO3–SO42–ClO4-) on granular iron (Fe-0) reactivity. *Environmental Science and Technology*, 43, 5975–5981.

Borda, M. J., Venkatakrishnan, R. & Gheorghiu, F. 2009. Status of nZVI technology. In: *Environmental Applications of Nanoscale and Microscale Reactive Metal Particles*, C. L. Geiger and K. M. Carvalho-Knighton, Eds. American Chemical Society, Washington, DC, pp. 219–232.

Busenberg, E., Plummer, L. N., Doughten, M. W., Widman, P. K. & Bartholomay, R. C. 2000. Chemical and isotopic composition and gas concentrations of ground water and surface water from selected sites at and near the Idaho

National Engineering and Environmental Laboratory, Idaho, 1994-97. Technical Report, DOE/ID-22164, TRN: AH200118%%529, U.S. Department of Energy.

Bystrzejewski, M. 2011. Synthesis of carbon-encapsulated iron nanoparticles via solid state reduction of iron oxide nanoparticles. *Journal of Solid State Chemistry*, 184, 1492–1498.

Cao, G. & Wang, Y. 2011. *Nanostructures and Nanomaterials: Synthesis, Properties, and Applications*. World Scientific, Washington, DC.

Carpenter, E. E., Calvin, S., Stroud, R. M. & Harris, V. G. 2003. Passivated iron as core–shell nanoparticles. *Chemistry of Materials*, 15, 3245–3246.

Comba, S. & Sethi, R. 2009. Stabilization of highly concentrated suspensions of iron nanoparticles using shear-thinning gels of xanthan gum. *Water Research*, 43, 3717–3726.

Crane, R. A. & Noubactep, C. 2012. Elemental metals for environmental remediation: Lessons from hydrometallurgy. *Fresenius Environmental Bulletin*, 21, 1192–1196.

Crane, R. A. & Scott, T. B. 2012. Nanoscale zero-valent iron: Future prospects for an emerging water treatment technology. *Journal of Hazardous Materials*, 211–212, 112–125.

Dalla Vecchia, E., Luna, M. & Sethi, R. 2009. Transport in porous media of highly concentrated iron micro- and nanoparticles in the presence of Xanthan Gum. *Environmental Science and Technology*, 43, 8942–8947.

Del Bianco, L., Hernado, A., Navarro, E., Bonetti, E. & Pasquin, L.. 1998. Structural configuration and magnetic effects in as-milled and annealed nanocrystalline iron. *Journal de physique. IV*, 08, 107–110.

Devlin, J. F. & Allin, K. O. 2005. Major anion effects on the kinetics and reactivity of granular iron in glass-encased magnet batch reactor experiments. *Environmental Science and Technology*, 39, 1868–1874.

Dong, H., Hou, K., Qiao, W., Cheng, Y., Zhang, L., Wang, B., Li, L., Wang, Y., Ning, Q. & Zeng, G. 2019. Insights into enhanced removal of TCE utilizing sulfide-modified nanoscale zero-valent iron activated persulfate. *Chemical Engineering Journal*, 359, 1046–1055.

Dong, J., Ding, L. J., Wen, C. Y., Hong, M. & Zhao, Y. S. 2013. Effects of geochemical constituents on the zero-valent iron reductive removal of nitrobenzene in groundwater. *Water and Environment Journal*, 27, 20–28.

Dong, J., Zhao, Y. S., Zhao, R. & Zhou, R. 2010. Effects of pH and particle size on kinetics of nitrobenzene reduction by zero-valent iron. *Journal of Environmental Sciences*, 22, 1741–1747.

El-Temsah, Y. S. & Joner, E. J. 2012. Impact of Fe and Ag nanoparticles on seed germination and differences in bioavailability during exposure in aqueous suspension and soil. *Environmental Toxicology*, 27, 42–49.

Eljamal, O., Mokete, R., Matsunaga, N. & Sugihara, Y. 2018. Chemical pathways of nanoscale zero-valent iron (nZVI) during its transformation in aqueous solutions. *Journal of Environmental Chemical Engineering*, 6, 6207–6220.

Elliott, D. W. & Zhang, W.-X. 2001. Field assessment of nanoscale bimetallic particles for groundwater treatment. *Environmental Science and Technology*, 35, 4922–4926.

Fan, J., Guo, Y. H., Wang, J. J. & Fan, M. H. 2009. Rapid decolorization of azo dye methyl orange in aqueous solution by nanoscale zerovalent iron particles. *Journal of Hazardous Materials*, 166, 904–910.

Filip, J., Zboril, R., Schneeweiss, O., Zeman, J., Cernik, M., Kvapil, P. & Otyepka, M. 2007. Environmental applications of chemically pure natural ferrihydrite. *Environmental Science and Technology*, 41, 4367–4374.

Ghosh, I., Mukherjee, A. & Mukherjee, A. 2017. In planta genotoxicity of nZVI: Influence of colloidal stability on uptake, DNA damage, oxidative stress and cell death. *Mutagenesis*, 32, 371–387.

Gillham, R. W. 1994. Enhanced degradation of halogenated aliphatics by zero-valent iron. *Ground Water*, 32, 958–967.

Glavee, G. N., Klabunde, K. J., Sorensen, C. M. & Hadjipanayis, G. C. 1995. Chemistry of borohydride reduction of iron(II) and iron(III) ions in aqueous and nonaqueous media. Formation of nanoscale Fe, FeB, and Fe2B powders. *Inorganic Chemistry*, 34, 28–35.

Greenlee, L. F., Torrey, J. D., Amaro, R. L. & Shaw, J. M. 2012. Kinetics of zero valent iron nanoparticle oxidation in oxygenated water. *Environmental Science and Technology*, 46, 12913–12920.

Guan, X. H., Jiang, X., Qiao, J. L. & Zhou, G. M. 2015. Decomplexation and subsequent reductive removal of EDTA-chelated Cu-II by zero-valent iron coupled with a weak magnetic field: Performances and mechanisms. *Journal of Hazardous Materials*, 300, 688–694.

Guha, T., Ravikumar, K. V. G., Mukherjee, A., Mukherjee, A. & Kundu, R. 2018. Nanopriming with zero valent iron (nZVI) enhances germination and growth in aromatic rice cultivar (Oryza sativa cv. Gobindabhog L.). *Plant Physiology and Biochemistry*, 127, 403–413.

He, F. & Zhao, D. Y. 2005. Preparation and characterization of a new class of starch-stabilized bimetallic nanoparticles for degradation of chlorinated hydrocarbons in water. *Environmental Science and Technology*, 39, 3314–3320.

He, F., Zhao, D. & Paul, C. 2010. Field assessment of carboxymethyl cellulose stabilized iron nanoparticles for in situ destruction of chlorinated solvents in source zones. *Water Research*, 44, 2360–2370.

He, F., Zhao, D. Y., Liu, J. C. & Roberts, C. B. 2007. Stabilization of Fe-Pd nanoparticles with sodium carboxymethyl cellulose for enhanced transport and dechlorination of trichloroethylene in soil and groundwater. *Industrial and Engineering Chemistry Research*, 46, 29–34.

Hernandez, R., Zappi, M. & Kuo, C. H. 2004. Chloride effect on TNT degradation by zerovalent iron or zinc during water treatment. *Environmental Science and Technology*, 38, 5157–5163.

Hoag, G. E., Collins, J. B., Holcomb, J. L., Hoag, J. R., Nadagouda, M. N. & Varma, R. S. 2009. Degradation of bromothymol blue by 'greener' nano-scale zero-valent iron synthesized using tea polyphenols. *Journal of Materials Chemistry*, 19, 8671–8677.

Hoch, L. B., Mack, E. J., Hydutsky, B. W., Hershman, J. M., Skluzacek, I. M. & Mallouk, T. E. 2008. Carbothermal synthesis of carbon-supported nanoscale zero-valent iron particles for the remediation of hexavalent chromium. *Environmental Science and Technology*, 42, 2600–2605.

Honetschlägerová, L., Janouškovcová, P. & Kubal, M. 2016. Enhanced transport of Si-coated nanoscale zero-valent iron particles in porous media. *Environmental Technology*, 37, 1530–1538.

Honetschlägerová, L., Janouskovcova, P., Kubal, M. & Sofer, Z. 2015. Enhanced colloidal stability of nanoscale zero valent iron particles in the presence of sodium silicate water glass. *Environmental Technology*, 36, 358–365.

Honetschlägerová, L., Janouškovcová, P. & Špaček, P. 2010. In situ dehalogenation of chlorinated hydrocarbons using zero valent nanoiron. In: *Inovativní sanační technologie ve výzkumu a praxi III,* M. Kubal, Ed. Vodní zdroje Ekomonitor spol. s. r. o, Beroun, pp. 18–22.

Honetschlägerová, L., Janouškovcová, P., Velimirovic, M., Kubal, M. & Bastiaens, L. 2018a. Using silica coated nanoscale zerovalent particles for the reduction of chlorinated ethylenes. *Silicon*, 10, 2593–2601.

Honetschlägerová, L., Škarohlíd, R., Martinec, M., Šír, M. & Luciano, V. 2018b. Interactions of nanoscale zero valent iron and iron reducing bacteria in remediation of trichloroethene. *International Biodeterioration and Biodegradation*, 127, 241–246.

Hua, Y., Wang, W., Huang, X., Gu, T., Ding, D., Ling, L. & Zhang, W.-X. 2018. Effect of bicarbonate on aging and reactivity of nanoscale zerovalent iron (nZVI) toward uranium removal. *Chemosphere*, 201, 603–611.

Huang, X. F., Shen, Y., Wang, Y. H., Liu, J. N., Peng, K. M., Lu, L. J. & Liu, J. 2016. Enhanced biodiesel production from glucose-fed activated sludge microbial cultures by addition of nZVI and $FeCl_3$. *RSC Advances*, 6, 88727–88735.

Huang, Y. H. & Zhang, T. C. 2005. Effects of dissolved oxygen on formation of corrosion products and concomitant oxygen and nitrate reduction in zero-valent iron systems with or without aqueous Fe^{2+}. *Water Research*, 39, 1751–1760.

Hwang, Y., Kim, D. & Shin, H. S. 2015. Inhibition of nitrate reduction by NaCl adsorption on a nano-zero-valent iron surface during a concentrate treatment for water reuse. *Environmental Technology*, 36, 1178–1187.

Hydutsky, B. W., Mack, E. J., Beckerman, B. B., Skluzacek, J. M. & Mallouk, T. E. 2007. Optimization of nano- and microiron transport through sand columns using polyelectrolyte mixtures. *Environmental Science and Technology*, 41, 6418–6424.

Chatterjee, S., Lim, S. R. & Woo, S. H. 2010. Removal of reactive black 5 by zero-valent iron modified with various surfactants. *Chemical Engineering Journal*, 160, 27–32.

Chen, S.-S., Hsu, H.-D. & Li, C.-W. J. J. O. N. R. 2004. A new method to produce nanoscale iron for nitrate removal. *Journal of Nanoparticle Research* 6, 639–647.

Chen, Z. X., Jin, X. Y., Chen, Z. L., Megharaj, M. & Naidu, R. 2011. Removal of methyl orange from aqueous solution using bentonite-supported nanoscale zero-valent iron. *Journal of Colloid and Interface Science*, 363, 601–607.

Choe, S., Chang, Y.-Y., Hwang, K.-Y. & Khim, J. 2000. Kinetics of reductive denitrification by nanoscale zero-valent iron. *Chemosphere*, 41, 1307–1311.

Choi, H., Al-Abed, S. R., Agarwal, S. & Dionysiou, D. D. 2008. Synthesis of reactive nano-Fe/Pd bimetallic system-impregnated activated carbon for the simultaneous adsorption and dechlorination of PCBs. *Chemistry of Materials*, 20, 3649–3655.

Iler, R. K. 1959. Product comprising a skin of dense, hydrated amorphous silica bound upon a core of another solid material and process of making same. USA Patent No. 2 885 366.

Im, J. K., Son, H. S. & Zoh, K. D. 2011. Perchlorate removal in Fe-0/H2O systems: Impact of oxygen availability and UV radiation. *Journal of Hazardous Materials*, 192, 457–464.

Jamei, M. R., Khosravi, M. R. & Anvaripour, B. 2013. Investigation of ultrasonic effect on synthesis of nano zero valent iron particles and comparison with conventional method. *Asia-Pacific Journal of Chemical Engineering*, 8, 767–774.

Jamei, M. R., Khosravi, M. R. & Anvaripour, B. 2014. A novel ultrasound assisted method in synthesis of nZVI particles. *Ultrasonics Sonochemistry*, 21, 226–233.

Jiang, X., Qiao, J. L., Lo, I. M. C., Wang, L., Guan, X. H., Lu, Z. P., Zhou, G. M. & Xu, C. H. 2015. Enhanced paramagnetic Cu^{2+} ions removal by coupling a weak magnetic field with zero valent iron. *Journal of Hazardous Materials*, 283, 880–887.

Jiemvarangkul, P., Zhang, W. X. & Lien, H. L. 2011. Enhanced transport of polyelectrolyte stabilized nanoscale zero-valent iron (nZVI) in porous media. *Chemical Engineering Journal*, 170, 482–491.

Johnson, R. L., Johnson, G. O., Nurmi, J. T. & Tratnyek, P. G. 2009. Natural organic matter enhanced mobility of nano zerovalent iron. *Environmental Science and Technology*, 43, 5455–5460.

Johnson, T. L., Scherer, M. M. & Tratnyek, P. G. 1996. Kinetics of halogenated organic compound degradation by iron metal. *Environmental Science and Technology*, 30, 2634–2640.

Kanel, R. J., Greneche, J.-M. & Choi, H. 2006. Arsenic(V) removal from groundwater using nano scale zero-valent iron as a colloidal reactive barrier material. *Environmental Science and Technology*, 40, 2045–2050.

Kanel, S., Nepal, D., Manning, B. & Choi, H. 2007. Transport of surface-modified iron nanoparticle in porous media and application to arsenic(III) remediation. *Journal of Nanoparticle Research*, 9, 725–735.

Kanel, S. R. & Choi, H. 2007. Transport characteristics of surface-modified nanoscale zero-valent iron in porous media. *Water Science and Technology*, 55, 157–162.

Kanel, S. R., Goswami, R. R., Clement, T. P., Barnett, M. O. & Zhao, D. 2008. Two dimensional transport characteristics of surface stabilized zero-valent iron nanoparticles in porous media. *Environmental Science and Technology*, 42, 896–900.

Kanel, S. R., Manning, B., Charlet, L. & Choi, H. 2005. Removal of arsenic(III) from groundwater by nanoscale zero-valent iron. *Environmental Science and Technology*, 39, 1291–1298.

Karn, B., Kuiken, T. & Otto, M. 2009. Nanotechnology and in situ remediation: A review of the benefits and potential risks. *Environ Health Perspect*, 117, 1813–1831.

Kim, S. A., Kamala-Kannan, S., Lee, K. J., Park, Y. J., Shea, P. J., Lee, W. H., Kim, H. M. & Oh, B. T. 2013. Removal of Pb(II) from aqueous solution by a zeolite-nanoscale zero-valent iron composite. *Chemical Engineering Journal*, 217, 54–60.

Klausen, J., Ranke, J. & Schwarzenbach, R. P. 2001. Influence of solution composition and column aging on the reduction of nitroaromatic compounds by zero-valent iron. *Chemosphere*, 44, 511–517.

Kober, R., Schlicker, O., Ebert, M. & Dahmke, A. 2002. Degradation of chlorinated ethylenes by Fe-0: Inhibition processes and mineral precipitation. *Environmental Geology*, 41, 644–652.

Kocur, C. M., O'Carroll, D. M. & Sleep, B. E. 2013. Impact of nZVI stability on mobility in porous media. *Journal of Contaminant Hydrology*, 145, 17–25.

Koch, A. M., Reynolds, F., Merkle, H. P., Weissleder, R. & Josephson, L. 2005. Transport of surface-modified nanoparticles through cell monolayers. *ChemBioChem*, 6, 337–345.

Kotchaplai, P., Khan, E. & Vangnai, A. S. 2017. Membrane alterations in pseudomonas putida F1 exposed to nanoscale zerovalent iron: Effects of short-term and repetitive nZVI exposure. *Environmental Science and Technology*, 51, 7804–7813.

Kristiana, I., Joll, C. & Heitz, A. 2011. Powdered activated carbon coupled with enhanced coagulation for natural organic matter removal and disinfection by-product control: Application in a Western Australian water treatment plant. *Chemosphere*, 83, 661–667.

Kuang, Y., Wang, Q., Chen, Z., Megharaj, M. & Naidu, R. 2013. Heterogeneous fenton-like oxidation of monochlorobenzene using green synthesis of iron nanoparticles. *Journal of Colloid and Interface Science*, 410, 67–73.

Kumar, N., Auffan, M., Gattacceca, J., Rose, J., Olivi, L., Borschneck, D., Kvapil, P., Jublot, M., Kaifas, D., Malleret, L., Doumenq, P. & Bottero, J. Y. 2014a. Molecular insights of oxidation process of iron nanoparticles: Spectroscopic, magnetic, and microscopic evidence. *Environmental Science and Technology*, 48, 13888–13894.

Kumar, N., Omoregie, E. O., Rose, J., Masion, A., Lloyd, J. R., Diels, L. & Bastiaens, L. 2014b. Inhibition of sulfate reducing bacteria in aquifer sediment by iron nanoparticles. *Water Research*, 51, 64–72.

Lefevre, E., Bossa, N., Wiesner, M. R. & Gunsch, C. K. 2016. A review of the environmental implications of in situ remediation by nanoscale zero valent iron (nZVI): Behavior, transport and impacts on microbial communities. *Science of the Total Environment*, 565, 889–901.

Li, F., Vipulanandan, C. & Mohanty, K. K. 2003. Microemulsion and solution approaches to nanoparticle iron production for degradation of trichloroethylene. *Colloids and Surfaces A-Physicochemical and Engineering Aspects*, 223, 103–112.

Li, J., Chen, C. L., Zhu, K. R. & Wang, X. K. 2016. Nanoscale zero-valent iron particles modified on reduced graphene oxides using a plasma technique for Cd(II) removal. *Journal of the Taiwan Institute of Chemical Engineers*, 59, 389–394.

Li, L., Fan, M. H., Brown, R. C., Van Leeuwen, J. H., Wang, J. J., Wang, W. H., Song, Y. H. & Zhang, P. Y. 2006a. Synthesis, properties, and environmental applications of nanoscale iron-based materials: A review. *Critical Reviews in Environmental Science and Technology*, 36, 405–431.

Li, S., Yan, W. & Zhang, W.-X. 2009. Solvent-free production of nanoscale zero-valent iron (nZVI) with precision milling. *Green Chemistry*, 11, 1618–1626.

Li, S. L., Wang, W., Liang, F. P. & Zhang, W. X. 2017. Heavy metal removal using nanoscale zero-valent iron (nZVI): Theory and application. *Journal of Hazardous Materials*, 322, 163–171.

Li, X. Q., Elliott, D. W. & Zhang, W. X. 2006b. Zero-valent iron nanoparticles for abatement of environmental pollutants: Materials and engineering aspects. *Critical Reviews in Solid State and Materials Sciences*, 31, 111–122.

Li, X. Q. & Zhang, W. X. 2006. Iron nanoparticles: The core-shell structure and unique properties for Ni(II) sequestration. *Langmuir*, 22, 4638–4642.

Li, X. Q. & Zhang, W. X. 2007. Sequestration of metal cations with zerovalent iron nanoparticles: A study with high resolution X-ray photoelectron spectroscopy (HR-XPS). *Journal of Physical Chemistry C*, 111, 6939–6946.

Li, Y. M., Cheng, W., Sheng, G. D., Li, J. F., Dong, H. P., Chen, Y. & Zhu, L. Z. 2015a. Synergetic effect of a pillared bentonite support on SE(VI) removal by nanoscale zero valent iron. *Applied Catalysis B-Environmental*, 174, 329–335.

Li, Z. J., Wang, L., Yuan, L. Y., Xiao, C. L., Mei, L., Zheng, L. R., Zhang, J., Yang, J. H., Zhao, Y. L., Zhu, Z. T., Chai, Z. F. & Shi, W. Q. 2015b. Efficient removal of uranium from aqueous solution by zero-valent iron nanoparticle and its graphene composite. *Journal of Hazardous Materials*, 290, 26–33.

Lien, H. L. & Zhang, W. X. 2001. Nanoscale iron particles for complete reduction of chlorinated ethenes. *Colloids and Surfaces A-Physicochemical and Engineering Aspects*, 191, 97–105.

Lien, H. L. & Zhang, W. X. 2007. Nanoscale Pd/Fe bimetallic particles: Catalytic effects of palladium on hydrodechlorination. *Applied Catalysis B-Environmental*, 77, 110–116.

Lim, T. T. & Zhu, B. W. 2008. Effects of anions on the kinetics and reactivity of nanoscale Pd/Fe in trichlorobenzene dechlorination. *Chemosphere*, 73, 1471–1477.

Lin, Y. T., Weng, C. H. & Chen, F. Y. 2008. Effective removal of AB24 dye by nano/micro-size zero-valent iron. *Separation and Purification Technology*, 64, 26–30.

Liu, M. H., Wang, Y. H., Chen, L. T., Zhang, Y. & Lin, Z. 2015. Mg(OH)(2) supported nanoscale zero valent iron enhancing the removal of Pb(II) from aqueous solution. *ACS Applied Materials and Interfaces*, 7, 7961–7969.

Liu, Y. & Lowry, G. V. 2006. Effect of particle age (Fe^0 content) and solution pH on nZVI reactivity: H_2 evolution and TCE dechlorination. *Environmental Science and Technology*, 40, 6085–6090.

Liu, Y., Majetich, S. A., Tilton, R. D., Sholl, D. S. & Lowry, G. V. 2005a. TCE dechlorination rates, pathways, and efficiency of nanoscale iron particles with different properties. *Environmental Science and Technology*, 39, 1338–1345.

Liu, Y., Phenrat, T. & Lowry, G. V. 2007. Effect of TCE concentration and dissolved groundwater solutes on nZVI-Promoted TCE dechlorination and H-2 evolution. *Environmental Science and Technology*, 41, 7881–7887.

Liu, Y. Q., Choi, H., Dionysiou, D. & Lowry, G. V. 2005b. Trichloroethene hydrodechlorination in water by highly disordered monometallic nanoiron. *Chemistry of Materials*, 17, 5315–5322.

Luo, S., Qin, P. F., Shao, J. H., Peng, L., Zeng, Q. R. & Gu, J. D. 2013. Synthesis of reactive nanoscale zero valent iron using rectorite supports and its application for Orange II removal. *Chemical Engineering Journal*, 223, 1–7.

Ma, X. M., Gurung, A. & Deng, Y. 2013. Phytotoxicity and uptake of nanoscale zero-valent iron (nZVI) by two plant species. *Science of the Total Environment*, 443, 844–849.

Macé, C., Desrocher, S., Gheorghiu, F., Kane, A., Pupeza, M., Cernik, M., Kvapil, P., Venkatakrishnan, R. & Zhang, W.-X. 2006. Nanotechnology and groundwater remediation: A step forward in technology understanding. *Remediation Journal*, 16, 23–33.

Mak, M. S. H., Rao, P. H. & Lo, I. M. C. 2009. Effects of hardness and alkalinity on the removal of arsenic(V) from humic acid-deficient and humic acid-rich groundwater by zero-valent iron. *Water Research*, 43, 4296–4304.

Malow, T. R., Koch, C. C., Miraglia, P. Q. & Murty, K. L. 1998. Compressive mechanical behavior of nanocrystalline Fe investigated with an automated ball indentation technique. *Materials Science and Engineering: A*, 252, 36–43.

Masciangioli, T. & Zhang, W.-X. 2003. Peer reviewed: Environmental technologies at the nanoscale. *Environmental Science and Technology*, 37, 102A–108A.

Matheson, L. J. & Tratnyek, P. G. 1994. Reductive dehalogenation of chlorinated methanes by iron metal. *Environmental Science and Technology*, 28, 2045–2053.

Meyer, D. E., Wood, K., Bachas, L. G. & Bhattacharyya, D. 2004. Degradation of chlorinated organics by membrane-immobilized nanosized metals. *Environmental Progress*, 23, 232–242.

Moore, A. M., De Leon, C. H. & Young, T. M. 2003. Rate and extent of aqueous perchlorate removal by iron surfaces. *Environmental Science and Technology*, 37, 3189–3198.

Moradi, M., Naeej, O. B., Azaria, A., Bandpei, A. M., Jafari, A. J., Esrafili, A. & Kalantary, R. R. 2017. A comparative study of nitrate removal from aqueous solutions using zeolite, nZVI-zeolite, nZVI and iron powder adsorbents. *Desalination and Water Treatment*, 74, 278–288.

Mueller, N. C., Braun, J., Bruns, J., Černík, M., Rissing, P., Rickerby, D. & Nowack, B. 2012. Application of nanoscale zero valent iron (nZVI) for groundwater remediation in Europe. *Environmental Science and Pollution Research*, 19, 550–558.

Naja, G., Halasz, A., Thiboutot, S., Ampleman, G. & Hawari, J. 2008. Degradation of hexahydro-1,3,5-trinitro-1,15-triazine (RDX) using zerovalent iron nanoparticles. *Environmental Science and Technology*, 42, 4364–4370.

NANOIRON Ltd. 2018. Zero-valent iron nanoparticles (nZVI) [Online]. Available: http://nanoiron.cz/en/products/zero-valent-iron-nanoparticles [Accessed 2018].

Nemecek, J., Lhotsky, O. & Cajthaml, T. 2014. Nanoscale zero-valent iron application for in situ reduction of hexavalent chromium and its effects on indigenous microorganism populations. *Science of the Total Environment*, 485, 739–747.

Noubactep, C. 2008. A critical review on the process of contaminant removal in Fe-0-H2O systems. *Environmental Technology*, 29, 909–920.

Noubactep, C., Care, S. & Crane, R. 2012. Nanoscale metallic iron for environmental remediation: Prospects and limitations. *Water Air and Soil Pollution*, 223, 1363–1382.

Nurmi, J. T., Tratnyek, P. G., Sarathy, V., Baer, D. R., Amonette, J. E., Pecher, K., Wang, C., Linehan, J. C., Matson, D. W., Penn, R. L. & Driessen, M. D. 2005. Characterization and properties of metallic iron nanoparticles: Spectroscopy, electrochemistry, and kinetics. *Environmental Science and Technology*, 39, 1221–1230.

O'Carroll, D., Sleep, B., Krol, M., Boparai, H. & Kocur, C. 2013. Nanoscale zero valent iron and bimetallic particles for contaminated site remediation. *Advances in Water Resources*, 51, 104–122.

Park, Y.-D., Park, C.-S. & Park, J.-W. 2010. Interaction between iron reducing bacteria and nano-scale zero valent iron. *Sustainable Environment Research*, 20, 233–238.

Phenrat, T., Fagerlund, F., Illangasekare, T., Lowry, G. V. & Tilton, R. D. 2011. Polymer-modified Fe-0 nanoparticles target entrapped NAPL in two dimensional porous media: Effect of particle concentration, NAPL saturation, and injection strategy. *Environmental Science and Technology*, 45, 6102–6109.

Phenrat, T., Kim, H. J., Fagerlund, F., Illangasekare, T., Tilton, R. D. & Lowry, G. V. 2009a. Particle size distribution, concentration, and magnetic attraction affect transport of polymer-modified Fe-0 nanoparticles in sand columns. *Environmental Science and Technology*, 43, 5079–5085.

Phenrat, T., Liu, Y. Q., Tilton, R. D. & Lowry, G. V. 2009b. Adsorbed polyelectrolyte coatings decrease Fe-0 nanoparticle reactivity with TCE in water: Conceptual model and mechanisms. *Environmental Science and Technology*, 43, 1507–1514.

Phenrat, T., Long, T. C., Lowry, G. V. & Veronesi, B. 2009c. Partial oxidation ("aging") and surface modification decrease the toxicity of nanosized zerovalent iron. *Environmental Science and Technology*, 43, 195–200.

Phenrat, T., Saleh, N., Sirk, K., Kim, H.-J., Tilton, R. D. & Lowry, G. V. 2008. Stabilization of aqueous nanoscale zerovalent iron dispersions by anionic polyelectrolytes: Adsorbed anionic polyelectrolyte layer properties and their effect on aggregation and sedimentation. *Journal of Nanoparticle Research*, 10, 795–814.

Phenrat, T., Saleh, N., Sirk, K., Tilton, R. D. & Lowry, G. V. 2007. Aggregation and sedimentation of aqueous nanoscale zerovalent iron dispersions. *Environmental Science and Technology*, 41, 284–290.

Philippini, V., Naveau, A., Catalette, H. & Leclercq, S. 2006. Sorption of silicon on magnetite and other corrosion products of iron. *Journal of Nuclear Materials*, 348, 60–69.

Ponder, S. M., Darab, J. G., Bucher, J., Caulder, D., Craig, I., Davis, L., Edelstein, N., Lukens, W., Nitsche, H., Rao, L., Shuh, D. K. & Mallouk, T. E. 2001. Surface chemistry and electrochemistry of supported zerovalent iron nanoparticles in the remediation of aqueous metal contaminants. *Chemistry of Materials*, 13, 479–486.

Ponder, S. M., Darab, J. G. & Mallouk, T. E. 2000. Remediation of Cr(VI) and Pb(II) aqueous solutions using supported, nanoscale zero-valent iron. *Environmental Science and Technology*, 34, 2564–2569.

Qian, L., Shang, X., Zhang, B., Zhang, W., Su, A., Chen, Y., Ouyang, D., Han, L., Yan, J. & Chen, M. 2019. Enhanced removal of Cr(VI) by silicon rich biochar-supported nanoscale zero-valent iron. *Chemosphere*, 215, 739–745.

Quinn, J., Geiger, C., Clausen, C., Brooks, K., Coon, C., O'Hara, S., Krug, T., Major, D., Yoon, W. S., Gavaskar, A. & Holdsworth, T. 2005. Field demonstration of DNAPL dehalogenation using emulsified zero-valent iron. *Environmental Science and Technology*, 39, 1309–1318.

Rathor, G., Chopra, N. & Adhikari, T. 2017. Remediation of nickel ion from soil and water using nano particles of zero-valent iron (nZVI). *Oriental Journal of Chemistry*, 33, 1025–1029.

Reinsch, B. C., Forsberg, B., Penn, R. L., Kim, C. S. & Lowry, G. V. 2010. Chemical transformations during aging of zerovalent iron nanoparticles in the presence of common groundwater dissolved constituents. *Environmental Science and Technology*, 44, 3455–3461.

Roduner, E. 2006. Size matters: Why nanomaterials are different. *Chemical Society Reviews*, 35, 583–592.

Sakulchaicharoen, N., O'Carroll, D. M. & Herrera, J. E. 2010. Enhanced stability and dechlorination activity of pre-synthesis stabilized nanoscale FePd particles. *Journal of Contaminant Hydrology*, 118, 117–127.

Saleh, N., Kim, H. J., Phenrat, T., Matyjaszewski, K., Tilton, R. D. & Lowry, G. V. 2008. Ionic strength and composition affect the mobility of surface-modified Fe-0 nanoparticles in water-saturated sand columns. *Environmental Science and Technology*, 42, 3349–3355.

Saleh, N., Phenrat, T., Sirk, K., Dufour, B., Ok, J., Sarbu, T., Matyiaszewski, K., Tilton, R. D. & Lowry, G. V. 2005. Adsorbed triblock copolymers deliver reactive iron nanoparticles to the oil/water interface. *Nano Letters*, 5, 2489–2494.

Saleh, N., Sirk, K., Liu, Y. Q., Phenrat, T., Dufour, B., Matyjaszewski, K., Tilton, R. D. & Lowry, G. V. 2007. Surface modifications enhance nanoiron transport and NAPL targeting in saturated porous media. *Environmental Engineering Science*, 24, 45–57.

Segura, Y., Martinez, F., Melero, J. A., Molina, R., Chand, R. & Bremner, D. H. 2012. Enhancement of the advanced Fenton process (Fe-0/H_2O_2) by ultrasound for the mineralization of phenol. *Applied Catalysis B-Environmental*, 113, 100–106.

Sevcu, A., El-Temsah, Y. S., Joner, E. J. & Cernik, M. 2012. Oxidative stress induced in microorganisms by zero-valent iron nanoparticles (vol 26, pg 271, 2011). *Microbes and Environments*, 27, 215–215.

Shan, G., Yan, S., Tyagi, R. D., Surampalli, R. Y. & Zhang, T. C. 2009. Applications of nanomaterials in environmental science and engineering: Review. *The Practice Periodical of Hazardous Toxic and Radioactive Waste Management*, 13, 110–119.

Sheng, G. D., Alsaedi, A., Shammakh, W., Monaquel, S., Sheng, J., Wang, X. K., Li, H. & Huang, Y. Y. 2016. Enhanced sequestration of selenite in water by nanoscale zero valent iron immobilization on carbon nanotubes by a combined batch, XPS and XAFS investigation. *Carbon*, 99, 123–130.

Shi, C. J., Wei, J., Jin, Y., Kniel, K. E. & Chiu, P. C. 2012. Removal of viruses and bacteriophages from drinking water using zero-valent iron. *Separation and Purification Technology*, 84, 72–78.

Shi, L. N., Zhang, X. & Chen, Z. L. 2011. Removal of chromium (VI) from wastewater using bentonite-supported nanoscale zero-valent iron. *Water Research*, 45, 886–892.

Shimizu, A., Tokumura, M., Nakajima, K. & Kawase, Y. 2012. Phenol removal using zero-valent iron powder in the presence of dissolved oxygen: Roles of decomposition by the Fenton reaction and adsorption/precipitation. *Journal of Hazardous Materials*, 201, 60–67.

Shin, K. H. & Cha, D. K. 2008. Microbial reduction of nitrate in the presence of nanoscale zero-valent iron. *Chemosphere*, 72, 257–262.

Shirin, S. & Balakrishnan, V. K. 2011. Using chemical reactivity to provide insights into environmental transformations of priority organic substances: The Fe-0-mediated reduction of acid blue 129. *Environmental Science and Technology*, 45, 10369–10377.

Schrick, B., Hydutsky, B. W., Blough, J. L. & Mallouk, T. E. 2004. Delivery vehicles for zerovalent metal nanoparticles in soil and groundwater. *Chemistry of Materials*, 16, 2187–2193.

Siciliano, A. 2015. Use of nanoscale zero-valent iron (nZVI) particles for chemical denitrification under different operating conditions. *Metals*, 5, 1507–1519.

Simonet, B. M. & Valcárcel, M. 2008. Monitoring nanoparticles in the environment. *Analytical and Bioanalytical Chemistry*, 393, 17.

Sirk, K. M., Saleh, N. B., Phenrat, T., Kim, H.-J., Dufour, B., Ok, J., Golas, P. L., Matyjaszewski, K., Lowry, G. V. & Tilton, R. D. 2009. Effect of adsorbed polyelectrolytes on nanoscale zero valent iron particle attachment to soil surface models. *Environmental Science and Technology*, 43, 3803–3808.

Sohn, K., Kang, S. W., Ahn, S., Woo, M. & Yang, S.-K. 2006. Fe(0) nanoparticles for nitrate reduction: Stability, reactivity, and transformation. *Environmental Science and Technology*, 40, 5514–5519.

Song, H. & Carraway, E. R. 2005. Reduction of chlorinated ethanes by nanosized zero-valent iron: Kinetics, pathways, and effects of reaction conditions. *Environmental Science and Technology*, 39, 6237–6245.

Song, S. K., Su, M. M., Adeleye, A. S., Zhang, Y. L. & Zhou, X. F. 2017. Optimal design and characterization of sulfide-modified nanoscale zerovalent iron for diclofenac removal. *Applied Catalysis B-Environmental*, 201, 211–220.

Stefaniuk, M., Oleszczuk, P. & Ok, Y. S. 2016. Review on nano zerovalent iron (nZVI): From synthesis to environmental applications. *Chemical Engineering Journal*, 287, 618–632.

Stohs, S. J. & Bagchi, D. 1995. Oxidative mechanisms in the toxicity of metal-ions. *Free Radical Biology and Medicine*, 18, 321–336.

Stumm, W. & Morgan, J. J. 1996. *Aquatic Chemistry: Chemical Equilibria and Rates in Natural Waters*. Wiley, New York.

Stumm, W. & Morgan, J. J. 2013. *Aquatic Chemistry: Chemical Equilibria and Rates in Natural Waters*. Wiley, New York.

Su, C. M. & Puls, R. W. 2001. Arsenate and arsenite removal by zerovalent iron: Effects of phosphate, silicate, carbonate, borate, sulfate, chromate, molybdate, and nitrate, relative to chloride. *Environmental Science and Technology*, 35, 4562–4568.

Su, L. H., Liu, C. W., Liang, K. K., Chen, Y. D., Zhang, L. J., Li, X. L., Han, Z. H., Zhen, G. Y., Chai, X. L. & Sun, X. 2018. Performance evaluation of zero-valent iron nanoparticles (nZVI) for high-concentration H2S removal from biogas at different temperatures. *RSC Advances*, 8, 13798–13805.

Su, Y. M., Adeleye, A. S., Zhou, X. F., Dai, C. M., Zhang, W. X., Keller, A. A. & Zhang, Y. L. 2014. Effects of nitrate on the treatment of lead contaminated groundwater by nanoscale zerovalent iron. *Journal of Hazardous Materials*, 280, 504–513.

Sun, Y. K., Li, J. X., Huang, T. L. & Guan, X. H. 2016. The influences of iron characteristics, operating conditions and solution chemistry on contaminants removal by zero-valent iron: A review. *Water Research*, 100, 277–295.

Sun, Y.-P., Li, X.-Q., Cao, J., Zhang, W.-X. & Wang, H. P. 2006. Characterization of zero-valent iron nanoparticles. *Advances in Colloid and Interface Science*, 120, 47–56.

Sun, Y.-P., Li, X.-Q., Zhang, W.-X. & Wang, H. P. 2007. A method for the preparation of stable dispersion of zero-valent iron nanoparticles. *Colloids and Surfaces A: Physicochemical and Engineering Aspects*, 308, 60–66.

Tang, C. L., Huang, Y. H., Zeng, H. & Zhang, Z. Q. 2014. Promotion effect of Mn^{2+} and Co^{2+} on selenate reduction by zero-valent iron. *Chemical Engineering Journal*, 244, 97–104.

Tao, N. R., Sui, M. L., Lu, J. & Lua, K. 1999. Surface nanocrystallization of iron induced by ultrasonic shot peening. *Nanostructured Materials*, 11, 433–440.

Tiraferri, A., Chen, K. L., Sethi, R. & Elimelech, M. 2008. Reduced aggregation and sedimentation of zero-valent iron nanoparticles in the presence of guar gum. *Journal of Colloid and Interface Science*, 324, 71–79.

Tiraferri, A. & Sethi, R. 2009. Enhanced transport of zerovalent iron nanoparticles in saturated porous media by guar gum. *Journal of Nanoparticle Research*, 11, 635–645.

Tratnyek, P. G. & Johnson, R. L. 2006. Nanotechnologies for environmental cleanup. *Nano Today*, 1, 44–48.

Tratnyek, P. G., Salter-Blanc, A. J., Nurmi, J. T., Amonette, J. E., Liu, J., Wang, C. M., Dohnalkova, A. & Baer, D. R. 2011. Reactivity of zerovalent metals in aquatic media: Effects of organic surface coatings. *Aquatic Redox Chemistry*, 1071, 381–406.

Tratnyek, P. G., Scherer, M. M., Deng, B. L. & Hu, S. D. 2001. Effects of natural organic matter, anthropogenic surfactants, and model quinones on the reduction of contaminants by zero-valent iron. *Water Research*, 35, 4435–4443.

Tsang, D. C. W., Graham, N. J. D. & Lo, I. M. C. 2009. Humic acid aggregation in zero-valent iron systems and its effects on trichloroethylene removal. *Chemosphere*, 75, 1338–1343.

Ulucan-Altuntas, K., Debik, E., Yoruk, I. I. & Kozal, D. 2017. Single and binary adsorption of copper and nickel metal ions on nano zero valent iron (nZVI): A kinetic approach. *Desalination and Water Treatment*, 93, 274–279.

U. S. Navy. 2007. Nanoscale zero valent iron [Online]. Available: https://portal.navfac.navy.mil/portal/page/portal/navfac/navfac_ww_pp/navfac_nfesc_pp/environmental/erb/nzvi [Accessed 28.3. 2013].

Uzum, C., Shahwan, T., Eroglu, A. E., Hallam, K. R., Scott, T. B. & Lieberwirth, I. 2009. Synthesis and characterization of kaolinite-supported zero-valent iron nanoparticles and their application for the removal of aqueous Cu^{2+} and Co^{2+} ions. *Applied Clay Science*, 43, 172–181.

Vecchia, E. D., Luna, M. & Sethi, R. 2009. Transport in porous media of highly concentrated iron micro- and nanoparticles in the presence of Xanthan Gum. *Environmental Science and Technology*, 43, 8942–8947.

Wang, C.-B. & Zhang, W.-X. 1997. Synthesizing nanoscale iron particles for rapid and complete dechlorination of TCE and PCBs. *Environmental Science and Technology*, 31, 2154–2156.

Wang, H. Y., Zhang, S. L., Chen, D. & He, Q. L. 2016. Nitrate removal from groundwater by nanoscale zero-valent iron (nZVI) coupling autohydrogenotrophic denitrification. *Proceedings of the 2016 International Conference on Education, Management, Computer and Society*, 37, 1346–1349.

Wang, J., Liu, G. J., Zhou, C. C., Li, T. F. & Liu, J. J. 2014. Synthesis, characterization and aging study of kaolinite-supported zero-valent iron nanoparticles and its application for Ni(II) adsorption. *Materials Research Bulletin*, 60, 421–432.

Wang, Q., Lee, S. & Choi, H. 2010a. Aging study on the structure of Fe-0-nanoparticles: Stabilization, characterization, and reactivity. *Journal of Physical Chemistry C*, 114, 2027–2033.

Wang, Q., Snyder, S., Kim, J. & Choi, H. 2009. Aqueous ethanol modified nanoscale zerovalent iron in bromate reduction: Synthesis, characterization, and reactivity. *Environmental Science and Technology*, 43, 3292–3299.

Wang, W., Zhou, M. H., Mao, Q. O., Yue, J. J. & Wang, X. 2010b. Novel NaY zeolite-supported nanoscale zero-valent iron as an efficient heterogeneous Fenton catalyst. *Catalysis Communications*, 11, 937–941.

Wei, Y. T., Wu, S. C., Chou, C. M., Che, C. H., Tsai, S. M. & Lien, H. L. 2010. Influence of nanoscale zero-valent iron on geochemical properties of groundwater and vinyl chloride degradation: A field case study. *Water Research*, 44, 131–140.

Wen, Z. P., Zhang, Y. L. & Dai, C. M. 2014. Removal of phosphate from aqueous solution using nanoscale zerovalent iron (nZVI). *Colloids and Surfaces A-Physicochemical and Engineering Aspects*, 457, 433–440.

Wiesner, M. R. & Bottero, J.-Y. 2007. *Environmental Nanotechnology: Applications and Impacts of Nanomaterials*. McGraw-Hill, New York.

Wu, L., Shamsuzzoha, M. & Ritchie, S. M. C. 2005. Preparation of cellulose acetate supported zero-valent iron nanoparticles for the dechlorination of trichloroethylene in water. *Journal of Nanoparticle Research*, 7, 469–476.

Wu, Y. W., Yue, Q. Y., Ren, Z. F. & Gao, B. Y. 2018. Immobilization of nanoscale zero-valent iron particles (nZVI) with synthesized activated carbon for the adsorption and degradation of chloramphenicol (CAP). *Journal of Molecular Liquids*, 262, 19–28.

Xiao, J. N., Gao, B. Y., Yue, Q. Y., Sun, Y. Y., Kong, J. J., Gao, Y. & Li, Q. 2015. Characterization of nanoscale zero-valent iron supported on granular activated carbon and its application in removal of acrylonitrile from aqueous solution. *Journal of the Taiwan Institute of Chemical Engineers*, 55, 152–158.

Xie, L. & Shang, C. 2005. Role of humic acid and quinone model compounds in bromate reduction by zerovalent iron. *Environmental Science and Technology*, 39, 1092–1100.

Xie, Y. K., Dong, H. R., Zeng, G. M., Tang, L., Jiang, Z., Zhang, C., Deng, J. M., Zhang, L. H. & Zhang, Y. 2017. The interactions between nanoscale zero-valent iron and microbes in the subsurface environment: A review. *Journal of Hazardous Materials*, 321, 390–407.

Xiu, Z. M., Jin, Z. H., Li, T. L., Mahendra, S., Lowry, G. V. & Alvarez, P. J. J. 2010. Effects of nano-scale zero-valent iron particles on a mixed culture dechlorinating trichloroethylene. *Bioresource Technology*, 101, 1141–1146.

Xu, J., Dozier, A. & Bhattacharyya, D. 2005. Synthesis of nanoscale bimetallic particles in polyelectrolyte membrane matrix for reductive transformation of halogenated organic compounds. *Journal of Nanoparticle Research*, 7, 449–467.

Xu, J., Hao, Z. W., Xie, C. S., Lv, X. S., Yang, Y. P. & Xu, X. H. 2012. Promotion effect of Fe^{2+} and Fe_3O_4 on nitrate reduction using zero-valent iron. *Desalination*, 284, 9–13.

Yan, W., Herzing, A. A., Li, X.-Q., Kiely, C. J. & Zhang, W.-X. 2010. Structural evolution of Pd-doped nanoscale zero-valent iron (nZVI) in aqueous media and implications for particle aging and reactivity. *Environmental Science and Technology*, 44, 4288–4294.

Yan, W. L., Lien, H. L., Koel, B. E. & Zhang, W. X. 2013. Iron nanoparticles for environmental clean-up: Recent developments and future outlook. *Environmental Science-Processes and Impacts*, 15, 63–77.

Yang, G. C. C., Tu, H. C. & Hung, C. H. 2007. Stability of nanoiron slurries and their transport in the subsurface environment. *Separation and Purification Technology*, 58, 166–172.

Yin, W. Z., Wu, J. H., Li, P., Wang, X. D., Zhu, N. W., Wu, P. X. & Yang, B. 2012. Experimental study of zero-valent iron induced nitrobenzene reduction in groundwater: The effects of pH, iron dosage, oxygen and common dissolved anions. *Chemical Engineering Journal*, 184, 198–204.

Yoo, B. Y., Hernandez, S. C., Koo, B., Rheem, Y. & Myung, N. V. 2007. Electrochemically fabricated zero-valent iron, iron-nickel, and iron-palladium nanowires for environmental remediation applications. *Water Science and Technology*, 55, 149–156.

You, Y. W., Han, J., Chiu, P. C. & Jin, Y. 2005. Removal and inactivation of waterhorne viruses using zerovalent iron. *Environmental Science and Technology*, 39, 9263–9269.

Yu, J., Liu, W. X., Zeng, A. B., Guan, B. H. & Xu, X. H. 2013. Effect of SO on 1,1,1-trichloroethane degradation by Fe^0 in aqueous solution. *Ground Water*, 51, 286–292.

Zhan, J. J., Kolesnichenko, I., Sunkara, B., He, J. B., Mcpherson, G. L., Piringer, G. & John, V. T. 2011. Multifunctional iron-carbon nanocomposites through an aerosol-based process for the in situ remediation of chlorinated hydrocarbons. *Environmental Science and Technology*, 45, 1949–1954.

Zhan, J. J., Zheng, T. H., Piringer, G., Day, C., Mcpherson, G. L., Lu, Y. F., Papadopoulos, K. & John, V. T. 2008. Transport characteristics of nanoscale functional zerovalent iron/silica composites for in situ remediation of trichloroethylene. *Environmental Science and Technology*, 42, 8871–8876.

Zhang, Q., Liu, H. B., Chen, T. H., Chen, D., Li, M. X. & Chen, C. 2017. The synthesis of nZVI and its application to the removal of phosphate from aqueous solutions. *Water Air and Soil Pollution*, 228.

Zhang, W.-X. 2003. Nanoscale iron particles for environmental remediation: An overview. *Journal of Nanoparticle Research*, 5, 323–332.

Zhang, X., Lin, Y. M. & Chen, Z. L. 2009. 2,4,6-Trinitrotoluene reduction kinetics in aqueous solution using nanoscale zero-valent iron. *Journal of Hazardous Materials*, 165, 923–927.

Zhang, X., Lin, Y. M., Shan, X. Q. & Chen, Z. L. 2010. Degradation of 2,4,6-trinitrotoluene (TNT) from explosive wastewater using nanoscale zero-valent iron. *Chemical Engineering Journal*, 158, 566–570.

Zhang, Y., Li, Y. M., Li, J. F., Hu, L. J. & Zheng, X. M. 2011. Enhanced removal of nitrate by a novel composite: Nanoscale zero valent iron supported on pillared clay. *Chemical Engineering Journal*, 171, 526–531.

Zhu, B. W., Lim, T. T. & Feng, J. 2006. Reductive dechlorination of 1,2,4-trichlorobenzene with palladized nanoscale Fe-0 particles supported on chitosan and silica. *Chemosphere*, 65, 1137–1145.

Zou, Y. D., Wang, X. X., Khan, A., Wang, P. Y., Liu, Y. H., Alsaedi, A., Hayat, T. & Wang, X. K. 2016. Environmental remediation and application of nanoscale zero-valent iron and its composites for the removal of heavy metal ions: A review. *Environmental Science and Technology*, 50, 7290–7304.

16

Carbon Nanostructures and Their Application to Water Purification

Stefano Bellucci
INFN Laboratori Nazionali di Frascati

16.1 Introduction

16.1.1 Water Pollution

Water pollution occurs when a body of water (lake, river, ocean, groundwater, etc.) is negatively affected by the addition of large quantities of material in the water itself. This form of environmental degradation occurs when pollutants are discharged directly or indirectly into water without having undergone an appropriate process that removes harmful compounds. In the first case, we talk about point sources, identifiable with a single and localizable source; a pipe that releases toxic chemicals directly into a river is an example. In a second case, there is a leak of pollutants in a stream, for example, when a field fertilizer is transported to a stream due to a leak from the surface; in this case, the sources are not point-like.

Water pollution affects the whole hydrosphere, i.e., plants and organisms living in water bodies. In most cases, the effects do concern not only a single species or population, but all the natural biological communities. As it is obvious, all kinds of natural waters (underground waters, surface waters, etc.) in their path cross the atmosphere and soil, enriching themselves with gases, minerals, salts, ions, and microorganisms. What matters to characterize water is not the water itself, but what it contains. In particular, it is necessary to consider drinking water, destined for human and animal consumption: it must be devoid of substances that could be hazardous to health. For drinking use, water must be colorless, tasteless, odorless, free of chemically pure suspended particles (devoid of toxic substances in harmful amounts for the organism), and bacteriologically pure (without pathogenic bacteria). At the same time, however, it is good to remember that several elements dissolved in the waters are part of the "trace elements" indispensable for our body, such as calcium, magnesium, and potassium.

It follows that one of the most important problems that the world today faces is the contamination of the environment and more specifically of surface and underground waters by toxic substances such as heavy metals and organic pollutants. Increasing the population, combined with the exploitation of water resources, has led to a shortage of supplying potable water in many parts of the world, as an increasing number of contaminants including organic dyes, heavy metal ions, and salts of light metals are introduced into water reserves due to human activities. The gradual decline in water has therefore become one of the world's worst concerns as it affects the health and life of humans: these substances can come to humans through food, drinking water, and even air, and then tend to bioaccumulate. For these reasons, it is necessary to increase the water reserves. In addition to what can be derived from the hydrologic cycle, processes such as salty water desalination or decontamination of wastewater must be carried out in order to guarantee the population the necessary water supply [1,2].

Heavy Metals

The metals are a group of 66 elements of the periodic table characterized by high thermal and electrical conductivity, gloss, hardness, ductility and malleability, as well as the peculiar metallic bond that joins their atoms. Also "heavy" is defined as those metals having a density greater than 5.0 g/cm^3 and are toxic in low concentrations. These are naturally present in the earth's crust and cannot be degraded or destroyed. Some of them (copper, zinc, selenium) in infinitesimal quantity are essential for the metabolism of the human body, but they become detrimental to higher concentrations. The presence of heavy metal ions in water has been one of the major concerns of recent years due to their toxicity to aquatic life, plants, animals, humans, and the environment. Heavy metals are commonly found in unpurified water or treated drinking water when accessing the distribution system. Such ions could even be in tap water as a result of corrosion of pipes and systems by "aggressive" water; among the sources of heavy metal contamination, there are tubes of old hydraulic systems as well as brass or bronze systems, which commonly contain lead. Since metals do not degrade biologically as organic pollutants, their presence on water is a problem for public health due to their absorption, and hence accumulation, in organisms. In addition, many metal ions interact with enzymes that contain sulfide groups due to their high affinity for sulfur, causing a conformational variation in the enzyme structure and thus becoming a true inhibitor of catalytic activity. In addition, many metals form bonds with the carboxylic and amide groups of amino acids precipitated with phosphorous compounds or may also bind to the cell membrane. This topic has gained groundbreaking growth in the industries: wastewater from metallurgy tanning, chemical, mining, and battery manufacturing, which contain several toxic heavy metal ions, which must be removed before they are released into the environment [3]. Generally, when the pH of the water decreases, the metal solubility increases and the particles become more mobile; consequently, metals are more toxic in freshwater. If high levels of concentration exist in the body, such ions may be immediately poisonous or lead to health problems in longer times. The most common pollutants formed by heavy metals are arsenic, cadmium, chromium, copper, nickel, lead, and mercury.

Due to their wide distribution, mercury and lead are today two of the most important environmental pollutants, especially for the question of tap water. The presence of these two pollutants depends on several factors, including the age and quantity of materials susceptible to corrosion present in the plants, the contact time between the pipes and the water, and the corrosive potential of the latter (influenced in turn by factors such as pH, hardness, and basicity). Since the first two factors can vary within the same building, the levels of mercury and lead can be very different between one tap and another in the same structure [4]. Despite intensive efforts to prevent the spread of these metals, they continue, together with other metals and semimetals such as arsenic, to pose a health risk. In this work, the operation of decontamination of heavy metal water was carried out about lead. In recent years, there has been an increasing interest in removing Pb^{2+} ions from drinking water, due to their extreme toxicity to our health. In fact, ingesting these ions for a long time, even at very low concentrations, can lead to various problems such as nausea, convulsions, coma, kidney problems, cancer, and effects on the metabolism and nervous system [5].

Organic Pollutants

As natural organic matter (NOM), we mean the large set of carbon-based compounds found both in the natural, terrestrial, and aquatic environment, and in those artificially synthesized. It is one of the major pollutants in the water; it can lead to the formation of carcinogenic by-products and complexing with metal ions. As a definition, NOM is a mixture of polyelectrolyte chemical complexes of various molecular weight, produced mainly by the decomposition of plant and animal residues [6]. It plays a fundamental role in water retention on the planet's surface and in the transport of nutrients into the environment.

Liquids in non-aqueous phase are hydrocarbons which in contact with water represent a separate, immiscible phase. The differences in physical and chemical properties between water and these substances result in the formation of a physical interface between the two liquids that prevents them from mixing. Liquids in non-aqueous phase are divided into Light non-aqueous phase liquids (LNAPLs), which have a lower density of water, and Dense non-aqueous phase liquids (DNAPLs), with a higher density than the latter.

These are characterized, in addition to the heaviness, by the low water solubility: the two physical properties mean that when released into the environment, in sufficient quantities, they can move through the soil and water until they meet a sufficiently resistant layer that will prevent, and they further vertical mass movement and allow it to accumulate. Depending on the nature of their release, the movement beneath the land surface can be quite complex, as the liquid follows the path that includes the least resistance. For example, homogeneous soils often have small differences in stratification that lead the DNAPLs to flow and then descend several times, to create a complex of vertical and horizontal clusters. Both the residuals of dense liquids in the non-aqueous phase in the soils, and the accumulations, over time become sources of contamination of groundwater and vapors from the ground. An example of DNAPL are halogenated hydrocarbons and chlorinated solvents.[1]

LNAPLs on the contrary, once they have penetrated the ground, stop at the water level because of their lower density. The costs for locating and removing these substances are in fact lower than those for DNAPLs thanks to this physical characteristic. However, they are sources of contamination of groundwater at many sites. They are usually released through petroleum products or organic mixtures of many chemical components with varying degrees of water solubility. Some additives are highly soluble (like alcohols), other components (such as benzene and toluene) are poorly soluble, and many components have relatively low water solubility under ideal conditions. Examples of LNAPLs are petrol, toluene, xylene, and other hydrocarbons.[2]

We will focus on the toxicity of phenols, in particular resorcinol, because in the work described below, experiments were carried out regarding the removal of this compound from water. Phenolic compounds are considered as environmental pollutants by the U.S. Environmental Protection Agency, EPA, as well as by the European Union, as they

[1] www.clu-in.org.

[2] www.epa.gov/sites/production/files/2015-06/documents/lnapl.pdf.

seriously threaten human health [7]. They are used in many types of industries, from tanning to cosmetics, from pharmaceuticals to dyes, and for this reason they are easily introduced into the environment through wastewater. Resorcinol, or 1,3-dihydroxybenzen, is a type of phenol with high toxicity. It can be easily absorbed into the gastric tract and from human skin, and can therefore cause dermatitis, phlegm, convulsions, poisoning, and even death [7,8].

16.2 Graphene

Graphene is a two-dimensional carbon nanomaterial organized in a hexagonal cell crystal structure, made of carbon atoms firmly bound together (the bonding distance is about 0.142 nm). Carbon has an electronic configuration ($1s^2$ $2s^2$ $2p^2$) in which the electrons in the orbital 1s are practically inert and do not contribute to the chemical bond. What affects the carbon–carbon bond stability is the presence of orbitals 2s, $2p_x$, and $2p_y$; in fact, after hybridization, they form a planar orbital, sp^2, which generates the covalent bond σ. The strength of this bond is responsible for the extraordinary mechanical stability of graphene. Also, the orbitals sp^2 form angles of 120° in the plane in which they are contained and then determine the hexagonal shape of the reticular structure. The orbital $2p_z$, not involved in hybridization, contains the fourth electron of each carbon atom and is positioned perpendicular to the plane in which the hybrid orbitals lie. Then, binding itself with an orbital $2p_z$ of another atom (parallel to it) constitutes the π bond and the electronic cloud extends above and below the plane identified by the carbon atoms. This whole system provides graphene with its great electrical properties.

It is therefore a monolayer with a thickness of about 340 pm and corresponds to one of the layers of graphite, taken individually and separated from the others [9]. It constitutes therefore the basic element for the formation of various carbon allotropes such as graphite, nanotubes, and others, which can be seen as graphene derivatives. The single graphene sheet can in fact be "folded" into a spherical shape to obtain a spherical fullerene C60, generally referred to as a zero-dimensional structure, "rolled up" in a cylindrical shape in a one-dimensional carbon nanotube (CNT), which depending on the number of layers of graphene is distinguished in single-walled nanotube (single-walled carbon nanotube, SWCNT) or multiple wall (multi-walled carbon nanotube, MWCNT); furthermore, more sheets of graphene can be "stacked" in a three-dimensional graphite structure. This structure generally consists of eight or more graphene sheets, separated by a distance of 0.335 nm and held together by weak intermolecular attractive forces [10].

The graphene nanoparticles (graphene nanoplatelets, GNP) are a graphene variation characterized by a number of sheets between 2 and 11, stacked on one another, which are often called double-, few-, or multilayered graphene sheets. They are different both from the graphene, composed of a single layer of carbon, and from the graphite, which implies a structure of eight or more sheets of graphene; they can therefore be considered an intermediate phase, with distinct properties that vary according to the number of layers, until the structure of the graphite is reached. GNPs are known as quasi-graphene, and precisely because their properties vary according to the number of sheets, it is good to specify the type of nanoplacchette with which you are working (i.e., to indicate the number of layers from which they are composed). Once the eight layers have been overcome, the electronic properties of the structure change, being more similar to those of graphite, so in these cases, the nanostructures are considered as such [9]. Another form of graphene commonly used in literature is graphene oxide (GO), which has a series of reactive functional groups containing oxygen. Depending on the manufacturing method used to synthesize it, its structure varies significantly in terms of the oxidized species present and their quantities. The adsorption capacity and the high surface area were highlighted in the GO, with a theoretical value of 2,620 m^2/g [11]. Graphene has been the subject of in-depth studies since 2004, despite its existence since the 1960s, given its large surface area (about 100–1,000 times greater than a typical organic molecule) and its excellent thermal, electrical, and mechanical. In fact, it has a high thermal conductivity and a conduction band that involves valence electrons. The mechanical properties are well represented by the definition of graphene as "the most resistant material ever discovered so far", a quality certainly linked to its thickness. As mentioned, this mechanical stiffness of the sheet is guaranteed by the strength of the C–C bond of type σ which is established between the hybrid sp^2 orbitals, while the interplanar bond of type π is very weak. This is what makes the separation of the graphite surfaces easily possible and is at the origin of the mechanical micro-exfoliation technique. Graphene is also very reactive on both sides, and the type of reactions that may occur or not is strictly related to its solubilization. Pure graphene is in fact hydrophobic and not soluble in many solvents, but by characterizing the carbon monolayer with various organic groups, it is possible to make it soluble in different solvents. Among the most characteristic properties of graphene, there is the possibility of observing it even with an optical microscope. Obviously, this visualization technique can be combined with other complementary ones, such as atomic force microscopy (AFM) or scanning electron microscopy (SEM) or transmission electron microscopy (TEM).

The preparation methods are not so immediate, given the high strength of graphene to fracture and deformation, and are still a widely studied topic. In fact, it is a highly coveted method that can generate reproducible sheets of high-quality monolayer graphene with a large surface area and large production volumes. There are several methods by which graphene was produced: mechanically, exfoliating the individual layers of graphite; by chemical vapor deposition (CVD); unrolling CNTs, electrochemically, chemically, or physically; by the chemical reduction of GO; and with the sonication of colloidal suspensions of graphite oxide [10].

As we will see below, structures derived from graphene have also been used for water decontamination. However, the micro- and nanoelectrical industry is the sector in which graphene has attracted more interest, given the presence in a single material of the characteristics of an ideal conductor: low resistivity value, high current density that can flow and high value of thermal conductivity.

16.2.1 Environmental Applications

As mentioned above, water pollution is largely due to incorrect wastewater disposal; the issue of decontaminating them from harmful substances, especially heavy metals and organic pollutants, becomes fundamental. As a result, an effective method is needed to remove such pollutants from wastewater before they are released into the environment, or directly from surface water. We have already mentioned in the above the toxicity of these substances, even in low concentrations, and their tendency to bioaccumulate. Over the years, different methods have been tested, since the concern increased, and there were increasingly strict regulatory criteria on the quality of drinking water. The main techniques used to guarantee the removal of pollutants are chemical precipitation, evaporation, ion exchange, adsorption, electrodialysis and reverse osmosis. Taking into account costs and efficiency, adsorption is without doubt one of the most promising methods; for this reason, it has been widely applied, it is easy to implement and economically advantageous [12–14].

Furthermore, in the adsorption on solid substrates, the materials used can be easily regenerated and undoubtedly lend themselves to automated analytical procedures for the preconcentration and determination of pollutants in natural waters [14].

There are several materials that have reported adsorbing capacity, especially against heavy metal ions: activated carbon, iron and manganese oxides, naturally condensed tannins and resins are just a few examples.

However, the adsorption capacity and the efficiency of decontamination did not turn out to be so high. For many researchers, it is therefore still important to find new materials that can act effectively by adsorbents.

Particular attention was paid to carbonaceous materials, as they proved to be excellent candidates for the removal of both heavy metals and their complexes. The way in which the internal structure of the pores is developed, the large surface area and the possibility of functionalizing them with different groups guarantee their adsorption capacity. Activated carbon (AC), as already mentioned, is a common adsorbent and consists of graphene sheets randomly substituted with hetero-atoms, and its characteristics depend on the precursor used and the activation technique, as well as the manufacturing process. Studies on carbon-encapsulated magnetic nanoparticles (CEMNP) have demonstrated high efficiency and adsorption capacity under acid conditions for metal ions such as copper and cobalt [14].

However, in the large carbon family, the materials that have been studied more extensively for environmental applications, based on their promising properties, are CNTs and graphene.

Carbon Nanotubes

In 2004, the U.S. Environmental Protection Agency (EPA) expressed the need to explore the possible applications of CNTs in terms of pollution. Naturally, the reclamation of the water was one of the biggest problems to be considered. Since then, the CNTs have obtained an increasing appreciation for their adsorption capacity, in addition to the already known mechanical and electrical. This is mainly due to their extremely small size, uniform pore distribution and a large specific surface area. Moreover, their structure is a hollow and multilayer structure (Figure 2.6), which makes them excellent candidates for the attraction of other substances. In particular, they showed a high efficiency of removal, better than AC, against organic toxic substances such as dioxins and inorganic ions such as fluorides [15,16].

An even more important application, given the huge consequences of the presence of metal in the environment, lies in the ability of nanotubes to adsorb lead ions from aqueous solutions, in particular after undergoing an oxidation process by nitric acid. The demonstrated adsorption capacity is 15.6 mg/g at a metal equilibrium concentration of 2.7 mg/g [17]. A positive value of the standard enthalpy variation suggests that the interaction between lead and nanotubes is endothermic, and at the same time, the spontaneity of the process is confirmed by the negative value of the standard free energy variation and by the positive value of the standard entropy variation. The lead can also be easily desorbed by the CNTs by modifying the pH value of the solution. Both the removal efficiency achieved and the capacity of adsorption make CNTs valuable materials for the decontamination of wastewater [5,17].

A study was also conducted in which the ability to adsorb the ions of various heavy metals by the nanotubes, in particular CNT sheets treated under oxidative conditions, was compared (in this way, the adsorption proved to be more significant than to pristine nanotubes, i.e., untreated). From the results, it is noted that as the initial concentration of the salts increases, the adsorbing capacities increase and the order of preference of adsorption on the sheets of oxidized nanotubes is $Pb^{2+} > Cd^{2+} > Co^{2+} > Zn^{2+} > Cu^{2+}$. It was therefore possible to treat the water relatively cheaply and without the occurrence of a CNT leak into the environment, which would have further increased the degree of contamination, given the toxicity of these materials [3].

We have spoken in the above of the various pollutants present in surface and ground waters, describing not only the toxicity of heavy metals, but also the dangers of organic natural matter (NOM) and in general other organic pollutants such as phenols. Also, in this case the nanotubes showed extraordinary abilities.

The organic natural material generally has a negative charge under natural conditions, due to the presence of carboxylic and phenolic groups distributed on the molecules. These physical and chemical characteristics are probably related to the interaction mechanism between NOM and the adsorbent substance. While the mechanism of interaction with AC is known, given the extensive use of this material over the years as an adsorbent for the treatment of contaminated water, that with CNTs should be studied keeping in mind that between NOM and AC. Obviously, the capacity and strength of adsorption depend on the type of organic matter and AC. Some factors influencing the interaction are the magnitude and chemical characteristics of NOM, the porous structure and the chemically AC surface. Moreover, given the so varied nature of organic matter, different types of NOM tend to have different degrees of interaction with the adsorbent substance. For example, the fraction of organic matter that tends to be more strongly adsorbed results in more favorable interactions with a lower AC dose. Other factors not to be underestimated are the parameters of the water under examination: ionic strength and pH clearly influence the charge and the configuration of the NOM, conditioning the adsorption process. In particular, when the organic substance is negatively charged, it is more adsorbed on the AC surface as ionic strength increases and pH decreases. The differences between carbon and AC nanotubes are considerable. First of all, the structure should be considered: AC is composed of micropores of different sizes that provide adsorption sites; the nanotubes instead contain sites for the interaction only along the surface of a cylindrical structure. We must then consider the differences in terms of chemical composition between AC and CNT. The first contains coals with different degrees of saturation and oxidation states, and functional groups formed during the activation process; the nanotubes consist instead of only globally conjugated unsaturated carbons in a three-dimensional arrangement. Despite these differences, experiments on both materials showed very similar adsorption, except for the fact that high-molecular-weight NOMs exhibited stronger interactions with multi-walled nanotubes and therefore greater adsorption on their surface [6].

A nanotube-based material that has exhibited higher adsorption to NOM than traditional ones, such as AC, is the nanotube buckypaper. It is nothing but a structure of tangled CNTs that form a self-supporting, chemically and physically stable architecture. The buckypaper has good thermal, electronic and mechanical properties; consequently, it is used in various fields, such as in the construction of artificial muscles, electrodes, and precisely water purification. It is traditionally prepared by the filtration of CNT dispersions in a solvent or in solutions with surfactants. The properties of the buckypaper as the best adsorbent are due to its porous structure and to the lower negative charge present on the surface of the nanotubes. Furthermore, the π–π interactions between the aromatic rings of nanotubes and organic matter contribute to a better adsorption. As proof of this, the buckypaper has shown a removal of more than 93% of humic acid, a natural substance that is formed as a result of the microbial biodegradation of organic matter. High efficiency has been attributed to the incorporation of hydrophilic functional groups on the purified nanotube buckypaper, improving its hydrophilicity and allowing a better and faster contact between the humic acid in solution and the surface of the carbonaceous material [18].

The adsorbing properties of CNTs were also studied for phenolics, whose toxicity was described in Section 16.1.1. In particular, it has been shown that even with a short contact time, multi-walled CNTs have a high potential as adsorbents for phenols. The ability of CNTs decreased after being treated with acid, probably due to a greater electrostatic repulsion and the possible formation of carboxylic groups that weakened the interaction and therefore the adsorption of water. The number and the position of hydroxy substituents in the aromatic ring also had an important influence: substances with more –OH groups showed a better ability to be adsorbed, and among the dihydroxybenzenes, the metaisomer (corresponding to resorcinol) was the one with the best results. The adsorption of phenolic derivatives can be useful not only in the environmental field but also in the field of catalytic supports and chemical sensors [13].

Graphene-Based Pollutants Removal

In the above, we discussed the various properties of graphene, which make it in effect one of the most interesting materials among carbon derivatives. As for the nanotubes, possible applications of this material have also been explored as adsorbent in the field of water pollution, in the form of GO or as functionalized structures based on graphene.

GO is a valid candidate for the absorption and removal of pollutants in water. Recently, several articles have shown that synthetic GO has high adsorbing capacities towards dyes, antibiotics and heavy metals [11].

In particular, heavy metal ions such as cobalt and cadmium were considered in a study, showing that GO is the material that has the highest adsorption capacity among those used so far. This is probably due to its large surface area and the presence of different functional groups containing oxygen on the surface, which make each atom of this element available to chelate the metal ion. Moreover, while CNTs require special oxidation processes to introduce hydrophilic groups and increase the adsorption of heavy metal ions, the preparation of GO with Hummer method is a simple technique that introduces groups such as carboxylic acid on the surface of the graphene. Carbonyl and hydroxyl are essential for the adsorption of metals.

The GO nano-sheets also act as Lewis base thanks to the delocalized π electron system of each graphene layer and form electron donor–acceptor complexes with the metal ions, in which the latter behave as Lewis acid. Therefore, a strong complexation occurs through an acid–base interaction of Lewis, which obviously contributes to the adsorption of ions on the GO sheets. The obtained thermodynamic data also suggest that the adsorption of cadmium and cobalt ions is a spontaneous and endothermic process. The values of the standard enthalpy variation are indeed positive, indicating that the process absorbs heat, as well as those of the variation of standard entropy, to indicate the spontaneity of adsorption. This is confirmed by the negative values of the standard free energy variation, and in particular, it is noted that this becomes more negative by increasing the temperature, to indicate a better adsorption. In fact, at higher temperatures, ions are easily dehydrated, and absorption becomes enhanced [19].

Further experiments were conducted with zinc ions, hypothesizing that the main force of absorption is ion exchange, while electrostatic interaction could also influence the process. In fact, the results show that the absorption depends strongly on the pH of the solution and slightly on the presence of foreign ions and the ionic strength of the solution. Also in this case, the thermodynamic parameters indicate a spontaneous and endothermic absorption [11].

In addition to GO, three-dimensional functionalized nanostructures with graphene as a substrate have attracted the attention of researchers. These have a large surface-to-volume ratio and a unique morphology, which justify their different applications. They have in fact been used for the detection of polluting agents, for the realization of superhydrophobic surfaces, of transparent conductors, batteries, etc. The three-dimensional structures produced up until now consisted of growing metallic nanomaterials on two-dimensional sheets of graphene. However, the specific values of these structures are relatively low, owing to the low specific surface area due to the metal tubular shapes present. In order to maximize the surface area of such materials, a solution consists in growing one-dimensional CNTs on the graphene (two-dimensional) sheet and subsequently applying metal nanoparticles, which are linked to both the CNT and the graphene.

This system has been tested in the decontamination of water from arsenic ions, a highly toxic metal that can lead to skin or lung cancers. In particular, the metal nanoparticles used are iron oxide. What we get then is a structure made up of nanotubes grown vertically on the surface of graphene sheets with iron oxide nanoparticles that are well distributed on both carbon nanostructures (G-CNT-Fe). The iron oxide nanoparticles are not only magnetically active but have also demonstrated a decisive adsorption and a consequent removal of the arsenic from the contaminated water. This three-dimensional system has a great capacity for adsorption thanks to the increased surface-to-volume ratio and the path that the 3D structures can undertake in water [20].

Acknowledgment

The participation of Antonino Cataldo and Arianna Filippini to the early stages of this work is gratefully acknowledged.

References

1. S. Chuanuwatanakul, W. Dungchai, O. Chailapakul, S. Motomizu, Determination of trace heavy metals by sequential injection-anodic stripping voltammetry using bismuth film screen-printed printed carbon electrode, *Anal. Sci.*, vol. 24, n. 5, pp. 589–594, 2008.
2. Z. Sui et al., Green synthesis of carbon nanotube-graphene hybrid aerogels and their use as versatile agents for water purification, *J. Mater. Chem.*, vol. 22, n. 18, pp. 8767–8771, 2012.
3. M. A. Tofighy, T. Mohammadi, Adsorption of divalent heavy metal ions from water using carbon nanotube sheets, *J. Hazard. Mater.*, vol. 185, n. 1, pp. 140–147, 2011.
4. A. Mandil, L. Idrissi, A. Amine, Stripping voltammetric determination of mercury (II) and lead (II) using screen-printed electrodes modified with gold films, and metal ion preconcentration with thiol-modified magnetic particles, *Microchim. Acta*, vol. 170, n. 3–4, pp. 299–305, 2010.
5. Y.-H. Li, Z. Di, J. Ding, D. Wu, Z. Luan, Y. Zhu, Adsorption thermodynamic, kinetic and desorption studies of Pb^{2+} on carbon nanotubes, *Water Res.*, vol. 39, n. 4, pp. 605–609, 2005.
6. H. Hyung, J.-H. Kim, Natural organic matter (NOM) adsorption to multi-walled carbon nanotubes: Effect of NOM characteristics and water quality parameters, *Environ. Sci. Technol.*, vol. 42, n. 12, pp. 4416–4421, 2008.
7. A. J. Ahammad, S. Sarker, M. A. Rahman, J.-J. Lee, Simultaneous determination of hydroquinone and catechol at an activated glassy carbon electrode, *Electroanalysis*, vol. 22, n. 6, pp. 694–700, 2010.
8. S. M. Ghoreishi, M. Behpour, E. Hajisadeghian, M. Golestaneh, Voltammetric determination of resorcinol on the surface of a glassy carbon electrode modified with multi-walled carbon nanotube, *Arab. J. Chem.*, vol. 9, pp. S1563–S1568, 2016.
9. A. C. Ferrari et al., Raman spectrum of graphene and graphene layers, *Phys. Rev. Lett.*, vol. 97, n. 18, p. 187401, 2006.
10. D. A. Brownson, C. E. Banks, *The Handbook of Graphene Electrochemistry*. Springer, London, 2014.
11. H. Wang et al., Adsorption characteristics and behaviors of graphene oxide for Zn (II) removal from aqueous solution, *Appl. Surf. Sci.*, vol. 279, pp. 432–440, 2013.
12. Y.-H. Li et al., Competitive adsorption of Pb^{2+}, Cu^{2+} and Cd^{2+} ions from aqueous solutions by multi-walled carbon nanotubes, *Carbon*, vol. 41, n. 14, pp. 2787–2792, 2003.
13. Q. Liao, J. Sun, L. Gao, The adsorption of resorcinol from water using multi-walled carbon nanotubes, *Colloids Surf. A*, vol. 312, n. 2, pp. 160–165, 2008.

14. K. Pyrzyńska, M. Bystrzejewski, Comparative study of heavy metal ions sorption onto activated carbon, carbon nanotubes, and carbon-encapsulated magnetic nanoparticles, *Colloids Surf. A*, vol. 362, n. 1, pp. 102–109, 2010.
15. R. Q. Long, R. T. Yang, Carbon nanotubes as superior sorbent for dioxin removal, *J. Am. Chem. Soc.*, vol. 123, n. 9, pp. 2058–2059, 2001.
16. Y. H. Li et al., Removal of fluoride from water by carbon nanotube supported alumina, *Environ. Technol.*, vol. 24, n. 3, pp. 391–398, 2003.
17. Y.-H. Li et al., Lead adsorption on carbon nanotubes, *Chem. Phys. Lett.*, vol. 357, n. 3, pp. 263–266, 2002.
18. X. Yang et al., Removal of natural organic matter in water using functionalised carbon nanotube buckypaper, *Carbon*, vol. 59, pp. 160–166, 2013.
19. G. Zhao, J. Li, X. Ren, C. Chen, X. Wang, Few-layered graphene oxide nanosheets as superior sorbents for heavy metal ion pollution management, *Environ. Sci. Technol.*, vol. 45, n. 24, pp. 10454–10462, 2011.
20. S. Vadahanambi, S.-H. Lee, W.-J. Kim, I.-K. Oh, Arsenic removal from contaminated water using three-dimensional graphene-carbon nanotube-iron oxide nanostructures, *Environ. Sci. Technol.*, vol. 47, n. 18, pp. 10510–10517, 2013.

17

Detection of Pesticides Using Cantilever Nanobiosensors

Janine Martinazzo, Alexandra Nava Brezolin, Clarice Steffens, and Juliana Steffens
URI—Campus of Erechim

17.1 Pesticides

The use of pesticides in agriculture has intensified due to the growing demand for food with the increase in world population (Muenchen et al. 2016). The wide use of pesticides has become a health concern worldwide, due to its persistence in the environment, leading to water, soil as well as air contamination (Bonansea, Ame and Wunderlin 2013; Sekhon 2014; Martinazzo et al. 2018). According to Sekhon (2014), the world population will increase to around 8 billion of people in 2025 and 9 billion in 2050, and is widely recognized that global agricultural productivity must also increase to keep up with this increase in population.

Agrochemicals, such as pesticides, are used to treat various types of pests, including insects, fungi and weeds (Xia and Leidy 2002; Borsoi et al. 2014; Karasali, Marousopoulou and Machera 2016). Although the use of pesticides has become essential for modern agriculture, constituting an agricultural defense strategy (Waichman 2012), their destination in the environment can cause contamination (Yusa, Coscolla and Millet 2014; Ccanccapa et al. 2016; Sun et al. 2016).

Pesticides are defined as any substance or mixture of substances intended for prevent, control or destroy any pest, including vectors of human or animal disease, undesirable plant or animal species, causing any kind of damage in the production, processing, storage, transport or trade of food, agricultural commodities, wood and wood products, or that may be administered to animals for the control of insects, arachnids or other parasites, as defined by Food and Agriculture Organization of the United Nations (FAO 2007). The term includes substances intended for use as a plant growth regulator, defoliants, desiccants or an agent for thinning fruits or preventing premature fruit falling, and applied in the crops (before or after harvesting) to protect against deterioration during storage and transport (Steffens et al. 2017).

There are numerous varieties of pesticides that can be classified according to three approaches: mode of action, pests control and chemical group to which they belong. According to the mode of action, they may be systemic or non-systemic. The first enters in the vascular system of the plant, whereas non-systemic ones do not, thus affecting the place which come into contact. Chemical structures are classified as carbamates, pyrethroids, organophosphates, organochlorines and triazines. According to their action to the pests are classified as follows: herbicides, for weeds control; fungicides, for fungus control; insecticides, for insect control; acaricides, and for the mites control, among others (Zacharia 2011).

These products cause undesirable effects in animals and humans when exposed directly or indirectly with the substances. According to studies carried out with the exposition of rats, fish and insects to herbicides, they cause changes in the development (cholinergic, immune and reproductive systems), causing changes in organs such as the brain, liver, kidney and lung, among others (Inouye et al. 2014; Lin et al. 2014; Vogel et al. 2015; Xing et al. 2015).

The main adverse health effects of pesticide exposure are cardiovascular and respiratory diseases, causing asthma and chronic obstructive pulmonary disease (COPD), endocrine disorders, neurological diseases such as Parkinson's, Alzheimer's and amyotrophic lateral sclerosis (ALS), as well as reproductive and cancer (Collotta, Bertazzi and Bollati 2013; Hernández et al. 2013; Mostafalou and Abdollahi 2013; Vakonaki et al. 2013; Blair et al. 2015). The exposure of pregnant women to pesticides can induce early changes in glucose metabolism of the newborn (Debost-Legrand et al. 2016).

17.1.1 Conventional and Alternative Detection Methods

The need to detect contaminants in the environment in very small amounts makes sense to look for alternatives with technologies of last generation that allow detecting a great number of compounds of fast and reliable form.

Conventional analytical methods for pesticide detection are used with high sensitivity and selectivity, at low detection limits. These include gas chromatography (El-Gawad 2016; Saitta et al. 2017), high-performance liquid chromatography (Botero-Coy et al. 2012; Jia et al. 2016), capillary electrophoresis (Hsu and Whang 2009; Bol'shakov and Amelin, 2016) and enzyme-linked immunosorbent assays—ELISA (Rekha, Thakur and Karanth 2000; Barchanska et al. 2012).

Generally, these detection methods are reliable and have high sensitivity and selectivity, but require expensive and specific equipment, and highly trained and specialized technicians, and involve the complex and time-consuming sample preparation, all of which are associated with high costs (Chauhan and Pundir 2012; Wang, Lu and Chen 2014; Lin et al. 2018; Madianos et al. 2018). There is a growing need for rapid, cost-effective, reusable, reliable and sensitive detection systems to identify the presence of environmental pollutants such as pesticides (Gupta et al. 2018).

The development of advanced methods has become a promising tool for simple and fast use, low cost and suitability for in situ monitoring (Samsidar, Siddiquee and Shaarani 2018). The determination of pesticides by advanced methods has increased significantly, based on the development of biosensors. This device can provide qualitative and quantitative information with minimal sample preparation and in real time (Pichetsurnthorn et al. 2012; Songa and Okonkwo 2016).

Sensor is defined as a device that identifies and transforms an information into measurable signal to be used (Hulanicki, Glab and Ingman 1991). A biosensor integrated device is capable of providing specific quantitative or semi-quantitative analytical information using a biological recognition element (biochemical receptor) which is in direct spatial contact with a transduction element (Thévenot et al. 1996). When the sensitivity and selectivity of biosensor are investigated at the nanoscale (1–100 nm), these devices are called nanobiosensors (Malik et al. 2013; Sang et al. 2013).

Second, in the study by Korotkaya (2014), a biosensor consists of three basic components: biological recognition element, which selectively recognize one or more analytes from a large number of other substances; physical transducer, which processes the signal produced by the interaction between the recognition element and the analyte of interest in a measurable signal and electronic system for signal amplification and data recording. Figure 17.1 presents a schematic diagram of biosensor operation.

Biosensors can be classified according to the type of interaction between the sensing layer and the analyte, methods used to detect desired interactions, the nature of

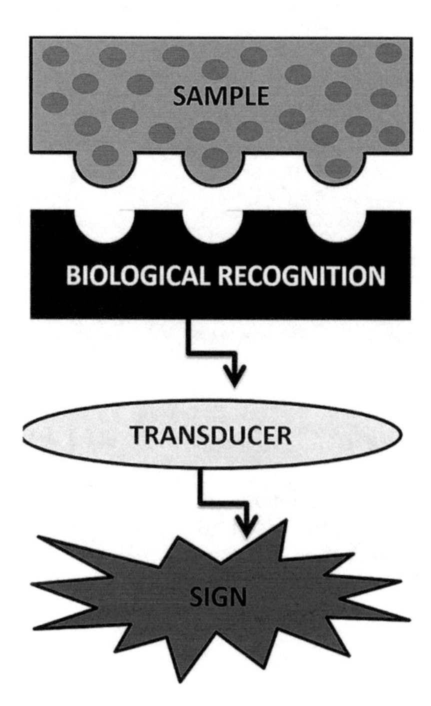

FIGURE 17.1 Schematic diagram of biosensor operating (Muenchen et al. 2016). (Reprinted from an open-access source.)

the recognition elements (sensitive layer) and transducing system (Rumayor et al. 2005). In relation to the transduction system, biosensors can be classified as

- Electrochemical biosensors: contain potentiometric, amperometric or conductometric cells as detectors that generate signals of electrochemical reactions, always based on the interaction between the analyte and the recognition element (sensitive layer) (Das et al. 2016).
- Optic biosensors: optical detection is performed by the interaction of the optical field with a biological recognition element. The goal is to produce a signal that is proportional to the concentration of the measured substance (Damborský, Švitel and Katrlík 2016).
- Piezoelectric biosensors: represent a suitable transducer for simple and fast determinations of viruses, bacteria and small molecules (drugs, hormones and pesticides) (Skládal 2016). They are based on the resonant frequency measurement due to changes in mass and/or microviscosity (Thévenot et al. 2001).
- Thermometric biosensors: convert thermal energy into electrical energy, and the output voltage is proportional to the temperature difference between the measurement and the reference (Nestorova et al. 2016).
- Biomechanical biosensors: the adsorption of molecules on a biosensor surface causes a mechanical response, such as deflection (Raiteri et al. 2001).

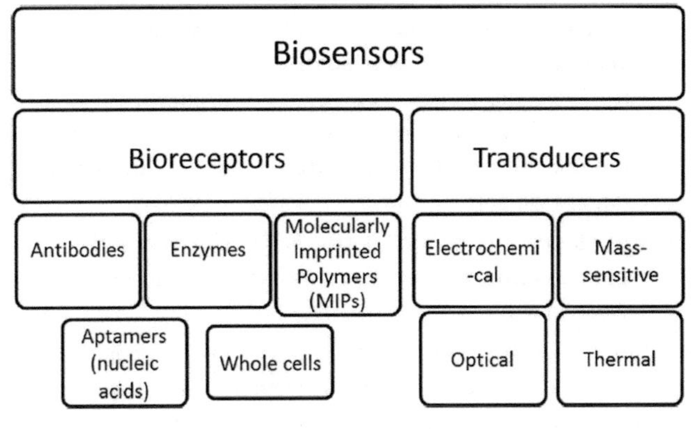

FIGURE 17.2 Different types of biosensors, classified by bioreceptor type or according to the transducer (Gaudin 2017). (Reprinted from an open-access source.)

New detection technologies: use small sensors and advanced materials with specific biological recognition molecules, and have been developed to produce sensors with low manufacturing cost, portability and simple operation (Madianos et al. 2018).

Among the several types of biosensors, the nanomechanical sensors based on cantilevers are used on atomic force microscopy (AFM) and could be applied to detect many specific substances (Steffens et al. 2017) (Figure 17.2).

17.2 Nanotechnology

Nanotechnology dominates research for development and finds several applications in agriculture, especially when it comes to sustainable agriculture, and can be the key to the survival of humanity, where they are inserted from the development of modified plants to the targeted application of chemicals (Poddar et al. 2018).

The new technologies have led to the development of potent but high-risk pesticides, including at the molecular level, and some molecules would be capable of causing harm to human health, showing the importance of developing technologies that can detect these pesticides. In this sense, nanotechnology can be a great ally, because it involves the manipulation of atoms and molecules for the fabrication/development of nanometer-sized materials (Tharmavaram et al. 2017; Rawtani et al. 2018).

The sensors developed at the nanoscale have differentiated physical and chemical properties (Zhao and Jiang 2010; Durán and Marcato 2013) and, in addition, increase the specific surface area (Malik et al. 2013), making them suitable and promising devices to detect specific analytes.

17.2.1 Cantilevers and Microscopy Force Atomic (AFM)

The invention of the AFM equipment in 1986 introduced the possibility of the creation of a new sensing tool for the detection of several analytes by means of cantilever functionalization, which then operate as sensors (Sang, Zhang and Zhao 2013). In this perspective, the cantilever acts as a mechanical transducer, translating the recognition of the analyte through the sensor layer (Carrascosa et al. 2006).

Cantilevers can be used as an extremely sensitive sensor to detect molecules of interest at the nanoscale and consist of a device that can act as a physical, chemical or biological sensor, detecting changes in flexion or vibrational frequency. They are composed of thin and flexible rods that in their lower part contain a needle, where the tip vertex radius has a dimension of some nanometers. One or more silicon rods or silicon nitride rods are used for construction, usually have a length and thickness in the order of micrometers, and may undergo deformations when a specific mass of analyte is adsorbed on its surface (Vashist 2007). Detection of molecules by cantilevers makes it possible to obtain mass resolutions in picograms (Okan and Duman 2018).

The AFM principle (Figure 17.3) is based on cantilevers with adequate spring constant, besides it is possible to measure interaction forces between the tip and atoms of the sample, resulting in a cantilever bending. The laser diode is focused at the free end of the cantilever, and the reflected laser beam is monitored using a position-sensitive photodetector. A feedback system adjusts the distance between the tip and the sample while maintaining the force applied to the surface. All this process is controlled by computer (Berger et al. 1997).

In this way, the elastic properties of the cantilever play a key role in the analysis of the laser beam on the detector signal (AFM) with high resolution, which can also be applied in cantilevers sensors.

The molecules adsorbed on the cantilever cause changes in vibrational frequency or deflection, and depending on the nature of the molecule chemical bind, the deformation can be up or down. Sensors with mechanical detection systems may be used, for example, silicon cantilevers coated with gold layer and a given receptor in one side; after analyte

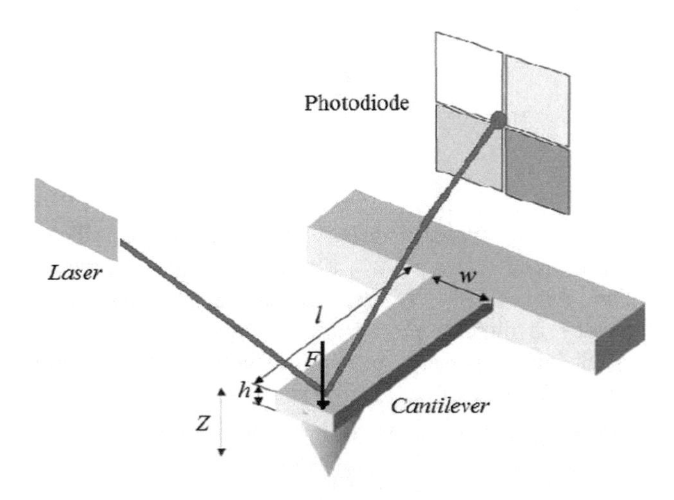

FIGURE 17.3 Illustration of the displacement of the cantilever (h, w and l, thickness, width and length of the cantilever, respectively) (Steffens et al. 2012). (Reprinted from NCBI with permission.)

binding with the receiver, the surface it is tensioned or relaxed. This deviation, typically in nanometers, can be measured using optical techniques, where the deflection is proportional to the concentration of analyte (Kapczyńska et al. 2012).

Some aspects in the choice of the cantilevers sensors to be considered: material with a low internal damping (Young's modulus) and geometry must provide a high-quality factor (Q). The cantilever surface is reflective enough to allow the position-sensing laser light to reflect with high quality be read optically and lower roughness for not disperse the light in all directions, and the coating (functionalization) should be specific to the desired application (Butt, Cappella and Kappl 2005; Zavala 2007).

The basic mechanical parameters of a cantilever beam are the spring constant and resonance frequency. The spring constant (K) is a proportionality factor between the force "F" (in Newton) applied and the deflection (displacement in nm) (ΔS), where this ratio is called Hook's law according to Eq. 17.1 (Boisen et al. 2011):

$$F = K \cdot \Delta S \tag{17.1}$$

The K denotes its elastic properties produce stiffness. The K expression is a function of geometrical dimensions and material parameters as given in Eq. 17.2:

$$K = \frac{Ewt^3}{4l^3} \tag{17.2}$$

where E is the modulus of elasticity for the material composing the cantilever ($E = 1.3 \times 10^{11}$ N/m^2 in <100> plan of silicon crystal structure), and w, t and l are the width, thickness and length respectively.

The Young's modulus is related to the K, which depends on the material properties, defining the sensitivity (Naddeo et al. 2012). Both K and mass (m), the resonance frequency (f_{res}) in the absence of damping, can be approximated as Eq. 17.3.

$$f_{\text{res}} = \frac{1}{2\pi}\sqrt{\frac{K}{m}} \tag{17.3}$$

The mass (kg) of the cantilever can be expressed as Eq. 17.4

$$m = \rho t l w \tag{17.4}$$

where ρ is the material density (2,300 kg/m^3 for silicon material).

The relation shows that increasing the resonance frequency is directly proportional to increase the spring constant and inversely to mass reduction (Wilson and Baietto 2009; Brattoli et al. 2011).

The nanomechanical sensors operation is based on the molecules adsorption in the surface, and three mechanical quantities are influenced by biomolecular adsorption: surface stresses, effective Young's modulus and viscoelasticity (Berger et al. 1997; Tamayo et al. 2013). All these changes can affect the sensor response including the following output signals:

a. Voltage nanomechanical system: usually caused by forces acting on a flat surface and result in a curvature (bending stress). The change in curvature is monitored in real time due to the interaction of the molecules with the sensitive layer of nanomechanical sensor;
b. Resonance frequency: depends on the geometry, size and configuration, and suffers a reduction with adsorbed mass;
c. Quality factor: is directly related to the precision of resonance frequency measurements, because it determines the slope of amplitude and resonance phase curves.

The stress effect on the cantilever surface is related to the amount of work per unit of area required to cause a bending or a frequency change in a preexisting surface, and adsorption on the cantilever surface can cause surface stresses as a result of adsorption/desorption interactions (Lakshmoji et al. 2014).

Cantilever sensors are most used in two modes: depending on the deflection or resonance frequency operation, these can be classified into the static mode and the dynamic mode (Figure 17.4). The operation mode is associated with the transduction mechanism. In static mode (deflection), the sensing biomolecular interactions or binding-induced changes are caused by differential surface stress. Cantilever sensors operating in the dynamic mode (resonant) surface biochemical binding-induced changes in resonant frequency caused by attached mass as well as viscoelastic properties of the medium. The adsorption of analyte molecules on a resonating cantilever decreases its frequency due to the increased suspended mass (Zavala 2007).

Other cantilever detection techniques include optical, capacitive, single-electron transistors, piezoresistive, piezoelectric, magneto motive, hard contact, direct tunneling and field emission, among others. The advantage of microelectromechanic system (MEMS), which included the AFM microcantilever, is the ability to adjust the size and structure of the device (Carrascosa et al. 2006). Also the resonance frequency and deformations shifts measured show information about the relations between the transducers and the environment, being that they can operate in liquids, gases and vacuum.

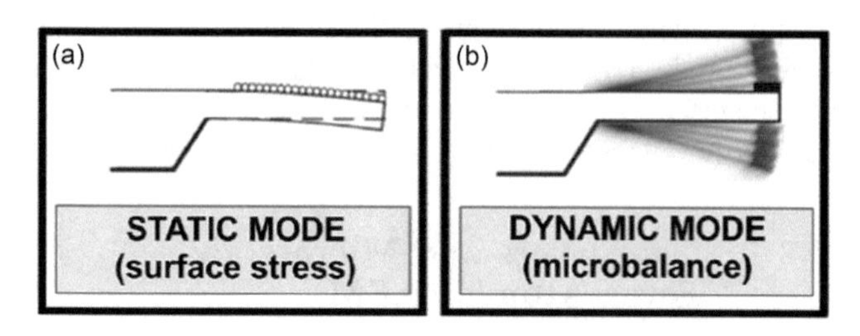

FIGURE 17.4 Schematic drawings of the two possible modes of operation of cantilever-based sensor: static mode (a) and dynamic mode (b) (Huber et al. 2015). (Reprinted from an open-access source.)

TABLE 17.1 Biosensors Developed for the Detection of Different Pesticides

Pesticide	Matrix	Biologic Material	Nanomaterial Exploited	Transduction	References
Dichlorvos	Cabbage juice	Acetylcholinesterase	Graphene oxide	Electrochemical	Cui et al. (2018)
Malathion	Real samples	Acetylcholinesterase	Nanocomposite chitosan-$Ti_3C_2T_x$ (T represents the terminating groups; x represents the number of these terminating groups)	Amperometric	Zhou et al. (2017)
Malathion	Fruits and vegetables	Acetylcholinesterase	Palladium-copper nanowires	Electrochemical	Song et al. (2017)
Carbamate	Real samples	Acetylcholinesterase	Sensing interface of citrate-capped gold nanoparticles (AuNPs)/(3-mercaptopropyl)-trimethoxysilane (MPS)/gold electrode (Au)	Electrochemical	Song et al. (2016)
Paraoxon and dimethoate	Water	Acetylcholinesterase	Gold nanorods	Amperometric	Lang et al. (2016)
Carbofuran, oxamyl, methomyl and carbaryl	Fruits and vegetables	Acetylcholinesterase	Gold nanoparticles	Optical	Kestwal et al. (2015)
Paraoxon	Vegetables	Acetylcholinesterase	Carbon nanotubes	Electrochemical	Yu et al. (2015)
Paraoxon, parathion and chlorfenvinphos	Water	Acetylcholinesterase	Carbon nanotubes	Amperometric	Kesik et al. (2014)
Methyl parathion	Water and soil	Acetylcholinesterase	Nanocomposites of multiple-walled carbon nanotubes and chitosan	Amperometric	Dong et al. (2013)
Chlorpyrifos-oxon, ethyl paraoxon and malaoxon	Milk	Acetylcholinesterase	–	Amperometric	Mishra et al. (2012)
Malathion, chlorpyrifos, monocrotophos and endosulfan	Soil and food	Acetylcholinesterase	Fe_3O_4 nanoparticle/multi-walled carbon nanotube	Amperometric	Chauhan and Pundir (2012)
Chlorpyrifos-oxon, chlorfenvinphos and azinphos-methyl-oxon	Water	Acetylcholinesterase	–	Amperometric	Alonso et al. (2012)
Organophosphate	Water, phosphate buffer, food or vegetable extracts	Acetylcholinesterase	–	Amperometric	Crew et al. (2011)
Carbaryl	Water	Acetylcholinesterase	Single-walled carbon nanotubes	Amperometric	Firdoz et al. (2010)
Organophosphate	Water	Acetylcholinesterase and choline oxidase	Graphene quantum dots	Photoluminescence	Sahub et al. (2018)
Chlorpyrifos, malathion and methyl parathion	Cabbage	Acetylcholinesterase and choline oxidase	Bimetallic nanoparticles Pt-Au and carbon nanotubes	Electrochemiluminescence	Miao et al. (2016)
Paraoxon and carbaryl	Apple	Acetylcholinesterase and organophosphate hydrolase	Multiple-walled carbon nanotubes	Electrochemical and optical	Zhang et al. (2015)
Imazaquin, metsulfuron-methyl and atrazine	Pesticide solution	Acetolactate synthase and anti-atrazine	–	Mechanic	Deda et al. (2013)
Atrazine	Water	Tyrosinase	–	Optic	Martinazzo et al. (2018)
Atrazine	Water	Tyrosinase	Graphene and multiple-walled carbon nanotubes	Amperometric	Tortolini et al. (2016)
Chlortoluron	Water	Tyrosinase	Zinc oxide nanoparticles	Electrochemical	Haddaoui and Raouafi (2015)
Acetamiprid	Soil	Aptameter (ABA)	Gold nanoparticle	Colorimetric	Shi et al. (2013)
Glyphosate and glufosinate	Soil and human serum samples	Amino acid	Gold nanoparticle and multiple-walled carbon nanotubes	Electrochemical	Prasad, Jauhari and Tiwari (2014)
Paraoxon	Water and milk	Butyrylcholinesterase	Silver nanowires	Amperometric	Turan et al. (2016)

Selective coating provides affinity of targeted analytes to the sensor active area. The adsorption of analytes on cantilever sensitive layer causes a surface tension, resulting in bending or changes in resonance frequency (Steffens et al. 2012; Etayash et al. 2016).

Cantilevers can be applied in gas or liquid phases. In the liquid environment, some aspects must be observed as prevention of leakage and of bubbles in microsystems, mixing liquids. Also, the biomolecules must be compatible with liquid phase.

The ability to adapt the size and the structure associated with the properties of micro- and nanomaterials indicates excellent prospects for the construction and application of new detection systems. These sensors have several advantages over conventional analytical techniques in terms of high sensitivity, simple, low requirement of analytes, non-hazardous procedures and rapid response (Bates and Lu 2016).

Another technology that has been gaining ground in the detection of pesticides using nanotechnology is nanobiosensors, working on improved analytical performance, and may assist in agriculture since the detection of chemicals, assisting in decisions and increasing yields (Antonacci et al. 2018).

Nanobiosensors

In recent decades, the use of pesticides has been intensified to help increase productivity and meet food demand because of the increasing population, and although of fundamental importance in agriculture, are dangerous compounds that are not directed towards the ecosystem and human health, and may cause harm. The nanobiosensors have several advantages: high sensitivity due to surface-to-volume ratio, fast response time, high stability and long shelf life, when compared to non-nanostructured biosensors (Scognamiglio 2013).

The development of nanobiosensors that can identify small-scale interactions or analytes with precision and maximum sensitivity is extremely important. Basically, they have three main components: the bioreactor, the transducer and the detector, with the main function of detecting specific materials, present in small quantities in the environment. The bioreactors can be antibodies, proteins, enzymes and immunological molecules, among others. These nanobiosensors have transduction mechanisms, which are the key elements, converting the responses of the interactions in a way that can be identifiable and reproducible, and at that point, the nanomaterials have a great advantage, due to the great surface area and excellent electromechanical characteristics, improving the characteristics of the sensor (Malik et al. 2013).

Table 17.1 presents biosensors developed for the detection of different pesticides, many of them using nanomaterials to improve detection characteristics.

The use of cantilever sensors associated with the miniaturization and use of specific recognition elements are presented as a promising technology with the possibility of application in the agriculture field, but still require further studies to optimize the response of these sensors.

17.3 Conclusion

The use of pesticides is of utmost importance for agriculture, increasing yields and meeting food demand, due to the prospect of increasing population. Along with the development of new pesticides, it is important to seek the development of new technologies to detect these, since minimal amounts can cause harm to human health and ecosystem.

In the search for these new technologies, nanobiosensors, which use biological elements for recognition associated with the benefits of nanotechnology, are very promising options. Among them, the optical and electromechanical sensors, developed with cantilevers used in AFM, present advantages, since they improve the sensitivity and limit of detection, being able to detect in different matrices, with fast response time and repeatability.

References

Alonso, G. A., Istamboulie, G., Noguer, T., Marty, J.-L. and Muñoz, R. 2012. Rapid determination of pesticide mixtures using disposable biosensors based on genetically modified enzymes and artificial neural networks. *Sensors and Actuators B: Chemical* 164: 22–8.

Antonacci, A., Arduini, F., Moscone, D., Palleschi, G. and Scognamiglio, V. 2018. Nanostructured (Bio)sensors for smart agriculture. *Trends in Analytical Chemistry* 98: 95–103.

Barchanska, H., Jodo, E., Price, R. G., Baranowska, I. and Abuknesha, R. 2012. Monitoringof atrazine in milk using a rapid tube-based ELISA and validation with HPLC. *Chemosphere* 87: 1330–34.

Bates, K. E. and Lu, H. 2016. Optics integrated microfluidic platforms for biomolecular analyses. *Biophysical Journal* 110: 1684–97.

Berger, R., Gerber, C. H., Lang, H. P. and Gimzewski, J. K. 1997. Micromechanics: A toolbox for femtoscale science: "Towards a laboratory on a tip". *Microelectronic Engineering* 35(1–4): 373–9.

Blair, A., Ritz, B., Wesseling, C. and Freeman, L. B. 2015. Pesticides and human health. *Occupational and Environmental Medicine* 72(2): 81–2.

Boisen, A., Dohn, S., Keller, S. S., Schmid, S. and Tenje, M. 2011. Cantilever-like micromechanical sensors. *Reports on Progress in Physics* 74(3): 1–30.

Bol'shakov, D. S. and Amelin, V. G. 2016. Determination of pesticides in environmental materials and food products by capillary electrophoresis. *Journal of Analytical Chemistry* 71(10): 965–1013.

Bonansea, R. I., Ame, M. V. and Wunderlin, D. A. 2013. Determination of priority pesticides in water samples combining SPE and SPME coupled to GC-MS. A case study: Suquía River basin (Argentina). *Chemosphere* 90(6): 1860–9.

Borsoi, A., Santos, P. R. R., Taffarel, L. E. and Gonçalves Júnior, A. C. 2014. Agrotóxicos: Histórico, atualidades e meio ambiente. *Acta Iguazu* 3(1): 86–100.

Botero-Coy, A. M., Marin, J. M., Ibanez, M., Sancho, J. V. and Hernandez, F. 2012. Multi-residue determination of pesticides in tropical fruits using liquid chromatography/tandem mass spectrometry. *Analytical and Bioanalytical Chemistry* 402(7): 2287–300.

Brattoli, M., Gennaro, G., Pinto, V., Loitile, A. D., Lovascio, S. and Penza, M. 2011. Odour detection methods: Olfactometry and chemical sensors. *Sensors* 11: 5290–322.

Butt, H.-J., Cappella, B. and Kappl, M. 2005. Force measurements with the atomic force microscope: Technique, interpretation and applications. *Surface Science Reports* 59: 1–152.

Carrascosa, L. G., Moreno, M., Álvarez, M. and Lechuga, L. M. 2006. Nanomechanical biosensors: A new sensing tool. *Trends in Analytical Chemistry* 25(3): 196–206.

Ccanccapa, A., Masiá, A., Andreu, V. and Picó, Y. 2016. Spatio-temporal patterns of pesticide residues in the Turia and Júcar Rivers (Spain). *Science of the Total Environment* 540: 200–10.

Chauhan, N. and Pundir, C. S. 2012. An amperometric acetylcholinesterase sensor based on Fe_3O_4 nanoparticle/multi-walled carbon nanotube-modified ITO-coated glass plate for the detection of pesticides. *Electrochimica Acta* 67: 79–86.

Collotta, M., Bertazzi, P. A. and Bollati, V. 2013. Epigenetics and pesticides. *Toxicology* 307: 35–41.

Crew, A. D., Lonsdale, D., Byrd, N., Pittson, R. and Hart, J. P. 2011. A screen-printed, amperometric biosensor array incorporated into a novel automated system for the simultaneous determination of organophosphate pesticides. *Biosensors and Bioelectronics* 26: 2847–51.

Cui, H.-F., Wu, W.-W., Li, M.-M., Song, X., Lv, Y and Zhang, T.-T. 2018. A highly stable acetylcholinesterase biosensor based on chitosan-TiO_2-graphene nanocomposites for detection of organophosphate pesticides. *Biosensors and Bioelectronics* 99: 223–9.

Damborský, P., Švitel, J. and Katrlík, J. 2016. Optical biosensors. *Essays in Biochemistry* 60: 91–100.

Das, P., Das, M., Madhuri, D., Chinnadayyala, S. R., Singha, I. M. and Goswami, P. 2016. Recent advances on developing 3rd generation enzyme electrode for biosensor applications. *Biosensors and Bioelectronics* 79: 386–97.

Debost-Legrand, A., Warembourg, C., Massart, C. et al. 2016. Prenatal exposure to persistent organic pollutants and organophosphate pesticides, and markers of glucose metabolism at birth. *Environmental Research* 146: 207–17.

Deda, D. K., Pereira, B. B. S., Bueno, C. C. et al. 2013. The use of functionalized AFM tips as molecular sensors in the detection of pesticides. *Materials Research* 16(3): 683–7.

Dong, J., Fan, X., Qiao, F., Ai, S. and Xin, H. 2013. A novel protocol for ultra-trace detection of pesticides: Combined electrochemical reduction of Ellman's reagent with acetylcholinesterase inhibition. *Analytica Chimica Acta* 761: 78–83.

Durán, N. and Marcato, P. D. 2013. Nanobiotechnology perspectives. Role of nanotechnology in the food industry: A review. *International Journal of Food Science and Technology* 48: 1127–34.

El-Gawad, H. A. 2016. Validation method of organochlorine pesticides residues in water using gas chromatography–quadruple mass. *Water Science* 30: 96–107.

Etayash, H., Khan, M. F., Kaur, K. and Thundat, T. 2016. Microfluidic cantilever detects bacteria and measures their susceptibility to antibiotics in small confined volumes. *Nature Communications* 7(12947): 1–9.

FAO. 2007. Coping with water scarcity: Challenge of the twenty-first century, 1–29.

Firdoz, S., Ma, F., Yue, X., Dai, Z., Kumar, A. and Jiang, B. 2010. A novel amperometric biosensor based on single walled carbon nanotubes with acetylcholine esterase for the detection of carbaryl pesticide in water. *Talanta* 83: 269–73.

Gaudin, V. 2017. Advances in biosensor development for the screening of antibiotic residues in food products of animal origin: A comprehensive review. *Biosensors and Bioelectronics* 90: 363–77.

Gupta, S., Murthy, C. N. and Ratna Prabhaa, C. 2018. Recent advances in carbon nanotube based electrochemical biosensors. *International Journal of Biological Macromolecules* 108: 687–703.

Haddaoui, M. and Raouafi, N. 2015. Chlortoluron-induced enzymatic activity inhibition in tyrosinase/ZnONPs/SPCE biosensor for the detection of ppb levels of herbicide. *Sensors and Actuators B: Chemical* 219: 171–8.

Hernández, A. F., Parrón, T., Tsatsakis, A. M., Requena, M., Alarcón, R. and López- Guarnido, O. 2013. Toxic effects of pesticide mixtures at a molecular level: Their relevance to human health. *Toxicology* 307: 136–45.

Hsu, C. C. and Whang, C. W. 2009. Microscale solid phase extraction of glyphosate and aminomethylphosphonic acid in water and guava fruit extract using alumina-coated iron oxide nanoparticles followed by capillary electrophoresis and electrochemiluminescence detection. *Journal of Chromatography A* 1216: 8575–80.

Huber, F., Lang, H. P., Zhang, J., Rimoldi, D. and Gerber, C. 2015. Nanosensors for cancer detection. *Swiss Medical Weekly* 145: 1–8.

Hulanicki, A., Glab, S. and Ingman, F. 1991. Chemical sensors definitions and classification. *Pure and Applied Chemistry* 63(9): 1247–50.

Inouye, L. A., Fernandez, L. M., Carneiro, L. F. S., Germano, J. J., and Crisci, A. R. 2014. Avaliação morfologica do fígado e do pulmão pós intoxicação por organofosforado em ratos Wistar. *Uniciências* 18(2): 103–9.

Jia, L., Su, M., Wu, X. and Sun, H. 2016. Rapid selective accelerated solvent extraction and simultaneous determination of herbicide atrazine and its metabolites in fruit by ultra high performance liquid chromatography. *Journal of Separation Science* 39: 4512–19.

Kapczyńska, K., Gamian, A., Nieradka, K., Gotszalk, T. and Rybka, J. 2012. Targeted protein immobilization on Si/Au surfaces for selective functionalization of Si/SiO_2microcantilevers with Au layer. *Procedia Engineering* 47: 1378–81.

Karasali, H., Marousopoulou, A. and Machera, K. 2016. Pesticide residue concentration in soil following conventional and low-input crop management in a mediterranean agro-ecosystem, in Central Greece. *Science of the Total Environment*, 541: 130–42.

Kesik, M., Kanik, F. E., Turan, J. et al. 2014. An acetylcholinesterase biosensor based on a conducting polymer using multiwalled carbon nanotubes for amperometric detection of organophosphorous pesticides. *Sensors and Actuators B: Chemical* 205: 39–49.

Kestwal, R. M., Bagal-Kestwal, D. and Chiang, B.-H. 2015. Fenugreek hydrogeleagarose composite entrapped gold nanoparticles for acetylcholinesterase based biosensor for carbamates detection. *Analytica Chimica Acta* 886: 143–50.

Korotkaya, E. V. 2014. Biosensors: Design, classification, and applications in the food industry. *Foods and Raw Materials* 2(2): 161–71.

Lakshmoji, K., Prabakar, K., Tripura, S. S., Jayapandian, J. and Sundar, C. S. 2014. Effect of surface stress on microcantilever resonance frequency water adsorption: Influence of microcantilevers dimensions. *Ultramicroscopy* 146: 79–82.

Lang, Q., Han, L., Hou, C., Wang, F. and Liu, A. 2016. A sensitive acetylcholinesterase biosensor based on gold nanorods modified electrode for detection of organophosphate pesticide. *Talanta* 156–157: 34–41.

Lin, B., Yan, Y., Guo, M. et al. 2018. Modification-free carbon dots as turn-on fluorescence probe for detection of organophosphorus pesticides. *Food Chemistry* 245: 1176–82.

Lin, Z., Roede, J. R., He, C., Jones, D. P. and Filipov, N. M. 2014. Short-term oral atrazine exposure alters the plasma metabolome of male C57BL/6 mice and disrupts α-linolenate, tryptophan, tyrosine and other major metabolic pathways. *Toxicology* 326: 130–41.

Madianos, L., Skotadis, E., Tsekenis, G., Patsiouras, L., Tsigkourakos, M. and Tsoukalas, D. 2018. Impedimetric nanoparticle aptasensor for selective and label free pesticide detection. *Microelectronic Engineering* 189: 39–45.

Malik, P., Katyal, V., Malik, V., Asatkar, A., Inwati, G. and Mukherjee, T. K. 2013. Nanobiosensors: Concepts and variations. *International Scholarly Research Notices* 2013: 1–9.

Martinazzo, J., Muenchen, D. K., Brezolin, A. N. et al. 2018. Cantilever nanobiosensor using tyrosinase to detect atrazine in liquid medium. *Journal of Environmental Science and Health, Part B: Pesticides, Food Contaminants, and Agricultural Wastes* 53(4): 229–36.

Miao, S. S., Wu, M. S., Ma, L. Y., He, X. J. and Yang, H. 2016. Electrochemiluminescence biosensor for determination of organophosphorous pesticides based on bimetallic Pt-Au/multi-walled carbon nanotubes modified electrode. *Talanta* 158: 142–51.

Mishra, R. K., Dominguez, R. B., Bhand, S., Munozc, R. and Martya, J.-L. 2012. A novel automated flow-based biosensor for the determination of organophosphate pesticides in milk. *Biosensors and Bioelectronics* 32: 56–61.

Mostafalou, S. and Abdollahi, M. 2013. Pesticides and human chronic diseases: Evidences, mechanisms, and perspectives. *Toxicology and Applied Pharmacology* 268(2): 157–77.

Muenchen, D. K., Martinazzo, J., de Cezaro, A. M. et al. 2016. Pesticide detection in soil using biosensors and nanobiosensors. *Biointerface Research in Applied Chemistry* 6(6): 1659–75.

Naddeo, V., Zarra, T., Giuliani, S. and Belgiorno, V. 2012. Odour impact assessment in industrial areas. *Chemical Engineering Transactions* 30: 85–90.

Nestorova, G. G., Adapa, B. S., Kopparthy, V. L. and Guilbeau, E. J. 2016. Lab-on-a-chip thermoelectric DNA biosensor for label-free detection of nucleic acid sequences. *Sensors and Actuators B: Chemical* 225: 174–80.

Okan, M. and Duman, M. 2018. Functional polymeric nanoparticle decorated microcantilever sensor for specific detection of erythromycin. *Sensors and Actuators B: Chemical* 256: 325–33.

Pichetsurnthorn, P., Vattipalli, K. and Prasad, S. 2012. Nanoporous impedemetric biosensor for detection of trace atrazine from water samples. *Biosensors and Bioelectronics* 32: 155–62.

Poddar, K., Vijayan, J., Ray, S. and Adak, T. 2018. Nanotechnology for sustainable agriculture. In: *Biotechnology for Sustainable Agriculture. Emmerging Approaches and Strategies*, eds. N. Maragioglio and K. Miller, 281–303. Cambridge: Academic Press.

Prasad, B. B., Jauhari, D. and Tiwari, M. P. 2014. Doubly imprinted polymer nanofilm-modified electrochemical sensor for ultra-trace simultaneous analysis of glyphosate and glufosinate. *Biosensors and Bioelectronics* 59: 81–8.

Raiteri, R., Grattarola, M., Butt, H.-J. and Skládal, P., 2001. Micromechanical cantilever-based biosensors. *Sensors and Actuators B: Chemical* 79(2–3): 115–26.

Rawtani, D., Khatri, N., Tyagi, S. and Pandey, G. 2018. Nanotechnology-based recent approaches for sensing and remediation of pesticides. *Journal of Environmental Management* 206: 749–62.

Rekha, K., Thakur, M. S. and Karanth, N. G. 2000. Biosensors for the detection of organophosphorous pesticides. *Critical Reviews in Biotechnology* 20: 213–35.

Rumayor, V. G., Iglesias, E. G., Galán, O. R. and Cabezas, L. G. 2005. *Aplicaciones de biosensores en la indústria agroalimentaria*. Madri: Comunidad de Madrid y la Universidad Complutense de Madrid.

Sahub, C., Tuntulani, T., Nhujak, T. and Tomapatanaget, B. 2018. Effective biosensor based on graphene quantum dots via enzymatic reaction for directly photoluminescence detection oforganophosphate pesticide. *Sensors and Actuators B: Chemical* 258: 88–97.

Saitta, M., Di Bella, G., Fede, M. R. et al. 2017. Gas chromatography-tandem mass spectrometry multiresidual analysis of contaminants in Italian honey samples. *Food Additives and Contaminants: Part A* 34: 800–8.

Samsidar, A., Siddiquee, S. and Shaarani, S. M. 2018. A review of extraction, analytical and advanced methods for determination of pesticides in environment and foodstuffs. *Trends in Food Science and Technology* 71: 188–201.

Sang, S., Zhang, W. and Zhao, Y. 2013. Review on the design art of biosensors. In: *State of the Art in Biosensors: General Aspects*, ed. T. Rinken, 89–110. Croácia: InTech.

Scognamiglio, V. 2013. Nanotechnology in glucose monitoring: Advances and challenges in the last 10 years. *Biosensors and Bioelectronics* 47: 12–25.

Sekhon, B. S. 2014. Nanotechnology in agri-food production: An overview. *Nanotechnology, Science and Applications* 7(2): 31–53.

Shi, H., Zhao, G., Liu, M., Fan, L. and Cao, T. 2013. Aptamer-based colorimetric sensing of acetamiprid in soil samples: Sensitivity, selectivity and mechanism. *Journal of Hazardous Materials* 260: 754–61.

Skládal, P. 2016. Piezoelectric biosensors. *Trends in Analytical Chemistry* 79: 127–33.

Song, D., Li, Y., Lu, X. et al. 2017. Palladium-copper nanowires-based biosensor for the ultrasensitive detection of organophosphate pesticides. *Analytica Chimica Acta* 982: 168–75.

Song, Y., Chen, J., Sun, M. et al. 2016. A simple electrochemical biosensor based on AuNPs/MPS/Au electrode sensing layer for monitoring carbamate pesticides in real samples. *Journal of Hazardous Materials* 304: 103–9.

Songa, E. A. and Okonkwo, J. O. 2016. Recent approaches to improving selectivity and sensitivity of enzyme-based biosensors for organophosphorus pesticides: A review. *Talanta* 155: 289–304.

Steffens, C., Leite, F. L., Bueno, C. C., Manzoli, A. and Herrmann, P. S. P. 2012. Atomic force microscopy as a tolls applied to nano/biosensors. *Sensors* 12: 8278–300.

Steffens, C., Steffens, J., Graboski, A. M., Manzoli, A. and Leite, F. L. 2017. Nanosensors for detection of pesticides in water. In: *New Pesticides and Soil Sensors*, ed. A. M. Grumezescu, 595–635. Cambridge: Academic Press.

Sun, J., Pan, L., Zhan, Y. et al. 2016. Contamination of phthalate esters, organochlorine pesticides and polybrominated diphenyl ethers in agricultural soils from the Yangtze River Delta of China. *Science of the Total Environment* 544: 670–76.

Tamayo, J., Kosaka, P. M., Ruz, J. J., San Paulo, A. and Calleja, M. 2013. Biosensors based on nanomechanical systems. *Chemical Society Reviews* 42: 1287–311.

Tharmavaram, M., Rawtani, D. and Pandey, G. 2017. Fabrication routes for one-dimensional nanostructures via block copolymers. *Nano Convergence* 4: 12–24.

Thévenot, D. R., Toth, K., Durst, R. A. and Wilson, G. S. 1996. International union of pure and applied chemistry physical chemistry division, steering committee on biophysical chemistry analytical chemistry division, commission V.5 (electroanalytical chemistry) electrochemical biosensors: Proposed definitions and classification synopsis of the report. *Sensors and Actuators B: Chemical* 30(1): 1.

Thévenot, D. R., Toth, K., Durst, R. A. and Wilson, G. S. 2001. Electrochemical biosensors: Recommended definitions and classification. *Biosensors and Bioelectronics* 16(1–2): 121–31.

Tortolini, C., Bollella, P., Antiochia, R., Favero, G. and Mazzei, F. 2016. Inhibition-based biosensor for atrazine detection. *Sensors and Actuators B: Chemical* 224: 552–8.

Turan, J., Kesik, M., Soylemez, S. et al. 2016. An effective surface design based on a conjugated polymer and silver nanowires for the detection of paraoxon in tap water and milk. *Sensors and Actuators B: Chemical* 228: 278–86.

Vakonaki, E., Androutsopoulos, V. P., Liesivuori, J., Tsatsakis, A. M. and Spandidos, D. A. 2013. Pesticides and oncogenic modulation. *Toxicology* 307: 42–5.

Vashist, S. K. 2007. A review of microcantilevers for sensing applications. *AZojono Journal of Nanotechnology Online* 3: 1–15.

Vogel, A., Jocque, H., Sirot, L. K. and Fiumera, A. C. 2015. Effects of atrazine exposure on male reproductive performance in *Drosophila melanogaster*. *Journal of Insect Physiology* 72: 14–21.

Waichman, A. V. 2012. A problemática do uso de agrotóxicos no Brasil: A necessidade de construção de uma visão compartilhada por todos os atores sociais. *Revista Brasileira de Saúde Ocupacional* 37(125): 17–50.

Wang, X., Lu, X. and Chen, J. 2014. Development of biosensor technologies for analysis of environmental contaminants. *Trends in Environmental Analytical Chemistry* 2: 25–32.

Wilson, A. and Baietto, M. 2009. Applications and advances in electronic-nose technologies. *Sensors* 9: 5099–148.

Xia, X.-R. and Leidy, B. 2002. Simplified liquid-solid extraction technique for the analyses of pesticide residues in soil samples. *Environmental Monitoring and Assessment* 73: 179–90.

Xing, H., Wang, Z., Wu, H. et al. 2015. Assessment of pesticide residues and gene expression in common carp exposed to atrazine and chlorpyrifos: Health risk assessments. *Ecotoxicology and Environmental Safety* 113: 491–8.

Yu, G., Wu, W., Zhao, Q., Wei, X. and Lu, Q. 2015. Efficient immobilization of acetylcholinesterase onto amino functionalized carbon nanotubes for the fabrication of high sensitive organophosphorus pesticides biosensors. *Biosensors and Bioelectronics* 68: 288–94.

Yusa, V., Coscolla, C. and Millet, M. 2014. New screening approach for risk assessment of pesticides in ambient air. *Atmospheric Environment* 96: 322–30.

Zacharia, J. T. 2011. Identity, physical and chemical properties of pesticides. In: *Pesticides in the Modern World: Trends in Pesticides Analysis*, ed. M. Stoytcheva, 1–19. Croácia: InTech.

Zavala, G. 2007. Atomic force microscopy, a tool for characterization, synthesis and chemical processes. *Colloid and Polymer Science* 286: 85–95.

Zhang, Y., Arugula, M. A., Wales, M., Wild, J. and Simonian, A. L. 2015. A novel layer-by-layer assembled multi-enzyme/CNT biosensor for discriminative detection between organophosphorus and non-organophosphrus pesticides. *Biosensors and Bioelectronics* 67: 287–95.

Zhao, Z. and Jiang, H. 2010. Enzyme-based electrochemical biosensors. In: *Biosensors*, ed. P. A. Serra, 1–23. Croácia: InTech.

Zhou, L., Zhang, X., Ma, L., Gao, J. and Jiang, Y. 2017. Acetylcholinesterase/chitosan-transition metal carbides nanocomposites-based biosensor for the organophosphate pesticides detection. *Biochemical Engineering Journal* 128: 243–9.

18

Zinc-Based Nanomaterials: From Synthesis and Characterization to Environmental Applications

Jimei Qi, Jiwei Hu, Yu Hou, and Yiqiu Xiang
Guizhou Normal University

Xionghui Wei
Peking University

18.1 Introduction

The rapid industrialization associated with the population explosion and urbanization has led to serious environmental pollution [1]. Currently, pollutants of concern include dyes, heavy metals, toxic gases, nutrients, hydrocarbons and persistent organic pollutants (POPs) [2–6]. Dyes have been extensively used in textile, leather tanning, cosmetics, pigment and many other industries [7–11].

Dyes can cause many health problems, because of their toxic, mutagenic and carcinogenic nature [12,13]. Heavy metals pollution does not only affects the production and quality of crops, but also influences the air quality and water bodies, and threatens the health and life of animals and humans through the food chain. The best known toxic gases are carbon monoxide (CO), methane (CH_4), nitrogen dioxide (NO_2) and phosgene, etc. [14]. In confined environments, high concentrations of these toxic gases can create serious harms for the human health and should be continuously monitored and controlled [15]. Most of the nutrients were from publicly owned treatment works (POTWs) and combined sewer overflows (CSOs). They can cause water quality degradation; hence, the biodiversity was reduced [16]. These pollutants cause a serious hazard to human health and ecological environment. Therefore, it is an important problem for people to effectively eliminate pollutants prior to their final discharge to the environment.

Faced with the problem of environmental pollution, a broad range of nanomaterials have been used for the treatment of pollutants. The nanomaterials have wide attention and studies because they possess small particle size, large specific surface area, high hardness and unique properties.

Wide-bandgap semiconductors, such as TiO_2 (3.2 eV), GaN (3.5 eV) and ZnS (3.67 eV), have come to the forefront in the past decades because of an increasing need for short-wavelength photonic devices and high-power, high-frequency electronic devices, and the wide-bandgap semiconductors [17]. The nanoscale zerovalent zinc is a wide-gap semiconductor with a high photocatalysis activity, which has been widely used in the photodegradation of organic pollutants in wastewater [18,19]. The semiconductor zinc oxide (ZnO) has gained a substantial interest in the research community in part because this material is classified as a semiconductor in groups II–VI and an important direct wide-bandgap semiconductor. ZnO has an energy bandgap of 3.37 eV and the excitation binding energy of 60 meV, which is a promising semiconductor material [20]. Zinc-based nanomaterials show special electrical, magnetic, optical and chemical properties that are different from those of zinc-based materials. The zinc-based nanomaterials are an excellent ultraviolet absorber, and it is expected to develop various light-emitting devices such as blue light, blue-green light and ultraviolet light, which are widely used in solar cells, liquid crystal displays, gas sensors, piezoelectric devices and catalysis [21–24] (Figure 18.1 and Table 18.1).

Response surface methodology (RSM) is a collection of mathematical and statistical techniques based on the fit of a polynomial equation to the experimental data, which can describe the behavior of a data set with the objective of making statistical provisions [25–27]. Artificial intelligence technologies (e.g., artificial neural networks, particle swarm optimization and genetic algorithm) have gained a tremendous advance in recent years, which are widely used in big data, autonomous driving, robots, intelligent search, computer, medicine, zero sales and fingerprint identification [28]. The RSM and artificial intelligence (AI) technologies were applied to find the optimum process variables and maximum removal efficiency of pollutants.

Therefore, the RSM and AI technologies were used for facilitating the applications of zinc-based nanomaterials in environmental monitoring and remediation. In this chapter, we have presented a summary on preparation, characterization and properties of zinc-based nanomaterials, and their applications in environmental monitoring and remediation.

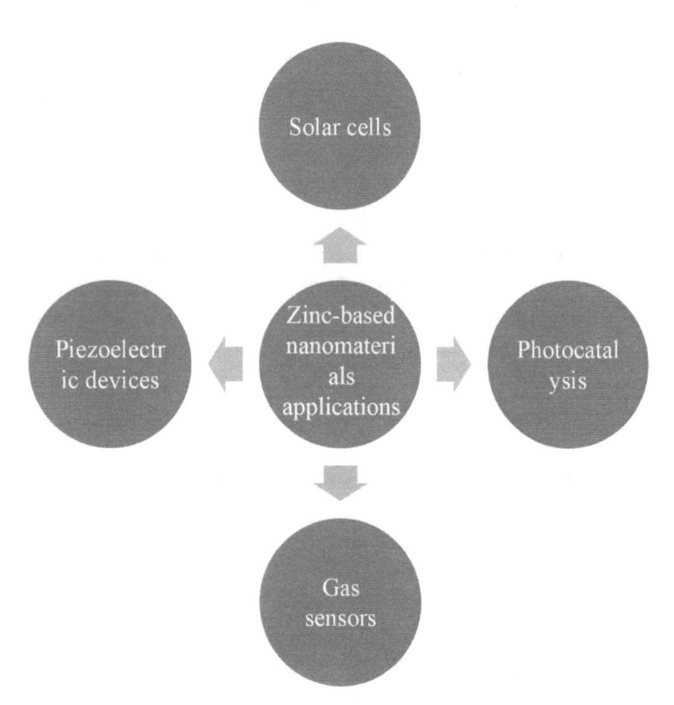

FIGURE 18.1 Industrial applications of zinc-based nanomaterials.

TABLE 18.1 Properties of Zinc Oxide

Zinc Oxide Properties		
1	Molecular weight	81.38 g/mol
2	Appearance	White solid
3	Odor	Odorless
4	Band gap	3.37 eV
5	The exciton binding energy	60 meV
6	Crystal structure	Wurtzite
7	Density	5.606 g/cm^3
8	Refractive index μ_d	2.0041

18.2 Nano-zinc Oxide

18.2.1 Zinc Oxide Nanomaterials

In general, zinc oxide crystals have two main forms: hexagonal wurtzite and cubic blende. The wurtzite is the stabler form, thus mostly observed under ambient conditions. The zinc blende form can be stabilized by growing ZnO on substrates with cubic lattice structure [29]. In the both forms, the zinc and oxide centers are tetrahedral, which is the most characteristic structural geometry for zinc. The hexagonal wurtzite and cubic zinc blende have a point group 6 mm in Hermann–Mauguin notation or C6vin Schoenflies notation, and the reported space group is generally P63mc or C6v4. However, this abnormality of ZnO is not completely defined.

18.2.2 Preparation of Nano-zinc Oxide

The nano-zinc oxide occurs in a rich variety of structures and possesses a wide range of properties. Hitherto, there are many methods for the synthesis of nano-zinc oxide, such as chemical vapor deposition, precipitation form microemulsions, mechanochemical processes, sol–gel method and hydrothermal method. Different processing methods can synthesize the nano-zinc oxide with particles differing in shape, size and structure.

Chemical Vapor Deposition Method Synthesis of Nano-ZnO

ZnO nanorods were grown in a two-temperature-zone furnace. Zinc acetylacetonate hydrate ($Zn(C_5H_7O_2)_2 \cdot H_2O$), which was used to grow ZnO whiskers and films, was vaporized at 130°C ± 140°C in a furnace. The vapor was carried by a N_2/O_2 flow into the higher temperature zone of the furnace in which substrates were located [30].

Sol–Gel Method Synthesis of Nano-ZnO

$Zn(NO_3)_2 \cdot 6H_2O$ and $C_6H_8O_7$ were selected as starting materials. Polyethylene glycol 2,000, ethylene glycol and polyvinyl alcohol (PVA, average polymerization degree 1,750 ± 50) were used as surfactants [31]. Zhang et al. reported that $Zn(NO_3)_2 \cdot 6H_2O$, citric acid and surfactants were dissolved in distilled water under vigorous magnetic stirring at room temperature. The pH of the formed solution was further adjusted from 1 to 7 using $NH_3 \cdot H_2O$. The solution was then immersed in a 353 K water bath for 4 h and aged at room temperature for another 4 h. The solution was then gelated at 358 K under magnetic stirring and was further dried at 383 K in vacuum. The as-obtained ZnO nanoparticles were annealed for 4 h in air between

773 and 1,473 K. The results indicated ZnO nanoparticles with a pure wurtzite structure were obtained after calcination at 773 K.

Hydrothermal Method Synthesis of Nano-ZnO

Analytical grade zinc acetate was adopted as the source material for zinc species. Zinc acetate was dissolved in deionized water for preparing zinc cation solution. Deionized water and potassium hydroxide (0.025–0.20 mol/L) were used as the solvents. In a typical procedure, the solvent (26 mL) was added to a Teflon-lined autoclave of 40 ml capacity filled with $Zn(CH_3COO)_2$ solution (1.0 mol/L, 6.5 mL) while stirring vigorously. The autoclave was maintained at 200°C for 2 h and then air-cooled to room temperature. The precipitate was dispersed and washed with deionized water. The white products were dried in air at 120°C for 30 min [32].

18.2.3 Properties of Nano-ZnO

Surface Effect of Nano-ZnO

The surface effect refers to a change in properties caused by a sharp increase in the ratio of the number of atoms on the surface of the nanoparticles to the total number of atoms as the particle size becomes smaller [33]. Since the particle size decreases significantly, the surface area, surface energy and surface bond of nanoparticles increase, and they have unsaturated properties, thus showing high chemical activity.

Small Size Effects of Nano-ZnO

The small size effect means that the size of the nanoparticle is equal to or smaller than that of light wave wavelength, De Broglie wavelength and the coherence length of superconducting state, and the periodic boundary conditions will be destroyed. Many novel phenomena would occur in melting point, magnetic properties, light absorption, thermal resistance and chemical activity [34]. The catalytic phase has a great change compared to ordinary particles, and this effect has opened a vast new field for its applications.

Catalytic Properties of Nano-ZnO

The nanoparticle crystal has small size, large specific surface area, large surface activity and high surface activity. This leads to its much higher catalytic activity and selectivity than those of conventional catalysts. The nanocatalyst has no pores which can avoid the effects of diffusion in the reactant phase pores caused by the use of conventional catalysts [35]. The nanomaterials can directly enter the liquid phase reaction system without relying on inert carriers.

Other Properties of Nano-ZnO

In addition to the aforementioned characteristics, the nano-ZnO also has quantum size effect, macroscopic quantum tunneling effect, chemical reaction property, optical property, hardness, strong plasticity, high specific heat, thermal expansion, low sintering temperature, large sintering shrinkage ratio, high electrical conductivity and diffusion [36,37]. These properties open a broad prospect for their applications (Figure 18.2).

18.2.4 Characterization of Nano-ZnO

Thermogravimetric-Differential Thermal (TG-DTA) Characterization of Nano-ZnO

The precursor obtained by the precipitation reaction was subjected to TG-DTA analysis to understand the thermal decomposition process and determine the minimum temperature required. Liu et al. reported that the condition of thermal decomposition of zinc carbonate at 350°C and 3.0 h can get nano-ZnO particles with smaller size and more uniform distribution.

X-Ray Diffraction (XRD) Characterization of Nano-ZnO

The XRD analysis was used to determine the phase structure and purity of the as-synthesized ZnO powder. Akhoon et al. indicated that the XRD data analysis clearly reveals the crystalline nature of ZnO nanopencils. The diffraction

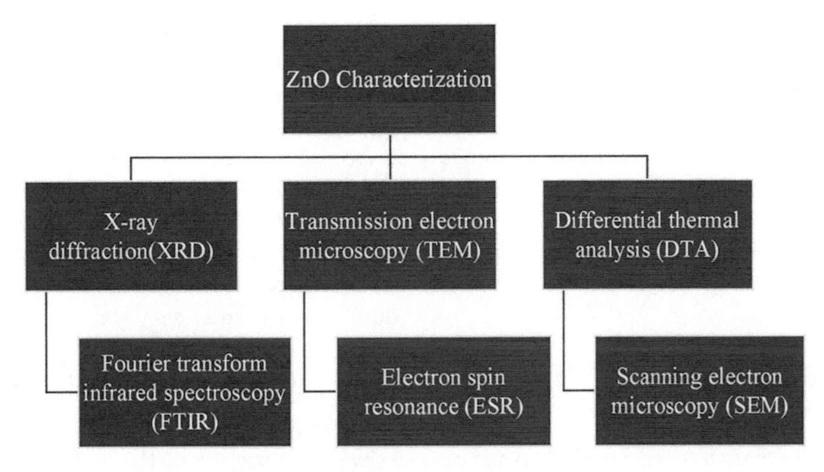

FIGURE 18.2 Nano-ZnO characterized by using different methods.

peaks and relative intensity matched well with hexagonal wurtzite phase without having any additional peaks, except for the peaks characteristic of ZnO [38]. The ZnO characteristic peaks at $2\theta = 31.77°, 34.47°, 36.23°, 47.53°, 56.62°, 62.90°, 66.40°$ and $67.90°$ were assigned to (100), (002), (101), (102), (110), (103), (200), (112), (201) and (004) reflection planes of hexagonal wurtzite ZnO, having lattice constants of $a = b = 0.3249$ nm, $c = 0.5206$ nm.

Transmission Electron Microscopy (TEM) Characterization of Nano-ZnO

The surface morphology of the sample was observed by TEM, and the primary particle diameter was measured [39]. TEM is an electron optical instrument with high-resolution power and high magnification using an electron beam. It also can directly observe the morphology and structure of zinc-based nanomaterials and get intuitive information.

Fourier Transform Infrared Spectroscopy (FTIR) Characterization of Nano-ZnO

Xiong et al. studied the FTIR spectra of unannealed ZnO samples; a series of absorption peaks from 1,000 to 4,000 cm^{-1} can be found, corresponding to the carboxylate and hydroxyl impurities in nanomaterials [40]. To be more specific, a broadband at 3,500 cm^{-1} is assigned to the O–H stretching mode of hydroxyl group. Peaks between 2,830 and 3,000 cm^{-1} are due to C–H stretching vibration of alkane groups. The peaks observed at 1,630 and 1,384 cm^{-1} are due to the asymmetrical and symmetrical stretching of the zinc carboxylate, respectively. As the size of the nanoparticles increases, the content of the carboxylate (COO–) and hydroxyl (–OH) groups in the samples decreased. The carboxylate probably comes from reactive carbon containing plasma species during synthesis and the hydroxyl results from the hygroscopic nature of ZnO. This suggests that these FTIR-identified impurities mainly exist near ZnO surfaces.

Electron Spin Resonance Spectroscopy (ESR) Characterization of Nano-ZnO

To obtain the concentration of paramagnetic centers, the ESR spectra were compared with a silicon spin standard. All ESR measurements were performed in the dark and at temperature of $T = 5$ K. Gluba et al. showed two ESR spectra of an as-grown and a post-hydrogenated sample. Both spectra exhibit a striking feature at $g = 1.957 \pm 0.001$ which is intensified by hydrogenation by a factor of six. Furthermore, the introduction of hydrogen triggers the observation of a variety of new lines over a large magnetic field scale [41].

Scanning Electron Microscopy (SEM) Characterization of Nano-ZnO

The morphology of ZnO is characterized by SEM. Amizam et al. summarized that the structures of ZnO films become dense and continuous when higher annealing temperatures and the resistivity of ZnO thin films decreased with the higher density of the ZnO structure [42].

18.2.5 Surface Modification Technology of Nano-ZnO

The large surface area and high specific surface energy responsible for the beneficial effects of zinc oxide nanomaterials cause agglomeration of particles. However, nano-zinc oxide has strong surface area and is not easy to be uniformly dispersed in organic medium, which greatly limits its nano-effect. Therefore, the dispersion and surface modification of the zinc oxide nanomaterials become the necessary treatment means for the nanomaterial before applications in the matrix. The so-called nanodispersion refers to the use of various principles, methods and means in a specific liquid medium (such as water), and the various forms of agglomerates composed of dried nanoparticles are reduced to primary particles and stabilized and evenly distributed in the medium [43]. The surface modification of nanopowder is based on the expansion and extension of nanodispersion technology. According to the needs of the applications, the surface of the dispersed nanoparticles is coated with a film of appropriate substance, or the nanoparticles are dispersed in a soluble solid-phase carrier. The surface-modified nanopowder has a series of surface properties such as adsorption, wetting and dispersion. This can be automatically or easily dispersed in a specific medium; therefore, it is very convenient to use. In general, there are three methods for modifying nanoparticles: (i) uniformly coating a film of other substances on the surface of the particles, thereby changing the surface properties of the particles. (ii) The surface of the nanoparticles is chemically adsorbed or reacted by using a charge transfer complex (such as a coupling agent, silane or titanate, stearic acid or silicone) as a surface modifier. (iii) The surface of the nanoparticles is modified by high energy means such as corona discharge, ultraviolet light, plasma and radiation.

Surface Coating Modification of Nano-ZnO

Surface coating is the main modification method of inorganic fillers or pigments. The surfactant is applied to the surface of the particles to impart new properties to the surface of the particles. Common surface modifiers include silane coupling agents, titanate coupling agents, stearic acid and silicones. In addition to utilizing surface functional groups, this method also involves surface modification using free radical reactions, chelation reactions, sol adsorption and coupling agent treatment [44].

Mechanochemical Modification of Nano-ZnO

Mechanochemical modification uses mechanical stress to activate the surface of the particle by means of pulverization, friction, etc. to change its structure [45]. This method causes the molecular lattice to be displaced, and the internal energy is increased. Under the action of external force, the surface of the active powder reacts with other substances to adhere to the surface modification.

Outer Membrane Modification of Nano-ZnO

Outer membrane modification is uniformly coating a layer of other substances on the surface of the particles to change the surface properties of the particles, and modifying them by precipitation reaction using an organic or inorganic substance to deposit a coating on the surface of the particles to change its surface properties.

18.3 Zinc Selenide (ZnSe) Nanomaterials

ZnSe is an important wide-bandgap (2.8 eV) material with a broad range of potential applications [46]. This exhibits an interesting photophysical phenomenon involving green-blue light emission, high transmission at the infrared and the visible spectrum. Nanoparticles of ZnSe, via quantum confinement, shape, size and surface dependent effects, hold a promise for tuning the optical properties and for assembling the particles in nanoscale structures [47].

18.3.1 Synthesis of ZnSe Nanomaterials

Hydrothermal Method Synthesis of Nano-ZnSe

Liu et al. found that in hydrothermal synthesis of ZnSe hollow nanospheres, $Zn(NO_3)_2{\cdot}6H_2O$ and Se-containing ionic liquid 1-*n*-butyl-3-methylimidazolium methylselenite ([BMIm][$SeO_2(OCH_3)$]) were dissolved into deionized water under vigorous magnetic stirring to form a clear solution, followed by the addition of $N_2H_4{\cdot}3H_2O$ (80%) [48]. Then, the mixture was transferred into a Teflon-lined stainless steel autoclave and maintained at 150°C for 12 h. The resulted yellow powders were centrifuged, washed with deionized water and anhydrous ethanol several times, and finally dried at 60°C for 4 h under vacuum.

Sol–Gel Method Synthesis of Nano-ZnSe

The sol–gel method is simple and an inexpensive. Hutagalung et al. indicated that the sol–gel method was found to play a key role in obtaining nano-sized ZnSe [49]. The sol of SiO_2 was at first prepared from tetraethylorthosilicate (TEOS) mixed with deionized water under ammonium hydroxide catalyst. Then, acetic acid was added into the solution to adjust pH of the solution. After that, zinc acetate dihydrate and selenic acid were added to the solution as zinc and selenium sources, respectively. The thin films were deposited on the glass substrate using dip coating method at room temperature, followed by being annealed under argon atmosphere at 500°C for 1 h to obtain the ZnSe nanoparticles/SiO_2 thin-film composites.

Solvothermal Method Synthesis of Nano-ZnSe

Solvothermal processing allows many inorganic nanomaterials to be prepared at temperatures substantially below those required by traditional solid-state reactions. Solvothermal synthesis in mixed solvents, which were composed of two or more components (including water acting as one of the components), in some cases, was much more effective than that in a single solvent. Du et al. reported that the Se powder was added to a conical flask containing 30 mL of hydrazine hydrate under stirring. After it was stirred for 30 min, anhydrous $ZnCl_2$ was added into the solution under continuous stirring, and the precipitate was formed. Then, the mixture was transferred into a 60 mL Teflon-lined stainless autoclave with ethylenediamine. The autoclave was sealed and maintained at 120°C for 10 h, and then, it was cooled to room temperature naturally. The resulted orange solid product was collected by filtration, washed with large amounts of absolute ethanol and distilled water, and finally dried in a vacuum box at 60°C for 4 h. The decomposition of the molecular precursor to ZnSe was conducted by carefully heating up to 300°C in rate of 30°C/h and maintaining for 0.5 h in N_2 in a horizontal furnace [50].

18.3.2 Characterization of ZnSe Nanomaterials

The synthesized ZnSe nanomaterials were characterized by XRD using the Cu Ka radiation, TEM, SEM, atomic force microscope (AFM), micro-Raman scattering (MRS) and photoluminescence (PL) spectroscopy.

XRD Characterization of Nano-ZnSe

Zeng et al. found that the ZnSe diffraction peaks (111), (200), (220), (311) and (222) are consistent with the standard values in Joint Committee on Powder Diffraction Standards (JCPDS) File no.37-1463. Very narrow and sharp diffraction peaks indicate a good crystallinity for the ZnSe samples [51].

TEM Characterization of Nano-ZnSe

Panda et al. reported an average diameter of 1.3 ± 0.05 nm and average length of 4.5 ± 0.05 nm for the ZnSe nanorods by examining the TEM image. The nanorods self-assemble into large regions of highly ordered 2D supercrystals. The width and length of nanorods can also reach 200 nm at longer reaction times. The wires are also self-assembled into strictly parallel conductor regions nearly with micrometer in size. The transverse distance of the wire rod is the same as that of the rod [47].

MRS Characterization of Nano-ZnSe

The excitation laser was focused on different locations to select either the nanowires or nanoribbons on the silicon substrate via an optical microscope. It was found that different regions containing primarily nanoribbons or nanowires had the same Raman spectrum. Yang et al. investigated a typical room-temperature Raman spectrum of the ZnSe nanostructures. The Raman peaks at 203 and 250 cm^{-1} are attributed to the transverse optic (TO) and longitudinal optic (LO) phonon modes of ZnSe, respectively. For ZnSe polycrystalline nanoparticles, the TO and LO phonon frequencies are 210 and 255 cm^{-1}, respectively, and both give a broad Raman peak due to the high surface-to-volume ratio of the small particles. Compared to these results, the LO and TO phonon peaks of the ZnSe nanostructures are both shifted toward lower frequency, which is probably due to the effects of the small size and high surface area [46].

AFM and PL Spectroscopy Characterization of Nano-ZnSe

The shape of ZnSe was confirmed by PL and AFM. Kishino et al. examined the structural and optical characterization for the lapped and polished (100) surface of ZnSe bulk crystals. They investigated the structural and photoluminescence characterization of mirror-polished ZnSe bulk crystals using AFM and low-temperature PL measurements; 8/1 μm diamond particle can be used to control the flatness of 7.64 nm and the surface roughness of 0.8 nm, corresponding to the surface actuation of a few atomic layers [52].

18.4 Zinc Sulfide (ZnS) Nanomaterials

ZnS is one of the first semiconductors discovered, and it has shown remarkable versatility of fundamental properties and a promise for novel diverse applications, including light-emitting diodes (LEDs), electroluminescence, flat panel displays, infrared windows, sensors, lasers and biodevices [53,54]. The structure and chemical properties of ZnS are comparable to more popular and widely known ZnO. However, certain properties pertaining to ZnS are unique and advantageous compared to ZnO.

For example, ZnS has a larger bandgap of 3.72 and 3.77 eV for cubic zinc blende (ZB) and hexagonal wurtzite (WZ) ZnS, respectively than ZnO (3.4 eV). Therefore, it is more suitable for visible-blind ultraviolet (UV)-light-based devices such as sensors/photodetectors [55]. ZnS is traditionally the most suitable candidate for electroluminescence devices [56]. However, the nanostructures of ZnS have not been investigated in detail relatively to ZnO nanostructures.

18.4.1 ZnS Nanomaterials

It is well known that ZnS has two structures: cubic ZB and hexagonal wurtzite. The former is stabler than the latter. Yeh et al. reported that the former is the stable low-temperature phase, and the latter is the high-temperature polymorph that can be formed at temperature higher than 1,296 K. The wurtzite phase has a higher bandgap of 3.77 eV than the ZB structure of 3.72 eV [57].

18.4.2 Synthesis of ZnS Nanomaterials

Sol–Gel Synthesis of Nano-ZnS

Sol–gel method is a widely used wet method to prepare nanomaterials. Recently, silica-modified ZnS nanoparticles were prepared by sol–gel method. Li et al. synthesized the ZnS nanomaterial by sol–gel method. $Zn(CH_3COO)_2$ and tetraethylorthosilicate (TEOS) were used as Zn and Si sources, respectively. First, TEOS was added to HCl solution and stirred to form a clear sol at 50°C. Second, $Zn(CH_3COO)_2$ and ethylene diamine tetraacetic acid (EDTA) were added to the above sol under stirring. $NH_3 \cdot H_2O$ was added dropwise to the above solution until the pH of solution was 6, and a transparent solution was formed. Third, Na_2S was added to the above solution. Finally, the hydrothermal product was filtrated and dried, and the silica-modified ZnS nanomaterial was obtained [58].

Hydrothermal Method Synthesis of Nano-ZnS

The hydrothermal processing represents an alternative to the calcination for the crystallization of ZnS at mild temperatures. In the hydrothermal treatment, grain size, particle morphology, crystalline phase and surface chemistry can be controlled via processing variables. Salavati-Niasari et al. prepared the ZnS nanoparticles by the hydrothermal method. $Zn(CH_3COO)_2 \cdot 2H_2O$ powder was dissolved in distilled water and thioglycolic acid (TGA) were mixed slowly under stirring. After stirring, the reactants were put into a Teflon-lined autoclave of 800 mL capacity. The autoclave was maintained at 80°C–140°C for 5–22 h and then cooled down to room temperature naturally. The reactants were centrifuged, washed with alcohol and distilled water for several times, and dried in the air [59].

18.4.3 Characterization of ZnS Nanomaterials

UV-vis Diffuse Reflectance Spectroscopy Characterization of Nano-ZnS

Zhang et al. analyzed ZnS nanoparticles by the UV-vis diffuse reflectance spectra using the Shimadzu UV-3101 equipped with an integrating sphere, using $BaSO_4$ as the reference. The results indicated that the two as-prepared ZnS samples show strong absorption, and both of them show an onset of absorption at ≈350 nm, which is shifted about 60 nm to a shorter wavelength as compared to that of large ZnS particles [60].

XRD Characterization of Nano-ZnS

Crystalline phase structure of ZnS was analyzed by the XRD. Furthermore, the crystalline grain size of ZnS was estimated according to Scherrer formula $L = K\lambda/(\beta\cdot\cos\theta)$, where λ is the wavelength of the X-ray radiation, K is a constant taken as 0.89, β is the line width at half maximum height and θ is the diffracting angle. Salavati-Niasari et al. observed the ZnS diffraction peaks corresponding to wurtzite (JCPDS card no. 36-1450) and no characteristic diffraction peaks from the possible impurity phases (such as TGA, S, ZnO). This shows that the phase-pure wurtzite ZnS sample could be prepared by the hydrothermal method [59].

N_2 Absorption Characterization of Nano-ZnS

Brunner–Emmet–Teller (BET) and Barrett–Joyner–Halenda (BJH) analyses were used to determine the total specific surface area (SBET) and the pore size distribution. N_2 adsorption–desorption isotherms were obtained on a Micromeritics TriStar 3,000 at 77 K under continuous adsorption conditions. Zhang et al. showed that the ZnS was mainly formed and retained in the channels of the MCM-41 host, and its growth was controlled by the channels. In contrast, the amount of ZnS on the external surface is much smaller [60].

18.5 Applications of Nanomaterials in Environmental Monitoring Remediation

Given the increasing production of nanomaterials of all types, they are not only in the forefront of the hottest fundamental research at present but also gradually intruded into our daily life [61]. As a "green" technology, semiconductor photocatalysis has been widely utilized for the treatment of water polluted by organic dyes and heavy metals [62]. Zinc-based nanomaterials have obtained an extensive attention due to the wide-bandgap, large surface area, good economic benefits and high yielding.

These state-of-the-art nanomaterials have been shown to efficiently remove heavy metals, and volatile hazardous substances, photodegrade persistent organic pollutants and other compounds, and inactivate bacteria. Such properties have enabled the use of these nanomaterials for wastewater treatment, soil remediation, air purification and substance monitoring.

The use of advanced oxidation processes (AOPs) for the effective oxidation of a wide variety of organics and dyes has gained special attention in recent years, due to the generation of the highly reactive hydroxyl radicals. AOPs include UV/H_2O_2, Fenton and photo-Fenton processes (Fe^{2+}/H_2O_2 and $Fe^{2+}/H_2O_2/UV$), the ultrasonic process, photocatalysis and electrochemical processes. Among them, photocatalytic degradation efficiency is high, and energy consumption is low. There are many types of photocatalysts such as titanium dioxide, zinc oxide, perovskites and titanates. Zinc-based nanomaterials have attracted a significant attention because its photodegradation mechanism is similar to that of nano-titanium dioxide.

18.5.1 Environmental Monitoring

Gas Sensors

In the recent years, there is a strong interest in the development of lightweight gas sensors capable of parts per million (ppm) range sensitivity and low-power levels operations. All experimental results demonstrate that zinc-based nanomaterials have a large surface area and the potential for detecting NO_2, NH_3, Cu^{2+}, Ni^{2+}, etc.

Mokhtar et al. revealed that ZnO nanoparticles annealed at high temperature (700°C) were found sensitive to CO, while they displayed negligible response to NO_2. The opposite behavior has been registered for the one-dimensional ZnO nanofibers annealed at medium temperature (400°C) [63].

Zhang et al. investigated ZnSe nanoparticles, which were modified with mercaptoacetic acid (MAA). This worked as novel fluorescence sensors for the quantitative determination of copper(II) and nickel(II). The results indicated that the tolerance limits of Zn^{2+}, K^+, Na^+, Mg^{2+} and Ca^{2+} were more than 10^{-2} g/L. Consequently, this method is suitable for the analysis of Cu^{2+} and Ni^{2+} [64].

18.5.2 Environmental Remediation

Dyes Removal

Dyes, such as sunset yellow, malachite green, rhodamine B and reactive red, are generally toxic and environmentally destructive even at very low concentrations, and were found as pollutants in groundwater and wastewater. The unique features of ZnO nanomaterials have given them the ability to effectively adsorb or photodegrade large quantities of these pollutants (Table 18.2).

Heavy Metals Removal

Toxic heavy metals of particular concern, in treatment of industrial wastewater, include lead, chromium, cadmium, mercury, arsenic, nickel and copper. A number of nanomaterials have been used for the removal of heavy metals from water. The zinc-based nanomaterials have the ability to effective removal of the toxic heavy metals (Table 18.3).

Solar Cells

Solar cells represent a very promising renewable energy technology because they provide reducing our dependence on fossil oil and our impact on the environment [85]. The zinc-based nanomaterials can synthesize the dye-sensitized solar cells, which are very efficient, inexpensive, large-scale solar energy conversion.

TABLE 18.2 Dyes Removal by Different Nanomaterials

Materials	Preparation/Modification	Adsorbate	Efficiency	References
γ-Fe_2O_3 nanoparticles	Co-precipitation method	Coomassie brilliant blue R-250	98%	[65]
CTF/Fe_2O_3 composites	Microwave synthesis	Methyl orange	90%	[66]
G-Fe_2O_3/ZnO nanocomposites	Hydrothermal method	Methylene blue	99.5%	[67]
Fe_3O_4@SDBS @LDHs composites	Hydrothermal method	Brilliant green (BG)	329.1 mg/g	[68]
Fe_3O_4 nanomaterials	Silica based-cyclodextrin immobilized on magnetic nanoparticles	Direct blue 15	98%	[69]
Zinc oxide/polypyrrole (ZnO/PPy) nanocomposites	In situ polymerization method	Brilliant green	140.8 mg/g	[70]
MgO nanoparticles	Activated carbon immobilized on MgO nanoparticles	Rhodamine B	16.2 mg/g	[71]
ZnS: Mn-NPs-AC	–	Malachite green (MG) and methylene blue (MB)	99.87% and 98.56%	[72]
ZnSe-graphene nanocomposites	Hydrothermal method	Methylene blue	99.6%	[73]
ZnS: Ni-NP-AC	–	Sunset yellow	98.7%	[74]
		Erythrosine	99.6%	
ZnS nanoparticles	Co-precipitation method	Direct blue 14	88.26%	[75]
ZnO nanoparticles	Evaporation–condensation method	Methyl orange	98%	[76]

TABLE 18.3 Heavy Metals Removal by Different Nanomaterials

Materials	Preparation	Adsorbate	Efficiency	References
Sodium dodecyl sulfate-coated Fe_3O_4 nanoparticles	Chemical co-precipitation method	Copper	24.3 mol/g	[77]
		Nickel	41.2 mol/g	
		Zinc	59.2 mol/g	
Montmorillonite/Fe_3O_4/humic acid (MFH) nanocomposites	Hydrothermal method	Cr(VI)	82.3%	[78]
Titanate nanomaterials	Hydrothermal Method	Cr(VI)	83.91 mg/g	[79]
Ag-$CoFe_2O_4$-GO nanocomposite	Solvothermal method	Pb(II)	99.9%	[80]
Fe_3O_4@DAPF CSFMNRs	–	Pb(II)	83.3 mg/g	[81]
Nanophase Fe_2O_3	Precipitation method	As(V)	4,600 mg/kg	[82]
Ga-doped ZnO (GZO) nanoparticles	Sol–gel method	Cr(VI)	220.7 mg/g	[83]
ZnS/graphene nanocomposites	Hydrothermal method	Cd(II)	97%	[84]

18.6 Conclusion and Future Prospects

This chapter highlights the zinc-based nanomaterials synthesis, characterization and environmental applications. Basic properties of zinc-based nanomaterials are outlined based on previous experiments. The properties of zinc-based nanomaterials, applications in removal dyes, heavy metals, gas sensors and solar cells are analyzed. Zinc-based nanomaterials have attracted a considerable interest because its properties are similar to that of nano-titanium dioxide. Therefore, it seems very plausible that zinc-based nanomaterials may find wide commercial applications in wastewater treatment in the near future. With a direct wide-bandgap semiconductor, excellent transport properties, good thermal stability and high electronic mobility, zinc-based nanomaterials are expected to play a key role in developing novel photovoltaic solar cells and creating renewable energy.

Although zinc-based nanomaterials have presented the ability for environmental monitoring and remediation, there are still some issues that need to be addressed. It is important to investigate the ecotoxicity of zinc-based nanomaterials in aquatic systems before their widespread applications. Finally, more research work should be focused on the cost-effective and feasible method of the zinc-based nanomaterials regeneration.

Acknowledgments

This work was supported by the National Natural Science Foundation of China under Grant No. 21667012, and the Government of Guizhou Province (Project No. [2017]5726-42).

References

1. Rose, M., Fernandes, A. *Persistent Organic Pollutants and Toxic Metals in Foods*. The Food and Environment Research Agency, Sand Hutton, 2013, 476–486.
2. Robinson, T., Mcmullan, G., Marchant, R. et al. Remediation of dyes in textile effluent: A critical review on current treatment technologies with a proposed alternative. *Bioresource Technology*, 2001, 77 (3), 247–255.
3. Ngah, W.S.W., Hanafiah, M.A.K.M. Removal of heavy metal ions by chemically modified plant wastes as adsorbents: A review. *Bioresource Technology*, 2008, 99 (10), 3935–3948.
4. Bessac, B.F., Jordt, S.E. Sensory detection and responses to toxic gases: Mechanisms, health effects, and countermeasures. *Proceedings of the American Thoracic Society*, 2010, 7 (4), 269.

5. Ahumadarudolph, R., Pozo, K., Harner, T. et al. Survey of persistent organic pollutants (POPs) and polycyclic aromatic hydrocarbons (PAHs) in the atmosphere of rural, urban and industrial areas of Concepción, Chile, using passive air samplers. *Atmospheric Pollution Research*, 2012, 3 (4), 426–434.
6. Mohamed, R.M., Algheethi, A.A., Aznin, S.S. et al. Removal of nutrients and organic pollutants from household greywater by phycoremediation for safe disposal. *International Journal of Energy and Environmental Engineering*, 2017, 8 (3), 259–272.
7. KabdaLi, I., Tünay, O., Orhon, D. Wastewater control and management in a leather tanning district. *Water Science and Technology*, 1999, 40 (1), 261–267.
8. Ahmad, A., Mohdsetapar, S.H., Chuong, C.S. et al. Recent advances in new generation dye removal technologies: Novel search for approaches to reprocess wastewater. *Cheminform*, 2015, 5 (39), 30801–30818.
9. Rajkumar, D., Kim, J.G. Oxidation of various reactive dyes with in situ electro-generated active chlorine for textile dyeing industry wastewater treatment. *Journal of Hazardous Materials*, 2006, 136 (2), 203–212.
10. Gómez Alvarez, A., Mezafigueroa, D., Valenzuela García-Jesús, L. Behavior of metals under different seasonal conditions: Effects on the quality of a mexico-USA border river. *Water Air and Soil Pollution*, 2014, 225 (10), 1–13.
11. Anastopoulos, I., Kyzas, G.Z. Agricultural peels for dye adsorption: A review of recent literature. *Journal of Molecular Liquids*, 2014, 200, 381–389.
12. Yagub, M.T., Sen, T.K., Afroze, S. et al. Dye and its removal from aqueous solution by adsorption: A review. *Advances in Colloid and Interface Science*, 2014, 209, 172–184.
13. Wong, Y.C., Szeto, Y.S., Cheung, W.H., Mckay, G. Adsorption of acid dyes on chitosan-equilibrium isotherm analyses. *Process Biochemistry*, 2004, 39 (6), 695–704.
14. Grant, W.B., Kagann, R.H., Mcclenny, W.A. Optical remote measurement of toxic gases. *Air Repair*, 1992, 42 (1), 18–30.
15. Neri, G., Bonavita, A., Micali, G. et al. Resistive CO gas sensors based on In_2O_3 and $InSnO_x$ nanopowders synthesized via starch-aided sol-gel process for automotive applications. *Sensors Actuators B*, 2008, 132 (1), 224–233.
16. Crawford, D.W., Bonnevie, N.L., Wenning, R.J. Sources of pollution and sediment contamination in Newark Bay, New Jersey. *Ecotoxicology Environmental Safety*, 1995, 30 (1), 85–100.
17. Look, D.C. Recent advances in ZnO materials and devices. *Materials Science & Engineering B*, 2001, 80, (1–3), 383–387.
18. Mir, N., Salavati-Niasari, M., Davar, F. Preparation of ZnO nano-flowers and Zn glycerolate nanoplates usinginorganic precursors via a convenient rout and application in dye sensitized solar cells. *Chemical Engineering Journal*, 2012, 181–182, 779–789.
19. Zhu, L.P., Wang, L.F., Xue, F. et al. Piezo-phototronic effect enhanced flexible solar cells based on n-ZnO/p-SnS core-shell nanowire array. *Advanced Science*, 2017, 4, 1600185.
20. Ozgur, U., Alivov, Y.I., Liu, C. et al. A comprehensive review of ZnO materials and devices. *Journal of Applied Physics*, 2005, 98 (4), 41301.
21. Liu, C., Yun, F., Morkoç, H. Ferromagnetism of ZnO and GaN: A review. *Journal of Materials Science: Materials in Electronics*, 2005, 16 (9), 555.
22. Wenas, W.W., Yamada, A., Takahashi, K. et al. Electrical and optical properties of boron - doped ZnO thin films for solar cells grown by metalorganic chemical vapor deposition. *Journal of Applied Physics*, 1991, 70 (11): 7119–7123.
23. Shanmugam, N.R., Muthukumar, S., Prasad, S. A review on ZnO-based electrical biosensors for cardiac biomarker detection. *Future Science OA*, 2017, 3 (4), FSO196.
24. Lee, K.M., Lai, C.W., Ngai, K.S. et al. Recent developments of zinc oxide based photocatalyst in water treatment technology: A review. *Water Research*, 2016, 88, 428–448.
25. Hanrahan, G., Garza, C., Garcia, E. et al. Experimental design and response surface modeling: A method development application for the determination of reduced inorganic species in environmental samples. *Journal of Environmental Informatics*, 2007, 9, 71–79.
26. Zhang, X.Y., Zhou, J.Y., Fu, W. et al. Response surface methodology used for statistical optimization of jiean-peptide production by Bacillus subtilis. *Electronic Journal of Biotechnology*, 2010, 13. doi: 10.2225/vol13-issue4-fulltext-5.
27. Chaves, L.L., Vieira, A.C., Ferreira, D. et al. Rational and precise development of amorphous polymeric systems with dapsone by response surface methodology. *International Journal of Biological Macromolecules*, 2015, 81, 662–671.
28. Fan, M.Y., Hu, J.W., Cao, R.S. et al. A review on experimental design for pollutants removal in water treatment with the aid of artificial intelligence. *Chemsphere*, 2018, 2, 111.
29. Uikey, P., Vishwakarma, K. Review of zinc oxide (ZnO) nanoparticles applications and properties. *Journal of Applied Physics*, 2005, 98, 041301.
30. Wu, J.J., Liu, S.C. Chem inform abstract: Low-temperature growth of well-aligned ZnO nanorods by chemical vapor deposition. *Cheminform*, 2010, 33 (15), 215–218.
31. Zhang, Y.L., Yang, Y., Zhao, J.H. et al. Preparation of ZnO nanoparticles by a surfactant-assisted

complex sol-gel method using zinc nitrate. *Journal of Sol-Gel Science and Technology*, 2009, 51 (2), 198–203.
32. Xu, H.Y., Wang, H., Zhang, Y.C. et al. Hydrothermal synthesis of zinc oxide powders with controllable morphology. *Ceramics International*, 2004, 30 (1), 93–97.
33. Dai, S., Park, H.S. Surface effects on the piezoelectricity of ZnO nanowires. *Journal of the Mechanics and Physics of Solids*, 2013, 61 (2), 385–397.
34. Erhart, P., Albe, K., Klein, A. First-principles study of intrinsic point defects in ZnO: Role of band structure, volume relaxation, and finite-size effects. *Physical Review B*, 2006, 73 (20), 205203.
35. Wang, Y., Li, X., Lu, G. Synthesis and photo-catalytic degradation property of nanostructured-ZnO with different morphology. *Materials Letters*, 2008, 62 (15), 2359–2362.
36. Viswanatha, R., Sapra, S., Satpati, B. Understanding the quantum size effects in ZnO nanocrystals. *Journal of Materials Chemistry*, 2004, 14 (4), 661–668.
37. Singh, S., Thiyagarajan, P., Mohan Kant, K. Structure, microstructure and physical properties of ZnO based materials in various forms: Bulk, thin film and nano. *Journal of Physics D: Applied Physics*, 2007, 40 (20), 6312–6327.
38. Akhoon, S.A., Rubab, S., Shah, M.A. A benign hydrothermal synthesis of nanopencils-like zinc oxide nanoflowers. International Nano Letters, 2015, 5 (1), 9–13.
39. Kołodziejczak-Radzimska, A., Jesionowski, T. Zinc oxide-from synthesis to application: A review. *Materials*, 2014, 7 (4), 2833.
40. Xiong, G., Pal, U., Serrano, J.G. Photoluminesence and FTIR study of ZnO nanoparticles: The impurity and defect perspective. *Physica Status Solidi*, 2011, 3 (10), 3577–3581.
41. Gluba, M.A., Friedrich, F., Lips, K. et al. ESR investigations on hydrogen-induced hyperfine splitting features in ZnO. *Superlattices and Microstructures*, 2008, 43 (1), 24–27.
42. Amizam, S., Abdullah, N., Rafaie, H.A. et al. SEM and XRD characterization of ZnO nanostructured thin films prepared by sol-gel method with various annealing temperatures. *AIP Conference Proceedings*, 2010, 1217 (1), 37–41.
43. Tang, E.J., Cheng, G.X., Ma, X.L. et al. Surface modification of zinc oxide nanoparticle by PMAA and its dispersion in aqueous system. *Applied Surface Science*, 2006, 252 (14), 5227–5232.
44. Peh, C.K.N., Ke, L., Ho, G.W. Modification of ZnO nanorods through Au nanoparticles surface coating for dye-sensitized solar cells applications. *Materials Letters*, 2010, 64 (12), 1372–1375.
45. Wei, W., Wang, M., Wang, L.J. One-step synthesis of surface modification ZnO nanopowders by mechanochemical reaction. *Chinese Journal of Materials Research*, 2010, 24 (3), 289–293.
46. Jiang, Y., Meng, X.M., Yiu, W.C. et al. Zinc selenide nanoribbons and nanowires. *Journal of Physical Chemistry B*, 2004, 108 (9), 2784–2787.
47. Panda, A.B., Acharya, S., Efrima, A. Ultranarrow ZnSe nanorods and nanowires: Structure, spectroscopy, and one-dimensional properties. *Advanced Materials*, 2010, 17 (20), 2471–2474.
48. Liu, X., Ma, J., Peng, P. et al. One-pot hydrothermal synthesis of ZnSe hollow nanospheres from an ionic liquid precursor. *Langmuir*, 2010, 26 (12), 9968–9973.
49. Hutagalung, S.D., Loo, S.C. Zinc selenide (ZnSe) nanoparticles prepared by sol-gel method. *IEEE Conference on Nanotechnology*, Hong Kong, China, 2007.
50. Du, J., Xu, L., Zou, G. et al. A solvothermal method to novel metastable ZnSe nanoflakes. *Materials Chemistry and Physics*, 2007, 103 (2–3), 441–445.
51. Zeng, Q.Z., Xue, S.L., Wu, S.X. et al. Synthesis, field emission and optical properties of ZnSe nanobelts, nanorods and nanocones by hydrothermal method. *Materials Science in Semiconductor Processing*, 2015, 31 (31), 189–194.
52. Kishino, M., Taguchi, T. AFM and photoluminescence characterization of defects in the mirror-polished ZnSe bulk crystals and MBE-grown homoepitaxial layers. *Journal of Crystal Growth*, 2000, 210 (1), 230–233.
53. Davidson, W.L. X-ray diffraction evidence for ZnS formation in zinc activated rubber vulcanizates. *Physical Review*, 1948, 74, 116–117.
54. Yeh, C.Y., Wei, S.H., Zunger, A. Relationship between the band-gaps of the zinc blende and the wurtzite modifications of semiconductors. *Physical Review B*, 1994, 50, 2715–2718.
55. Chen, H., Shi, D., Qi, J. et al. The stability and electronic properties of wurtzite and zinc-blende ZnS nanowires. *Physical Letter A*, 2009, 373, 371–375.
56. Tran, T.K., Park, W., Tong, W. et al. Photoluminescence properties of ZnS epilayers. *Journal Applied Physics*, 1997, 81, 2803.
57. Yeh, C.Y., Lu, Z.W., Froyen, S. et al. Zinc blende wurtzite polytypism in semiconductors. *Physical Review B*, 1992, 46, 10086.
58. Li, Z., Shen, W., Fang, L. et al. Synthesis and characteristics of silica-modified ZnS nanoparticles by sol-gel-hydrothermal method. *Journal of Alloys and Compounds*, 2008, 463 (1–2), 0–133.
59. Salavati-Niasari, M., Loghman-Estarki, M.R., Davar, F. Controllable synthesis of wurtzite ZnS nanorods through simple hydrothermal method in the presence of thioglycolic acid. *Journal of Alloys and Compounds*, 2009, 475 (1–2), 0–788.
60. Zhang, W., Shi, J., Chen, H. et al. Synthesis and characterization of nanosized ZnS confined in ordered

mesoporous silica. *Chemistry of Materials*, 2001, 13 (2), 648–654.
61. Feng, J., Wang, Y.T., Zou L.Y. et al. Improved visible-light photocatalytic properties of $ZnFe_2O_4$ synthesized via sol-gel method combined with a microwave treatment. *Chemical Research in Chinese Universities*, 2015, 31 (3), 439–442.
62. Hwang, D.K., Oh, M.S., Lim, J.H. et al. ZnO thin films and light-emitting diodes. *Journal of Physics D: Applied Physics*, 2007, 40 (22), R387–R412.
63. Hjiri, M., Mir, L.E., Leonardi, S.G. et al. CO and NO_2 selective monitoring by ZnO-based sensors. *Nanomaterials*, 2013, 3 (3), 357.
64. Zhang, F., Li, L., Ding, Y. et al. Application of functionalized ZnSe nanoparticles to determinate heavy metal ions. *Journal of Fluorescence*, 2010, 20 (4), 837–842.
65. Chaudhary, G.R., Saharan, P., Kaur, G. et al. Removal of coomassie brilliant blue R-250 dye from water using γ-Fe_2O_3 nanoparticles. *Journal of Nanoengineering and Nanomanufacturing*, 2012, 2 (3), 304–308.
66. Zhang, W., Liang, F., Li, C. et al. Microwave-enhanced synthesis of magnetic porous covalent triazine-based framework composites for fast separation of organic dye from aqueous solution. *Journal of Hazardous Materials*, 2011, 186 (2–3), 984–990.
67. Kumar, S.V., Huang, N.M., Yusoff, N. et al. High performance magnetically separable graphene/zinc oxide nanocomposite. *Materials Letters*, 2013, 93 (1), 411–414.
68. Zhang, D., Zhu, M.Y., Jin-Gang, Y.U. et al. Effective removal of brilliant green from aqueous solution with magnetic Fe_3O_4@SDBS@LDHs composites. *Transactions of Nonferrous Metals Society of China*, 2017, 27 (12), 2673–2681.
69. Zhang, X., Zhang, P., Wu, Z. et al. Adsorption of methylene blue onto humic acid-coated Fe_3O_4, nanoparticle. *Colloids and Surfaces A*, 2013, 435 (9), 85–90.
70. Zhang, M., Chang, L., Zhao, Y. et al. Fabrication of zinc oxide/polypyrrole nanocomposites for brilliant green removal from aqueous phase. *Arabian Journal for Science and Engineering*, 2019, 44, 111–121.
71. Daniel, S., Shoba, U.S. Synthesis, characterization and adsorption behaviour of MgO nano particles on Rhodamine B dye. *Journal Chemical Pharmaceutical Research*, 2015, 7, 713–723.
72. Asfaram, A., Ghaedi, M., Yousefi, F. et al. Experimental design and modeling of ultrasound assisted simultaneous adsorption of cationic dyes onto ZnS: Mn-NPs-AC from binary mixture. *Ultrasonics Sonochemistry*, 2016, 33, 77–89.
73. Hsieh, S.H., Chen, W.J., Yeh, T.H. Degradation of methylene blue using ZnSe–graphene nanocomposites under visible-light irradiation. *Ceramics International*, 2015, 41 (10), 13759–13766.
74. Roosta, M., Ghaedi, M., Daneshfar, A. et al. Simultaneous ultrasound-assisted removal of sunset yellow and erythrosine by ZnS: Ni nanoparticles loaded on activated carbon: Optimization by central composite design. *Ultrasonics Sonochemistry*, 2014, 21 (4), 1441–1450.
75. Mehrizad, A., Gharbani, P. Optimization of operational variables and kinetic modeling for photocatalytic removal of direct blue 14 from aqueous media by ZnS nanoparticles. *Journal of Water and Health.* doi: 10.2166/wh.2017.269.
76. Wang, H., Xie, C. The effects of oxygen partial pressure on the microstructures and photocatalytic property of ZnO nanoparticles. *Physica E: Low-Dimensional Systems and Nanostructures*, 2008, 40 (8), 2724–2729.
77. Adeli, M., Yamini, Y., Faraji, M. Removal of copper, nickel and zinc by sodium dodecyl sulphate coated magnetite nanoparticles from water and wastewater samples. *Arabian Journal of Chemistry*, 2012, 304, (S1), S514–S521.
78. Lu, H., Wang, J., Li, F. et al. Highly efficient and reusable montmorillonite/Fe_3O_4/humic acid nanocomposites for simultaneous removal of Cr(VI) and aniline. *Nanomaterials*, 2018, 8 (7), 537.
79. Fan, G., Lin, R., Su, Z. et al. Removal of Cr (VI) from aqueous solutions by titanate nanomaterials synthesized via hydrothermal method. *Canadian Journal of Chemical Engineering*, 2017, 95 (4), 717–723.
80. Ma, S., Zhan, S., Jia, Y. et al. Highly efficient antibacterial and Pb(II) removal effects of Ag-$CoFe_2O_4$-GO nanocomposite. *ACS Applied Materials and Interfaces*, 2015, 7 (19), 10576–10586.
81. Sada, V., Minyoung, Y. Core-shell ferromagnetic nanorod based amine polymer composite (Fe_3O_4 @DAPF) for fast removal of Pb(II) from aqueous solutions. *ACS Applied Materials Interfaces*, 2015, 7 (45), 25362–25372.
82. Luther, S., Borgfeld, N., Kim, J. et al. Removal of arsenic from aqueous solution: A study of the effects of pH and interfering ions using iron oxide nanomaterials. *Microchemical Journal*, 2012, 101 (3), 30–36.
83. Ghiloufi, I., Ghoul, J.E., Modwi, A. et al. Ga-doped ZnO for adsorption of heavy metals from aqueous solution. *Materials Science in Semiconductor Processing*, 2016, 42, 102–106.
84. Sahoo, A.K., Srivastava, S.K., Raul, P.K. et al. Graphene nanocomposites of CdS and ZnS in effective water purification. *Journal of Nanoparticle Research*, 2014, 16 (7), 1–17.
85. Yeul, D.M., Dhote, D.S. A review on applications of zinc oxide nanostructures. *International Journal of Innovative Science, Engineering and Technology*, 2015, 2 (11), 354–384.

19

Nanomaterials in Foodstuffs: Toxicological Properties and Risk Assessment

Holger Sieg, Linda Böhmert, Albert Braeuning, and Alfonso Lampen
German Federal Institute for Risk Assessment (BfR)

In recent years, it has been revealed that nanoparticles, for example, titanium dioxide particles, can be detected in different foodstuffs. A nanomaterial is a material that has at least one dimension in nanometer size. More stringent definitions set a range of 1–100 nm. Nanomaterials can be categorized according to their shape into nanoparticles, nano-rods or -wires, nano-films, or structures made out of nanopore-containing materials. Another way of categorizing nanomaterials is by their chemical nature consisting of organic and inorganic compounds. Nanoscaled structures can also be differentiated by their physicochemical properties and the way of formation.

Small particles in the nano range possess altered physicochemical properties compared to the respective bulk materials, which is the underlying cause for a suspected increased toxicity of such compounds. A lot of research has been performed in the field of nanotoxicology in the past 15 years and is still ongoing. With respect to nanomaterials in foodstuff, the oral uptake route is of major relevance. On their way into the human body, the nanoparticles subsequently pass various conditions of biological environment which have an influence on their properties and behavior. Nanoparticles interact with the food matrix, saliva, gastric juice and intestinal juice before they come into contact with intestinal epithelial cells, which control the further fate of the particles, namely the uptake into the human body or the excretion via feces.

Only intentionally produced nanomaterials have to be labeled as "nano" in the European Union. However, nanoparticles might also be unintendedly formed in foodstuff by natural processes or during the gastrointestinal passage. Numerous naturally occurring structures in food exhibit a size in the nano range and might therefore be regarded as nanostructures.

In this work, we will reflect current applications of nanoparticles in the food sector and focus on toxicological and analytical issues with regard to the oral ingestion of food-derived nanomaterials.

19.1 Occurrence

In recent years, it is getting more and more obvious that nanoparticles like titanium dioxide are detectable in foodstuffs (Agir pour L'Environnement, 2016; Yang et al., 2014; Faust et al., 2016), even though proper methods to detect nanoparticles in complex matrices like foodstuffs are still under development. So far, there are just few method combinations published that have been demonstrated to be suitable to detect lead nanoparticles in meat, silver nanoparticles in chicken and titanium dioxide nanoparticles in sweets (Kollander et al., 2017; Peters et al., 2014; Weigel et al., 2017). Anyhow, many food products contain organic or inorganic particles whose dimensions lie at least partially in the nanometer range, as detailed below. These nanoparticles can occur naturally in food, but they can also be added deliberately or find their way into foods by accident through contamination and additives or through packaging and manufacturing stages.

Natural organic nanostructures in food can be found, for example, in milk. Casein micelles have a size in the range of a few tenths of a nanometer up to several hundred nanometers (Holt et al., 2003; Livney, 2010). Ferritin, as another example, is a protein found in foods and in the human body which serves mainly to store iron. It has a size of roughly 12 nm in diameter and contains a nucleus made of hydrated iron oxide (Fe(III)O·OH) (Theil, 1987). In addition to organic nanoparticles, inorganic nanoparticles also occur naturally and unintentionally. It is known that particles of this size are released during explosive volcanic eruptions, dust storms and forest fires and become a part of the fine dust which reaches the human environment (Gieré and Querol, 2010). Nanoscaled metallic particles can also occur in the food production process through mechanical abrasion, for instance, and remain in the food (Beltrami et al., 2011). Several food additives can contain some nanoscaled particles, such as titanium dioxide (E171) and silicon dioxide (E551) (Dekkers et al., 2011; van Kesteren et al., 2015; Weir et al., 2012). Data on exposure is only available for a few particle species. According to a current re-evaluation of titanium dioxide (E171) and silicon dioxide (E551) published by the European Food Safety Authority (EFSA), the absorption of orally administered titanium dioxide is extremely low and the low bioavailability of titanium dioxide appears to be independent of particle size. A small amount (maximum of 0.1%) of orally ingested titanium dioxide was absorbed by the gut-associated lymphoid tissue (GALT) and subsequently distributed to various organs. Elimination rates from these organs were variable. In its latest assessment of E171 for the maximum exposure scenario, the EFSA calculated mean exposure levels of 0.4 mg of titanium dioxide per kilogram body weight and day (nanoscaled and non-nanoscaled) for adults, while the corresponding values for children of 10.4 mg/kg body weight and day are significantly higher (EFSA, 2016a). Depending on the used detection method, up to roughly one third of the particles in E171 are nanoscaled in their number size distribution. In relation to the mass proportion distribution, the share of nanoscaled particles in E171 reaches to a maximum of 3.2% (EFSA, 2016a). By contrast, the food additive silicon dioxide (E551) is a material comprised of aggregated nanosized primary particles. These aggregates can further agglomerate to form larger structures. The sizes of the aggregates and agglomerates are normally bigger than 100 nm. However, depending on the starting material and/or on the manufacturing process, it cannot be excluded that some aggregates of primary particles are smaller than 100 nm in size (EFSA, 2018). In principle, however, metallic particles can also originate from ionic, detached metal compounds. This could be shown for silver as an example (Hansen and Thünemann, 2016; Juling et al., 2016, 2018). The occurrence of nanoparticles in food containing detached ionic metal compounds cannot, therefore, be excluded.

Nanoparticles are also produced intentionally and deliberately added to foods. Accordingly, synthesized or isolated vitamins (e.g. vitamins D and E), minerals (e.g. iron and calcium) and secondary plant materials (e.g. curcumin and quercetin) are enriched in foods in nanoscaled form or bonded to nanoscaled structures (Oehlke et al., 2014). These food supplements and "nutraceuticals" are advertised with various health promises. The nano- or micro-formulations here are supposed to enhance the solubility of bioactive substances, have an effect on the durability of the substance or its behavior during the production process, reduce undesired sensory properties, or increase bioavailability. This topic is dealt with in detail in the overview studies by Oehlke et al. (2014) and Livney (2015). Hardly any *in vivo* data has been available up to now, however, to verify the possible positive health effects of nano-formulations of vitamins, minerals or secondary plant materials of this kind or which could eliminate the possibility of negative effects as a result of a possible overdose of bioactive substances caused by increased bioavailability.

19.2 Legal Situation

Legal definitions of nanomaterials exist which differ slightly from one another depending on the manner in which they are used. Some of them stipulate threshold values for the number size distribution of the nanoparticulate substances, while others make a distinction between intentionally produced and naturally occurring substances of similar magnitude. Relevant definitions of nanomaterials are contained in DIN CEN ISO/TS Norm 27687 2008, the European cosmetics regulation (EC) No. 1223/2009, the European food information regulation (EU) No. 1169/2011 and European Commission Recommendation 2011/696/EU. According to the latter, European Commission recommendation, at least 50% of the particles in the number size distribution must have measurements in the range of 1–100 nm in one or more dimensions. A detailed summary of the various regulatory definitions of nanoparticles inside and outside the EU is to be found in the overview study by Amenta et al. (2015). None of the definitions mentioned orientates around any proven health risks posed by the particles included in or excluded from each respective definition.

The definition relevant to consumers in the area of food labeling is to be found in the food information regulation (EU) No. 1169/2011 which was recently aligned with the new novel food regulation (EU) No. 2015/2283. Accordingly, a "technically manufactured nanomaterial" is an intentionally produced material which has a size in the range of 100 nm or less in one or more of its dimensions or whose inner structure or surface consists of individual functional parts, mainly in the size range of 100 nm or less in one or more of their dimensions, including structures, agglomerates and aggregates which may have a size above the order of 100 nm but retain properties which are characteristic of the nanoscale. Properties of the nanoscale include those characteristic connected with the large specific surface of each material and/or special physicochemical properties which differ from those of the same material in non-nanoscale form.

Since December 2014, all nanoparticles in food which are covered by this definition must be preceded by the prefix "nano" to inform consumers of the nanoscale characteristics of the ingredients. However, the stipulation of the "intentional production" of the contained particles, as well as the requirements for number size distribution (Commission recommendation 2011/696/EU) and the combination of upper size limits in the valid particle definition with the method used to determine particle size severely restrict the particle species which are in actual practice subject to labeling requirements: the proportion of nanoparticles in a food or food ingredient usually lies well below the legally specified threshold value of the number size distribution of 50% (Commission recommendation 2011/696/EU). Furthermore, natural nanoparticles are expressly not covered by the valid definition and are therefore not subject to any labeling requirements. Suitable detection methods for the use of nanomaterials in complex matrices, such as foods, are currently under development. In addition to this, the point in time in the life cycle of the nanomaterial at which the analysis is made also has a considerable influence on the results (Abassi et al., 2009). From 2018, the authorization of novel foods consisting of technically manufactured nanomaterials as defined above, which were not used for human consumption to any appreciable extent in the European Union prior to 15 May 1997, will be in compliance with the provisions of the new novel food regulation (EU) No. 2015/2283.

19.3 Gastrointestinal Uptake and Toxicology of Nanoparticles

Oral uptake of nanoparticles can be divided into four main parts, as schematically delineated in Figure 19.1: (i) interaction of nanoparticles with the food matrix, (ii) interaction of nanoparticles with intestinal fluids, (iii) intestinal uptake and bioavailability of nanoparticles and (iv) toxicity and metabolism. The intestinal epithelium plays a decisive role in the possible uptake of nanoparticles to the body while simultaneously constituting a primary target for the potentially toxic effects of the particles. Irrespective of whether they occur naturally or are consciously manufactured and used, people ingest nanoparticles every day with their food. Prior to possible resorption in the gut and contact with the cells of the intestinal mucosa, these particles must at first pass through the human digestive system consisting of the oral cavity, esophagus, stomach and intestine. On their way through the gastrointestinal tract, the nanoparticles are exposed to alternating conditions which can influence the behavior of the particles through changes to properties such as size, form and charge which in turn have an influence on whether they remain in the body or not. Particles can dissolve, for instance, be resorbed in the small intestine or

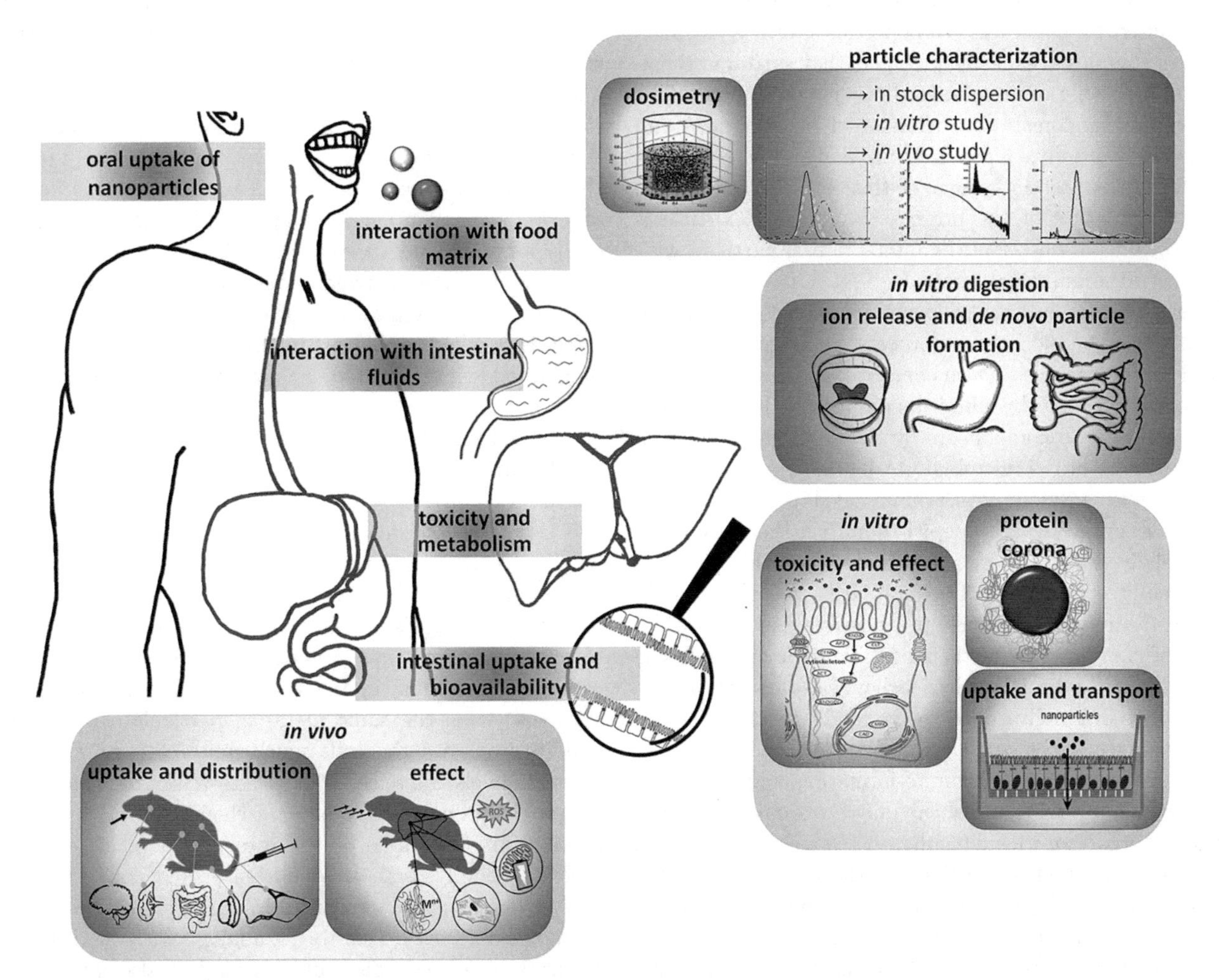

FIGURE 19.1 Overview of aspects of interest to the oral uptake of nanoparticles and of corresponding research fields.

reach the large intestine, interact with the microbial gut microbiota or be excreted undigested.

Gastric juices with different compositions are to be found in the various sections of the digestive system, with the pH value and ionic strength fluctuating significantly. In addition to this, surface-active substances, such as bile acids, phospholipids, fatty acids and proteins, support the decomposition of food and, depending on the material properties of each material, potentially that of nanoscaled structures too. Owing to the chemical environment of the gastrointestinal tract, the complete dissolution of inorganic nanoparticles or their aggregation to larger particles can occur, depending on the individual physicochemical nature of the observed nanoparticle (Böhmert, 2014; Kästner et al., 2017; McCracken et al., 2013; Sieg et al., 2017). A re-formation of nanoscaled structures from dissolved compounds in the gastrointestinal tract was also shown at the same time (Walczak et al., 2012). Moreover, different enzymes, such as glycosidases, proteinases and lipases, which can lead in particular to changes in organic nanoparticles, are also present in gastric juice.

Biopolymers consisting of proteins and mucins ensure the low-friction transport of food through the gastrointestinal tract and the protection of the epithelia from mechanical influences emanating from the complex flowing and shearing forces caused by peristalsis. Nanoparticles form a so-called corona in complex organic matrices, i.e. they surround themselves in a coating of proteins and other biomolecules. The stability and agglomeration behavior of nanoparticles can change substantially, depending on changes to or the re-formation of a protein corona of this kind (Treuel et al., 2015). The nature of the protein corona of a particle is an important parameter in the interaction of particles with epithelia of the intestinal barrier.

Conversely, not only the properties of particles are altered and influenced by the digestive process, a change of digestion caused by the particles is also conceivable in principle. This could occur through the inhibition of digestive enzymes, the absorption of food components by the particles or through the influencing of the microbial gut biota.

The precise form in which nanoscaled particles in the gut get in contact with the epithelium can therefore differ substantially from the chemical nature of the particles as originally contained in the food. It is therefore barely possible to draw conclusions on a possible gastrointestinal uptake of nanoparticles or their possible toxic effects in the gut solely on the basis of physicochemical data relating to the particles originally added to the food.

The changes to the particle structure outlined above, i.e. partial or complete dissolution, re-formation, aggregation or formation of a protein corona, can only be examined directly in a living organism with difficulty. Instead, artificial models of the digestive tract which simulate the various sections of the gastrointestinal passage and their chemical conditions, such as saliva in the oral cavity, gastric juice or intestinal liquid, are mostly used to study these phenomena. This enables the examination of the particle properties through the substages of digestion, also under consideration of the simultaneous presence of various food components (Böhmert et al., 2014; Kästner et al., 2017; Lichtenstein et al., 2015; McCracken et al., 2013; Peters et al., 2012; Walczak et al., 2012).

Due to their altered properties compared to the same material in a larger form, nanoparticles are not only of interest for various technical applications, they have also been the focus of toxicological examinations for many years. These altered properties can be caused, for example, by the enlarged surface-to-volume ratio that results from their small size. The latter leads to a stronger chemical reactivity of the nanoparticles than is to be observed with larger particles of comparable material (Zhang et al., 2010). This altered behavior in the nanoscale can be of advantage in technical and other applications, but it is also the basis of the speculation that nanoparticles have an altered and possibly increased toxicological potential compared to dissolved substances or larger particles.

It is also being discussed and examined in this regard to which degree nanoparticles may possibly tend to overcome physiological barriers due to their small size, so that they can be ingested into cells or tissue. According to a study in mice, the overall absorption of orally administered silica particles was as low as <4% and <3% for nano silica and bulk silica, respectively. Furthermore, particle size did not affect solubility, tissue distribution patterns and excretion kinetics (Lee et al., 2017). A 28-day *in vivo* study on the oral uptake of silver nanoparticles showed that the excreted silver in the feces was estimated at >99% of the intake, implying that only a very minor fraction was absorbed (van der Zande et al., 2012). The absorption of orally administered titanium dioxide particles (micro- and nanosized) in the gastrointestinal tract is described to be negligible, estimated at most as 0.02%–0.1% of the administered dose, whereas the vast majority of an oral dose of titanium dioxide is eliminated unchanged with the feces (EFSA, 2016a).

Detailed knowledge on a possibly increased nano-specific cellular uptake is currently of major interest for research, because this process would enable particles to infiltrate cells and tissue with toxic quantities of a material like released ions which is only ingested in small quantities in other forms. A mechanism of this kind would be possible if, for example, the intake of a dissolved substance, such as a potential toxic metallic ion, into the cell by means of the transport or channel proteins in the cell membrane was only very slight; whereas the same metal in particulate form could find its way into the cells much more efficiently through other channels and release increased levels of metal ions from the particles due to the dominant chemical conditions, such as an acidic pH value or the presence of reactive oxygen species in the lysosomes. This is suspected, for instance, of silver nanoparticles in murine microglial cells and astrocytes (Hsiao et al., 2015). Due to the analogy with the historical occurrence, a mechanism of this kind is also known as a "Trojan horse". Effects of this kind would not appear to be the rule for nanoparticles of arbitrary chemical composition. However, they would seem to occur only in specific instances, where an interplay of several factors comprising the chemical nature of the nanoparticles, their individual capability to produce ions, as well as the specific willingness of each examined cell population to ingest the relevant nanoparticles, should be regarded as being decisive for any toxicity which may be observed. A toxicity mechanism of the "Trojan horse" type cannot therefore be concluded at the moment purely from the possibility of ion release by a particle.

Another instance of a "Trojan horse" would be particles which are not toxic themselves and do not release any toxic components but which bind other substances, such as certain environmental contaminants, before releasing these contaminants into the body or cells after the particles have been ingested. In cases of this kind, the type of each harmful effect would in principle comply with the toxicological profile of each contaminant, even though the quantity of this substance ingested into the cell or the speed of ingestion could be altered by the presence of the particles.

Apart from possible nano-specific toxicological properties, which could be proven in individual instances as outlined above, the toxic effect of nanoscaled metallic nanoparticles would appear in many cases to be conveyed essentially by ions released by the particles and not by nano-specific effects. If differences between the effects of particles and ions were to be established here, the specific kinetics of the release of ions from the particles could explain at least some of the differences. It is also of great relevance in these comparisons to establish the doses on the basis of which the observed effect and its differences were compared with one another; concentration, mass or surface has frequently been used as the basis for the dose up to now (Oberdörster et al., 2007; Wittmaack, 2007). It should be considered here, however, that while the constitution of the nanoparticles allows them to diffuse much more slowly than dissolved ions, especially in *in vivo* experiments, it also allows them to settle much more slowly than larger particles. Various calculation models (Cohen et al., 2014, 2015; Hinderliter et al., 2010; Teeguarden et al., 2007) exist to determine the quantity of nanoparticles which actually have contact *in vitro* with the cellular test system used (the so-called effective dose), but they are not yet applied consistently.

In addition to *in vitro* experiments, all the abovementioned studies conducted *in vivo* traditionally play the decisive role in risk assessment. In its guideline document on the risk assessment of nanotechnology applications in the food and feed chain, EFSA recommends among other things that the following procedure be used to enable the risk assessment of nanomaterials. Extensive physicochemical characterization of the nanomaterials used is required in the first stage. As the properties of the nanomaterials can change in different environments, EFSA recommends that the following parameters are examined: the conditions under which the nanomaterial was manufactured, the condition in which the nanomaterial is used or occurs in foods and feeds, the condition in which the nanomaterial occurs in the toxicological examinations and the condition of the nanomaterials in biological liquids and tissues. If the properties and behavior of the nanomaterials are known and exposure is to be expected, possible hazards should be identified. To do so, *in vitro* genotoxicity examinations, toxicokinetic analyses and a 90-day study on rodents with repeated oral administration should be used as a minimum requirement (EFSA, 2011).

19.4 Risk Assessment

General statements along the lines that nanoscale particles can fundamentally be regarded as safe or unsafe where consumer health protection is concerned cannot be made on the basis of the available data. Despite the availability of a large number of experimental studies on various aspects of the toxicology of many different nanoparticles, only a few risk assessments of nanomaterials are currently on hand, for example, a publication by van Kesteren et al. (2015): their estimations of the silicon concentration in human liver for average-to-worst-case dietary exposure to synthetic amorphous silica (SAS) predict concentrations in humans at a similar level as the measured or estimated liver concentrations in animal studies in which adverse effects were found (van Kesteren et al., 2015).

Within the scope of the re-evaluation of authorized food additives, EFSA has also taken a look at percentages of nanoparticles in several of its scientific opinions (EFSA, 2015, 2016b,c), some of which concluded that the data situation is currently insufficient to make a final risk assessment, as in the case of E172 (yellow and red iron oxide) (EFSA, 2015). Overall, due to the many differences that exist in the properties of the many different particle types, the current assessment of the possible health risks of orally ingested nanoparticles is based essentially on the observation of individual cases. A transition of risk assessment towards the possible grouping of similar nanomaterials which can be jointly assessed would appear to be worthwhile for the future and is the subject of current research projects.

Whereas the focus of nanotoxicological research in the area of food safety has been on inorganic nanoparticles up to now, it is expected that the growing importance will be placed on the field of organic nanoparticles in future. In addition to questions about the toxicological properties of orally ingested nanoparticles, recent findings of plastic microparticles – a topic that was mainly discussed up to now in relation to environmental toxicology and marine organisms – in foods such as beer and honey have met with considerable public interest (Liebezeit and Liebezeit, 2014, 2015). As it is not to be expected that the weathering of plastic in the environment that leads to the occurrence of plastic microparticles will end in the micrometer range, it would seem conceivable that the possible effects of nanoscaled plastic particles, which can find their way from the environment into the human body via the food chain, could also become the focus of interest. Moreover, due to the many different technical uses of nanoparticles, the occurrence of nanomaterials of this kind as contaminants in food would appear to be merely a question of time. There is also a great need for research in the development of sensitive analytical methods for detecting organic nanoparticles, in particular, in complex matrices such as foods.

Outside the food area, nanotechnology and nanomaterials are used mainly in areas such as electronics, materials technology and medicine. Apart from the technical advantages that materials in this small dimension can produce, above all, their general acceptance by the public is an important factor which will decide on the future application and distribution of the entire technology. Whereas the technical applications of nanomaterials usually meet with a positive response from consumers, the acceptance of nanoparticles in food is significantly lower (Zimmer et al., 2009). The future will show here whether and to what extent the use of deliberately manufactured nanomaterials as additives in the food area will assert itself. From a scientific point of view, meticulous examination of the possible toxicological potential of nanoscale materials and the comprehensible and transparent communication of the results can help to ensure consumer safety. In addition to this, understandable marking and labeling regulations will play a role in safeguarding consumers' freedom of choice.

References

Abassi, Y. A., Xi, B., Zhang, W., Ye, P., Kirstein, S. L., Gaylord, M. R., Feinstein, S. C., Wang, X. & Xu, X. 2009. Kinetic cell-based morphological screening: Prediction of mechanism of compound action and off-target effects. *Chemical Biology*, 16, 712–723.

Agir pour L'Environnement. 2016. Les nanoparticules dans l'alimentation: Dangereuses, inutileset incontrôlées Un moratoire s'impose! [Online]. www.agirpourlenvironnement.org/sites/default/files/communiques_presses/160613_Dossier_de_presse_Enquete_Nano.pdf. [Accessed].

Amenta, V., Aschberger, K., Arena, M., Bouwmeester, H., Botelho Moniz, F., Brandhoff, P., Gottardo, S., Marvin, H. J. P., Mech, A., Quiros Pesudo, L., Rauscher, H., Schoonjans, R., Vettori, M. V., Weigel, S. & Peters, R. J. 2015. Regulatory aspects of nanotechnology in the agri/feed/food sector in EU and non-EU countries. *Regulatory Toxicology and Pharmacology*, 73, 463–476.

Beltrami, D., Calestani, D., Maffini, M., Suman, M., Melegari, B., Zappettini, A., Zanotti, L., Casellato, U., Careri, M. & Mangia, A. 2011. Development of a combined SEM and ICP-MS approach for the qualitative and quantitative analyses of metal microparticles and sub-microparticles in food products. *Analytical and Bioanalytical Chemistry*, 401, 1401–1409.

Böhmert, L. 2014. Aufnahme und Toxizität von Silbernanopartikeln im Intestinalmodell Caco-2. www.depositonce.tu-berlin.de/handle/11303/4234

Böhmert, L., Girod, M., Hansen, U., Maul, R., Knappe, P., Niemann, B., Weidner, S., Thünemann, A. F. & Lampen, A. 2014. Analytically monitored digestion of silver nanoparticles and their toxicity for human intestinal Cells. *Nanotoxicology*, 42, 8959–8964.

Cohen, J. M., Deloid, G. M. & Demokritou, P. 2015. A critical review of in vitro dosimetry for engineered nanomaterials. *Nanomedicine (London)*, 10, 3015–3032.

Cohen, J. M., Teeguarden, J. G. & Demokritou, P. 2014. An integrated approach for the in vitro dosimetry of engineered nanomaterials. *Particle and Fibre Toxicology*, 11, 20.

Dekkers, S., Krystek, P., Peters, R. J., Lankveld, D. P., Bokkers, B. G., van Hoeven-Arentzen, P. H., Bouwmeester, H. & Oomen, A. G. 2011. Presence and risks of nanosilica in food products. *Nanotoxicology*, 5, 393–405.

EFSA. 2011. Guidance on the risk assessment of the application of nanoscience and nanotechnologies in the food and feed chain. *EFSA Journal*, 9, 2140.

EFSA. 2015. Scientific opinion on the re-evaluation of iron oxides and hydroxides (E172) as food additives. *EFSA Journal*, 13, 4317.

EFSA. 2016a. Re-evaluation of titanium dioxide (E171) as a food additive. *EFSA Journal*, 14, e04545.

EFSA. 2016b. Scientific opinion on the re-evaluation of gold (E175) as a food additive. *EFSA Journal*, 14, 4362.

EFSA. 2016c. Scientific opinion on the re-evaluation of silver (E174) as food additive. *EFSA Journal*, 14, 4364.

EFSA. 2018. Re-evaluation of silicon dioxide (E551) as a food additive. *EFSA Journal*, 16, 5088.

Faust, J. J., Doudrick, K., Yang, Y., Capco, D. G. & Westerhoff, P. 2016. A facile method for separating and enriching nano and submicron particles from titanium dioxide found in food and pharmaceutical products. *PLoS One*, 11, e0164712.

Gieré, R. & Querol, X. 2010. Solid particulate matter in the atmosphere. *Elements*, 6, 215–222.

Hansen, U. & Thünemann, A. F. 2016. Considerations using silver nitrate as a reference for in vitro tests with silver nanoparticles. *Toxicol in Vitro*, 34, 120–122.

Hinderliter, P. M., Minard, K. R., Orr, G., Chrisler, W. B., Thrall, B. D., Pounds, J. G. & Teeguarden, J. G. 2010. ISDD: A computational model of particle sedimentation, diffusion and target cell dosimetry for in vitro toxicity studies. *Part Fibre Toxicol*, 7, 36.

Holt, C., De Kruif, C., Tuinier, R. & Timmins, P. 2003. Substructure of bovine casein micelles by small-angle X-ray and neutron scattering. *Colloids and Surfaces A: Physicochemical and Engineering Aspects*, 213, 275–284.

Hsiao, I.-L., Hsieh, Y.-K., Wang, C.-F., Chen, I.-C. & Huang, Y.-J. 2015. Trojan-horse mechanism in the cellular uptake of silver nanoparticles verified by direct intra-and extracellular silver speciation analysis. *Environmental Science and Technology*, 49, 3813–3821.

Juling, S., Bachler, G., Von Götz, N., Lichtenstein, D., Böhmert, L., Niedzwiecka, A., Selve, S., Braeuning, A. & Lampen, A. 2016. In vivo distribution of nanosilver in the rat: The role of ions and de novo-formed secondary particles. *Food and Chemical Toxicology*, 97, 327–335.

Juling, S., Böhmert, L., Lichtenstein, D., Oberemm, A., Creutzenberg, O., Thunemann, A. F., Braeuning, A. & Lampen, A. 2018. Comparative proteomic analysis of hepatic effects induced by nanosilver, silver ions and nanoparticle coating in rats. *Food and Chemical Toxicology*, 113, 255–266.

Kästner, C., Lichtenstein, D., Lampen, A. & Thünemann, A. F. 2017. Monitoring the fate of small silver nanoparticles during artificial digestion. *Colloids and Surfaces A: Physicochemical and Engineering Aspects*, 526, 76–81.

Kollander, B., Widemo, F., Agren, E., Larsen, E. H. & Loeschner, K. 2017. Detection of lead nanoparticles in game meat by single particle ICP-MS following use of lead-containing bullets. *Analytical and Bioanalytical Chemistry*, 409, 1877–1885.

Lee, J. A., Kim, M. K., Song, J. H., Jo, M. R., Yu, J., Kim, K. M., Kim, Y. R., Oh, J. M. & Choi, S. J. 2017. Biokinetics of food additive silica nanoparticles and their interactions with food components. *Colloids and Surfaces B: Biointerfaces*, 150, 384–392.

Lichtenstein, D., Ebmeyer, J., Knappe, P., Juling, S., Böhmert, L., Selve, S., Niemann, B., Braeuning, A., Thünemann, A. F. & Lampen, A. 2015. Impact of food components during in vitro digestion of silver nanoparticles on cellular uptake and cytotoxicity in intestinal cells. *Biological Chemistry*, 396, 1255–1264.

Liebezeit, G. & Liebezeit, E. 2014. Synthetic particles as contaminants in German beers. *Food Additives and Contaminants: Part A*, 31, 1574–1578.

Liebezeit, G. & Liebezeit, E. 2015. Origin of synthetic particles in honeys. *Polish Journal of Food and Nutrition Sciences*, 65, 143–147.

Livney, Y. D. 2010. Milk proteins as vehicles for bioactives. *Current Opinion in Colloid and Interface Science*, 15, 73–83.

Livney, Y. D. 2015. Nanostructured delivery systems in food: Latest developments and potential future directions. *Current Opinion in Food Science*, 3, 125–135.

Mccracken, C., Zane, A., Knight, D. A., Dutta, P. K. & Waldman, W. J. 2013. Minimal intestinal epithelial cell toxicity in response to short-and long-term food-relevant inorganic nanoparticle exposure. *Chemical Research in Toxicology*, 26, 1514–1525.

Oberdörster, G., Oberdörster, E. & Oberdörster, J. 2007. Concepts of nanoparticle dose metric and response metric. *Environmental Health Perspectives*, 115, A290–A293.

Oehlke, K., Adamiuk, M., Behsnilian, D., Gräf, V., Mayer-Miebach, E., Walz, E. & Greiner, R. 2014. Potential bioavailability enhancement of bioactive compounds using food-grade engineered nanomaterials: A review of the existing evidence. *Food and Function*, 5, 1341–1359.

Peters, R., Kramer, E., Oomen, A. G., Herrera Rivera, Z. E., Oegema, G., Tromp, P. C., Fokkink, R., Rietveld, A., Marvin, H. J. P., Weigel, S., Peijnenburg, A. A. C. M. & Bouwmeester, H. 2012. Presence of nano-sized silica during in vitro digestion of foods containing silica as a food additive. *ACS Nano*, 6, 2441–2451.

Peters, R. J., van Bemmel, G., Herrera-Rivera, Z., Helsper, H. P., Marvin, H. J., Weigel, S., Tromp, P. C., Oomen, A. G., Rietveld, A. G. & Bouwmeester, H. 2014. Characterization of titanium dioxide nanoparticles in food products: Analytical methods to define nanoparticles. *Journal of Agricultural and Food Chemistry*, 62, 6285–6293.

Sieg, H., Kästner, C., Krause, B., Meyer, T., Burel, A., Böhmert, L., Lichtenstein, D., Jungnickel, H., Tentschert, J., Laux, P., Braeuning, A., Estrela-Lopis, I., Gauffre, F., Fessard, V., Meijer, J., Luch, A., Thünemann, A. F. & Lampen, A. 2017. Impact of an artificial digestion procedure on aluminum-containing nanomaterials. *Langmuir*, 33, 10726–10735.

Teeguarden, J. G., Hinderliter, P. M., Orr, G., Thrall, B. D. & Pounds, J. G. 2007. Particokinetics in vitro: Dosimetry considerations for in vitro nanoparticle toxicity assessments. *Toxicological Sciences*, 95, 300–312.

Theil, E. C. 1987. Ferritin: Structure, gene regulation, and cellular function in animals, plants, and microorganisms. *Annual Review of Biochemistry*, 56, 289–315.

Treuel, L., Docter, D., Maskos, M. & Stauber, R. H. 2015. Protein corona: From molecular adsorption to physiological complexity. *Beilstein Journal of Nanotechnology*, 6, 857–873.

van der Zande, M., Vandebriel, R. J., van Doren, E., Kramer, E., Herrera Rivera, Z., Serrano-Rojero, C. S., Gremmer, E. R., Mast, J., Peters, R. J. B., Hollman, P. C. H., Hendriksen, P. J. M., Marvin, H. J. P., Peijnenburg, A. A. C. M. & Bouwmeester, H. 2012. Distribution, elimination, and toxicity of silver nanoparticles and silver ions in rats after 28-day oral exposure. *ACS Nano*, 6, 7427–7442.

van Kesteren, P. C., Cubadda, F., Bouwmeester, H., Van Eijkeren, J. C., Dekkers, S., De Jong, W. H. & Oomen, A. G. 2015. Novel insights into the risk assessment of the nanomaterial synthetic amorphous silica, additive E551, in food. *Nanotoxicology*, 9, 442–452.

Walczak, A. P., Fokkink, R., Peters, R., Tromp, P., Herrera Rivera, Z. E., Rietjens, I. M. C. M., Hendriksen, P. J. M. & Bouwmeester, H. 2012. Behaviour of silver nanoparticles and silver ions in an in vitro human gastrointestinal digestion model. *Nanotoxicology*, 7, 1198–1210.

Weigel, S., Peters, R., Loeschner, K., Grombe, R. & Linsinger, T. P. J. 2017. Results of an interlaboratory method performance study for the size determination and quantification of silver nanoparticles in chicken meat by single-particle inductively coupled plasma mass spectrometry (sp-ICP-MS). *Analytical and Bioanalytical Chemistry*, 409, 4839–4848.

Weir, A., Westerhoff, P., Fabricius, L., Hristovski, K. & Von Goetz, N. 2012. Titanium dioxide nanoparticles in food and personal care products. *Environmental Science and Technology*, 46, 2242–2250.

Wittmaack, K. 2007. In search of the most relevant parameter for quantifying lung inflammatory response to nanoparticle exposure: Particle number, surface area, or what? *Environmental Health Perspectives*, 115, 187–194.

Yang, Y., Doudrick, K., Bi, X., Hristovski, K., Herckes, P., Westerhoff, P. & Kaegi, R. 2014. Characterization of food-grade titanium dioxide: The presence of nanosized particles. *Environmental Science and Technolog*, 48, 6391–400.

Zhang, J., Su, D. S., Blume, R., Schlögl, R., Wang, R., Yang, X. & Gajović, A. 2010. Surface chemistry and catalytic reactivity of a nanodiamond in the steam-free dehydrogenation of ethylbenzene. *Angewandte Chemie International Edition*, 49, 8640–8644.

Zimmer, R., Hertel, R., & Böl, G.-F. 2009. *BfR-Delphi-Studie zur Nanotechnologie: Expertenbefragung zum Einsatz von Nanomaterialien in Lebensmitteln und Verbraucherprodukten.* BfR. www.mobil.bfr.bund.de/cm/350/bfr_delphi_studie_zur_nanotechnologie.pdf.

20

Nano-Minerals as Livestock Feed Additives

N. B. Singh
Sharda University Greater Noida, India

Chris U. Onuegbu
Sharda University
Bioresources Engineering Federal Polytechnic
OKO, Nigeria

20.1 Introduction

In strict sense, the term livestock refers especially to animals for the production of milk, meat, work and wool. Such animals include beef and dairy cattle, swine, sheep, horses and goats. Poultry also come in this category. For the present purpose, all domesticated animals that are of economic importance to man have been considered. Livestock species are of high economic, social and cultural importance to rural communities, in particular, and nations at large. This is on account of the roles, functions or parts they play in improving the income and well-being of not only the rural people but also that of a nation in general. The functions and contributions of livestock to human well-being cannot be overstressed. As a component of the agricultural sector, livestock contributes significantly to food supply and security, family nutrition, family income, asset savings, soil productivity, livelihoods, transport, agricultural traction, agricultural diversification and sustainable agricultural production, family and community employment, ritual purposes and social status (Tazhibaev et al., 2014; Benttencourt et al., 2015). For convenience, the whole spectrum of livestock functions can be classified and studied under two broad categories. The first category is based on the kind of output produced. The second classification is based on their economic roles, such as source of cash income, means of savings accumulation, direct food use for family subsistence, farm input such as fertilizer and animal draught and capacity to comply with a set of social rules obligations.

Livestock production plays a complementary role with crop production. Livestock and their by-products, manure, are important in crop production. Livestock can act as source of energy, providing draught and animal power for a wide range of farm operations, while its manure improves soil structure, fertility and water retention capacity (Kaur et al., 2017). Both uses are environmentally friendly, as they improve energy and nutrient cycle. Livestock is also used to transport agricultural inputs and outputs and human labor. The key to effective performance of livestock function is animal health which can directly affect both production and productivity of livestock. In summary, livestock functions include provision of food, contribution to the general agriculture, as a means of transport, source of raw materials, source of foreign exchange and contribution to gross domestic product (GDP).

Despite these obvious direct and indirect, immediate and remote, short- and long-term impacts of livestock on human life, its production is far from meeting present demands as its development is challenged by several factors. Increasing human population and industrialization have given rise to serious reduction in agricultural grazing lands. Consequently, manufactured feeds are filling the gaps created by dwindling field feeds. But the cost of manufactured feeds is so high that it makes the business of livestock farming nonprofitable. This has given rise to studies on how best to increase the quality and productivity of conventional livestock feeds.

Minerals play very vital roles in the growth and health of animals. Although they are required in minute amounts in animal body, their deficiency can be of catastrophic consequences on the life of the animal, while their availability in appropriate amounts promotes fast and healthy growth in animals. Basic feed ingredients do not contain enough minerals to meet animal requirements, hence, there is a need to incorporate them as additives or supplements in animal feeds (Bruggar and Windisch, 2015). However, the efficiency of utilization and actions of these minerals depend largely on their solubility in water and bioavailability in the animal body. These, in turn, are dependent on particle size of the minerals. In their bulk forms, some of the minerals have either low solubility and/or bioavailability. This makes it difficult for animals to assimilate enough levels of minerals from their bulk forms.

Nanotechnology is the trending technology of the 21st century and involves the study, design, synthesis, manipulation and handling of materials at the nanometer (0–100 nm) range. Hence, nanomaterials are defined as materials which have at least one of their dimensions in the nanometer range. Studies have shown that materials in the nanometer range possess novel properties completely different from, and for which they find wider areas of applications than, their bulk (macro) counterparts. For instance, they are reported to be more reactive, on account of their small particle size, surface chemistry and extended surface area, than their bulk counterpart. The performance of nanomaterials particularly in animal food depends on several factors including particle size, shape, chemical and structural properties (Ramachandraiah et al., 2018). Nanoparticles (NPs) added to feed can provide a platform to incorporate various compounds, such as vaccines and nutrient supplements, due to large surface area-to-volume ratio and high absorption in the body. This can enable direct transportation of compounds to targeted organs or systems while avoiding fast degradability often seen with antibiotics and can encourage multiple health benefits. They are known to pass through the gastrointestinal tracts (GITs) and cross the membranes of animal cells more easily than their counterparts in the bulk forms. Currently research attention in animal nutrition is involving the use of nano forms of minerals to improve nutrient utilization and, consequently, enhance growth and general health in livestock (Rajendran, 2013). In the present chapter, the effects of different mineral nanoparticles, as feed additives, on the growth performance of different animals have been discussed.

20.2 Nanomaterials

The term 'nano' is derived from the Greek word *nanos,* meaning dwarf and the size is <100 nm. Nanomaterials may be *naturally occurring* (as caseins in milk) or they may be man-made. When they are deliberately designed and produced, they are called *engineered* nanomaterials. Nanomaterials which are not deliberately designed and produced but which accidentally result from anthropogenic activities are called *incidental* nanomaterials.

20.3 Synthesis of Nanomaterials

Basically there are two methods for the synthesis of nanomaterials as given in Figure 20.1. Chemical methods also involve green route.

Metal NPs can be prepared by green route. The production of metal NPs by metal salt reduction through plant extracts is a comparatively facile ambient atmosphere activity. The working process is very simple; the plant extract and the solution of metal salt are mixed well at

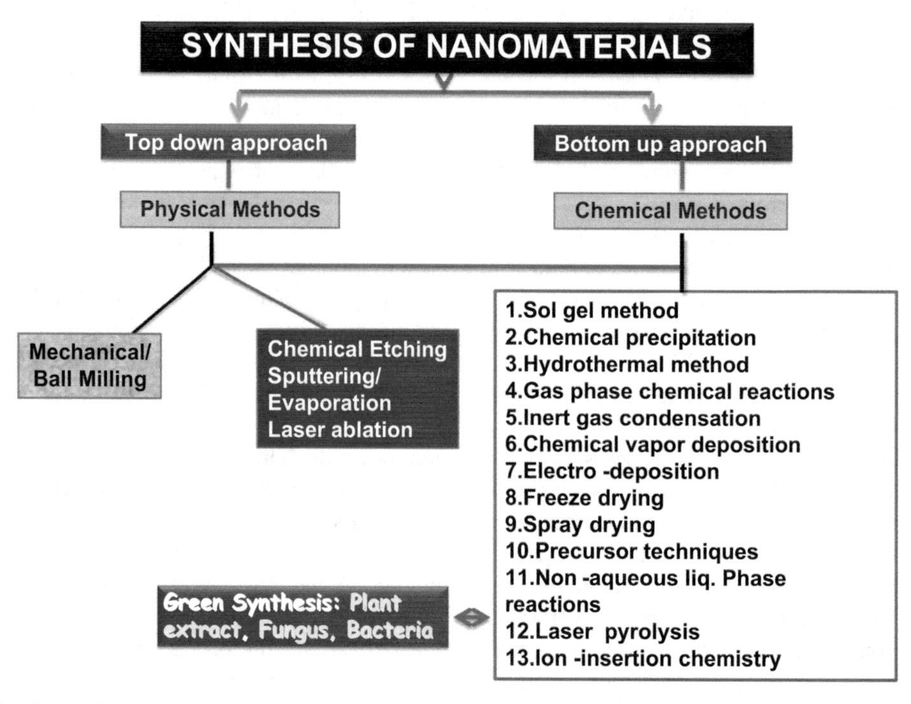

FIGURE 20.1 Synthesis of nanomaterials.

room temperature (Sharma et al., 2018). The reaction is completed within few minutes. Flowchart for green biosynthesis of metal NPs is shown in Figure 20.2 (Sharma et al., 2018).

Nano ZnO (nZnO) can also be prepared by green route i.e. by mixing zinc salt solution with plant extract (Figure 20.3) (Sharma et al., 2018).

In general, a number of bioreductants are found in plant extracts for reducing metal ions to nano metals (Figure 20.4) (Shamaila et al., 2016).

Different microbes are also used for green synthesis of nanomaterials (Figure 20.5) (Shamaila et al., 2016).

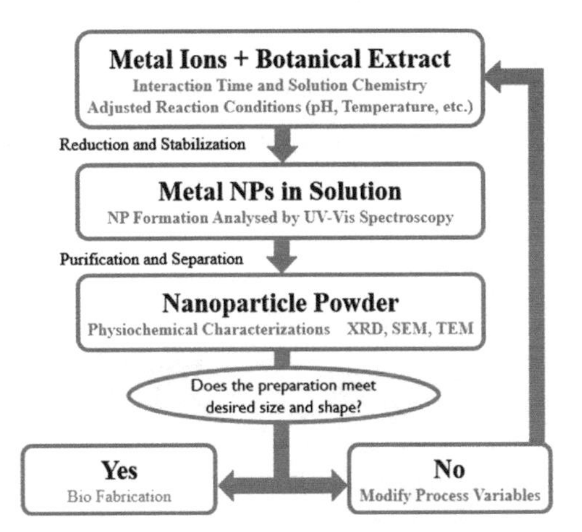

FIGURE 20.2 Flowchart for green biosynthesis of metal NPs (Sharma et al., 2018). (Reproduced with the permission of Elsevier.)

Some nanomaterials such as ZnO can be synthesized by precipitation method. Zinc sulfate is dissolved in water, and dilute solution of ammonium hydroxide is added till complete precipitation of $Zn(OH)_2$. The precipitate was filtered and washed several times with hot water and then heated at 400°C for 3 h to have nZnO.

20.4 Characterization of Nanomaterials

Basically there are four techniques to characterize nanomaterials. The techniques are scanning electron microscopy, transmission electron microscopy (TEM), atomic force microscopy (AFM) and powder X-ray diffraction (XRD). SEM is an electron microscope that images the sample surface by scanning it with a high energy beam of electrons and the instrument is shown in Figure 20.6a. SEM picture of nZnO is shown in Figure 20.6b (Zhang et al., 2014). TEM is a microscopic technique whereby beam of electrons is transmitted through an ultrathin specimen and interacts as it passes through the sample (Figure 20.6c). An electron beam illuminates the sample, and transmitted beam is imaged. TEM picture of nZnO is shown in Figure 20.6d (Varaprasad et al., 2016). Morphology of nanomaterials is examined by AFM technique (Figure 20.6e), and the AFM picture of nZnO is shown in Figure 20.6f (Sabry and Azeez, 2014). Using Scherrer formula (Eq. 20.1), the crystallite size is calculated by XRD technique (Figure 20.6g).

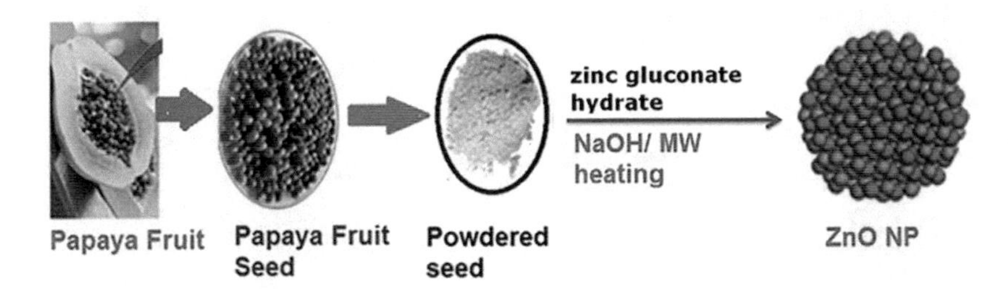

FIGURE 20.3 Green route for preparation of nZnO (Sharma et al., 2018). (Reproduced with the permission of Elsevier.)

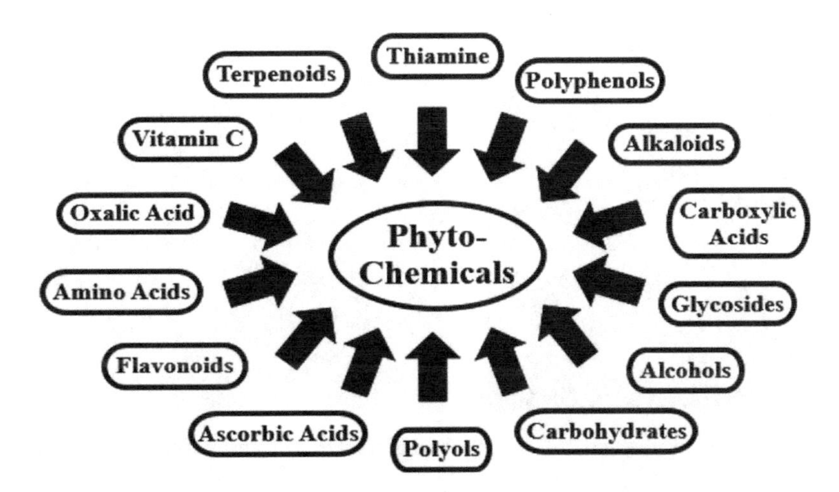

FIGURE 20.4 Important bioreductants found in plant extracts (Shamaila et al., 2016). (Reproduced with the permission of Elsevier.)

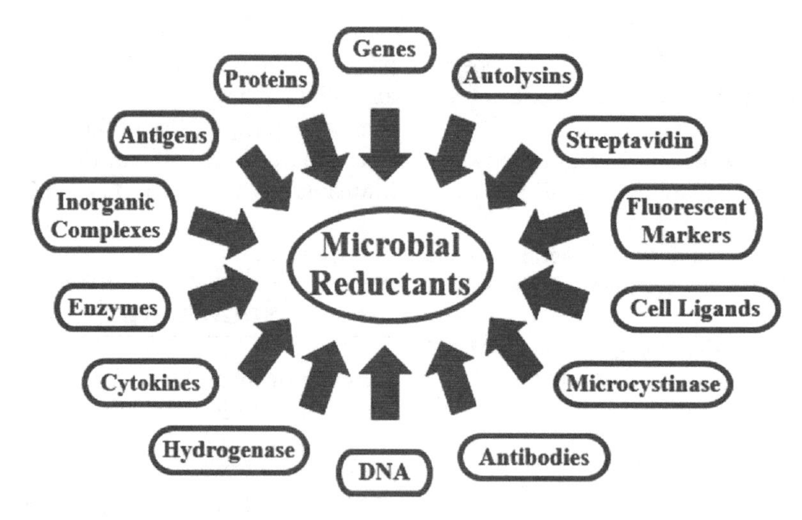

FIGURE 20.5 Reducing agents involved in microbial synthesis (Shamaila et al., 2016). (Reproduced with the permission of Elsevier.)

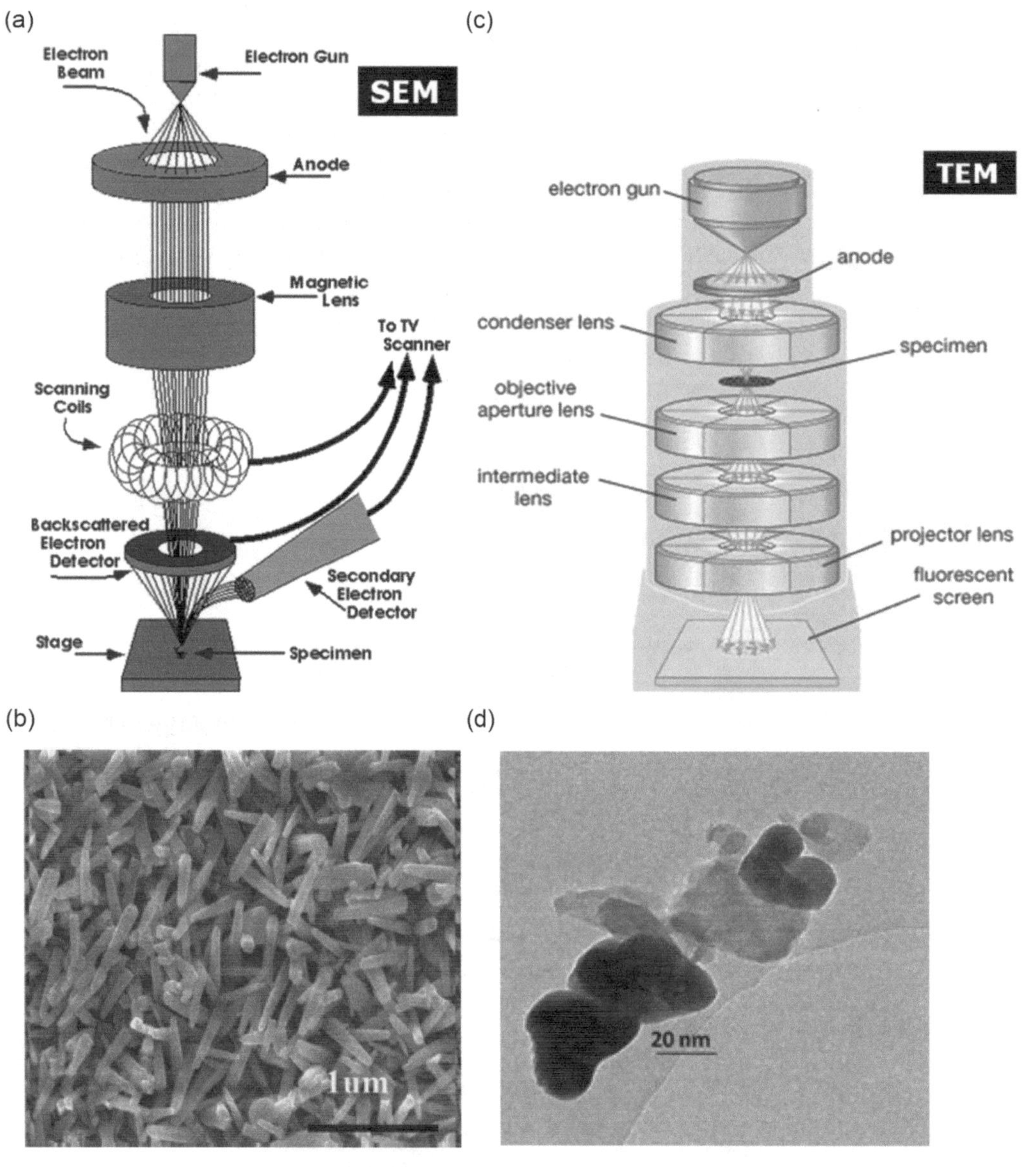

FIGURE 20.6 (a) Scanning electron microscope, (b) SEM picture of nZnO, (c) transmission electron microscope, (d) TEM picture of nZnO.

(Continued)

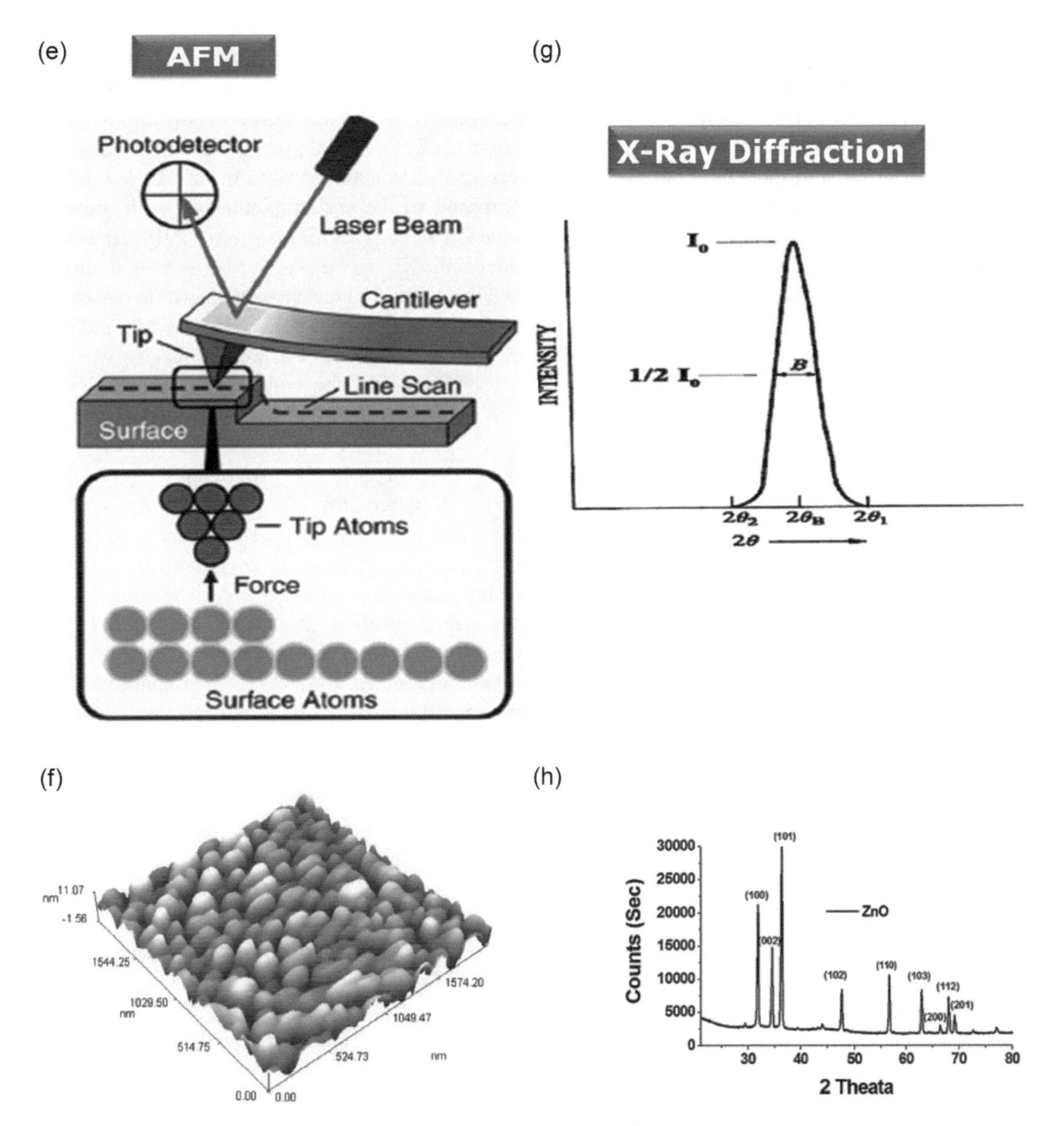

FIGURE 20.6 (CONTINUED) (e) atomic force microscope, (f) AFM picture of nZnO, (g) P. Scherrer method of crystallite size determination by XRD and (h) XRD of nZnO. (Reproduced with the permission of Elsevier.)

$$t = \frac{K\lambda}{B \cdot \cos\theta_B} \quad (20.1)$$

where t = thickness of crystallite, K = constant dependent on crystallite shape (0.89), λ = x-ray wavelength, B = FWHM (full width at half maxima) or integral breadth in radians and θ_B = Bragg angle. XRD of nZnO is shown in Figure 20.6h (Varaprasad et al., 2016).

20.5 Effect of Nano-Mineral Feeding on Livestock Production

Minerals refer to valuable chemical substances formed naturally in rocks and in the earth. Minerals are very important parts of animals as they are found in almost all cells, organs and tissues of the body, playing roles as either structural components or as cofactor of many enzymes and proteins. Minerals in animal body are categorized into macro and micro forms. Macro minerals (such as calcium, magnesium, potassium, and sodium) refer to those which are required and are found in large quantities in the animal body, while micro (or trace) minerals are those required by the animals, and are found in their systems, in very small amounts. Some trace minerals are so important to animal health that their deficiency leads to health abnormalities and even death. They are called essential micro minerals and include, in order of importance, iron (Fe), zinc (Zn), copper (Cu), selenium (Se), molybdenum (Mo), nickel (Ni), manganese (Mn), boron (Bo), chlorine (Cl) and iodine (I).

Despite their importance in animal physiology and metabolism, essential trace mineral requirements of animals are hardly met via conventional feeds. Hence, dietary supplementation is an imperative. The bioavailability of many minerals is low. The direct implication of this is the administration of a larger quantity of material than is usually required. The advent of nanotechnology has ushered in the use of nanostructured forms of essential trace elements as supplements in livestock feed, and there have been reports of positive results in terms of efficacy of action, efficiency of material use, cost-effectiveness and environmental friendliness. Nanomaterials can enter very easily in different parts of animal body, giving lots of advantages as schematically shown in Figure 20.7.

The effects of dietary administration of NPs of some essential trace minerals in livestock are discussed below.

20.5.1 Zinc

Zinc (Zn), the second most abundant trace element in animal body, can't be stored in the body and requires regular dietary intake to meet the physiological needs. As a component of numerous enzymes and hormones, Zn is necessary for the proper physiological functioning. In the animal body, nano-minerals interact more effectively with organic and inorganic substances due to their larger surface area (Swain et al., 2016). When given in appropriate amounts, zinc promotes growth, acts as antimicrobial agent and modulates both immune and reproductive systems of animals. The essentiality of Zn in different animal species is given in Table 20.1 (Swain et al., 2016).

However, health indicator has been observed to be more significantly enhanced in early weaned piglets (Yang and Sun, 2006) and layer chicken (Misra et al., 2014) fed nZnO supplemented diets compared to those fed the basal diet and conventional bulk zinc oxide (ZnO) supplemented diets. Two main symptoms of subclinical mastitis in cows are elevated somatic cell counts and depressed milk production, both of which can now be remedied by nZnO supplementation diets of the affected animals (Rajendran, 2013). Reproductive abnormalities in cow such as abortion can now be diagnosed by means of nano-sensors which are now available, just as placenta retention after calving can be prevented and general fertility in the animal improved by means of nano antioxidants (Rajendran, 2013). Similarly, supplementation of nZnO at 0.06 ppm to basal feed of broiler birds has been shown to significantly improve their immune response compared to dietary supplemented with organic and inorganic Zn at 15 ppm (Sahoo et al., 2014). It is reported that when ZnO-NP was given to hen as a food supplement, egg shell thickness and egg shell strength increased as compared to control ($P < 0.05$) (Abedini et al., 2017). The bone breaking strength and ash weight also increased in the presence of ZnO-NP. The same source has also suggested that nZnO supplementation in the rations of ruminants could cause changes in their fermentation kinetics as well as bring about alterations in the proportions of volatile fatty acids produced. There have also been indications that the growth of ruminal microorganisms, the synthesis of ruminal proteins as well as the efficiency of energy utilization in early incubation phase in ruminants could be improved through *in vitro* nZnO supplementation (Zhisheng, 2011). Recently it is found that growth indices such as final weight (FW), percent weight gain (% WG), feed conversion ratio (FCR) and hematological parameters were significantly ($P < 0.05$) improved in African catfish (*Clarias gariepinus*) fingerlings fed diets supplemented with nZnO, against those fed the basal and macro-ZnO supplemented diets (Onuegbu et al., 2018). Similar improvements on the biology, physiology and immune responses of the freshwater prawn, *Macrobrachium rosenbergii* through dietary nano-zinc supplementation, have also been reported (Muralisankar et al., 2014).

The performance of juvenile grass carp fish (*Ctenopharyngodon idella*) was studied in the presence of different sources of dietary Zn (ZnO, $ZnSO_4$ and ZnO-NPs), supplemented at low (30 mg/kg) and high (60 mg/kg) doses to a basal feed, using growth and blood parameters as indices to assess performance. Results showed that the best performance in terms of WG, growth rate and feed conversion ratio was from fish fed feed supplemented with nZnO at

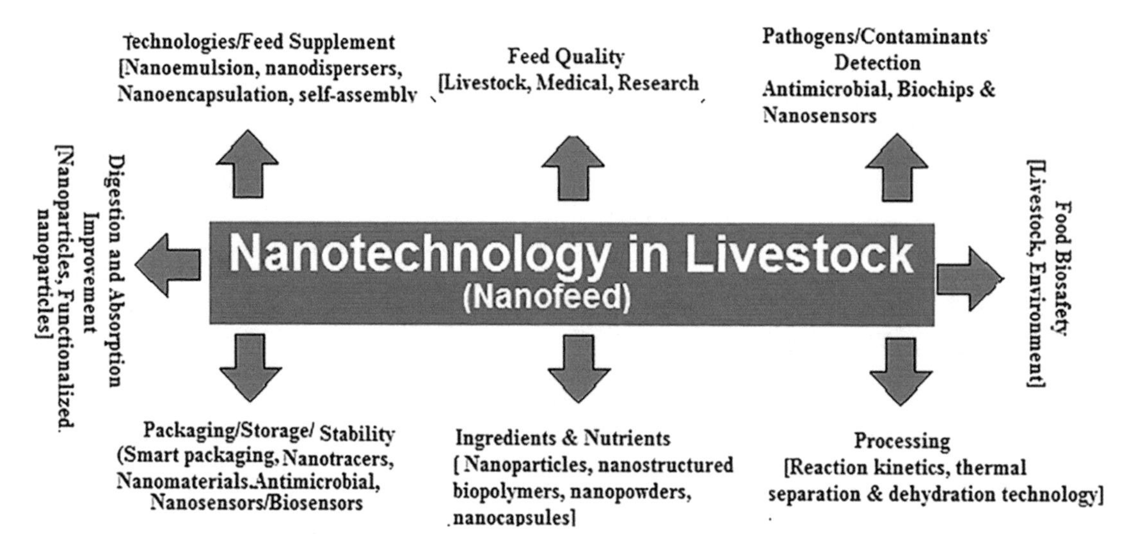

FIGURE 20.7 Nanotechnology in livestock.

TABLE 20.1 Essentiality of Zinc in Animal Species

Species	Systems under Investigation	Effects and Conclusions
Poultry	Growth, carcass trait, meat quality	Increased ADFI, ADG, DM and intramuscular fat contents of the breast muscle, percentage of eviscerated yield, redness value in breast muscle and pH values in thigh muscle and decreased shear force in thigh muscle, drip loss in breast and thigh muscle
Mice	Epithelial cells	Enhances proliferation of the cells and does not injure the cells at lower concentrations; impacts on epithelial cell integrity of the animals
Mice	Vision	Accumulation of inclusion bodies in the retinal pigment epithelium cause its alterations due to Zn-deficient diets
Mice	Immunity	Alcoholism reduces Zn transporter gene expression and, thus, reduces immunity compared with normal subjects
Rabbit	Reproduction	Increased semen volume, total live sperm concentration, percent sperm motility, conception rate in heat-stressed rabbits
Ruminants	Reproduction	Higher incidence of abortions and stillbirths in Zn-deficient ewes
In vivo and *in vitro*	Antioxidant	Exhibits antioxidant-like effects *in vitro*. At pharmacological doses *in vivo*, Zn has a protective effect against pro-oxidants and dietary Zn deficiency predisposes to oxidative damage in cells by protection of sulfhydryl groups and inhibits production of reactive oxygen species (ROS) by transition metals
Goats	Immunity	Enhanced resistance to udder stress in dairy goats to Zn supplementation
Human, lab animals	Immunity	Reduced immune responses and disease resistance in Zn-deficient subjects

Source: Swain et al. (2016)

low concentration (30 mg/kg), followed by those fed same supplement at high concentration (60 mg/kg). Also, best hematological parameters were obtained in fish fed diet supplemented with low dose of nZnO. The high dose of the supplement was reported to have caused significant decrease ($P < 0.05$) in the parameters. This result was interpreted to mean preference for ZnO-NPs as dietary zinc source for the fish (Faiz et al., 2015). Other effects of ZnO-NPs on animal well-being are given in Table 20.2.

20.5.2 Selenium

Selenium is an essential trace element for animals on account of its wide range of biological functions (Zhou et al., 2017). It is considered very essential in the physiological processes of growth, fertility, immune system modulation, hormone metabolism and antioxidant defense in animals and humans. It is also a very important structural component of many enzymes and proteins in animal tissues, particularly glutathione peroxidase (GSH-PX), which is the enzyme that destroys and inhibits the actions of naturally occurring peroxides that cause cell damage. Many physiological disorders in poultry such as oxidative diathesis, pancreatic dystrophy and nutritional muscle dystrophy of the gizzard, heart and skeletal muscle, as well as lowered fertility/laying capacity and decreased feathering have been associated with Se deficiency. Hence, dietary Se supplementation in animal feed is important to guard against deficiency and promote healthy growth and productivity. Several studies have confirmed the need for dietary selenium supplementation for the enhancement of growth and physiological activities in both man and farm animals (Cao et al., 2017; Zhang et al., 2014). However, excess dietary selenium is reported to be poisonous (Zhou et al., 2017). The efficacy of conventional dietary selenium sources (inorganic selenium as sodium selenate or selenite) is limited by several factor which include low intestinal absorption, interaction and antagonism with other feed minerals and nutrients, loss in storage, narrow safety margin, poor supply and maintenance of body selenium reserve and nonspecific binding pattern to tissue proteins. The use of nano-selenium (nSe) as an alternative source of dietary selenium for supplementation in livestock has increased. A couple of studies have been carried out with this novel source of Se for a number of farm animals including chicken (Visha et al., 2017; Mohapatra et al., 2014), swine (pig) and cow (Horky, 2015; Sordillo, 2016), with improvement in different aspects of their physiology, health and growth. Similarly, reports (Mohapatra et al., 2014) have demonstrated that layer grower birds showed significantly ($P < 0.05$) higher body weight when fed diets supplemented with nSe up to a concentration of 0.3 mg/kg diet, compared to birds fed the basal, and sodium selenite supplemented, diets. Generally, the report showed that the body weight, antioxidant status, feed consumption ratio, tissue Se deposition and immunity of grower birds were significantly ($P < 0.05$) improved by dietary nSe

TABLE 20.2 Effect of Nano Zinc oxide on Animal Performance

Species	Effects	Remarks
Pig	Immunity	Diarrhea incidence reduced
Poultry (Broilers)	Growth; feed consumption	Improves in growth performance, FCR and dressing performance; decreases in the cost of production
Cattle (Holstein Friesian)	Milk production	Reduce somatic cell count in subclinical mastitis. Increase in milk production
Poultry	Growth	Improves growth performance and FCR
In vitro supplementation of 100 and 200 mg/kg of ZnO NP at the 6th and 12th h of incubation	Fermentation	Improved concentration of volatile fatty acid and microbial crude protein production and fermentation of organic matters. Concentration of ammonia nitrogen and the ratio of acetate to propionate are adversely affected
Sheep	Reproduction	High incidence of abortions and stillbirths in the ewes in ZnO NP-deficient diets

Source: Swain et al. (2016)

supplementation compared to bulk sodium selenite supplementation. In male Boer goats, dietary elemental nSe supplementation has been used to improve testis ultrastructure, glutathione peroxidase activity and semen quality (Shi et al., 2010); while improvement in growth performance, blood parameters and tissue concentrations of selenium has been achieved in growing male goats through the same dietary nSe supplementation (Shi et al., 2011). This implied more effective utilization of selenium in the nano, against its macro organic and inorganic forms in animal systems. nSe has reportedly (Xun and Shi 2012) improved ruminal fermentation and feed conversion efficiency in sheep. It was therefore suggested by the researchers that the material (nSe) is preferably used as dietary (nutritional) selenium source not just for sheep but for ruminants at large. Similarly, studies (Sadeghian et al. 2012) have found that nSe, as dietary selenium supplement in sheep feed, was less toxic and more bioactive and was recommended for that purpose against other forms of selenium. Feeding broiler chicken with nSe supplemented diet at 0.1875 mg/kg was reported to bring about an improvement in the gene, GSHP $\times$ 1 mRNA, expression to the tune of 430% against birds fed diet containing conventional bulk selenium particles supplemented at 0.15 mg Se kg^{-1} dry feed (Aparna et al., 2017). Similarly, interesting improvements in growth (measured by specific growth rate, feed conversion efficiency, % WG and FCR); serum growth hormone concentration and hematological parameters (red blood cells count, hemoglobin level and hematocrit value) were recorded in fish, *Tor putitora*, by a nutrient synergism mediated by a combination of vitamin C (100 mg/kg) and nSe (0.68 mg/kg) supplemented in a basal diet against the control diet (Khan et al., 2017).

20.5.3 Iron

Iron is very important in the physiological functioning in all higher animals. It is critical in such processes as oxygen transport, cellular respiration and lipid peroxidation. It is also very essential in immune system modulation, body defense against infections, steroid synthesis, DNA synthesis, drug metabolism and electron transfer (Crichton, 1991). Iron is a constituent of many proteins (such as hemoproteins: hemoglobin, myoglobin and non-hemoproteins: ferritin, transferrin) and is a very important cofactor for a wide range of iron-dependent enzymes. Physiological disorders such as vulnerability to diseases, changes in hematological parameters, microcytic anemia, poor feed conversion, growth depression and immune suppression in some animals are associated with iron deficiency. Dietary iron supplementation is therefore considered very important for healthy growth and performance in livestock. Iron supplementation in animal feed in two oxidation states, and redox interconversions forms, ferrous (Fe(II)) and ferric (Fe(III)), to meet dietary requirements have been known for long and are central to the biological properties of this trace element (Ponka et al., 2015). But most chemical forms of iron (with the exception of ferrous fumarate ($C_2H_2FeO_4$) and ferrous sulfate ($FeSO_4{\cdot}7H_2O$)) have low bioavailability while others induce offensive sensory changes (such as color and/or taste) to feed, in addition to being poorly absorbed. Iron oxides have been reported to give better sensory effects to feed but have poor solubility and bioavailability. Improving iron bioavailability and its imparted sensory appeals improves livestock feed quality and guarantees higher feed intake and better performance. Solubility and bioavailability are particle size-dependent phenomena, and particle size reduction of otherwise poorly soluble iron compounds to the micrometer range has been known to significantly improve both their solubility and bioavailability. Consequently, several health indices in animals such as growth, hematological and immunological parameters are now better improved by the use of nano, rather than bulk, forms of iron as dietary supplement. For instance, significant improvements in these parameters have been reported in fish, *Labeo rohita* H, fed diets supplemented with low levels of ferric oxide in nano form (Behera et al. 2014). Similarly, in a study (Srinivasan et al., 2016) on the effects of dietary iron oxide NPs on the growth performance, biochemical constituents and physiological stress response of the giant prawn (*Macrobrachium rosenbergii*), not only was the carcass iron content significantly elevated in fish fed nano-iron supplemented diet against those fed the control, but contents of other minerals (Cu, Ca, Mg, Zn, k and Na) were also significantly elevated. This implied that the uptake and body incorporation of the supplemented iron oxide NPs facilitated the absorption of other minerals also. Also, increased growth rates of 30% and 24%, respectively, were achieved in fish, young carp (*Carassius auratus*) and sturgeon (*Acipenser gueldenstaedtii*) through dietary supplementation of iron NPs.

20.5.4 Copper

Copper is one of the major trace elements that play vital role in animal metabolism, but its delivery via oral pathway is challenged by limited bioavailability, as in conventional nutrient delivery. The importance of copper in animal nutrition derives from its critical roles in diverse physiological processes such as hemoglobin synthesis, bone formation and myelin maintenance in the nervous system. It is also a key structural component of important enzymes such as cytochrome oxidase, dopamine hydroxylase ferroxidase and tyrosinase, and it is also a growth promoter (Mohseni et al., 2014). Like most trace minerals, copper requirements of animals are hardly met from conventional feed sources, thereby making dietary supplementation a necessity.

However, the chemical form in which Cu is supplemented in feed is a big factor in determining its bioavailability. Recently, the discovery and production of materials with nano-range dimensions, and their novel properties, have opened a new gateway for investigations on their possible effects, as alternatives to their bulk counterparts, as supplements in livestock feeds (El-Basuini et al., 2016).

20.5.5 Silver

Silver NP is generally used in colloidal form in poultry. Silver NPs can directly target a particular cell and interact. Growth and development of muscle cells are enhanced in presence of Ag NPs. As a result, body weight gain due to the expansion of breast muscle can be increased. NPs are studied as nutritional supplements in diets for the improvement of broiler health and performance since they are able to carry nutrients directly to cells. Broilers given diets containing Ag NPs (2, 4, 6, 8 and 10 ppm concentrations) showed increased body weight gain and total serum antioxidant with decreased serum total protein and decreased cholesterol (Gangadoo et al., 2016).

20.6 Miscellaneous Role of Nanoparticles to Livestock

Besides the specific effects produced in animals by oral administration of NPs as discussed so far, there are several effects/benefits which are achievable or derivable via oral administration of NPs to livestock. In all animals, NPs have found applications in medical imaging, diagnosis and as microbial therapies. On account of their small particle size, NPs can be used to encapsulate nutrients for easy, safe and timely delivery to the blood stream. Also certain quantum effects of some NPs can be utilized to deliver drugs not only to specific/target sites but to target cells in animals, for more effective treatment of some diseases. In addition to these, they act as nutraceuticals with better intestinal interactions and higher bioavailability in animals, compared to their bulk counterparts (Hill and Li, 2017).

20.7 Digestion and Absorption of Nanoparticles

Digestion refers to the breakdown of complex food materials, consumed by animals, into simpler substances usable by the animal body. It is a complex biochemical processes involving some nutrients as reactants and natural organic catalysts called enzymes. Several nutrients (carbohydrates, proteins, fats minerals and vitamins) are needed in the right amounts and availability in animal body to achieve healthy growth and reproduction. When nutrients, especially minerals, play complimentary/synergic roles and enhance the effectiveness of one another, they are called agonists; if their roles oppose the effectiveness of one another, they are antagonists. Mineral synergism and antagonism have been known to influence extents and rates of digestion in animals. Useful end products of digestion must cross the domain of digestion (the GIT) and enter the blood stream from where they are distributed to all cells, tissues and organs of the body for the cell/tissue building and repair. Any undigested and unassimilated material passes out of the GIT as a waste, not biologically available for anabolic processes. The more materials digested and assimilated, the better the production and productivity of the animal involved. Hence, success in farm animal production depends on the efficiencies of feed digestion, assimilation and conversion to body tissues to attain fast growth rate and high fecundity. Hence an underlying issue in increasing the efficiency of livestock production is an understanding of basic scientific principles such as genetics, nutrition, utilization of cellulose by ruminants among others (Scanes, 2018).

The effects of nanomaterials in the growth and health of animals have been studied along their roles in the metabolic processes of nutrient digestion, assimilation, distribution and cellular synthesis. In these regards, the following have been suggested as possible roles of NPs:

I. They may act as catalysts and speed up otherwise slow catabolic and anabolic reactions.
II. They may be useful as functional materials required in anabolism (Figure 5a); or they may act as carriers for products of digestion from the GIT to the blood stream, tissues and cells either by conjugation or by encapsulation. In these cases NPs pass through GIT, and other barriers, to reach reaction sites easily, safely and more rapidly, while the enhanced exposed surface areas and surface chemistry increase their reaction kinetics and make reactions proceed at faster rates. Consequently, more nutrients, conjugated or encapsulated, can be delivered for cellular reactions before being degraded, thus enhancing bioavailability and performance. Other novel properties such as magnetic and photo effects can also be utilized in controlled nutrient delivery, while same nano size can confer on materials catalytic properties which may be absent in their bulk forms.

20.8 Excretion of Nanoparticles in Livestock

Materials that enter animal body are either taken up and utilized to the benefit of the animal or rejected and excreted from the body as wastes. Excretion pathways are; the GIT, as undigested/unassimilated materials and the urinary tract/the skin as by-product of renal (liver) filtration which, according to Singh (2014), permit hydrophilic particles <5 nm diameter to pass into the urine for the maintenance of body homeostasis. However, NP excretion is not always an indication of an immediate toxicity but that of removal of potential (long-term) health risks. A clear perception of excretion pathways of NPs is critical for their safe application in biomedicine; but this is still not very clearly understood. A recent insight into this problem involving direct live examination of the actions of activated carbon NPs has revealed a novel excretion pathway; 30–200 nm activated carbon nanoparticles (ACNP) microinjected into zebrafish yolk sac were observed to be directly excreted into, and through, the gut lumen (intestinal tract) without connecting the hepato-biliary (hap-bile) system. Similarly, organ distribution of orally administered silica NPs in organs (kidney, liver, lung and spleen) of animals such as rat is known to be particle-size independent. The particles tend to retain their particulate forms, but with higher decomposition of the lower particle size group taking place in the kidney. Excretion of these particles from the body is however particle size dependent, being more rapid among smaller, compared to the larger, particle size ranges, and occurring principally by the urinary and fecal excretion pathways.

20.9 Encapsulated Food Components and Edible Supplements

Nutritional supplements and nutraceuticals having nanosized ingredients such as vitamins, antimicrobials, antioxidants, and preservatives are now available for enhanced taste, absorption, and bioavailability. The ingredients when encapsulated have several advantages, such as a longer shelf life, enhanced stability, consecutive delivery of multiple active ingredients, and pH-triggered controlled release. Functional nutrients such as vitamins, antioxidants, probiotics, carotenoids, preservatives, omega fatty acids, proteins, peptides, and lipids as well as carbohydrates are incorporated into a nano delivery system. The functionality and stability of these foods are increased, as they are not used in their pure form. Lipid based nanoencapsulation can potentially improve the solubility, stability, and bioavailability of foods, thus preventing unwanted interactions with other food components. Nanoencapsulated probiotics can be selectively delivered to certain parts of the GIT, where they have the capability to modulate immune responses (Pathakoti et al., 2017).

20.10 Nanotoxicity in Livestock

Nanotoxicity refers to all forms of adverse effects induced in a system as a result of the presence of nanomaterials. Nanotoxicity has attracted a lot of research interests because of the potential dangers posed by the increasing synthesis and applications of nanomaterials on humans, animals and the environment. Toxicity of nanomaterials is rooted on the same properties which have endeared them to general acceptance and wide applications; i.e. small particle size, large exposed surface area and quantum effects among others. It has also been shown that nanotoxicity is related to exposure time, exposure dose/concentration, the biological model involved as well as the reference protocol (Christen and Fent, 2012). Studies have indicated that nanomaterials can be harmful to the animal and human systems as well as the environment if inappropriately handled and/or administered. In rainbow trout, for instance, feeding diet supplemented with ZnO NPs (20–30 nm) at 1,000 mg/kg feed for 10 days, reportedly induced oxidative stress in the liver, indicating the potentiality of ZnO NPs to interfere with cytochrome metabolic processes in the animal (Connoly et al., 2016). Similarly, rats exposed to diet supplemented with ZnO NPs at 2,000 mg/kg feed were reported to have suffered severe hemato-biochemical abnormalities after 48 h of exposure (Srivastav et al., 2016). Copper oxide NP is viewed as the most toxic of metal oxide NPs. In a study exposing red sea bream (*Pagrus major*) to diets supplemented with Cu-NPs, depressed tolerance to low-salinity stress was reported in fish fed Cu-NPs concentrations above 4 mg/kg feed (El-Basuini et al., 2016). In a comparative study of the toxicity of three different metal oxide NPs (CuO-NPs, CdO-NPs and TiO_2-NPs), it was shown (Zhu et al., 2013) that the most potent of the three in cytotoxicity and DNA damage was CuO-NPs, which gave rise to 8-hydoxy-2-dyoxysuanosine (8-OHdG) formation. TiO_2-NPs showed least toxicity among the three and induced no significant level of 8-OHdG. Similarly, investigation involving comparative study of cytotoxicity of four metal oxide NPs (CuO-NPs, TiO_2-NPs, ZnO-NPs and Co_3O_4-NPs) was conducted using primary hepatocytes of catfish and human HepG2 cells as models. Report showed that induced toxicity in both cells followed the trend, $TiO_2 < Co_3O_4 < ZnO < CuO$; with higher levels of toxicity occurring in the human cells against the primary hepatocytes of fish (Wang et al., 2011).

Despite its excellent bioactivities (like anti-cancer and immunomodulation properties), nSe is feared to be more or less toxic than the other forms depending on the type and surface stabilization of the nSe involved. A recent study (Shi et al., 2018) reported that nSe showed similar toxicity compared to the other forms, in a feeding trial with Japanese medaka fish (*Oryzias latipes*).

20.11 Conclusions and Future Prospects

Nanostructured forms of minerals hold great promises for application as feed supplements in the livestock industry. Even at low concentrations, they have shown better effectiveness than their conventional (organic and inorganic) bulk counterparts. Nanotechnology has been demonstrated to have the capacity and promise to impact livestock nutrition through feed quality improvement, enhancement of nutrient bioavailability, improvement of growth and production performances as well as enhancement of animal immune status. Studies have shown that nano-applications have the potential to provide smarter solutions for various applications in the livestock production systems, which can help in reducing costs and enhancing the final product quality. However, nanoscience is at its initial stages in the field of mineral nutrition, and further work is required to understand the effect of nano-minerals, their site of absorption, mechanism of absorption, molecular basis of distribution and mode of action. The expected risks and hazards related to nano-application in livestock production systems that can affect animal, human and environment should also be studied in detail. Further, research should be directed to find the optimum levels of nanomaterials that can provide better performance and economic benefits. Extensive risk assessments should be conducted to ensure the safety of the nano-products before making them available for animal or human use.

References

Abedini M., Shariatmadari F., Torshizi M.A.K., and Ahmadi H., 2017. Effects of a dietary supplementation with zinc oxide nanoparticles, compared to zinc oxide and zinc methionine, on performance, egg quality, and zinc status of laying hens. *Livestock Science*, 203: 30–36.

Aparna N., Karunakaran R., and Parthiban M., 2017. Effect of selenium nano particles on glutathione peroxidase mRNA gene expression in broiler chicken. *Indian Journal of Science and Technology*, 10(32):0974–6846.

Behera T., Swain P., Rangacharulu P.V., and Samanta M., 2014. Nano-Fe as feed additive improves the hematological and immunological parameters of fish, *Labeo rohita* H. *Applied Nanoscience*, 4:687–694.

Benttencourt E.M.V., Tilman M., Narcisco V., Calvalho M.L.S., and Henriques P.D.S., 2015. The livestock roles in the wellbeing of rural communities of timor-leste. *Revista de Economia e Sociologia Rural*, 53. doi:10.1590/1234-56781806-94790053s01005.

Bruggar D. and Windisch W.M., 2015. Environmental responsibilities of livestock feeding using trace mineral supplements. *Animal Nutrition*, 1(3):113–116.

Cao L., Zhang L., Zeng H., Wu R.T, Wu T.-L., and Cheng W.-H., 2017. Analyses of selenotranscriptomes and selenium concentrations in response to dietary selenium deficiency and age reveal common and distinct patterns by tissue and sex in telomere-dysfunctional mice. *The Journal of Nutrition*, 147(10):1858–1866.

Christen V. and Fent K., 2012. Silica nanoparticles and silver-doped silica nanoparticles induce endoplasmatic reticulum stress response and alter cytochrome P4501A activity. *Chemosphere*, 87:423–434.

Connoly M., Fernandez M., Conde E., Torrent F., Navas J.M., and Fernandez-Crez M.L., 2016. Tissue distribution of Zn and subtle oxidative stress effects of after dietary administration of ZnO nanoparticles to rainbow trout. *Total Environment*, 551–552:334–343.

Crichton R. R., 1991. *Inorganic Biochemistry of Iron Metabolism*. Ellis Horwood, West Sussex.

El-Basuini M.F., El-Hais A.M., Dawood M.A.O., Abou-Zeid A.E.-S., EL-Damrawy S.Z., EL-Sayed Khalafalla M.M. et al. 2016. Effect of different levels of dietary copper nanoparticles and copper sulfate on growth performance, blood biochemical profiles, antioxidant status and immune response of red sea bream (*Pagrus major*). *Aquaculture*, 455:32–40.

Faiz H., Zuberi A., Nasir S., and Rauf M., 2015. Zinc oxide, zinc sulfate and zinc oxide nanoparticles as source of dietary zinc: Comparative effects on growth and hematological indices of juvenile grass carp (Ctenopharyngodon idella). *International Journal of Agriculture and Biology*, 17:568–574.

Gangadoo S., Stanley D., Hughes R.J., Moore R.J., and Chapman J., 2016. Nanoparticles in feed: Progress and prospects in poultry research. *Trends in Food Science and Technology*, 58:115–126.

Hill E.K. and Li J., 2017. Current and future prospects for nanotechnology in animal production. *Journal of Animal Science and Biotechnology*, 8:26.

Horky P., 2015. Effect of selenium on its content in milk and performance of dairy cows in ecological farming. *Potravinarstvo*, 9:324–329.

Kaur G., Brar Y.S., and Kothari D.P., 2017. Potential of livestock generated biomass: Untapped energy source in India. *Energies*, 10(7):847.

Khan K.U., Zuberi A., Nazir S., Ullah I., Jamil Z., and Sarwar H., 2017. Synergistic effects of dietary nano selenium and vitamin C on growth, feeding, and physiological parameters of mahseer fish (*Tor putitora*). *Aquaculture Reports*, 5:70–75.

Misra A., Swain R.K., Misra S.K., Panda N., and Sethy K., 2014. Growth performance and serum biochemical parameters as affected by nano zinc supplementation in layer chicks. *Indian Journal of Animal Nutrition*, 31(4):384–388.

Mohapatra P., Swain R.K., Mishra S.K., Behera T., Swain P., Behura N.C. et al., 2014. Effects of dietary nano-selenium supplementation on the performance of layer grower birds. *Asian Journal of Animal and Veterinary Advances*, 9:641–652.

Mohseni M., Pourkazemi M., and Bai S.C., 2014. Effects of dietary inorganic copper on growth performance and immune responses of juvenile beluga, Huso huso. *Aquaculture Nutrition*, 20:547–556.

Muralisankar T., Bhavan P.S., Radhakrishnan S., Seenivasan C., Manickam N., and Srinivasan V., 2014. Dietary supplementation of zinc nanoparticles and its influence on biology, physiology and immune responses of the freshwater prawn, Macrobrachium rosenbergii. *Biological Trace Element Research*, 160:56–66.

Onuegbu C.U., Agarwal A., and Singh N.B., 2018. ZnO nanoparticles as feed supplement on growth performance of cultured African Catfish Fingerlings. *Journal of Scientific and Industrial Research*, 77:213–218.

Pathakoti K., Manubolu M., and Hwang H.-M., 2017. Nanostructures: Current uses and future applications in food science. *Journal of Food and Drug Analysis*, 25: 245–253.

Ponka R., Abdou Bouba A., Fokou E., Beaucher E., Piot M., Leonil J. et al., 2015. Nutritional composition of five varieties of pap commonly consumed in Maroua (Far-North, Cameroon). *Polish Journal of Food and Nutrition Sciences*, 65:183–190.

Rajendran D., 2013. Application of nano minerals in animal production system. *Research Journal of BioTechnology*, 8(3):1–3.

Ramachandraiah K., Choi M.-J., and Hong G.-P., 2018. Micro- and nano-scaled materials for strategy-based applications in innovative livestock products: A review, *Trends in Food Science and Technology*, 71:25–35.

Sabry R.S. and Azeez O.A., 2014. Hydrothermal growth of ZnO nano rods without catalysts in a single step. *Manufacturing Letters*, 2:69–73.

Sadeghian S., Kojouri G.A., and Mohebbi A., 2012. Nanoparticles of selenium as species with stronger physiological effects in sheep in comparison with sodium selenite. *Biological Trace Element Research*, 146(3): 302–308.

Sahoo A., Swain R.K., and Mishra S.K., 2014. Effect of inorganic, organic and nano zinc supplemented diets on bioavailability and immunity status of broilers. *International Journal of Advanced Research*, 2(11): 828–837.

Scanes C., 2018. Animal agriculture: Livestock, poultry, and fish aquaculture. In: *Animals and Human Society*, Scanes C. and Toukhsati S. (eds.). Academic Press, Elsevier, London, 133–179.

Shamaila S., Sajjad K.A.L., Ryma N.-A., Farooqia S.A., Jabeen N., Majeed S. et al., 2016. Advancements in nanoparticle fabrication by hazard free eco-friendly greenroutes. *Applied Materials Today*, 5:150–199.

Sharma D., Sabela M.I., Kanchi S., Bisetty K., Skelton A.A., and Honarparvar B., 2018. Green synthesis, characterization and electromechanical sensing of silymarin by ZnO nanoparticles: Experimental and DFT studies. *Journal of Electroanalytical Chemistry*, 808:160–172.

Shi L.G., Xun W., Yue W., Zhang C., Ren Y., Liu Q. et al., 2011. Effect of elemental nano-selenium on feed digestibility, rumen fermentation, and purine derivatives in sheep. *Animal Feed Science and Technology*, 163: 136–142.

Shi L.G., Yang R.J., Yue W.B., Xun W.J., Zhang C.X., and Ren Y.S., 2010. Effect of elemental nano-selenium on semen quality, glutathione peroxidase activity and testis ultrastructure in male Boer goats. *Animal Reproduction Science*, 118:248–254.

Shi M., Zhang C., Xia I.F., Cheung S.T., Wong K.S., Wong K.-H. et al., 2018. Maternal dietary exposure to selenium nanoparticles led to malformation in offspring. *Ecotoxicology and Environmental Safety*, 156: 34–40.

Singh A.K., 2014. Nanoparticle pharmacokinetics and toxicokinetics. In: *Engineered Nanoparticles: Structures, Properties and Mechanisms of Toxicity*. Available at https://books.google.co.in/books?isbn=012801492X.

Sordillo L.M., 2016. Nutritional strategies to optimize dairy cattle immunity. *Journal of Dairy Science*, 99:1–16.

Srinivasan V., Saravana Bhavan P., Rajkumar G., Satgurunathan T., and Muralisankar T., 2016. Effects of dietary iron oxide nanoparticles on the growth performance, biochemical constituents and physiological stress responses of the giant freshwater prawn Macrobrachium rosenbergii post-larvae. *International Journal of Fisheries and Aquatic Studies*, 4(2):170–182.

Srivastav A.K., Kumar M., Ansari N.G., Jain A.K., Shankar J., Arjaria N. et al., 2016. A comprehensive toxicity study of zinc oxide nanoparticles versus their bulk in wistar rats: Toxicity study of zinc oxide nanoparticles. *Human and Experimental Toxicology*, 35(12): 1286–1304.

Swain P.S., Rao S.B.N., Rajendran D., Dominic G., and Selvaraju S., 2016. Nano zinc, an alternative to conventional zinc as animal feed supplement: A review. *Animal Nutrition*, 2:134–141.

Tazhibaev S., Musabekov K., and Yesbolova A., 2014. Issues in the development of livestock sector in Kazakhstan. *Procedia-Social and Behavioral Sciences*, 143:610–614.

Varaprasad K., Raghavendra G.M., Jayaramudu T., and Seo J., 2016. Nano zinc oxide–sodium alginate antibacterial cellulose fibres. *Carbohydrate Polymers*, 135:349–355.

Visha P., Nanjappan K., Selvaraj P., Jayachandran S. and Thavasiappan V., 2017. Influence of dietary nanoselenium supplementation on the meat characteristics of broiler chickens. *IJCMAS*, 6(5):340–347.

Wang Y., Aker W.G., and Hwang H.M. 2011. A study of the mechanism of in vitro cytotoxicity of metal oxide nanoparticles using catfish primary hepatocytes and human HepG2 cells. *Science of the Total Environment*, 409(22):4753–4762.

Xun W. and Shi L., 2012. Effect of high-dose nano-selenium and selenium-yeast on feed digestibility, rumen fermentation, and purine derivatives in sheep. *Biological Trace Element Research*, 150(1–3):130–136.

Yang Z.P. and Sun L.P., 2006. Effects of nanometre ZnO on growth performance of early weaned piglets. *Journal of Shanxi Agricultural Sciences*, 3:024.

Zhang J.-L., Zhang Z.-W., Shan A.-S., and Xu S.-W., 2014. Effects of dietary selenium deficiency or excess on gene expression of selenoprotein in chicken muscle tissues. *Biological Trace Element Research*, 157:234–241.

Zhisheng C.J., 2011. Effect of nano-zinc oxide supplementation on rumen fermentation *in vitro*. *Chinese Journal of Animal Nutrition*, 8:023.

Zhou J.-C., Zheng S., Mo J., Liang X., Xu Y., Zhang H. et al., 2017. Dietary selenium deficiency or excess reduces sperm quality or testicular mRNA abundance of nuclear Glutathione Peroxidase 4 in rats. *The Journal of Nutrition*, 10:1947–1953.

Zhu X., Hondroulis E., Liu W., and Li C.-Z. 2013. Biosensing approaches for rapid genotoxicity and cytotoxicity asseys upon nanoparticles exposure. *Small*, 9:1821–1830.

21

Nanobiosensors and Its Application in Agriculture and Food

Mohammed Nadim Sardoiwala, Anup K. Srivastava, Subhasree Roy Choudhury, and Surajit Karmakar
Institute of Nano Science and Technology

21.1 Introduction

Reduction of fertile agriculture land and resources of farming with a continuous increment of heavy metals, herbicides and pest control chemicals in agricultural land are very sensitive issues today. Economic growth of any country also depends on the fruitfulness of agriculture and food sectors. Detection and removal of heavy metals, herbicides, pesticides and other toxic compounds from agricultural and food resources demand newer technologies and sensing approaches. Conventional technologies have biotechnological approaches to enhance shelf life of agriproducts and sensing techniques based on pH, temperature and chemical reactivity to monitor and enhance the quality of food and agricultural outputs (Heldman and Lund, 2011). Limitations of these conventional technologies include the use of hazardous chemicals, cost of techniques and requirement of trained hands. Hence, there is need of new technology to enhance the shelf life of foods and effective monitoring of food quality to protect humans from adverse effects of ruined foods. In demand of 21st century, nanotechnology emerges as a new technology with the advancement of earlier conventional techniques to make a revolution in food and agriculture industries (Neethirajan and Jayas, 2011). Nanoscience and technology have already been applied in wide areas such as electronics and communication, energy production, health and medicine, diagnostic and medical examination and computer. There are various nanotechnology-based commercial products such as long-lasting varnishes and paints, stain and odor repulsive materials, dirt protective coating, transparent sunscreens, moisturizers and light scattering cosmetics. Nanotechnologies are beneficial on several aspects due to different physical, chemical and biological features of nanomaterials as compared to their macroscopic structure (Silva, 2004). Research advancement in different areas of chemistry, biology and physics expedites the growth and development of the nanoscience and technology. In the impact of this development, some industries such as pharmaceutics, electronics and communications, cosmetic and aerospace have started manufacturing of nanotechnology-based products, and food and agricultural industries have also begun to understand features of nanomaterials to explore applications (Ravichandran, 2010). Hence, nanotechnology is viewed to boost the economy by enhancing the quality of agriproducts, reducing farm cost and promoting higher yield production (Sekhon, 2014). In correspondence to that one of the nanotechnology-based sensing, an application is recognized to monitor the quality of agricultural raw materials and products (Fraceto et al., 2016). In this sensing application, a modified version of a biosensor is used in which biological entity-derived sensitizing element is linked with a transducer, which is called as nanobiosensor (Turner, 2000). The characteristic feature of a biological entity to respond specific substrate is linked to the design of nanobiosensors in which nanomaterial-based transducers are incorporated. Nanobiosensors are more beneficiary over biosensor due to quick response, easy to use with enhanced sensitivity and specificity (Bellan et al., 2011). The present chapter elaborates various nanobiosensors, their principles and applications in the food and agricultural sectors. Nanoparticles such as gold nanoparticles, magnetic nanoparticles, upconversion nanoparticles, single- and multi-walled carbon nanotubes (SWNTs and MWNTs) and quantum dots are utilized in designing of nanobiosensors. Nanobiosensors have potential application in the food and agriculture sector as in monitoring quality and shelf life of the product, microbial and pathogen contamination, mycotoxin, pesticide and herbicide residual contamination. The major application in the era of agriculture is the assessment of soil quality in terms of pH, microbial and residual contamination, humidity, etc. (Rai et al., 2012). The marketizations of the nanobiosensors are crucial for the empowerment of the food and agricultural sectors. In food technology, nanomaterials are used as food additives and food packaging materials (Sozer and Kokini, 2009). Nanoparticles as food additives play role in the detection of pathogen, assurance food quality and enhancement of food quality. In food packaging, nanomaterials indicate spoilage of food by preventing gas flow to food and increase product quality and shelf life (Kulkarni et al., 2016). However, any food products with the involvement of nanotechnology are considered as novel products and that require appropriate assessment of nanoparticle associated toxicity and adverse effects (Tiede et al., 2008). Such concerns should be taken into consideration because the success of nanotechnology-derived food and agricultural consumable products will be a relay on consumer acceptance. The recent studies are showing a good impact of nanotechnology on the food and agricultural sectors. This chapter focuses on nanomaterial- based biosensors, focuses on their applications in agriculture and food sector and comprises future developmental aspects.

21.2 Nanobiosensor: Concept and Classification

In general, a biosensor is a sensor to recognize any biological product. A biosensor is a device that combines a biological recognition element with a physical or chemical transducer to detect a biological product. In more technical term, a biosensor is a probe that couples biological element with an electronic transducer that generates a measurable signal as recognition of biological product. The biosensor is an interesting research topic for researchers from biology, chemistry, physic and interdisciplinary applied science. In result of its wide research interest, several biosensors have been under development for various applications such as food, agriculture, environment, pharmaceutical and medical. Researchers always emphasize to produce better and more reliable system for enhancement of accuracy. In way of enhancement of potential and reliability of biosensor, researchers integrate nanomaterials in the construction of biosensor. Nanotechnology is the most promising and emerging area of science that deals with the production, modulation and fabrication of nanometer size of materials. Hence, nanobiosensors were designed and developed in the 21st century based on the concept of nanotechnology. Submicron size sensors with biological recognition element specific for particular target molecules are called as nanobiosensor that provides extraordinary sensing specificity, sensitivity, and reliability with high spatial and temporal resolution. Nanomaterial-based biosensor-related research is gaining interest in which nanomaterial is used to either transducer or biological recognition system or in both to boost their multi-detection specificity, reliability and sensitivity (Jianrong et al., 2004). Nanomaterials that are being used in the production of biosensor would be nanoparticles, nanowires, nanotubes and quantum dots having various properties such as submicron size, less power and voltage consumption with fast reactivity. These properties of nanomaterials have revolutionized the field of biosensor in sensing approach of multiple targets (Guo and Wang, 2007) and also facilitate real-time biosensing that was difficult in the past (Erickson et al., 2008; Health Quality Ontario, 2006; Mahato et al., 2018; Shi et al., 2007). Advanced biosensors are applied to agriculture and food sectors especially in the detection of the foodborne pathogen, metabolites and pesticides (Chaudhry et al., 2018).

21.2.1 Features of Ideal Nanobiosensors

- Higher specificity toward target analytes.
- Good stability under room storage condition.
- Biological interaction process should not be affected by nanobiosensor under alteration of physical parameters.
- Faster reactivity with higher reliability, accuracy, and sensitivity for a wide range of analytes.
- It should be small and portable without any electronic noise.
- It should be nontoxic, non-immunogenic and biocompatible.
- It should be cost-effective and easy to operate.

21.2.2 Components of Nanobiosensors

Nanobiosensors also comprise three components system: (i) biologically sensitizing element as a probe (ii) transducer and (iii) signal processor (detector) (Liu et al., 2018). Basically, there are five steps involved in sensing process of a nanobiosensor (Figure 21.1): (i) biologically sensitizing element (probe) that sense the specific form of a sample. Here, a probe is submicron size material immobilized with an enzyme, antibody, tissue, bacteria or any biologically derived chemical that acts as nanoreceptor. Nanoreceptors interact with target samples and transfer generated product or signal to the transducer. (ii) An electrochemical interface made up of nanomaterials which provide platform to specific biological recognition processes. (iii) A transducer that converts the output energy of biological recognition process to electrical signal. (iv) A signal processor that gives meaningful physical parameters to generate electrical signals, and at last, (v) data is displayed at a computational system to interpret and store results.

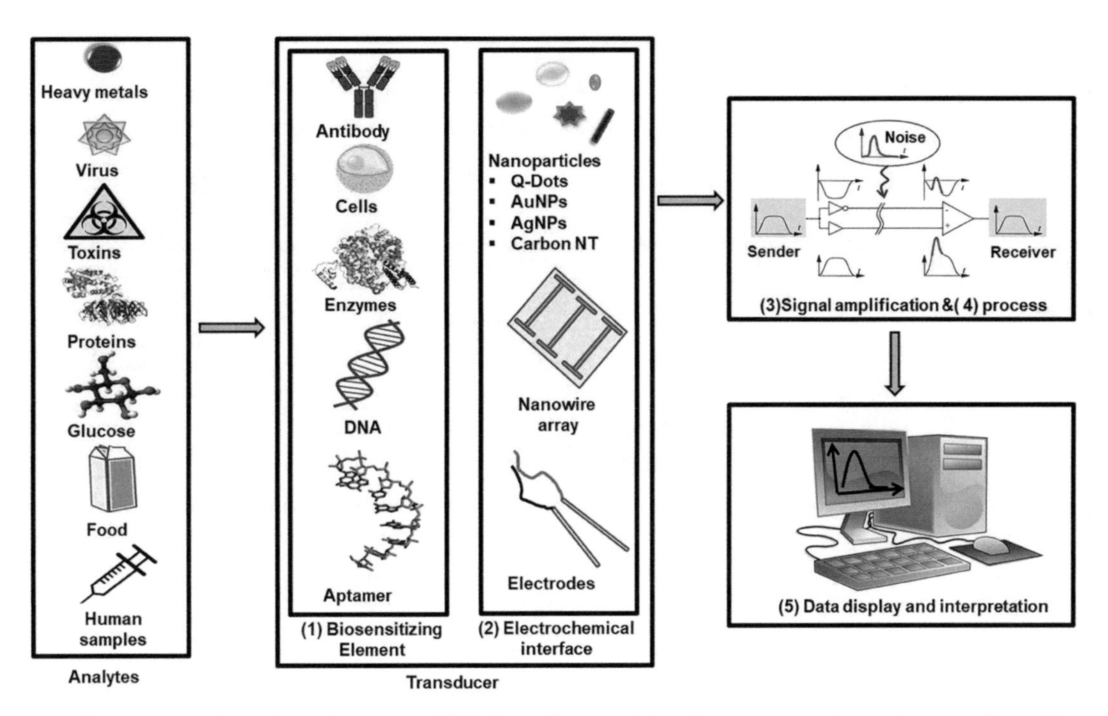

FIGURE 21.1 Sensing procedure of nanobiosensor: (1) biologically sensitizing elements recognize analyte samples and generate signal transfer to a transducer, (2) electrochemical interfaces allow specific biological recognition processes, (3) transducers convert the output energy to electrical signal, (4) signal processors amplify and provide data and (5) data is interpreted and saved.

21.2.3 Usefulness of Nanobiosensor over Other Biosensors

- Nanobiosensors are highly specific with a better limit of detection (LOD) that enables ultralow concentration of a harmful substance.
- Applicable to more surface area due to submicron size.
- The high speed of detection with extraordinary efficiency.

21.2.4 Limitation of Nanobiosensors

- Nanobiosensors are ultrasensitive that have accuracy at some level.
- Nanobiosensors are under the developmental stage that cannot be considered as a biocompatible and environmentally friendly system.

21.2.5 Classification of Nanobiosensors

A classification of nanobiosensors is not simple because types of biosensors overlap according to the criteria of classification. There are two major criteria of classification: (i) Signal transduction mechanism applied and (ii) nanomaterial used. According to the signal transduction mechanism, the main types of biosensors are electrochemical, optical, acoustic, magnetic and mechanical. Each class is then subcategorized according to the difference in sensing approach of the respective class of biosensors. For example, electrochemical biosensors are subcategorized to amperometric, voltammetric and potentiometric biosensors. Same as optical biosensors, they also comprise an optical fiber, surface plasma resonance (SPR), surface-enhanced Raman scattering and fluorescence-based biosensors (Srivastava et al., 2017). In the category of nanomaterial-based nanobiosensors, nanotube, nanowire, nanoshell, ion channel switch, viral and PEBBLE (Photonic Explorer for Bioanalysis with Biologically Localized Embedding)-based nanobiosensors are included. Here, various nanomaterials are being used to develop more efficient, specific and signal-enhancing nanobiosensor.

Signal Transduction Mechanism-Based Nanobiosensors

In the sensing process, the mechanism of signal transduction differs according to the approach of nanobiosensors. These approaches with respect to signal transduction mechanism are illustrated in Figure 21.2. Signal transduction mechanism-based nanobiosensors are described here.

Electrochemical Nanobiosensors

Electrochemical nanobiosensors generate and measure electrochemical signal as an output of biomolecular interaction. These nanobiosensors are majorly made up of metallic nanoparticles. Fabrication of nanoparticles and immobilization of one of the reactive biomolecules on the surface of nanoparticles facilitate specific, faster and efficient biomolecular reaction to generate an appropriate electrochemical signal. The electrochemical signal can be measured as current, voltage or ion potential. According to signal transduction mechanism, electrochemical nanobiosensors

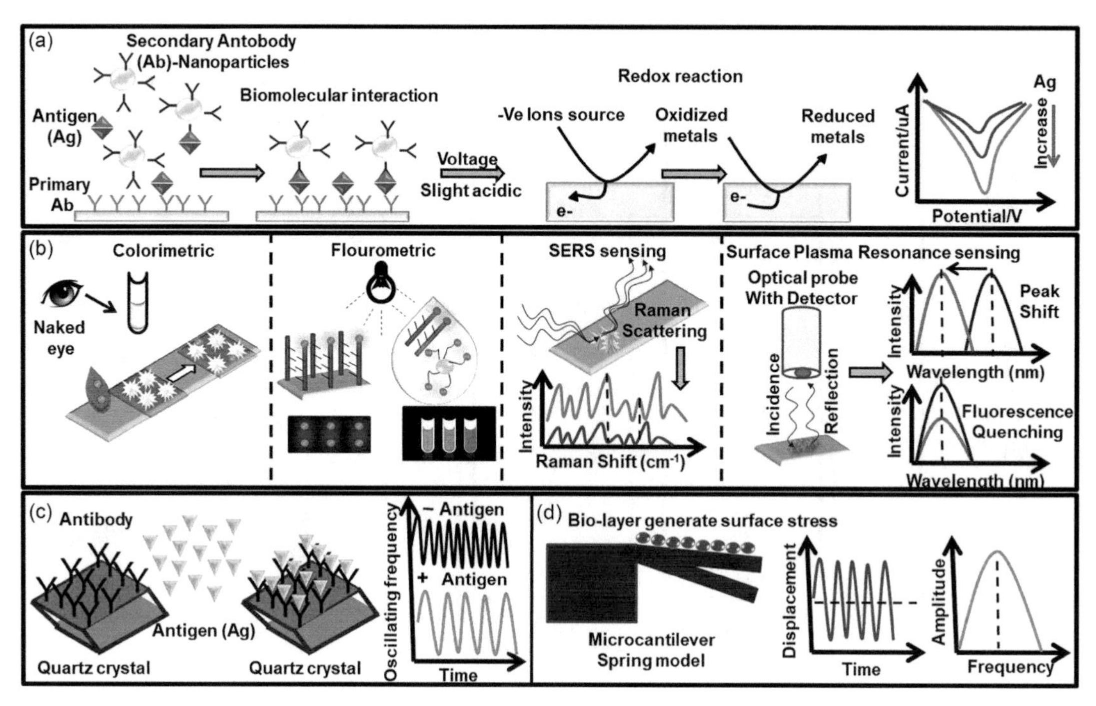

FIGURE 21.2 Major classes and principle of nanobiosensors: (a) electrochemical nanobiosensors generate electrical signals as an output signal of biomolecular interaction. (b) Optical nanobiosensors are subcategorized into colorimetric (output is a visualization of color), fluorometric (fluorescence intensity), surface-enhanced Raman spectroscopy (SERS) sensing (Raman peak intensity and shift) and SPR, where an output signal is based on fluorescence. Acoustic nanobiosensors detects antigenic compound on the basis of changes of the oscillation frequency (c). Microcantilever nanobiosensors measures surface stress level and diagnose biological entity according to changes of displacement and amplitude of it (d).

are subcategorized into amperometric, voltammetry and potentiometric nanobiosensors.

The amperometric nanobiosensor works on transduction of electrochemical signal as ampere current that generates as a biochemical interaction of reactants. Measured current at constant voltage indicates the amount of analytes in solution (Chaubey and Malhotra, 2002). In this reference, majorly gold nanoparticles have been used with immobilization of DNA, enzymes and antibodies to identify glucose, xanthine, plant metabolites and hydrogen peroxide (Saei et al., 2013; Zhao et al., 1996; Xiaoxing et al., 2003). Uses of nanobiosensors over conventional sensors provide much efficient and significant result with lowering LOD (Zhang et al., 2011). Amperometric nanobiosensors were also applied to agriculture and food sector to identify toxic ractopamine and salbutamol (Lin et al., 2013), hydrogen peroxide (Nasirizadeh et al., 2014), fructose (Antiochia et al., 2013), glucose (Saei et al., 2013) and xanthine (Zhao et al., 1996).

Voltammetry nanobiosensor principle is based on the measurement of current with controlled variation of applied potential that shows the amount of analytes reacted during a biochemical reaction. Displayed data as the format of results shows a graph between current and voltage called a voltammogram. This nanobiosensor has been developed to detect pathogen (Chen and Park, 2016), heavy metal (Yavuz et al., 2015) and carbosulfan (Nesakumar et al., 2016).

Potentiometric nanobiosensors sense ionic potential at ion-selective electrodes with respect to a reference electrode. Ion-sensing electrode detects various ions such as sodium, potassium, ammonium, calcium, hydrogen or other ions that are present in a biological sample (Radomska et al., 2004; De Marco et al., 2007). In similar ways, photoelectrochemical nanobiosensor was also developed that was made up of nanosized semiconductor crystals. Curri et al. (2002) immobilized formaldehyde dehydrogenase to CdS nanocrystalline semiconductor to detect formaldehyde. Recently, Kalita et al. (2016) have developed graphene quantum dot-based ion-selective nanobiosensor that selectively measures soil moisture content by analyzing ion conductivity. Hence, electrochemical nanobiosensors have remarkable efficiency, specificity with low power and fast reactivity that make them useful nanobiosensors to be used in a wide range of applications.

Optical Nanobiosensors

Optical nanobiosensors comprise an arrangement of optics in that circulation of light beam occurs, and on exposure of analytes to this path, change in resonance frequency is measured. There are various optical nanobiosensors categorized based on optical sources. Luminescence-based nanobiosensors' measurement relies on the emitted light as a result of released energy from excite state electron of analytes when it achieves a ground state. Majorly lasers are used to excite electron of reactants that help to monitor and quantify biomolecular interaction (Vo-Dinh, 2005). Luminescence nanobiosensors are highly sensitive, reactive and also provide useful tools for imaging methodology to make it more easy and adaptable system. Fluorescent nanoparticles such as gold, silver, quantum dots and upconversion nanoparticles have inherent fluorescence property. Biomolecular interactions of analytes with nanomaterials either enhance or quench the fluorescence of nanomaterials, and changes in parameters of fluorescence help to measure a concentration of analytes. Another type of optical biosensor is SPR nanobiosensors that depend on the optical-electrical interactive phenomenon. Incident photons of light transfer energy to free oscillating electrons at the surface of metal lead to charge density oscillation of electron that is called surface plasma polaritons. Surface plasma polariton measured in respect to generation of electrical field and ultrasmall fluctuation in surrounding medium of a resonating or oscillating electron can be measured effectively. Therefore, SPR nanobiosensors are highly in demand as they enable very less concentration of analytes. Majorly conductive materials such as metallic nanoparticles, graphene-based nanomaterials and magnetic or engineered nanoparticles are applied in the construction of SPR nanobiosensors (Haes et al., 2004). Using the principle of SPR, surface-enhanced Raman scattering techniques were developed in that interaction of light to a resonating electron at the surface of metals to observe an enhancement of SPR in respect to applied laser field. Surface-enhanced Raman scattering has lowered the LOD with highest detection sensitivity. Immobilized analytes to nanomaterials can be detected with enhancement of plasma resonance under applied electromagnetic field generated through specific wavelength of a laser. Some optical nanobiosensors are developed that cover application in agriculture and food industry such as detection of foodborne pathogen (Radhakrishnan and Poltronieri, 2017), cyanotoxin (Shi et al., 2013), abscisic acid (Wang et al., 2017), palytoxins (Zamolo et al., 2012), free radicals (Prow et al., 2013) and aflatoxin (Karczmarczyk et al., 2016) that are further discussed in this chapter.

Acoustic Nanobiosensor

An acoustic biosensor is a topic of research from a long time to develop a sensing platform for biological and chemical analytes. These nanobiosensors comprise acoustic wave system whose physical properties such as resonance frequency and velocity are modulated on exposure of analytes and that is useful for the measurement of amount of analytes. These nanobiosensors would be widely accepted due to their cost, micron size and feasible integrated construction with microfluidic and electronic systems. Smaller size makes it compatible for multiplex detection platform in that several hundred nano-arrays are embedded to a single device that enhance sensitivity, efficiency and reliability with less error at higher spatiotemporal resolution. Nanoparticles generally recommended for preparing acoustic wave nanoparticles are gold, titanium oxide and cadmium sulfide (Su et al., 2000; Heydari and Haghayegh, 2014). Recently, Karczmarczyk et al. (2017) have developed acoustic nanobiosensors that have application in food analysis to detect ochratoxin A in red wine.

Mechanical Nanobiosensors

The principle of mechanical nanobiosensor relies on mechanical forces between biomolecules that provide the measurable outcome of biomolecular interaction (Cheng et al., 2006). In a construction of mechanical nanobiosensor, cantilever such as AFM (atomic force microscopy) probe, optical laser and signal processor are incorporated. Microscale cantilever is bent due to biomolecular interaction at a surface that can be optically measured as a difference of deflection of cantilever laser beam. From the measurement of a difference in deflection level experienced by cantilever laser beam in response to biomolecular interaction, the amount of target analyte can be accessed. A high sensitivity of mechanical nanobiosensors has been developed as a powerful platform for resolving time and label-free sensing (Backmann et al., 2010). Mechanical nanobiosensor follows three mechanisms due to asymmetric chemisorption of analytes to microcantilever (Ziegler, 2004): (i) microcantilever bending as a response to analyte induced surface stress. This mechanism is also referred to as static mode in that molecular interaction at the surface of microcantilever due to electronic charge distribution and lateral electrostatic interaction of analyte to microcantilever generates stress. (ii) The difference in the resonance frequency of oscillating microcantilever due to analyte mass that is called as dynamic mode (Fritz, 2008) and (iii) microcantilever bending in respect to temperature change. Mechanical nanobiosensors are highly mass sensitive and they gain the interest of several researchers due to its transduction ability over wide signal domains. For example, any change in physical parameters such as temperature, mass and heat leads to changes in resonance frequency or bending of microcantilever that can be measured with high spatial and temporal resolution. Recently, Kim et al. (2015) developed highly sensitive microcantilever nanobiosensor for detection of early human papilloma virus infection. They generated sensing area on microcantilever with a coating of gold nanoparticles and gold binding proteins having amine and carboxyl functional group. These reactive functional groups have interacted on the introduction of nanobiosensor to HPV DNA-conjugated beads and a generated signal was 1,000 times more amplified as compared to nonconjugated beads. Hence, mechanical nanobiosensor is one of a good choice for development of nanobiosensor for agricultural and food product analysis.

Nanoparticle-Based Biosensor

A class of biosensor that is categorized on the basis of nanomaterials or properties of nanomaterials has been used in the construction of nanobiosensor like carbon nanotube-based nanobiosensors are called as nanotube biosensor and magnetic nanoparticle-based biosensors are termed as magnetic nanobiosensor. In similar fashion, nanowire, nanoshell biosensors are enlisted and discussed.

Magnetic Nanobiosensor

Magnetic nanomaterials are interesting materials to work with biosensing approach due to the presence of negligible magnetization in a biological system. Modulation and engineering of magnetic nanomaterials facilitate a wide range of biomolecular sensing. (Koh and Josephson, 2009). Magnetic nanoparticles are mainly based on iron oxides that enable iron binding metals or magnetically sensible materials. Incorporation of magnetic nanoparticles to conventional biosensors reforms them as more sensitive and efficient biosensor. Metals having an unpaired electron in d-orbitals are being used for the construction of magnetic nanobiosensors due to their greater magnetization. The magnetometer is used to identify magnetically labeled analytes. Superconducting quantum interference devices, one of magnetometer device, have been used for the screening of magnetically labeled biomolecules from a sample (Tian et al., 2018). Recently in 2018, Alaa et al. developed magnetic nanobiosensor to detect *Serratia marcescens*, a food, and hospital origin pathogen. Magnetic nanoparticle-conjugated viral part has been used to develop magnetic nanobiosensor for clinically important virus detection (Shelby et al., 2017).

Nanotube-Based Biosensors

Carbon nanotubes are widely used and popular nanomaterials in the area of material science as it has extraordinary characteristics such as electronic conductivity, flexible geometry, better physicomechanical and electrical property and functionalization ability. These properties make carbon nanotubes more important nanomaterials in the development of versatile nanobiosensor. Both SWNTs and MWNTs are employed in designing a biosensor for the betterment its performance and utility (Davis et al., 2003). Carbon nanotube-based biosensors initially developed in the detection of glucose from various sources of samples as a glucose sensor (glucose oxidase) is embedded in matric of the carbon nanotube. Performance of nanotube-based glucose biosensor is reported to be far better as compared to conventional biosensor because it can detect a lower concentration of glucose from saliva or tear samples (Lin et al., 2004). Here, a combination of higher enzyme's presence as a result of immobilization and better electrical conductivity of nanotube lowers the LOD with better efficiency (Anthony et al., 2002). This major breakthrough in the area of nanobiosensors also flexes in agricultural and food sector as a result of the friendliness of nanotubes to detect toxins, food pathogen and plant metabolites that are further elaborated in application part (Marques et al., 2017; Wang et al., 2009; Malhotra et al., 2015; Akanbi et al., 2017).

Nanowire-Based Biosensor

Nanowire biosensor is composite of detector and transmitter system in which DNA or fabricated or decorated DNA works as a detector, and electrical conductive nanomaterials like carbon nanotubes serve as a transmitter. Nanowires are several millimeters in length, and the diameter of this one-dimensional cylindrical fibril-like structure is in nanorange. The nanowire-based biosensor is considered as a more powerful tool for the development of nanobiosensor compared to nanotube-based biosensor due to its highly flexible surface modification on exposure of biological ligands and effective electron transduction. Generation of the electrical signal as a result of the binding process at the surface of nanowires can be measurable in real time and extremely sensitive. However, synthesis of nanowire-based biosensor is not easy due to the complex and crucial synthesis process of a nanowire without the loss of electrical conductive property (Umar et al., 2009). In contrast, some reports enlighten research development in nanowire-based biosensor and also prove their ability to detect a lower concentration of analytes (Namdari et al., 2016; Cui and Lieber, 2001). Recently, Tien et al. (2018) have developed a gold nanowire-based biosensor to design very sensitive and efficient probing system for the detection of oral cancer cells. The nanoconfined dimension of nanowire-based biosensors gain interest in the study of real-time *in vivo* detection, and such kind of nanobiosensor has been developed by Cullum et al. (2000) to detect toxicants' intracellular level. In a similar way, the recent development of nanowire-based biosensors in an area of pathogen and plant diseases' detection has an open door of a nanowire-based biosensor in an application of agriculture and food analysis (Vikesland and Wigginton, 2010; Ariffin et al., 2014). In recent 2018, Ali et al. demonstrated silver nanowire-based disposable all-printed biosensor to detect and classify pathogen in which they showed detection of *Salmonella typhimurium* and classified the *Escherichia coli* strains JM109 and DH5-α based on impedance analysis. Research has been in the process to utilize the benefit of more morphological changes of nanobiosensor such as nanodots, nanobelt and nanoribbon to enhance more specificity and efficiency (Ding et al., 2004).

21.3 Role of Nanobiosensor in Agriculture

Nanobiosensor has greater potential over conventional biosensor that makes them useful for wide range of application in diagnostic, medical, environmental, agriculture and food technology. This chapter is focused on the role of nanobiosensor in agriculture and food sector (Figure 21.3).

In the agricultural sector, nanobiosensors have applications to detect soil and water quality and pathogens, metabolites, plant diseases and toxins that are described in Table 21.1. Some of the innovations are elaborated below. Kalita et al. (2016) have developed graphene quantum dots to sense moisture content of soil in that they have used interdigitated electrodes, and graphene dots work as a channel between these electrodes to sense moisture content in two types of soils inluding white clay and bentonite clay. They found conductance changes from $0.06 \times (10\text{–}61)/\Omega$ to $0.68 \times (10\text{–}61)/\Omega$ in white clay as the gravimetric moisture content varies from 4% to 45%. In case of

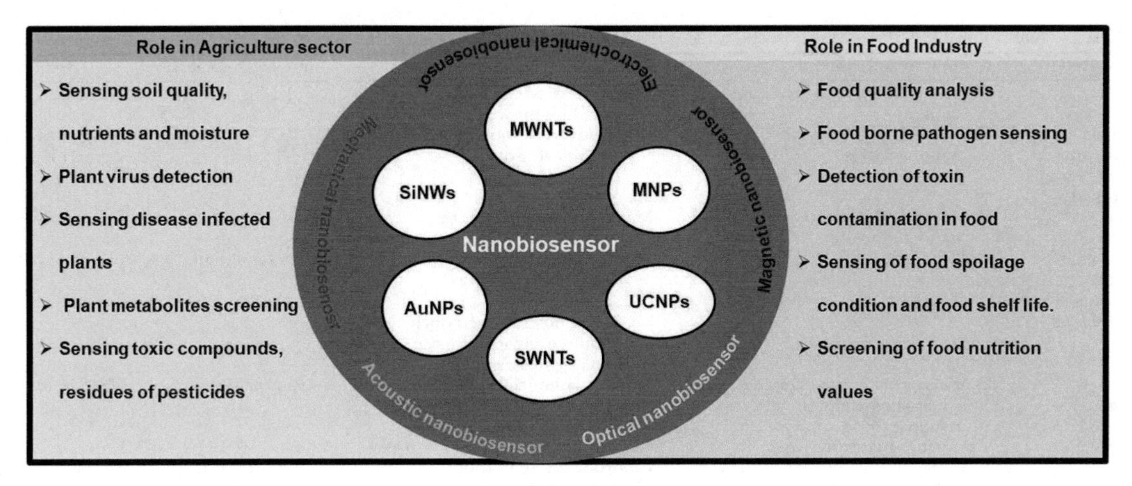

FIGURE 21.3 Role of nanobiosensors in agriculture and food sectors: widely utilized nanomaterials SWNTs, MWNTs, AuNPs (gold nanoparticles), MNPs (Magnetic nanoparticles), SiNWs (silicon nanowires) and UCNPs (upconversion nanoparticles) potentiate biosensors to be used for various described applications in agriculture and food industry.

bentonite clay, they found changes in conductance from 0.06 × (10–61)/Ω to 0.48 × (10–61)/Ω in respect to changes in the gravimetric moisture from 11% to 90%. Nutrient detection of soil is also facilitated by nanobiosensor to test the quality of the soil. Mura et al. (2015) have developed optic bionanosensor by using gold nanoparticles to detect nitrate in which gold nanoparticles were modified with cysteamine to sense nitrate with better efficiency and changes in color within a range of 35 ppm can be visualized with naked eyes. In another study, Shi et al. (2013) have developed optical nanobiosensor to detect cyanotoxin from water in which they immobilized microcystin-LR ovalbumin on to biochip surface to detect cyanotoxin residues (microcystin-LR). They followed indirect competitive approach, premixed microcystin-LR with fluorescent-labeled specific monoclonal antibody and exposed them to the surface of biochips. In this system, less fluorescence is generated with high percent contamination of cyanotoxin residue in water. They quantified microcystin-LR amount was 0.2–4 μg/L, with an LOD of 0.09 μg/L. Hence, it is helpful to sense water from cyanotoxin, a hazardous toxin of cyanobacteria, whose acute dose exposure causes inflammation and cell death. Wang et al. (2017) have also demonstrated SPR-based optical nanobiosensor to detect abscisic acid, an important abiotic stress-responsive plant hormone. They have utilized aptamer functionalized gold nanoparticles. The principle of abscisic acid nanobiosensor is based on conformational and stability changes of gold nanoparticles on the binding of an aptamer to abscisic acid against NaCl-induced aggregation that leads to changes in SPR spectra to quantify the presence of abscisic acid. They quantify a good linear range of abscisic acid concentration from 5×10^{-7} M to 5×10^{-5} M with 0.33 μM LOD. In similar ways, other plant hormones can also be detected by changing hormone-specific aptamer. In the detection of plant metabolites, Bagheri et al. (2016) have also developed FRET (fluorescence resonance energy transfer)-based optical nanobiosensor to detect scopolamine, a tropane alkaloid as well as anticholinergic agent found in hairy root extraction of medicinal Atropa belladonna plant. They utilized thioglycolic acid-conjugated cadmium telluride quantum dots (Cd/Te) with functionalization of an M_2 muscarinic receptor (Cd/Te QDs-M_2R) and rhodamine123 tagged scopolamine that respectively functions as donor and acceptor pairs to transfer charge. The nanobiosensor response was found linear over a range of 0.01–4 μmol/L of scopolamine hydrochloride concentration with LOD of 0.001 μmol/L. They have successfully used nanobiosensors to demonstrate in vitro recognition of scopolamine in investigated plant extracts. In similar ways, detecting plant disease condition and health of a plant is possible by sensing virus itself or metabolites generated during disease condition. In an area of that, Akanbi et al. (2017) have developed carbon nanotube-based nanobiosensors with decoration of gold and chitosan nanoparticles to a screen-printed electrode of a MWNT. This nanobiosensor is able to detect quinolone that generates in *Ganoderma boninense* infectious plant. Measurement of quinolone electrooxidation was performed with cyclic voltammetry and linear sweep voltammetry that shows outstanding detection ability of nanobiosensor in a linear range of detection between 0.0004 and 1.0 μM, with a detection limit of 3.75 nM. Here, sensitivity and the lower detection limit are attributed to salient characteristics and synergistic effects of gold, chitosan nanoparticles, and multiwalled carbon tubes. In another study, Ariffin et al. (2014) have developed a nanowire-based sensor to detect plant virus-like cucumber mosaic virus and papaya ringspot virus that viruses are fatal to plant and early detection of an infectious plant can be curable. In a study, they used silicon-based nanowire and functionalized them with bioreceptors that are specific to grab plant diseases like cucumber mosaic virus and papaya ringspot virus. On exposure of virus, bioreceptor interaction process elicits electrical impulse that can be monitored to quantify virus load.

Furthermore, developments of nanobiosensor useful in agriculture comprise nanobiosensors that have the ability to

TABLE 21.1 Nanobiosensors have been Used in the Detection of Various Contaminants and Agricultural Assessment

Target Analyte	Utilized Nanomaterial	Class of Nanobiosensor	Limit of Detection	References
Deltamethrin	Cadmium telluride-silica quantum dots	Fluorometric-based optical nanobiosensor	0.16 g/mL	Ge et al. (2011)
Parathion-methyl, monocrotophos, dimethoate and acetamiprid	Gold and UCNP hybrid nanoparticles	Fluorometric-based optical nanobiosensor	0.67, 23, 67 and 3.2 nM	Long et al (2015) and Hu et al. (2016)
Dimethoate	Graphene oxide-magnetic nanocomposites	Magnetic nanobiosensor	1–20 μM	Liang et al. (2013)
Cymbidium mosaic virus and *Odontoglossum ringspot virus*	Gold nanorods	Surface plasma resonance (SPR)-based optical nanobiosensor, fluorometric-based optical nanobiosensor, acoustic nanobiosensors	48 and 42 pg/mL, 1 ng/mL	Lin et al. (2014)
Bacillus thuringiensis (Bt) gene	Oligonucleotide-conjugated silica-encapsulated gold nanoparticles	Surface-enhanced Raman spectroscopy (SERS)-based optical nanobiosensors	0.1 pg/mL	Chen et al. (2012)
Soil moisture	Graphene quantum dots	Electrochemical nanobiosensor	0.06 × (10–61)/Ω	Kalita et al. (2016)
Cyanotoxin	Fluorescent antibody tagged to biochip surface	Fluorometric-based optical nanobiosensor	0.09 μg/mL	Shi et al. (2013)
Abscisic acid	Aptamer functionalized gold nanoparticles	SPR-based optical nanobiosensor	0.33 μM	Wang et al. (2017)
Scopolamine	Thioglycolic acid-conjugated cadmium telluride quantum dots	Fluorometric-based optical nanobiosensor	0.001 μM	Bagheri et al. (2017)
Quinolone	Carbon nanotubes functionalized with chitosan and gold nanoparticles	Electrochemical nanobiosensor	3.75 nM	Akanbi et al. (2017)

detect pesticides. Pesticidal exposures to environment especially organophosphorus pesticides are harmful due to their inhibitory effect of acetylcholinesterase enzyme. Liposomal nanobiosensor has been developed to sense organophosphorus pesticides such as dichlorvos and paraoxon at very less concentration in the construction of liposome nanobiosensors, enzyme acetylcholinesterase, and pH-sensitive fluorescent dye pyranine which is effectively immobilized in the internal environment of the liposome. Enzyme activity has been found to be decreasing as exposure to pesticides that have been indirectly measured with decreasing fluorescence of pH-sensitive dye. Liposome nanobiosensor's LOD was reported as 10^{-10} M (Vamvakaki and Chaniotakis, 2007). Hence, use of nanobiosensor in the field of agriculture has proved itself as a boon for farmers, environment and health.

21.4 Application of Nanobiosensors in the Food Sector

District shape, size and morphology carrying nanomaterials have the potential to generate fast and very sensitive electron transfer and they also acquire mechanical, optical, chemical, electrical and other versatile properties that can be also useful to design nanobiosensor having application in food industry to detect food pathogen, food contaminants and assessment of food quality such as vitamin analysis and food spoilage. Applications of nanobiosensors are listed in Table 21.2, and some innovative applications are explained. Recently, Cdse and ZnS quantum dot-based optical nanobiosensor was prepared with surface modification of Q-dots with silane and methylacrylate group to detect dicyandiamide in the milk product. The measured concentration of dicyandiamide is linearly correlated with quenching of fluorescent Q-dots with LOD of 2.7 μmol/L (Liu et al., 2016). An inflammatory action of dicyandiamide acute dose is harmful to liver, kidney and blood. Hence, it is good to detect it in food prior to consumption. Similarly, Karczmarczyk et al. (2016) have reported the detection of aflatoxin, a carcinogenic toxin in milk, by using SPR-based optical nanobiosensor in which low molecular weight mycotoxin, aflatoxin, was measured using indirect competitive immunoassays. Gold nanoparticle-conjugated secondary antibodies help to amplify signals generated from immunoassay. They reported this nanobiosensor as much as effective to sense 18 pgm/L LOD. In a further report in 2017, Karczmarczyk has demonstrated piezoelectric nanobiosensor to detect ochratoxin A in red wine. Ochratoxin A is a highly toxic mycotoxin that is referred to as a carcinogen that contaminates a wide variety of agricultural food. Hence, ochratoxin A-specific secondary antibody-functionalized gold nanoparticle-based quartz crystal microbalance with dissipation monitoring (QCM-D) acoustic nanobiosensors was developed in demand of its feasible detection in food. They combined indirect competitive immunoassay with QCM-D device to generate a specific piezoelectric signal in correspondence to frequency or dissipation change in an interaction of analytical mass to the surface of an electrode. They proposed a better system to detect ochratoxin A in food as compared to existing nanobiosensor having poor LOD due to a very small molecular weight of ochratoxin A. They measured the response of proposed nanobiosensor in a linear detection range of 0.2–40 ngm/L with outstanding LOD of 0.16 ngm/L of ochratoxin A in red wine. Here, they pre-treated red wine with 3% poly (vinylpyrrolidone) to eliminate the nonspecific interaction of polyphenols at the sensor surface. In the same area of food analysis application, Zamolo et al. (2012) have reported the development of carbon

TABLE 21.2 Nanobiosensors have been Used in the Detection of Various Contaminants and Analysis of Food Products

Target Analyte	Utilized Nanomaterial	Class of Nanobiosensor	Limit of Detection	References
Brilliant blue FCF	L-Cysteine capped cadmium sulfide quantum dots	Fluorometric-based optical nanobiosensor	4.0–4.5 μM	Sivasankaran et al. (2017)
Melamine	Chondroitin sulfate reduced gold nanoparticles, UCNPs (sodium yttrium fluoride doped lanthanide)	SPR-based optical nanobiosensor, fluorometric-based optical nanobiosensor	12.6 ppb, 32–500 nM	Noh et al. (2013) and Wu et al. (2015)
Tobramycin	Polymer film-gold nanoparticles	SPR-based optical nanobiosensor	5.7 pM	Yola et al. (2014)
Cu ions	2–4,6-di-tert-butylphenol doped nano-fibrous (polymeric ethyl cellulose) films	Fluorometric-based optical nanobiosensor	33 pM	Kacmaz et al. (2015)
Mercury	Lanthanide, europium, terbium tagged silica nanoparticles	Fluorometric-based optical nanobiosensor	7.7 nM	Tan et al. (2015)
Polyphenol	Iron oxide and tannic acid hybrid nanoparticle	Electrochemical nanobiosensor	8.57 μM	Magro et al. (2016)
Dicyandiamide	Cdse and ZnS hybrid quantum dots	Fluorometric-based optical nanobiosensor	2.7 μM	Liu et al. (2016)
Aflatoxin	Antibody-conjugated gold nanoparticles	SPR-based optical nanobiosensor	18 pM	Karczmarczyk et al. (2016)
Ochratoxin A	Gold nanoparticle-based quartz crystal	Acoustic nanobiosensor	0.2–40 nM	Karczmarczyk et al. (2017)
Palytoxin	Carbon nanotube	Nanotube-based nanobiosensors	22 μg/kg	Zamolo et al. (2012)
E. coli (O157: H7)	22mer oligonucleotide functionalized silica nanoparticles	Acoustic nanobiosensor	1.8 fM	Ten et al. (2017)
Bacillus cereus	DNA functionalized gold nanoparticles	Electrochemical nanobiosensor	9.4 pM	Izadi et al. (2016)
Food allergen (Arah1)	Q-dots aptamer graphene oxide nanoparticles	Fluorometric-based optical nanobiosensor	56 ng/mL	Weng and Neethirajan (2016)
Glucose	Magnetite – Prussian blue nanocomposites	Fluorometric-based optical nanobiosensor	0.5 μM	Jomma and Ding (2016)

nanotube-based nanobiosensor having electrochemiluminescence phenomena to detect palytoxin, a potent marine toxin having the ability to cause cardiac arrest. Contamination of palytoxin generally present in seafood can be sensed with developed ultrasensitive nanobiosensor which utilizes anti-palytoxin antibody to detect palytoxin, and this biomolecular interaction can be measured by electric signal conduction through carbon nanotubes and luminescence generation in correspondence to a concentration of palytoxin present in mussel meat. They quantified palytoxin in mussel meat with an excellent LOD of 2.2 μg/kg that is sensitive more than twofold magnitudes compared to LC-MS/MS. A food-borne pathogen is one of the indications of food spoilage and nanobiosensor having the efficiency to detect the food-borne pathogen. Eser et al. (2015) have developed SPR-based optical nanobiosensor that is sensitive in the detection of *Salmonella enteritidis*. In a construction of biosensor, they utilized anti-*S. enteritidis* antibody-immobilized gold nanoparticles to detect various concentrations of bacteria. The changes in SPR spectrum in response to a different amount of bacteria were measured. They also used *Listeria monocytogenes* to confirm the selectivity of nanobiosensor, and they detected *S. enteritidis* bacteria in milk samples also. Hence, nanobiosensors those are commercialized and are under development are rapid, highly sensitive, selective, reliable and cost-effective to be utilized in field of agriculture and food sector.

21.5 Challenges and Future Aspects

In recent decades, applications of nanobiosensors facilitate new dimension in the development of novel biosensing technologies. Nanomaterials have the potential to improve electrical, mechanical, optical and magnetic properties of biosensors that leads to the development of single molecular sensor and high throughput sensor with high accuracy. However, like any other emerging area, nanobiosensors also face many challenges (Neethirajan et al., 2018) like functionalization of biomolecules to nanoparticles because biological molecules are structure specific and their functionality also relies on specific structure and conformation. Hence, this is a crucial task to develop nanobiosensor without affecting the functionality of biomolecules. Indeed, multifactorial and multifunctional nanobiosensors are still a great challenge. The site-specific interactions of biomolecules to nanoparticles are still not well understood to apply it as a principle to develop further multifunctional nanoarrays or nanofilm system. Besides that, interference of nonspecific moieties at nanoparticles' site is also a major concern in real-time observation. Crucial processing, characterization of nanomaterials, availability of highly purified nanomaterials and exact mechanism that drive the interaction of nanoreceptors to a surface of the electrode are also considered as challenges in existing biosensing techniques. Enhancement of signal-to-noise ration and signal transduction plus amplification in real-time assessment are still considered as a research problem. In spite of that, a major challenge is relying with less knowledge of toxicity and biocompatibility of nanomaterials to the environment and human health. Future development in an area of nanobiosensor should be focused on understanding the mechanism of interaction between biological moiety and nanomaterials as well as in between bioreceptor-conjugated nanoparticles and electrode surface that allow developing process of multifactorial and multifunctional nanobiosensors to detect several biological analytes with single device and in elaboration and exploration of knowledge about toxicity of nanomaterials. Nevertheless, nanobiosensors have emerged as a better solution of sensing approach that will be widely accepted and applied in diagnostic, food processing and analysis, agriculture and environmental monitoring in the upcoming future.

21.6 Conclusion

Research thrust and development in the areas of innovation of nanobiosensors are rapidly enhanced in the last 10 years. Review of this chapter gives idea and understanding about nanobiosensors, their classes and their potential application in the field of agriculture and food sector. Majorly, quantum dots, carbon nanotubes, gold nanoparticles and silicon nanowires have been preferred in the construction of nanobiosensors in which bioreceptor-conjugated or functionalized nanoparticles have a diverse role in the enhancement of nanobiosensor specificity. Recent developments in nanobiosensors based on novel signal transduction processes are extensively described. In application of nanobiosensors in agriculture and food sectors, distinct and highly sensitive detection mechanism of toxic compounds in food/soil/water and verification of virus-infected plants, identification of pesticides residues and isolation of medically important plant metabolite are discussed. In an enhancement of nanobiosensors applications to agriculture and food industry, further research and novel sensors developments are expected.

Acknowledgments

Authors acknowledge Council of Scientific and Industrial Research for providing fellowship to Mr. Mohammed Nadim and Mr. Anup K. Srivastava and also appreciate the financial support of Department of Biotechnology grant BT/PR27008/NNT/28/1525/2017.

References

Akanbi, F., N. Yusof, J. Abdullah, Y. Sulaiman, and R. Hushiarian. Detection of quinoline in G. boninense-infected plants using functionalized multi-walled carbon nanotubes: A field study. *Sensors* 17, no. 7 (2017): 1538. www.mdpi.com/1424-8220/17/7/1538.

Alaa, A. A. A., H. Emad, A. Omar, Z. Mazhar Al, A. Bahaa, A. Khaled, and A. A. Al-Razaq Mutaz. Rapid magnetic nanobiosensor for the detection of serratia marcescen. *IOP Conference Series: Materials Science and Engineering* 305, no. 1 (2018): 012005. http://stacks.iop.org/1757-899X/305/i=1/a=012005.

Ali, S., A. Hassan, G. Hassan, C.-H. Eun, J. Bae, C. H. Lee, and I.-J. Kim. Disposable all-printed electronic biosensor for instantaneous detection and classification of pathogens. *Scientific Reports* 8, no. 1 (2018): 5920. doi: 10.1038/s41598-018-24208-2.

Anthony, G.-E., L. Chenghong, and H. B. Ray. Direct electron transfer of glucose oxidase on carbon nanotubes. *Nanotechnology* 13, no. 5 (2002): 559. http://stacks.iop.org/0957-4484/13/i=5/a=303.

Antiochia, R., G. Vinci, and L. Gorton. Rapid and direct determination of fructose in food: A new osmium-polymer mediated biosensor. *Food Chemistry* 140, no. 4 (2013): 742–47. doi: 10.1016/j.foodchem.2012.11.023, www.sciencedirect.com/science/article/pii/S0308814612017505.

Ariffin, S. A. B., T. Adam, U. Hashim, S. Faridah Sfaridah, I. Zamri, and M. N. Aiman Uda. Plant diseases detection using nanowire as biosensor transducer. *Advanced Materials Research* 832 (2014): 113–17. www.scientific.net/AMR.832.113.

Backmann, N., N. Kappeler, T. Braun, F. Huber, H. P. Lang, C. Gerber, and R. Y. Lim. Sensing surface pegylation with microcantilevers. *Beilstein Journal of Nanotechnology* 1 (2010): 3–13. doi: 10.3762/bjnano.1.2, www.ncbi.nlm.nih.gov/pubmed/21977390.

Bagheri, F., Piri, K., Mohsenifar, A., and Ghaderi, S. FRET-based nanobiosensor for detection of scopolamine in hairy root extraction of Atropa belladonna. *Talanta* (2016). 164. doi:10.1016/j.talanta.2016.12.013.

Bellan, L. M., D. Wu, and R. S. Langer. Current trends in nanobiosensor technology. *Wiley Interdisciplinary Reviews: Nanomedicine and Nanobiotechnology* 3, no. 3 (2011): 229–46. doi: 10.1002/wnan.136, www.ncbi.nlm.nih.gov/pubmed/21391305.

Chaubey, A. and B. D. Malhotra. Mediated biosensors. *Biosensors and Bioelectronics* 17, no. 6–7 (2002): 441–56. www.ncbi.nlm.nih.gov/pubmed/11959464.

Chaudhry, N., S. Dwivedi, V. Chaudhry, A. Singh, Q. Saquib, A. Azam, and J. Musarrat. Bio-inspired nanomaterials in agriculture and food: Current status, foreseen applications, and challenges. *Microbial Pathogenesis* 123 (2018): 196–200. doi: 10.1016/j.micpath.2018.07.013, www.ncbi.nlm.nih.gov/pubmed/30009970.

Chen, J., and B. Park. Recent advancements in nanobioassays and nanobiosensors for foodborne pathogenic bacteria detection. *Journal of Food Protection* 79, no. 6 (2016): 1055–69. doi: 10.4315/0362-028X.JFP-15-516, www.ncbi.nlm.nih.gov/pubmed/27296612.

Chen, K., H. Han, Z. Luo, Y. Wang, and X. Wang. A practicable detection system for genetically modified rice by sers-barcoded nanosensors. *Biosensors and Bioelectronics* 34, no. 1 (2012): 118–24. doi: 10.1016/j.bios.2012.01.029, www.ncbi.nlm.nih.gov/pubmed/22342698.

Cheng, M. M., G. Cuda, Y. L. Bunimovich, M. Gaspari, J. R. Heath, H. D. Hill, C. A. Mirkin, et al. Nanotechnologies for biomolecular detection and medical diagnostics. *Current Opinion in Chemical Biology* 10, no. 1 (2006): 11–9. doi: 10.1016/j.cbpa.2006.01.006, www.ncbi.nlm.nih.gov/pubmed/16418011.

Cui, Y., and C. M. Lieber. Functional nanoscale electronic devices assembled using silicon nanowire building blocks. *Science* 291, no. 5505 (2001): 851–3. doi: 10.1126/science.291.5505.851, www.ncbi.nlm.nih.gov/pubmed/11157160.

Cullum, B. M., G. D. Griffin, G. H. Miller, and T. Vo-Dinh. Intracellular measurements in mammary carcinoma cells using fiber-optic nanosensors. *Analytical Biochemistry* 277, no. 1 (2000): 25–32. doi: 10.1006/abio.1999.4341, www.ncbi.nlm.nih.gov/pubmed/10610686.

Curri, M. L., A. Agostiano, G. Leo, A. Mallardi, P. Cosma, and M. Della Monica. Development of a novel enzyme/semiconductor nanoparticles system for biosensor application. *Materials Science and Engineering: C* 22, no. 2 (2002): 449–52. doi: 10.1016/S0928-4931(02)00191-1, www.sciencedirect.com/science/article/pii/S0928493102001911.

Davis, J. J., K. S. Coleman, B. R. Azamian, C. B. Bagshaw, and M. L. Green. Chemical and biochemical sensing with modified single-walled carbon nanotubes. *Chemistry* 9, no. 16 (2003): 3732–9. doi: 10.1002/chem.200304872, www.ncbi.nlm.nih.gov/pubmed/12916096.

De Marco, R., G. Clarke, and B. Pejcic. Ion-selective electrode potentiometry in environmental analysis. *Electroanalysis* 19, no. 19–20 (2007): 1987–2001. doi: 10.1002/elan.200703916, https://onlinelibrary.wiley.com/doi/abs/10.1002/elan.200703916.

Ding, Y., P. X. Gao, and Z. L. Wang. Catalyst-nanostructure interfacial lattice mismatch in determining the shape of Vls grown nanowires and nanobelts: A case of Sn/ZnO. *Journal of the American Chemical Society* 126, no. 7 (2004): 2066–72. doi: 10.1021/ja039354r, www.ncbi.nlm.nih.gov/pubmed/14971941.

Erickson, D., S. Mandal, A. H. Yang, and B. Cordovez. Nanobiosensors: Optofluidic, electrical, and mechanical approaches to biomolecular detection at the nanoscale. *Microfluid Nanofluidics* 4, no. 1–2 (2008): 33–52. doi: 10.1007/s10404-007-0198-8, www.ncbi.nlm.nih.gov/pubmed/18806888.

Eser, E., O. Ö. Ekiz, H. Çelik, S. Sülek, A. Dana, H. İ. Ekiz. Rapid detection of foodborne pathogens by surface plasmon resonance biosensors. *International Journal of Bioscience, Biochemistry and Bioinformatics* 5, no. 6 (2015). doi: 10.17706/ijbbb.2015.5.6.329-335.

Fraceto, L. F., R. Grillo, G. A. de Medeiros, V. Scognamiglio, G. Rea, and C. Bartolucci. Nanotechnology in agriculture: Which innovation potential does it have? [In English]. Perspective. *Frontiers in Environmental Science* 4, no. 20 (2016). doi: 10.3389/fenvs.2016.00020, www.frontiersin.org/article/10.3389/fenvs.2016.00020.

Fritz, J. Cantilever biosensors. *Analyst* 133, no. 7 (2008): 855–63. doi: 10.1039/b718174d, www.ncbi.nlm.nih.gov/pubmed/18575634.

Ge, S., J. Lu, L. Ge, M. Yan, and J. Yu. Development of a novel deltamethrin sensor based on molecularly imprinted silica nanospheres embedded CdTe quantum dots. *Spectrochimica Acta Part A: Molecular and Biomolecular Spectroscopy* 79, no. 5 (2011): 1704–9. doi: 10.1016/j.saa.2011.05.040, www.ncbi.nlm.nih.gov/pubmed/21684806.

Guo, S., and E. Wang. Synthesis and electrochemical applications of gold nanoparticles. *Analytica Chimica Acta* 598, no. 2 (2007): 181–92. doi: 10.1016/j.aca.2007.07.054, www.ncbi.nlm.nih.gov/pubmed/17719891.

Haes, A. J. and R. P. Van Duyne. Preliminary studies and potential applications of localized surface plasmon resonance spectroscopy in medical diagnostics. *Expert Review of Molecular Diagnostics* 4, no. 4 (2004): 527–37. doi: 10.1586/14737159.4.4.527.

Health Quality Ontario. Nanotechnology: An evidence-based analysis. *Ontario Health Technology Assessment Series* 6, no. 19 (2006): 1–43. www.ncbi.nlm.nih.gov/pubmed/23074489.

Heldman, D. R. and D. B. Lund. *The Beginning, Current, and Future of Food Engineering: A Perspective.* Springer, New York (2011).

Heydari, S. and G. H. Haghayegh. Application of nanoparticles in quartz crystal microbalance biosensors. *Journal of Sensor Technology* 04, no. 02 (2014): 81–100. doi: 10.4236/jst.2014.42009.

Hu, W., Q. Chen, H. Li, Q. Ouyang, and J. Zhao. Fabricating a novel label-free aptasensor for acetamiprid by fluorescence resonance energy transfer between Nh2-Nayf4: Yb, Ho@Sio2 and Au nanoparticles. *Biosensors and Bioelectronics* 80 (2016): 398–404. doi: 10.1016/j.bios.2016.02.001, www.ncbi.nlm.nih.gov/pubmed/26874106.

Izadi, Z., M. Sheikh-Zeinoddin, A. A. Ensafi, and S. Soleimanian-Zad. Fabrication of an electrochemical DNA-based biosensor for bacillus cereus detection in milk and infant formula. *Biosensors and Bioelectronics* 80 (2016): 582–89. doi: 10.1016/j.bios.2016.02.032, www.ncbi.nlm.nih.gov/pubmed/26896793.

Jianrong, C., M. Yuqing, H. Nongyue, W. Xiaohua, and L. Sijiao. Nanotechnology and biosensors. *Biotechnology Advances* 22, no. 7 (2004): 505–18. doi: 10.1016/j.biotechadv.2004.03.004, www.ncbi.nlm.nih.gov/pubmed/15262314.

Jomma, E. Y. and S. N. Ding. One-pot hydrothermal synthesis of magnetite prussian blue nano-composites and their application to fabricate glucose biosensor. *Sensors (Basel)* 16, no. 2 (2016): 243. doi: 10.3390/s16020243, www.ncbi.nlm.nih.gov/pubmed/26901204.

Kacmaz, S., K. Ertekin, D. Mercan, O. Oter, E. Cetinkaya, and E. Celik. An ultra-sensitive fluorescent nanosensor for detection of ionic copper. *Spectrochimica Acta, Part A: Molecular and Biomolecular Spectroscopy* 135 (2015): 551–9. doi: 10.1016/j.saa.2014.07.056, www.ncbi.nlm.nih.gov/pubmed/25123945.

Kalita, H., V. S. Palaparthy, M. S. Baghini, and M. Aslam. Graphene quantum dot soil moisture sensor. *Sensors and Actuators B: Chemical* 233 (2016): 582–90. doi: 10.1016/j.snb.2016.04.131, www.sciencedirect.com/science/article/pii/S092540051630613X.

Karczmarczyk, A., K. Haupt, and K. H. Feller. Development of a QCM-D biosensor for ochratoxin a detection in red wine. *Talanta* 166 (2017): 193–97. doi: 10.1016/j.talanta.2017.01.054, www.ncbi.nlm.nih.gov/pubmed/28213222.

Karczmarczyk, A., M. Dubiak-Szepietowska, M. Vorobii, C. Rodriguez-Emmenegger, J. Dostalek, and K. H. Feller. Sensitive and rapid detection of aflatoxin M1 in milk utilizing enhanced SPR and P(HEMA) brushes. *Biosensors and Bioelectronics* 81 (2016): 159–65. doi: 10.1016/j.bios.2016.02.061, www.ncbi.nlm.nih.gov/pubmed/26945182.

Kim, H. H., H. J. Jeon, H. K. Cho, J. H. Cheong, H. S. Moon, and J. S. Go. Highly sensitive microcantilever biosensors with enhanced sensitivity for detection of human papilloma virus infection. *Sensors and Actuators B: Chemical* 221 (2015): 1372–83. doi: 10.1016/j.snb.2015.08.014, www.sciencedirect.com/science/article/pii/S0925400515301921.

Koh, I. and L. Josephson. Magnetic nanoparticle sensors. *Sensors (Basel)* 9, no. 10 (2009): 8130–45. doi: 10.3390/s91008130, www.ncbi.nlm.nih.gov/pubmed/22408498.

Kulkarni, A. S., P. S. Ghugre, and S. A. Udipi. 15 - Applications of nanotechnology in nutrition: Potential and safety issues. In: *Novel Approaches of Nanotechnology in Food*, edited by A. M. Grumezescu, 509–54. Academic Press Elsevier, Amsterdam, Netherlands (2016).

Liang, R. P., X. N. Wang, C. M. Liu, X. Y. Meng, and J. D. Qiu. Construction of graphene oxide magnetic nanocomposites-based on-chip enzymatic microreactor for ultrasensitive pesticide detection. *Journal of Chromatography A* 1315 (2013): 28–35. doi: 10.1016/j.chroma.2013.09.046, www.ncbi.nlm.nih.gov/pubmed/24084001.

Lin, H. Y., C. H. Huang, S. H. Lu, I. T. Kuo, and L. K. Chau. Direct detection of orchid viruses using nanorod-based fiber optic particle plasmon resonance immunosensor. *Biosensors and Bioelectronics* 51 (2014): 371–8. doi: 10.1016/j.bios.2013.08.009, www.ncbi.nlm.nih.gov/pubmed/24001513.

Lin, K.-C., C.-P. Hong, and S.-M. Chen. Simultaneous determination for toxic ractopamine and salbutamol in pork sample using hybrid carbon nanotubes. *Sensors and Actuators B: Chemical* 177 (2013): 428–36. doi: 10.1016/j.snb.2012.11.052, www.sciencedirect.com/science/article/pii/S0925400512012488.

Lin, Y., F. Lu, Y. Tu, and Z. Ren. Glucose biosensors based on carbon nanotube nanoelectrode ensembles. *Nano Letters* 4, no. 2 (2004): 191–95. doi: 10.1021/nl0347233.

Liu, Y., S. Kumar, and R. E. Taylor. Mix-and-match nanobiosensor design: Logical and spatial programming of biosensors using self-assembled dna nanostructures. *Wiley Interdisciplinary Reviews: Nanomedicine and Nanobiotechnology* 10, no. 6 (2018): e1518. doi: 10.1002/wnan.1518, www.ncbi.nlm.nih.gov/pubmed/29633568.

Liu, H., Zhou, K, Wu, D, Wang, J, and Sun, B. A novel quantum dots-labeled on the surface of molecularly imprinted polymer for turn-off optosensing of dicyandiamide in dairy products. *Biosensors and Bioelectronics* 77 (2016): 512–17, ISSN 0956-5663, doi:10.1016/j.bios.2015.10.007. www.sciencedirect.com/science/article/pii/S0956566315304723.

Long, Q., H. Li, Y. Zhang, and S. Yao. Upconversion nanoparticle-based fluorescence resonance energy transfer assay for organophosphorus pesticides. *Biosensors and Bioelectronics* 68 (2015): 168–74. doi: 10.1016/j.bios.2014.12.046, www.ncbi.nlm.nih.gov/pubmed/25569873.

Magro, M., E. Bonaiuto, D. Baratella, J. de Almeida Roger, P. Jakubec, V. Corraducci, J. Tucek, et al. Electrocatalytic nanostructured ferric tannates: Characterization and application of a polyphenol nanosensor. *Chemphyschem* 17, no. 20 (2016): 3196–203. doi: 10.1002/cphc.201600718, www.ncbi.nlm.nih.gov/pubmed/27464765.

Mahato, K., P. K. Maurya, and P. Chandra. Fundamentals and commercial aspects of nanobiosensors in point-of-care clinical diagnostics. *3 Biotech* 8, no. 3 (2018): 149. doi: 10.1007/s13205-018-1148-8, www.ncbi.nlm.nih.gov/pubmed/29487778.

Malhotra, B., S. Srivastava, and S. Augustine. Biosensors for food toxin detection: Carbon nanotubes and graphene. *Symposium I – Emerging 1D and 2D Nanomaterials in Health Care* 1725 (2015). doi: 10.1557/opl.2015.165.

Marques, I., J. P. da Costa, C. Justino, P. Santos, K. Duarte, A. Freitas, S. Cardoso, A. Duarte, and T. Rocha-Santos. Carbon nanotube field effect transistor biosensor for the detection of toxins in seawater. *International Journal of Environmental Analytical Chemistry* 97, no. 7 (2017): 597–605. doi: 10.1080/03067319.2017.1334056.

Mura, S., G. F. Greppi, P. P. Roggero, E. Musu, D. Pittalis, A. Carletti, G. Ghiglieri, and J. Irudayaraj. Functionalized gold nanoparticles for the detection of nitrates in water. *International Journal of Environmental Science and Technology* 12 (2015). doi:10.1007/s13762-013-0494-7.

Namdari, P., H. Daraee, and A. Eatemadi. Recent advances in silicon nanowire biosensors: Synthesis methods, properties, and applications. *Nanoscale Research Letters* 11, no. 1 (2016): 406. doi: 10.1186/s11671-016-1618-z, www.ncbi.nlm.nih.gov/pubmed/27639579.

Nasirizadeh, N., S. Hajihosseini, Z. Shekari, and M. Ghaani. A novel electrochemical biosensor based on a modified gold electrode for hydrogen peroxide determination in different beverage samples. *Food Analytical Methods* 8, no. 6 (2014): 1546–55. doi: 10.1007/s12161-014-0041-2.

Neethirajan, S. and D. S. Jayas. Nanotechnology for the food and bioprocessing industries. *Food and Bioprocess Technology* 4, no. 1 (2011): 39–47. doi: 10.1007/s11947-010-0328-2.

Neethirajan, S., V. Ragavan, X. Weng, and R. Chand. Biosensors for sustainable food engineering: Challenges and perspectives. *Biosensors (Basel)* 8, no. 1 (2018). doi: 10.3390/bios8010023, www.ncbi.nlm.nih.gov/pubmed/29534552.

Nesakumar, N., S. Sethuraman, U. M. Krishnan, and J. B. B. Rayappan. Electrochemical acetylcholinesterase biosensor based on Zno nanocuboids modified platinum electrode for the detection of carbosulfan in rice. *Biosensors and Bioelectronics* 77 (2016): 1070–77. doi: 10.1016/j.bios.2015.11.010, www.sciencedirect.com/science/article/pii/S0956566315305613.

Noh, H. J., H. S. Kim, S. Cho, and Y. Park. Melamine nanosensing with chondroitin sulfate-reduced gold nanoparticles. *Journal of Nanoscience and Nanotechnology* 13, no. 12 (2013): 8229–38, www.ncbi.nlm.nih.gov/pubmed/24266218.

Prow, T. W., D. Sundh, and G. A. Lutty. Nanoscale biosensor for detection of reactive oxygen species. *Methods in Molecular Biology* 1028 (2013): 3–14. doi: 10.1007/978-1-62703-475-3_1, www.ncbi.nlm.nih.gov/pubmed/23740110.

Radhakrishnan, R. and P. Poltronieri. Fluorescence-free biosensor methods in detection of food pathogens with a special focus on listeria monocytogenes. *Biosensors (Basel)* 7, no. 4 (2017). doi: 10.3390/bios7040063, www.ncbi.nlm.nih.gov/pubmed/29261134.

Radomska, A., E. Bodenszac, S. Głb, and R. Koncki. Creatinine biosensor based on ammonium ion selective electrode and its application in flow-injection analysis. *Talanta* 64, no. 3 (2004): 603–08. doi: 10.1016/j.talanta.2004.03.033, www.sciencedirect.com/science/article/pii/S0039914004001535.

Rai, V., S. Acharya, and N. Dey. Implications of nanobiosensors in agriculture. *Journal of Biomaterials and Nanobiotechnology* 03, no. 02 (2012): 315–324. doi: 10.4236/jbnb.2012.322039, www.scirp.org/journal/PaperInformation.aspx?PaperID=18992.

Ravichandran, R. Nanotechnology applications in food and food processing: Innovative green approaches, opportunities, and uncertainties for global market. *International Journal of Green Nanotechnology: Physics and Chemistry* 1, no. 2 (2010): P72–P96. doi: 10.1080/19430871003684440.

Saei, A. A., J. E. N. Dolatabadi, P. Najafi-Marandi, A. Abhari, and M. de la Guardia. Electrochemical biosensors for glucose based on metal nanoparticles. *TrAC Trends in Analytical Chemistry* 42 (2013): 216–27. doi: 10.1016/j.trac.2012.09.011, www.sciencedirect.com/science/article/pii/S016599361200283X.

Sekhon, B. S. Nanotechnology in agri-food production: An overview. *Nanotechnology, Science and Applications* 7 (2014): 31–53. doi: 10.2147/NSA.S39406, www.ncbi.nlm.nih.gov/pubmed/24966671.

Shelby, T., T. Banerjee, I. Zegar, and S. Santra. Highly sensitive, engineered magnetic nanosensors to investigate the ambiguous activity of Zika virus and binding receptors. *Scientific Reports* 7, no. 1 (2017): 7377. doi: 10.1038/s41598-017-07620-y.

Shi, H., T. Xia, A. E. Nel, and J. I. Yeh. Part II: Coordinated biosensors: Development of enhanced nanobiosensors for biological and medical applications. *Nanomedicine (London)* 2, no. 5 (2007): 599–614. doi: 10.2217/17435889.2.5.599, www.ncbi.nlm.nih.gov/pubmed/17976023.

Shi, H. C., B. D. Song, F. Long, X. H. Zhou, M. He, Q. Lv, and H. Y. Yang. Automated online optical biosensing system for continuous real-time determination of microcystin-Lr with high sensitivity and specificity: Early warning for cyanotoxin risk in drinking water sources. *Environmental Science and Technology* 47, no. 9 (2013): 4434–41. doi: 10.1021/es305196f, www.ncbi.nlm.nih.gov/pubmed/23514076.

Silva, G. A. Introduction to nanotechnology and its applications to medicine. *Surgical Neurology* 61, no. 3 (2004): 216–20. doi: 10.1016/j.surneu.2003.09.036, www.ncbi.nlm.nih.gov/pubmed/14984987.

Sivasankaran, U., S. T. Cyriac, S. Menon, and K. G. Kumar. Fluorescence turn off sensor for brilliant blue FCF: An approach based on inner filter effect. *Journal of Fluorescence* 27, no. 1 (2017): 69–77. doi: 10.1007/s10895-016-1935-8, www.ncbi.nlm.nih.gov/pubmed/27639570.

Sozer, N. and J. L. Kokini. Nanotechnology and its applications in the food sector. *Trends in Biotechnology* 27, no. 2 (2009): 82–9. doi: 10.1016/j.tibtech.2008.10.010, www.ncbi.nlm.nih.gov/pubmed/19135747.

Srivastava, A. K., A. Dev, and S. Karmakar. Nanosensors and nanobiosensors in food and agriculture. *Environmental Chemistry Letters* 16, no. 1 (2017): 161–82. doi: 10.1007/s10311-017-0674-7.

Su, X., F. T. Chew, and S. F. Y. Li. Design and application of piezoelectric quartz crystal-based immunoassay. *Analytical Sciences* 16, no. 2 (2000): 107–14. doi: 10.2116/analsci.16.107.

Tan, H., Q. Li, C. Ma, Y. Song, F. Xu, S. Chen, and L. Wang. Lanthanide based dual-emission fluorescent probe for detection of mercury (II) in milk. *Biosensors and Bioelectronics* 63 (2015): 566–71. doi: 10.1016/j.bios.2014.08.015, www.ncbi.nlm.nih.gov/pubmed/25168765.

Ten, S. T., U. Hashim, S. C. B. Gopinath, W. W. Liu, K. L. Foo, S. T. Sam, S. F. A. Rahman, C. H. Voon, and A. N. Nordin. Highly sensitive escherichia coli shear horizontal surface acoustic wave biosensor with silicon dioxide nanostructures. *Biosensors and Bioelectronics* 93 (2017): 146–54. doi: 10.1016/j.bios.2016.09.035, www.ncbi.nlm.nih.gov/pubmed/27660016.

Tian, B., X. Liao, P. Svedlindh, M. Strömberg, and E. Wetterskog. Ferromagnetic resonance biosensor for homogeneous and volumetric detection of DNA. *ACS Sensors* 3, no. 6 (2018): 1093–101. doi: 10.1021/acssensors.8b00048.

Tiede, K., A. B. Boxall, S. P. Tear, J. Lewis, H. David, and M. Hassellov. Detection and characterization of engineered nanoparticles in food and the environment. *Food Additives and Contaminants Part A: Chemistry, Analysis, Control, Exposure and Risk Assessment* 25, no. 7 (2008): 795–821. doi: 10.1080/02652030802007553, www.ncbi.nlm.nih.gov/pubmed/18569000.

Tien, C. H., H. V. Linh, P. X. T. Tung, and L. Van Hieu. Improving gold nanowire-based biosensor sensitivity by changing probe design. *International Journal of Nanotechnology* 15, no. 1–3 (2018): 199–209. doi: 10.1504/IJNT.2018.089568, www.inderscienceonline.com/doi/abs/10.1504/IJNT.2018.089568.

Turner, A. P. F. Biosensors: Sense and sensitivity. *Science* 290, no. 5495 (2000): 1315. doi: 10.1126/science.290.5495.1315, http://science.sciencemag.org/content/290/5495/1315.abstract.

Umar, A., M. M. Rahman, A. Al-Hajry, and Y. B. Hahn. Highly-sensitive cholesterol biosensor based on well-crystallized flower-shaped zno nanostructures. *Talanta* 78, no. 1 (2009): 284–9. doi: 10.1016/j.talanta.2008.11.018, www.ncbi.nlm.nih.gov/pubmed/19174239.

Vamvakaki, V. and N. A. Chaniotakis. Pesticide detection with a liposome-based nano-biosensor. *Biosensors and Bioelectronics* 22, no. 12 (2007): 2848–53. doi: 10.1016/j.bios.2006.11.024, www.sciencedirect.com/science/article/pii/S0956566306005641.

Vikesland, P. J. and K. R. Wigginton. Nanomaterial enabled biosensors for pathogen monitoring: A review. *Environmental Science and Technology* 44, no. 10 (2010): 3656–69. doi: 10.1021/es903704z.

Vo-Dinh, T. Optical nanosensors for detecting proteins and biomarkers in individual living cells. *Methods in Molecular Biology* 300 (2005): 383–401. doi: 10.1385/1-59259-858-7:383, www.ncbi.nlm.nih.gov/pubmed/15657493.

Wang, L., W. Chen, D. Xu, B. S. Shim, Y. Zhu, F. Sun, L. Liu, et al. Simple, rapid, sensitive, and versatile swnt-paper sensor for environmental toxin detection competitive with ELISA. *Nano Letters* 9, no. 12 (2009): 4147–52. doi: 10.1021/nl902368r, www.ncbi.nlm.nih.gov/pubmed/19928776.

Wang, S., W. Li, K. Chang, J. Liu, Q. Guo, H. Sun, M. Jiang, et al. Localized surface plasmon resonance-based abscisic acid biosensor using aptamer-functionalized gold nanoparticles. *PLoS One* 12, no. 9 (2017): e0185530. doi: 10.1371/journal.pone.0185530, www.ncbi.nlm.nih.gov/pubmed/28953934.

Weng, X. and S. Neethirajan. A microfluidic biosensor using graphene oxide and aptamer-functionalized quantum dots for peanut allergen detection. *Biosensors and Bioelectronics* 85 (2016): 649–56. doi: 10.1016/j.bios.2016.05.072, www.ncbi.nlm.nih.gov/pubmed/27240012.

Wu, Q., Q. Long, H. Li, Y. Zhang, and S. Yao. An upconversion fluorescence resonance energy transfer nanosensor for one step detection of melamine in raw milk. *Talanta* 136 (2015): 47–53. doi: 10.1016/j.talanta.2015.01.005, www.ncbi.nlm.nih.gov/pubmed/25702984.

Xiaoxing, X., S. Liu, and J. Huangxian. A novel hydrogen peroxide sensor via the direct electrochemistry of horseradish peroxidase immobilized on colloidal gold modified screen-printed electrode. *Sensors,* 3 (2003). doi:10.3390/s30900350.

Yavuz, S., A. Erkal, İ. A. Kariper, A. O. Solak, S. Jeon, İ. E. Mülazımoğlu, and Z. Üstündağ. Carbonaceous materials-12: A novel highly sensitive graphene oxide-based carbon electrode: Preparation, characterization, and heavy metal analysis in food samples. *Food Analytical Methods* 9, no. 2 (2015): 322–31. doi: 10.1007/s12161-015-0198-3.

Yola, M. L., L. Uzun, N. Ozaltin, and A. Denizli. Development of molecular imprinted nanosensor for determination of tobramycin in pharmaceuticals and foods. *Talanta* 120 (2014): 318–24. doi: 10.1016/j.talanta.2013.10.064, www.ncbi.nlm.nih.gov/pubmed/24468376.

Zamolo, V. A., G. Valenti, E. Venturelli, O. Chaloin, M. Marcaccio, S. Boscolo, V. Castagnola, et al. Highly sensitive electrochemiluminescent nanobiosensor for the detection of palytoxin. *ACS Nano* 6, no. 9 (2012): 7989–97. doi: 10.1021/nn302573c, www.ncbi.nlm.nih.gov/pubmed/22913785.

Zhang, Y., Z. Wang, Y. Wang, L. Huang, W. Jiang, and M. Wang. Electrochemical detection of sequence-specific DNA with the amplification of gold nanoparticles. *International Journal of Electrochemistry* 2011 (2011): 1–7. doi: 10.4061/2011/619782.

Zhao, J., J. P. O'Daly, R. W. Henkens, J. Stonehuerner, and A. L. Crumbliss. A xanthine oxidase/colloidal gold enzyme electrode for amperometric biosensor applications. *Biosensors and Bioelectronics* 11, no. 5 (1996): 493–502. doi: 10.1016/0956-5663(96)86786-8, www.sciencedirect.com/science/article/pii/0956566396867868.

Ziegler, C. Cantilever-based biosensors. *Analytical and Bioanalytical Chemistry* 379 (2004). doi: 10.1007/s00216-004-2694-y.

Index

A

B

D

G

H

I

K

L

M

N

O

P

Q

R

S

T

U

V

W

X

Y

Z

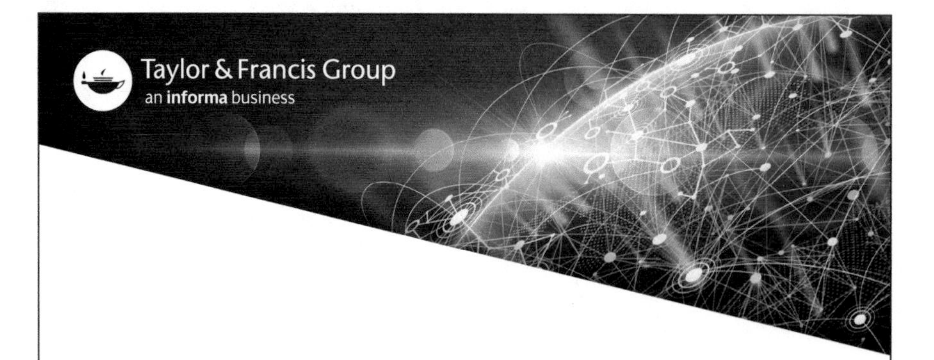

Taylor & Francis eBooks

www.taylorfrancis.com

A single destination for eBooks from Taylor & Francis with increased functionality and an improved user experience to meet the needs of our customers.

90,000+ eBooks of award-winning academic content in Humanities, Social Science, Science, Technology, Engineering, and Medical written by a global network of editors and authors.

TAYLOR & FRANCIS EBOOKS OFFERS:

- A streamlined experience for our library customers
- A single point of discovery for all of our eBook content
- Improved search and discovery of content at both book and chapter level

REQUEST A FREE TRIAL

support@taylorfrancis.com